5th International Conference on Green Composite Materials and Nanotechnology (GCMN 2021)

Journal of Physics: Conference Series Volume 2011

Nanjing, China
23 - 25 July 2021

ISBN: 978-1-7138-4982-7
ISSN: 1742-6588

Printed from e-media with permission by:

Curran Associates, Inc.
57 Morehouse Lane
Red Hook, NY 12571

Some format issues inherent in the e-media version may also appear in this print version.

Institute of Physics
Dirac House, Temple Back
Bristol BS1 6BE UK

Phone: 44 1 17 929 7481
Fax: 44 1 17 920 0979

techtracking@iop.org

Additional copies of this publication are available from:

Curran Associates, Inc.
57 Morehouse Lane
Red Hook, NY 12571 USA
Phone: 845-758-0400
Fax: 845-758-2633
Email: curran@proceedings.com
Web: www.proceedings.com

TABLE OF CONTENTS

Preface

The 2021 5th International Conference on Green Composite Materials and Nanotechnology (GCMN 2021) was planned to be held on July 23-25, 2021 in Nanjing, China. Taking into account the COVID-19 pandemic, the GCMN 2021 conference was held virtually as agencies around the world are now issuing restrictions on travel, gatherings, and meetings in an effort to limit and slow the spread of this pandemic. The health and safety of our participants and members of our research community is of top priority to the Organizing Committee. Therefore, the GCMN 2021 conference was held online through Zoom software.

A key aspect of this conference is the strong mixture of academia and industry. This allows for the free exchange of ideas and challenges faced by these two key stakeholders and encourage future collaboration between members of these groups. GCMN 2021 is to bring together innovative academics and industrial experts in the field of Green Composite Materials and Nanotechnology to a common forum. The primary goal of the conference is to promote research and developmental activities in Green Composite Materials and Nanotechnology and another goal is to promote scientific information interchange between researchers, developers, engineers, students, and practitioners working all around the world.

More than 60 participants attended the online conference, they were from China, Singapore, Malaysia and more. We were greatly honor to have invited Prof. Guanjun Qiao, from Jiangsu University and Prof. SHAHID HUSSAIN, from Jiangsu University to as our Conference Chairs.

During the conference, the conference model was divided into three sessions, including oral presentations, keynote speeches, and online Q&A discussion. In the first part, some scholars, whose submissions were selected as the excellent papers, were given about 5-10 minutes to perform their oral presentations one by one. Then in the second part, we invited eight professors as our keynote speakers. They were Prof. Guanjun Qiao, our beloved Conference Chairmen; Prof.SHAHID HUSSAIN, from Jiangsu University; Prof. Mingsong Wang, from Jiangsu University; A.Prof Muhammad Bilal Tahir, from Khwaja Fareed University of Engineering and information technology, KFUEIT, RYK Pakistan; Prof. Taiqi Liu, from Beijing Institute of Petrochemical Technology; Prof. Yasin Orooji, from Nanjing Forestry University; A.Prof. Sayyar Ali Shah, from Jiangsu University of science and technology; A.Prof. Ziwei Xu, from Jiangsu University. Their insightful speeches had triggered heated discussion in the third session of the conference. Every participant praised this conference for disseminating useful and insightful knowledge.

The proceedings are a compilation of the accepted papers and represent an interesting outcome of the conference. Topics include but are not limited to the following areas: Green Composite Materials, Nanotechnology and Other related topics. All the papers have been through rigorous review and process to meet the requirements of International publication standard.

This year, we have all worked and lived with the daily threat of Covid-19, the need for social distancing and relative isolation has impacted upon the familiar and fundamental nature of conferences. Regardless, such interactions are the lifeblood of our respective communities, and so we made the decision to proceed with the Workshop, bring participants together from around the world through a digital platform. We would like to acknowledge all of those who supported GCMN 2021. The help and contribution of each individual and institution was instrumental in the success of the conference. In particular, we would like to thank the organizing committee for its valuable inputs in shaping the conference program and reviewing the submitted papers.

The Committee of GCMN 2021

Journal of Physics: Conference Series **2011** (2021) 011001 doi:10.1088/1742-6596/2011/1/011001

Committee member

Conference Chairman
Prof. Guanjun Qiao, Jiangsu University, China
Prof. SHAHID HUSSAIN, Jiangsu University, China

Program Committees
Prof. Ziad Moumni, ENSTA Paris Tech, France
Prof. Weihua Dan, Sichuan University, China
Prof. Zujin Zhao, South China university of Technology, China
Prof. Zongqiong Lin, Northwest Polytechnic University, China
Prof. Seeram Ramakrishna, National University of Singapore, Singapore
A. Prof. Rongyun Zhang, Anhui Polytechnic University, China
A. Prof. Wei Wei, Xi'an University of Technology, China
A. Prof. LIU Erjia, National University of Singapore, Singapore

Technical Program Committees
Prof. Wenjing Tian, Jilin University, China
Prof. Xiaotao Hao, Shandong University, China
Prof. Ming-Yang Chen, Jiangsu University, China
Prof. Ming-Yang Chen, Jiangsu University, China
Prof. Yuwu Zhong, Chinese Academy of Sciences, China
Prof. Ir. Dr. Ts. Abdul Talib bin Din, Universiti Teknikal, Malaysia

Journal of Physics: Conference Series **2011** (2021) 011002 doi:10.1088/1742-6596/2011/1/011002

Peer review declaration

All papers published in this volume of Journal of Physics: Conference Series have been peer reviewed through processes administered by the Editors. Reviews were conducted by expert referees to the professional and scientific standards expected of a proceedings journal published by IOP Publishing.

- **Type of peer review: Double-blind**
- **Conference submission management system: AI Scholar System**
- **Number of submissions received: 195**
- **Number of submissions sent for review: 189**
- **Number of submissions accepted: 113**
- **Acceptance Rate (Number of Submissions Accepted / Number of Submissions Received X 100): 57.9%**
- **Average number of reviews per paper: 2**
- **Total number of reviewers involved: 50**
- **Any additional info on review process:**
- **Step 1. Each of selected paper will be reviewed by two/three professional experts in the related subject area.**
- **Step 2. Review Reports received from the experts will be judged by one of the editors either Review Reports are logical or not?**
- **Step 3. If not logical, then editor can assign new reviewer or can also judge at his/her own.**
- **Step 4. If logical, then Review Reports will be sent to authors to modify the manuscript accordingly.**
- **Step 5. Authors will be required to revise their papers according to the points raised.**
- **Step 6. Revised version will then be evaluated by the editor for the incorporation of the points raised by the reviewers.**
- **Step 7. Then the editor will send the revised manuscript to the reviewers again for re-evaluation.**
- **Step 8. If the reviewers approve the revise version of the manuscript, then will be accepted for publication.**
- **Contact person for queries:**
  Xuexia Ye
  publication@keoaeic.org
  AEIC Academic Exchange Information Centre

Study on Tensile Properties of Reinforced Composite Laminates with Open Holes

Ding Cong

Shanghai Aircraft Design and Research Institute, Shanghai 201200, China

Email: dingcong@comac.cc

Abstract. The performance of open-hole composite laminates with different sizes and fillets under tensile load was analyzed by experiments and finite element software of ABAQUS. The effects of different hole sizes and fillet radius on the tensile properties of laminates were discussed. The results showed that the tensile failure load decreased with the increase of the hole size, and the trend was obvious. As the fillet radius of the hole increased, the failure load tended to increase.

1. Introduction

Carbon fiber reinforced composites have been widely used in aircraft structures since the 1970s due to their high specific strength, high specific stiffness, strong designability, corrosion resistance, and good fatigue properties [1]. In the design of composite structure, it is inevitable to consider the stress concentration and the reduction of bearing capacity caused by the open holes [2]. Therefore, certain reinforcement measures must be taken. In order to obtain a lightweight and high-efficiency composite with open-hole reinforcement, it is necessary to study the mechanical properties and parameter selection of the typical open area reinforcement.

Many studies have been carried out at home and abroad. O'Neill analyzed the influence of asymmetric reinforcement on the strength of laminates and found that the strength of the reinforced laminates could be increased by 5% to 12% [2]. Luo et al. conducted experimental research and numerical analysis on different aperture diameters in open-hole reinforcement and found that there was a reinforcement aperture with the highest reinforcement efficiency [3]. The above studies have done a lot of analysis on the influence of strength by reinforcement form, reinforcement aperture, and so on. However, there is no in-depth study on the reinforcement size of open-hole composite laminates and the relationship between the reinforcement thickness and the laminates thickness.

In this paper, the strength values of composite laminates with different reinforcement diameters and thicknesses (2.68mm in thickness) were obtained through tensile tests. At the same time, the stress change and damage process of the open-hole laminates were obtained by finite element simulation using ABAQUS software.

2. Experiments and Methods

A composite laminate with a length of 1000 mm and a width of 600 mm was designed, with a circular reinforcement opening in the center of the web. The materials were 27 pieces of CYCOM977-2/12K HTS unidirectional tape. The typical size of the test piece was shown in Figure 1, the laying order was shown in Table 1, and the test piece matrix was shown in Table 2.Each test piece was attached with 15 pieces of reinforcing patches in total.

Journal of Physics: Conference Series **2011** (2021) 012001 doi:10.1088/1742-6596/2011/1/012001

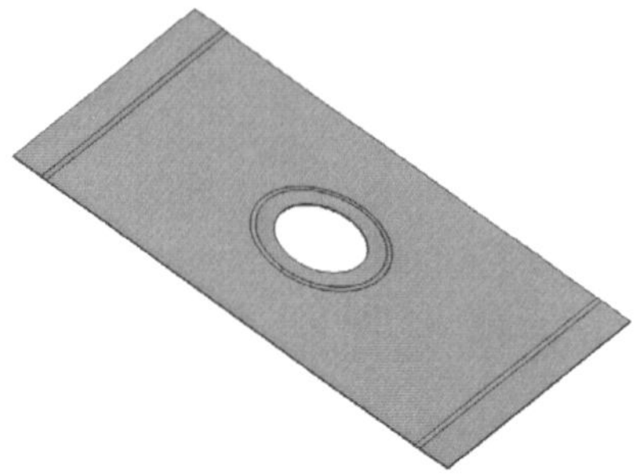

Figure 1 Dimension drawing of the typical test piece

Table 1 Ply parameters of test laminates

Ply angle	Nominal thickness (mm)
$[0/\pm18/\pm36/\pm54/\pm72/90]_S$	2.680

Note: the ply with brackets is bi-directional fabric.

Table 2 Matrix of test piece

Test piece number	Reinforcement parameters	Number of test pieces
1	D2/D1=1.4; t2/t1=1.4	3
2	D2/D1=1.5; t2/t1=1.4	3
3	D2/D1=1.6; t2/t1=1.4	3
4	D2/D1=1.4; t2/t1=1.5	3
5	D2/D1=1.4; t2/t1=1.5	3
6	D2/D1=1.6; t2/t1=1.5	3
7	D2/D1=1.4; t2/t1=1.6	3
8	D2/D1=1.5; t2/t1=1.6	3
9	D2/D1=1.5; t2/t1=1.6	3

Note: D2 is the diameter of reinforcing patch and D1 is the diameter of open hole.
t2 is the thickness of reinforcing patch and t1 is the thickness without reinforcement.

According to ASTM D 5766/D 5766M-02 standard, 200×100 reinforcing patches with a thickness of 5 cm were glued on both sides of the two ends of the test piece. The chucks at the upper and lower ends of the testing machine were clamped on the reinforcing patches at the upper and lower ends of the test piece respectively. The WDW-1000E static testing machine was used for displacement loading, with a loading speed of 2 mm/min. The installation of the test piece was shown in Figure 2.

Figure 2 Installation diagram of test piece

3. Results and Discussion

There were three typical failure modes in the tensile test of the open-hole test piece: brittle fracture, shear failure, and mixed failure. The three typical failure modes were shown in Figure 3.

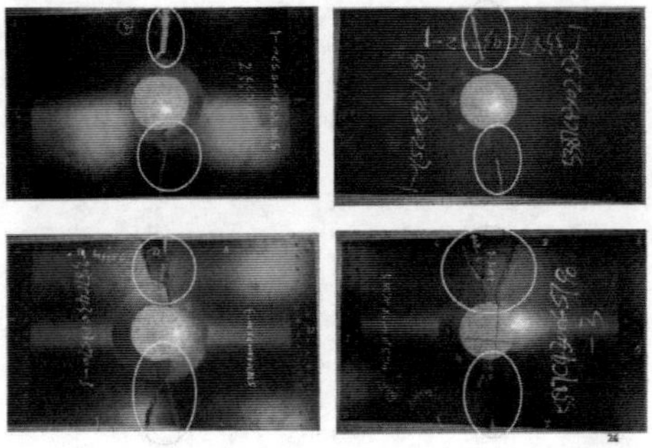

Figure 3 Typical failure modes of test piece

3.1. Results of tensile failure load test

There were 27 open-hole tensile test pieces in total, and there were 3 pieces per group. The failure modes of all the test pieces were in accordance with ASTM standards and there was no abnormal test data. The statistical results of failure load and average load of each test piece were shown in Table 3 and Figure 4

Table 3 Failure load table of test piece (KN)

Hole (mm) / Fillet (mm)	50×50	40×40	30×30
R10	101.67	114.97	111.13
R15	100.27	108.72	103.68
R20	91.12	107.80	-
R25	97.53	-	-

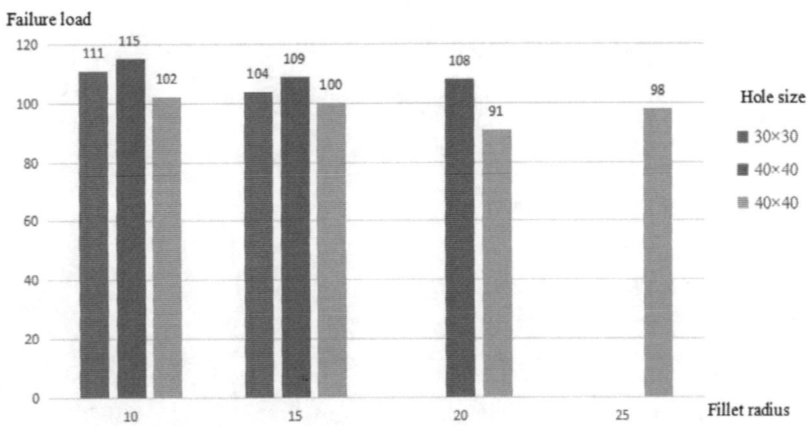

Figure 4 Histogram of failure load in tensile test of test piece

The time-strain curve, the load-displacement curve, and the strain-load curve of the test piece with a central hole of 50 mm × 50 mm and a fillet of R20 were shown in Figures 5 to 7.

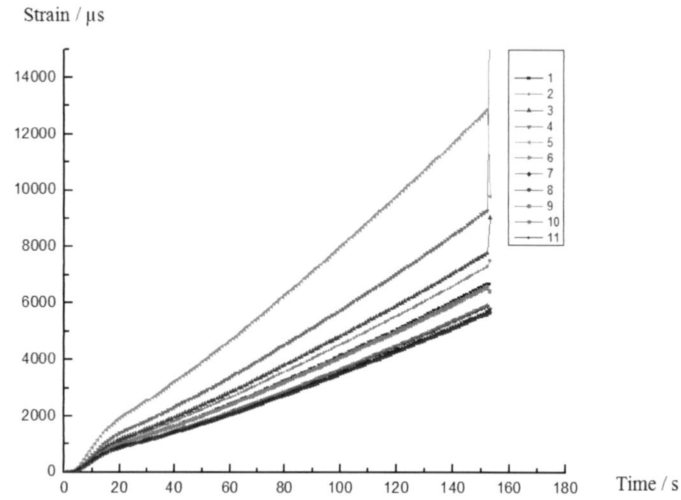

Figure 5 The time-strain curve

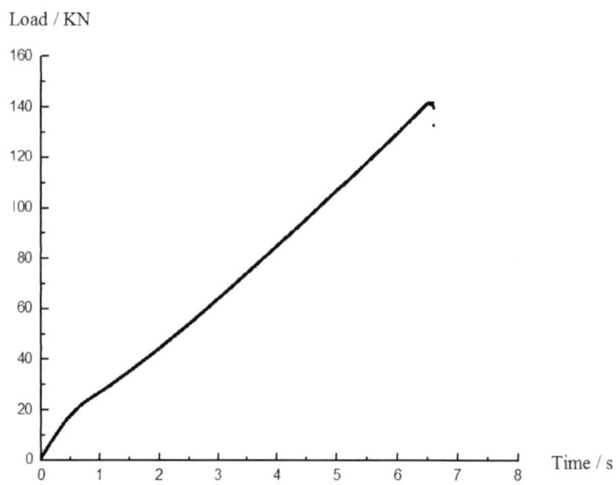

Figure 6 The load-displacement curve

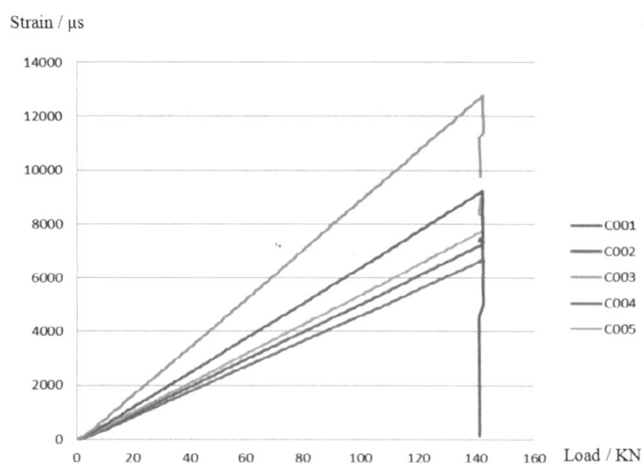

Figure 7 The strain-load curve

3.2. Analysis of tensile failure load test

The model adopted a test piece with a hole size of 30 mm×30 mm and a fillet radius of 10 mm. The finite element model was shown in Figure 8. The boundary condition was that one end was fixed, the other end was applied with a displacement load (1000με) of 1 mm, and both sides were free. In the simulation process, the 8-node secondary shell element S8R was adopted, and the analysis step was linear static general analysis step [4].

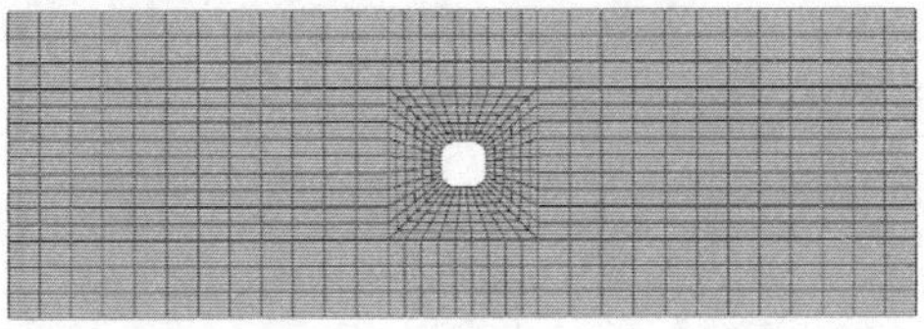

Figure 8 The finite element model

It can be seen that the stress concentration factor was 3.615 and the maximum tensile strain occurred at the end of the fillet of the minimum section of the structure. At this time, the reaction force of the loading point was 31.12 K. Assuming that the failure occurred when the strain reaches 10000 μ ε and the load displacement changed linearly, the failure load was estimated to be 111.2 KN. Compared with the actual failure load of 111.13 KN, the error was within 1%, which proved that the analysis method was effective.

The stress concentration factors and estimated failure loads of other test pieces were shown in Table 4 and Table 5.

Table 4 Stress concentration factor of test piece

Opening (mm) / Fillet (mm)	50×50	40×40	30×30
R10	4.089	3.819	3.615
R15	3.963	3.898	4.255
R20	3.985	4.452	-
R25	4.588	-	-

Table 5 Estimated failure load of test piece (KN)

Hole(mm) / Fillet (mm)	50×50	40×40	30×30
R10	98.8	105.8	111.2
R15	99.9	106.6	111.7
R20	100.5	107.6	-
R25	102.0	-	-

4. Conclusion

The test uses ASTM D 5766/D standard to obtain the failure load and strain statistics of Reinforced Composite Laminates with Open Holes.The stress concentration factors were calculated to be used in the design of similar structures.According to the test results and finite element calculations, the tensile failure load decreased obviously with the increase of the hole size of the composite. Besides, the failure load increased with the increase of the fillet radius. Therefore, circular openings should be used as far as possible.

References

[1] Niu M C Y. Airframe structural design. Aviation Industry Press; 2008.
[2] Yang NB, Zhang YN. Structural design of composite aircraft.Aviation Industry Press; 2002.

[3] Shen Z, Yang SC. Property requirement of composite systems applicable to aircraft structures. Journal of Materials Engineering. 2007; S1: 248-252.

[4] Ye CJ, Jiao JQ, Wang B, Huang T. Investigation on tensile performance of open-holed thin laminates with bi-directional cloth lamina. Journal of Mechanical Strength. 2012; 34: 767-771.

[5] Zeng L, Guan ZD, He W, Liu DB. Damage analysis of open-hole tension laminates. Acta Materiae Compositae Sinica. 2012; 1: 169-175.

Analysis on the Conductivity of Graphene-based Composite Material

Dalei Yang

College of Engineering, University of New South Wales, Sydney, NSW, Australia

Z5192309@ad.unsw.edu.au

Abstract. Electric energy is the main energy source in production and life. How to make efficient transmission and utilization of electric energy is a problem that requires long-term research. Graphene, as a material with a single-layer two-dimensional honeycomb lattice carbon chain structure, has good electrical conductivity. Its composite with conductive polymer or inorganic metal materials can effectively improve its electrical conductivity and obtain more outstanding electrochemical performance. In order to understand the application of graphene's electrical conductivity in life in more detail, this article first reviews the electrical properties of graphene, and then reviews and analyzes the research progress of graphene polymer electrical conductivity based on different composite materials and processing methods. The author also points out that the biggest problem facing graphene composites now is that the expensive price of graphene nanosheets and its unavailableness in large-scale applications and industrial production, as well as proposes that the future of graphene composites is still through scientific research in a short period of time. Continuously improving its electrical conductivity is necessary, yet the long-term strategic goal should be reducing the cost of industrialized production of graphene products and find the preparation conditions of graphene composite materials that are more suitable for industrial production.

1. Introduction

Since the discovery of graphene in 2004, British scientists successfully exfoliated flakes composed of only one layer of carbon atoms from highly oriented pyrolytic graphite. Graphene has important application prospects in materials science, micro-nano processing, energy, biomedicine, and drug delivery due to its excellent optical, electrical and mechanical properties, and is considered to be a revolutionary material in the future [1], because pure graphene is difficult to manufacture and expensive to manufacture. Although graphene has good chemical properties, it has good toughness and is easy to bend. Therefore, it is difficult to apply graphene in production and life so scientists have determined the direction of its research and application.

In order to analyze how graphene plays a higher role in daily production in detail, which is based on the electrochemical properties of graphene as a starting point, this article reviews its electrochemical properties and summarizes the analysis history of the electrical conductivity of composite materials with graphene as the main raw material, according to different preparation methods and different substrates. This article briefly discusses the properties of each different composite material and focuses on the analysis of the current problems facing graphene composite materials and the future development direction.

2. The electrical properties of graphene

A carbon atom has four valence electrons so that each carbon atom contributes an unbonded π electron. These π electrons can form an orbital perpendicular to the plane. The π electrons can move freely in the crystal. Graphene is a kind of carbon atoms connected by sp2 hybridization that are closely packed into a single-layer two-dimensional honeycomb lattice structure. The theoretical specific surface area is as high as $2630m^2/g$. The large area ofπ-π conjugate determines its excellent electrochemical performance [2]. The carrier mobility of graphene at room temperature is about $15000cm^2/(V{\cdot}s)$, and the conduction rate of electrons in it can reach 1/300 of the speed of light. It is 10 times that of silicon material and more than twice that of indium antimonide (InSb), which is the substance with the highest carrier mobility currently known. The reason why graphene has local superconductivity and high carrier mobility is that the carriers in graphene follow a special quantum tunnelling effect, which does not produce backscatter when encountering impurities [3].

In addition, graphene has an energy band structure with a zero bandgap. The resistivity is $10^{-8}\Omega{\cdot}m$, which is lower than the resistivity of copper/silver. As shown in figure 1, graphene is a zero-distance semiconductor because its conduction and valence band meet at the Dirac point. The Brillouin division at the edge of the momentum space at the six positions of the Dirac point is divided into two sets of equivalent triplicates. In contrast, the main point of traditional semiconductors is usually Γ, with zero momentum [4].

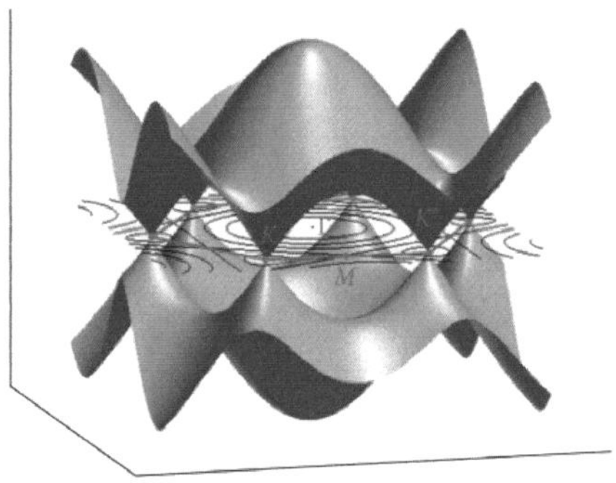

Figure 1 Brillouin band and band structure of graphene [4]

3. Graphene composite

Graphene itself is thinner and lighter than paper, and its flexibility is twice as hard as diamond, but it is expensive to manufacture. Therefore, it cannot be widely used in production and life. For this reason, graphene composite materials have emerged, and graphene composite materials have various forms according to their composites with other materials. It can form a multi-phase material with metals, inorganic non-metal materials and organic polymer materials, which can not only have the good conductivity of graphene itself but also have the good properties of the matrix itself.

3.1. Graphene/polymer composite

3.1.1. Graphene filled composite

The composite of graphene and polymer is to break the van der Waals force between graphene molecules in a covalent manner. Through melt blending or in-situ polymerization to mix graphene with organic polymers can effectively improve the conductivity of the polymer itself.

Zhang Haobin prepared PET graphene nanocomposites by using a 285°C melt blending method. In the experiment, they compared the conductivity of the filling rate of the graphite composite material with the conductivity of the filling rate of the graphene composite material and obtained the relational image shown in figure 2 [5].

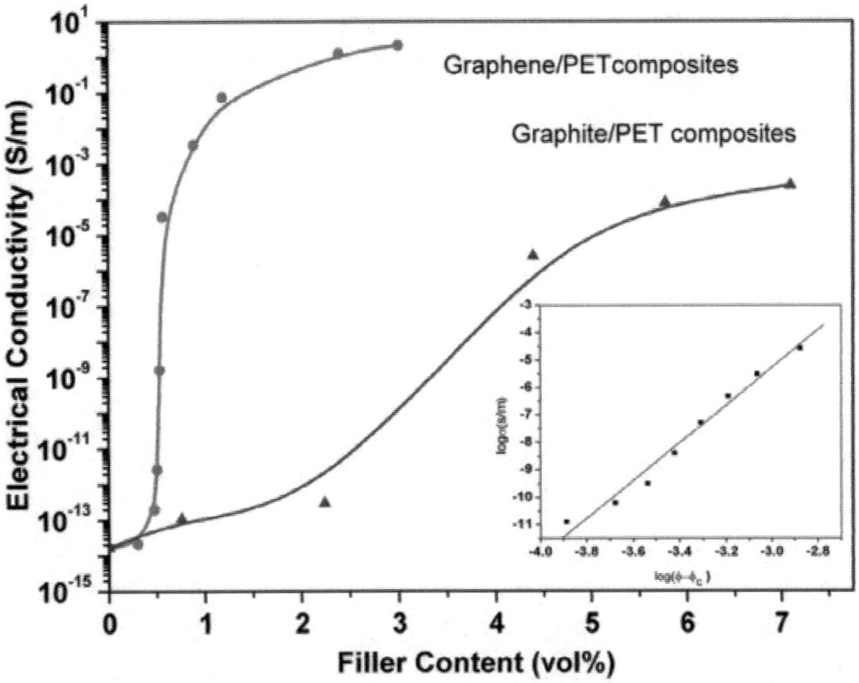

Figure 2 Graphene composite material and graphite composite material filling rate and electrical conductivity relationship [5]

As can be seen from the above figure, when the permeation threshold is determined to be 0.47%, when graphene is used instead of graphite to participate in the production of composite materials. The composite material can achieve an ultra-high conductivity of 2.11 S/m at a graphene content of 0.3 vol%.

3.1.2. Functionalized graphene polymer composite

Graphene and its derivatives can also be covalently or non-covalently functionalized by polymer modification to form functionalized graphene polymer composites. Non-covalent functionalization is to modify the surface of graphene without changing its chemical structure through van der Waals force, electrostatic interaction and π-π stacking. This method can improve the electrical properties of composite materials.

Yang Ying-Kui used NaBH4 as a reducing agent and successfully non-covalently functionalized graphene oxide with an imidazole ionic liquid (Imi-IL) with active vinyl benzyl under alkaline conditions. Afterwards, the uniform dispersion of the graphene sheets and the strong interface bonding with the poly (methyl methacrylate) (PMMA) matrix were successfully achieved, and the conductivity images of the composite materials containing different volume fractions of graphene were obtained as shown in figure 3 [6].

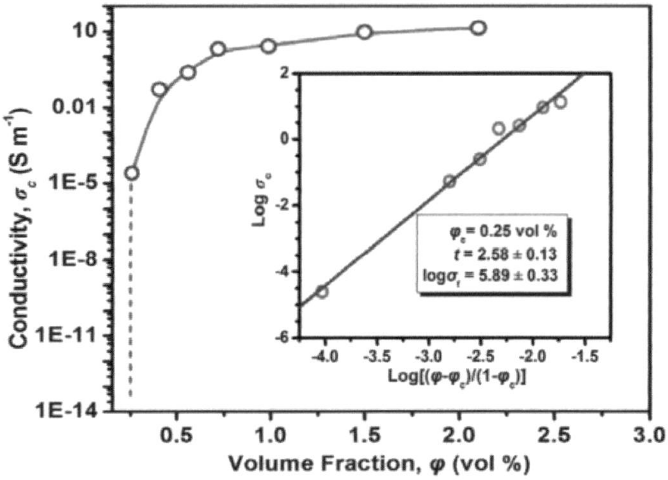

Figure 3 Graphene/PMMA composite material volume fraction-conductivity image [6]

It can be seen from the above figure that when the low permeability threshold is 0.25 vol.%, the composite material achieves a high conductivity of 13.3S/m with a low content of 2.08vol.%. This proves that the functional composite can significantly improve the electrical properties of the material.

3.2. Graphene/inorganic composite material

Dispersing inorganic materials (metal nanomaterials, semiconductors and insulating nanomaterials) on the surface of the graphene nanolayer can synthesize graphene-based inorganic nanocomposites. Inorganic nanoparticles can reduce the interaction between graphene sheets, and the combination of graphene and specific nanoparticles makes this type of composite material have a wide range of applications in the fields of catalysts, electricity, etc.

3.2.1. Chemical reduction method

The reducing agent and organic solvent used in the chemical reduction method will reduce the activity of the graphene-nanoparticle binding interface, thereby reducing the performance of the composite material. The electrochemical deposition of inorganic nanomaterials directly on the graphene matrix is a green method for preparing graphene composite films. Environmentally friendly and efficient method.

Li Fenghua successfully used $SnCl_2$ to reduce graphene oxide to graphene flakes using sodium chloride and urea [7], and then combined SnO_2 particles with graphene flakes perfectly. Then he compared graphene/Sn composite (GS), graphite oxide (GO) and pure graphene (CCG) synthesized by the chemical method in sulfuric acid solution. Nyquist plots as shown in the figure 4 are obtained through the summary of the experiment.

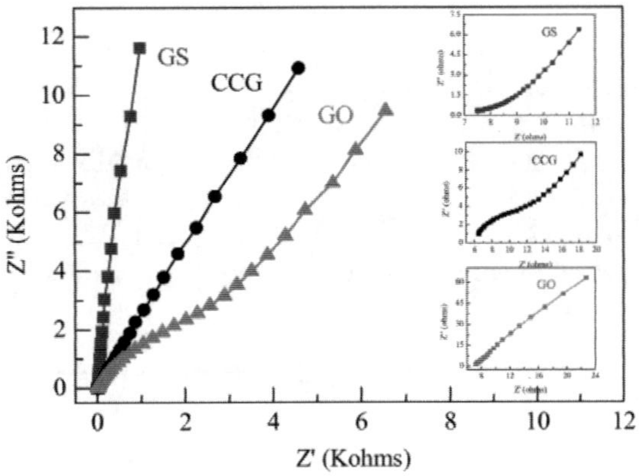

(a)

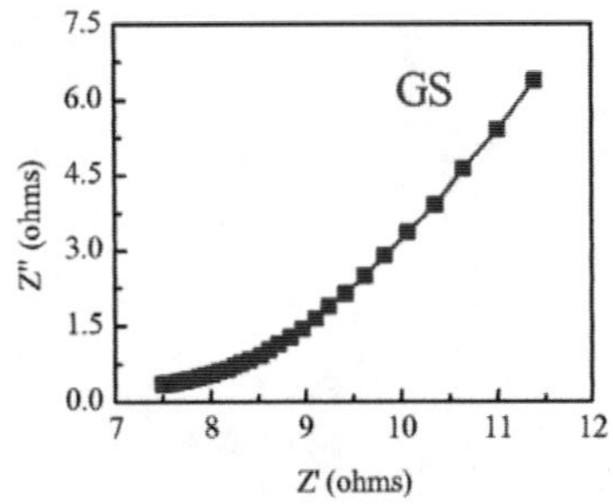

(b)

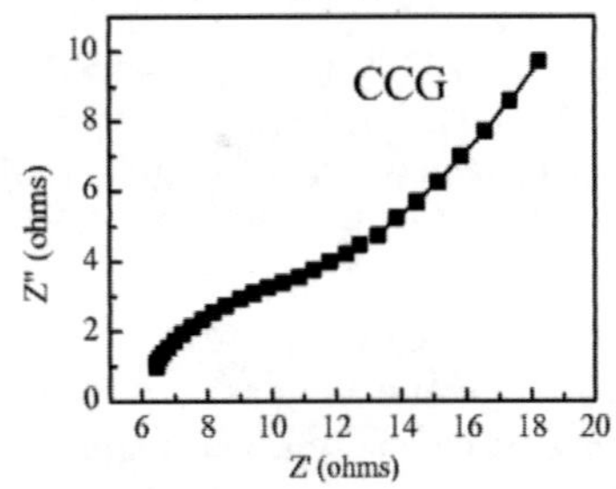

(c)

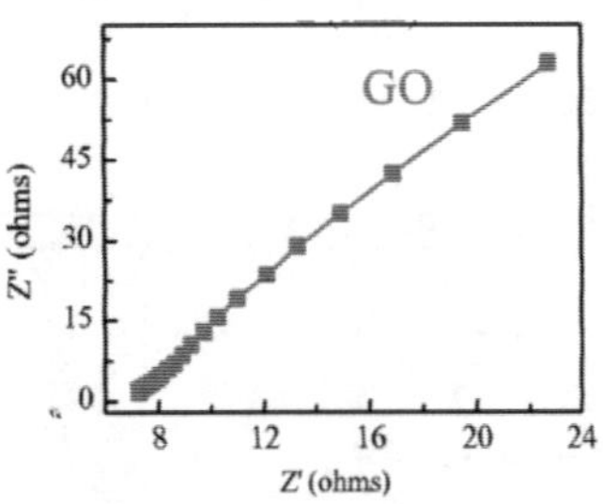

(d)

Figure 4 Nyquist plots of GS, CCG and GO [7]

Figure 4 a is the summary of the three composite materials, b, c, d are the data of the three composite materials themselves in a [7]. As can be seen from the above figure, compared with CCG and GO, the conductivity of graphene/Sn composites is much stronger than the former two. In addition, GS composite materials have broader application prospects in the fields of batteries, supercapacitors and sensors.

3.2.2. Electrochemical deposition

The hydrothermal method is a simple method that can generate high pressure at a high temperature and a fixed volume to prepare inorganic nanocomposites.

Wang Huanwen used a two-step method to prepare graphene-bismuth oxide composite materials [8]. First, the graphite oxide and $Bi(NO_3)_2 \cdot 5H_2O$ are mixed ultrasonically and crushed, and then heated with high-pressure steam at 180°C for 12 hours. Next, the mixture is thoroughly mixed and reacted with hot air heated to 300° C for 4 hours to obtain a composite material. Finally, the resistance value of the composite material at different potentials was tested accordingly and the corresponding curve as shown in the figure 5 below was obtained.

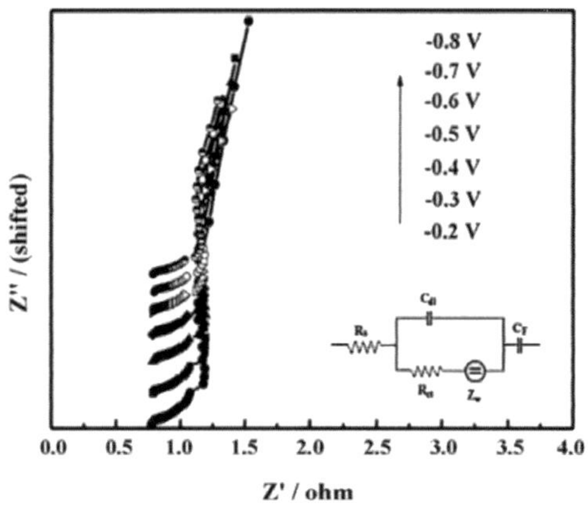

Figure 5 EIS analysis of graphene–Bi_2O_3composite at the different potentials [8]

As can be seen from the above figure, at low frequencies, the resistance of the graphene-bismuth oxide composite material tends to a vertical line, which indicates that the composite material has very good capacitance. This composite material has very good potential for supercapacitors [9].

4. Discussion

Although graphene composite materials have developed a large number of new and improved materials through continuous research, 80% of graphene composite materials are now used in scientific research. Only a small part will be used in actual production and manufacturing because all the improvements mentioned in the previous article are done in the laboratory. Although the performance has been significantly improved, it does not conform to the principle of low cost in industrial production. According to table 1, which shows the global price of graphite products in 2018, provided by China Foresight Industry Research Institute, the price of graphene nanomaterials far exceeds that of graphite products of the same level.

Table 1 Carbon product material price (2018)

Material	Market price(USD/kg)
Natural graphite	0.5
Graphite product (+50mesh)	3-10
Graphene nanosheets (2-10nm)	3000-30000
Graphene nanosheets (1-2nm)	500000
Single wall carbon nanotube	300000
Multi-walled carbon nanotubes	30000
Silicon/carbon composite materia	15000

In industrial production, there are far more problems that need to be solved than in laboratory environments. For example, hydrothermal processing methods are very expensive in industrial environments. Simultaneously, industrial production also needs to ensure the controllability of the product. That is, the structure of the graphene composite material can be controlled relatively easily, which is still a problem to be solved today.

5. Conclusion

Since the birth of graphene, it has been a research hotspot because of its excellent properties. Although there are many preparation methods, it is still difficult to prepare high-quality graphene in large quantities and at a low cost. Therefore, graphene composite materials are still the focus of current research. At present, the application of graphene composite materials in electricity is mainly concentrated in electronic materials, thin-film materials, energy storage materials and liquid crystal materials. However, there are relatively few studies on bulk graphene metal matrix composites. The excellent reinforcement effect of graphene and its unique interface with the metal matrix will make this type of composite material a research hotspot for graphene composites in the future. Moreover, large-scale synthesis and industrial applications still face a large number of problems and challenges. The short-term goal should still be to explore the new properties of graphene conductive polymer functionalized composites so as to continue to broaden the new fields of composite applications. However, the long-term strategic goal should be to reduce graphene production costs and environmental requirements for production.

Acknowledgment

The author is grateful for the help of the University of New South Wales Library and the guidance of professor Erik Luijten. Thanks to Jane for critically reading an initial draft of this manuscript, as well as Wang Ting and Haitong for their help and guidance in selecting the topic.

References

[1] A. K. Geim, "Graphene : Status and Prospects," vol. 324, no. June, pp. 1530–1535, 2009.

[2] H. Chen, M. B. Müller, K. J. Gilmore, G. G. Wallace, and D. Li, "Mechanically strong, electrically conductive, and biocompatible graphene paper," Adv. Mater., vol. 20, no. 18, pp. 3557–3561, 2008, doi: 10.1002/adma.200800757.

[3] K. S. Novoselov et al., "Two-dimensional gas of massless Dirac fermions in graphene," vol. 438, no. November, pp. 197–200, 2005, doi: 10.1038/nature04233.

[4] D. R. Cooper et al., "Experimental Review of Graphene," vol. 2012, 2012, doi: 10.5402/2012/501686.

[5] H. Zhang et al., "Electrically conductive polyethylene terephthalate / graphene nanocomposites prepared by melt compounding," Polymer (Guildf)., vol. 51, no. 5, pp. 1191–1196, 2010, doi: 10.1016/j.polymer.2010.01.027.

[6] Y. K. Yang et al., "Non-covalently modified graphene sheets by imidazolium ionic liquids for multifunctional polymer nanocomposites," J. Mater. Chem., vol. 22, no. 12, pp. 5666–5675, 2012, doi: 10.1039/c2jm16006d.

[7] F. Li et al., "One-step synthesis of graphene/SnO2 nanocomposites and its application in electrochemical supercapacitors," Nanotechnology, vol. 20, no. 45, 2009, doi: 10.1088/0957-4484/20/45/455602.

[8] H. W. Wang et al., "Facile solvothermal synthesis of a graphene nanosheet-bismuth oxide composite and its electrochemical characteristics," Electrochim. Acta, vol. 55, no. 28, pp. 8974–8980, 2010, doi: 10.1016/j.electacta.2010.08.048.

[9] M. D. Stoller, S. Park, Y. Zhu, J. An, and R. S. Ruoff, "Graphene-Based Ultracapacitors," pp. 6–10, 2008.

Study on the application properties of high hydrophobic paper yarn base paper

Zhang Bianzhi[1,a], Yang Qian[1,b], Cheng Jinru[1,c], Yuan Lin[2,d], Liu Zhong[1,e*]

[1]Tianjin Key Laboratory of Pulp and Paper, College of Papermaking Science and Technology, Tianjin University of Science and Technology, Tianjin, China

[2]Shandong Marriott Paper Group Co. Ltd, Shandong, China

[a]E-mail: 1090687031@qq.com, [b]E-mail: beyondmop@163.com, [c]E-mail: 15822539439@163.com , [d]E-mail: 1355772338@163.com

*Corresponding author： [e]email: mglz@tust.edu.cn

Abstract. As a new kind of paper-based textile materials and thanks to its healthiness and environmental protection, the paper yarn and its production will have a good future. This study mainly explores the papermaking process as follows: the beating degree of hemp pulp is 75°SR, the addition amount of PVA fiber, polyamide polyamine epichlorohydrin (PAE) and alkenone dimer (AKD) is 2.0%, 2.5% and , 0.2%, respectively. In addition, the pressure light pressure is 15 N/mm. The high hydrophobic paper yarn base paper has the wear resistance, water resistance, acid and alkali resistance. The analysis shows that the paper yarn base paper can be used as raw material in textile and weaving industries, thus reducing the use of viscose, polyester, plastic, etc.

1. Introduction

Paper yarn base paper, which is made from natural hemp pulp at specific conditions by papermaking machine, can be tailored into required width for various application(such as Fig.1(a)). And also, yarn base paper can be composited with thin nylon(such as Fig.1(b)), then twisted into string, by which enlarged its application in textiles and packing industrials [1].

(a) (b)

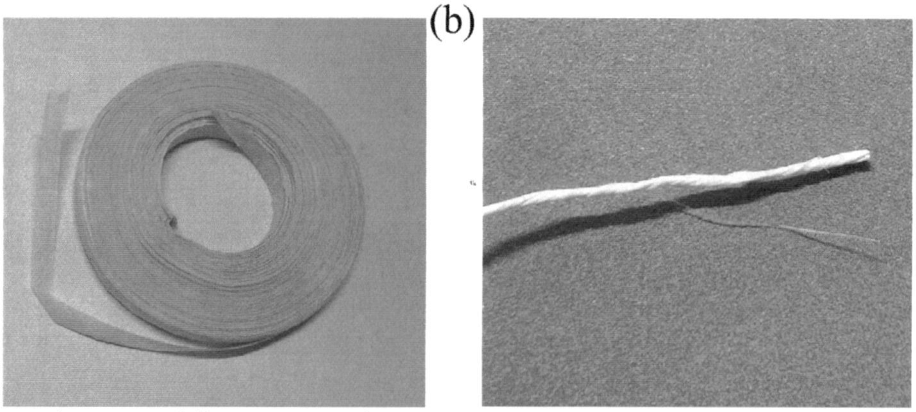

Fig.1 Paper yarn base paper follow-up processing. a) The appearance of a yarn base paper after cutting strips. b) Yarn base paper twisted with nylon.

Nowadays, braided twine, bedroom accessories, ornaments, wearable and different types of furniture (armchairs, sofas, beds, air purification filters etc.) can be made from paper yarn, carpets, upholstery fabric, and the blinds [2-4]. Besides, the blinds also can be made of paper or combined with cotton and wool yarn, yarn paper reduces the development of the viscose and the use of leather products, etc. At the same time, yarn base paper can also be used to make straw, paper bags and other to replace plastic products. Although, these products are becoming more and more popular, the disadvantage of the high manufacturing cost, high technical content [5,6]. Therefore, it seems that the above products only seem to be aimed at the elite market. On the contrary, the domestic research is very few, so it is necessary to explore and develop this technology to develop paper yarn base paper products suitable for public use.

The paper yarn not only had low basis weight (around 20 g/m^2), but also affected by complex forces such as stretching and torsion in the process of cutting into a narrow series of paper, paper line. So that its paper performance put forward higher requirements. The yarn base paper was required to have higher strength and elongation. In addition, due to the reusable characteristics of paper yarn products, it was also required to have good water resistance, friction resistance and acid and alkali resistance [7].

2. Materials and Methods

2.1. Materials
The manila hemp, with a length of 2.8 mm and a width of 21.3 μm, was provided by Shandong Longde Composites Technology Co., Ltd. PVA fiber was provided by Jinan Chida Technology and Trade Co., Ltd, and its length is 6.0 mm, water solubility temperature is 60 °C. The experimental regents included PAE was bought from Nantong Huachuan Chemical Industry Co., Ltd. The AKD, sodium hydroxide and hydrogen chloride were bought from Solis Shanghai Chemical Industry Co., Ltd and Sinopharmaceutical Chemical Reagents Co., Ltd, respectively.

2.2. Methods

2.2.1. Preparation of handsheets
The hemp pulp fibers were beaten by Valley for 20 min and by PFI grinding to 75 °SR. Then, 2.0% PVA fiber was added to disperse together, and 2.5% PAE, 0.05% AMPAM and 0.2% AKD were added successively after stirring for a period of time. Finally, take a certain amount of paste to make a quantitative 20g/m^2 of handwritten sheet. The handwritten sheet was wet pressed, dried at 85°C , and pressed with a light pressure of 15 N/mm. Handsheets produced under these conditions would be used for further research.

2.2.2. Determination of wear resistance of handsheets
Ground the surface of the handsheets several times until the internal fibers were exposed. The contact Angle of the surface was observed with a contact Angle contact. The wear test schematic diagram is shown in Fig. 2.

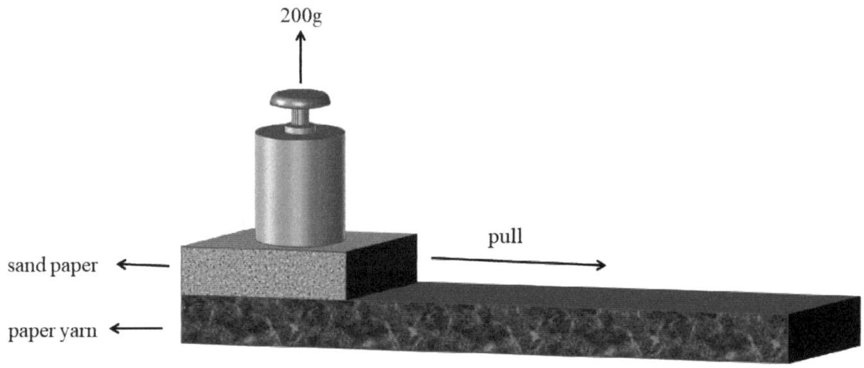

Fig. 2 Experimental schematic diagram of wear of handsheets

2.2.3. Determination of water resistance of handsheets
Several handsheets were cut into strips of 15 mm wide and put into a beaker filled with water. The paper was stirred in a magnetic stirrer (200 r/min) for a period of time to test the wet strength of the paper pattern. Spread out the mixed pattern on the press cloth and re-dry to test its tensile strength.

2.2.4. Determination of acid and alkali resistance of handsheets
Hydrochloric acid and sodium hydroxide were prepared into solutions with different pH values. A number of handsheets were cut into strips of 15 mm wide and immersed in the solution for 30 min. The changes before and after treatment of morphology of fibers were observed under a scanning electron microscope (SEM). The tensile strength and elongation were measured and the rate of decrease in tensile strength and elongation was calculated. The calculation formula is as follows:

$$\text{Decreasing rate} = \frac{X_0 - X}{X_0} \times 100\% \quad (1)$$

X_0: targeted value before treatment
X: targeted value after treatment

3. Results & Discussion

3.1. Hydrophobic properties of handsheets
Due to the low quantity and thickness of the yarn base paper, AKD was added into the pulp to make the paper have hydrophobic property [8]. The addition of AKD would affect the wear-resistant and washer-resistant properties of the paper. Meanwhile, the change of contact Angle was one of the methods to characterize the wear-resistant and washer-resistant degree of the paper.

As can be seen from Fig. 4 (a) and Fig. 4 (b), the contact Angle of the handwritten sheet was larger and the hydrophobic performance was greatly improved. As can be seen from the comparison between Fig. 4(c) and Fig. 4(d), with the addition of AKD, AKD molecules were spread on the surface of fibers. AKD had hydrophobicity due to its aliphatic hydrocarbon long chain functional groups. After adding AKD to the dreged slurry, during the drying process, hydrophobic groups were oriented outward, and AKD was oriented toward the hydroxyl group and active functional group of the fibers to react and form β-ketone ester bond, which was fixed on the fibers. The alkyl group with hydrophobicity was outward, resulting in the hydrophobicity of the paper. At the same time, AKD also reacted with water molecules to produce unstable β-ketoic acid, which was decarboxylated to form ketones [9-14], as shown in Fig. 3.

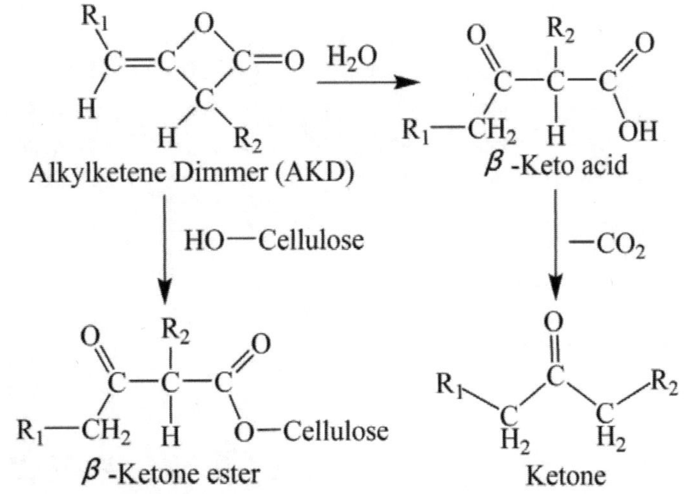

Fig. 3 Mechanism of action of AKD with cellulose and water

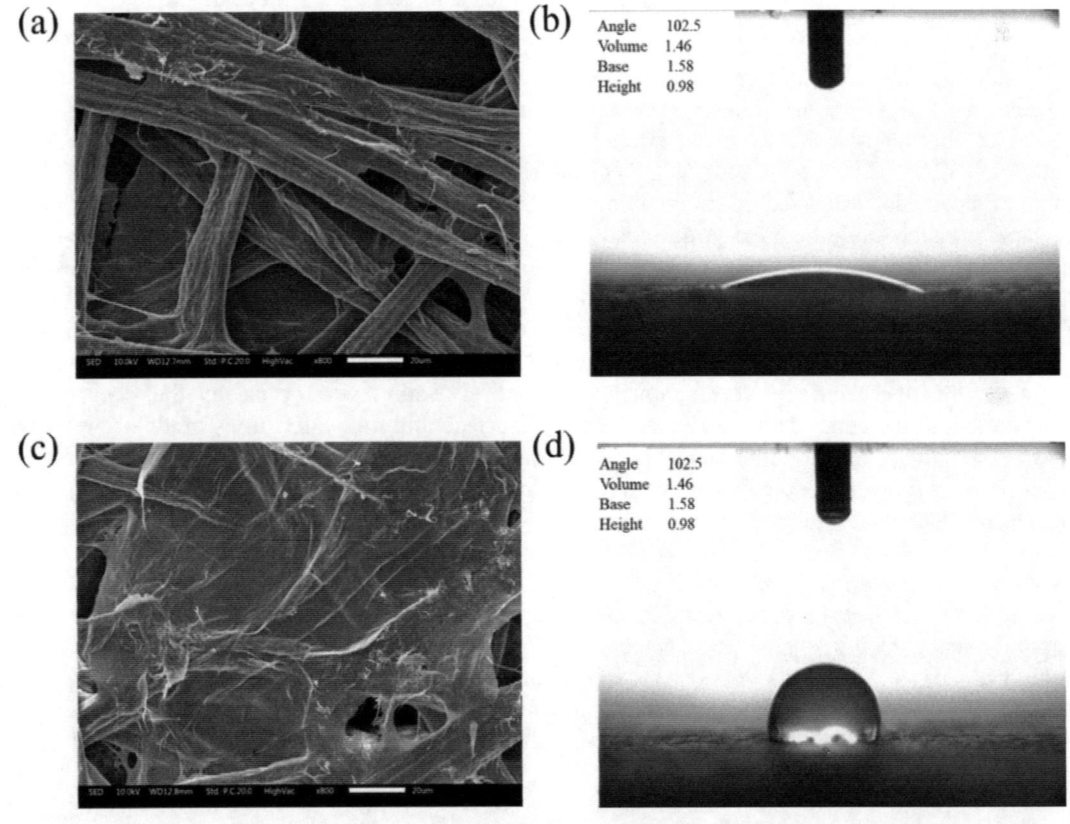

Fig.4 The wettability of handshheets changed before and after adding AKD. a) SEM image of handshheets surface before adding AKD. b) Contact Angle diagram of handshheets surface before adding AKD. c) SEM image of handshheets with AKD content of 0.2%. d) Surface contact Angle of handshheets with 0.2% AKD added.

3.2. The wear resistance of handsheets

Due to the subsequent processing process of paper yarn base paper needed to suffer twisting, weaving and other processes, and paper yarn products were often rubbed, so it needed good wear resistance.

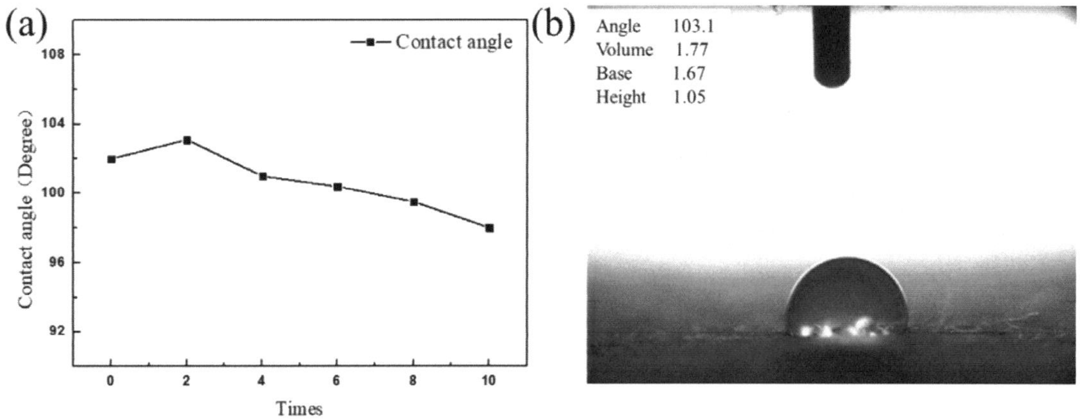

Fig. 5 Surface wettability of worn handshheets. a) Influence of friction times on wettability of handshheets. b) Contact Angle diagram of handshheets after 2 times of friction.

As can be seen from Fig. 5 (a), with the increase of the number of friction, the static contact Angle between the handsheets and water increased first and then decreased. When the water drops on the surface of the handshheets were not friction, the static contact Angle of the water drops on the handshheet was 102.0°, and when the paper handshheets was friction for 2 times, the static contact Angle of the handshheets reached the maximum value of 103.1° in Fig. 5 (b). As the friction continues, the static contact Angle begined to decrease. This was because the paper surface was slight friction handshheets, fibers slightly damaged, form a tiny nap, in the density of handshheets surface fibers cluster, made the water droplets and formed a layer of air layer on the surface of handshheets, had the effect to hold up water. Water and paper yarn on the surface of the handshheets contact area was reduced, eventually the contact Angle was increased performance for yarn handshheets. As the friction continues, the fibers surface were seriously damaged. The air layer formed by the fibers clusters disappears, and the contact area between the fibers surface and the water drops gradually increased, making the contact Angle of the handshheets decreased [15-17]. After 10 times of friction, handshheets still had strong hydrophobic property, and the contact Angle was 98.0°, which indicated that handshheets had good wear resistance.

3.3. Washable property of handsheets

In order to ensure that the paper yarn base paper products could still be reused after washing, it was required to have good washable performance.

As can be seen from Fig. 6 (a), the wet strength of handsheets decreased gradually with the increase of mixing time, while the redrying tensile index changed little with the increase of mixing time. This was because with the increase of washing time, water slowly permeates into the fibers of handsheets, and finally completely permeates, resulting in the decrease of wet strength [18,19]. After washing for 30 min, the tensile index of handsheets decreased from 81.1 N·m/g to 79.6 N·m/g, which was only 1.85% lower than that of handsheets without washing, and the contact Angle between handsheets and water decreased from 102.0° to 97.2° after washing for 30 min. The contact Angle of handsheets are 4.71% lower than that of handsheets without washing. The contact Angle is shown in Fig. 6 (b), indicating that handsheets of paper yarn had strong washable performance.

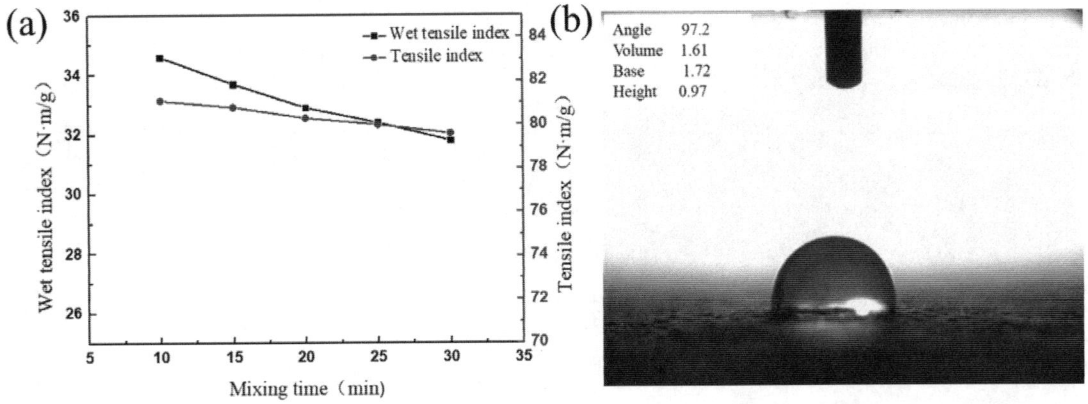

Fig.6 The mechanical properties of handsheets after washing and the surface wettability of handsheets after drying again. a) Effect of stirring time in water on strength of handsheets. b) Contact Angle of handsheets after 30 min washing.

3.4. Acid and alkali resistance of handsheets

In order to store paper yarn base paper for a longer time and increase the service life of paper yarn products, it is necessary to explore the acid and alkaline resistance of paper yarn base paper and minimize the damage caused by the external environment to the fibers.

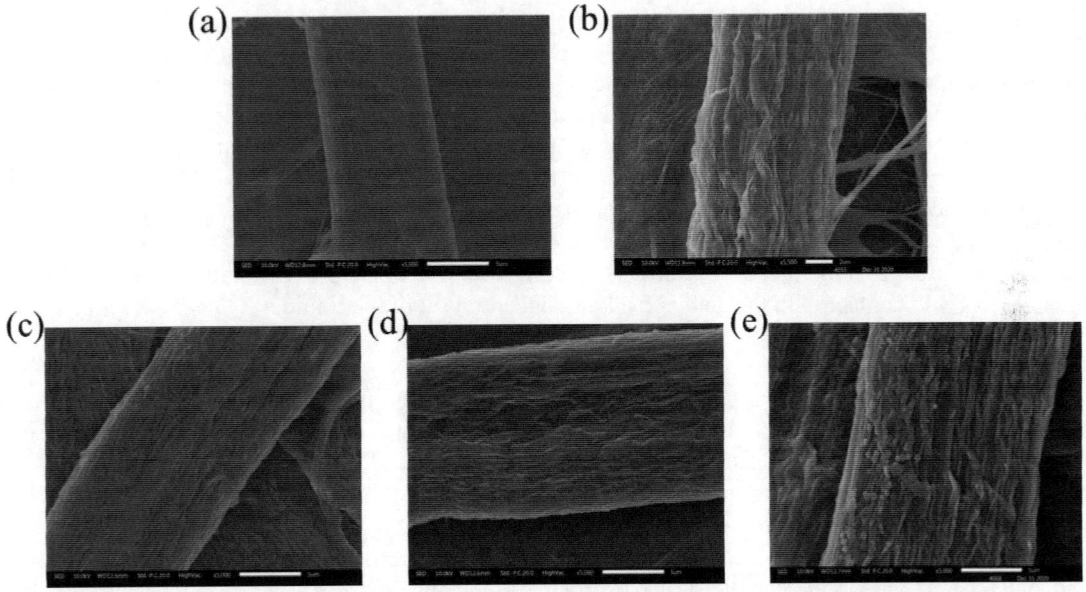

Fig.7 SEM image of fibers surface morphology after solution treatment with different pH values. a) SEM image of fibers surface without acid-base treatment. b) SEM image of fibers surface treated with solution of pH 1. c) SEM image of fibers surface treated with solution of pH 5. d) SEM image of fibers surface treated with solution of pH 9. e) SEM image of fibers surface treated with solution of pH 13.

As can be seen from Fig. 7 (a), without acid-base treatment of the fibers surface of handsheets was smooth, with almost no gullies. Fig. 7 (c) and Fig. 7 (d) show that after handsheets were treated with a lower concentration of acid-base solution, the fibers surface became rough and there were fewer ravines. As can be seen from Fig. 7 (b) and Fig. 7 (e), when the acid-base concentration was high, the surface of the fibers had more gullies, and the depth and width of the gullies became larger. However, by comparing the fibers on the surface of handsheets treated with the same concentration of acid and alkali solution, it could be seen that the surface of fibers treated with acid was more and the depth of the gullies was larger than that treated with alkali [20].

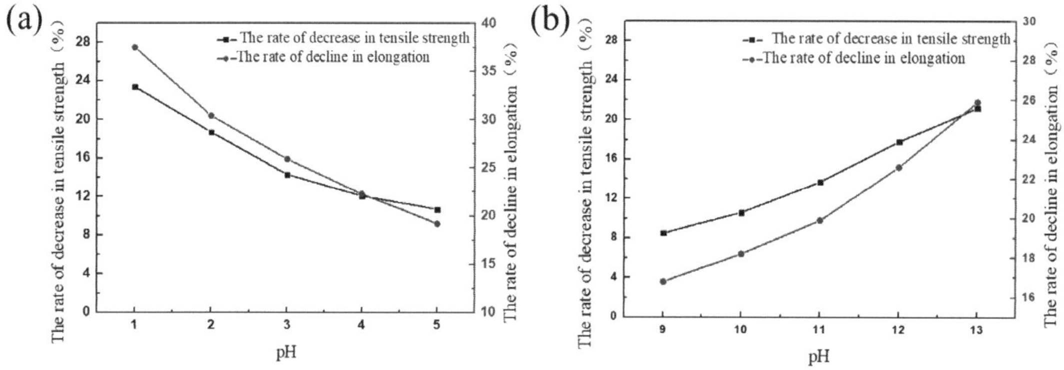

Fig.8 Effect of acid-base treatment on physical properties of handsheets. a) Effect of acid solution treatment at different pH on strength and elongation of handsheets. b) Effect of alkali solution treatment at different pH on strength and elongation of handsheets.

As can be seen from Fig. 8 (a), as the concentration of acid solution decreases, the decrease rates of tensile strength and elongation of handsheets decrease as the concentration of acid solution decreases. As can be seen from Fig. 8 (b), the decrease rate of tensile strength and elongation of handsheets increase with the increase of alkali solution concentration. This is because with the increase of the concentration of acid and base solution, the more severe the damage to the fibers structure, the strength of the fibers itself decreases, resulting in the reduction of the strength of the handsheets [21]. However, the decrease rate of tensile strength and elongation of handsheets treated with the same concentration of alkali solution was lower than that treated with acid solution. This is because the alkali resistance of hemp pulp fibers are better than that of acid solution and the damage degree of the fibers treated with alkali are lower than that treated with acid [22, 23]. In the actual production and use, attention should be paid to the control of the concentration of acid and alkali solution, as far as possible to avoid contact with the acid and alkali concentration of the liquid.

4. Conclusions

The handsheet prepared in this study has good hydrophobic property. After 10 times of wear, the contact Angle of the handsheet is 98°, which only decreases by 5.77%, indicating that the handsheet has good wear resistance. After the handsheet is stirred in water for 30 min, the tensile index of the handsheet is 79.6 N·m/g after drying, which only decreases by 1.85%, which proves that the handsheet has excellent washable property. After the handsheet is treated with different concentration of acid and alkali, the fibers are damaged, and its strength and elongation are decreased. Therefore, the handsheet and other paper yarn products should be avoided from being placed in the environment with high concentration of acid and alkali for a long time.

Acknowledgments

This paper was supported by the Key Research Program of the Ministry of Science and Technology (2017YFB307901)

References

[1] DANG Y H, WANG Z J, LI H K. Study on Base Paper of Paper Yarn for Textile[J]. Transactions of China Pulp and Paper, 2006，21(4):56-61.

[2] Wang H, Liu Z, Hui L , et al. Utilization of Xylan-rich Steam Explosion Liquid from Processing of Poplar for Hydrogel Synthesis[J]. Bio Resources, 2020, 15(2), 2525-2539.

[3] Chonsakorn S, Piromthamsiri K, Sirikasemlert C. Physical Properties of Woven Fabric from Paper Mulberry Yarn[J]. Advanced Materials Research, 2014, 849: 157-161.
[4] SUN LH. Paper cones of high precision manufactured of Majer plant [J].Textile Review, 1993, 3(22): 41-41.
[5] Takasaki M , Ogura R , Morikawa H , et al. Preparation and Properties of Paper Yarn from Mulberry[J]. Advanced Materials Research, 2011, 175: 575-579.
[6] Peterson J, Eckard A, Hjelm J, et al. Mechanical-Property-Based Comparison of Paper Yarn with Cotton, Viscose, and Polyester Yarns[J]. Journal of Natural Fibers, 2019(3):1-10.
[7] SUN H X, XU H, ZHU Z F. Study on properties of paper yarn and its ply yarn. Shanghai Textile Science and Technology, 2011, 5(15): 23-25.
[8] ZHOU L J, Chen F S, HU H R. Mechanism of AKD sizing and its effect on friction properties of paper [J]. Heilongjiang Paper, 2005, 2(15): 34-35. (in Chinese)
[9] Seppänen R, Tiberg F, Valignat MP, Mechanism of internal sizing by alkyl ketene dimers (AKD): The role of the spreading monolayer precursor and autophobicity[J]. Nordic Pulp & Paper Research Journal, 2000, 15(5): 452-458.
[10] Garnier G, Bertin M, Smrckova M. Wetting dynamics of alkyl ketene dimer on cellulosic model surfaces[J]. Langmuir, 1999, 15(22): 7863-7869.
[11] Kumar S, Chauhan VS, Chakrabarti SK. Separation and analysis techniques for bound and unbound alkyl ketene dimer (AKD) in paper: A review[J]. Arabian Journal of Chemistry, 2016, 9(2): 1636-1642.
[12] Huang LQ, Song XM, Cui LQ, et a. Preparation and characterization of AKD sizing agent by "one-pot cooking"[J]. Nordic Pulp & Paper Research Journal, 2018, 33(2): 317-326.
[13] Bildik AE, Hubbe MA, Gurboy KB. Alkyl ketene dimer (AKD) sizing of paper under simplified treatment conditions[J]. Tappi Journal, 2016, 15(8): 545-552.
[14] Seppanen R, Tiberg F, Valignat MP. Mechanism of internal sizing by alkyl ketene dimers (AKD): The role of the spreading monolayer precursor and autophobicity[J]. Nordic Pulp & Paper Research Journal, 2000, 15(5): 452-458.
[15] ZHENG G L. Study on Preparation of Superhydrophobic/Oil-wear-resistant Cotton Fabric by Polydopamine and Its Properties [D]. Jiangnan University, 2020.
[16] Ap A, Mk A, Mr B, et al. Magnesium stearate-coated superhydrophobic sponge for oil/water separation: Synthesis, properties, application[J]. Separation and Purification Technology, 2020(251): 117105-117114.
[17] ZHAO D D. Preparation and Properties of Inorganic/Organic Composite Superhydrophobic Wear Resistant Coating [D]. North University of China, 2018.
[18] Wei Y, Cao R, Zhang X, et al. Air-Permeable and washable paper-based triboelectric nanogenerator based on highly flexible and robust paper electrodes[J]. Advanced Materials Technologies, 2018, 3(11): 178-187.
[19] SUN H Y, YAN Y S, SHEN Z J, et al. Preparation and washing application of Gleditsia sinensis extract [J]. Dyeing and Finishing Technology, 2019, 41(12): 32-36. (in Chinese)
[20] SHI S W, SHEN Y, WANG L M, et al. Study on the resistance of Parser fiber to acid and alkali [J]. Journal of Textile Science and Technology of Shanghai, 2014, 42(1): 15-19.
[21]TANG Y Y, QIAN J F, Yu C W. Study on acid and alkaline resistance of domestic and Bangladesh jute [J]. Guangxi Textile Science and Technology, 2009, 38(1): 25-28.
[22] Baley C, Le Duigou A, Morvan C, et al. Tensile properties of flax fibers, Handbook of Properties of Textile and Technical Fibres[M]. 2018: 275-300.
[23] Baley C, Gomina M, Breard J, et al. Specific features of flax fibres used to manufacture composite materials[J]. International Journal of Material Forming, 2018, 12(6): 1023-1052.

Study on the Aggregation Performance of Nitrogen and Oxygen Gas Based on Permanent Magnet Gradient Magnetic

Zunhai Man, Keli Zhao*, Yuqi Liu and Aiwei Yu

School of Mechanical and Aerospace Engineering, Jilin University, China

Email: zhaokl@jiu.edu.cn

Abstract. Oxygen and nitrogen in the air have different magnetic properties in a magnetic field. Oxygen and nitrogen are paramagnetic gases and diamagnetic gases, and the magnetizing force received in a gradient magnetic field is in opposite directions. Based on the magnetization theory and simulation experiment analysis, this paper uses the difference in the direction and magnitude of the magnetization force of oxygen and nitrogen in a gradient magnetic field to study the force state of oxygen and nitrogen molecules in the input air in a gradient magnetic field and the oxygen mass fraction of the output gas to explore effective ways to obtain oxygen-enriched gas. The results of this study can provide oxygen-enriched gas for operations such as oxygen-enriched combustion technology and sewage treatment engineering, and improve operational efficiency.

1. Introduction

High-gradient magnetization separation technology is widely used in ore screening, separation of magnetic immobilized enzymes, and separation of red blood cells. The use of high-gradient magnetization separation technology to enrich oxygen based on different magnetic properties of oxygen and nitrogen is the main direction of current research. With the advancement of science and technology, the application of oxygen-enriched gas is now more widely used, and oxygen-enriched combustion technology is becoming more and more mature. The installation of magnetized oxygen-enriching mechanisms on internal combustion engines of automobiles and ships has been applied abroad. In addition, oxygen-enriched gas is also widely used in sewage treatment, indoor oxygenation and refreshing [1].

2. Mechanism of the Magnetizing Force of Gas Molecules in a Gradient Magnetic Field

According to the Helmholtz free energy, combined with Landau's derivation [2], the magnetic field force experienced by a unit volume of gas molecules F can be expressed as:

$$F = \frac{1}{2}\nabla\left[H^2\rho(\frac{\partial\mu}{\partial\rho})\right] - \frac{1}{2}H^2\nabla\mu + \mu\left[j \times H\right] \quad (1)$$

In the formula, H is the magnetic field strength, ρ is the molecular density, μ is the magnetic permeability of the magnetic molecule, and j is the current density.

According to the magnetization law and electromagnetic knowledge of magnetic media, the magnetic permeability μ and volume susceptibility χ can be expressed as:

$$\mu = \mu_0 \cdot \mu_r, \mu_r = 1 + \chi, \chi = \rho\chi_m \quad (2)$$

In the formula, μ_0 is the vacuum permeability, μ_r is the relative permeability of the magnetic molecule, χ is the volume susceptibility of the magnetic molecule, and χ_m is the mass susceptibility.

The Lorentz force experienced by gas molecules per unit volume is zero. Substituting equation (2) into equation (1), the magnetic field force experienced by gas molecules per unit volume is:

$$F = \frac{1}{2}\mu_0 \rho \chi_m H \nabla H \tag{3}$$

The magnetic field intensity and the magnetic induction intensity have the following relationship: B=μH, where B is the magnetic induction intensity generated by the magnetic field. Using the relationship between the two, the magnetizing force received by a single gas molecule can be obtained as formula (4):

$$f = \frac{1}{2n\mu_0}\rho \chi_m B \nabla B \tag{4}$$

In the formula, $B\nabla B$ is the magnetic induction intensity gradient of the non-uniform magnetic field, ***n*** is the number of molecules per unit volume

3. Force Analysis of Gas Molecules in Gradient Magnetic Field

3.1. Performance Analysis of Gradient Magnetic Field Magnetic Induction Intensity between Permanent Magnets

In this paper, the gradient magnetic field generated by two permanent magnets is used to explore the gas magnetization performance. Two permanent magnets are magnetized in the same direction, with opposite poles facing each other. The arrangement of two permanent magnets is shown in figure 1. Where L is the length of the permanent magnet, W is the width of the permanent magnet, h is the thickness of the permanent magnet, δ is the air gap between the two permanent magnets, L=78mm, W=38mm, h=30mm,δ=3mm.

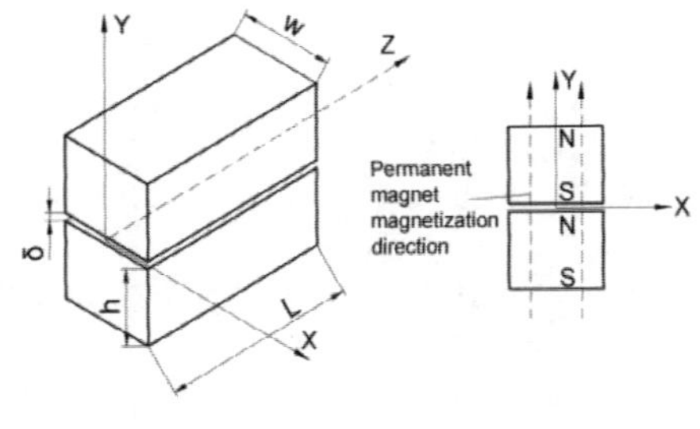

Figure 1. Arrangement of two permanent magnets.

Using Maxwell software to simulate and simulate, the distribution of magnetic induction intensity (vector diagram) generated around the permanent magnet is shown in figure 2. It can be seen from figure 2 that the magnetic flux density in the gap between the two permanent magnets is changing. In order to understand its internal changes more clearly, the XZ plane magnetic induction intensity (scalar) cloud diagram is obtained by further simulation as shown in figure 3. It can be seen that the middle area is a uniform magnetic field. Due to the edge effect, the magnetic induction intensity of the peripheral area decreases, forming a gradient magnetic field.

On the XZ plane, X = 0mm, 16mm, 17mm, 18mm, respectively, and the change curve of the magnetic induction intensity corresponding to the change in the Z direction is shown in figure 4. In the interval of 8mm<Z<70mm, the magnetic induction intensity B is almost unchanged; in the interval of Z≤8mm and Z≥70mm, the magnetic induction intensity B shows a gradient change. On the XZ plane, when X takes different values, the magnetic induction intensity change curve corresponding to the change in the Z direction has similar changes. Therefore, a gradient magnetic field is formed in this area [X=-19mm~19mm, Y=0mm, Z=0mm~8mm]. In the same way, the other three areas around the area of figure 3 are also gradient magnetic fields. Due to the different directions of the magnetic induction intensity gradient of the gradient magnetic field in the four regions, in order to facilitate the distinction, we named the four regions as gradient magnetic fields 1, 2, 3, and 4, as shown in figure 3.

3.2. Analysis of the Magnetizing Force of Oxygen and Nitrogen Molecules in a Gradient Magnetic Field

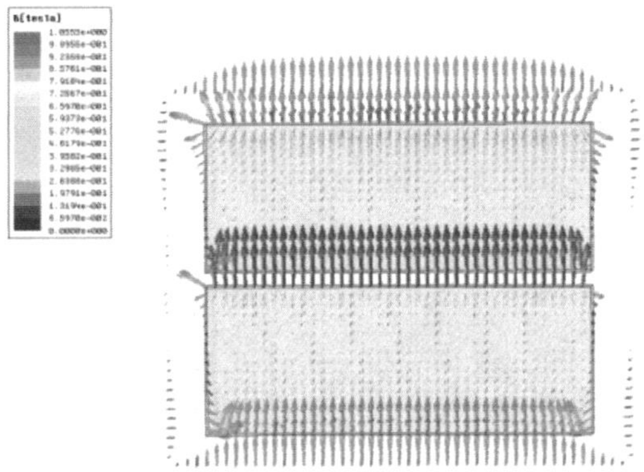

Figure 2. Magnetic induction intensity produced by permanent magnets (vector diagram).

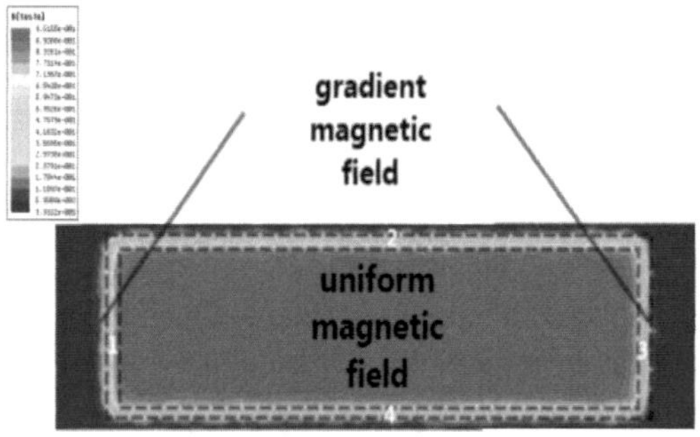

Figure 3. The XZ plane magnetic induction intensity.

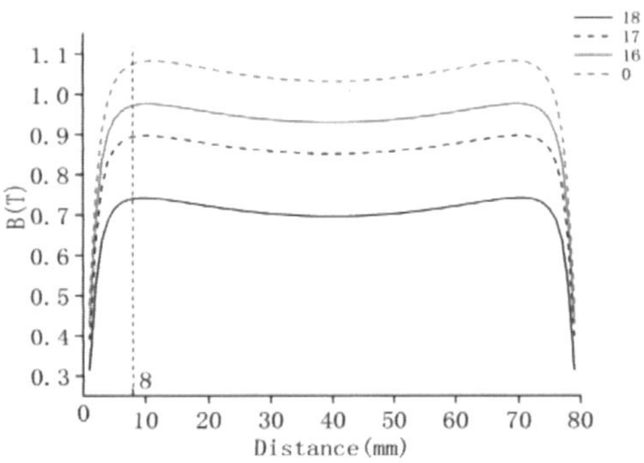

Figure 4. The change curve of magnetic induction intensity corresponding to the change of Y=0mm, X takes different values and changes in the Z direction.

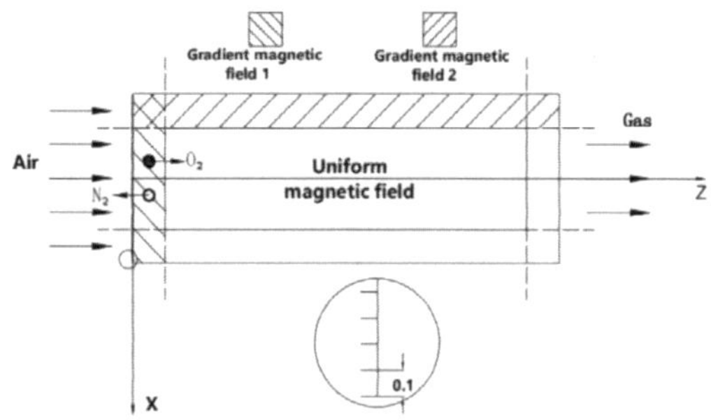

Figure 5. Schematic diagram of the force of oxygen and nitrogen molecules in a gradient magnetic field.

Oxygen molecules and nitrogen molecules have different magnetic properties: oxygen molecules are paramagnetic, and their mass magnetic susceptibility $\chi_m>0$, the direction of the magnetizing force received by oxygen molecules is the direction of increased magnetic induction; nitrogen molecules are diamagnetic, and their mass magnetic susceptibility $\chi_m< 0$, the direction of the magnetizing force received by the nitrogen molecules is the direction in which the magnetic induction intensity decreases. When air is input from the position shown in figure 5, the force of oxygen molecules and nitrogen molecules in the gradient magnetic field in the edge area is shown in figure 5.

Usually a gradient magnetic field formed by two permanent magnets is used, the magnetic induction intensity is usually between 0.1-2T, and the magnetic induction intensity gradient is 0-200T/m. Select the magnetic induction intensity as 1T and the magnetic induction intensity gradient as 100T/m as the prerequisites for the calculation. The mass magnetic susceptibility of oxygen and nitrogen are $+43.43\times10^{-6}m^3/kg$,$-15.08\times10^{-8}m^3/kg$ respectively[3], combined with formula

(4), we can get, the final calculation is that the oxygen and nitrogen molecules are subjected to The magnetizing force is 1.835×10^{-22}N, 5.581×10^{-25}N .

When the air passes through the gap between the two permanent magnets, the oxygen and nitrogen molecules in the air will not only be magnetized, but also be affected by their own gravity, buoyancy, and other forces. After calculation, it can be obtained that the gravity and buoyancy of oxygen molecules is 5.22×10^{-25}N and 4.72×10^{-25}N respectively; the gravity and buoyancy of nitrogen molecules is 4.56×10^{-25}N and 4.89×10^{-25}N respectively. The gravity and buoyancy of oxygen and nitrogen molecules are almost equal, and their combined force is much lower than the magnetizing force they are subjected to[4].

4. Oxygen and Nitrogen Molecules Pass Performance Analysis in a Gradient Magnetic Field

4.1. Analysis on the Movement Characteristics of Oxygen and Nitrogen Molecules in Gradient Magnetic Field

In solving the magnetization movement of oxygen and nitrogen molecules, the micro-element method is used for processing. In a short time, the product of the magnetic induction intensity and the magnetic induction intensity gradient can seem to be constant.

The magnetic induction intensity gradient around the permanent magnet is not uniformly distributed, but it can be approximately regarded as a uniform gradient magnetic field in the instantaneous region of each micro-element, and the trajectory of the gas molecule movement can be obtained as follows [5]:

$$\begin{cases} \Delta x = u_t \cdot \Delta t + \dfrac{1}{2} a_x \cdot \Delta t^2, u_{t+\Delta t} = u_t + a_x \cdot \Delta t \\ \Delta y = v_t \cdot \Delta t + \dfrac{1}{2} a_y \cdot \Delta t^2, v_{t+\Delta t} = v_t + a_y \cdot \Delta t \\ \Delta z = w_t \cdot \Delta t + \dfrac{1}{2} a_z \cdot \Delta t^2, w_{t+\Delta t} = w_t + a_z \cdot \Delta t \end{cases} \tag{5}$$

Since the influence of gas buoyancy and gravity are ignored, only the magnetizing force is considered, and the influencing factors of the molecular motion only need to consider the magnitude of the magnetizing force in different directions. Through simulation and theoretical analysis, as shown in figure 1, the magnetizing force in the Y direction is very small and can be ignored, that is, the magnetizing force has no effect on the movement of the molecules in the Y direction, and the magnetizing force is mainly reflected in the Z and X directions. When the air is input from the position shown in figure 5, it is mainly affected by the gradient magnetic field 1, that is, it is mainly affected by the magnetizing force in the Z direction. For the convenience of research, we set the initial velocity of the out flowing molecules to be along the Z direction, that is, the gas molecules in the gradient magnetic field 1 region can be seen as the superimposed effect of uniformly accelerating linear motions in multiple micro-element intervals. In the region of gradient magnetic field 1, the magnetizing force of nitrogen molecules is opposite to the speed of nitrogen molecules in the Z direction. The nitrogen molecules will decelerate, and the speed of some molecules will drop to zero, and then the speed will be discharged in the reverse direction, making it along the Z axis. The content of oxygen in the output gas increases to form an oxygen-rich gas.

It can be seen from figure 4 that at different positions of the magnetic field, the magnetic induction intensity is different. Figure 5 shows the B∇B change curve of the magnetic induction intensity and the magnetic induction intensity gradient product B∇B corresponding to changes in the Z direction with Y=0mm and X being 0mm, 16mm, 17mm, and 18mm. It can be seen from figure 9 that when X takes different values, the product B·∇B of the magnetic induction intensity and the magnetic induction intensity gradient changes differently. X=0mm, the value of B·∇B is larger. When nitrogen molecules are input from positions with different X values on the left side of the permanent magnet, the magnetizing force they receive is different. When entering from the position of X=0mm, the magnetizing force of nitrogen molecules is greater.

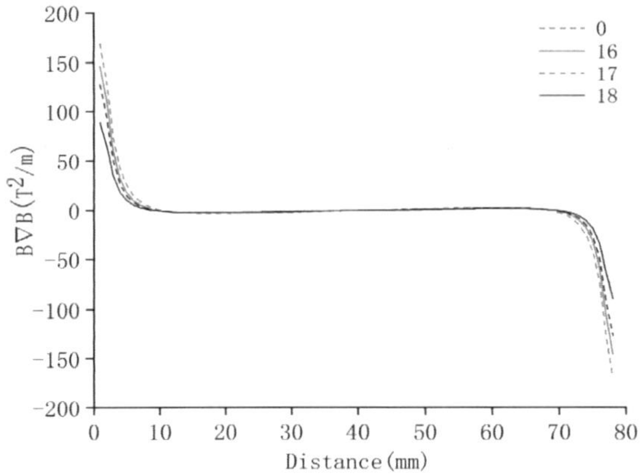

Figure 6. Y=0mm, X takes different values, the $B\nabla B$ change curve corresponding to changes in the Z direction.

4.2. Using MATLAB Software to Analyze the Oxygen Mass Fraction of Oxygen-Enriched Gas

In order to analyze the passing rate of nitrogen and oxygen gas molecules in the gradient magnetic field, set the gas input position to [X=-19mm~+19mm (interval 0.1, 381 analysis points in total), Y=0mm, Z=0mm], initial velocity v0=8m/s (the direction is along the positive Z axis) as shown in figure 5. Based on the ratio of velocity to magnetic field value change λ($\lambda = \frac{v}{B\nabla B}$,the unit of λ is $m^2 \cdot s/T^2$). Using MATLAB software to calculate the oxygen mass fraction of the output oxygen-enriched gas in figure 5 (oxygen mass fraction: the percentage of the oxygen mass to the total mass of the gas).

The analysis shows that the nitrogen molecules at different positions along the edge of the gas introduced into the magnetic block show different states due to the difference in the product of the magnetic induction intensity and the magnetic induction intensity gradient of the gradient magnetic field. In the area of [X=15.8mm~19mm and X=-15.8mm~-19mm, Y=0mm, Z=0mm~8mm], although the gradient magnetic field hinders the effect, the nitrogen molecules all pass through this area; and in [X =-15.8mm~15.8mm, y=0mm, Z=0mm~8mm], the nitrogen molecules are discharged out of the permanent magnet because the magnetizing force it receives is greater than its inertial force, without entering the gap between the magnet blocks. The result of calculation and analysis is that the nitrogen molecules that have not entered the gap between the magnetic blocks account for 81.1% of the total nitrogen molecules introduced into the gas.

When oxygen molecules pass through the gradient magnetic field formed by the permanent magnet, they accelerate first, then at a constant speed, and then decelerate. Due to the symmetry of the magnetic induction intensity, the gradient magnetic field does not promote or hinder the passage of oxygen molecules. The gradient magnetic field has an obstructive effect on nitrogen molecules, and has a propelling effect on oxygen molecules, so that the nitrogen content in the output gas is reduced, the oxygen content is increased, and oxygen-enriched air is formed. Based on the above-mentioned velocity and magnetic field change ratio λ value, it is calculated that the mass fraction of oxygen in the oxygen-enriched gas output after the gradient magnetic field is increased to 59.02%.

Using the same principle to obtain the value of v_0=5m/s~9.5m/s with an interval of 0.5m/s, the results of the mass fraction of oxygen in the output gas are shown in table 1.

Table 1. Statistics table of oxygen mass fraction of output gas at different initial speeds.

Round	V_0 (m/s)	$B\cdot\nabla B(T^2/m)$	$\lambda(10^{-2}\cdot m^2\ s/T^2)$	Oxygen mass fraction (%)
1	4.58	150	3.05	100
2	5.0	150	3.33	88.81
3	5.5	150	3.67	83.03
4	6.0	150	4.0	77.95
5	6.5	150	4.33	73.46
6	7.0	150	4.67	69.46
7	7.5	150	5.0	64.22
8	8.0	150	5.33	59.02
9	8.5	150	5.67	53.45
10	9.0	150	6.0	46.67
11	9.5	150	6.33	37.11
12	9.84	150	6.56	21

Figure 10 is a schematic diagram of the oxygen mass fraction in the output gas changing with λ.

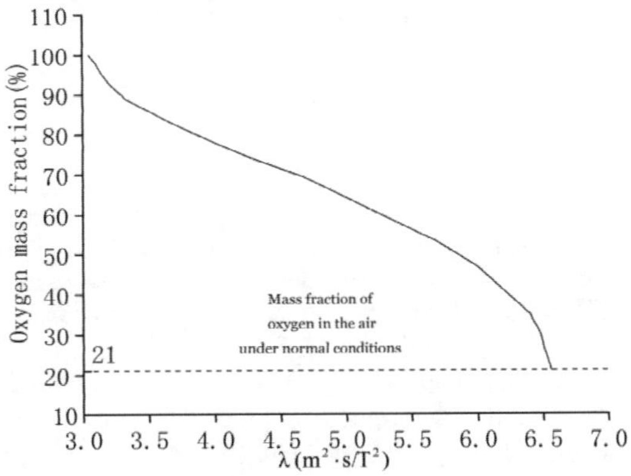

Figure7. Schematic diagram of the oxygen mass fraction in the output gas changing with λ.

5. Conclusion

(1) When $\lambda \leq 3.05$, the oxygen mass fraction in the output oxygen-rich gas after the gradient magnetic field magnetization is 100%; when $\lambda \geq 6.56$, the output gas the oxygen mass fraction is 21%.When $3.05 < \lambda < 6.56$, as λ increases, the oxygen mass fraction of the output oxygen-rich gas after being magnetized by the gradient magnetic field will decrease.

(2) In order to increase the mass fraction of oxygen in the output gas , the goal can be achieved by increasing $B\cdot\Delta B$ or reducing the initial velocity v_0 of gas molecules.

References

[1] Zhu Ch G 1996 The effect of transverse non-uniform magnetic field On atoms *Journal of Huaibei Coal Teachers College (Natural Science Edition)* (01).

[2] Landau L D, Lifshitz E M 1960 *Electrodynamics of Continuous Media* Oxford: Pergamon.

[3] Wang X K, Chen Zh J, Hong G H, Meng Zh J 2005 Research on the mechanism and experimental scheme of high gradient and strong magnetic field for oxygen accumulation and nitrogen accumulation Journal of Shenyang Institute of Technology (Natural Science Edition) (Z1).

[4] Li F Ch, Wang L, Wu P, Zhang Sh P 2011 Study of flow and diffusion behavior of oxygen in gradient magnetic field *Journal of Engineering Thermo Physics* (11).

[5] Asako Y, Suzuki Y 2007 Oxygen separation/enrichment from atmospheric air using magnetizing force *Journal of Fluids Engineering* 129(4): 438-445.

Simulation Analysis of Thermoplastic Composites on Suspension Brackets of Electric Commercial Vehicles

Haijun Zhao [1] *, Haixia Wang[2] *, Naibin Zhai [1], Hongliang Li [3], Zhiqiang Wang [1], Feng Wang [1]

[1]College of Automobile and Transportation, Tianjin University of Technology and Education, Tianjin, 300222, China

[2]Changshu Institute of Technology, Changshu, Jiangsu, 215500

[3]Automotive Engineering Research Institute, China Automotive technology & Research Center, Tianjin, 300300, China

*Corresponding author's e-mail:hjzhaoly@lit.edu.cn

Abstract. Lightweight is more important in electric commercial vehicles, the static simulation analysis is first carried out on the suspension system bracket using the material of steel under three working conditions in this paper. Taking the suspension system bracket connection plate as an example, the bracket connecting plate model of the suspension system is established with thermoplastic composite materials, which is nine-layer three-dimensional model. The pre-processing, solving and post-processing procedures of the stress analysis of the part are introduced, and the stress distribution of different layers is obtained. Results show that the maximum stress of the steel bracket of the suspension system under three working conditions of triple gravity, acceleration of gravity adding steering inertia, acceleration of gravity adding braking inertia are 189.22, 92.48, and 156.9Mpa for steel, which are all less than the ultimate stress of 235Mpa. The maximum stress of each layer of the thermoplastic composite connecting plate meets the design requirements. In the case of the same ply angle, the bottom layer will bear less and less stress, the base layer will bear very little stress, and the main stress will be borne by the reinforcing material, support is provided for the application of thermoplastic composite materials on automobile chassis.

1. Introduction

Body weight of traditional energy vehicles consumes about 70% of fuel. If vehicle weight is reduced by 10%, fuel consumption can be reduced by 6%-8% [1]. Lightweight is an important method to improve fuel economy and reduce pollutants and carbon emissions. Driven by national policies and regulations and the fierce market competition environment, lightweight design of automobile has become an inevitable trend in automobile development. The body weight reduction design can be realized through three dimensions, which are structural optimization design, the use of advanced manufacturing technology and new lightweight materials. It is very limited for automotive structural design to reduce weight through structural optimization, and weight reduction effect is significant using composite material body [2]. Thermoplastic composite materials not only have the advantages of light weight and high strength, composite materials can be integrated, which greatly reduces the number of parts and fasteners. Thermoplastic composite materials are used in automobile bodies to improve collision energy absorption, vibration damping and fatigue resistance [3]. Future vehicles are green cars that adapt to environmental protection. Therefore, it is inevitable to mention the environmental protection awareness of composite materials. Thermoplastic composite materials have characteristics, low emissions during the production process, and in addition to the advantages of general new composite materials. It has the advantage of being recyclable, so it is in line with the future development of environmentally friendly cars [4]. Due to the pressure of environmental protection, motor vehicles in almost all cities, including counties, are restricted to driving only in some time. At present, the distribution of goods in cities is mainly realized by electric commercial vehicles. The cruising range of electric vehicles is directly related to the quality of the whole vehicle. Therefore, the meaning is greater.

In order to reduce the weight and maintain the performance of electric commercial vehicles, the static simulation analysis is first carried out in this paper on the suspension system bracket using steel material under three working conditions. Taking the suspension system bracket connection plate as an example, and the system support connecting plate is modelled using the thermoplastic composite material for suspension, and the processes are introduced in detail on the pre-processing, solving, and post-processing of the three-dimensional finite element simulation analysis, and results are obtained on the stress distribution of different layers.

2. Model establishment and working conditions

2.1. Building a model

In the field of finite element, Hypermesh is a convenient meshing software, and Ansys Workbench is a multi-physics coupling analysis software with excellent performance. Its powerful calculation accuracy and multi-physics coupling are favoured by most automakers. The combination of the strong and the strong is currently the mainstream analysis method for most CAE analysis. Import the suspension bracket digital model into Hypermesh for mesh division. The mesh is required to use a hexahedral mesh, try to avoid mesh distortion, and the part of the mesh that involves a large curvature change is appropriately refined, and the divided mesh is imported The three-dimensional finite element model shown in the figure 1 in Ansys Workbench has an average mesh quality of 0.94 and a full score of 1, which meets the requirements for meshing quality.

Figure 1. Three-dimensional finite element model after meshing.

3. Static analysis of steel materials

3.1. Building a model

The static analysis of the mounting system bracket simulates and calculates three working conditions respectively. The first working condition simulates the force when 3 times the acceleration of gravity, the negative direction of the Z axis in the vehicle coordinate system, and the second working condition simulates the force when the car is turning. That is, acceleration of gravity adding steering inertia, the third working condition simulates the force situation when the car is braking, that is, acceleration of gravity adding braking inertia. The first working condition, a uniform load of 350×3 =1029N perpendicular to the body downward is applied at A and B respectively to simulate the weight of the motor under 3 times the gravitational acceleration, and 3 times the gravitational acceleration is applied. That is, 9806.6mm/s^2×3=29420mm/s^2, fixed with 8 bolt holes.

The second working condition is that applying 350N uniformly distributed loads perpendicular to the vehicle body downwards at A and B to simulate the weight of the motor, respectively apply a gravity acceleration of 9806.6 mm/s^2 and 9806.6 in the Y direction of the vehicle coordinate system. The acceleration of mm/s^2 is used to simulate the steering inertia, and the fixed 8 bolt holes are similar to the point A in the figure 2. The third working condition is that applying 350N uniformly distributed loads perpendicular to the vehicle body downwards at A and B to simulate the weight of the motor, respectively apply a gravity acceleration of 9806.6 mm/s^2 and 9806.6 in the X direction of the vehicle coordinate system. The acceleration of mm/s^2 is used to simulate the braking inertia, and the fixed 8 bolt holes are similar to the point A.

4. Static analysis of steel materials

4.1. Static analysis on the first working condition

After finishing editing the boundary conditions, perform static analysis through Workbench. The stress cloud diagram under the first working condition is shown in the figure 2. The maximum stress is 189.22MPa, which is located at the mounting hole of the body connecting plate. MAX shows the position in the figure. The maximum stress of the position is less than the yield limit of the selected material 235MPa, which meets the design requirements.

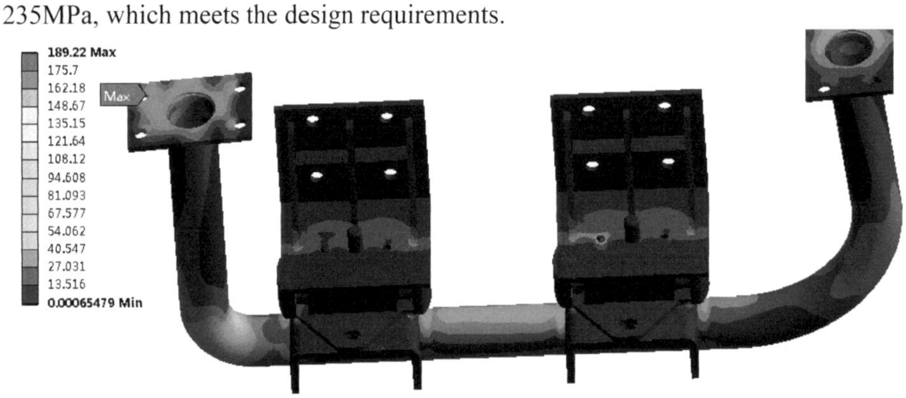

Figure 2. Static analysis cloud diagram under the first working condition

4.2. Static analysis on the second working condition

The stress cloud diagram of the suspension bracket under the second working condition is shown in the figure 3, where the maximum stress is 92.483MPa, which is located at the mounting hole of the body connecting plate, and the maximum stress at the dangerous position is less than the yield limit of the selected material 235MPa, which meets the design requirements.

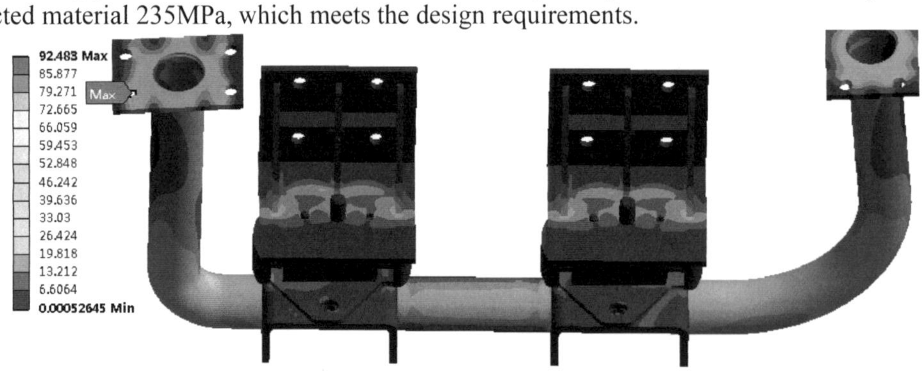

Figure 3. Static analysis cloud diagram under the second working condition

4.3. Static analysis on the third working condition

The stress cloud diagram of the suspension bracket under the third working condition is shown in the figure 4, where the maximum stress is 156.9MPa, which is located at the mounting hole of the body connecting plate, and the maximum stress at the dangerous position is less than the yield limit of the selected material 235MPa, which meets the design requirements.

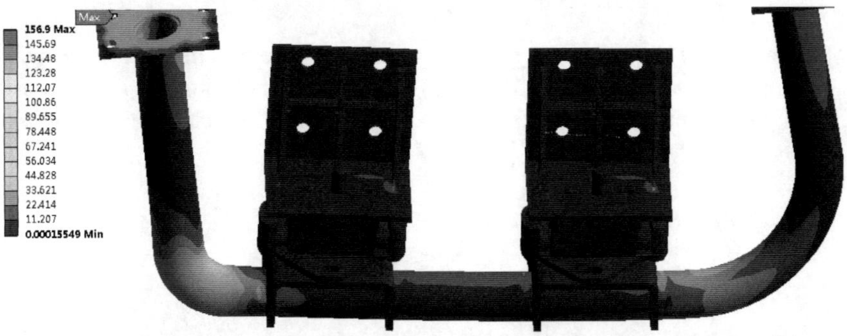

Figure 4. Static analysis cloud diagram under the third working condition

Results show that the maximum stresses of the suspension system support steel in the three working conditions of triple gravity, gravitational acceleration adding steering inertia, gravitational acceleration adding braking inertia are 189.22, 92.48, and 156.9Mpa respectively, which are all less than The ultimate stress is 235Mpa, and the margin is relatively large. The currently selected materials meet the strength design requirements. In order to ensure driving safety, higher-grade bolts can be used for the installation of the motor lower beam, or to enhance the strength and precision of the body link plate. Methods such as the design of higher-strength composite materials.

5. Simulation analysis for composite material

Through static analysis, it can be concluded that under the three working conditions, the maximum stress of the suspension system is concentrated at the mounting hole of the body connecting plate. In order to improve the safety factor and ensure driving safety, the use of composite materials can meet the requirements of high strength and light weight. The purpose of quantification.

5.1. Analysis program

The 3D simulation analysis process of the thermoplastic composite material of the bracket connecting plate of the suspension system is the same as other materials, including the pre-processing part, solution, and post-processing to view the results. Due to the material processing properties, the pre-processing and post-processing processes are much more complicated. The pre-processing is roughly divided into 23 steps, the solution is simple, click the solve button to solve it directly, and post-processing is also more complicated, and it includes 8 steps.

5.2. Composite plies

The lay-up method of thermoplastic composite materials is basically processed in accordance with the processing process, and at the same time according to certain rules [5], it is divided into nine layers, the sixth layer is the matrix with a thickness of 2mm, and the rest are fibre reinforced layers with a thickness of 0.2mm, due to the properties of the material, the maximum tensile stress and maximum compressive stress of the matrix and the fibre reinforced layer are very different. Due to the different layup angles, the maximum tensile stress and maximum compressive stress of the fibre-reinforced layer with the same thickness are also different.

5.3. Boundary conditions

Through static analysis, the magnitude of the force on the connecting plate in all directions can be obtained. Import the connecting plate model into Ansys ACP, as shown in Figure 9, apply a force of -133.3N in the X direction, 399.06N in the Y direction, and -1218.4N in the Z direction to constrain all the degrees of freedom of the four mounting holes.

5.4. Result analysis

The stress cloud diagram of the sixth layer is shown in the figure 5. This layer is a matrix material, where the maximum tensile stress is 0.34Mpa less than the maximum allowable tensile stress of 1.6MPa, and the maximum compressive stress is 0.64Mpa less than the maximum allowable compressive stress of 1.1MPa. Meet the design requirements.

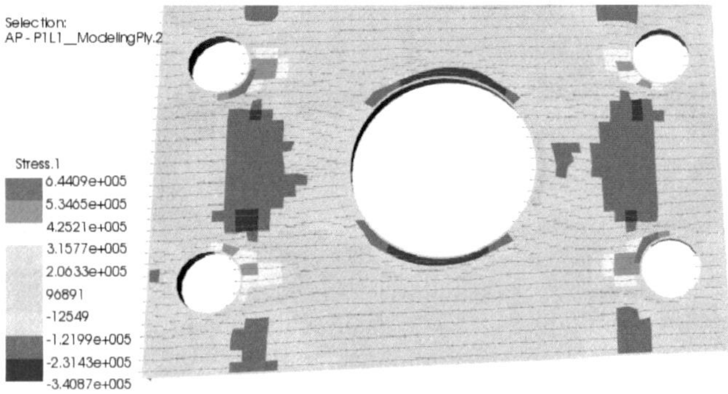

Figure 5. Stress cloud map of the sixth layer

The first layer of stress cloud is shown in the figure 6. This layer is made of UD-T700 laminated at -45°, and the direction of the layer is shown by the arrow in the figure. The maximum tensile stress is 993.19Mpa which is less than the allowable tensile. The maximum stress is 1500MPa, and the maximum compressive stress is 363.69Mpa, which is less than the maximum allowable compressive stress of 700MPa, which meets the design requirements. The second layer is made of UD-T700 layered at 45°. The maximum tensile stress is 892.81 MPa, which is less than the maximum allowable tensile stress of 1500 MPa, and the maximum compressive stress is 327.99 MPa, which is less than the maximum allowable compressive stress of 700 MPa, which meets the design requirements. The third layer is made of UD-T700 layered at 90°. The maximum tensile stress is 405.12MPa, which is less than the maximum allowable tensile stress of 1500MPa, and the maximum compressive stress is 255.66MPa, which is less than the maximum allowable compressive stress of 700MPa, which meets the design requirements. The fourth layer is made of UD-T700 layered at -45°. The direction of the layering is as shown by the arrow in the figure. The maximum tensile stress is 448.53MPa, which is less than the allowable maximum tensile stress of 1500MPa, and the maximum compressive stress is 231.73 MPa. The maximum compressive stress is less than 700 MPa, which meets the design requirements. The other four lays all meet the design requirements too.

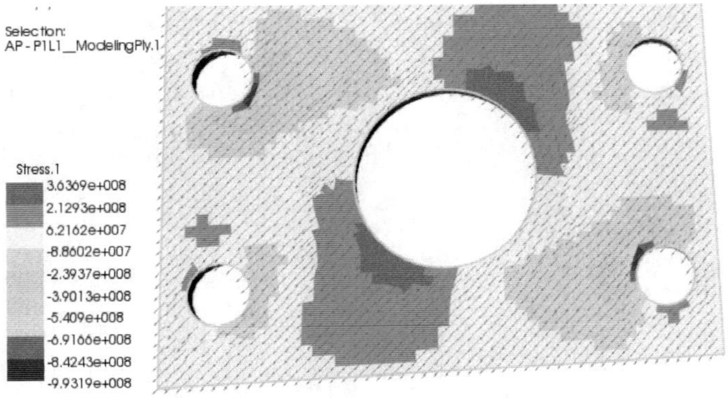

Figure 6. Stress cloud of the first layer

The analysis conclusions are as follows:

(1) The maximum stress of each layer of the composite material meets the design requirements.

(2) Under the action of working condition 1, the paving angle of 90 degrees (the third, seventh, eighth, and ninth floors) has better stress and lower stress.

(3) By comparing the static analysis cloud images of the seventh, eighth, and ninth floors, under the same ply angle, the lower the bottom level bears less and less stress.

(4) The stress borne by the base layer is very small, and the main stress is borne by the reinforcing material.

6. Conclusions

(1) The maximum stress of the suspension system bracket steel under three working conditions of triple gravity, gravitational acceleration adding steering inertia, and gravitational acceleration adding braking inertia are 189.22, 92.48, and 156.9Mpa respectively, which are all less than the ultimate stress of 235Mpa.

(2) A nine-layer three-dimensional model of the thermoplastic composite material suspension connecting plate is established, and the pre-processing, solving and post-processing processes of the thermoplastic composite material simulation analysis are introduced.

(3) The maximum stress of each layer of the thermoplastic composite material meets the design requirements. In the case of the same ply angle, the lower the layer is subjected to less and less stress.

(4) The stress borne by the thermoplastic composite matrix layer is very small, and the main stress is borne by the reinforcing material.

Acknowledgments

This research was financially supported by National Natural Science Foundation of China (U1604141) and China Scholarship Council (CSC). The authors would like to acknowledge suggestion from the colleges in the past laboratory and several electric car companies.

References

[1] Xiong F., Wang D.F., etal. Lightweight optimization of the side structure of automobile body using combined grey relational and principal compnent analysis. Struct Multidisc Optim. 2018, 57: 441-461.

[2] Tetsuya Yamamoto, Katsumasa Uematsu, Toshihira Irisawa, Yasuhiro Tanabe. A polymer colloidal technique for enhancing bending properties of carbon fiber-reinforced thermoplastics using nylon modifier[J]. Composites Parts, 2018, (112):250-254.

[3] Mayer D, Bein T, Buff H, etal. Enhanced lightweight design by composites-results of the EU project ENLIGHT[J]. Journal of Reinforced Plastics and Composites. 2018, 37(9):1217-1224.

[4] Kang Q., Gu P.Y., Zuo S.G.. A research on design principles for electric drive assembly mounts of pure electric vehicles. Automotive Engineering. 2019, 41(11):1236-1242.
[5] Zhao H.J., Zhu W.C., Li H.L., etal. (2018) Determination and performance analysis on miniature and electric commercial vehicle frames of carbon fiber materials in the laying mode. 4th International Conference on Energy Equipment Science and Engineering. Xi'an. pp. 1-6.

Investigations on oxidation resistance of phase change materials for heat storage

Wei Zhang[1,a], Xiao Meng[2,b], Wenjie Liu[3,c], Baorang Li[4,d*]

School of Energy, Power and Mechanical Engineering, North China Electric Power University, Beijing, China

[a]email: 982953131@qq.com, [b]email: 1257409643@qq.com, [c]email: 1806863473@qq.com

[d]email: libr@ncepu.edu.cn

Abstract. Antioxidant graphite was prepared from $Al(H_2PO_4)_3$ solution by solution impregnation method. The as-prepared antioxidant graphite was further compounded with phase change materials for heat storage depending upon the form-stable technique. SEM with EDS analyses were used to characterize both the morphology and microstructure of the impregnated graphite while DSC were used to measure thermal properties of the composite. The results showed that the best impregnation mass ratio of $Al(H_2PO_4)_3$to graphite should be 2:1~3:1. In this case, the obtained composite heat storage material can realize certain oxidation resistance and obtain good heat storage performance after thermal cycling.

1. Introduction

Research on heat storage materials is one of the important strategies to solve energy problems. Graphite processes high thermal conductivity and is regarded as good additive, which is often used to compound with phase change materials for heat storage to improve their thermal performance. But it is well known that the additive graphite will be oxidized at the temperature higher than 450°C, which greatly limits its applications in high temperature heat storage materials [1-2]. Many researchers have been reported their investigations on the oxidation resistance of graphite materials, among which the solution impregnation is regarded as an effective way to improve the oxidation resistance of graphite materials at high temperature. For example, in 1997, T. DuRKI, A. PER proved the mixed acid solution of phenolic resin and boron can increase the oxidation stability of the sample surface [3]; Manocha et. al adopted the silica sol and SiOC sol as medium to impregnate C/C composites. After effective impregnation and heat treatment, C/C composites can show good antioxidant effect at temperatures above 1000°C [4]; In 2002, Wei-ming Lu et. al pretreated graphite materials with ozone and then impregnated them with phosphoric acid and dissolved aluminum hydroxide for the first time. Their results suggested the oxidation resistance of carbon can be improved significantly [5]. In this study, the additive graphite was first treated by solution impregnation method and then mixed with phase change materials using form-stable technique for improving high-temperature oxidation resistance of phase change materials for heat storage. It is no doubt that the related research has far reaching significance in the application of high temperature phase change materials. Our studies suggested graphite treated by solution impregnation technique has certain practical significance for the high temperature application of phase change materials.

2. Experiment

2.1. Experimental material

Materials used in the experiment include graphite（purity≥99%,flake size≤30μm）, Na_2CO_3（GR, purity≥99.9%）, K_2CO_3（GR, purity≥99.5%）, MgO（AR, purity≥99.9%）, H_3BO_3（AR, purity≥99.5%）, $Al(H_2PO_4)_3$（AR,50%w/w）. The composition of the immersion liquid is shown in Table 1 while the composition of each component in the obtained composite phase change heat storage material is shown in Table 2.

Table 1 Composition of impregnating solution

The number of Impregnation solution	solvent	solute	the amount of solvent	The amount of solute
1	H_2O	H_3BO_3	80ml	5g
2	$Al(H_2PO_4)_3$+H_2O	H3BO3	40ml+40ml	5g
3	$Al(H_2PO_4)_3$+H_2O	-	40ml+40ml	-
4	$Al(H_2PO_4)_3$+H_2O	H_3BO_3	40ml+40ml	20g
5（Initial sample）	-	-	-	-

Table 2 Chemical composition of heat storage material（wt%）

Na_2CO_3	K_2CO_3	MgO	MG（Modified graphite）
28.6	26.4	35	10

2.2. Sample preparation

Graphite with a weight of 4g were first added to the impregnation solutions presented in table 1, then the obtained solutions were mixed and heated at the impregnated temperature of 70°C. After effective mixture, the solutions were finally dried in an oven at 120°C. Before mixing with Na_2CO_3-K_2CO_3 eutectic salt and MgO, these dried samples were put into alumina crucible and heat treated in muffle furnace and then cooled to room temperature. The corresponding calcinations technique is described in Fig.1(a). Fig.1(b) is the sintering procedure for fabricating phase transition heat storage materials. A typical preparation route for heat storage materials is all the materials including eutectic salt, the impregnated products and MgO etc. were mixed together and pressed into a die with a diameter of 12.6mm, and further sintered according to techniques shown in Fig.1(b).

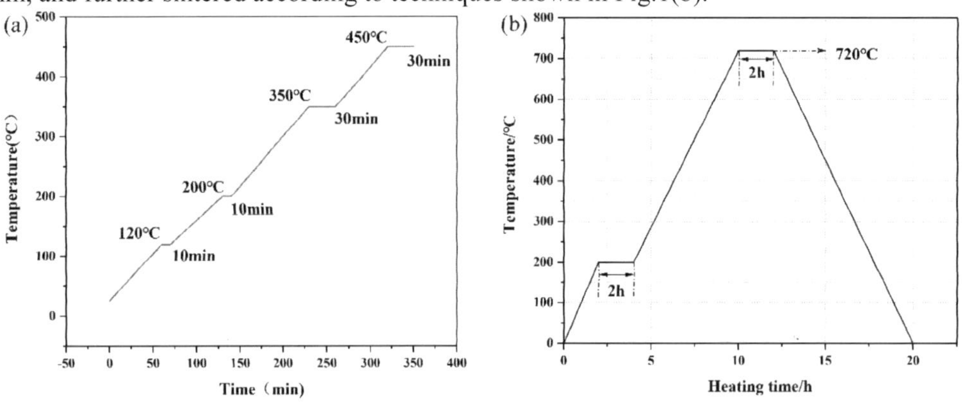

Figure 1 Heat treatment procedure and sintering procedure

2.3. Tests and characterization methods

The oxidation weight loss of the as-prepared heat storage materials was measured using an electronic balance (METTLERTOLEDO ML204). SEM combined with EDS (SU1080) were used to observe and analyze the possible microstructure evolutions of both the composite and coated graphite. DSC (STA 449) were used to measure the cycled thermal stability of heat storage materials.

2.4. Results and discussion

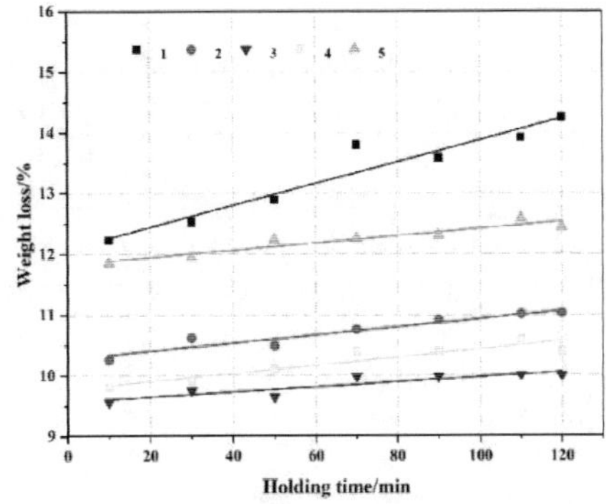

Figure 2 weight loss curves of the samples (the numbers are in Table 1) obtained under various conditions

Dependence of the oxidation weight losses obtained under various conditions on the holding times is shown in Fig. 2. It was reported that boric acid could be changed into B_2O_3 after dehydration. The formed B_2O_3 would become glass melt with low volatility, high viscosity and strong adhesion at high temperature. Therefore, it can adhere to the active sites of graphite to protect graphite from oxidation [6]. However, Fig. 2 suggests that the mass loss rate of the sample impregnated with high concentration boric acid solution alone is obviously higher than the untreated sample. The increased weight loss rates might be due to the following two reasons:

1) Few H_3BO_3 molecules and acid ions in the solution resulting from the lower solubility of H_3BO_3 cannot be inserted into the graphite layer effectively.

2) Part of the substance in the antioxidants will be lost due to the condensation reaction occurrence or volatilization.

The results in Fig.2 also indicate that the weight loss rates of all the samples obtained by impregnation with aluminum dihydrogen phosphate solution became decreased obviously. Aluminum dihydrogen phosphate can be converted to $Al(PO)_3$ at 300 °C, and further polymerized into aluminum metaphosphate, which usually becomes mesh-like shape and further plays a role of protection for graphite from oxidation[7]. Obviously, the best impregnation result can be achieved in aluminum dihydrogen phosphate. SEM observations on the samples with different impregnation ratios of aluminum dihydrogen phosphate and graphite are given in Fig.3.

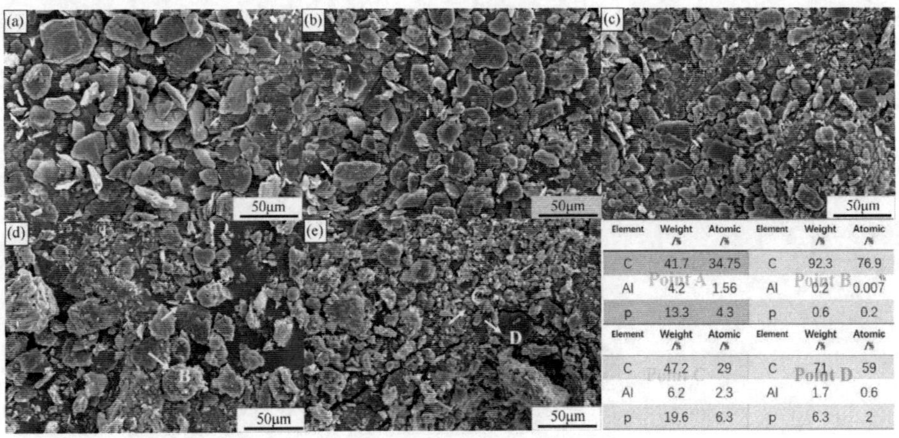

Figure 3 the impregnated samples obtained under various conditions (a)Initial graphite (b)$Al(H_2PO_4)_3$:C=0.5:1； (c) $Al(H_2PO_4)_3$:C =1:1； (d) $Al(H_2PO_4)_3$:C =2:1； (e) $Al(H_2PO_4)_3$:C =3:1

It can be seen from Fig.3 that the graphite flakes tend to become thicker due to the presence of the coating layer with the increased aluminum dihydrogen phosphate ratios. The increased thickness of the coating layer can reach the maximum value between 2:1 and 3:1. Therefore, it is reasonably concluded that the optimum ratio of aluminum dihydrogen phosphate to graphite should be between 2:1 and 3:1. For further proving this, the point scanning were carried out and the results were given in Fig.3.The point scanning data in Fig.3 revealed only part of the graphite in the sample is effectively coated in the case of the impregnation ratio of 2:1, as marked by point A shown in Fig.3 (d). The other part of the graphite isn't coated successfully, as shown in point B in figure (d). For the samples with the impregnation ratio of 3:1, the point scanning results suggested these samples could be effectively coated, as marked by points C and D shown in Fig.3 (e). These results further confirmed that the optimum impregnation ratio of aluminum dihydrogen phosphate to graphite is 2:1 ~ 3:1.

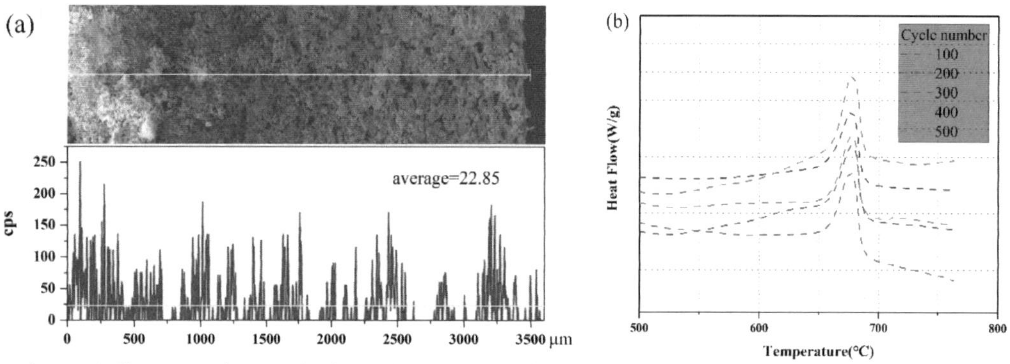

Figure 4 line scanning analysis on the cross-section of the sample (a) and DSC curves of the samples with various thermal cycle numbers(b)

Fig. 4 (a) shows the line scanning results on the cross-section of composite phase change material fabricated using the impregnated sample 3 (as shown in Fig.2) and Fig. 4 (b) is the DSC curves of the same sample with various thermal cycles. It can be seen that the average strength of C on the fracture surface of the sample doped with impregnated sample 3 is 22.85, which indicates good oxidation resistance. Meanwhile, almost unchanged values in latent heat suggest the stable thermal performance of the samples.

3. Conclusion

In this study, aluminum dihydrogen phosphate was used to impregnate flake graphite samples for the first time, and the optimum impregnation ratio was obtained as follows: the mass ratio of aluminum dihydrogen phosphate to graphite was 2:1 to 3:1. The impregnated graphite was combined with phase change material for achieving better oxidation resistance. The results show that the sample has good oxidation resistance and excellent thermal cycle stability.

Acknowledgments

This research was supported by 'the Fundamental Research Funds for the Central Universities' (NO.2021MS012).

References

[1] H L Zhang, X D Fang. (2014) Research progress of phase change heat storage materials and technologies [J]. Materials Guide, 28(13):26-32.
[2] L M Xue, T Jin. (2018) Research progress on oxidation resistance of graphite materials [J]. Carbon, (04):42-45.
[3] T. Ďurkić,A ,Perić,M. (1997) Boron and phosphorus doped glassy carbon: I. Surface properties[J]. Carbon, 35(10):1567–1572.

[4] L.M. Manocha,S, Manocha,Ketan B,Patel,P. Glogar. (2000) Oxidation behaviour of Carbon/Carbon Composites impregnated with siilicon oxycarbide[J]. Carbon, 38(10) :1481-1491.
[5] Welica and Siming Lu, D.D.L. Chung. (2002) Oxidation protection of carbon materials by acid phosphate impregnation[J]. Carbon, 40(8):1249-1254
[6] H Y Chen, T Zhang. (2015) Effect of boric acid complex salt on oxidation resistance of artificial graphite electrode [J]. carbon technology, 34(02):27-30.
[7] Y X Zhao. (2014) Effect of additive-phosphoric acid on properties of flexible graphite materials [J]. Nonmetallic Minerals, 37(06):17-18+25.

Effect of aging of foam rubber on properties and constitutive models

ZhaoQun Shao[1,a], Min Zhu[1,b*], HaoYu Wang[2,a], MuYi Li[2,a]

[1]College of Nuclear Science and Technology, Naval University of Engineering, Wuhan 430033

[2]Marine Equipment Major Project Center, Beijing, China, 100071

[a]email: 2297089907@qq.com

*Corresponding author: [b]email: zljg03@aeutlook.com

Abstract: This article reviews the effects of aging on properties and constitutive models of foam rubber materials. Firstly the change of aging indicators such as compression set and stress relaxation with time and temperature, and the influence of relative density, hardness of materials, the size and shape of the cell on the aging indicators are introduced in this article. Secondly, this article also introduces the strengths and application of commonly used hyper-elastic material constitutive models such as Arruda-Boyce model, Van der Waals model, Neo-Hookean model, Mooney-Rivlin model, Yeoh model, and Odgen model, and introduced the influence of aging on model parameters. Finally, the influence of aging on the model parameters is introduced.

1. Introduction

Foam rubber is a hyper-elastic material obtained from a rubber matrix through a foaming method. It has good shock absorption, heat insulation, sound insulation, and sealing properties. It is often used for positioning, shock absorption, anti-rotation, and other working conditions between important parts[1]. Foam rubber is subjected to tensile or compressive stress for a long time during use, and it may be exposed to temperature and humidity, which can lead to chemical and structural changes[2]-[3]. These changes will lead to the degradation of the macro properties of foam rubber and affect the safety of the equipment.

The performance of foam rubber under compression and tension is very different. This article pays more attention to the change in performance of foam rubber during the aging process under the condition of long-term compression load. Indexes such as compression set and stress relaxation are commonly used to measure the degree of performance of compressed foam rubber. In the aging process of foam rubber, the compression set and stress relaxation are related to the change of the stress-strain curve[4]. Therefore, choosing a constitutive model that can accurately describe the stress-strain response of the material, and studying the changes of model parameters during the aging are also important means to study the aging of foam rubber.

This article aims to overview the changes regularities of compression set and stress relaxation of compressed foam rubber during the aging, and study the reasons and influencing factors of these changes. In order to find a suitable constitutive model for foam rubber, the common constitutive models and their application of Hyper-elastic Materials are introduced. Finally, the changes of model parameters during aging are introduced.

2. The changes regularities of aging index

Accelerated aging tests are often used to study changes in foam rubber properties. Zhang[5] and Yan[6] used the accelerated aging method to study the performance changes of silicon foam and found that the compression set of silicon foam gradually increased with aging time, and the higher the temperature, the faster the compression change. Zhang[5] also found that the displacement-load curve of silicon foam gradually shifts to the left and becomes steeper with the aging time. Zhang analyzed that the reason for these changes is that the rubber matrix undergoes oxidative degradation during the aging process, and the internal cells will undergo permanent deformation, which will cause the material to become hard and brittle, resulting in changes in the macroscopic properties of the silicon foam. Although the research of Zhang revealed the changing trend and change mechanism of foam rubber properties, it failed to give a specific function description.

Huang's research[7] found that the stress relaxation of silicon foam has a linear relationship with the natural logarithm of time. Liu[8] found that the relationship between the change rate of load and temperature can be described using the Arrhenius model for polyurethane foam. Their research results show that the aging process of foam rubber conforms to the principle of time-temperature equivalence.

The research results of Huang[7] and Ma[9] show that the stress relaxation of Silicon foam reduces at first and increases then with the increase of the initial compression amount. This means that the relationship between the rate of stress relaxation degradation and compression is not monotonous, which is contrary to common sense, this paper holds that the main reason for this phenomenon is that the stress relaxation data obtained in the literature are not for the same compression. Although the research results in the literature have strong engineering guiding significance, in order to show the effect of different compression on the stress relaxation clearly, the initial load value before aging should be tested at the specified compression (such as 20%), and then measure the load at this compression (20%) after aging to determine the stress relaxation level, rather test the initial load value under varies compression(such as 20%, 30%, 40%), and then test the value under varies compression (such as 20%, 30%, 40%) after aging.

The aging characteristics of silicon foam are also related to the properties of the material. Shi[1] found that the stress relaxation of silicon foam increases with density, and the relationship between the two is approximately linear. The cell types and sizes also have an impact on the aging performance of silicon foam. Foams with high open cell content, uniform cell size, and circular cell, helps reduce the stress relaxation of foamed silicon[10], and the cell diameter has a smaller effect on the stress relaxation rate[12]. The hardness of silicon foam also affects the aging characteristics of the material. Zhang[13] found that the hardness and stress relaxation are inversely proportional. The increase of hardness is helpful to maintain stress during aging.

3. The constitutive model of common hyper-elastic materials and the change of model parameters with aging time

3.1. Common constitutive models of hyper-elastic materials

The constitutive models of hyper-elastic materials are divided into two categories: one is the continuum constitutive model based on phenomenological theory, and the other is the constitutive model based on thermodynamic statistical methods[14].

The thermodynamic constitutive model has the advantages of few parameters and clear physical meaning. Commonly used constitutive models based on thermodynamic statistical methods include the Arruda-Boyce model[15] and Van der Waals model[16] are commonly used constitutive models for this method. The Arruda-Boyce model has two parameters, which parameters can only change the stress-strain ratio but cannot change the curve shape, so the data fitting effect is not well when as a foam rubber constitutive model. The Van der Waals model has 4 parameters, compared with the Arruda-Boyce model, the Van der Waals model can change the shape of the curve and has better data fitting capabilities.

The constitutive models based on phenomenological theory include Neo-Hookean[17], Mooney-Rivlin[18], Yeoh[19] and Odgen[20]. The Mooney-Rivlin, Neo-Hookean, and Yeoh model mainly uses strain invariants to describe the constitutive relationship of materials. These three models are all deformed based on formula (1).

$$W = \sum_{i+j=1}^{N} C_{ij}\left(I_1 - 3\right)^i \left(I_2 - 3\right)^j \tag{1}$$

Where I_1 represents the first strain invariant, I_2 represents the second strain invariant, C_{ij} represents the model coefficient.

When formula (1) keeps only the first term, the Neo-Hookean model is obtained

$$W = C_{10}\left(I_1 - 3\right) \tag{2}$$

When formula (2) keeps the first two terms, the Mooney-Rivlin model is obtained

$$W = C_{10}\left(I_1 - 3\right) + C_{01}\left(I_2 - 3\right) \tag{3}$$

The Neo-Hookean model and Mooney-Rivlin model are simple and convenient to apply. However, because there are not enough expansion items and the shear modulus does not change with strain, the two models have a better fitting effect for the small deformation stage, but there will be deviations in the large deformation stage[14], [21].

As shown in equation (4), the Yeoh model discards the second strain invariant and retains the higher-order terms of the first strain invariant. The Yeoh model is simple in form, but because it has a higher-order term of strain invariant, it can better reflect the behavior of foam rubber in the large deformation stage. However, because the model discards the second strain the model appears "softer" when predicting the biaxial tensile strain behavior[22].

$$W = C_{10}(I_1 - 3) + C_{20}(I_1 - 3)^2 + C_{30}(I_1 - 3)^3 \tag{4}$$

The Odgen model mainly uses three main elongation rates λ_1, λ_2, λ_3 to describe the material constitutive relationship. Compared with formula (1), the Odgen model considered that the three principal elongations are asymmetric[23], so the Odgen model is more flexible, adapted to a wide range of deformation, and is widely used. The deformed Hyper-foam model of the Odgen model is also often used as a constitutive model of foam rubber materials. The Hyper-foam model expression is shown in formula (5)

$$U\left(\lambda_1,\lambda_2,\lambda_3\right) = \sum_{i=1}^{N} \frac{2\mu_i}{\alpha_i^2}\left\{\lambda_1^{\alpha_i} + \lambda_2^{\alpha_i} + \lambda_3^{\alpha_i} - 3 + \frac{1}{\beta_i}\left[\left(\lambda_1\lambda_2\lambda_3\right)^{-\alpha_i\beta_i} - 1\right]\right\} \tag{5}$$

Where, the λ_1, λ_2, λ_3 are the main deformation, μ_i and, α_i and β_i are the material parameters.

3.2. The constitutive model parameters of foam rubber change with aging time

During the aging process, the stress-strain curve of the foam rubber will change. The slope of the curve increases with the aging time[24], the parameters of the constitutive model also change accordingly. Xie[25] used the Odgen model as a constitutive model of a filled rubber and found that the parameter μ_i changes significantly with aging, and the change α_i is not obvious. By assuming α_i constant, Xie found that the parameter μ_i and the aging temperature meet the Arrhenius model. Jin[26] used the Hyper-foam model as the constitutive model of silicon foam, by using the Prony series expansion to fit the model parameters, Jin started from the perspective of uncertainty quantification in the fitting process[27] considers that there is a correlation between the parameters μ_i and α_i. Maiti[4] used the Ogden Hyper-foam + Two-network Tobolsky model as the constitutive model of a certain silicon foam cushion, by considering that α_i is unchanged, the parameter μ_i during the aging process is fitted. Maiti used the constitutive model to predict the compression set and load retention of the silicon foam cushion and found that the prediction results were consistent with the experimental structure. It can be analyzed from the literature that at present, numerical fitting methods are mainly used to study the parameter changes of the constitutive model, to converge the calculation results, the above methods all limit the model parameters.

4. Conclusions

This article reviews the effects of foam rubber on material properties and constitutive models during the aging process. In the aging process, the oxidative degradation of the rubber matrix and the damage of the cell microstructure will cause irreversible changes in the permanent of the foam rubber materials. The higher the temperature, the longer the aging time, the more obvious the material performance degradation. The performance degradation of foam rubber is also affected by the characteristics of the material itself. The higher the open-cell content of the material, the more uniform the cell distribution, the greater the hardness, the less stress relaxation during the aging process. As the relative density of the material increases, the stress relaxation will increases. The size of the cell has no obvious effect on the stress relaxation of the material.

This article also introduces common constitutive models of hyper-elastic materials and the influence of aging on model parameters. Constitutive models can be divided into thermodynamic statistical constitutive models and phenomenological constitutive models. Commonly used thermodynamic statistical constitutive models include Arruda-Boyce model and Van der Waals model. Commonly used phenomenological constitutive models include Neo-Hookean model, Mooney-Rivlin model, Yeoh model, and Odgen model. Because the situation is flexible and can be applied to large deformation strains, the Odgen Hyper-foam model is often used as a constitutive model of foam rubber materials. At present, numerical fitting methods are mainly used to study the changes of constitutive model parameters with aging time. The research on the parameter changes of the foam rubber constitutive model needs to be further in-depth.

Acknowledgments
Key academies and key disciplines and major construction projects in the military (430169, 430183), Naval academies and major disciplines major construction projects (430618)

References

[1] Shi Y, Zhang C, Zhao Q, Ding G, Luo S. (2007) Study on the effects on compression stress relaxation of HTV silicone rubber-foams.J. New Chemical Materials. 21-22.
[2] Patel, M., Morrel, P. R., Murphy, J. (2005) Continuous and intermittent stress relaxation studies on foamed polysiloxane rubber. J. Polym. Degrad. Stab. 87, 201–206.
[3] Coons, J. E., McKay, M. D. , Hamada, M. S. (2006) A Bayesian analysis of the compression set and stress–strain behavior in a thermally aged silicone foam. J. Polym. Degrad. Stab. 91, 1824–1836.
[4] Maiti, A. , Small, W. , Lewicki, J. P. , Chinn, S. C. , Saab, A. P. . (2019). Age-aware constitutive materials model for a 3d printed polymeric foam. J. Scientific Reports. 9(1): 15923.
[5] ZHANG K, FAN J, WU J, MA Y. (2007) Study on Ther maloxydative Ageing Mechanisms of Silicone Rubber Foam Materials. J. Synthetic Materials Aging and Application. (03):18-21.
[6] Yan X, Zhou Y, Wen M, Li J, Zhou X, Wang P. (2009) Study on the effects of multi-factor accelerating ageing on the mechanical properties of the silicone rubber foam materials. J. New Chemical Materials. 37(05): 66-68.
[7] Huang Y, Zhang F, Hu W. (2009) The research of the stress relaxation of cellular silicone materials. J. New Chemical Materials. 37(08): 85-87.
[8] Liu Y, Wu L, He C, Deng J, Zhang T, Yi K, Sun S. (2005) Research on Prediction of Storage Life of Rigid Polyurethane Foam. J. China Plastics. (05):87-90.
[9] Ma Y, Wang F. (2019) Study on the compressive stress relaxation properties of foamed silicone rubber. J. Organic Silicone Material. 33(04): 296-301.
[10] Ding G, He C, Shi Y, Luo S. (2008) Study on the pertinences between the compress stress relaxation characteristics and the pore structure of the silicone rubber foam. J. New Chemical Materials. (06): 47-49.
[11] Fan Z, Chen C, Hu W, Wan Q. (2015) EFFECTS OF MICR OSTRUCTURE ON THE LARGE COMPRESSION BEHAVIOR OF RUBBER FOAMS. J. Mechanical Strength. 37(05): 892-

897.
[12] Sha Y, Zhang C, Li J, Luo S. (2013) The influence of cell structure on the properties of silicone rubber foam materials. J. Materials for Mechanical Engineering. 37(02):25-28.
[13] Zhang C, Luo S, Shi Y, Lu A, Ding G. (2008) Study on the connection between rigidity sand compression properties o fsilicone rubber foam. J. New Chemical Materials. (08):77-78.
[14] Xu L, Wu G. (1999) Some forms of rubber strain energy function in finite element analysis. J. China Rubber Industry. (12): 707-711
[15]]Ellen, M, Arruda, et al. (1993) A three-dimensional constitutive model for the large stretch behavior of rubber elastic materials. J. Journal of the Mechanics & Physics of Solids.
[16] GUO Z, SLUYS L J. (2008) Constitutive Modelling of Hyperelastic Rubber-like Materials. J. Delft University of Technology. 55(3) : 109-132.
[17] Treloar L R G. (1943) The elasticity of a network of long-chain molecules I. J. Transactions of the Faraday Society. 39: 36-41.
[18] DONG L, LAKES R S. (2011) Frequency Dependence of Poisson's Ratio of Viscoelastic lastomer Foam. J. Cellular Poly- mers. 30(6) : 277-286.
[19] Yeoh O H. (1993) Some forms of the strain energy function for rubber. J. Rubber Chemistry nd Technology. 65(5): 754—771.
[20] Ogden R W. (1972) Large deformation isotropic elasticity-On the correlation of theory and experiment for incompressible rubberlike solids. J. Proceedings of the Royal Society of London. A326: 565—584.
[21] Wei Y, Fang Q, Jin Z, Feng X. (2014) Research progress on the constitutive model of filled rubber. J. Chin Polym Bull. (05): 15-21.
[22] Li X, Wei Y. (2016) AN IMPROVED YEOH CONSTITUTIVE MODEL FOR HYPERELASTIC MATERIAL. J. Engineering Mechanics. 33(12): 38-43.
[23] GREGORY M J. (1975) The Stress-strain Behavior of Filled Rubbers at Moderate Strains. J. Plastics and Rubbers Materials and Applications. 4(4) : 184-188.
[24] ANHTH, VU-KHANHT. (2005) Effects of thermal aging on fracture performance of polychloroprene. J. Mater Sci., 40(19) :5243 -5248.
[25] Xie Z, Wang Y, Wan Z, Gu Q. (2008) Thermal aging of filled rubber constitutive relationship. J. Journal of Harbin Institute of Technology. (09): 1404-1407.
[26] Jin F, Xiao S, He Q, Jia D, Fang Y. (2018) Uncertainty characterization of super-elastic compression and stress relaxation of silicon foam. J. Chin J Appl Mech. (06): 1200-1206 +1414.
[27] ZHANG Y, NIU W, LIU X. (2013) Selection of static constitutive models for polyurethane based on Bayesian posterior probability. J. Chinese journal of solid mechanics. 33(S): 242-246.

Microstructural simulation of initial particle-matrix interfacial defect in HTPB solid propellants

Yufei Hou[1], Jinsheng Xu[1*], Changsheng Zhou[1], Xiong Chen[1]

[1]School of Mechanical Engineering, Nanjing University of Science and Technology, Nanjing 210094, PR China

Corresponding author: email: xujinsheng@njust.edu.cn.

Abstract: Impregnation capability at particle-matrix interfacial deteriorates during the preparation of composite solid propellant due to manufacturing and environmental factors, which leads to the initial defects in the HTPB solid propellant. In the present work, the effects of initial particle-matrix interfacial defects on the debonding, nucleation, and crack propagation of the composite solid propellant is investigated experimentally and numerically. A random packing model is developed based on the molecular dynamics method by embedding AP particles in the HTPB propellant. The strength of particle-matrix interface obeys Weibull function. Moreover, a zero-thickness cohesive element is used to simulate the particle-matrix interfacial debonding and matrix rupture. According to the simulation and experiment results, the initial interface defects makes a marked effect on the initial modulus, limit strength and elongation at fracture of the composite solid propellant. The initial modulus, limit strength and elongation at fracture of the composite solid propellant decrease with the increase of the initial interface defects.

1. Introduction

Solid propellants are extensively used in rocketry systems due to their high energy density, safety and long shelf life. Usually, the solid propellants are composed of ammonium perchlorate (AP) particles, metal powder and matrix which lead the mechanical properties of the solid propellants depend on the mechanical properties of the particles, the matrix and the particle/matrix interface [1]. Van Ramshorst et al. [2] studied the relationship between the microscopic morphology and macroscopic deformation and failure of the solid propellants based on scanning electron microscope (SEM), and found that the particle/matrix interface debonding and cracks grow along the particle/matrix interface under loading. Therefore, it is necessary to analyze the microscopic damage of the solid propellants.

Matouš and Geubelle [3] simulate the particle/matrix interface debonding process with small deformation theory. They proved that the interface debonding lead a non-homogeneous stress. H. Arora [4] developed 2D circles, 2D polygons and 3D spheres to compare the effects of the particle geometries on the damage evolution of particulate reinforced materials. Li Gaochun [5] found that the strength of the interface is the key factors to enhance the mechanical properties of the solid propellants. To describe the viscoelastic mechanical properties of the solid propellants, Cui [6] proposed a novel time-dependent cohesive zone model (CZM) based on the Maxwell box to simulate relaxation responses. However, above studies have ignored the performance defects of the solid propellants in production which causing errors in the simulate result and experimental result. Balzer [7] observed the particle/matrix interface by using SEM, and found that there were so many micro holes in the solid propellant, especially in the particle/matrix interface. This is because the material properties of the particles and the matrix are different during the curing process of the solid propellant, which causing gaps between particle and matrix, resulting in the existence of randomly distributed initial interface defects in the solid propellants. The initial interface defect will change the failure path of the solid propellants which accelerated damage. Wang [8] found that the effect of the initial defects on the macro-mechanical properties of the particulate reinforced materials cannot be ignored.

The present research investigat the mechanical properties and fracture behavior of the HTPB propellant through microstructure-based numerical simulations. The strength density function of the particle/matrix interface obey Weibull distribution to analyze the influence of initial defect on the mechanical properties of the HTPB propellants.

2. Microstructure-based numerical model

2.1. RVE model generation

A two-dimensional (2D) representative volume element (RVE) model is developed by using Fortran language, as shown in Fig. 1. It is considered as effective when the RVE model size is 3 times the maximum particle size [9]. Therefore, the $500\ \mu m \times 500\ \mu m$ edge-sized RVE model is selected.

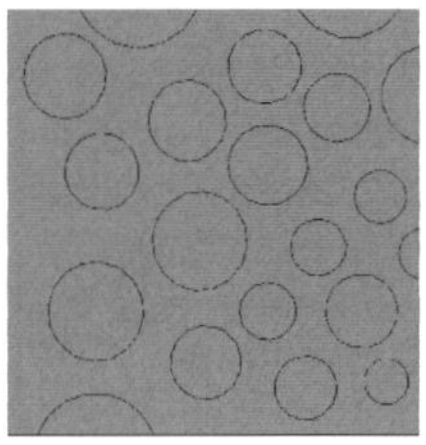

Figure 1 Microscopic particle filled model

2.2. Cohesive zone model

In order to capture the particle/matrix interfacial debonding and matrix rupture, the zero-thickness cohesive elements are embedded in the packing model by using Python scripting language (Fig. 2). Fig. 2(a) shows the initial mesh of the particle and the matrix and the mesh type is CPE4R. Fig. 2(b) shows the cohesive element of the interface, the matrix and the mesh type is COH2D4.

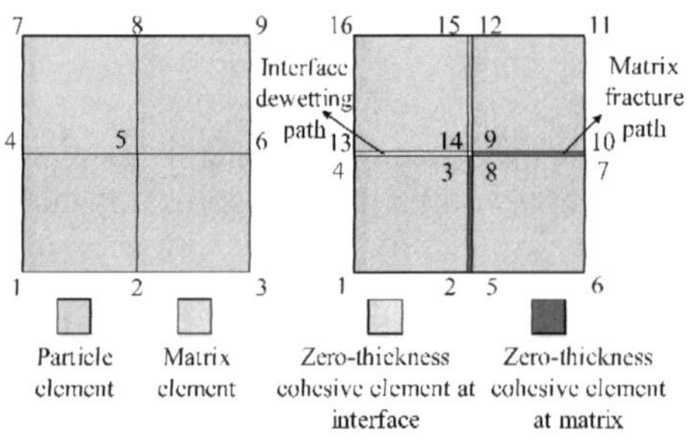

(a) solid element (b) cohesive element

Figure 2 Embedded cohesive element diagram

The cohesive zone model was first proposed by Dugdale [10] and Barenblatt [11]. A fracture process area at the crack tip is proposed which is determined by the normal traction and tangential traction. The cohesive zone models are divided into bilinear, trapezoidal and exponential, etc. Every cohesive zone model has its scope of application. According to the damage characteristics of the HTPB propellant, the exponential cohesive zone model is used to describe the damage evolution process of the particle-matrix interface, and the trapezoidal cohesive zone model is used to describe the rupture of the matrix.

The exponential cohesive zone model and trapezoidal cohesive zone model have been obtained by using Fortran language. The fracture energy $\phi(t)$ of the exponential cohesive zone model is:

$$\phi(t)=\phi_{\mathrm{n}}+\phi_{\mathrm{n}}\exp\left(-\frac{t_{\mathrm{n}}}{\delta_{\mathrm{n}}}\right)\left\{\left(1-r+\frac{t_{\mathrm{n}}}{\delta_{\mathrm{n}}}\right)\frac{1-q}{r-1}-\left[q+\left(\frac{r-q}{r-1}\right)\frac{t_{\mathrm{n}}}{\delta_{\mathrm{n}}}\right]\exp\left(-\frac{t_{\mathrm{t}}^2}{\delta_{\mathrm{t}}^2}\right)\right\}, \tag{1}$$

where δ_{n} and δ_{t} are the characteristic length of the normal and tangential. The cohesive element normal traction and tangential traction are obtained by the following equation.

$$T_{\mathrm{n}}=-\frac{\phi_{\mathrm{n}}}{\delta_{\mathrm{n}}}\exp\left(-\frac{t_{\mathrm{n}}}{\delta_{\mathrm{n}}}\right)\left\{\frac{t_{\mathrm{n}}}{\delta_{\mathrm{n}}}\exp\left(-\frac{t_{\mathrm{t}}^2}{\delta_{\mathrm{t}}^2}\right)+\frac{1-q}{r-1}\left[1-\exp\left(-\frac{t_{\mathrm{t}}^2}{\delta_{\mathrm{t}}^2}\right)\right]\left(r-\frac{t_{\mathrm{n}}}{\delta_{\mathrm{n}}}\right)\right\}, \tag{2}$$

$$T_{\mathrm{t}}=-2\frac{\phi_{\mathrm{n}}}{\delta_{\mathrm{t}}}\cdot\frac{t_{\mathrm{t}}}{\delta_{\mathrm{t}}}\left\{q+\left(\frac{r-q}{r-1}\right)\frac{t_{\mathrm{n}}}{\delta_{\mathrm{n}}}\right\}\exp\left(-\frac{t_{\mathrm{n}}}{\delta_{\mathrm{n}}}-\frac{t_{\mathrm{t}}^2}{\delta_{\mathrm{t}}^2}\right), \tag{3}$$

The fracture energy $\phi(t)$ of the trapezoidal cohesive zone model is:

$$\phi(t)=\frac{1}{2}\sigma_{\max}\left(\delta_f+\delta_2-\delta_1\right) \tag{4}$$

The cohesive element normal traction and tangential traction are obtained by the following equation.

$$T_n,T_t=\begin{cases}\dfrac{\sigma_{\max}}{\delta_1}\delta_{n,t} & \delta<\delta_1\\ \sigma_{\max} & \delta_1\le\delta\le\delta_2\\ \dfrac{\sigma_{\max}}{\delta_f-\delta_2}\left(\delta_f-\delta_{n,t}\right) & \delta_1\le\delta\le\delta_f\end{cases} \tag{5}$$

During the mixing and curing of the HTPB solid propellant, the bonding of the particle-matrix interface will decrease, and voids will be generated between the particle and the matrix interface, as shown in Fig. 3. The initial defects causing the instability mechanical propose of the HTPB propellant. Therefore, it is necessary to consider the initial defects between the particle and the matrix interface when predicting the mechanical properties of the propellant. In this paper, the density function of the bonding strength between the particle and the matrix is set as the Weibull function. The mesoscopic numerical model with initial defects is developed. The Weibull function is:

$$f(x)=\frac{m}{\lambda}\left(\frac{x}{\lambda}\right)^{m-1}\exp\left[-\left(\frac{x}{\lambda}\right)^m\right], \tag{6}$$

where x represents a random function. m is a shape parameter. λ is a proportional parameter. The Weibull function with different shape parameters m is shown in Fig. 4.

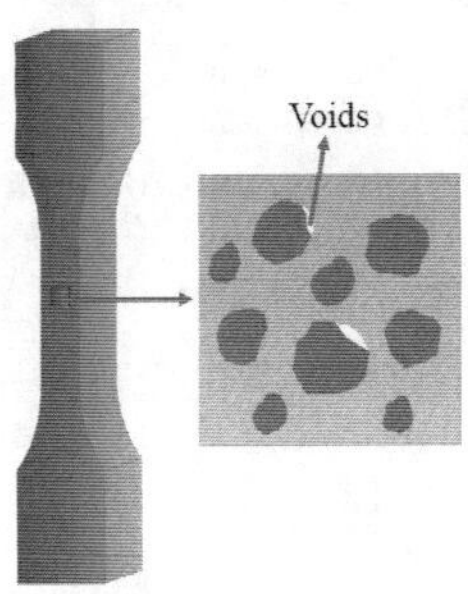

Figure 3 Initial defect of solid propellant

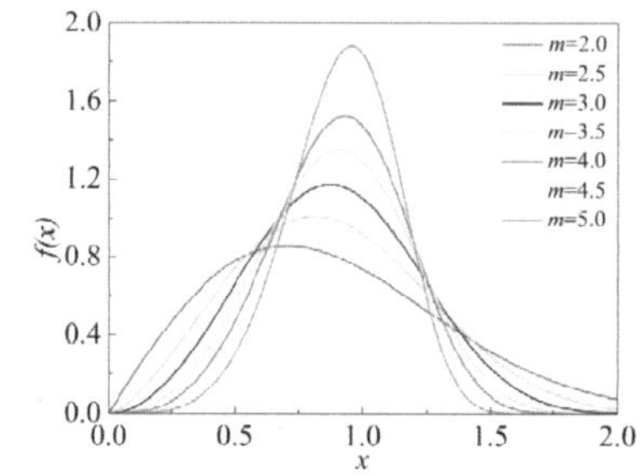

Figure 4 Weibull distribution density function

3. Results and discussion

3.1. Damage behavior

Under uniaxial tension, the deformation process of the packing model was divided into four stages-I, II, III, and IV, as shown in Fig. 5. In stage I, the particle and the matrix experienced elastic deformation and interface between the particle and matrix is bonded well. However, the rigidity of the particles and the matrix is different, causing the extremely uneven stress distribution within the model, and the particle stress is significantly higher than the matrix stress. In stage II, due to the non-uniformity of the interfacial bonding strength distribution, the location where the particle and the matrix interface damage content is higher will be the first to debond. In stage III, the deformation of the matrix increases, and the amount of separation displacement of the particles from the matrix increase, causing the particles to be exposed form the holes. In stage IV, as the strain continues to increase, the pores after debonding of the particles and the matrix converge, and the microcracks formed inside the matrix.

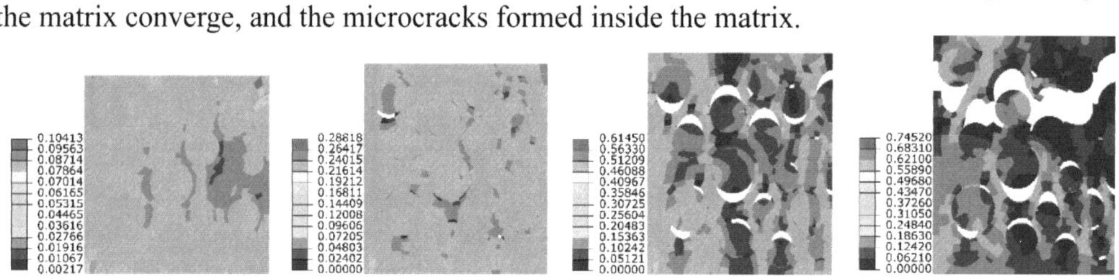

Figure 5 von Mises stress nephograms of initial damage model

3.2. Influence of initial defects on HTPB solid propellant

The effect of initial defects on the mechanical property of the HTPB solid propellant is examined by varying the m values. With the increase of the m values, the mechanical property of the HTPB solid propellant changed significantly, as shown in Fig. 6. The initial modulus, limit strength and elongation increase with the increase of the m values. This is because the m value determines the distribution law of the bonding strength between the particle and the matrix. The smaller of the m value, the more dispersed the bonding strength, the more initial defects inside the RVE model. As a result, the interface between the particle and the matrix is more likely to form debonding damage. Moreover, the load transfer capacity of the interface decreases, resulting in a decrease in the overall strength of the model. Finally, the initial modulus, the limit strength and elongation decreases. On the contrary, the larger of the m value, the more concentrated the bonding strength distribution, and the higher the initial modulus, the limit strength and elongation obtained from the RVE model.

The initial defect models better explain the discreteness of the HTPB solid propellant experimental results. It also proved the randomness of the HTPB solid propellant fracture position. Therefore, the initial defect model is more coincident with the real situations.

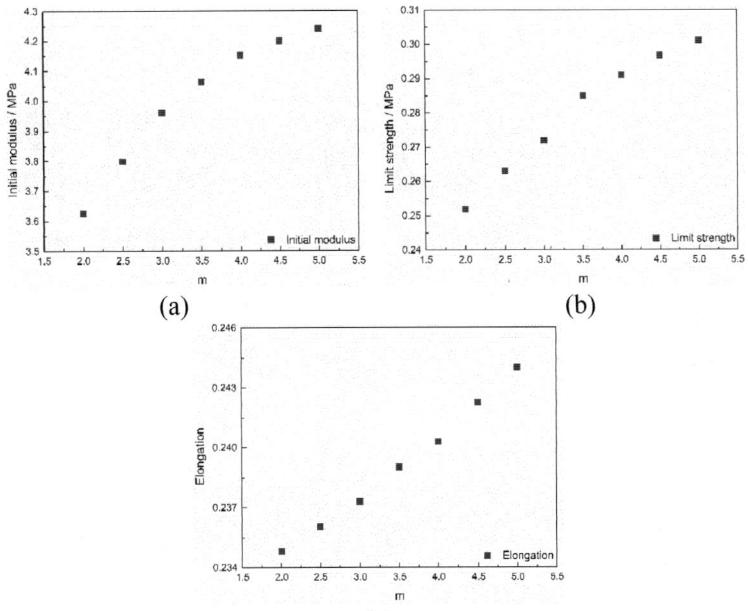

Figure 6 (a) Initial modulus of initial damage model (b) Limit strength of initial damage model (c) Elongation of initial damage model

4. Conclusion

A RVE model of HTPB solid propellant is developed to explore the influences of initial interfacial defect on the mechanical properties. The cohesive elements are presented to simulate the particle-matrix interfacial debonding and matrix rupture. It is found that the damage evolution of the HTPB solid propellant from the particle-matrix interfacial debonding to the matrix rupture. Moreover, the initial defect model also presents that the initial modulus, the limit strength and the elongation decreases with the increases of the initial interfacial defect.

Reference

[1] Francqueville, J., Diani, J., Gilormini, P,. et al. (2021) Use of a micromechanical approach to understand the mechanical behavior of solid propellants[J]. Mech Mater, 153: 103656.

[2] Vanramshorst, M.C.J., Benedetto G.L., Duvalois, W,. et al. (2016) Investigation of the Failure Mechanism of HTPB/AP/Al Propellant by In-situ Uniaxial Tensile Experimentation in SEM[J]. Propell, Explos, Pyrot, 41(4): 700-708.

[3] Matouš, K., Geubelle, P.H. (2006) Finite element formulation for modeling particle debonding in reinforced elastomers subjected to finite deformations[J]. Comput Method Appl M, 196(1-3): 620-633.

[4] Arora, H., Tarleton, E., Li-Mayer, J., et al. (2015) Modelling the damage and deformation process in a plastic bonded explosive microstructure under tension using the finite element method[J]. Comp Mater Sci, 110: 91-101.

[5] Li, G., Wang, Y., Jiang, A., et al. (2018) Micromechanical investigation of debonding processes in composite solid propellants[J]. Propell, Explos, Pyrot,, 43(7): 642-649.

[6] Cui, H., Shen, Z., Li, H. (2018) A novel time dependent cohesive zone model for the debonding interface between solid propellant and insulation[J]. Meccanica, 53(14): 3527-3544.

[7] Balzer, J.E., Siviour, C.R., Walley, S.M., et al. (2004) Behaviour of ammonium perchlorate–based propellants and a polymer–bonded explosive under impact loading[J]. Proceedings of the Royal Society of London. Series A: Mathematical, Physical and Engineering Sciences, 460(2043): 781-806.

[8] Wang, J., Qiang, H., Wang, Z., et al. (2020) Numerical Study of Mechanical Properties of Composite Solid Propellant with Initial Defects[C]//Journal of Physics: Conference Series. IOP Publishing, 1634(1): 012146.
[9] Bernard, F., Kamali-Bernard, S., Prince, W. (2008) 3D multi-scale modelling of mechanical behaviour of sound and leached mortar[J]. Cement Concrete Res, 38(4): 449-458.
[10] Dugdale, D.S. (1960) Yielding of steel sheets containing slits[J]. J Mech Phys of Solids, 8(2): 100-104.
[11] Barenblatt, G. I. (1962) The mathematical theory of equilibrium cracks in brittle fracture[M]. Adv Appl Mech, 7: 55-129.

Study on sedimentary microfacies and sand body distribution of Chang 8 oil reservoir group in Xunyi area, southern Ordos

Wang Yin[1]*, Zhang Tianlong[1], Zhang Ningning[2], Song Xiaogang[3]

[1]Research Institute of Shaanxi Yanchang Petroleum (Group) Co., Ltd., Xi'an, Shaanxi, 710000, China

[2]Shaanxi Energy Vocational and Technical College, Xianyang, Shaanxi, 710000,China

[3]Fuxian Oil Production Plant, Shaanxi Yanchang Petroleum (Group) Co., Ltd., Yan'an, Shaanxi, 727500, China

*Corresponding author's e-mail: wangyin@sxycpc.com

Abstract. Geological surveys were conducted in this area as early as the 1960s. Only a few wells were drilled in the 1970s, and there will be no large-scale exploration in the future. It is currently difficult to meet the need for further exploration. The provenance direction of the Chang 8 reservoir in this area is still controversial, the sedimentary microfacies characteristics and sand body distribution are unclear, and the development law is unclear. Therefore, the subsequent exploration and development process of Chang 8 reservoir in this area is greatly affected. Therefore, it is urgent to carry out research on sedimentary microfacies and sand body distribution. This paper studies the distribution of sedimentary microfacies and sand bodies of Chang 81 and Chang 82 oil layers, which provides an important basis for future reservoir research and analysis of accumulation factors, ascertaiN the distribution of oil reservoirs, predicting favorable zones, and evaluating and optimizing exploration targets.

1. Introduction

The study area is located in the southern Ordos Basin. across Shaanxi and Gansu provinces, with an area of about 12,000 km^2, starting from Zhangjiawan in the north, Liulin in the south, Panke in the west, and Huangling in the east. The structure is located in two tectonic units of the northern Shaanxi slope and the Weibei uplift [1.2].This paper studies the sedimentary microfacies and sand body distribution characteristics of Chang 8 reservoir in Xunyi area.Another section of your paper.

2. Regional sedimentary background

The Chang 8 sedimentary period of the Ordos Basin was controlled by four major provenance systems in the southwest, northwest, west and northeast. Among them, the northwest, northeast, and southwest are important source systems, and the west is a secondary source system. The Chang 8 sedimentary period is in the middle of the Chang 9 Lake flooding period and the Chang 7 Lake transgression period. The lake basin changes rapidly, and delta accumulation and degeneration occur in a short period of time. Therefore, the Chang 8 oil layer is divided into Chang 81 and Chang 82. Chang 82 is the process of base-level descent after the flooding period of Chang 9 Lake, and delta proliferation occurs [3-4]. Chang 81 is the period of base-level rise before the transgression of Lake Chang 7 and most of it is the period of delta regression.

3. Sedimentary facies signs

Most of Chang 8 in this area is light gray, gray subangular fine sandstone, silty mudstone intercalated with dark mudstone, and medium sandstone is seen locally. There are obvious oblique bedding, progressive positive bedding, homogeneous bedding and bottom conglomerate in medium and fine sandstone, and wave bedding and sand-grain bedding in siltstone. Wormholes and carbonized plants are common in Well Chang 8 and it is determined that the Chang 8 sedimentary period in this area should be a shallow delta sedimentary environment.

4. Granularity characteristics

The particle size probability curve of length 8 can reflect that the debris particles are between 1.90 and 5.10. The cumulative probability curve shows a "three-level" pattern of traction flow samples, and the jump is mainly divided into "two stages". The particle size characteristics indicate the effect of two-way flow (Figure 1). Generally speaking, the underwater distributary channel at the front edge of the shallow delta has such hydrodynamic characteristics.

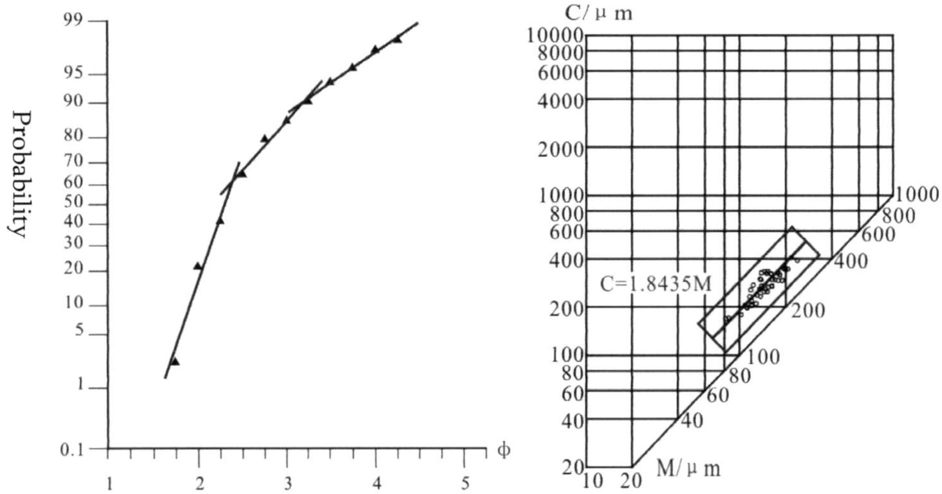

a. Well Z 97, length 82, 1966.68m b. C-M map of Chang 8 reservoir

Figure 1 Grain size probability curve and C-M diagram of Chang 8 sandstone

The analysis of the CM map of the Chang 8 oil layer group in this area (Figure 2) shows that the C value is between 168 and 404 μm, the M value is between 82 and 237 μm, and the average value of C/M is 1.85. The formation of sampling points is roughly parallel to C=m-line bars. The gradual suspended sedimentation of traction flow.

5. Logging facies

The Chang 8 logging curves of 67 wells in this area have been studied. The results show that the SP and GR curves of Chang 8 wells are mainly shaped like a toothed bell, and the thickness is generally between 4 to 14m; the second is a toothed box shape. The thickness is usually 11~21m, and the shape and scale of the sand body are easily formed on the front edge of the shallow lake delta (Figure 2).

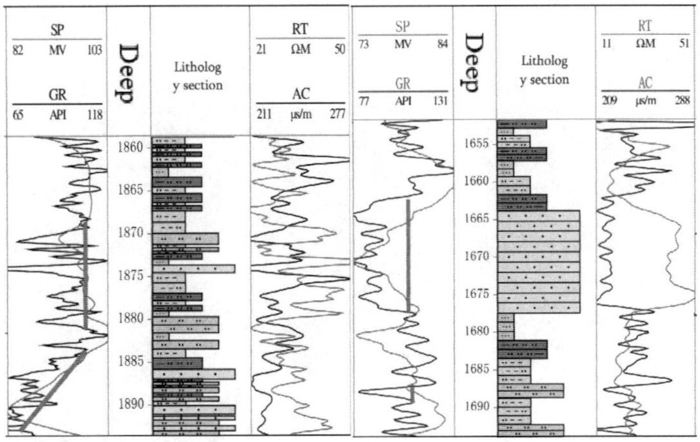

a. Well N20, length 81, length 82 Toothed bell-shaped b Well Z64, length 81 peaked toothed box type

Figure 2 The characteristics of the Chang 8 logging curve in the study area

6. Sedimentary microfacies and distribution characteristics of sand bodies

Based on the analysis of the color, lithology, sedimentary structure, grain size characteristics and logging curve shape of the Chang 8 oil layer group, Chang 8 is determined to be a delta-lacustrine sedimentary system, and the delta front is the main one. Two types of sedimentary facies, three types of sedimentary subfacies, and seven types of sedimentary microfacies are identified from the Chang 8 reservoir group, as shown in Table 1.

Table 1 Sedimentary facies types and characteristics of Chang 8 in the study area

phase	Subphase	Microphase	Developmental characteristics	Identification mark
Delt a- lake	Plain	Chang 8 is basically underdeveloped		
	The front edge of the delta	Underwater diversion channel	There are multiple sub-diversion and pooling functions	The delta front deposits are composed of medium to fine sandstone, fine sandstone and siltstone, with mudstone interbedded, with trough, plate-like cross bedding, parallel bedding and sand-line bedding, often with sliding deformation bedding and enveloping bedding And horizontal bedding, including ostracods, leaf limbs, lobular branches, fish fossils, and plant fossil fragments
		Estuary dam	Underdeveloped, often scoured and reformed by the submerged diversion channel in the later period	
		Far sand dam	The end of the delta front extending to the lake is mostly siltstone with a small thickness	
		Diverging channel flanks	Mostly submerged embankment flood deposits	
		Diversion bay	Often separated by a network of underwater diversion channels	
	Front Delta-Shallow Lake	Shallow lake mud /Front Delta Mud	Deposited with horizontal bedding dark to dark gray mudstone, silty mudstone or argillaceous siltstone mainly	The lithology is fine, dominated by mudstone, and the color is mostly gray and dark gray. There are often fishes, lobulins, leaf limbs and ostracods, and plant fossil fragments are also seen. The natural gamma curve is basically low amplitude, tooth-like, with low spikes

6.1. Chang 82 sedimentary microfacies and sand body distribution

As shown in Figure 3, it is jointly controlled by the southern provenance and the southwestern provenance. There are 6 underwater distributary channels entering the study area: Xunyi-N 71 well area, 0.94-12.87km wide and 14.00-22.40m sand thickness; Xunyi Z 65 well has a width of 0.22-9.52km and a well thickness of 14.00-24.0. m; the Luoshan area of N 29 is 0.86～4.58km wide, and the sand thickness is 14.00～32.70m. The inflow area of the two underwater distributary channels is about the intersection of N 28 and N 88 well.

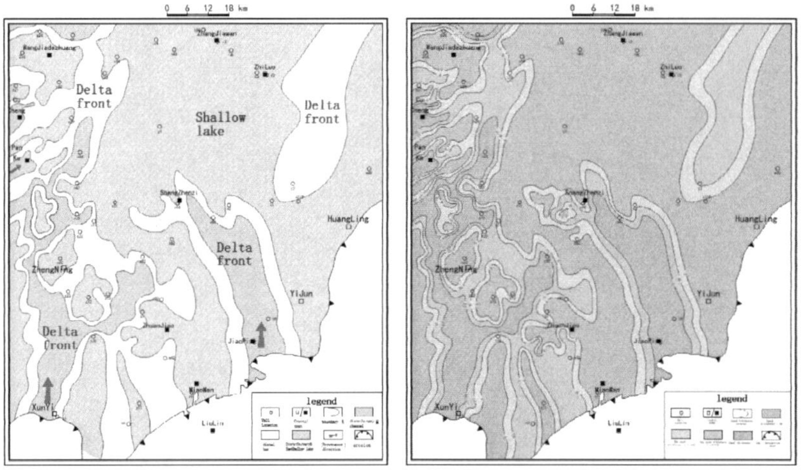

Figure 3 Distribution of Chang 82 sedimentary facies and sand bodies

There are 10 irregular diversion bays, of which 3 are larger, with an area larger than 220km^2, and the remaiN 7 have a larger area of about 11-24km^2.

The underwater distributary channel flanks are distributed between the underwater distributary channel microfacies and the interdistributed bay microfacies, mostly in the southwest and west of the area, with a width of about 1.51-17.27km and a sand body thickness of 6-15.9m.

6.2. Chang 81 sedimentary microfacies and sand body distribution

It is controlled by provenance in the south, southwest and northeast. There are 4 underwater diversion channels entering the study area: N 201 Zheng 71 well area, 1.65 ~ 14.53 km wide and 15 ~ 23.90 m thick; N 51 well and N 41 well area, well width 1.74 ~ 3.49 km, sand thickness 15 ～17.70m; T1-N 73 well area, the well width is 1.1-19.3km, the sand thickness is 15-22.5m; the width of the T 19 to the north of Luoshanfu is 4.29-22.50km, and the sand thickness is 15.00-22.50m. There are two underwater distributary channels that enter the study area, which converge in the area from Well N 73 to Well N 39 north of Luoshan Prefecture. There are 5 irregular diversion bays, of which 3 are larger than 270km^2 and 2 are smaller, with 106.74km^2 and 68.53km^2 respectively.

The side of the underwater distributary channel is distributed between the underwater distributary channel microfacies and the interdistributed bay microfacies. The most developed area in the north-central part of the study area is 1.4～20.74km wide and 5～14.6m sand thickness.

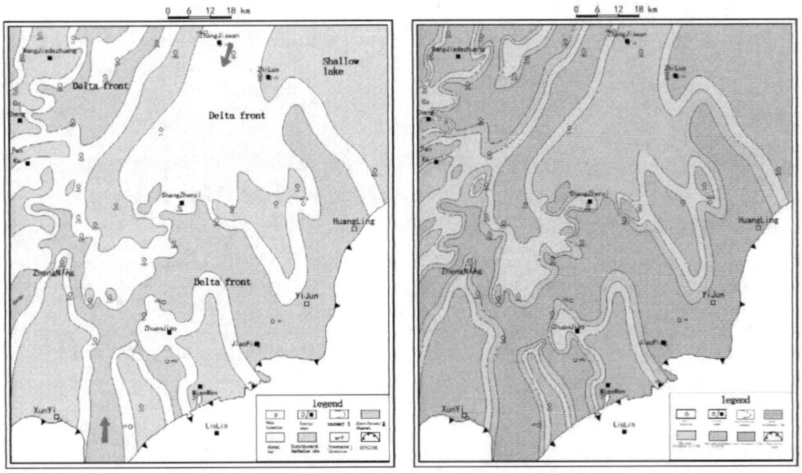

Figure 4 Distribution of Chang 81 sedimentary facies and sand bodies

The distribution of sedimentary facies and sand bodies of the Chang 81 reservoir in this area is generally distributed in an irregular band. The sand rate of Chang 81 is 0.01 to 0.62, with an average of 0.27. The number of wells with a sand ratio greater than 0.4 accounted for 5.45%, the number of wells with a sand ratio between 0.2 and 0.4 accounted for 36.63%, and the number of wells with a sand ratio less than 0.2 accounted for 53.61%; the cumulative thickness of the sand body was 2-30m, with an average of 11.2m. The number of wells with a thickness of 20m accounted for 8.32%, the number of wells with a thickness of 13-18m accounted for 21.46%, the number of wells with a thickness of 11-16m accounted for 24.22%, and the number of wells with a thickness of 10m accounted for 42.51%; the thickness of a single sand body was 1.60m-15.30m, with an average of 3.5m. In general, the average value of the total thickness of the sand body is slightly smaller than that of the single sand body, and the average value of the total thickness of the sand body is basically the same as that of the single sand body (Table 2).

Table 2 Statistics of sand-to-ground ratio and sand thickness of each layer of Chang 8 in the study area

Stratum / Sand-to-ground ratio and sand thickness	Chang81	Chang82
Range of sand-to-ground ratio (m)	0.01 ~ 0.62	0.01 ~ 0.72
Average sand to land ratio (m)	0.27	0.20
Accumulated thickness range of sand body (m)	2.00 ~ 32.00	2.00 ~ 38.90
Average cumulative thickness of sand body (m)	11.2	8.60
Single sand body thickness range (m)	0.50 ~ 15.3	0.50 ~ 23.4
Average thickness of single sand body (m)	2.80	2.81

7. Conclusion

（1）From the lateral view of the Xunyi-Chang 8 reservoir group, the underwater distributary channel sand bodies and the remote sandbar sand bodies of the delta front interweave in the diversion bay, surrounded by shallow lake subfacies.

（2）On the plane of Chang 8 oil layer group in Xunyi area, there are delta front and shallow lake subfacies. Multiple underwater distributary rivers in the study area converge and diverge. The flanks of the underwater distributary channel are relatively developed. The diversion bay is irregular in shape and distributed around the diversion bay. Shallow lakes are distributed in the middle and northeast. The size of Chang 81 sand body is larger than that of Chang 82 sand body.

References

[1] Li Keyong.Study on the sedimentary system and reservoir characteristics of the Yanchang Formation in the rich yellow exploration area of the Ordos Basin[D]. Xi'an, Shaanxi: Northwest University, 2011.
[2] Li Xiangbo, Fu Jinhua, Chen Qilin, et al. The concept of sandy clastic flow and its application in deep-water sedimentation research of Yanchang Formation in Ordos Basin[J], Advances in Earth Science, 2011, 26(3): 287-293.
[3] Chu Meijuan, Guo Zhengquan, Bai Chang'e, et al. Sedimentary and evolution characteristics of Chang 8 oil layer group of Yanchang Formation in Ordos Basin[J], Journal of Oil and Gas Technology, 2012, 34(2): 13-18.
[4] Li Shixiang, Chu Meijuan, Guo Zhengquan, et al. Sandbody structure characteristics and genetic mechanism of Chang 8 reservoir of Yanchang Formation in Ordos Basin[J], Acta Petrolei Sinica, 2013, 34(3): 435-444.

Sedimentary facies characteristics of Yan 9 oil-bearing formation of Yan'an Formation in D area of northern Shaanxi

Yunhong Zhao[1], Yangjun Li[1], and Shasha Yang[2,*]

[1]Fuxian Oil Production Plant, Shaanxi Yanchang Petroleum（Group）Co. Ltd., Yan'an, Shaanxi, 727500, China

[2]Shaanxi Key Laboratory of Safety and Durability of Concrete Structures, XiJing University, Xi'an, Shaanxi, 710123, China

* Corresponding author: 20190100@xijing.edu.cn

Abstract. In order to promote the exploration and development of oil reservoirs in northern Shaanxi, the study of sedimentary facies characteristics in D area was carried out. Using core, logging and logging data, the Yan 9 oil layer group in the D area is divided into two small layers, Yan 9_{1+2} and Yan 9_3. Based on the core description, the analysis of single-well sedimentary facies, continuous-well sedimentary facies, and planar sedimentary facies is carried out. It is believed that the Yan 9 member of the study area is a delta plain deposit, which is mainly composed of distributary channels, natural embankment and inter-branch swamp microfacies. Vertically, the sand bodies in the Yan 9 oil-bearing group are discontinuous, with distributary channels and inter-branch swamps mainly developed. Horizontally, the channels of the Yan 9 oil layer group spread out in strips on the plane, extending in the northwest-south east direction. There are 6 main rivers in the study area of Yan9_3 stage and 7 main rivers in Yan9_{1+2} stage. Channel deposits constitute the main body of the sand body skeleton and are also the main oil and gas enrichment area.

1. Introduction

The Ordos Basin is the second largest basin in China. The Jurassic Yan'an Formation in the basin made a major breakthrough in the early stage of exploitation, and has been the main target layer for oil and gas exploration ever since [1,2]. The Yan'an Formation in the basin is a set of coal-bearing rock series, belonging to fluvial marsh facies deposits. It is a proliferation-type multi-cycle deposit composed of sandstone, mudstone, coal or carbonaceous mudstone in the vertical position. In the early stage of deposition, it is mostly a thick multi-stage braided river Coarse clastic rocks filled with sediments[3,4]. The rock and mineralogical characteristics are mainly gray-white and light gray feldspar quartz sandstone, and the rock structure is mature. Kaolin is cemented with voids, and more muscovite flakes can be seen, and sandstone has good oil storage properties[5].

Area D is located in the northern Shaanxi area of the Ordos Basin. It is a loess hilly and mountainous area with an altitude of 1500 ~ 1800m. The sedimentary system is located on the southwest bank of the lower reaches of the ancient Ningshaan river. This area is a favorable zone for oil and gas generation and storage. The Jurassic Yanchang Formation in the Ordos Basin consists of 10 oil layer groups. Among them, the Yan 9 oil layer group is the main oil-bearing interval in the study area. Sedimentary facies is a combination of the sedimentary environment and the characteristics of the sedimentary rocks formed in the environment. Sedimentary facies not only determine the production status of the reservoir, including the horizon, reservoir cap conditions, reservoir scale and distribution form, but also determine the internal structure and rock type of the reservoir, thereby controlling the distribution of oil and gas[6]. Therefore, the study of sedimentary facies is of great significance to the exploration and development of oil fields. In this paper, using core, logging and logging data, a detailed study of the sedimentary facies characteristics of the Yan 9 oil layer of the Yan'an Formation in the D area of northern Shaanxi is carried out to provide a theoretical basis for the further exploration and development of the oilfields in the study area.

2. Stratigraphic characteristics

For the D area of northern Shaanxi, the Yan 9 oil layer group was finely divided and compared using the marker layer method, the principle of equal thickness, and the sedimentary cycle analysis method. According to the results of the stratigraphic division, the plan view of the thickness of the small layer and the structure of the top of the strata were drawn. It is believed that the $Yan9_3$ and $Yan9_{1+2}$ stratums are relatively complete in the whole area and have not suffered denudation. The overall thickness of the formation does not vary widely, and the thickness of the two small layers is mainly about 20m. The nose-like structure at the top of Yan 9 is densely developed, forming multiple rows of nearly east-west nose-like structures, with a structural amplitude less than 25m, and the distribution of rhinoplasty has a certain difference. The distance between rhinoplasty is 3-8km. These nose structures provide favorable trap conditions for hydrocarbon accumulation. The study of the stratigraphic and structural characteristics of the Yan 9 oil-bearing group provides basic geological data for the later sedimentary facies and sand body distribution characteristics.

3. Sedimentary facies characteristics

3.1. Sedimentary facies type in the study area

Area D is located in the Yishan Slope, which is the broadest in the Ordos Basin, and is one of the main areas for oil exploration and development in the basin. According to previous research results and regional sedimentary background, Yan 9 oil layer of Yan'an Formation in area D is comprehensively analyzed as a delta plain subfacies, which is divided into three microfacies: distributary channel, natural embankment and interbranched swamp（Table 1）.

Table 1 Sedimentary facies classification of Yan 9 oil layer group in area D

Sedimentary facies	Sedimentary subfacies	Sedimentary microfacies
Delta	Delta plain	Diversion channel
		Branch swamp
		Natural embankment

(1) Diversion channel

The distributary channel is the main sand body deposition unit on the delta plain and the most important oil and gas reservoir. It is mainly composed of clastic sandstone, with extremely well-developed parallel bedding, including horizontal bedding (Figure 1) and oblique bedding (Figure 2). There are often interlayer erosion surfaces, and thin layers of fine conglomerate or gravel coarse sandstone lenses are often associated with the erosion surface, and the upward trend of grain size is not obvious. In the middle part, there is a large slab-like cross-bedded gravel-bearing coarse sandstone. The bottom surface and interlayers are common with scour-filling structures. There are often thin conglomerate lenses on the scour surface, which are the products of lateral sand dams. The electrical characteristics are characterized by rich response patterns of channel sedimentation logging curves. The most typical logging patterns are: the spontaneous potential and natural gamma are box-shaped with mid-to-high amplitude and micro-toothed box.

(2) Natural embankment

natural embankments are developed on both sides of the distributary channel, and are mainly characterized by fine sand (Figure 3), silt sand and thin muddy interbeds. Far away from the river bed, the sediments become finer and mud content increases. Common wavy bedding, wavy bedding and flowing water wave marks. Carbonized plant fragments can be seen.

(3) The swamp between branches

The branch swamp is located in the low-lying area between the distributary channels of the delta plain. Due to the luxuriant vegetation and poor drainage of the swamp, it is a stagnant and reductive environment, so its sediments are mainly dark organic clay, layered silt and carbonaceous Mezzanine (Figure 4). Often developed massive bedding and horizontal bedding. The biggest sedimentary feature of the delta plain is the wide distribution of marsh deposits. Therefore, the extensive and stable distribution of layered organic matter swamp deposits can be used as a marker layer for the stratigraphic contrast of the delta plain.

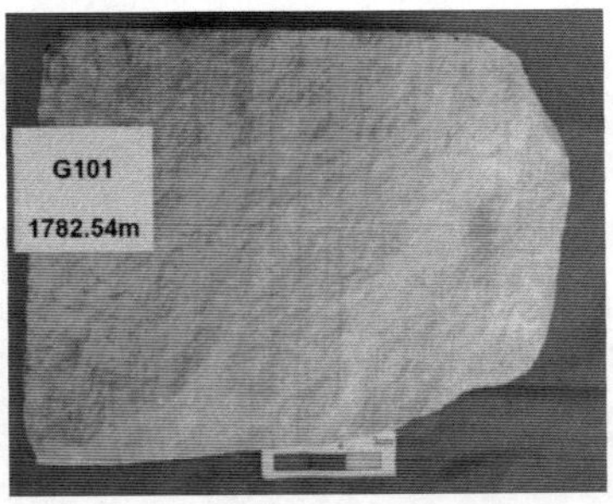

Figure 1 Well G101, Yan9, 1782.54m, horizontal bedding

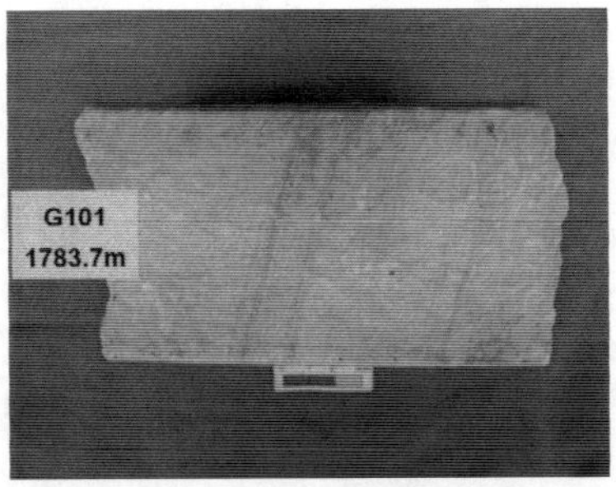

Figure 2 Well G101, Yan9, 1783.73m, inclined bedding

Figure 3 Well H8 , Yan9, 1833.36m, brown fine sandstone

Figure 4 Well H125 ,Yan9, 1937.70m, carbonaceous interlayer

3.2. Single-well facies analysis in the study area

Based on the classification of sedimentary facies and the study of sedimentary microfacies characteristics, in order to reveal the law of vertical evolution and planar distribution of sedimentary facies of Yan 9 oil layer group in the study area, single-well sedimentary facies analysis was carried out on two wells in the study area.

(1) Single-well facies analysis of Yan 9 oil layer group in Well G137

Well G137 is located in the middle of the study area. From the bottom to the upper delta plain, the distributary channels, inter-branch swamps and natural embankments frequently overlap each other, and the top coal seam of Yan 9 is developed. Yan9_3 mainly developed distributary channel microfacies, and Yan9_{1+2} developed distributary channel and inter-branch swamp microfacies. The sand body of Yan9_3 is thicker, and the curve is mostly box-shaped. In comparison, the Yan9_{1+2} sand body is slightly thinner.

(2) Single-well facies analysis of Yan 9 oil layer group in Well G213

Well G213 is located in the eastern part of the study area. The distributary channels, inter-branch swamps and natural embankments frequently overlap each other from the bottom to the upper delta plain. Coal seams are developed in the upper part of Yan 9_3 and the top of Yan 9_{1+2}. Yan9_3 developed distributary channel and inter-branch swamp microfacies, the lower part of Yan9_{1+2} was distributary channel, and the upper part was natural embankment microfacies. In comparison, the sand body of Yan9_3 is thicker, and the sand body of Yan9_{1+2} is thinner.

3.3. Analysis of the connected profile of the sand body in the study area

Profile facies analysis is based on the analysis of single well sedimentary facies, making full use of electrical logging data for comparison, establishing the phase sequence relationship between adjacent wells, and determining the distribution characteristics of sedimentary facies in two-dimensional space. Phase analysis method.

Combining the structural location and geological evolution background of the study area, based on a large number of exploration and development practices, three continuous well profiles that can reflect the sedimentary facies of the area have been established. The two sections show that the Yan 9 oil layer group has little undulation, and the sand bodies of Yan 9_3 and Yan 9_{1+2} are good, with a thickness of more than 10 m, and the sand bodies have good connectivity. On the whole, the Yan 9 oil-bearing group mainly develops distributary channels and inter-branch swamps. The river channel is mainly composed of medium and fine sandstone deposits, and the sand bodies are distributed in the river channel in a lens shape. The inter-branch swamp is mainly composed of fine sandstone, siltstone, and mudstone. The prominent feature in the vertical direction is the thin interbedded sand and mudstone.

3.4. Sedimentary microfacies and distribution characteristics of sand bodies

The study of the planar characteristics of sedimentary microfacies is an important geological basis for reservoir development and analysis. It is a key step in the study of reservoir heterogeneity and remaining oil distribution. It is also the main purpose of microfacies research. It is restricted by a series of factors, including Factors such as paleo-climatic conditions, source area direction and paleo-topographic features, as well as the development status and energy of river systems. On the basis of determining sedimentary facies, subfacies, and microfacies, following the point-line-plane research method, starting from the single-well facies diagram analysis, the well facies diagram analysis is connected again, and finally the sedimentary microfacies of the target layer in the study area is carried out. Accurate analysis and summary of phase plane characteristics.

(1) Yan9_3 sedimentary microfacies and sand body distribution characteristics

The Yan9_3 stage sedimentation was formed in the river-controlled delta plain environment. Drawing the Yan9_3 stage sand body thickness map (Figure 5) and the sedimentary microfacies plan (Figure 6) shows that the study area developed 6 river channels in this stage, which are strips on the plane. It is spread out in a shape of a northwest-south east direction. The widest part of the river is about 7.86km, and the narrowest part is about 0.73km.

In plane, these rivers flow in from the northwest direction and flow out from the southeast direction. The area where the sand ratio is between 0 and 0.2 develops inter-branch swamps, the area where the sand ratio is between 0.2 and 0.4 develops natural embankments, and the area where the sand ratio is greater than 0.4 develops channel deposits, and most of the study area has a well-sand ratio. Above 0.4, channel deposits constitute the main body of the sand body skeleton and are also the main favorable enrichment area.

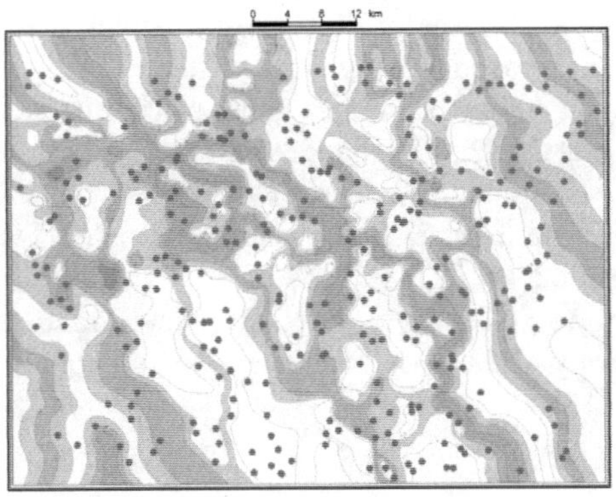

Figure 5 Plan of thickness of $Yan9_3$ sand body in area D

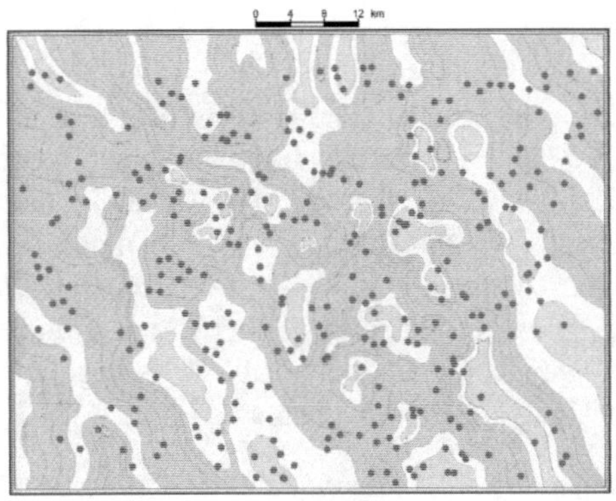

Figure 6 Plan of $Yan9_3$ sedimentary microfacies in area D

(2) $Yan9_{1+2}$ sedimentary microfacies and sand body distribution characteristics

The $Yan9_{1+2}$ stage deposits in area D were formed in the river-controlled delta plain environment. Drawing the $Yan9_{1+2}$ stage sand body thickness map (Figure 7) and the sedimentary microfacies plan (Figure 8) shows that there are 7 river channels in the study area during this period. It is distributed in strips on the plane, and the extension direction is northwest-southeast. The widest part of the river is about 7.78km, and the narrowest part is about 0.61km.

In plane, these rivers flow in from the northwest direction and flow out from the southeast direction. The area of the study area with a sand ratio of 0 to 0.2 develops inter-branch swamps, the area with a sand ratio of 0.2 to 0.4 develops natural embankments, and the area with a sand ratio of greater than 0.4 develops channel sedimentation, and most of the study area has a well-sand ratio of more than 0.2 , Channel sediments constitute the main body of the sand body skeleton and are also the main favorable enrichment area.

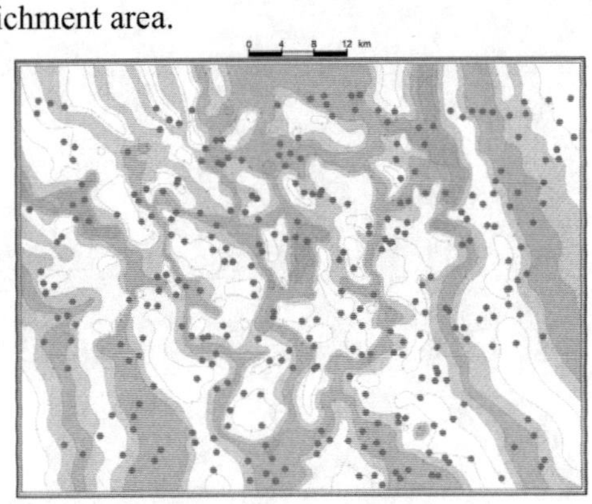

Figure 7 Plan of thickness of Yan91+2 sand body in area D

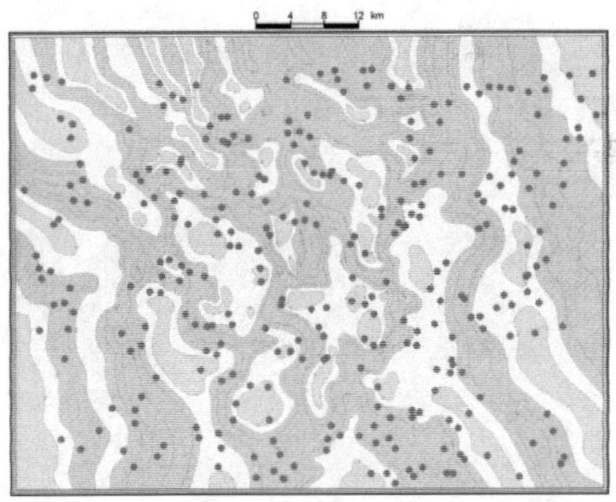

Figure 8 Plan of Yan91+2 sedimentary microfacies in area D

4. Conclusion

The Yan 9 oil-bearing group in the study area is developed with delta plain distributary channels as the dominant facies. The development of sand bodies is strictly controlled by sedimentary microfacies. The main channels are diversion channels, natural embankments and swamp microfacies between branches, and two to three main channels are developed in the two small layers of Yan 9 in the study area.

(2) Vertically, the two sedimentary cycles of $Yan9_3$ and $Yan9_{1+2}$ are mainly developed by distributing channels, and the sand body of $Yan9_3$ is more developed than $Yan9_{1+2}$. The thickness of the sand body can reach more than 10m, and the sand body has good connectivity. On the whole, the Yan 9 oil-bearing

group mainly develops distributary channels and inter-branch swamps. The river channel is mainly composed of medium and fine sandstone deposits, and the sand bodies are distributed in the river channel in a lens shape. The inter-branch swamp is mainly composed of fine sandstone, siltstone, and mudstone. The prominent feature in the vertical direction is the thin interbedded sand and mudstone.

(3) On the plane, the Yan 9 stage sedimentation was formed in the river-controlled delta plain environment. The sand body of Yan 9_3 is more developed than that of Yan 9_{1+2}, and the channel width is larger. The channels of the Yan 9 oil layer group spread out in strips on the plane, extending in the northwest-south east direction. There are 6 main rivers in the study area of Yan9_3 stage and 7 main rivers in Yan9_{1+2} stage. Channel deposits constitute the main body of the sand body skeleton and are also the main oil and gas enrichment area.

References

[1] Tang Jianyun. Study on the difference in oil accumulation conditions and main controlling factors between Yan'an Formation and Yanchang Formation in Dingbian area, Ordos Basin[D]. Northwest University, 2014.

[2] Bai Weiwei. Research on the Jurassic Yan'an Formation Sedimentary System in the Southern Ordos Basin[D]. Northwest University, 2007.

[3] Yang Zhiwen. Characteristics analysis of Yan 6 and Yan 9 oil reservoirs in Yan'an Formation in the southern east area of Luopangyuan in Dingbian Oilfield[D]. Northwest University, 2016.

[4] Liang Jiwei. The Jurassic Sedimentary System and Sequence Stratigraphy in Ordos Basin[D]. Northwest University, 2007.

[5] Yuan Zhen. Research on Sedimentary Facies and Hydrocarbon Accumulation Law of Jurassic Yan'an Formation in Ansai Oilfield, Ordos Basin[D].Northwest University,2007.

[6] Han Meimei,Li Zhongdong,Feng Yibo,Xie Shiwen.Sequence Stratigraphy and Sedimentary Facies Analysis of Yan'an Formation in Mahuangshan Area, Ordos Basin[J].Lithologic Reservoirs,2010,22(03):43-47.

Effect of fiber size of straw on properties of straw board

Li-guang Xiao[1,a] and Yan-bo Ding[2,b]

[1]School of Materials Science and Engineering, Jilin Jianzhu University， Changchun 130118)

[a]Corresponding authors's e-mail: xlg627@163.com @jlju.edu.cn

Abstract: The fiber and powdered reed straw fibers of different sizes were pretreated with 2% NaOH solution at 180℃ and 6MPa for 8min. The straw board without adhesive was prepared by hot pressing. The surface morphology, energy spectrum analysis, thermal stability and mechanical properties of the straw board were studied.The experimental results showed that the surface Si content of straw was significantly increased by alkali activation pretreatment, and the mechanical properties of straw board were significantly improved. The static bending strength (MOR), elastic modulus (MOE) and internal bonding strength (IB) were significantly increased, while the water absorption and expansion rate (TS) were significantly decreased.Compared with fibrous straw board, the fiber bond of straw board made from powdered straw is more compact, its mechanical properties are improved greatly, and its water absorption and expansion rate are reduced obviously.

1. Introduction

In recent years, with the development of infrastructure in various countries, people need a lot of wood and its products. As the main source of wood products, the number of trees has changed greatly. On July 21, 2020, the Food and Agriculture Organization of the United Nations released the Global Forest Resources Assessment 2020, which shows [1] that the total forest area of the world is only 4.06 billion hectares, accounting for 31% of the land area. And the world's forests continue to shrink. In most developing countries, deforestation and overfishing lead to insufficient forest resources [2]. In the context of decreasing wood resources and environmental protection, the use of agricultural and forestry wastes, such as corn straw [3], cotton straw [4] and other resources to replace wood resources, and the research on the preparation of environmentally friendly straw boards with excellent performance has become a hot research issue. As for the selection of raw materials, there are few reports on the preparation of glueless straw board from reed straw. The composition of reed straw is the same as that of wood, but the proportion is different.The main components are cellulose, hemicellulose, lignin and so on. Reed straw can replace wood to produce straw board with good performance. In this experiment, the alkali activation pretreatment method was used to treat the reed straw, and the straw fiber was self-cementing without any adhesive. After alkali solution pretreatment, the straw fibers expand and the surface becomes rough, removing the impurities on the straw surface and improving the interfacial properties of the straw [5]. It is conducive to the bonding of fibers. By using two different sizes of reed straw fiber to prepare straw board, the influence of straw fiber size on the performance of straw board was explored.

2. Materials and test methods

2.1. Materials and Instruments

Reed straw: common annual reed straw in Jilin Province, after drying and dusting, it is crushed by jaw crusher. One part is screened by a sieve with aperture of 1.18mm for reserve use, and the other part is crushed by a sieve with aperture of 0.6mm for powder reserve use. Sodium hydroxide: solid particles, analytical pure, Tianjin Jindong Tianzheng Fine Chemical Reagent Factory; The water used in this experiment: the water with sodium hydroxide solution is deionized water, and the other water is from laboratory tap water.ML-203 Electronic Analytical Balance (Metler Tolidor Instrument Co., Ltd., China), WGL-65 Electric Hot Blow Drying Oiler (Tianjin Tester Instrument Co., Ltd., China), DM-4L Jaw Crusher ((DM-4L, Nanjing Daran Technology Co., Ltd., Nanjing, China), TM3030 Scanning Electron Microscopic ((MIRA 3 LMH,TESCAN, USA), IRAffini-1 IR Spectrometer ((IrAffini-1, Shimadsu Manufacturing Institute, Japan), TGA-50 Thermogravimetric Analyzer (TGA-50, TA Instruments, Inc., Newcastle, USA), YAD-2000 Temperature Controlled Pressure Forming-System (Changchun Kexin Equipment Co., Ltd.,China).

Table 1 Analysis results of reed straw components

Composition	Crude fiber				ash	other
	cellulose	hemicellulose	lignin	horny		
percentage (%)	42.7	29.5	14.3	7.1	5.8	0.6

2.2. Preparation

Treat all ingredients at room temperature.NaOH solution with a mass fraction of 2% was prepared. Two kinds of reed straw fibers of different sizes were immersed in alkali solution for 40min. After the soaking, the two kinds of straw fibers of different sizes were washed with deionized water, respectively, until the pH value of the water was about 7. Put the cleaned straw into the oven and dry it until the moisture content is 30%.According to GB/T 27796-2011[6] , the design density of the straw board is 0.8g/cm3, the size is 100mm×100mm×10mm, and the size of the molding mold is 100mm×100mm×80mm.The treated reed stalks are weighed and put into the mold. After manual prepressing, they are put into the hot press. The pressure set by the hot press is 6MPa, the temperature is 180°C, and the time is 8min. The hot-pressed reed straw sheet is shown in Figure 1.At the end of the experiment, the straw board prepared by two different sizes of reed straw fibers was placed at about 24°C and relative humidity was about 60%. After stable for 24h, the test and analysis began.

(a) (b)

Figure 1 . Straw board prepared from reed straw fibers of different sizes.(a) Fibrous straw;(b) Powder straw.

2.3. Test

The profile microstructure of reed straw board and the energy spectrum analysis of reed straw before and after alkali treatment were tested and analyzed by MIRA 3 LMH scanning electron microscope and energy spectrum analyzer.Thermogravimetric analyzer was used to analyze the thermal stability of reed straw. The temperature was raised from the ambient temperature to 600°C, and the heating rate was

15°C/min. IRAffinity-1 IR spectrometer. According to GB/T 17657-2013[7], the physical performance test is to test the static bending strength (MOR), elastic modulus (MOE), internal bonding strength (IB), water absorption thickness and expansion rate (TS).The straw board made of two different sizes of reed straw fibers was tested with 3 specimens respectively, and the average value was taken.

3. Test results and Analysis

3.1. Microscopic morphology analysis
Fig. 2 shows the micromorphology of straw board prepared by reed straw fibers of different sizes. In Fig. 2 (a), after the fibrous straw was hot-pressed at 180°C, the microscopic morphology of the straw board showed that irregular cracks and irregular cracks existed obviously in the observed area, and the straw fibers were not tightly bonded.At the same time, the irregular cracks mean that the fiber distribution in the straw board is not uniform during the hot pressing process, which may lead to poor mechanical properties of the straw board. In Fig. 2 (b), the microscopic morphology of the straw board after the hot pressing of the powdered straw at 180°C shows that the tissues are tightly bonded and seamless.This represents superior mechanical performance in the later stage. It can also be observed that there are no holes and thermal decomposition in the microscopic planing surface of the straw board after hot pressing, which indicates that the hot pressing temperature has no effect on the mechanical properties of the straw board in the later stage.

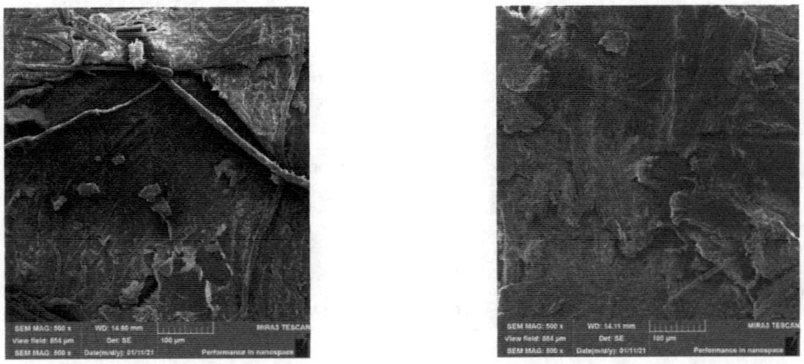
(a) (b)

Figure 2 Microstructure diagram of straw board prepared by reed straw fibers of different sizes.(a) Fibrous straw; (b) Powder straw.

3.2. Spectrum analysis
Fig. 3 shows the Si element distribution on the surface of straw before and after pretreatment and on the surface of straw board with different sizes of straw fibers after hot pressing. As shown in figure (a), (b), without pretreatment of straw surface can be found that there are a lot of Si elements per unit area, per unit area after alkali pretreatment of straw surface Si element within the mass fraction of greatly reduced, it shows that the pretreatment of straw using alkali solution can effectively remove the straw surface nonpolar siliceous, improve the wettability on the surface of the reed straw, Also makes the polar fiber exposed, conducive to the bonding between the fibers. As shown in Fig. (c) and (d), there is little difference in the Si element content per unit area on the surface of straw board prepared with two different sizes of straw fibers, which indicates that different straw sizes have little influence on the Si element content on the surface of straw. In addition, the literature [8] also showed that sodium hydroxide solution treatment of straw can dissolve the non-polar organic wax on the surface of straw, expose the internal polar fiber, and further promote the bonding between corn straw fibers.

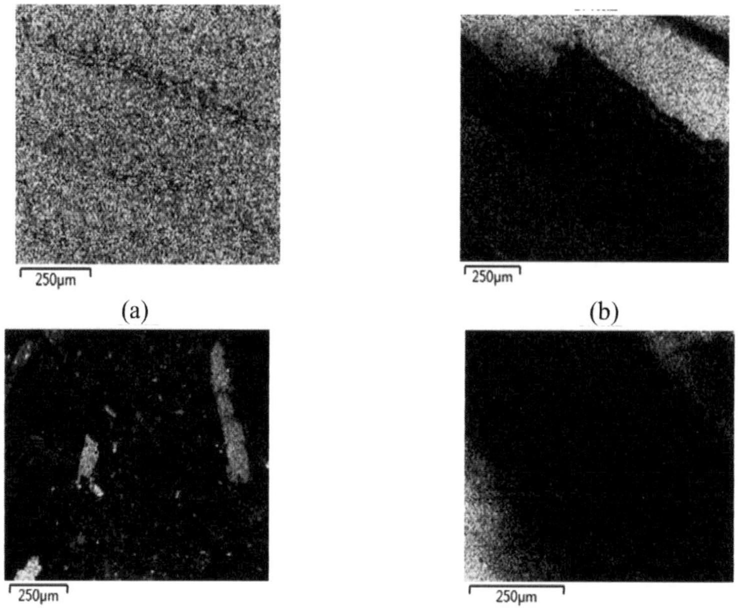

Figure 3 Si element distribution on the surface of straw before and after pretreatment and on the surface of straw board with different sizes of straw fibers after hot pressing.(a) Not processed;(b) NaOH treatment;(c) Fibrous straw;(d) Powder straw.

3.3. *Thermal stability analysis*

Fig. 4 shows the thermogravimetric analysis curve of reed straw.As can be seen from the figure, when the temperature is about 100°C, the weight of reed straw begins to decrease. The reason may be that the moisture in reed straw evaporates and diffuses into the air due to heating. After that, the weight does not change until the temperature rises to about 200°C and the weight begins to decrease. From 200°C to 450°C, the weight has been decreasing.Literature [9] reported that the thermal degradation sequence of plant macromolecules is hemicellulose (200-260°C), cellulose (240-350°C) and lignin (280-500°C). The thermal decomposition of hemicellulose may occur in the hot pressing of straw board at temperatures of 200°C and 230°C. At temperatures above 400 ° C, the remaining lignin gradually degrades until about 14% of the ash content is left. It can be seen that the straw fiber has good stability and no decomposition trend in the process of hot pressing at 180°C.

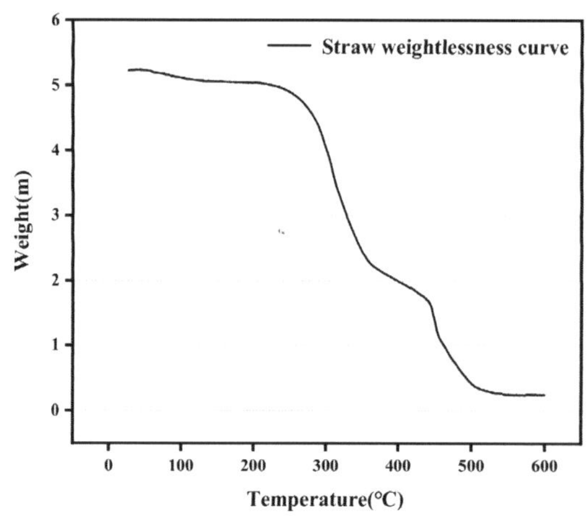

Figure 4 weightlessness curve of reed straw.

3.4. FT-IR

Figure 5 shows the infrared spectra of reed straw fibers of different sizes after alkali treatment and hot pressing. The -C=O asymmetrical vibration absorption peak of reed straw in 1610 ~ 1640cm^{-1} was obvious, which indicated that reed straw was modified with carboxyl group. 2900.2cm-1 and 1372cm^{-1} are respectively the expansion vibration absorption peak and bending vibration absorption peak of -CH;1636.0cm^{-1} is the vibration absorption peak of adsorbed water.1130~1170cm^{-1} is the C-O stretching vibration absorption peak of annular C-O-C [10]. The C-O vibration of cellulose and xylan is characterized by an absorption band with a wavelength of 1000-1200cm^{-1} [11].After hot pressing, the C-O carbonyl absorption bands shift at the 1000-1100cm-1 and 1739cm^{-1} bases, indicating more hydrolysis of hemicellulose and the formation of aldehyde compounds [10].Studies have shown that lignin has the same characteristics as polymer, which can improve the bonding property of biological boards, and then improve the mechanical properties of biological boards [11]. After hot pressing, a relatively obvious change at 1162cm^{-1} is considered to be the generation of new ether bonds, which is the reaction between the by-product of hemicellulose and lignin in the hot pressing process, i.e., lignin-carbohydrate complex (LCC)[12]. The enhancement at 900cm^{-1} can be attributed to the polymerization of glycoside components in low-molecular compounds [13]. All these reactions can improve the bond strength of reed straw board.

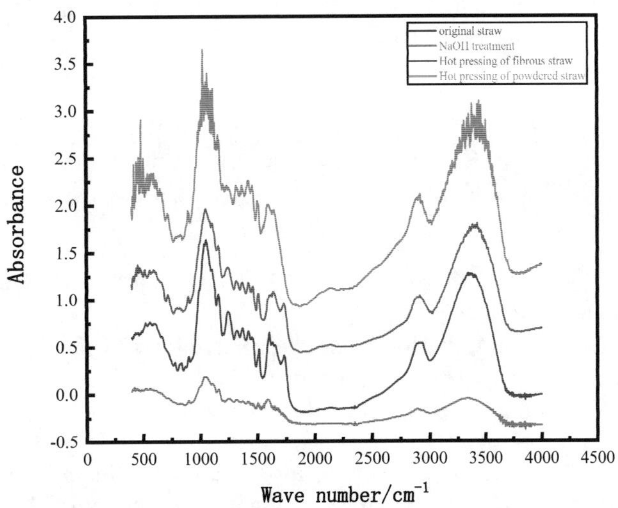

Figure 5 FT-IR images of reed straw of different sizes before and after alkali treatment and hot pressing

3.5. Mechanical property analysis

Table 2 shows the static bending strength (MOR), elastic modulus (MOE), internal bonding strength (IB) and water absorption and expansion rate (TS) of straw board prepared with reed straw fibers of different sizes. Group A is the mechanical properties of the straw board prepared by fibrous straw, and group B is the mechanical properties of the straw board prepared by powder straw. It can be seen from Table 1 that the mechanical properties of straw board can be changed by improving the size of straw fiber. Compared with the fibrous straw board, the mechanical properties of the powdered straw board were improved, including the static bending strength (MOR) increased by 40.5%, the elastic modulus (MOE) increased by 7%, the internal bonding strength (IB) increased by 28.6%, and the water absorption and expansion rate decreased by 25.6%. The reasons for the improvement of mechanical properties of straw board prepared by powder straw may be as follows:Compared with the straw board prepared by fibrous straw, the hydrothermal action of powdered straw is more intense in the process of hot pressing, and the specific surface area of powdered straw is larger, in which more hemicellulose and cellulose hydrolyze, and more low-molecular weight furfural hydrolytic products that can react with lignin are generated.More phenolic-like adhesives were produced that acted as adhesives [14]. The infrared spectrum also confirmed the results.

Table 2 MOR, MOE, IB and TS values of different pretreatment artificial straw boards

Code	MOR（MPa）	MOE（MPa）	IB（MPa）	TS（%）
A	17.3	2379	0.49	39
B	24.3	2563.3	0.63	29

4. Conclusion

In summary, two kinds of straw boards with different fiber sizes were prepared by alkaline solution pretreatment and hot pressing. The microscopic morphology of straw board prepared by powdered straw showed that the fiber in the straw board was compact and uniform, and there were no holes and signs of thermal decomposition in the microscopic section of straw board, which indicated that the hot pressing temperature had no effect on the mechanical properties of straw board in the later stage. At the same time, alkali solution pretreatment can reduce the Si element content on the straw surface, improve the wettability of the straw surface and expose the polar fiber on the straw surface, so that the straw fibers are easy to glue.After alkali treatment and hot pressing, reed straw produced reactions and products that were beneficial to self-cementing. Compared with the straw board prepared by fibrous straw with large size, the mechanical properties of straw board prepared by powdered straw were significantly improved.

Acknowledgments

This work was financially supported by China's 13th five-year national key research and development plan (No.2018YFD1101001) and the Key project of science and technology development plan of jilin province, China (No.No.20180312018ZX).

Reference

[1] Global Forest Resources Assessment 2020,United Nations Food and Agriculture Organization,Rome, Italy:2020(5).

[2] Azizi K , Tabarsa T , Ashori A . Performance characterizations of particleboards made with wheat straw and waste veneer splinters[J]. Composites Part B Engineering, 2011, 42(7):2085-2089.

[3] Wang D , Sun X S . Low density particleboard from wheat straw and corn pith[J]. Industrial Crops & Products, 2002, 15(1):43-50.

[4] A Z H , A H N C , A K T K , et al. Optimization and practical application of cottonseed meal-based wood adhesive formulations for small wood item bonding[J]. International Journal of Adhesion and Adhesives,2019(95):1147-1152.

[5] LIU F H . Properties of composites made of pretreated-corn stalk fibers and high-density polyethylene[D]. Harbin: Northeast Forestry University,2015.

[6] GB/T 27796-2011 Plant boards of straw for construction (2); Inspection and Quarantine of the People's Republic of China/Standardization Administration of the People's Republic of China: Beijing, China, 2013.

[7] GB/T 176577-2013 Test Methods of Evaluating the Properties of Wood-Based Panels and Surface Decorated Wood-Based Panels; Inspection and Quarantine of the People's Republic of China/Standardization Administration of the People's Republic of China: Beijing, China, 2013.

[8] F zhang,L Zhang，QI Chu-Sheng,et al.Effects of pretreatment methods on properties of corn straw board[J].Journal of Beijing Forestry University,2017,9(39):112-118.

[9] Macedo, J.S.; Otubo, L.; Ferreira, O.P.; de Fátima Gimenez, I.; Mazali, I.O.; Barreto, L.S. Biomorphic activated porous carbons with complex microstructures from lignocellulosic residues. Microporous Mesoporous Mat. 2008, 107, 276–285.

[10] CHEN J, YAN T, YAN S, et al.Experimental study on liquefied corn straw bonding powder coal [J]. Cellulose Science and Technology, 2019(3):8-14.

[11] YANG Z L, DU S R, MEI J Q, et al. Characterization of lignin content and distribution in corn straw treated with alkali by FTIR microscopic imaging [J].Transactions of the CSAE, 2019, 035(008):280-286.

[12] Jiajun, Wang, Bo, et al. Effect of Hot-Pressing Temperature on Characteristics of Straw-Based Binderless Fiberboards with Pulping Effluent.[J]. Materials, 2019,12(922):6-17.

[13] Dam J E G V , Oever M J A V D , Teunissen W , et al. Process for production of high density/high performance binderless boards from whole coconut husk: Part 1: Lignin as intrinsic thermosetting binder resin[J]. 2004, 19(3):207-216.

[14] Okuda N , Hori K , Sato M . Chemical changes of kenaf core binderless boards during hot pressing (I): influence of the pressing temperature condition[J]. Journal of Wood ence, 2006, 52(3):244-248.

Pore-network distribution laws of cementitious materials detected by high-energy X-ray computed tomography imaging system

Qifeng Lyu[1*], Yanwei Guo [1], Anguo Chen[1] and Jie Jia[1]

[1] Department of Civil Engineering, Changzhou University, Changzhou, Jiangsu Province, 213164, China

[*]Corresponding author's e-mail: lqf@cczu.edu.cn

Abstract. Pore networks of cementitious materials determine the materials' properties and engineering safety. In this work, four kinds of cementitious materials which are geopolymer cement, foamed cement, ordinary concrete and pervious concrete were prepared, and a high-energy X-ray computed tomography imaging system was used to detect the samples' inner structures. It was found that the pore network of such materials show similarity to social networks. However, the traditional mathematical degree which is the coordination number of a network is insufficient to characterize the pore networks, because such networks have volume limitations. Thus a new parameter, physical degree, was proposed in this work. By analyses of the pore networks of the samples, it was found the mathematical degrees follow Poisson distribution law whereas the physical degree can be fitted by power law. This indicates the pore networks are random and scale free. These findings may provide insight for future intelligent design of digital cementitious materials.

1. Introduction

Pore networks of cementitious materials determine the materials' strength[1], durability[2], permeability[3], and thermal resistivity[4] etc. Thus understanding the governing laws of pore-network distribution is crucial[5], especially from macroscopic perspective. On the other hand, with the rise of additive manufacturing, the 3D printing of cementitious materials [6] can also benefit from the thorough understanding of pore-network distributions.

The traditional methods to detect the pore network of cementitious materials are mainly indirect methods such as mercury intrusion porosimetry (MIP) [7] which is a destructive method that can give information about the connected porosity, pores volume and size distribution. Besides this method, the nuclear magnetic resonance (NMR) method [8] can also get the information of inner pore networks of cementitious materials by detecting hydrogen nuclear magnetization relaxation. This method is nondestructive and allows 3D image reconstructions. But it also needs the pores are permeable with detectable quantities of hydrogen nuclear, though the majority pores in cement are impermeable. Nevertheless, the X-ray computed tomography (CT) method has no such limitation, and thus it is widely used [9]. This method detects the X-rays attenuation when they pass through the materials and thus can provide detailed inner structure of the materials [10].

Usually, the CT image data are extremely huge whereas the sample size is very small. To correlate the sample's micro parameters with its macro properties, we may draw support from the network theories which are highly developed but lack enough application in the analysis of pore networks of cementitious materials. Indeed, pore networks are quite similar to social networks, cf. Figure 1 where two pore networks of cementitious materials and two social networks [11] are compared. Their degree distributions show similarity to each other. Here the degree is also named as coordination number in literatures of porous media. From the similarity, it can be inferred that pore networks may follow similar laws of social networks. This paper was planed to check this point. Here, four kinds of samples of cementitious materials were prepared, and an industrial X-ray computed tomography imaging system was used to detect the samples' inner structures. By data analysis, the pore-network distribution laws of such materials were concluded in this paper.

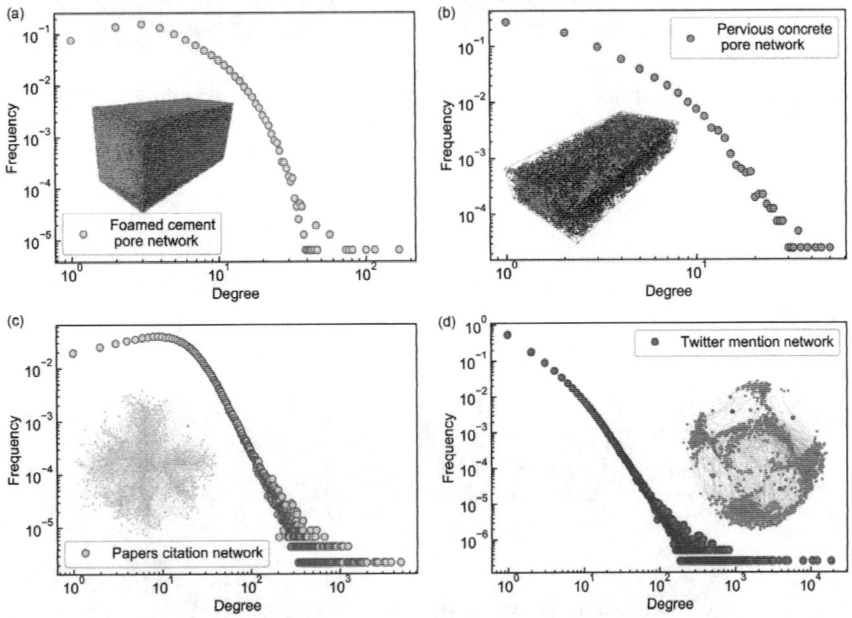

Figure 1. Degree distributions of four networks: (a) Pore network of foamed cement; (b) Pore network of pervious concrete; (c) Papers citation network of American Physical Society journals until 2005; (d) Twitter mention network during January 23rd and February 8th, 2011.

2. Materials and Methods

2.1. Samples

Four kinds of cementitious materials which are geopolymer cement, foamed cement, pervious concrete, and ordinary concrete were used in the experiment, cf. Figure 2. Specifically, Figure 2a is 3 pieces of geopolymer cement with the same dimensions of 50 mm × 50 mm × 50 mm. Figure 2b shows a piece of foamed cement with the dimensions of 78 mm × 86 mm × 200 mm. Figure 2c is a piece of ordinary concrete with the dimensions of 50 mm × 110 mm × 200 mm, which was made of ordinary Portland cement, water and aggregates. Figure 2d shows a piece of pervious concrete with the dimensions of 53 mm × 115 mm × 240 mm, which is made of Portland cement, water, slag and coarse recycled aggregates.

2.2. X-ray CT imaging system

All the samples were detected by a high-energy X-ray CT imaging system. The CT facilities mainly consist of three parts which are X-ray source, sample table and X-ray detector as shown in Figure 3. The maximum voltage of the X-ray source in this CT system can reach up to 450kV which can generate high-energy X rays that can penetrate large or dense samples.

Figure 2. Samples: (a) geopolymer cement, (b) foamed cement, (c) ordinary concrete and (d) pervious concrete.

In detection, the X-ray source of the CT facilities generate X-rays with an incident intensity I_0 to penetrate the samples on the sample table, and then the X-ray detector receives attenuated X-ray intensity I which follows the Beer-Lambert law. Indeed, the intensity reflects material densities varying in space. Through the histogram of intensities, air (pore) can be differentiated from solid (structure) by the Otsu method. All the CT images were processed in VG Studio and Avizo. Then the PNextract was used to extract the conventional pore networks from the processed image data. After that, the pore networks can be analyzed by OpenPNM and NetworkX which are specific Python packages for network analysis.

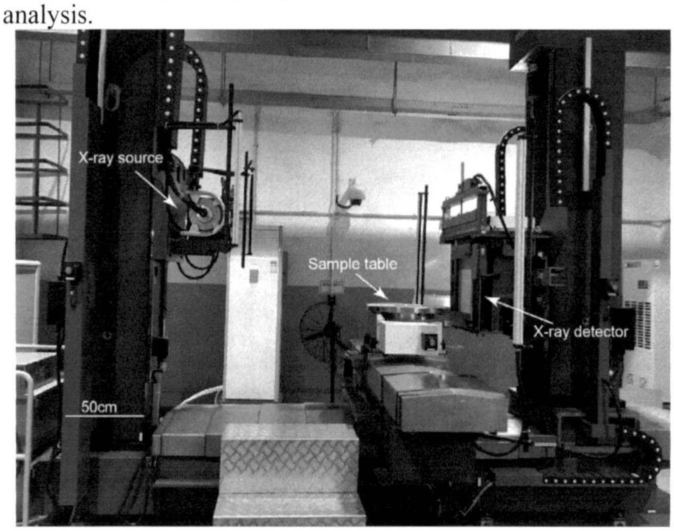

Figure 3. X-ray CT facilities.

2.3. Degrees

The degree, also named coordination number, means the number of links attached to a pore in a network, which plays crucial roles in network theories. However, the traditional degree is insufficient to describe the pore networks, because these networks, unlike social networks, have space dimensions. That is to say they are affected by their volumes. Thus a new parameter, physical degree d_p, was defined in this work, and the traditional degree is renamed as mathematical degree d_m. The physical degree d_{pi} for the ith pore in a pore network can be written as

$$d_{pi} = V_{pi} + \sum_{j=1}^{n} V_{tj}, \tag{1}$$

where V_{pi} represents the volume of the ith pore which has n throats attaching to it, and V_{tj} represents the volume of the jth throat of the ith pore.

2.4. Distribution laws

Two kinds of distribution law were considered in this work, which are power law and Poisson law. The power law is the only scale-free distribution law and also the main concern of this work, which can be written as

$$f = \gamma d^{-\alpha}, \tag{2}$$

where d represents the degree, f represents the frequency of degrees, γ and α are fitting coefficients. In logarithmic coordinates, the power law can be written in a form of linear function which is

$$y = -\alpha x + \beta, \tag{3}$$

where $y = \log f$, $\beta = \log \gamma$, and $x = \log d$. The Poisson law, which is also widely used to demonstrate the randomness of variables, can be written as

$$f = \frac{e^{-\lambda}\lambda^{d}}{d!}, \tag{4}$$

where e is the Euler's number, λ is a fitting coefficient.

The Levenburg-Marquardt algorithm in SciPy, which is a Python package, was used to fit the data, and the coefficient of determination was used to show the goodness of fit, which is

$$r^2 = 1 - \frac{\sum(y - \hat{y})^2}{\sum(y - \bar{y})^2}, \tag{5}$$

where y represents the sample data, $\hat{y}$ represents the fitted data, and y represents the mean of sample data. When r^2 approaches 1, the fitting result is the best.

3. Results and discussion

The Poisson distribution law can demonstrate the randomness of variables, which was used to fit the mathematical degrees. Results are shown in Figure 4. Basically, the Poisson distribution law catches the trends of mathematical degrees. The fitting formula as well as its related goodness of fit r^2 are also shown in each sub-figure of Figure 4. The expectation (λ) of mathematical degrees of each sample can be found in each fitting formula. Specifically, the λ of foamed cement in Figure 4b is 3.998 which is the highest expectation among the four pore networks. This indicates the majority degree of foamed-cement pore network is around 3.998 (i.e. 4). Besides the foamed cement, the λ of pervious concrete (λ =1.373, cf. Figure 4d) is also beyond 1. These higher expectations (i.e., higher λ) indicates the pores in the networks are well connected. On the contrary, the λs of geopolymer cement (λ = 0.036, cf. Figures 4a) and ordinary concrete (λ =0.274, cf. Figures 4c) are less than 1. That is to say the pores of such samples are mainly isolated, which indeed can be also conjectured from their CT images shown in each subfigures of Figure 4.

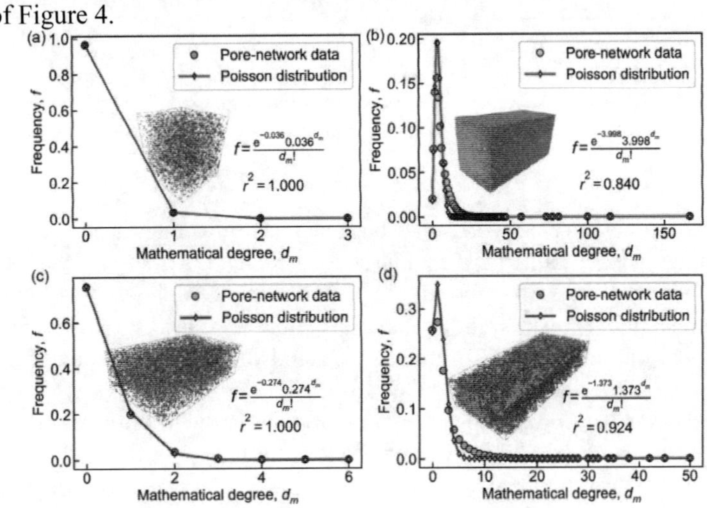

Figure 4. Fitting results of Poisson distribution law for the mathematical degree of (a) geopolymer cement, (b) foamed cement, (c) ordinary concrete and (d) pervious concrete.

It can be inferred from Figure 1 that mathematical degrees of cementitious materials do not perfectly follow the power law which should be a straight line in the logarithmic coordinates. From another point of view, the physical degree (Equation 1) proposed in this work was meant to demonstrate the preferential-attachment mechanism, and thus the physical degree may follow the power law. Indeed, the fitting results of physical degrees of each sample are shown in Figure 5, where it can be found the goodness of fit for each sample approaches 1. This indicates the fitting results are excellent and the physical degree follows the power law. Nevertheless, the data of foamed cement (Figure 5b) show slight deviation from the power law although the coefficient of determination r^2 is greater than 0.8 which is indeed a high value for fitting results. The reason for this deviation is due to the special pores of foamed cement, and these pores are not totally controlled by the preferential-attachment mechanism. Indeed, the pores in foamed cement are made by blowing agent and they are usually uniformed. This contributes to the slight deviation of the fitting result. For the parameters in the fitting formula, α is the slope of the fitting line, which satisfy $2 < \alpha < 3$. This is coincident with social networks [13]. β represents the intercept of the fitting line. In Figure 5, it can be found the absolute value of β for the geopolymer cement is the highest. This indicates the frequency of larger pores in geopolymer cement is the lowest. In other word, there are many more small pores in geopolymer cement whereas the pores in other samples are relatively larger.

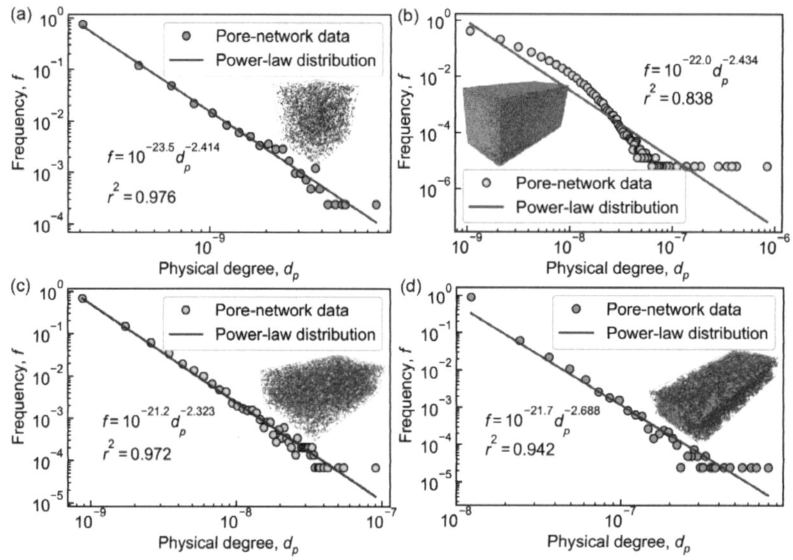

Figure 5. Fitting results of power law for the physical degree of (a) geopolymer cement, (b) foamed cement, (c) ordinary concrete and (d) pervious concrete.

4. Conclusions

To find the macroscopic pore-network distribution laws of cementitious materials, four kinds of cementitious samples were scanned by a high-energy X-ray CT imaging system. By analyzing the reconstructed CT pore structures of each sample, conclusions were drawn as below:

- The pore networks of cementitious materials show similarity to social networks. But the pore networks are limited by their volume space and the traditional degree distribution laws are insufficient to describe the particular properties of pore networks. Thus a new parameter, physical degree was proposed in cooperation with the traditional mathematical degree to characterize the pore network of cementitious materials.
- The Poisson distribution law is suitable to describe the mathematical degrees and showed excellent fitting agreement and reasonable parameters. For the physical degree, which was defined to represent the preferential-attachment mechanism that pervades a scale-free network,

the power law can describe the data trend, though the physical degrees of foamed cement deviate a little since their pore networks are also controlled by other mechanisms.

- the fitting parameters also show specific physical meanings. Larger λ of the Poisson distribution law indicates the pores in the sample are mainly connected whereas smaller ones, especially when $\lambda < 1$, indicate the pores are isolated. The slopes of power law for the physical degrees show $2 < \alpha < 3$. And the lower intercepts of the fitting lines (power law) indicates the frequency of larger pores in the samples are lower.

These findings may provide insights for the design of cementitious materials. Especially, a model could be proposed based on the findings. And such model could be developed and used in optimizing both the permeability and strength of cementitious materials in future.

Acknowledgments
This work was supported by the National Natural Science Foundation of China (51908075)

References

[1] T. T. Nguyen, H. H. Bui, T. D. Ngo, G. D. Nguyen, M. U. Kreher, F. Darve. (2019) A micromechanical investigation for the effects of pore size and its distribution on geopolymer foam concrete under uniaxial compression, Engineering Fracture Mechanics 209: 228-244.
[2] M. A. Vicente, D. C. Gonzlez, J. Mnguez, M. A. Tarifa, G. Ruiz, R. Hindi. (2018) Influence of the pore morphology of high strength concrete on its fatigue life, International Journal of Fatigue 112: 106-116.
[3] F. Yu, D. Sun, M. Hu, J. Wang, Study on the pores characteristics and permeability simulation of pervious concrete based on 2d/3d CT images. (2019) Construction and Building Materials 200: 687-702.
[4] F. L. Monte, R. Felicetti, M. J. Miah. (2019) The influence of pore pressure on fracture behaviour of normal-strength and high-performance concretes at high temperature, Cement and Concrete Composites 104: 103388.
[5] J. Zhang, Y. Huang, G. Ma, J. Sun, B. Nener. (2020) A metaheuristic-optimized multi-output model for predicting multiple properties of pervious concrete, Construction and Building Materials 249: 118803.
[6] M. Chen, L. Li, J. Wang, Y. Huang, S. Wang, P. Zhao, L. Lu, X. Cheng. (2020) Rheological parameters and building time of 3d printing sulphoaluminate cement paste modified by retarder and diatomite, Construction and Building Materials 234: 117391.
[7] J. Tatar, N. R. Brenkus, G. Subhash, C. R. Taylor, H. Hamilton. (2018) Characterization of adhesive interphase between epoxy and cement paste via Raman spectroscopy and mercury intrusion porosimetry, Cement and Concrete Composites 88 187-199.
[8] R. S. Holthausen, M. Raupach. (2019) A phenomenological approach on the influence of paramagnetic iron in cement stone on 2D t1-t2 relaxation in single-sided 1h nuclear magnetic resonance. Cement and Concrete Research 120: 279-293.
[9] S. Brisard, M. Serdar, P. J. Monteiro. (2020) Multiscale X-ray tomography of cementitious materials: A review, Cement and Concrete Research 128: 105824.
[10] P. Wang, H. Qiao, Y. Zhang, Y. Li, Q. Feng, K. Chen. (2020) Meso-damage evolution analysis of magnesium oxychloride cement concrete based on X-CT and grey-level co-occurrence matrix, Construction and Building Materials 255: 119373.
[11] S. Pei, L. Muchnik, J. S. Andrade Jr, Z. Zheng, H. A. Makse. (2014) Searching for superspreaders of information in real-world social media, Scientific Reports 4: 5547.

Finite Element Analysis on Hysteretic Behavior of H-shaped Columns of Thin-Walled Sections of 690 MPa High Strength Steel

Zhongqi Chen[1,*], Wei Feng[2], Ruipeng Guo[2], Chunfeng Liu[2]

[1]School of Civil Engineering, Qingdao University of Technology, Qingdao 266033, China

[2] Shandong Hi-Speed Engineering Construction Croup Co. Ltd., Jinan 250014, China

[*]Corresponding author's e-mail: zhongqi.chen2018@outlook.com

Abstract. High strength steel is widely used in engineering, especially in light steel structures. However, due to the material properties of high strength steel, its seismic performance has attracted wide attention. To study the hysteretic behavior of H-shaped members of thin-walled sections of Q690 high strength steel, a finite element model was built up by using the general finite element analysis software ABAQUS. Based on the finite element model, an extensive parametric analysis of H-shaped members of thin-walled sections of Q690 high strength steel was carried out. The results showed that the limiting flange width-thickness ratio and web height-thickness ratio for H-shaped members of Q690 high strength steel were recommended to be 11 and 65, respectively, in actual engineering. Also, an increase in web height-thickness ratio and axial compression ratio could lead to a significant decrease in the ductility of H-shaped members of thin-walled sections of Q690 high strength steel. The thinner and more flexible the section is, the larger the axial compression ratio is, further reducing the energy dissipation ability of the members.

1. Introduction

In recent years, with the widespread application of high-strength structural steels (with a yield strength of not less than 460 MPa, and hereinafter referred to as HSS) in construction projects, the seismic performance of high strength steels has been gaining prominent attention [1, 2]. Generally, in order to ensure the plastic rotation capacity of structural members under seismic actions, steel structures are usually designed as members with a small section width-thickness ratio. When the weight per unit length is certain, since HSS members with a larger section width-thickness ratio have greater bending stiffness, yield moment, and overall elastic stability than those with a smaller section width-thickness ratio, HSS members have good economic advantages and can be used in multi-story light steel structural systems [3]. However, due to their large width-thickness ratio, slabs usually are locally unstable, making it hard for the members to fully develop their plastic bending capacity. Besides, the members have low ductility and energy dissipation capacity, which greatly limits their applications in seismic design with high requirements for the plastic energy dissipation capacity of the members [4].

In the Chinese Code for Design of Steel Structures (GB50017-2017) [5], sections are divided into five classes, that is, S1, S2, S3, S4, and S5. Systematic studies on the hysteretic behavior of Q690 HSS H-shaped members with a section class of S1, S2, and S3 have been conducted. Chen et al [6-7] performed cyclic loading tests and finite element analysis for Q690D HSS welded H-section and box-section columns. The results showed that the members had good energy dissipation capacity and the maximum inter-story displacement angle met the requirements of the Code for Design of Seismic Design of Buildings (GB 50011-2010) [4]. Through the study on Q690D HSS compact H-sections, Hai et al [8, 9] found that Class 1 and Class 2 H-sections had good plastic rotation capacity and could be used in seismic steel framing systems according to Eurocode [10] (EC3). However, few studies have been conducted for H-shaped members of Q690 thin-walled sections with a relatively large section width-thickness ratio.

In order to study the hysteretic behavior of H-shaped members of Q690 thin-walled sections, this paper established a finite element analysis model using the general finite element analysis software ABAQUS and conducted a parametric analysis for the hysteretic behavior of cantilever columns under constant axial force and horizontal cyclic load. In addition, the effects of web width-thickness ratio, flange width-thickness ratio, and axial compression ratio on the hysteretic behavior of H-shaped columns of Q690 thin-walled sections were analyzed.

2. Establishment of Finite Element Model

2.1 Material Model

The Mises yield criterion and the kinematic hardening model were used, and the trifold model was adopted in the material constitutive model. In parametric analysis, the measured values of 10mm Q690D steels in tensile testing in the literature [11] were used as the material strength. The yield strength f_y of the steel plates was 823.7 MPa, the ultimate strength fu was 890.4 MPa, the elastic modulus E was 207 GPa, and the tangential modulus in the strengthening phase was taken as $E_h=1/100E$ with a Poisson's ratio of 0.3.

2.2 Boundary Conditions and Mesh Division

The S4R element was used for building the model with a uniform mesh division pattern as shown in Fig. 1. The boundary conditions of the finite element analysis were the same as that of the test, with the top being the free end and the bottom being completely rigidly connected. The boundary conditions of the finite element analysis were used as shown in Fig. 2.

2.3 Division Initial Defects and Residual Stresses

Through eigenvalue buckling analysis, the first-order buckling mode of the members under the combined actions of axial and horizontal loads was obtained, which could be used as the geometric initial defect mode. Then the mode was applied to the overall model. The size of the geometric defects was taken as B/100 and the minimum value was 2 mm [12], where B was the width of the flange of the H-shaped members. Considering the effects of residual stresses, the Q690 residual stress distribution model proposed by Li et al [13] was used and applied to the overall model.

3. Parametric Analysis

3.1 Parameter Setting

Based on the verified finite element model, an extensive parametric analysis was performed on the H-shaped members of Q690 HSS thin-walled sections. According to the classification of plate sections in GB 20017-2017 [5], the sections with a section class of S5 are called thin-walled sections ($M_u < M_y$). Therefore, the limiting width-thickness ratio of S5 sections was used for the parametric analysis (Table 1). For the parametric analysis, five groups of members were selected in terms of flange width-thickness ratio and web width-thickness ratio, that is, 11, 13, 15, 17, 19 and 55, 60, 65, 70, 75, respectively. The axial compression ratio was 0.2 and 0.35, and a total of 50 members was selected, and all the members were of the S5 section. The naming rule of the members is H-x-y-z, where H stands for H-shaped section member, x stands for flange width-thickness ratio, y stands for web height-thickness ratio, and z stands for axial compression ratio.

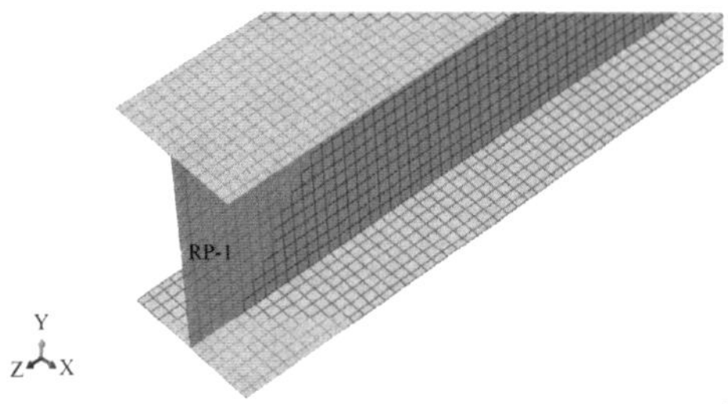

Fig. 1. Mesh Division Pattern

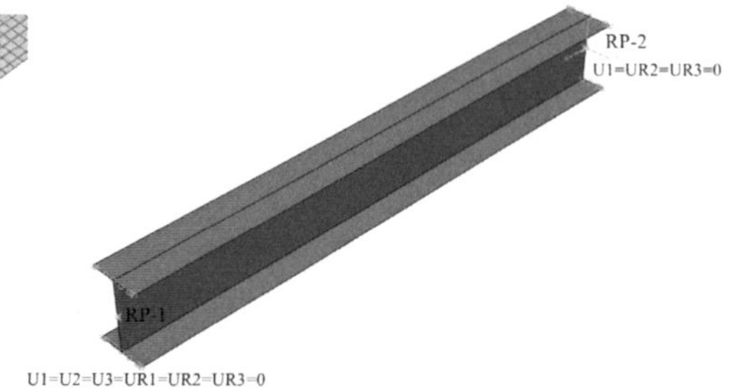

Fig. 2. Boundary Conditions

3.2 Ultimate Bearing Capacity

Table 2 shows the calculation results of the members through the parametric analysis. The ultimate flexural bearing capacity M_u can be obtained from the peak value of the M-θ curve. The nondimensionalized ultimate bearing capacity (M_u/M_p, M_u/M_y) reflects the plastic development capacity of the section to a certain extent, where M_p is the plastic moment of the full section with the influence of the axial force into consideration and M_y is the elastic yield moment calculated according to the elastic theory. The M_u/M_p value of all members is summarized in Fig. 3 and Fig. 4. Fig. 3 shows the effect of flange width-thickness ratio and axial compression ratio on M_u/M_p, from which it can be seen that M_u/M_p decreases with the increase of flange width-thickness ratio. As we can see from Fig. 4, the effect of web height-thickness ratio on M_u/M_p of the members is significantly smaller than that of the flange width-thickness ratio and shows a slowly decreasing trend, indicating that the flange width-thickness ratio is a decisive factor affecting the ultimate bearing capacity of the members. In addition, from Table 2, it can be seen that M_u/M_p is either greater than 1 or less than 1, which means that the members are in an elastic state or can only develop limited plastic capacity when reaching the ultimate bearing capacity. In this case, the members belong to an S4 section or S5 section, which is because the current code does not consider the interaction between the plates. Under the premise of considering the interaction between the plates, the limiting width-thickness ratio of the plates of the compression-bending members of Q690 HSS H-shaped sections can be extended appropriately. Based on the calculated data in Table 2, in order to ensure that the members can enter the plastic stage under the compression-bending load state and give full play to the plastic capacity of the sections, it is recommended that the limiting flange width-thickness ratio of Q690 HSS H-shaped members in engineering is 11, and the limiting web height-thickness ratio is 65, and the latter can be extended to 75 when the frame column is subjected to a relatively smaller axial force.

Table 1 Limiting Width-Thickness Ratio of Q690 HSS Compression-Bending Members

Section Class	S4	S5
Flange b/t_f	$15\varepsilon_k$	20
Web h/t_w	$(45+25\alpha_0^{1.66})\ \varepsilon_k$	250

Table 2 Calculation Results of Members

Member No.	M_u/M_y		M_u/M_p	
Axial Compression Ratio	0.2	0.35	0.2	0.35
H-11-55-0.2/0.35	1.15	1.10	0.84	0.67
H-13-55-0.2/0.35	1.05	0.97	0.77	0.60
H-15-55-0.2/0.35	0.97	0.84	0.72	0.54
H-17-55-0.2/0.35	0.89	0.80	0.67	0.51
H-19-55-0.2/0.35	0.84	0.73	0.64	0.47
H-11-60-0.2/0.35	1.16	1.06	0.83	0.64
H-13-60-0.2/0.35	1.04	0.95	0.76	0.58
H-15-60-0.2/0.35	0.95	0.83	0.70	0.52

H-17-60-0.2/0.35	0.87	0.76	0.65	0.48
H-19-60-0.2/0.35	0.82	0.72	0.62	0.45
H-11-65-0.2/0.35	1.11	0.99	0.79	0.59
H-13-65-0.2/0.35	1.01	0.89	0.73	0.54
H-15-65-0.2/0.35	0.92	0.80	0.65	0.49
H-17-65-0.2/0.35	0.86	0.73	0.64	0.46
H-19-65-0.2/0.35	0.81	0.68	0.61	0.43
H-11-70-0.2/0.35	1.04	0.95	0.74	0.56
H-13-70-0.2/0.35	0.98	0.85	0.70	0.51
H-15-70-0.2/0.35	0.91	0.76	0.66	0.47
H-17-70-0.2/0.35	0.83	0.70	0.61	0.43
H-19-70-0.2/0.35	0.80	0.66	0.59	0.41
H-11-75-0.2/0.35	1.04	0.89	0.74	0.53
H-13-75-0.2/0.35	0.95	0.82	0.68	0.49
H-15-75-0.2/0.35	0.87	0.73	0.63	0.44
H-17-75-0.2/0.35	0.82	0.67	0.60	0.41

3.3 Ductility

Ductility is an important characteristic of the seismic performance of the member, characterizing the plastic deformation capacity of the member under seismic loading. θ_u is used to describe the ductility of the member [10], where θ_u is the angle of rotation when the load drops to 0.85 times the ultimate load after the member reaches its limit. The θ_u value of all members is divided into four groups in terms of the flange width-thickness ratio, web height-thickness ratio, and axial compression ratio, as shown in Fig. 5 and Fig.6. By comparing Fig. 6 and Fig. 7, it can be seen that the web height-thickness ratio is the decisive factor affecting the ductility of the members. The ductility of the members decreases significantly with the increase of the web height-thickness ratio, with an average decrease rate of 25.3%, while that of the flange width-thickness ratio is only 6.4%. By comparing Fig. 5(a) and Fig. 5(b), Fig. 6(a) and Fig. 6(b), respectively, it can be seen that with the increase of the axial compression ratio, the ductility of the members decreases significantly, with an average decrease rate of 24.4% and 25.7%, respectively. In summary, increasing the web height-thickness ratio and axial compression ratio will cause a significant decrease in the ductility of the members, which is not conducive to the development of plastic deformation.

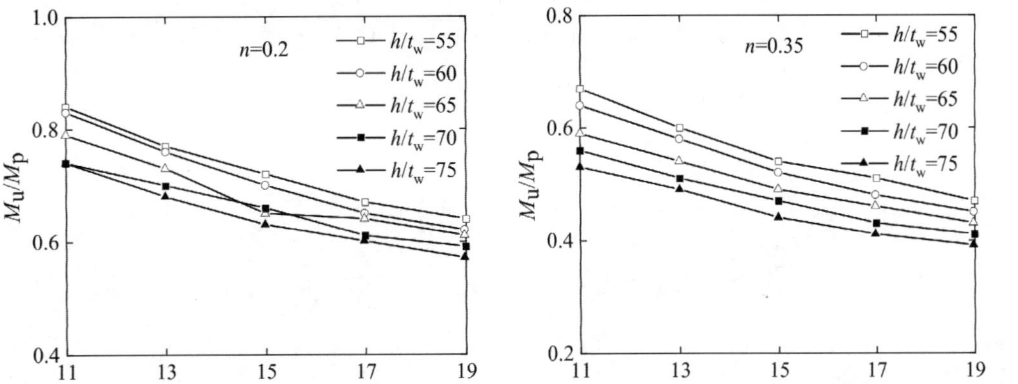

(a) Member with Axial Compression Ratio of 0.2 (b) Member with Axial Compression Ratio of 0.35

Fig. 3. Effect of Flange Width-Thickness Ratio and Axial Compression Ratio on M_u/M_p

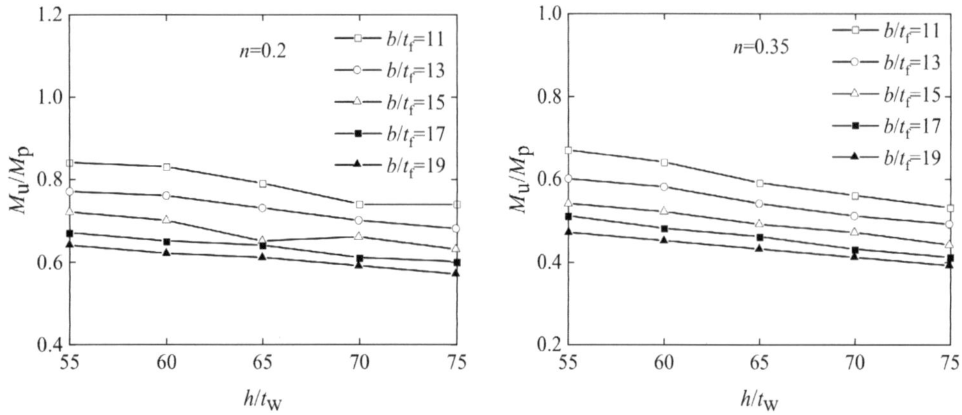

(a) Member with Axial Compression Ratio of 0.2 (b) Member with Axial Compression Ratio of 0.35

Fig. 4. Effect of Web Height-Thickness Ratio and Axial Compression Ratio on M_u/M_p

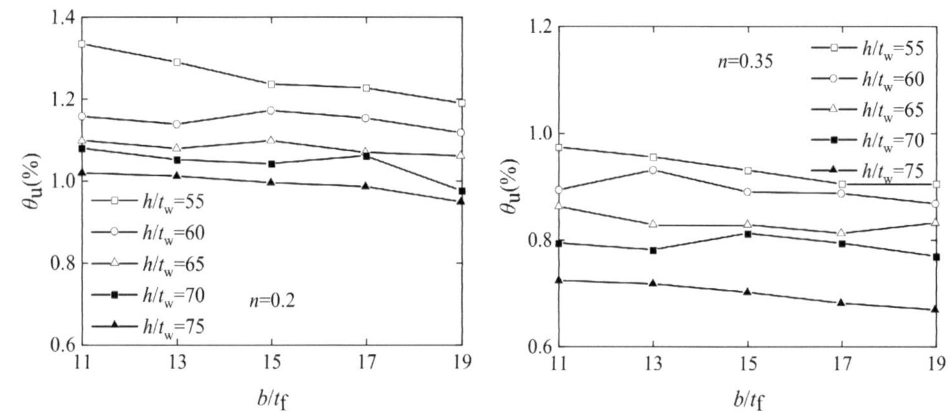

(a) Member with Axial Compression Ratio of 0.2 (b) Member with Axial Compression Ratio of 0.35

Fig. 5. Effect of Flange Width-Thickness Ratio and Axial Compression Ratio on Ductility

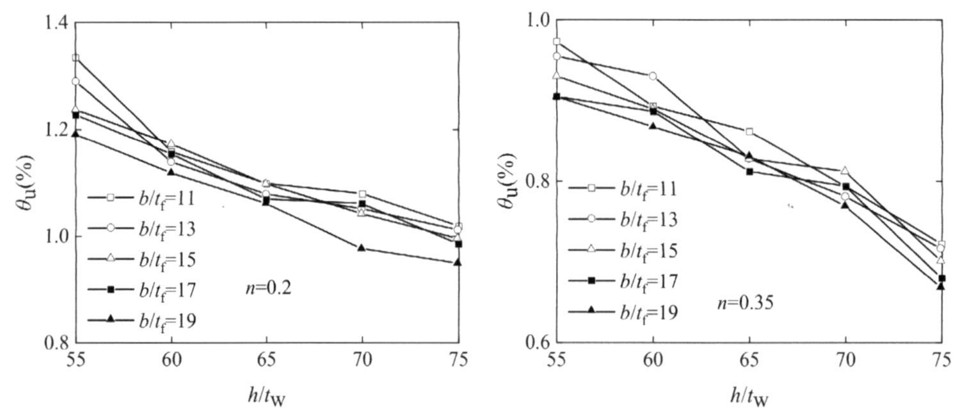

(a) Member with Axial Compression Ratio of 0.2 (b) Member with Axial Compression Ratio of 0.35

Fig. 6. Effect of Web Height-Thickness Ratio and Axial Compression Ratio on Ductility

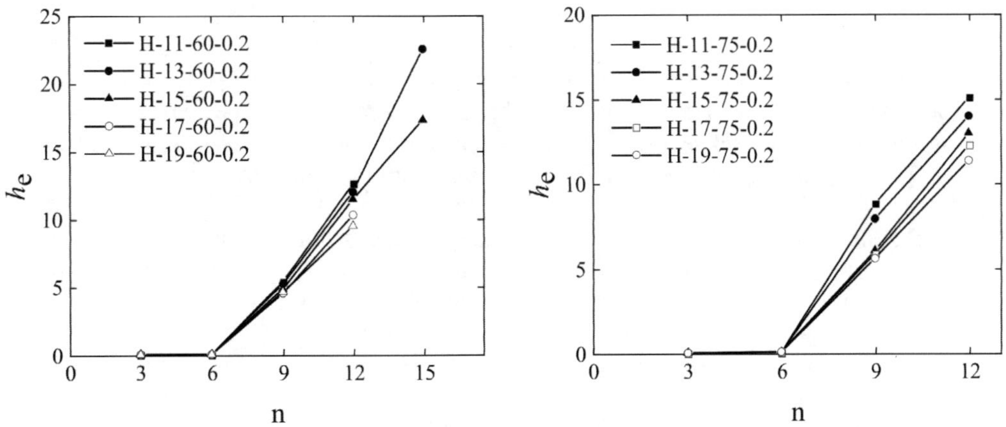

(a) Member with $n = 0.2$, $h/t_w = 60$ (b) Member with $n = 0.2$, $h/t_w = 60$

Fig. 7. Effect of Flange Width-Thickness Ratio on Energy Dissipation Coefficient

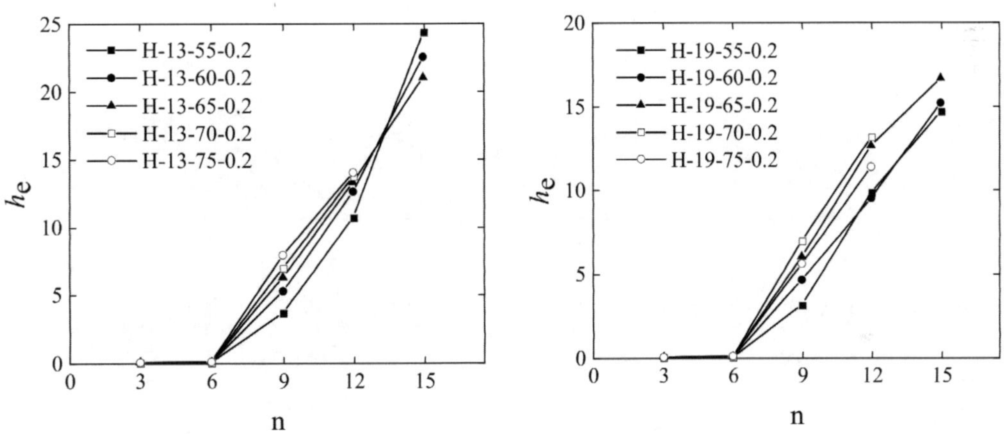

(a) Members with $n = 0.2$, $b/t_f = 13$ (b) Members with $n = 0.2$, $b/t_f = 19$

Fig. 8. Effect of Web Height-Thickness Ratio on Energy Dissipation Coefficient

3.4 Energy Dissipation Capacity

The energy dissip ation capacity is an important indicator for evaluating the seismic performance of the members. In this paper, the energy dissipation capacity of the members and their development trend are evaluated by using the whole-process nondimensionalized energy dissipation coefficient. The sum of the area enclosed by each hysteresis loop in the *M-θ* curve is nondimensionalized by the yield rotation angle and yield bending moment. The nondimensionalized energy dissipation coefficient he can be obtained as follows.

$$h_{e} = \frac{\sum_{i=1}^{n} E_i}{M_y \theta_y} \tag{1}$$

where E_i is the energy consumed by each cycle of the member.

The effects of flange width-thickness ratio, web height-thickness ratio, and axial compression ratio on the energy dissipation coefficient are shown in Figs. 7, 8, and 9, respectively. It can be found that the energy dissipation coefficient decreases with the increase of the flange width-thickness ratio. This is because the larger the flange width-thickness ratio is, the earlier the local buckling appears, and the more significantly the bearing capacity decreases. However, it can be seen from Fig. 8 that the increase of web height-thickness ratio has a favorable effect on the energy dissipation coefficient to some extent. But when the number of cycles continues to increase, the energy dissipation coefficient starts to decrease with the increase of the web height-thickness ratio. The reason is that with the increase of the web height-thickness ratio, the sectional area of the member will grow, thus increasing the ultimate bearing capacity of the member. Therefore, there is an increase in the energy dissipation coefficient during the first few cycles, but after exceeding the limit, the excessive width-thickness ratio will accelerate the degradation of the bearing capacity of the member, which will have a negative impact on the energy dissipation. The effect of axial compression ratio on the energy dissipation coefficient is shown in Fig. 9. The increase of axial compression ratio enables the member to enter the energy dissipation stage in advance, but with the increase of the number of cycles, after exceeding the limit, the excessive axial compression ratio will accelerate the degradation of the member, further leading to the loss of the bearing capacity so that the member cannot dissipate energy.

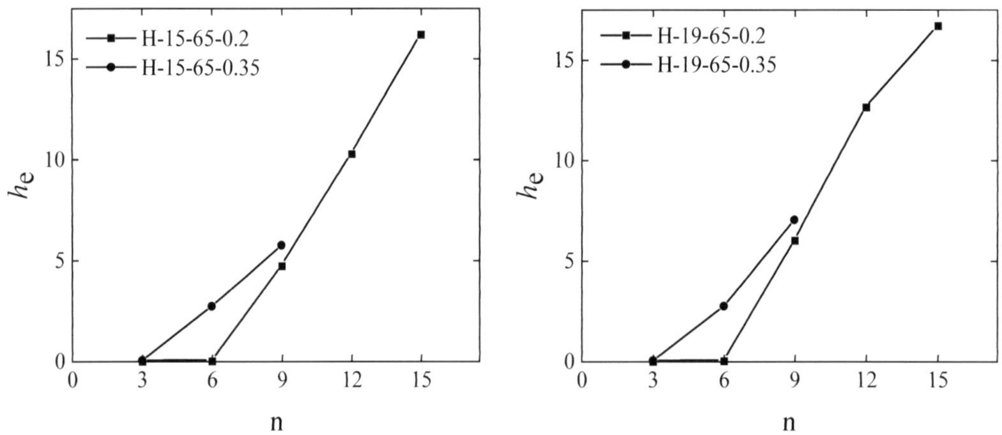

(a) Members with $b/t_f = 15$ and $h/t_w = 65$ (b) Members with $b/t_f = 19$ and $h/t_w = 60$

Fig. 9. Effect of Axial Compression Ratio on Energy Dissipation Coefficient

4. Conclusions

In order to fully understand the seismic performance of thin-walled high-strength steel members, based on the established finite element model, a parametric study of H-shaped members of Q690 HSS thin-walled sections was conducted to analysed the effects of flange width-thickness ratio, web height-thickness ratio, and axial compression ratio on the ultimate bearing capacity, ductility, and energy dissipation capacity. The main conclusions are as follows.

(1) Under the category of the S5 section, the flange width-thickness ratio is the decisive factor affecting the ultimate bearing capacity of Q690 HSS H-shaped members, and the impact of web height-thickness ratio on the ultimate bearing capacity is relatively weaker.

(2) In the actual engineering with regard to Q690 HSS H-shaped members, it is recommended that the limiting flange width-thickness ratio and web height-thickness ratio are 11 and 65, respectively. Further study on the interaction between flange and web of high strength steel members is needed to guide engineering practice.

(3) The increase of axial compression ratio and web height-thickness ratio will significantly reduce the ductility of the members. The thinner and more flexible the sections are, the larger the axial compression ratio is, and the more negative the impacts on the energy dissipation of the members are.

References

[1] Li G.Q., Wang Y.B., Chen S.W., et al. (2013) State-of-the-art on research of high strength structural steels and key issues of using high strength. J. Build. Struct., 34(01):1-13.

[2] Yin F., Yang L., Shi G., et al. (2020) Overview of research progress for seismic behavior of high strength steel structures. J. Steel. Constr., 35(03): 1-25.
[3] Chen Y.Y., Tong L.W., Yue C.Z., et al. (2004) Research on application of high frequency welding H-section steel in light steel structure system of multi-storied building. J. Hou. Sci., (04): 28-32.
[4] (2010) Code for seismic design of buildings. China Architecture & Building Press, Beijing.
[5] (2017) Code for design of steel structures. China Architecture & Building Press, Beijing.
[6] Chen S.W., Lu Z.l., Li G.Q., et al. (2014) Cyclic loading tests of Q690D high strength steel welded columns. J. Build. Struct., 35(12): 97-103.
[7] Chen S.W., Lu Z.L., Chen X., et al. (2015) Hysteretic beahvior of Q690D high strength steel Box-section beam-columns. J . Tongji. university., 43(08): 1135-43.
[8] Hai L.T., Li G.Q., Wang Y.B., et al. (2019) Experimental investigation on cyclic behavior of Q690D high strength steel H-section beam-columns about strong axis. J. Eng. Struct., 189: 157-73.
[9] Hai L.T.Wang Y.B., Li G.Q., et al. (2020) Numerical investigation on cyclic behavior of Q690 high strength steel beam-columns. J. Constr. Steel. Res., 167: 105814.
[10] (2005) Eurocode3: Design of steel structures-Part 1-1: General rules and rules for buildings. European Committee for Standardization, Brussels.
[11] Hai L.T., Sun F.F., Zhao C., et al. (2018) Experimental cyclic behavior and constitutive modeling of high strength structural steels. J. Constr. Build. Mater., 189: 1264-85.
[12] (2020) Code for acceptance of construction quality of steel structures. China Architecture & Building Press, Beijing.
[13] Li T.J., Li G.Q., Wang Y.B., (2015) Residual stress tests of welded Q690 high-strength steel box- and H-sections. J. Constr. Steel. Res., 115: 283-9

Study on shear fracture flow capacity of hard brittle rocks

Yetong Xie[1], Jing Li[1*], Ziyi Song[2], Chongzhi Wang[1], Kaiyu Wang[3], Wenyan Li[3] and Haoyang Wei[1]

[1]College of Pipeline and Civil Engineering in China University of Petroleum, Qingdao, Shandong, 266580, China

[2]CNOOC Research Institute Co. Ltd, Beijing 100028, China

[3]Research Institute of Exploration and Development, Tarim Oil field Company, PetroChina, Korla, Xinjiang, 841000, China

[*]Corresponding author's e-mail: jingli@upc.edu.cn

Abstract. It is of great significance for the construction and operation of high head pumped storage power station, high dam project and deep tunnel project to master the seepage characteristics and laws of rock mass shear fracture. In this paper, the seepage characteristics of marble shear fractures are summarized as follows. (1) In the radial seepage test, the seepage process conforms to darcy's law, and there is a linear relationship between head pressure and flow. (2) The characteristics of local tensile failure in the cracks generated under low normal force during the shear process hinder the seepage. (3) The normal force has a significant influence on the fracture seepage. In the process of the normal force increasing from 36kN to 48kN, the seepage characteristics show a sudden change, and the contact form of the shear fracture surface changes, indicating that part of the seepage channel in this stage appears a closed phenomenon with the increase of the normal force, and a large area of closure is formed in the fracture.

1. Introduction

The essence of engineering seepage instability is that water enters into the pores and fissures of surrounding rock, which causes the redistribution of stress field of surrounding rock and increases the opening of original pores and fissures of surrounding rock, even create new cracks. In turn, the deformation of surrounding rock and provides a more powerful seepage path, adding to the seepage influence on surrounding rock. Under the condition of high water pressure, the mutual promotion between seepage and surrounding rock may cause outburst and water inrush, resulting in engineering accidents. Therefore, it is increasingly important to study the flow capacity and permeability of rock fractures[1]. Iwai et al.[2] found that the influence of roughness of fracture surface on fracture flow law was mainly related to contact rate of fracture surface area; Morteza Javadi et al.[3] studied the effects of shear on critical Reynolds number and nonlinear flow characteristics, and established a quantitative criterion for nonlinear flow by combining fushi's law and Reynolds number; Zhou et al.[4] conducted the seepage test on the fractured granite fracture surface and the sandstone samples with slight dislocation in the fracture surface; Zou et al.[5] studied the influence of shear fracture structure change and flow condition change on nonlinear flow. Li et al.[6] used the micro CT scanning technology to scan the tight sandstone samples, established the microscopic model that can finely depict the rock samples, and simulated the seepage characteristics of fluid in the rock pores. Single fracture is the basic element that constitutes the fracture network of rock mass, and the seepage law of single fracture is the foundation of the seepage law of fractured rock mass. But at present, the study on seepage law of shear fracture in hard brittle rocks needs to be strengthened. The purpose of this paper is to study the flow capacity of shear fractures in hard brittle rocks. Therefore, taking marble as an example, the seepage characteristics of shear fractures in hard and brittle rocks are studied by controlling the infiltration flow, which provides a theoretical basis for the stability analysis of the surrounding rock of deep buried tunnel.

2. Experimental research

2.1. Test rock sample

In this experiment, samples were made of fine-grained marble, sample size for Φ50×100mm.The uniaxial compressive strength was about 125MPa by uniaxial compression test. The cylindrical sample was drilled with a diamond drill bit at a depth of 60mm and a bore diameter of 4mm.The finished sample is shown in figure 1. The longitudinal wave velocity of the cylindrical sample is between 2460 ($m \cdot s^{-1}$) and 2750 ($m \cdot s^{-1}$).

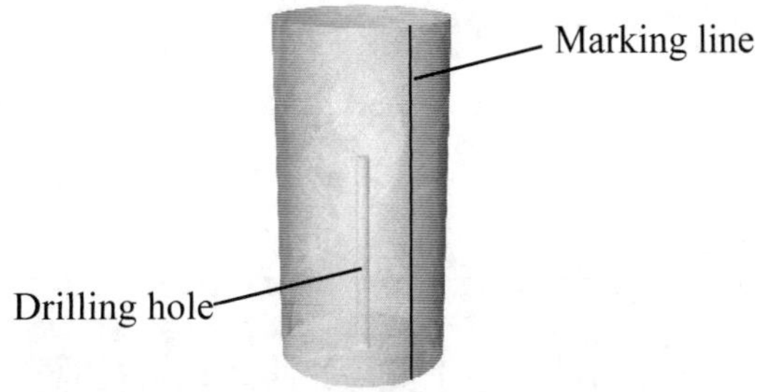

Figure 1. Schematic of drilled Marble sample

2.2. Direct shear test

Table 1 shows the values of normal stress and peak shear stress of samples J_1~J_4 in direct shear test. Normal stress σ_n and shear stress τ can be respectively as follows.

Table 1. Relationship between normal stress and peak shear stress

Sample	Normal force (kN)	Normal stress (MPa)	Peak shear force (kN)	Peak shear stress (MPa)
J_1	12.160	6.448	35.320	18.730
J_2	12.140	6.425	41.056	21.772
J_3	12.120	6.396	35.900	18.945
J_4	12.120	6.385	40.280	21.221

2.3. Fracture surface morphology analysis

The scanning of fracture surface is completed by Photographic 3d scanning system. The cloud data of crack fracture were obtained by this equipment, then import the cloud data into Geomagic Studio for mesh surface encapsulation. Finally, noise data outside the rock structure plane was deleted. Thus, the reconstructed digital image shown in figure 2 is completely consistent with the surface morphology of the fracture surface. From left to right, figure 2 shows the morphologies of shear sections formed by specimen J_1, J_2 and J_3 under the normal pressure of 12kN.The shear fracture surface of J_1 and J_2 is relatively flat compared with that of J_3. The local shear fracture surface of J_3 presents the characteristics of tensile failure, with a large range of protrusions and grooves. Such local tensile failure characteristics may have a certain impact on the seepage characteristics of the fracture surface.

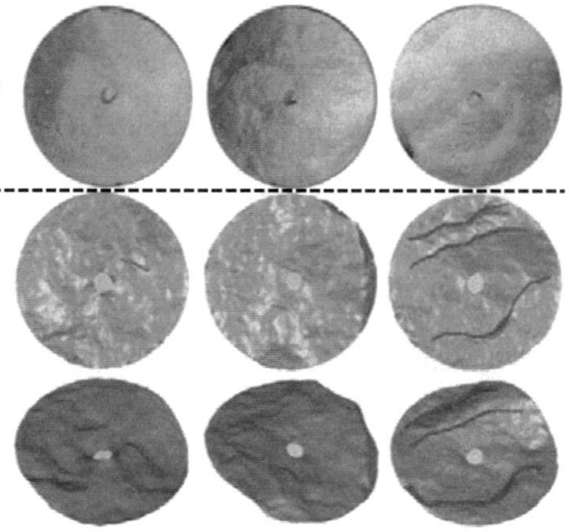

Figure 2. Shear surface morphology under the same normal force

The calculation formula of three-dimensional roughness coefficient Z_{2s}[7] is shown in equation (1).

$$Z_{2S}=\left\{\frac{1}{(N_x-1)(N_y-1)}\left[\sum_{i=1}^{N_x-1}\sum_{j=1}^{N_y-1}\frac{(z_{i+1,j+1}-z_{i,j+1})^2+(z_{i+1,j}-z_{i,j})^2}{2\Delta x^2}+\sum_{i=1}^{N_x-1}\sum_{j=1}^{N_y-1}\frac{(z_{i+1,j+1}-z_{i+1,j})^2+(z_{i,j+1}-z_{i,j})^2}{2\Delta y^2}\right]\right\}^{1/2} \tag{1}$$

Herein: Δx and Δy are the sampling interval in x direction and direction respectively; $Z_{i,j}$ Is the Z direction coordinate of the sampling point; N_x and N_y are the number of sampling points in X direction and Y direction respectively.

The parameters of 3d morphology calculated by Matlab are shown in table 2. It can be seen that under the constraint of the same normal force, the roughness of fracture surface obtained by shear failure has a certain difference. According to the three-dimensional roughness coefficient Z_{2s}, the roughness of fracture surface J_1, J_2 and J_3 increases successively. Secondly, the peak difference of fracture surface J_1, J_2 and J_3 also increased successively. Therefore, the Matlab calculation results are consistent with the macro features shown in the Geomagic Studio processed reconstruction image of the fracture surface, that is, the roughness of the fracture surface J_1, J_2 and J_3 increases successively.

Table 2. 3D Morphological parameter of shear fracture surface

Sample	Z_{2S}	R_S	Maximum height(mm)	Minimum height(mm)	Maximum height difference(mm)
J_1	0.2893	1.0371	1.7996	-1.6513	3.4509
J_2	0.3202	1.0475	3.9244	-1.6909	5.6154
J_3	0.3672	1.0562	4.0739	-1.8185	5.8925
J_4	0.3233	1.0462	2.8525	-2.9801	5.8327

2.4. Fracture seepage test

The seepage test was carried out by the seepage test instrument as shown in figure 3. Before the test, the fracture surface of the shear sample was closed along the mark line, and the side of the sample with water injection channel was bonded to the seal gasket with epoxy resin glue. At the beginning of the test, normal force is applied to the fractured sample with a loading rate of 1kN/s. When the normal force is loaded to a predetermined value, open the water valve between the supplementary water tank and the water pump, fill the water pump with water, and close the water valve between the water pump and the supplementary water tank when the water pump is filled. At the beginning of seepage test, open the water valve between the water pump and the rock sample and start the seepage test in the fracture.

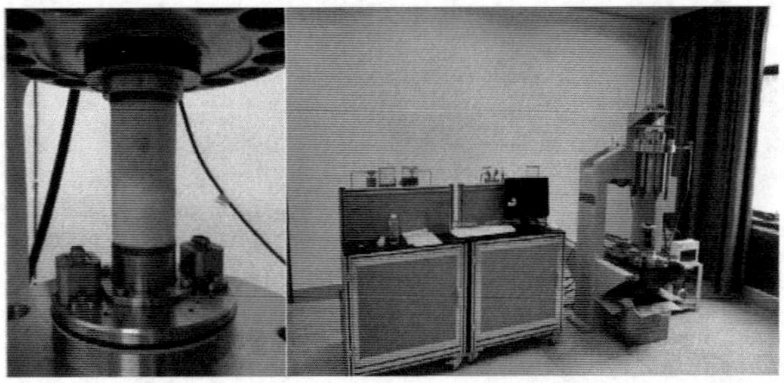

Figure 3. Fracture seepage test

3. Seepage characteristics of fracture surface under constant normal force

Constant normal force of 12kN was applied to shear samples J_1, J_2 and J_3. After the normal force was loaded to a predetermined value, fracture seepage was started. The flow rate started from 1ml/min and rose to 15ml/min successively, and then the hydraulic head pressure under a series of flows was measured. Under the normal force of 12kN, the relation curve of flow-infiltration water pressure on the crack surface of specimen J_1, J_2 and J_3 is shown in figure 4. It can be seen from the figure that in the seepage curve of specimen J_1, J_2 and J_3, the head pressure increases linearly with the increase of flow rate, and the seepage process roughly conforms to Darcy law. In the seepage curve, with the increase of flow rate, the pressure growth rate of J_1, J_2 and J_3 water heads was gradually accelerated. The seepage curves of J_1 and J_2 are relatively close, while the seepage curves of J_3 show that the hydraulic head pressure is significantly higher than that of J_1 and J_2 at the same flow rate. This means that under the same head pressure, the seepage process in J_1 and J_2 fractures is easier than that in J_3.

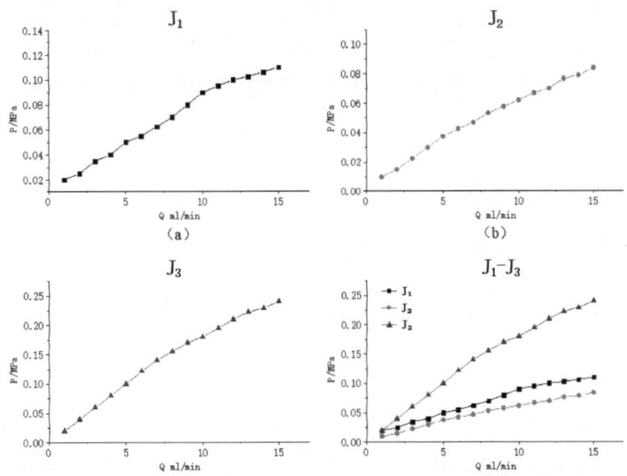

Figure 4. Seepage characteristic curves of fracture surface under same normal force

We find that there are two main reasons for this phenomenon: (1) as can be seen from figure 4, the roughness coefficients of J_1, J_2 and J_3, Z_{2s}, increase successively, indicating that the roughness degree of fracture has a certain influence on seepage, that is, the greater the roughness degree of fracture surface, the greater the obstruction effect on seepage. (2) figure 5 is a schematic diagram of radial seepage, in which the black arrow represents the direction of seepage. It can be seen from the figure 5 that the local tensile failure characteristics in the J_3 fracture surface hinder the progress of radial seepage in some directions, so as to create a seepage advantage channel. As a result, the seepage process is mainly parallel to the direction of protrusion and groove, while perpendicular to the direction of protrusion and groove, the seepage process is hindered. It can be inferred that the local tensile failure characteristics generated during the shear process of marble will hinder the seepage process.

This is consistent with the test results in literature[8] that the tensile crack surface is relatively smooth and the two sides can contact closely under the action of pressure, resulting in a sharp decrease in the hydraulic opening.

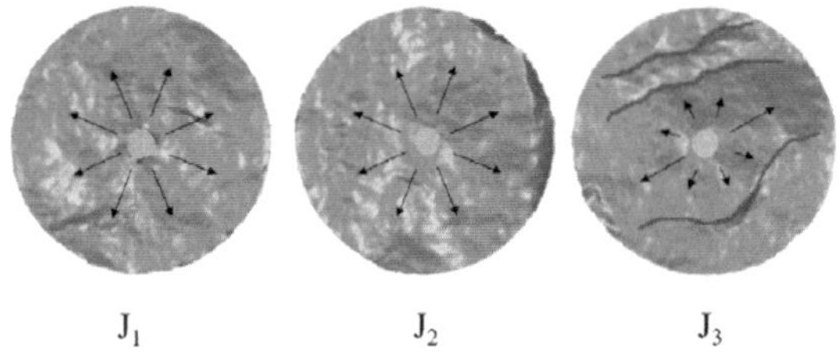

Figure 5. Schematic diagram of radial seepage

4. Conclusion

The purpose of this paper is to study the relationship between the seepage flow rate, the seepage head and the normal constraint force of the fracture in the process of marble shear fracture seepage. The conclusions are as follows, which have certain guiding significance for the construction and operation of high head pumped storage power station, high dam project and deep tunnel project.

(1) In the radial seepage test, when the flow rate is within the range of 0~15ml/min, the seepage process is characterized by linear darcy flow, that is, the head pressure increases linearly with the increase of the flow rate.

(2) The local tensile failure characteristics generated during the shear process of marble hinder the seepage of shear fractures.

(3) In the process of fracture seepage, the influence of normal force on seepage is more significant than that of fracture surface morphology.

(4) During the increase of normal force from 36kN to 48kN, the contact form of upper and lower crack surface changed. During the process, some seepage channels closed with the increase of normal force, and a large area of stagnant water was formed in the fracture surface.

Based on the test method adopted in this paper, the author will carry out seepage test research on marble fractures with different sizes, and further verify and improve the results obtained in this paper.

Acknowledgments

This work was financially supported by the National Natural Science Fund of China (41972138) and the National Great Science Fund of China (ZD2019-183-007).

References

[1] Zhou, H., Hu, D. W., Zang, K. (2014) Mechanism and theory of fluid solid coupling for porous and fractured rocks under high osmotic pressure. The Science Publishing Company, Beijing.
[2] Iwai, K. (1976) Fundamental Studies of Fluid Flow through a single fracture. Berkely: University of California.
[3] Javadi, M., Sharifzadeh, M., Shahriar, K. (2014) Critical Reynolds number for nonlinear flow through rough-walled fractures: The role of shear processes. J. Water Resources Research, 50: 1789–1804.
[4] Zhou, J.Q., Hu, S.H., Fang, S. (2015) Nonlinear flow behavior at low Reynolds numbers through rough-walled fractures subjected to normal compressive loading. J. Inter national Journal of Rock Mechanics & Mining Sciences, 80: 202-218.
[5] Zou, L., Jing, L., Cvetkovic, V. (2017) Shear-enhanced nonlinear flow in rough-walled rock fractures. J. International Journal of Rock Mechanics & Mining Sciences, 97: 33-45.
[6] Yang, F., Wang, H., Hang, B. (2019) Study on the stress sensitivity and seepage characteristics of tight sandstone based on CT scanning. J. Journal of Geomechanics, 2019,25(04):475-482.

[7] Zhang, P., Li, N., Chen, X.M. (2009) A new representation method for three-dimensional surface roughness of rock fracture. J. Chinese Journal of Rock Mechanics and Engineering, 28: 3477-3483.
[8] Zhang, F., Wang, L., Zhao, J. (2016) Evolution of permeability of granite with tensile and compressive-shear cracks. J. Rock and Soil Mechanics, 37: 2803-2809.

A novel reversible thermometric organic material for power transmission line detection application

Li Cheng*, Shang Ziwen, Wang Minzhen, Liu Li, Zhang Guangxin, Zhao Liyin, Ni Hongxia, Zhang Xiao, Zhang Jian, Qiao Dongliang

Changchun Institute of Technology, Changchun, Jilin Province, China

*Corresponding author's e-mail: li.cheng@ccit.edu.cn

Abstract. Due to the inaccuracy of the existing fault detection method caused by the overheating of the equipment of the long-spanning overhead transmission line, this paper creatively proposes a detection method of the reversible thermochromic organic thin film material HCBCC which can be deposited on the equipment or hardware of the transmission line based on the electron transfer mechanism. In this method, the film material made with five components in accordance with the optimal ratio was named HCBCC film, and the coloration performance of the material was determined by UV-Vis absorption spectroscopy, temperature variable infrared spectroscopy, differential scanning calorimetry and other characterization methods. Then, its durability was verified through environmental tests such as high and low temperature, rain and snow, wind and sand, and aging. HCBCC material has the advantages of easy observation, fast color rendering and wide temperature range in the temperature indicator, which not only greatly reduces the intensity of inspection work for staff and can find fault points in time, but also saves a lot of time for the emergency repair of power equipment and provides guarantee for the safe and stable operation of power grid.

1. Introduction

The construction of the transmission line in China has been in the stage of rapid development. The lightning strike or short circuit fault would lead to arc corrosion at the key nodes caused the local heating is difficult to be observed by the naked eye or instruments. According to the relevant regulations, the maximum allowable temperature of wire and gold metal crossing the overhead transmission lines is 90 °C. Therefore, if the temperature exceed limit, it will cause damage to the transmission equipment and affect the working life, and even cause major accidents. In order to find the location of the fault point in time, it is urgent to need an effective method to enter the heating parts Line detection in time, eliminate potential hidden danger of power transmission system safety.

Toru Katsumata(2014)[1] used SiO2 glass fiber doped with Eu and Al to make optical fiber thermometers, whose temperature measurement principle is based on the temperature dependence of the luminous and visible photothermal radiation properties of the glass fiber. The fiber thermometer greatly improves the measurement range thermal radiation temperature measurement, but with low accuracy. C.Wen (2002)[2][3] used the multispectroscopic radiation temperature measurement algorithm for aluminum alloy property research and temperature measurement, which determined the characteristics and temperature relationship of aluminum alloy. N.Renault (2010)[4] developed an infrared temperature measurement method for high temperature heat sources, however it had low sensitivity when temperture below 20 degrees. F. Girard(2014)[5] extended the detection range of multispectral temperature measurement theory, selected various metal materials and different emission models, and significantly reduced the temperature measurement error in the band range at 0.355μm-0.95μm. This indicates that the measurement range of multi-spectral radiation temperature measuring equipment is increased, and provides theoretical support and data analysis for temperature measurement in the ultraviolet region. Although some of these temperature measurement technologies have the advantages of high accuracy and wide range of temperature measurement, there are also defects such as high cost, tedious operation and poor stability, which make these methods difficult to popularize. This paper designs and manufactures a thermal chromic HCBCC material that can be used at low cost, high durability, good discoloration effect and wide discoloration temperature range, which can be applied to the surface of the monitored power equipment to show the long-term temperature change to achieve early warning or indication.

2. Analysis and preparation of HCBCC of reversible thermal chromic material

2.1 Material analysis and screening

2.1.1. Intramular proton transfer mechanism
At high temperatures, the molecular structure of the material changes as protons migrate, which results in the color change. The representative ones are salicylaldehyde aniline schiff bases. When the temperature rises, the hydrogen atoms on the phenolic hydroxyl group in the molecular structure will be transferred, so that the enol compound structure will change into ketone structure and form ketone-enol tautomer, thus causing the color change. Fig. 1 left shows changes in its molecular structure before and after the discoloration.

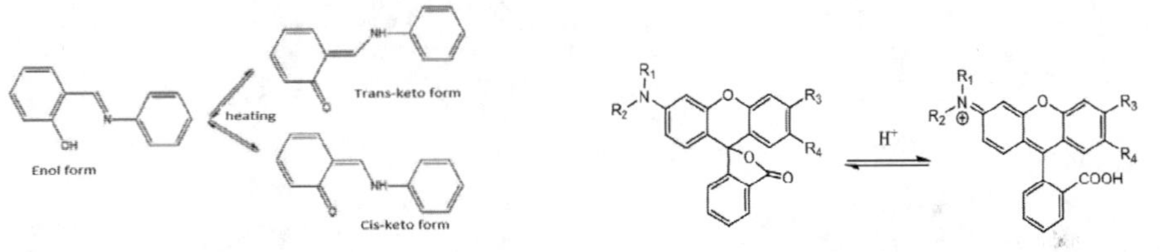

Figure 1 Changes of molecular structure before and after the alkali discoloration of aniline Changes of molecular structure before and after discoloration of fluorane.

2.1.2. Electronic transfer
When the temperature increases, the color changing agent and the color developer will constantly conduct electron transfer, resulting in the molecular structure of the color changing material with a fixed absorption of light at some wavelengths to other structures, resulting in the color change. The material with the discoloration mechanism as fluorane, and the discoloration mechanism is shown in right in Fig. 1.

Therefore, according to the most suitable materials are mainly selected from three aspects: coloration mechanism, coloration temperature, and color contrast before and after coloration. Crystalline violet lactone, bisphenol A, high impact polystyrene, xylene and di-butanone were finally selected as the core parts of reversible thermochromic film materials. Among them, crystal violet lactone and bisphenol A as the color and color agent, high impact polystyrene as, xylene and dibutanone as the solvent.

2.2 Material type selection and preparation
The quality ratio of high anti-resistant polystyrene (HIPS), crystalline violester, bisphenol A, methylene benzene, 2-butyl one will affect the performance of heat chromic films, so it is necessary to determine the best blending scheme and obtain heat chromic films with excellent effect. First, conduct the experiment strictly according to the complex scheme, and determine the discoloration temperature, speed and time of the complex compound. The steps are as follows:

(1) Determination of color interval: the test of color interval needs to call different ratios of color films of equal quality, and test the color temperature of reversible thermal color film respectively. During the test, the sample to be tested is divided into two parts, one for reference and the other in a petri dish and in a constant temperature water bath pot. Then, it gradually increased at the speed of 3°C/5 min, observed the color change of the color film, and tested the color film temperature in real time with an infrared thermometer to record the temperature when the color change was completely stopped, so as to determine the color interval. In order to reduce the experimental error, the test for the sample was been repeated four times to average of the test results.

(2) Determination of color time: adjust the temperature of the vacuum drying box to make it within the color temperature range of the changing film. The cooled petri dish was placed in the oven with discolouration from the film surface to record how long the sample needs to complete restoration.

(3) Determination of recolor time: place the film at room temperature and record the time from the initial color to the full return to the original color.

The test results are shown in Table 1.

Table 1 Color changing properties of reversible material under different mass ratios

Sample number	Proratio	The temperature interval for changing the color(°C)	Fully change the color time (T/s)	Restore the color time	The effect of changing the color	Rereversibility	Comprehensive evaluation
1	100:1: 3:600:500	74～126	73	64	Obviously	Good	Excellent
2	100:1: 3:600:400	73～125	73	62	Obviously	Good	Excellent
3	100:1: 3:500:500	71～129	65	56	Obviously	Excellent one	Excellent
4	100:1: 3:500:400	70～124	62	54	Very obvious	Excellent one	Perfect
5	100:1: 2:600:500	76～125	78	65	Obviously	Good	Excellent
6	100:1: 2:600:400	74～124	79	67	Obviously	Good	Excellent
7	100:1: 2:500:500	73～127	81	70	General rule	Good	Ordinary ones
8	100:1: 2:500:400	73～126	82	71	General rule	Good	Ordinary ones
9	100:1: 1:600:500	72～126	80	71	General rule	Good	Ordinary ones
10	100:1: 1:600:400	71～125	80	70	Obviously	Good	Ordinary ones
11	100:1: 1:500:500	72～128	77	65	Obviously	Good	Ordinary ones
12	100:1: 1:500:400	70～127	80	69	Obviously	Good	Ordinary ones

3. Test and analysis of HCBCC materials

3.1 Analysis of the discoloration mechanism of HC BCC

From the experimental data of Table 2, the dosage of the film determines the color; the dosage of bisphenol A, xylene and 2-butylone determines the depth of the color of the compound, but the dosage of crystalline purple polyester plays a leading role; the dosage of xylene and 2-butylone determines the discoloration temperature of the reversible thermochromic material. According to the comparison of the five components, the optimal ratio of the five components is 100: 1: 3: 500: 400, and when the use of the solvent is less, the thermal chromic film discoloration is slow, which shows that the appropriate increase of the solvent under the allowable conditions, will change the thermochromic film Performance has a beneficial effect. To facilitate the formulation of this reversible thermochromic material, the film material made by the five ingredients according to the optimal ratio was named HCBCC films. The after discoloration are shown in Fig. 2.

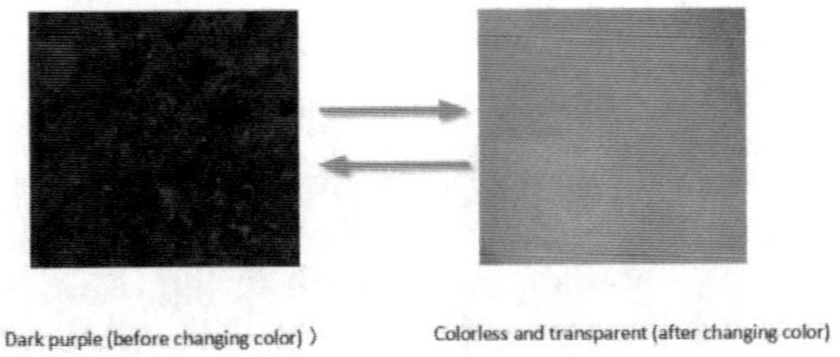

Figure. 2 Best color material before and after discoloration

3.2 Experimental test results analysis

3.2.1 Analysis of UV-visible absorption spectroscopy results

In the range of visible light, the optimal UV ratio of the visible absorption spectrum is shown in Fig. 3. At room temperature 25°C, there is a wider absorption peak in the wavelength range of 500-750nm, when the temperature rises from 25 °C to 70 °C is not obvious, the peak gradually decreases as the temperature continues to rise, each 10 °C for 70-90 °C, compared to the other temperature intervals.

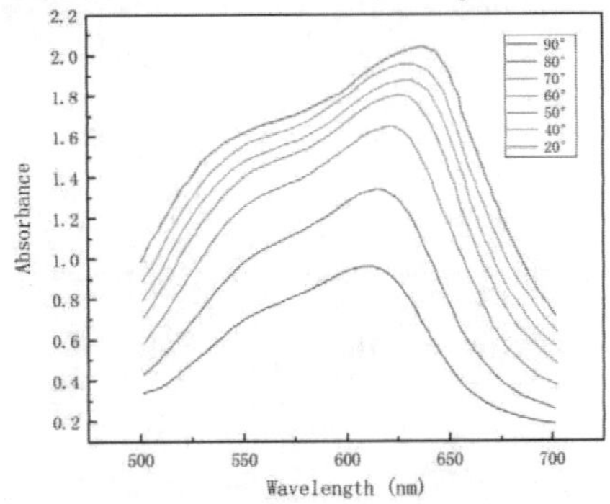

Figure. 3 UV-visible absorption spectra of HCBCC films

3.2.2. Analysis of infrared spectrum observation variable temperature

The temperature changing infrared spectrum of the HCBCC film and the temperature changing infrared spectrum after image processing are shown in Fig.4a and Fig.4b. It is seen from Fig.4a that the increase in temperature rise in the range of 2300-2800cm-1 leads to its reduced spectral band strength, indicating the reduced strength of substances such as carboxylic acid. When the reversible thermochromic film is in the range 1345-1380cm-1, its temperature-varying infrared spectrum is shown in Fig. 4b, whose spectral band strength decreases with temperature, which also shows that the characteristic peak of the carboxylate also gradually decreases with temperature.

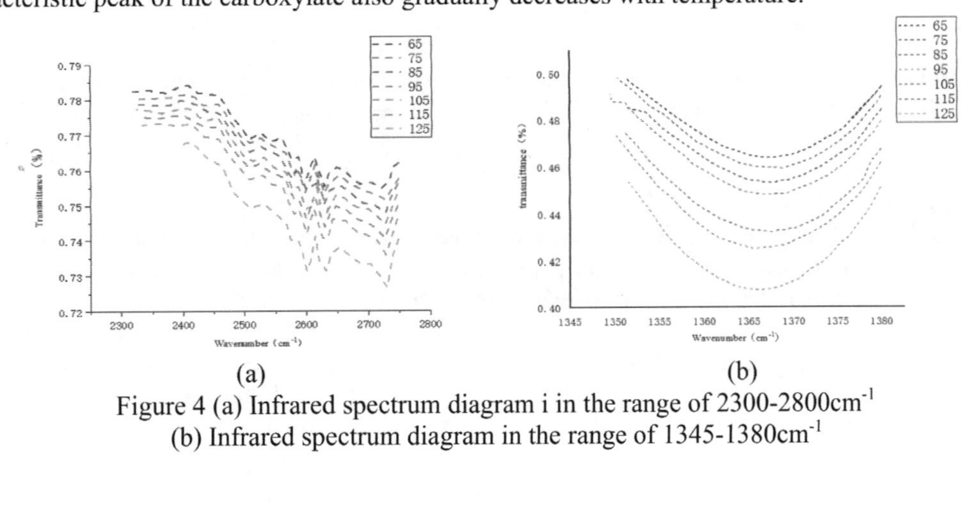

(a) (b)

Figure 4 (a) Infrared spectrum diagram i in the range of 2300-2800cm^{-1}

(b) Infrared spectrum diagram in the range of 1345-1380cm^{-1}

4.Study on online properties of HCBCC films

4.1 Line clamp coating process

In order to further verify the actual performance of HCBCC material online clamping, a series of experiments were carried out in this paper to verify its working effect in different external environments. In order to prevent the damage of HCBCC film by ultraviolet radiation, UV-531 ultraviolet absorber was used to enhance the anti-ultraviolet ability of HCBCC film. The UV absorber and epoxy resin were mixed at a ratio of 1：10 to make the anti-UV coating. The UV coating can absorb the UV light at the wavelength of 240 ~ 340 nm. In order to protect the performance of the wire clip and HCBCC film, this paper applies different layers of fire retardant paint, insulating paint and anti-ultraviolet resin varnish evenly on the wire clip.The line sandwich film structure is shown in Fig. 5 left part, and the applied finished product is shown in Fig. 5 right.

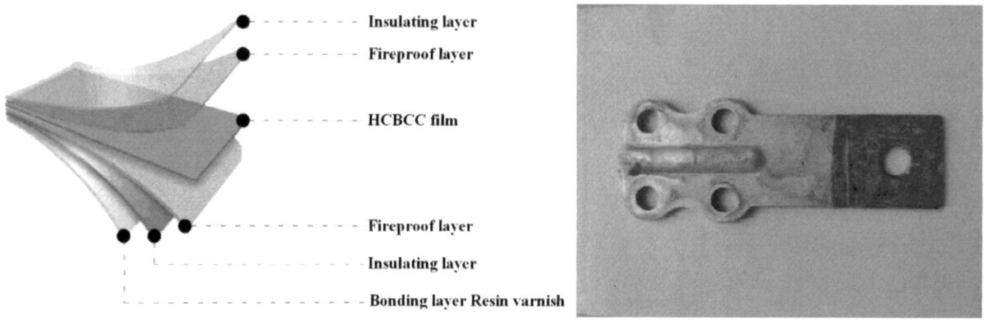

Figure 5 Film structure diagram of clips and the sample of HCBCC film smear on the wire clip

4.2 Performance test and aging test analysis in different environments

To simulate the complex environment after its actual installation, this paper place wire clips covered with thermochromic films in different environments as shown in Table 2 to detect their discoloration effect and performance.

Table 2 HCBCC discoloration performance in different environments

	Low Tempe (No sunlight)	Low Tempe (sunlight straight)	High Tempe (No sunlight)	High Tempe (sunlight straight)	Room Tempe (No sunlight)	Room Tempe (sunlight straight)
Start the discoloration temp(°C)	77	78	73	75	71	70
Color change time(s)	49	50	49	51	50	51
Full-colour Tempe(°C)	97	95	93	94	91	90
Fully color discoloration Time (s)	66	67	66	66	65	66

According to Table 2, the thermochromic material has good discoloration performance, good discoloration effect, and is suitable for different complex environments, with high comprehensive value.

5.Conclusions

According to the characteristics of coloration mechanism, coloration temperature and coloration time of reversible thermochromic material, this paper designed and manufactured a thermochromic HCBCC material which can be used on overhead transmission lines with good coloration effect and wide range of coloration temperature. According to the discoloration experiment of HCBCC material, the optimal ratio, discoloration mechanism and discoloration time were determined. The discoloration performance was determined by ultraviolet spectrum analysis, infrared spectrum analysis and differential scanning calorimetry. According to the analysis, the best wire laminating materials and structures of HCBCC films were obtained, and the properties were tested in different environments. The overall performance of HCBCC film is superior and the prospect is broad. In the future, it will have a very good development prospect in the overheat fault warning of power equipment.

Acknowledgments

This work was funded by Jilin Province Science and Technology Association, Project No. QT202014, and Changchun Institute of Technology, Project No. 04010192020024 and 320200022.

References

[1] Toru K, Shuji K, Hiroaki A, et al. Photoluminescence and thermal radiation from Eu and Al doped SiO2 for fiber-optic thermometer application[J]. Journal of Luminescence, 2014, 154 (5):549-552.

[2] C. Wen, I. Mudawar. Experimental investigation of emissivity of aluminum alloys and temperature determination using multispectral radiation thermometry (MRT) algorithms[J]. Journal of Materials Engineering and Performance, 2002, 11(5): 551-562.

[3] C. Wen. Emissivity characteristics of aluminum alloy surfaces and assessment of multispectral radiation thermoy1etry emissivity models[D]. Purdue University, 2005: 10-30.

[4] J. Travis, R. Matthew, R. Dale, et al. An experimental method for making spectral emittance and surface temperature measurements of opaque surfaces[J]. Journal of Quantitative Spectroscopy and Radiative Transfer, 2011, 112(7): 1191-1196.

[5] Girard F, Battuello M, Florio M. Multiwavelength Thermometry at High Temperature: Why It is Advantageous to Work in the Ultraviolet[J]. International Journal of Thermophysics, 2014, 35(6-7): 1401-1413.

Research on Porosity and Performance of Self-compacting Backfill Materials

Zhao Jian[1,a], Zhao Mingzhe[2,b], Tang Lei[2,c], Chen Penghui[2,d], Gao Peiwei[2*]

[1]Department of Civil Engineering, Nanjing University of Aeronautics and Astronautics, Nanjing 210016, China;

[2]Research Center on Levee Safety Disaster Prevention, MWR, Zhengzhou， 450003， China;

[a]zhao1062020@163.com(ZJ), [b]109483193@qq.com（ZMZ)

[c]470795032@qq.com（TL), [d]285169275@qq.com(CPH)

*gpw1963@nuaa.edu.cn(GPW)

Abstract: This paper researched and developed a new type of self-compacting backfill material behind the wall of underground pipelines, and compared and analyzed the relationship between the water-binder ratio, porosity and compressive strength of the two backfill materials. The average pore size of the foam backfill material increases with the water-binder ratio. While increasing, the porosity tends to decrease with the increase of water-binder ratio, and the compressive strength first increases and then decreases with the increase of water-binder ratio; self-compacting backfill material has excellent properties such as self-compacting, light weight and anti-vibration. It is a new type of building materials which can meet the requirements of fluidity and strength for backfilling after penetrating the yellow pipe wall.

1. introduction

In recent years, with the rapid development of national economic construction, there has been an increasing number of water tunnel and subway tunnel projects, and the number of projects crossing the Yellow River and Yangtze River dikes has gradually increased[1,2]. Dyke-crossing pipelines may affect the safety of the dikes during the construction and operation periods, and will have certain impacts on river flooding, river regime stability, dike safety, flood control, etc., especially if the backfill around the pipeline is not dense and the surrounding soil will be prone to settlement, deformation, water seepage and displacement, which not only causes stratum loss, but even causes major accidents such as the collapse of dikes and the collapse of nearby buildings[3-5].

Although the grouting material and raw soil are used for backfilling, it is not dense and requires multiple backfilling. At the same time, due to the long setting time, the construction period is affected, and embankment settlement and collapse may occur [6,7]. In the construction of large-scale yellow tunnels or water pipelines, there are a large number of backfill areas behind the pipe wall. The treatment of these areas not only needs to meet the mechanical performance requirements, but also it should meet the requirements of compactness, reduced settlement, and short backfilling time. This requires that the backfill materials have relatively high fluidity and volume stability, by adding appropriate stabilizing materials and admixtures, not only it can improve material strength and fluidity, reduce sedimentation[8,9], but it will also meet the requirements of sustainable development.

The self-compacting foam backfill material has outstanding characteristics in engineering, with low self-weight, good fluidity, and compressive strength that can be adjusted and controlled within a certain range[10,11]; it also has characteristics such as excellent construction performance, self-standing after curing, permeability and low water absorption performance; it has other properties such as heat preservation, earthquake resistance, frost resistance[12,13]. Foam backfilling materials are used in the backfilling after crossing the yellow pipeline wall, which can avoid the damage to the environment such as high filling and high excavation, and protect the embankment and the surrounding ecological environment[14-16].

The self-compacting foam backfill material has high requirements for the comprehensive performance of all aspects of the material due to long-term service in the underground humid environment[17,18]. In this study, self-compacting foam was prepared using the self-made FC-1 high-performance foam with lightweight filling materials, water-binder ratio, fly ash content, etc. as variables, and through a large number of experiments to determine the range of changes, through optimizing the ratio design, to explore its impact on the physical and mechanical properties of self-compacting foam filling materials, based on direct methods such as measurement and electronic scanning to test material pore structure parameters and analyze the influence mechanism of these macro factors on the micro-structure to provide a theoretical basis for its large-scale application in the backfill of underground pipeline walls of the Yellow River embankment.

2. Test raw materials and test methods

2.1 Test raw materials

This test research uses 42.5MPa ordinary Portland cement and Grade II fly ash, which come from Nanjing and Zhenjiang respectively. The performance indicators of cement and fly ash meet the requirements of "General Portland Cement" GB175-2007 and according to "Fly Ash Used in Cement and Concrete" GB/T1596-2005, the performance indicators of cement and fly ash are shown in Table 1.

The foaming agent used in the experiment is the FC-1 foaming agent developed by the research group. The dilution ratio of the foaming agent is 40, the bleeding rate in one hour is about 5% to 15%, and the settlement distance in one hour is 4 to 7 mm; The water-reducing agent used is a polycarboxylic acid-based water-reducing agent, from a water-reducing agent company in Nanjing, with a water-reducing rate of about 32%. In the process of designing the mix ratio of the self-compacting backfill material, by controlling the total amount of raw materials in the foam backfill material per unit volume, the quality of various raw materials in the material under different water-to-binder ratios is determined. The dry density and water glue ratio of the two self-compacting backfill materials used in the test are shown in Table 2.

Table 1 Physical properties of cement and fly ash

Cement		Fly ash	
Project	Test results	Project	Test results
Specific surface area(m^2/kg)	370	Specific surface area (m^2/kg)	315
SO3 (%)	2.14	SO3 (%)	0.35
Ignition loss (%)	1.80	Ignition loss (%)	3.87
MgO (%)	1.50	Bulk density (kg/m^3)	2.1
Chlorine content (%)	0.007	Fineness (45 μm)	18
Initial/final setting time (min)	120/ 205	Moisture content （%）	0.2
Standard consistency (%)	28	Water demand ratio （%）	94
3/28 days Flexural strength(MPa)	6.7/9.0	Strength activity index(%)	78
3/28 days Compressive strength(MPa)	30.0/50.0	SiO2+Al2O3+Fe2O3 (%)	89
Stability	Qualified	f-CaO (%)	0.35
Alkali content (%)	0.47	Alkali content (%)	0.40

Table 2 Dry density and water-binder ratio of self-compacting backfill materials

Specimen	Dry densitykg/m^3	Water-binder ratio%	Specimen	Dry densitykg/m^3	Water-binder ratio%
FC_{700}-1	734	0.40	FC_{900}-1	928	0.40
FC_{700}-2	723	0.50	FC_{900}-2	916	0.50
FC_{700}-3	720	0.55	FC_{900}-3	908	0.55
FC_{700}-4	736	0.60	FC_{900}-4	893	0.60
FC_{700}-5	741	0.65	FC_{900}-5	906	0.65

2.2 Test method

After the self-compacting backfill material is mixed according to the dry density and water-binder ratio described in Table 1, the mixed self-compacting backfill material is immediately poured into a mold of 100mm×100mm×100mm, vibrated, formed and flattened, and then placed in the room. (24±2) hours later, the mold will be removed, and the sample will be cured in a standard curing room for 28 days for performance testing and microanalysis.

The self-compacting backfill material specimens cured for 28 days were cut into 20mm thick slices, and the surface morphology characteristics were observed and collected by an electron microscope, and the pore structure of the material was analyzed using Image-J software to calculate the average pore size and pores of the self-compacting backfill material .The porosity rate I is directly measured by the mass-volume method[4], and calculated according to formula (1):

$$I=(\rho-\rho 0)/\rho \qquad (1)$$

In the formula: ρ-the dry density of the test piece, in kg/m^3; ρ0-the actual density of the test piece, in kg/m^3.

3 Test results and analysis

3.1 Research on the fluidity of self-compacting backfill materials with different water-to-binder ratios

Due to the low density and high volume utilization of self-compacting foam backfill materials, when used in different underground backfill projects, due to the surrounding environment of the pipeline, it is usually impossible to vibrate and compact. The fluidity of the backfill material determines the compactness of the backfill. The degree and different construction requirements are shown in Figures 1 and 2 as a schematic diagram of the fluidity of the self-compacting backfill material with a density of 700 kg/m^3 as a function of the water-binder ratio and the amount of fly ash.

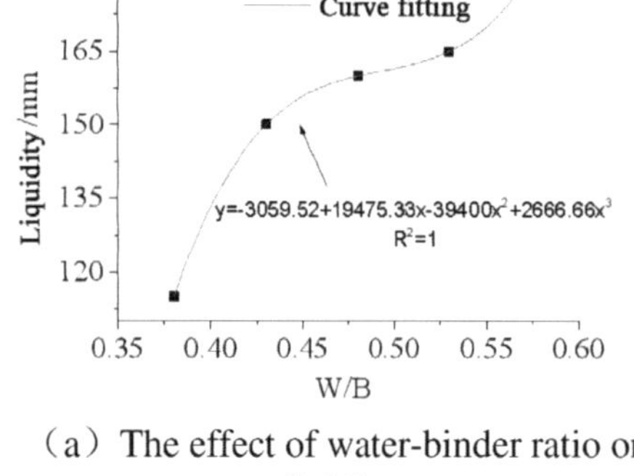

(a) The effect of water-binder ratio on fluidity

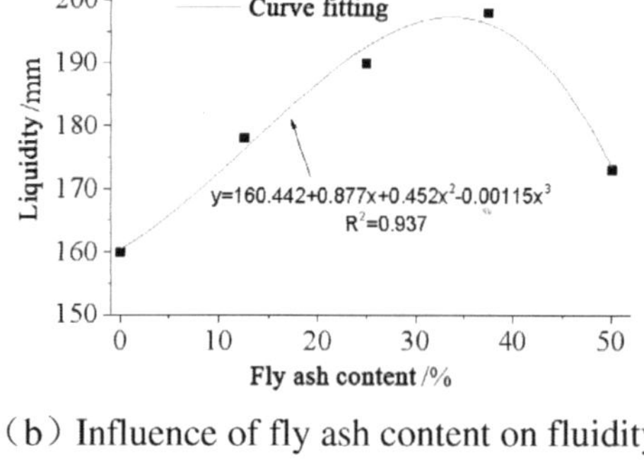

(b) Influence of fly ash content on fluidity

Figure 1 The influence of different components on the fluidity of backfill material slurry

It can be seen from Figure1(a) that the fluidity of the self-compacting backfill material slurry shows an upward trend with the increase of the water-to-binder ratio. When the water-to-binder ratio increases sequentially, the fluidity increases by 30.43%, 39.13%, 43.48% and 60.87%. The fluidity of the backfill material slurry increases with the increase of the water-to-binder ratio. The increase in the water-to-binder ratio increases the dispersion of foam and cement in the slurry, and the packing of the air bubbles by the cementitious material also becomes more uniform, which promotes greater fluidity.

The influence of fly ash content on the fluidity of backfill material slurry is shown in Figure 1(b). With the increase of fly ash content, it first increases and then decreases. It has a good fluidity when the content of fly ash is 35%~45%; when the fly ash content gradually increases, the fluidity of the backfill material slurry increases by 11.25%, 18.75%, 23.75% and 8.13%, respectively. This is because the addition of fly ash improves the dispersion of cement particles to a certain extent, which promotes the relative movement between bubbles, and improves the fluidity of the slurry; if the amount of fly ash is too large, it will affect the hydration of cement particles and reduce the degree of mobility[4]. Figure 2 shows the fluidity of backfill materials with different water-to-binder ratios. It can be clearly seen that the fluidity and surface morphology of the low water-to-binder ratio are not good. Therefore, the water-to-binder ratio used in the research is selected as 0.40～0.65.

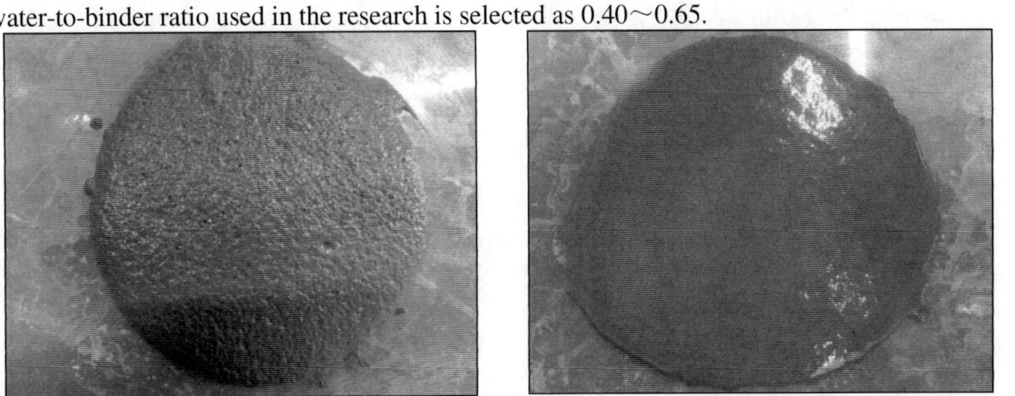

(a) W/B=0.38 (b) W/B=0.53

Figure 2 Backfill material slurry fluidity and surface state under different water-binder ratios

3.2 Compressive strength of backfill materials with different water-binder ratios

Figures 3 and 4 are schematic diagrams of the compressive strength of self-compacting backfill materials with density grades of 700 kg/m^3 and 900 kg/m^3 as a function of water-binder ratio.

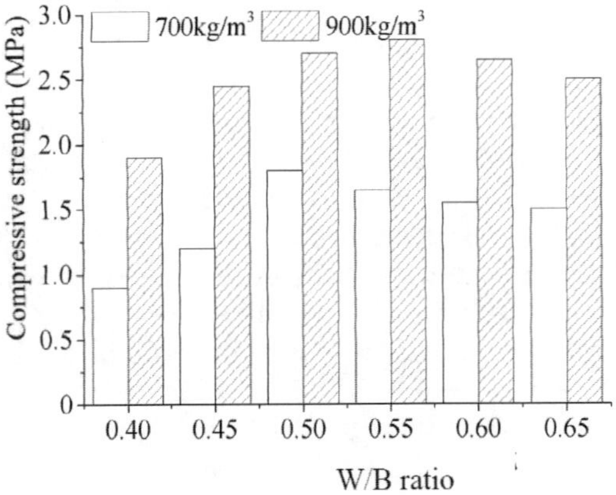

Figure 3 Compressive strength of backfill materials with different water-to-binder ratios

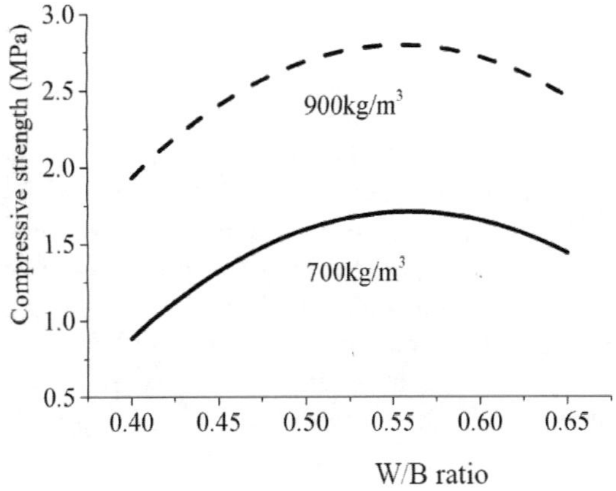

Figure 4 The compressive strength of backfill material changes with water-to-binder ratio

It can be seen from Figures 3 and 4 that the compressive strength of the self-compacting backfill material gradually rises first and then slightly decreases with the increase of the water-to-binder ratio. The compressive strength of the backfill material with a density of 700 kg/m^3 reaches the maximum when the water-to-binder ratio increases. The compressive strength of the backfill material with a density of 900 kg/m^3 reaches the maximum when the water-to-rubber ratio is 0.55, which is 2.73MPa, and basically exceeds 2MPa (except for the slightly lower water-to-rubber ratio of 0.40), and the compressive strength fully meets the design requirements of the backfill material for the Yellow River Crossing Project. This is because as the water-binder ratio in the self-compacting backfill material increases, the viscosity of the slurry decreases, the bubbles in the slurry merge with each other, and the volume of the bubbles in the slurry decreases; the flow of bubbles will promote the gelation in the backfill material, which makes the hydration reaction of the material more complete; the hydration product fills the gap between the pore walls, which reduces the porosity, and improves the compressive strength of the backfill material[19]; if the water-binder ratio increases again, the porosity of the backfill material decreases, however, the increase of the pore diameter makes the pores in the backfill material unevenly distributed in the material, and stress concentration is prone to occur, which reduces the compressive strength of the backfill material.

3.3 Research on the average pore size and porosity of backfill materials with different density levels

Figures 5 and 6 are the average pore size and porosity test results of self-compacting backfill materials with density grades of 700 kg/m^3 and 900 kg/m^3.

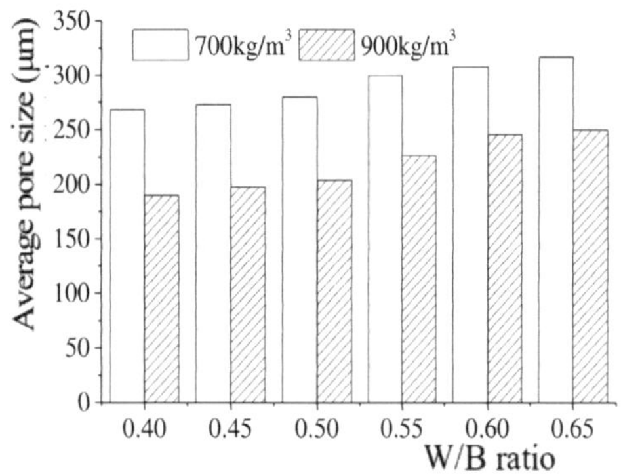

Figure 5 Average pore size of backfill materials with different densities

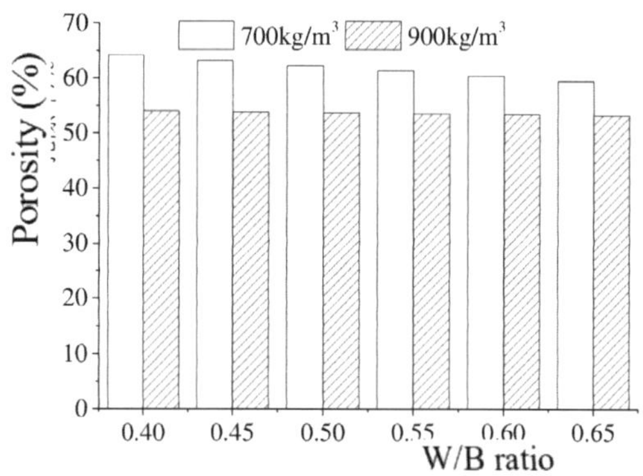

Figure 6 Porosity of backfill materials with different densities

It can be seen from Figure 5 and Figure 6 that as the water-to-binder ratio increases, the average pore size of the self-compacting backfill materials of different density grades is on the rise, and the average pore diameter of the backfill materials of the density grade of 900 kg/m^3 is smaller than 700 kg/m^3; As the water-binder ratio increases, the porosity of backfill materials of different density grades decreases slightly, and the porosity of backfill materials with density grades of 900 kg/m^3 is lower than 700 kg/m^3. This is because as the water-to-binder ratio increases, the bonding performance of the cementitious material gradually decreases, and the ability to wrap bubbles is correspondingly weakened. A large number of bubbles in the backfill material merge and communicate with each other, and the average pore size shows an upward trend[20]; If the water-binder ratio of the backfill material is further increased, the increase in the pore size of the bubbles due to the friction between the cementitious materials will be correspondingly weakened. For backfill materials with different density levels under the same water-to-binder ratio, the average pore size decreases with the increase of the density level of the backfill material. This is because as the density level of the backfill material increases, the cementitious material accounts for the proportion of the backfill material gradually increase, the resistance encountered during bubble formation and movement also increases, and its average pore size also decreases.

3.4 Research on the morphology of backfill materials with different density levels

Figure 7 is a schematic diagram of the morphology of the self-compacting backfill material with a density of 700 kg/m^3 and 900 kg/m^3 when the water-binder ratio is 0.50.

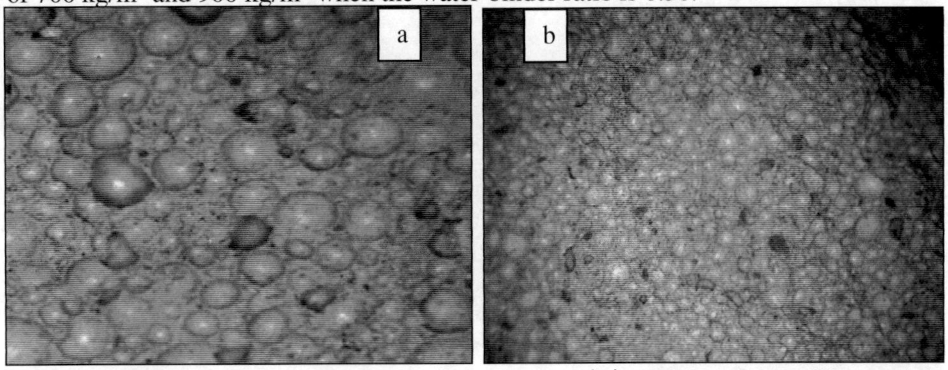

(a) 700 kg/m3 backfill materials (b) 900 kg/m3 backfill materials

Figure 7 Internal bubble morphology of backfill materials with different densities

It can be seen from Figure 7 that the water-binder ratio is 0.50, and the self-compacting backfill materials with different densities have different morphologies. As the density level increases, it is obvious that the bubble diameter in the backfill material gradually decreases. For example, the pore size of the backfill material with a density of 700 kg/m3 in Figure 7 (a) is significantly larger than that of 900 kg/m3 in Figure 7 (b). This is basically consistent with the average pore size test result in Figure 5; at the same time, the number of bubbles has increased significantly, the pore distribution is more uniform, the bubble spacing is also significantly reduced, and the porosity is reduced, which is basically consistent with the porosity test results in Figure 6. The porosity decreases, the compactness of the backfill material increases, the contact surface and adhesion of the cementitious material also increase, the rate and degree of cement hydration reaction increase, and the compressive strength of the self-compacting backfill material increases. This matched the result of compressive strength in Figure 3.

4. Conclusion

The physical and mechanical properties, pore structure and morphology of self-compacting backfill materials with densities of 700 kg/m^3 and 900 kg/m^3 are studied through experiments, and the following conclusions are drawn:

1) The fluidity of the self-compacting backfill material slurry increases with the increase of the water-binder ratio. The fluidity of the backfill material slurry is better when the water-binder ratio is greater than 0.4; the fluidity of the backfill material slurry increases with the amount of fly ash. First rises and then declines. When the content is 35%~45%, the fluidity is better.

2) The compressive strength of the self-compacting backfill material first gradually increases with the water-binder ratio and then slightly decreases, gradually increasing with the increase in density; the compressive strength of the backfill material with the water-binder ratio is greater than 0.4 and the density of 900 kg/m3 is greater than 2MPa, which meets the design requirements for backfill materials for the Yellow River Crossing Project.

3) The average pore size of the self-compacting backfill material increases slightly with the increase of the water-binder ratio, and decreases with the increase of the density; the porosity decreases slightly with the increase of the water-binder ratio, and decreases with the increase of the density.

4) Through the analysis of the microstructure and morphology, it can be observed that the morphology of the self-compacting backfill materials of different densities is different, the density increases, and the bubble diameter in the self-compacting backfill material decreases, indicating that the pore structure and strength have a certain correlation.

The project was support by the traffic construction in Tongling（20210023）and Jiangxi Province (2020H0047), Jiangsu Natural Science Foundation (BK20200429), Post-doctoral fund of China and Jiangsu (2020M671485、2020Z321)

Reference

[1] R Wang, P Gao, Y Dai, Experimental study on mechanical and waterproof performance of lightweight foamed concrete mixed with crumb rubber [J]. Construction and Building Materials, 2019, (209): 655-664.

[2] Wang Wuxiang, Liu Ning. Application of foam concrete in the backfilling of tunnels through the Yellow River Diversion Project [J]. Concrete and Cement Products, 2002, (4): 12-15.

[3] Sun Yafei, Gao Peiwei. Thermal conductivity and mechanical properties of porous concrete materials[J]. Material letter, 2017, (209): 349-352.

[4]Dai Yuchen. Preparation of high-performance foaming agent and its application in foam lightweight materials[D]. Nanjing University of Aeronautics and Astronautics, 2019.

[5] C. Bai, T. Ni, Q. Wang, et al. Porosity, mechanical and insulating properties of geopolymer foams using vegetable oil as the stabilizing agent[J]. Journal of the European Ceramic Society, 2018, 38: 799-805.

[6] A. Abd, S. Abd. Modelling the strength of lightweight foamed concrete using support vector machine [J]. Case Studies in Construction Materials, 2017, 6: 8-15.

[7] Bai Jing, Xie Chongbao, Wu Zhiqin. Research on antifreeze mode of shallow buried pipeline replacement in seasonal frozen soil area[J]. Journal of Hydraulic Engineering, 2018, 148: 140-147. [8] M. Siva, K. Ramamurthy, R. Dhamodharan. Development of a green foaming agent and its performance evaluation[J]. Cement & Concrete Composites，2017, 80: 245-257.

[9] J. Lee, K. Lee. Carbonation Behavior of Lightweight Foamed Concrete Using Coal Fly Ash[J]. Journal of the Korean Ceramic Society, 2016, 53(3): 354-361.

[10] Fan Qixiang, Huang Canxin, Jiang Xiaochun. Intelligent control method and system for cement grouting in hydropower engineering[J].Journal of Building Materials, 2019, 2: 165-173. .

[11] Gao P, LU X. Microstructure and pore structure of concrete mixed with superfine phosphorous slag and superplasticizer[J]. Construction and Building Materials, 2008, (22): 837-840.

[12] Tikalsky P, Pospisil J, Macdonald W. A method for assessment of the freeze-thaw resistance of preformed foam cellular concrete[J]. Cement & Concrete Research, 2004, 34(5): 889-893.

[13] Guan Lingyue. Study on the Characterization and Control Method of Foam Concrete Pore Structure and Its Performance[D]. Wuhan University of Technology, 2014.

[14] Yang K, Lee K. Tests on alkali-activated slag foamed concrete with various water-binder ratios and substitution levels of fly ash[J]. Journal of Building Construction and Planning Research, 2013, 49(5): 588-596.

[15] Y. Amran, R. Rashid, F. Hejazi, et al. Response of precast foamed concrete sandwich panels to flexural loading[J]. Journal of Building Engineering, 2016, 7: 143-158.

[16] H. Guo, W. Guo, Y. Shi. Computational modeling of the mechanical response of lightweight foamed concrete over a wide range of temperatures and strain rates[J]. Construction & Building Materials, 2015, 96: 622-631.

[17] Sang G, Zhu Y, Yang G. Influence of water-cement ratio on property of light weight cement based foam material[J]. Journal of Materials Science and Engineering, 2015, (3): 339-342.

[18] Z. Liu, K. Zhao, C. Hu, et al. Effect of Water-Cement Ratio on Pore Structure and Strength of Foam Concrete[J]. Advances in Materials Science and Engineering, 2016, 11: 1-9.

[19] A. Hilal, N. Thom, A. Dawson. On entrained pore size distribution of foamed concrete[J]. Construction & Building Materials, 2015, 75: 227-233.

[20] M. Liu, U. Alengaram, M. Jumaat, et al. Evaluation of thermal conductivity, mechanical and transport properties of lightweight aggregate foamed geopolymer concrete[J]. Energy & Buildings, 2014, 72: 238-245.

Mechanism experimental study on flooding control of Soft Microgel by low-field nuclear magnetic resonance

Yinzhu Ye[1,a], Xingcai Wu[1], Zhuowei Huang[2], Taifei Bi[3], Zhe Yang[1,2], Xiaocong Wang[1], Lipeng He[1], Yang Liu[1,4], Shichao Li[1]

[1] Research Inst. of Petroleum E&D (RIPED), CNPC.

[2] China University of Geosciences, Beijing .

[3]Oil Production Plant 1 of PetroChina Changqing Oilfield

[4] Research Institute of Exploration and Development, PetroChina Southwest Oil&Gasfield Company, Chengdu, Sichuan 610041, China.

[a]email: Yeyinzhu@petrochina.com.cn

Abstract: The research method of LF-NMR(low field nuclear magnetic resonance) combined with core displacement device has become important experiment was used to study the Water Flooding SCT(swept control technology by water flooding) of soft microgel. Besides, the transverse relaxation time spectrum and the "visual" evaluation method based on LF-NMR were used to investigate the microscopic mechanism of expended sweep volume of soft microgel in single core experiment. The displacement conformance effect of flexible microgel in parallel double-core and its displacement mechanism at different pore scales were analyzed by transverse relaxation time spectrum. The results of single core experiments show that the soft microgel can expand the sweep volume, mainly in the relatively large pore throat, with 15.43%. The results of parallel double-core experiments indicate that after SMG flooding, the recovery of low permeability core increased by 29.5%, high permeability core increased by 15.3%, comprehensive recovery increased by 22.0%, and the ratio of low permeability recovery increased by nearly 2 times that of high permeability core. The soft microgel flooding technology can effectively activate the remaining oil in different pore sizes of the core, among which the medium pore (12.1%) is more used than the large pore (5.6%) and the small pore (4.2%).

1. Introduction

In China's waterflooding oilfields, there are some problems, such as serious reservoir heterogeneity, ineffective injection water or inefficient circulation[1-2]. The technology of expanded water flooding with flexible microgel is an enhanced oil recovery technology developed in recent years[3]. It has the advantages of stable chemical properties, good environmental tolerance and simple injection process. The technology is widely used and has achieved good technical and economic results[4]. In order to study the influence of volume and profile control of flexible microgels, conventional core displacement, microscopic displacement models, CT imaging and low field NMR can be used. In comparison, low field NMR and imaging technology is more suitable for testing the fluid distribution in core[5-6]. In this paper, the low field NMR off-line measurement method is used to study the mechanism of the flexible micro gel flooding in parallel dual pipe displacement, which provides a reference for similar chemical system to study the mechanism of profile control and flooding.

2. Principle and method of low field NMR experiment

The object of nuclear magnetic resonance is the relaxation behavior of atomic nuclei (such as hydrogen nuclei) at different resonance frequencies. In the process of NMR experiment, the measured signal is the signal of hydrogen element in the fluid in the core, and the signal quantity measured by NMR reflects the fluid content in the core. NMR results show that the transverse relaxation time is proportional to the signal amplitude, and the transverse relaxation time is proportional to the pore throat size[7], That is:

$$r = 0.735 T_2 / C \tag{1}$$

Where:

r—— pore radius， μm;

T_2——NMR relaxation time， ms;

C——conversion coefficient， nearly 1.71 ms/μm.

According to the relaxation mechanism of NMR, the NMR signal with longer relaxation time on T2 spectrum corresponds to the fluid in larger pores, and the NMR signal with shorter relaxation time on T2 spectrum corresponds to the fluid in fine pores[20]. The relationship between transverse relaxation time and pore size is defined as that the pores corresponding to transverse relaxation time (T2) ≤10 ms are small pores with pore radius≤ 4.3μm. T2 between 10-100 ms, the corresponding pore is medium pore, and the pore radius is 4.3-43.0μm. T2≥100 ms is macropore with pore radius≥43.0μm. According to the principle of NMR technology, the longer the relaxation time is, the larger the pore size of simulated core is; the larger the integral area of signal amplitude is, the higher the water saturation of porous media is.

Magnetic resonance imaging is to apply three mutually vertical controllable linear gradient magnetic fields on the target object to achieve the spatial positioning of the magnetic resonance signal, collect the amplitude of the magnetic resonance signal and the corresponding spatial position information, and obtain the magnetic resonance image of the sample after certain processing[8].

3. Experimental process

3.1. Experimental materials

The chemical agent used in the experiment is micron grade SMG, and the median particle size is determined by laser particle size analyzer. The basic physical and chemical performance parameters are shown in Table 1.

Table 1 Physical and chemical properties of SMG

Performance parameter	Median of original particle size μm	Median particle size after swelling μm	Swelling ratio
Evaluation results	3.2	28.6	7.9

The experimental water is the formation water from an oil field in Bohai Sea, and the salinity is 9374.13 mg/L. the composition is shown in Table 2.

Table 2 Inorganic ion composition of simulated water in SZ36-1 Oilfield

Ionic composition	Na^+, K^+	Ca^{2+}	Mg^{2+}	CO_3^{2-}	HCO_3^-	SO_4^{2-}	Cl^-
content， mg/L	3091.96	276.17	158.68	14.21	311.48	85.29	5436.34

Conventional simulated oil or water can produce hydrogen signal. In order to distinguish T2 spectrum signal of transverse relaxation time, deuterium water is used to prepare bound water in single pipe extended sweep volume experiment, deuterium water is from isowater company, D content is 99.9%; fluorine oil is used to replace simulated oil in parallel double pipe profile control and displacement experiment, and fluorine oil is from 3M company.

Experimental core: artificial standard columnar core, with cross-sectional area of 4.91 cm^2. Other parameters are shown in Table 3

Table 3 Parameters of columnar core

Core number	length /cm	Gas logging permeability /mD	porosity /%
1-1(Single tube)	5.03	960	27.6
2-1(Hypertonic)	5.32	1965	26.5
2-2(Hypotonic)	5.28	485	23.8

3.2. Experimental equipment

The experimental instruments used in the low field NMR core displacement experiment include Nantong Huaxing constant pressure and constant speed double cylinder pump, pressure collection system, Nantong Huaxing mixing intermediate vessel, intermediate vessel, oven, measuring cylinder, oven, Suzhou newmai large aperture NMR imaging analyzer Macro MR12-150H-I (magnetic field intensity 0.5T, resonance frequency 21MHz, probe coil diameter 70mm), etc[9].

3.3. Flooding experiment steps

Two cores with different permeability were dried in oven at 80°C for 24 hours, then the cores were taken out and vacuumized for 12 hours to saturate the fluorine oil and calculate the porosity. The low field nuclear magnetic "off-line test" method was used to measure the initial signal of two cores, and the T_2 spectrum of the core was obtained. The experimental temperature of core displacement is 65°C.

A water flooding experiment: Put the high and low permeability cores in the core holders respectively, and carry out the parallel double-pipe oil flooding experiment. Simulated water was injected at an injection rate of 1.0 mL/min to record the fluid and oil production of high- and low-permeability cores. Until the water content of the production end of the high-permeability core reaches more than 98% (take three stable points). After that, the core was taken out, and the signal after water flooding of the high and low permeability core was measured under low-field nuclear magnetic conditions to obtain the T_2 spectrum of the core.

Flexible gel flooding and subsequent water flooding: injecting 1.0 PV of SMG chemical system at 1.0 mL/min displacement rate, recording injection pressure, and producing liquid and oil yield of core. Until the water content of core production end reaches above 98% (take three stable points). After that, the core is taken out, and the signal quantity of high and low permeability core after SMG profile control and displacement is measured by low field nuclear magnetic "off-line test" method, and the T_2 spectrum of the core is obtained.

4. Results and discussion

According to the relationship between the different water content in the control substance (glass bottle) and the intensity of the nuclear magnetic signal, the water marking line is obtained, as shown in Figure 1. Use this curve to calculate the change of the water content in the core during the oil displacement process, so as to calculate the change of the recovery factor during the oil displacement process. The NMR T2 test data of high-permeability and low-permeability cores in different displacement stages of parallel double pipe experiment are shown in Figure 2, the data processing results of oil recovery efficiency of different pores in high-permeability and low-permeability cores are shown in Figure 3, and the comprehensive oil recovery efficiency of parallel double pipe flooding is shown in Table 4.

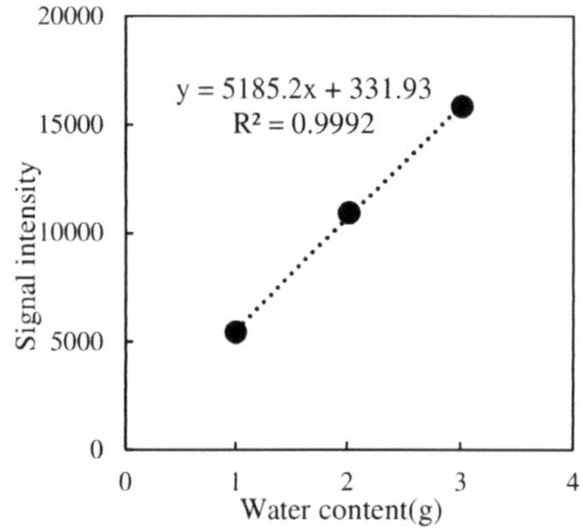

Figure.1 Calibration curve of NMR experiment

It can be seen from Fig. 5 and Fig. 6 that compared with water flooding, after SMG profile control and flooding, the follow-up water flooding T_2 spectrum curves of high permeability core and low permeability core move to the left and increase compared with water flooding curve, which is more obvious in low permeability core. The results show that after SMG profile control and flooding, both high permeability core and low permeability core use smaller pores in varying degrees, and the low permeability core is more obvious.

The T2 spectrum curve of low permeability core increases obviously in the range of less than 100ms (medium and small pores). In water flooding stage, the recovery factor of high permeability core mainly comes from macropores, which is 37.9%, and the cumulative recovery factor of water flooding is 51.2%; The recovery rate of low-permeability cores is mainly from large and medium pores, which is 20.1%, of which the cumulative recovery rate of water flooding is 23.8%. After SMG control and flooding, the production of small pores in high-permeability cores is still limited, only 8.7%, and the cumulative recovery rate of high-permeability cores is 15.3%; However, the production degree of small pores in low-permeability cores has increased significantly, reaching 25.1%, and the cumulative recovery rate of low-permeability cores has been increased by 29.5%. After comprehensive consideration of the experimental results of large and small pore throats of high and low permeability cores, it is found that after SMG profile control and flooding, the comprehensive macropore recovery of high and low permeability cores is increased by 5.6%, the medium pore by 12.1%, and the small pore by 4.2%.

It can be seen from Table 4 that in the parallel double-pipe experiment, after SMG profile control and flooding, the recovery rate of high-permeability cores increased by 15.3%, the recovery rate of low-permeability cores increased by 29.5%, and the comprehensive recovery rate increased by 22.0%.

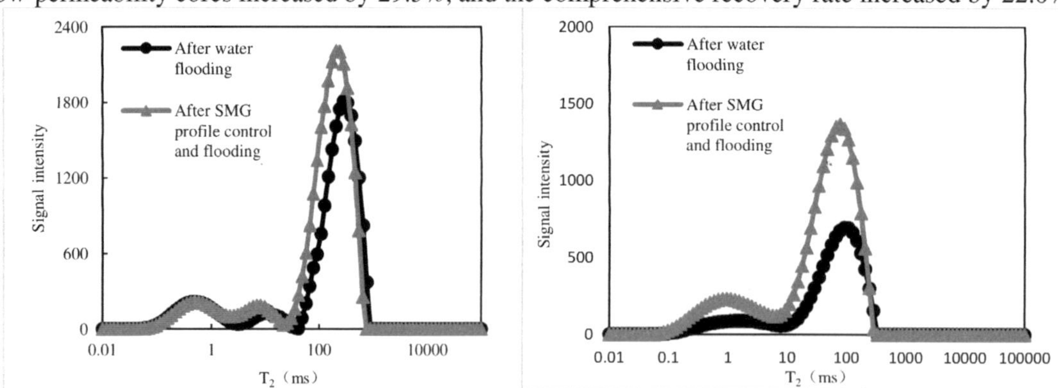

(1) High permeability core (2) Low permeability core

Figure.2 T_2spectrum of high and low permeability cores in parallel dual tube core flooding experiment

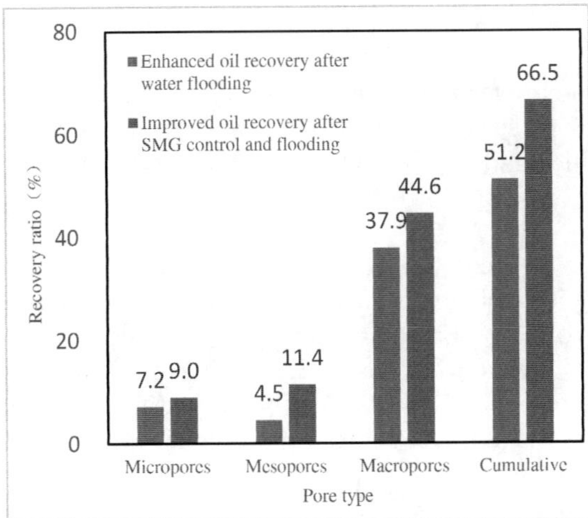

(1) High permeability core

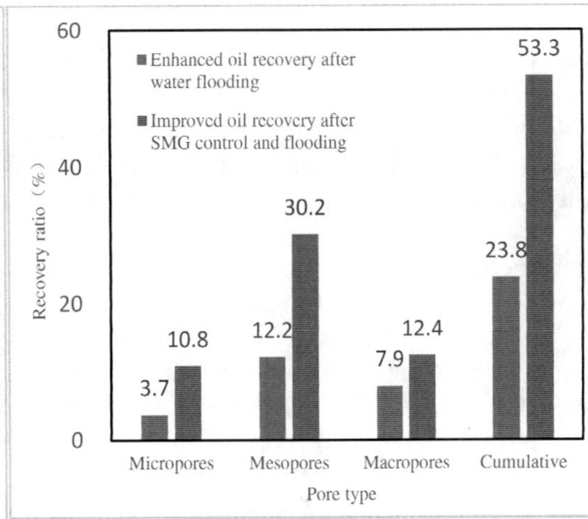

(2) Low permeability core

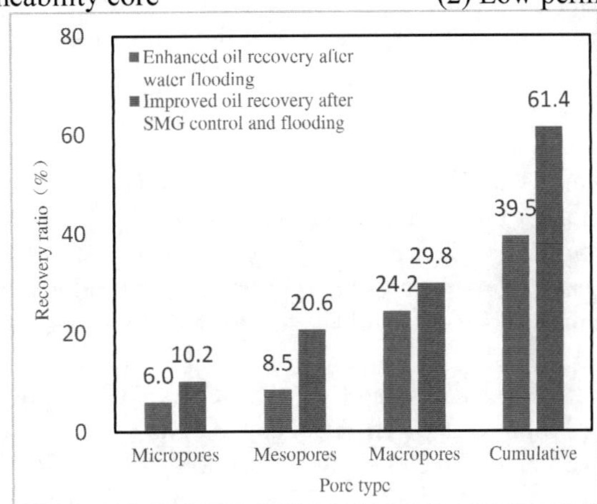

(3) Comprehensive porosity of high and low permeability core

Figure.3 Recovery degree of high and low permeability cores in parallel dual tube core flooding

Table 4 Corresponding recovery factor of parallel double pipe flooding experiment

Core type	Water drive /%	SMG profile control and flooding /%	Enhanced oil recovery /%	Comprehensive recovery factor /%
Hypertonic	51.2%	66.5%	15.3%	22.0%
Hypotonic	23.8%	53.3%	29.5%	

Combined with the above experiments, it can be found that the displacement fluid is mainly along the high permeability area in the water flooding stage. After SMG profile control and flooding, SMG aqueous dispersion is effectively retained in the high permeability core, and the subsequent injected SMG aqueous dispersion enters into the low permeability core and continuously migrates to the deep part of the core, and the medium and small pores with rich residual oil are activated. At the same time, the SMG water dispersing liquid entering into the high permeability core improves the producing degree of the remaining oil in the pores of the high permeability core (mainly Acting in the medium and large porosity), and finally greatly improves the comprehensive recovery. Therefore, SMG dispersion system has a good role in deep profile control and flooding, especially for heterogeneous reservoirs. In the SMG profile control and flooding stage, the injected SMG dispersion system will preferentially enter the high permeability layer, and play an effective role in plugging and deep profile control and flooding in the high permeability layer. At the same time, it will carry liquid into the medium and low permeability layer to improve water sweep efficiency and further improve water flooding effect[10]. Compared with the traditional core displacement experimental method, the low field NMR method can characterize the start-up position and profile control and displacement effect of remaining oil in different pore sizes.

5. Conclusion

(1) SMG can effectively inhibit the seepage capacity of high-permeability reservoirs and activate low-permeability reservoirs. The enhanced value of low-permeability recovery factor (29.5%) is greater than the enhanced value of high-permeability core recovery factor (15.3%), which is close to 2 times. Considering the comprehensive recovery factor of high- and low-permeability cores, the contribution of enhanced recovery in a waterflooding stage mainly comes from the large pores. The contribution of enhanced oil recovery after SMG flooding mainly comes from mesopores.

(2) The in-depth study of this experimental method is expected to further reveal the expanding sweeping volume mechanism of flexible microgels or similar chemical agent systems, the starting mechanism of remaining oil at different pore scales, and the evaluation of profile control and flooding effects.

References

[1] Wang Yupu, Liu Yikun, Deng Qingjun. Current situation and development strategy of the extra high water cut stage of continental facies sandstone oil fields in China[J]. Journal of Northeast Petroleum University, 2014. 38(1):1-9.
[2] Yuan Shiyi, Wang Qiang. New progress and prospect of oilfields development technologies in China[J]. Petroleum Exploration and Development, 2018, 4594); 637-667.
[3] Wu Xingcai, Han Dakuang, Lu Xiangguo, et al. Oil Displacing Mechanism of Soft Micro gel Particle Dispersion in Porous Media[J]. Earth science, 2017. 42(8):1348-1355.
[4] Wu Xingcai, Xiong Chunming, Han Dakuang, et al. A New IOR Method for Mature Waterflooding Reservoirs: "Sweep Control Technology"[R]. SPE-171485-MS, 2014.
[5] Wu Xingcai, Qu Debin, Xu Hanbing, et al. The Economic Analysis and Application Strategies of EOR Technology in Low-Oil-Price Period---8 Case Study of a New Polymerflooding Technology[R]. SPE-182145-MS, 2016.
[6] Yi Zhe. Study on Profile-control Flooding Mechanism of Polymer Microsphere[J]. Advances in fine petrochemicals, 2013. 14(6):1-4.
[7] Di Qinfeng, Zhang, Jingnan, Hua Shuai, et al. Visualization experiments on polymer-weak gel profile control and displacement by NMR technique[J]. Petroleum Exploration and Development, 2017, 44(2): 270-274.
[8] Lei Qun, Luo Jianhui, Peng Baoliang, et al. Mechanism of expanding swept volume by nano-sized oil-displacement agent[J]. Petroleum Exploration and Development, 2019, 46(5): 937-942.
[9] Zhang Jingnan, Di Qinfeng, Hua Shuai, et al. Nuclear magnetic resonance experiments on foam flooding and evaluation of foam dynamic stability[J]. Petroleum Exploration and Development, 2018, 45(5): 853-860.
[10] Cheng Yichong, Di Qinfeng, Gu Chunyuan, et al. Visualization study on fluid distribution and end effects in core flow experiments with low-field mri method[J]. Journal of Hydrodynamics, Ser. B, 2015, 27(2):187-194.

Journal of Physics: Conference Series **2011** (2021) 012018 doi:10.1088/1742-6596/2011/1/012018

Development of S22053+12MnNiVR rolled clad steel plate for multi-stage flash desalination evaporator shell

Ouyang Xin[1,a*], Hu Xinming[1], Sun Diandong[1], Wang Chu[1], Wang Yong[1], Xing Mengnan[1]

[1]Anshan Iron and Steel Group Steel Research Institute, Anshan, Liaoning, China

*Corresponding author: [a]email: 02004450@ansteel.com.cn

Abstract. This paper describes the development process of S22053+12MnNiVR composite steel plate using vacuum electron beam welding technology and double-sided symmetrical billet assembly method in ANSTEEL Co., Ltd., which has a product thickness of 2+12mm and is used in multi-stage flash seawater desalination evaporator shell. The stainless-steel clad steel plate has excellent comprehensive mechanical properties, good resistance to intergranular corrosion and seawater corrosion. The bonding rate of its composite interface meets the requirements of GB/T 8165-2008 Class I, and the bonding rate between atoms is 100%. As a new generation of green composite material, the successful development of S22053+12MnNiVR stainless steel-clad steel plate is of great significance in reducing the cost of seawater desalination, increasing the life of seawater desalination equipment, and realizing the localization of equipment, which provides an important way for China to solve the shortage of fresh water resources and lays a solid foundation for China to realize the green recycling of seawater resources.

1.Introduction

S22053 duplex stainless steel has excellent stress corrosion resistance, pitting corrosion resistance and crevice corrosion resistance under typical extreme environments, and can improve the corrosion resistance and service life of seawater desalination devices. Therefore, it becomes an ideal manufacturing material for evaporator shells [1]. However, the production cost of pure stainless-steel medium and heavy plates is relatively high, which is not conducive to the large-scale production of seawater desalination devices. ANSTEEL Co., Ltd. combined with the existing level of process equipment and production capacity, using vacuum electron beam welding technology and double-sided symmetrical billet assembly method, successfully developed a carbon steel with good mechanical properties of the same strength level and excellent corrosion resistance. S22053+12MnNiVR composite steel plate is used for the shell of the first-level flash desalination evaporator. Compared with traditional duplex stainless-steel medium and heavy plates, the price advantage of stainless-steel composite plates is obvious, which is of great significance for cost control and large-scale applications in seawater desalination devices.

2.Materials and Methods

2.1Chemical Composition

The chemical composition of 12MnNiVR shall meet the requirements of GB/T 19189-2011 and the chemical composition of S22053 shall meet the requirements of GB/T 4237-2015.

Table 1 Chemical composition of 12MnNiVR、S22053(Wt%)

	C	Si	Mn	P	S	Ni	Cr	Mo	Cu	V	B	N
12MnNiVR	≤ 0.150	0.15-0.40	1.20-1.60	≤ 0.020	≤ 0.010	0.15-0.40	≤ 0.30	≤ 0.30	≤ 0.25	0.02-0.06	≤ 0.0020	-
S22053	≤ 0.030	≤ 1.00	≤ 2.00	≤ 0.030	≤ 0.020	4.5-6.5	22.0-23.0	3.00-3.50	-	-	-	0.14-0.20

2.2 Mechanical Properties

The mechanical properties of composite steel plate shall meet the requirements of GB/T 8165-2008, the inspection direction shall be transverse, and the inspection position shall be the full thickness of steel plate.

Table 2 Mechanical properties of S22053+12MnNiVR

Tensile strength at room temperature			Impact at -20°C	Bending	Shear strength
R_{eL}/ MPa	R_m/ MPa	A/ %	KV_2/ J	b=2a 180°	τ/MPa
≥490	610~730	≥18	≥100	d=2a	≥210

2.3 Delivery status and inspection requirements

Solid solution heat treatment + pickling + passivation should be carried out before obtaining S22053 duplex stainless-steel billet. The upper and lower two sub-boards after the division are required to be ultrasonically inspected according to the NB/T47013.1-2015 standard, and the qualified level should be T I level.

2.4 Production process route

ANSTEEL Co., Ltd. adopts the symmetric billet method + vacuum electron beam welding technology to roll the S22053+12MnNiVR composite steel plate with a thickness of 2+12mm for the shell of the multistage flash seawater desalination evaporator. The production process is shown in Figure 1.

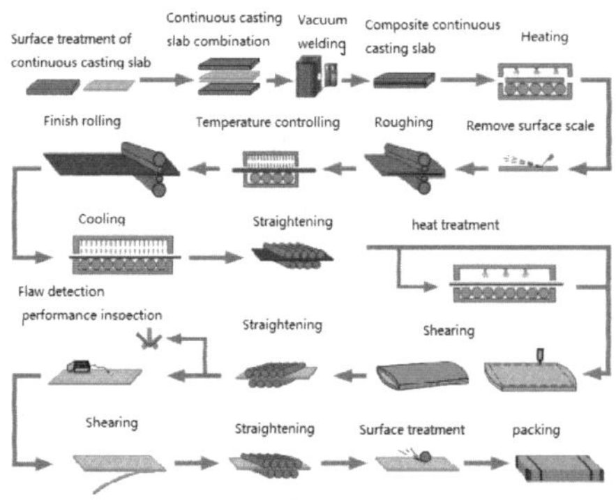

Fig 1 Production process flow chart

2.5 Symmetric billet assembly process

The symmetrical composite billet uses quenched and tempered high-strength steel 12MnNiVR as the base material, and duplex stainless steel S22053 as the composite material. After the size is determined, each layer of the board is finely polished with a drilling and milling machine to make the surface of the board have a uniform roughness, and this step also has a surface derusting effect. After polishing, the surface of the board is repeatedly cleaned with alcohol and acetone to remove oil stains, iron shavings and other attachments. Then, each layer of the plates is arranged in a four-layer symmetrical arrangement, and a high-temperature release agent is added between the two layers of S22053 to facilitate the cutting and separation of the plates after rolling and forming. The high-temperature release agent is evenly sprayed on the surface of the stainless-steel cladding, and center the top and bottom to complete the billet assembly [2].

2.6 Vacuum electron beam welding process

The composite continuous casting billet is welded in a vacuum chamber. The pressure in the vacuum chamber must reach a high vacuum state below 1Pa. By creating a high vacuum environment, the problem of interface oxidation pollution during the later hot rolling process can be effectively solved to ensure the interface bonding of the composite steel plate rate. The entire vacuum welding process needs to be carried out under argon protection. The two laser welding heads need to be welded in opposite directions along the two parallel sides to complete multiple passes until the four sides of the composite billet are welded.

2.7 Rolling process

After the billet is taken out of the furnace, it is descaled by descaling box, and then descaled by high-pressure water again before rolling, which purpose is to remove the primary and secondary scales on the surface of the slab to ensure the surface quality of the steel plate. TMCP two-stage rolling is adopted, the initial temperature is ≥1050°C. The first-stage adopts rapid high-reduction rolling, the single-pass reduction rate of the first three times is ≥15%. The austenite grain refining degree increases in this process, and the ferrite grain after transformation is also refined. At the same time, the roller has a higher temperature after the first three times of rolling, so in the subsequent rolling process, the plate loses less heat through contact with the roller. In addition, the heat generated by plastic deformation itself lengthens the duration of the composite interface in the high temperature rolling stage, providing sufficient energy for refining the stainless-steel grains and sufficient diffusion energy and time for the alloy elements at the interface to ensure the good bonding of the composite interface [3].Without rolling in the crystallization zone, the grain of the base carbon steel plate is further refined, so that the base steel plate has a good strength and toughness match, and can promote the remaining small unbonded zone at the composite interface to complete welding, to ensure that the bonding rate of the composite interface reaches 100%.

3.Results & Discussion

3.1Mechanical properties and analysis

Table 3 Mechanical properties of S22053+12MnNiVR

	Tensile strength at room temperature			Impact at -20°C			Bending b=2a 180°		Shear strength	Interface binding rate	Flaw detection
	R_{eL}/ MPa	R_m/ MPa	A/ %	KV_2/ J			Inward d=2a	outward d=2a	τ/MPa	100%	
Results	544	651	21.3	211	234	226	qualified	qualified	478	qualified	qualified
Standard	≥490	610~730	≥19	≥100			No crack	No crack	≥210	I	T I

It can be seen from Table 3 that the clad steel plate has good mechanical properties. Among them, the shear strength value, as an important index to characterize the bonding quality of the composite steel plate composite interface. A good interface bonding rate is the key to ensuring the shear strength and flaw detection results. During the tensile-shearing process of the shear specimen, the unbonded area at the interface will not cause stress concentration and induce the fracture morphology to appear as flat brittle cracks. But as the tensile-shearing process progresses, the shear force is uniformly transferred from the composite layer to the base layer through the composite interface, and the final shear fracture is located in the base layer. Before the fracture, the organization first undergoes the plastic deformation stage, and finally because the shear strength is greater than the tensile strength of the material itself, the grain around the fracture is elongated. This is caused by the material itself hindering the occurrence of deformation. A large amount of carbon compounds can be found at the fracture, indicating that the interface bonding strength of the composite plate is relatively high [4].

3.2Steel plate resistance to intergranular corrosion
The steel plate is sampled for sensitization treatment, the process is 675 °C. When the set temperature is reached, the sample is put into the furnace. After 1h of net heat preservation, the sample is cooled to room temperature with water, and then the steel plate is subjected to the intergranular corrosion test by ASTM A262-2015 standard E method. Observing the surface of the 180° curved sample under the microscope 10 times and observing the microstructure of the bending position under the metallographic microscope 50 times, no cracks and microscopic intergranular corrosion induced by intergranular corrosion were found, indicating that the intergranular corrosion resistance of the composite steel plate is good.

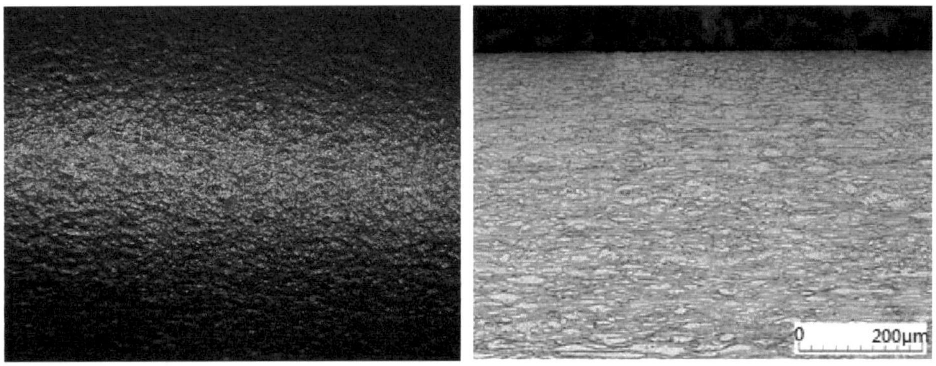

Stereomicroscope photos Metallographic photograph

Fig.2 Bending photos of intergranular corrosion specimens

3.3 Corrosion performance of steel plate with high temperature and high concentration seawater
Based on the JB/T 7901-2001 standard, the composite plate multi-layer S22053 stainless steel is subjected to high temperature and high concentration seawater corrosion performance evaluation test. The simulated test environments are room temperature, 70°C, and 90°C 10% NaCl aqueous solution, and the test results are shown in Figure 3 and Figure 4.

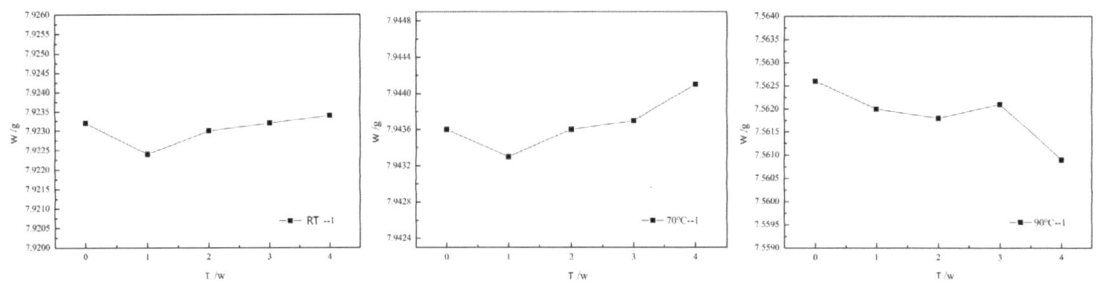

room temperature 70°C 90°C

Fig.3 Corrosion kinetics curve of etched sheet in 10% NaCl aqueous solution at different temperatures

room temperature 70°C 90°C

Fig.4 Surface photos of etched sheet in 10% NaCl aqueous solution for 672h at different temperatures

Figure 3 shows that in a 10% NaCl environment, at room temperature, 70°C, and 90°C, after 4W (672h) corrosion test period, the most serious weight loss is the corroded sheet in an environment of 90°C, and its weight reduction ratio before and after the corroded sheet is about 1‰. From the photo of the sample after 4W (672h) immersion corrosion in Figure 4, it can also be seen that there is no obvious corrosion product and corrosion crack formation on the surface of the corrosion sheet, and the surface is still relatively smooth and bright.

The reasons for duplex stainless steel has good corrosion resistance in seawater are as follows: (1) The cause of corrosion cracking is that the passive film on the surface of the stainless-steel slips and is destroyed, causing the corrosive medium to be directly connected to the substrate. contact. Compared with the single-phase austenitic stainless steel, the ferrite + austenite two-phase structure of duplex stainless steel has a higher yield strength, that is, under the same corrosive environment, the critical stress required for the surface slip of dual-phase stainless steel is higher, which is difficult to cause the coarse slip, and the surface corrosion cracking is difficult to form. (2) Ferrite has an electrochemical protective effect on austenite. Due to the different corrosion electrode potentials of the two-phase structure of duplex stainless steel, the expansion mechanism of corrosion cracking at different phases and phase boundaries is also different, which effectively prevents or inhibits the propagation of cracks, and the generation of corrosion cracks is very slow. (3) S22053 duplex stainless steel contains a higher content of chromium and molybdenum elements. The addition of these elements can make the stainless steel have better pitting corrosion resistance, inhibit the expansion of the connection between pitting points, and prevent the generation of corrosion crack source. (4) The two-phase structure of duplex stainless-steel acts as a mechanical barrier to each other during the propagation of corrosion cracks, thereby preventing the possibility of continuous crack propagation. In addition, due to the different crystal shapes and orientations of the two-phase structure, the cracks frequently change their directions during the propagation process, which greatly extends the growth period of corrosion cracks and is not easy to cause transcrystalline corrosion[5-6].

3.4 Macroscopic and metallographic structure of steel plate

The metallographic structure observation of the fusion line position of the composite plate was sampled for micro-observation, and the results are shown in Figure 5 .

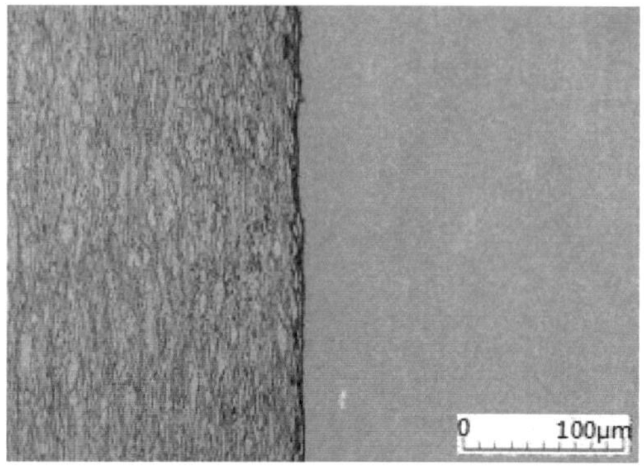

Fig.5 Metallographic photograph of composite interface

The metallographic photo shows that the metallographic structure of the composite layer S22053 is austenite + ferrite, and the metallographic structure of the base layer 12MnNiVR is mainly composed of recrystallized ferrite and aggregated cementite to form a bainite tempered structure. Compared with the position close to the center of the base layer, the carbide aggregation distribution in the composite interface position is more obvious, and there is a certain proportion of ferrite structure. This is because there is a gradient distribution of multiple elements at the bonding interface position of the composite plate sample. Under the action of element concentration difference, the interaction diffusion of C, Cr, Ni and other elements occurs at the interface. The carbon element in the base carbon steel diffuses to the composite layer, forming a decarburization layer in the composite interface base. Cr and Ni diffuses to the base layer, and combines with the C element in the base layer to form stable carbides. At the same time, a chromium-poor layer is formed in the lamination of the composite interface, and the interaction and diffusion of elements ensure the good combination of the composite interface[7-8].

4.Conclusions

(1) The S22053+12MnNiVR composite steel plate produced by ANSTEEL Co., Ltd. has a good match of strength and toughness, the bonding rate of the composite interface meets the requirements of GB/T 8165-2008 Class I, and the application performance meets the requirements of equipment manufacturing;

(2) The S22053+12MnNiVR composite steel plate produced by ANSTEEL Co., Ltd. has good intergranular corrosion resistance and seawater corrosion resistance, meeting the long-term service requirements as a shell material in the special environment of multi-stage flash desalination evaporators;

(3) The S22053+12MnNiVR composite steel plate produced by ANSTEEL Co., Ltd. has good overall performance and can be used as the raw material for the production of seawater desalination equipment to manufacture multi-stage flash desalination evaporators.

References

[1] Wang Dong. Research on the application of duplex stainless steel in seawater desalination[J]. Metallurgical Power, 2018,4(2):56-59.
[2] Yang Munan, Zuo Xiaoqing, Zhao Mingwei, et al. Research progress in the preparation of stainless-steel composite plates[J]. Hot Working Technology, 2012, 41(20): 93-96.
[3] Jin Herong, Zhang Lei. Thermal-mechanical coupling simulation and experimental study of multi-pass hot-rolled stainless steel-clad plate[J]. Chinese Journal of Plasticity Engineering, 2018, 25(05): 158-165.
[4] Wang S, Liu B X, Chen C X, et al. Microstructure, mecha Nical properties and interface bonding mecha Nism of hot-rolled stainless steel clad plates at different rolling reduction ratios[J]. Journal of alloys and compounds, 2018, 766(16): 517-526.

[5] Jin Duoren, Hongrui, Liu Fenzhao, et al. Research on seawater corrosion resistance of marine high-strength duplex stainless steel[J]. Ship Science and Technology. 2018, 8(20): 40-42.
[6] Zhou Bo, Research on the corrosion resistance of new seawater corrosion resistant stainless steel [D]. Gansu: Lanzhou University of Technology, 2009.
[7] Lu Zehua, et al. Study on the microstructure and properties of hot-rolled double-clad stainless steel/carbon steel composite plate[J]. Journal of Plasticity Engineering. 2020, 27(07): 30-34.
[8] Chen Shenglu, et al. The effect of reduction rate on the structure and bonding strength of stainless-steel composite plates[J]. Mechanical Design and Manufacturing. 2020, (04): 23-27.

The impact of the use of new environmentally friendly materials on the management of construction projects: Taking straw fiber materials as an example

Lin Chen[1*], and Mingyu Yang[1,2]

[1] Department of Civil Engineering, Xi'an Jiaotong-Liverpool University, Suzhou, Jiangsu, 215000, China

[2] Department of Civil Engineering, Xi'an Jiaotong-Liverpool University, Suzhou, Jiangsu, 215000, China

[*]Corresponding author's e-mail: Lin.Chen20@student.xjtlu.edu.cn

Abstract. With the rapid development of the construction industry, the social economy has experienced a golden period of development, and the construction industry has experienced an unprecedented high rate of growth. Because of its rapid development, the construction industry has contributed to the development of the world economy, but at the same time, there are other problems, such as the construction of houses, roads, and tunnels, which consume much non-renewable energy and release many harmful gases, leading to the greenhouse effect and the rise in global temperature. As non-renewable energy sources continue to be consumed, society's resources are decreasing, and the earth's ecological environment is gradually deteriorating. Therefore, in the future of the construction industry, not only its materials in terms of strength, durability, and other aspects of higher requirements, but also on the environment-friendly construction materials more attention, so the project cost control and management of materials play a more critical role. This paper takes straw fiber material as an example through the extraction and preparation of straw fiber and then analyses its material in quality, cost, and construction of the three aspects of the content, derived from the material in the quality, cost, and construction of the three sides of the impact, through the rational and effective use of new environmentally friendly building materials, thus to drive the world's new green building materials in the development of scientific and technological innovation, while promoting the world's low-carbon economy in the development of important the role of guidance.

1. Introduction

The rapid development of the construction industry has both advantages and disadvantages. The advantages are that it has boosted the country's economic development, and the disadvantages are that the construction of buildings has consumed a large amount of non-renewable energy, leading to a series of environmental problems and a lack of resources [1]. Therefore, the choice of a new sustainable and environmentally friendly material may solve the current problems faced. As a new environmentally friendly material, the straw fiber material is widely available and exists in most rural areas [2]. Waste straw is usually disposed of by burning, which not only pollutes the air but may also cause a series of problems such as fires [3]. Therefore, the study of the application of straw fiber materials in the construction market has some value, while turning the material into a treasure is in line with world's sustainable development strategy. In this paper, through reading a lot of literature and experimental analysis, the use of straw fiber materials in the quality, cost, construction of the three parties to analyses the impact, as shown in Figure 1 below, first of all, the quality of its materials in different straw types and different fiber additions in different solutions under treatment, learned that its bending and compressive strength value after treatment by a specific concentration of sodium hydroxide solution is higher than that of ordinary water treatment; then its cost is analyzed in terms of the impact of the purchase, extraction, transport and preparation of the straw fiber itself, suggesting that detailed data survey collection and analysis is still required; finally, this paper analyses the impact of the material in terms of construction, due to the emergence of new materials, causing specific difficulties for both construction companies and personnel, and the need to simulate the complete process of the material in the construction of future construction by a detailed construction manual is developed for the use of enterprises and personnel.

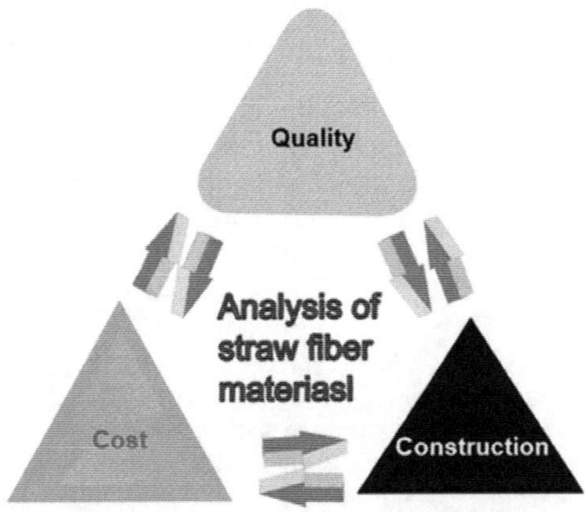

Figure 1 Analysis of straw fiber materials from three aspects

2. Methodology

The methodology of the paper is through experiment and reading a lot of literature and books, analysis research and summary of the impact of straw fiber materials on the management of construction projects, understanding to grasp the previous experts and scholars and other relevant inquiry on the new environmentally friendly materials, and then to analyze and summaries the impact of the material in three aspects: quality, cost, and construction.

3. Study of the impact on the quality of straw fiber materials

The composite material was made using straw fibers; the test was prepared using standard cold-water treatment and 4% sodium hydroxide treatment for numbers A and B, respectively. Then, at room temperature in a specific ratio, the prepared specimens were cured under standard conditions for seven days and 28 days to determine the specimens' compressive strength and flexural strength, respectively. The first is in accordance with the cement 1000g, the water-cement ratio of 0.60, 3000g of sand, straw added into 50g, 100g, 150g, 200g for sample one, and measured flexural strength compressive strength results are as follows Table 1. Secondly, in accordance with the cement 1300g, the water-cement ratio was 0.60, 520g of sand, straw fiber added 65g, 130g, 195g, 260g respectively for sample two, and measured flexural strength and compressive strength results are as follows Table 2.

Table 1 Test results of specimen one

Number	Cement/g	Water-cement ratio	Sand/g	Straw fiber/g	Flexural strength /MPa 7d	Flexural strength /MPa 28d	Compressive strength /MPa 7d	Compressive strength /MPa 28d
A1	1000	0.6	3000	50	1.16	1.95	4.36	7.33
A2	1000	0.6	3000	100	0.46	1.03	1.59	2.96
A3	1000	0.6	3000	150	0	0.47	0	1.12
A4	1000	0.6	3000	200	0	0	0	0
B1	1000	0.6	3000	50	1.57	2.63	5.94	9.78
B2	1000	0.6	3000	100	0.96	1.93	3.19	5.66
B3	1000	0.6	3000	150	0.49	0.92	1.51	2.14
B4	1000	0.6	3000	200	0	0.42	0	0.94

Table 2 Test results of specimen two

Number	Cement/g	Water-cement ratio	sand/g	Straw fiber/g	Flexural strength /MPa 7d	Flexural strength /MPa 28d	Compressive strength /MPa 7d	Compressive strength /MPa 28d
A1	1300	0.6	520	65	0	0.77	0	1.67
A2	1300	0.6	520	130	0	0.46	0	0.84
A3	1300	0.6	520	195	0	0.22	0	0
A4	1300	0.6	520	260	0	0	0	0
B1	1300	0.6	520	65	1.31	2.33	1.86	3.07
B2	1300	0.6	520	130	0.63	1.63	1.07	2.11
B3	1300	0.6	520	195	0.19	0.75	0.42	1.16
B4	1300	0.6	520	260	0	0.48	0.22	0.78

According to the comparison of the two different groups of tests, it can be seen from Tables 1 and 2 that the flexural and compressive strengths of the straw fiber material prepared after treatment with 4% sodium hydroxide solution are higher than those of the untreated specimens with the same amount of straw fiber added, from which it can be inferred that after treatment with this solution, the bond between the cement and the fiber is stronger, and with the curing time, at 28d, the flexural and compressive strengths of the specimens were significantly higher than those of the untreated specimens. Although the strength of the treated straw fiber material has changed significantly, the hydration of the cement is an extremely complex process and should be compared with ordinary cement without the addition of straw fiber material, and other properties and applications of different straw fiber contents are subject to further comparative testing.

4. Study on the impact of the cost of straw fiber materials

At present, straw fiber materials are still in their early stages of application regarding their specific dosage and use and are usually used as straw bricks in rural housing [4]. The performance and requirements of rural housing construction are not representative of all buildings in all respects, so the means of construction and the amount of straw fiber material added vary considerably from building to building. As a result, there are significant problems with the statistics of the amount of straw material, the mix ratio, and other indicators, which may lead to difficulties in controlling the cost of construction materials [5]. In addition, straw fiber materials are made from different crop species, and different materials are available in different sizes [6]. In some construction standards, it is necessary to use a specific straw fiber material because it has a higher quality performance, but it is also more expensive, which can increase the cost of construction using this material. Cost control of straw fiber materials is an essential part of the overall project cost management, both in construction and in the budget, to ensure that accounts and materials are uniforms [7]. However, the current specifications and grades of straw fiber materials are inconsistent, creating considerable difficulties in the final settlement of construction projects. If these new materials need to be used frequently, it is not conducive for construction companies to control the economic costs of new construction materials projects, and the straw fiber material currently utilized is a composite material formed by adding straw fibers to the original cement concrete, the composition of which is shown in Table 3. The cost of the cement concrete itself is added to the cost of the solution and other materials utilized in the above tests, which adds to the total cost of materials.

Table 3 Composition of different straw fiber materials

Type	Ingredients					
Ordinary cement concrete	Water	Cement	Sand	Additives	Admixtures	-
Wheat straw fiber material	Water	Cement	Sand	Additives	Admixtures, 4% sodium hydroxide solution	Wheat straw fiber
Hemp straw fiber material	Water	Cement	Sand	Additives	Admixtures, PVA emulsions	Hemp straw fiber
Nano straw fiber material	Water	Cement	Sand	Additives	Admixtures	Nano straw fiber

Table 3: Self-made (Partial references)

5. Study of the impact of straw fiber materials on construction

The straw fiber material has only appeared in the last decade in people's eyes and in the construction market, and for the previous builders or some of the current builders, they do not have much experience or skills in the application of this new straw fiber material [8]. In the design phase of the building, the personnel concerned also needs to detail the differences between straw fiber materials and ordinary building materials and then combine these differences to make design adjustments [8]. At the same time, the builders may not have a clear idea of how to apply the material effectively in the building, as they do not know much about the nature of straw fiber materials [9]. As a result, these shortcomings may prevent the straw fiber material from being used efficiently in construction and may even cause unprecedented construction problems in the process, affecting the progress of subsequent works and design contract disputes. These problems not only lead to higher costs but also create difficulties in maintaining the project at a later stage.

There are certain differences between straw fiber materials and ordinary construction materials. When using this material, building construction units need to effectively reorganize and integrate the project's construction organization design and other aspects of management according to the requirements for the use of the new material, and only after ensuring the timeliness of the management of these new materials is it possible to ensure the quality of construction. In addition, in the actual construction of straw fiber materials, as many construction personnel and units do not know enough to grasp the knowledge, so the construction site is still somewhat messy, and to some extent, also affects the construction quality of the building project [10, 11]. In the pre-construction phase of building construction, the project requires construction program design work, pre-construction budget, and other work, but these straw fiber materials touch on many new construction techniques and brand-new equipment and devices. Finally, there is no unified, standard construction plan for wheat straw materials, which has led to many construction units being in a state of ignorance and confusion when using these materials.

6. Conclusion

This paper mainly studies the impact of using straw fiber materials on the management of construction project. Through the study and analysis of its straw fiber materials in terms of quality, cost, and construction, its main conclusions are as follows.

(1) Two control tests were carried out on wheat straw, and hansa straw materials with the same water to ash ratio and different straw fiber additions; the material was also treated with 4% sodium hydroxide solution and normal water, and the flexural and compressive strength values were measured at seven

days and 28 days respectively. According to the comparative analysis of the test results, the flexural and compressive strength values of the straw fiber material treated with sodium hydroxide solution were higher than those of the straw fiber material treated with ordinary water. However, due to the chance nature of the test, it was necessary to carry out multiple comparative tests of the concentrations to arrive at the best concentration values for the high-strength straw fiber material. Because of the late start on straw fiber, it is also necessary to pass a large number of tests to derive the indicators of each performance so that the construction personnel or enterprises know these precise quality data so that they can effectively use the straw fiber material in the actual construction and control the quality of the project.

(2) The variety of straw fiber species, which can be extracted from wheat, hansom, and other crops, and the composition of its constituent parts, makes it difficult to control the economic cost of the project. Although the usual disposal method of waste straw materials is incinerated, the cost of raw material selection can be almost negligible. In practice, the extraction, preparation, transport, aging, and later maintenance of different straw fiber materials have not been further studied and analyzed. If the straw material is used in large quantities in the market, it will also raise the requirements of the purchasing staff and testers because not only do they have to master the nature of these straw fiber materials in a short period, but they also need to establish a new set of rules for the use of straw fiber materials.

(3) Straw fiber materials have only appeared in the construction market in the last decade. In the construction, the previous construction personnel and construction units using straw fiber material have not yet reached proficiency. This may cause certain defects in the project due to the construction personnel being unfamiliar with the application of the material, which may cause operational errors during construction. In the design phase, due to the use of new materials, the personnel concerned also have no way to reasonably control the construction and cannot efficiently utilize these materials in the construction. Therefore, it may lead to the whole construction process being affected to a certain extent, and in serious cases, it may also affect the subsequent construction progress.

Each figure should have a brief caption describing it and, if necessary, a key to interpret the various lines and symbols on the figure.

7. Future outlooks

This paper analyses the impact of straw fiber materials on construction projects in terms of quality, cost, and construction, and finds that the research on the three aspects of straw fiber materials is still scattered and at an early stage and that the application of some straw fiber materials is also at an early stage. At present, the application of straw fiber materials and exploration in the construction market is still at the beginning of the stage but also needs to the material from the preparation, procurement, construction, post-maintenance, and other aspects of the whole of a large number of simulations and tests. This paper has drawn some of the limiting conclusions through experimentation and reading a large amount of literature, and it is hoped that the following aspects can be studied and explored in-depth in the future.

(1) In this paper, only the flexural and compressive strength values of wheat straw and hansa straw were analyzed in 4% sodium hydroxide solution and water at room temperature, but not in different concentrations of sodium hydroxide solution, e.g., 5% and 6%. In addition, the experiment also requires a comparative analysis of different qualities of straw fibers and different additions of straw fibers, as different concentrations, different material qualities, and additions have a specific influence on the flexural and compressive strength values of the specimens, so a great deal of research is still required in this area.

(2) The cost impact analysis of straw fiber materials in this paper is only derived from a large amount of literature on the partial impact of many new materials, including straw fiber materials on the cost, for the straw fiber itself the purchase, presentation, transport, preparation, and other aspects have not been investigated in detail in the analysis of data. The cost of using materials as a whole is an integral part of the project management, so detailed analysis is needed in this area to form specific cost data tables.

(3) In the analysis of the impact of straw fiber material construction, as it is the emergence of new materials, its application is still in the early stages, so to the construction, no doubt to the construction

enterprises and personnel have caused certain difficulties. Therefore, in future research, it is also necessary to simulate the application of this material in construction to form a regular construction project and compile a specific straw fiber material construction manual for the use of construction enterprises and personnel.

Conflicts of Interest

The author declares that there is no conflict of interest regarding the publication of this article.

References

[1] Goosen, M.F.J.P.E., Environmental management and sustainable development. 2012. **33**: p. 6-13.
[2] Madurwar, M.V., et al., Application of agro-waste for sustainable construction materials: A review. 2013. **38**: p. 872-878.
[3] Ghaffar, S., Straw fibre-based construction materials, in Advanced High Strength Natural Fibre Composites in Construction. 2017, Elsevier. p. 257-283.
[4] Maraveas, C.J.M., Production of sustainable construction materials using agro-wastes. 2020. **13**(2): p. 262.
[5] Toguyeni, D.Y., et al., Study of the influence of roof insulation involving local materials on cooling loads of houses built of clay and straw. 2012. **50**: p. 74-80.
[6] Petrella, A., et al., Use of cellulose fibers from wheat straw for sustainable cement mortars. 2019. **8**(3): p. 161-179.
[7] Govindan, K., et al., Sustainable material selection for construction industry–A hybrid multi criteria decision making approach. 2016. **55**: p. 1274-1288.
[8] Xie, X., et al., Influence of pretreatment of rice straw on hydration of straw fiber filled cement based composites. 2016. **113**: p. 449-455.
[9] Bouasker, M., et al., Physical characterization of natural straw fibers as aggregates for construction materials applications. 2014. **7**(4): p. 3034-3048.
[10] Ashour, T., et al., The influence of natural reinforcement fibres on insulation values of earth plaster for straw bale buildings. 2010. **31**(10): p. 4676-4685.
[11] Quagliarini, E. and S.J.J.o.C.H. Lenci, The influence of natural stabilizers and natural fibres on the mechanical properties of ancient Roman adobe bricks. 2010. **11**(3): p. 309-314.

Numerical study of the thermo-fluid flow and weld appearance of 2195 Al-Li alloys in wire feed laser welding

Qihang Xv[1,a], Gan Tian[1,*], jinshu Chen[1,b], Zhengwei Yang[1,c]

[1] Xi'an Research Institute of High-Tech, Xi'an, China

[a]e-mail: xvqh@jcut.edu.cn , [b]e-mail: ensail@aliyun.com

[c]e-mail: yangzhengwei1136@163.com

[*]Corresponding Author e-mail: tiangan_2012@jcut.edu.cn

Abstract. The wire feed laser welding has significant merits in metal joining technology for the manufacturing industry, which including high energy efficiency, high welding speed, automated workflow and narrow heat-affected zone. During to the interaction between weld pool and welding wire, the welding process is more complicated. In this work, a numerical model to the wire feed laser welding for the liquid bridge transition was built. By constructing a movable welding wire entity accurately, the simulation result for flow and heat transfer inside weld pool and welding wire were acquired by the numerical solution. The temperature field, velocity field and weld appearance can be obtained by simulation results. The results shown that the rapid wire feed speed will increase the heat absorption of welding wire, which can improve the hardness due to higher cooling rate. Through the analysis of experimental and simulation results, the quantitative relationship between weld parameters and weld result can be obtained. The numerical simulation can guide the optimization of welding parameters and control the weld appearance.

1. Introduction

Wire feed laser welding need to not only melt the workpiece but also need to allocate energy to melt the welding wire by using high energy laser beam. In this process, it involves multiple physics changes, so it is a very complex physical process. For this reason, the welding process window is narrow. It is challenging to find and optimize a suitable welding technology, because it needs to optimize the welding parameters by a lot of experiments. The thermo-fluid flow field, weld morphology, cooling rate and other parameters in the welding process can be obtained by numerical simulation, which can provide quantitative guidance for the setting of welding process parameters. The numerical model of laser welding can not only simulate the macro heat transfer and fluid flow, but also can solve the transport equation of the wire filling process, which can obtain the element distribution in the weld [1,2]. The process of weld recrystallization also can be simulated by using cellular automata (CA) or phase field methods (PF), and the grain structure was obtained finally[3–5]. Although the numerical simulation of wire feed laser welding can be carried out at various scales, this numerical simulation always depends on the data of heat transfer and flow in the weld pool during the welding process. Therefore, the numerical simulation of fluid flow and heat transfer in welding process is a great significance for the optimization of welding parameters and microstructure simulation.

In this paper, the process of wire feed laser welding was simulated and corresponding experiments also be carried out. The velocity and temperature field, weld appearance and fusion line were obtained in this simulation. The accuracy of the numerical simulation can be verified accordingly by comparing calculated and experimental fusion line shapes.

2. Numerical model

Wire feed laser welding is a complicated multiple physics process, so the contributing factors in this process is taken as source term or boundary condition in numerical model, which includes melt, evaporation and solidification process in welding region. This numerical model was simulated by CFD solver, all the contributing factors were showed in Table1.

Table1. Numerical model contributing factors

	Contributing factor model
Flow model	Gravity
	Surface tension + Marangoni shear stress[6]
	Recoil pressure[7]
	Solidification damp force[8]
	Buoyancy
Heat transfer	Evaporation loss[9]
	Surface radiation
	Latent heat of melting
	Convective heat loss
	Solidification releases energy

The welding wire is reconstructed by multiphase method, so that the mass and energy transfer between wire and workpiece can be considered into the model. In this simulation, the heat transfer and multiphase flow in the whole region. Mass transfer in the flow process is calculated by discretized continuity equations, fluid flow is calculated by discretized momentum equations in multiphase region, and the change of energy is calculated by discretized energy equation in the whole region.

3. Materials and experiment method

The workpiece is 2195 Al-Li alloys plate, which thickness is 2 mm. laser source was produce by an Nd:YAG laser machine. The welding wire is fed in front of the laser focus on the workpiece, and their distance is 1 mm. The filler wire had 45° with the workpiece. Shielding gas of pure Argon is set behind the laser focus. The detailed technological parameters are illustrated in Table2. The experiment material is 2195-T8 Al-li alloys, its chemical composition is shown in Table3. The laser beam moved along the butt between two plates.

Table2. Experimental parameters

Welding parameter	Value
Power (W)	1300
Welding speed (mm/s)	3
Feeding wire speed (mm/s)	2
Shield gas flux (L/min)	20
Defocus distance (mm)	160

Table3. Chemical composition of workpiece and filler wire (%)

Workpiece	Cu	Li	Ag	Zr	Fe	Mg	Ti	Al
2195 Al-Li alloy	3.7	0.8	0.2	0.1	0.1	0.2	0.068	Balance

4. Result analysis

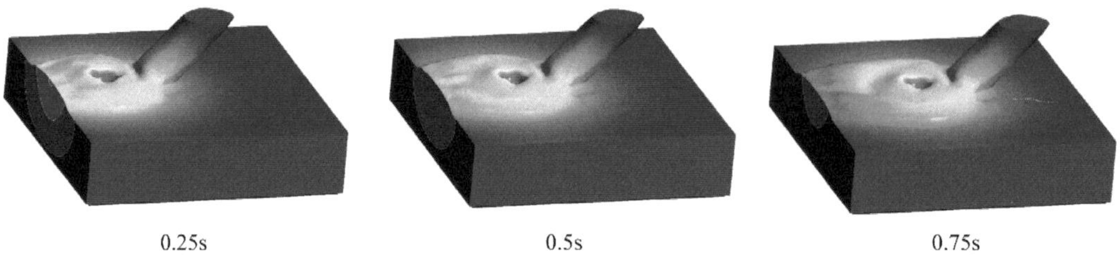

Fig.1. Simulation results of dynamic laser welding with filler wire process

Fig.1 show the simulation results in different. In this simulation result, the welding wire overlaid on the workpiece surface and form the weld appearance. The keyhole was emerged by recoil pressure, which due to metal evaporation. The keyhole wall emerges rapid fluctuations since the welding wire absorbs energy continuously. Fig.1. gives the curve of keyhole depth change in the simulation process. The maximal depth is about 1.3 mm and dynamic range is from 0.8 mm to 1.3 mm (except the overlap height). Keyhole has narrowed before depth becomes shallow. Fig.2. (a)-(d) illustrates this phenomenon, fluctuation emerges and the below part of keyhole wall becomes narrow. Afterwards, it will be expanded by recoil pressure again in this region due to laser leam irradiate.

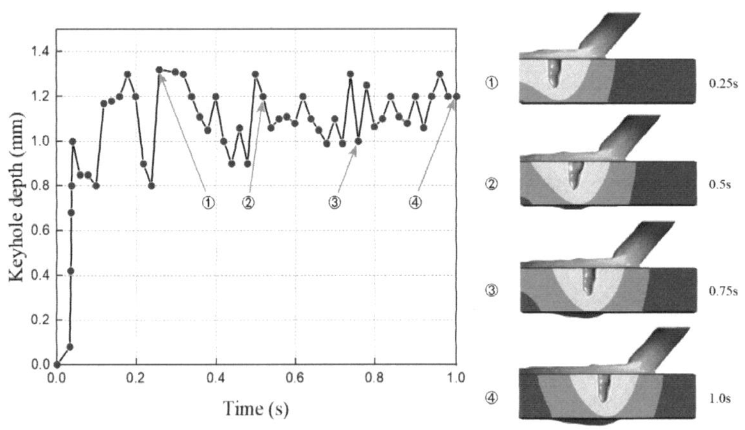

Fig.2. Keyhole depth evolution process with time.

The velocity and temperature field in simulation were shown in Fig.3. the velocity magnitude is around 2 mm/s-20 mm/s in weld pool. In order to clearly illustrate the flow field and the region location, the dotted line which can reflect the flow trend and the region which is circled by the yellow line are only marked in Fig.3. (d). Recoil pressure is the main driven force to push keyhole to expand and overcomes the contracted effect which produced by surface tension and gravity. With the combination of Fig.3. (a)-(d), the melt flow from bottom of the keyhole wall (region B) to top of the keyhole wall, then melt horizontally flows ahead and mixes with molten welding wire[10]. The molten wire hardly exits any flow trend in the opposite direction of laser movement, it flows downward and forms a vortex flow mixed with molten flow which comes from weld pool.

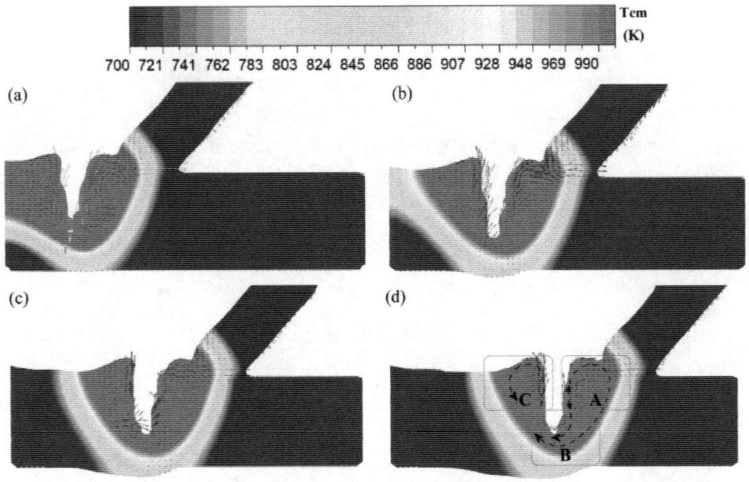

Fig.3. Temperature field and flow patterns in different time:

Fig.4. (a) is the experiment result and Fig.4. (b) is the temperature field in weld cross-section for simulation result. The vector arrow in Fig.4. (b) represents the temperature gradient vector. The simulation results are consistent with the growth of columnar crystals along the opposite direction of the temperature gradient and the growth from the boundary of the heat-affected zone. Comparing the heat-affected zone obtained by the experiment in Fig.4. (a), it can be seen that the temperature field results of the weld pool obtained by simulation in Fig.4. (b) are more consistent with the experimental results.

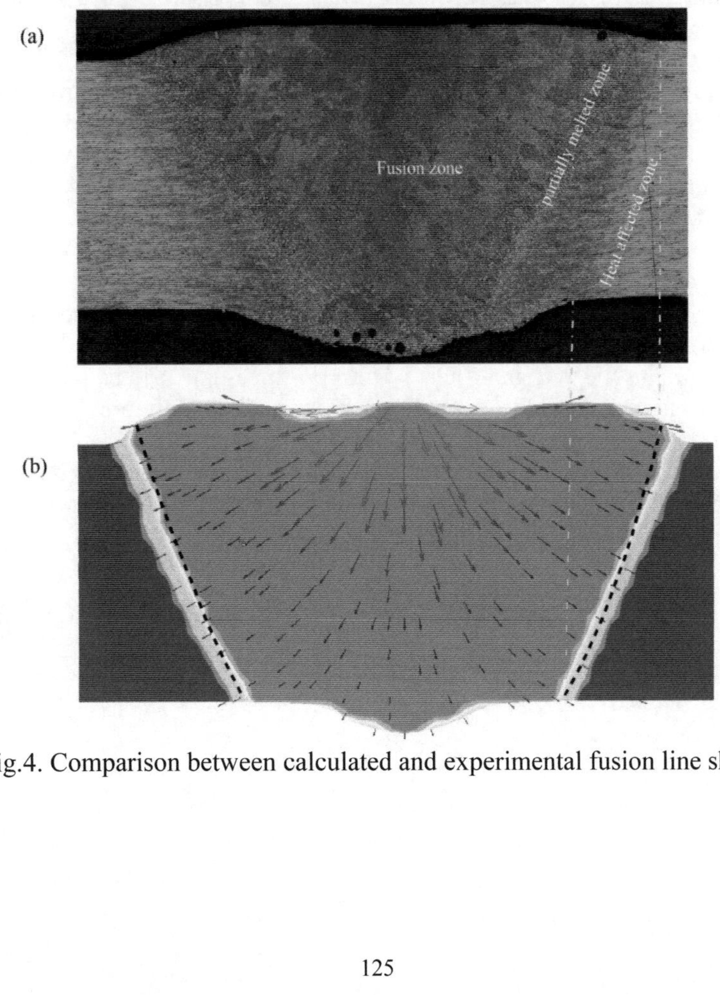

Fig.4. Comparison between calculated and experimental fusion line shape

5. Conclusion

In this work, a numerical simulation was carried out for wire feed laser welding of 2195 Al-li alloys. By using numerical simulation, the temperature field, velocity field, flow trend in weld pool and temperature gradient and fusion line were obtained quantitatively. This simulation result can correspond experiment result, which proves that the validity of this numerical model. Through compare experimental with simulation results, the quantitative relationship between weld parameters and weld result can be obtained. The numerical simulation can guide the optimization of welding parameters and control the weld appearance.

Acknowledgements

This work was supported by the National Natural Science Foundation of China (Grant Number:92060106).

References

[1] Meng X, Artinov A, Bachmann M and Rethmeier M (2020) Theoretical study of influence of electromagnetic stirring on transport phenomena in wire feed laser beam welding Journal of Laser Applications, 32: 022026

[2] Meng X, Artinov A, Bachmann M and Rethmeier M (2019) Numerical and experimental investigation of thermo-fluid flow and element transport in electromagnetic stirring enhanced wire feed laser beam welding International Journal of Heat and Mass Transfer, 144: 118663

[3] Tan W and Shin Y C (2015) Multi-scale modeling of solidification and microstructure development in laser keyhole welding process for austenitic stainless steel Computational Materials Science, 98: 446–58

[4] Fallah V, Amoorezaei M, Provatas N, Corbin S F and Khajepour A (2012) Phase-field simulation of solidification morphology in laser powder deposition of Ti–Nb alloys Acta Materialia, 60: 1633–46

[5] Yu F, Wei Y, Ji Y and Chen L-Q (2018) Phase field modeling of solidification microstructure evolution during welding Journal of Materials Processing Technology, 255: 285–93

[6] Brackbill J U, Kothe D B and Zemach C (1992) A continuum method for modeling surface tension Journal of Computational Physics, 100: 335–54

[7] Semak V and Matsunawa A (1997) The role of recoil pressure in energy balance during laser materials processing J. Phys. D: Appl. Phys., 30: 2541–52

[8] Rai R, Elmer J W, Palmer T A and DebRoy T (2007) Heat transfer and fluid flow during keyhole mode laser welding of tantalum, Ti–6Al–4V, 304L stainless steel and vanadium J. Phys. D: Appl. Phys., 40: 5753–66

[9] Pang S, Chen L, Zhou J, Yin Y and Chen T (2011) A three-dimensional sharp interface model for self-consistent keyhole and weld pool dynamics in deep penetration laser welding J. Phys. D: Appl. Phys., 44: 025301

[10] Peng J, Li L, Lin S, Zhang F, Pan Q and Katayama S (2016) High-Speed X-Ray Transmission and Numerical Study of Melt Flows inside the Molten Pool during Laser Welding of Aluminum Alloy Mathematical Problems in Engineering, 2016: 1–13

EMG-based Monitoring Muscle Contraction Force to Determine Most Effective Exercise

Jiawei Fu[1], Yichen He[2], Siming Jiang[3], Zhongqi Tao[4], Yifan Wen[5], Xingyu Yi[6]

[1]Beijing Jiaotong University, Nanomaterials and Technology, j72fu@uwaterloo.ca

[2]Beijing Jiaotong University, Nanomaterials and Technology, yichen20008@163.com

[3]Xidian University, Electronic Information Engineering, 1798024657@qq.com

[4]University of Nottingham, Electrical and Electronic Engineering, shyzt1@nottingham.ac.uk

[5]WuHan University, Microelectronic Science and Engineering, 417990013@qq.com

[6]University of Macao, Electromechanical Engineering, 1152131610@qq.com

Abstract—Electromyography (EMG) signals have been widely applied in fields of biomedical, clinical and human-machine interface [1]. And surface EMG (sEMG) technology could provide a non-invasive way to detect the stimulation degree of muscles and enable the control of artificial limbs. In this research, the main focus lies on one of the social problems—sedentariness. With the development of internet and electronic economy, computing time has undergone a steady increase around the globe, which extends the job time in long-term sitting [2]. Together with some non-ergonomic designs of chairs, many officers might experience an abnormal sitting posture and get injured in their muscles unconsciously. To scope the phenomenon in detail, a simulation is applied to reappear the condition of sedentariness of a common office worker. And a series of easily achieved movements were investigated to find out the corresponding effect in muscle exercise to compensate the negative influence caused by long-term sitting.

1. Introduction

Surface electromyography (sEMG) is a good medium for human-computer interaction, which can transmit real-time information of muscle function and all kinds of movement in non-injured state. Through the measurement and analysis of EMG, we can judge the health status of muscle and the underlying diseases. It has been widely applied in fields of biomedical, clinical and human-machine interface.

Combined with the knowledge learned in CIS, the EMG signal can monitor and reflect the muscle power status in real time. We will combine the EMG of measuring the symmetrical muscle power status to judge whether the muscle power is balanced and hope to be used to determine whether the exercise standard reaches the effective exercise in the process of fitness.

With the development of internet and electronic economy, computing time has undergone a steady increase around the globe, which extends the job time in long-term sitting. Together with some non-ergonomic designs of chairs, many officers might experience an abnormal sitting posture and get injured in their muscles unconsciously. Especially the young office workers suffer a lot from these diseases, such as humpback and cervical spondylopathy. It will be a long-term social problem in the future. Our group took note of this phenomenon and did research.

It has been shown that there is a strong linear correlation between force and the EMGs features MAV, RMS and WL. [3]. Strength training involves against resistance that gradually increase over time. The strain on muscles causes damage to muscle fibers by repeatedly challenging the muscles can cause them to adapt by growing in size and strength [4]. So, it can be a useful tool to monitor and reflect the muscle power status in real time.

1.1 Goals of Research
Two main goals of our research:

1.1.1. Sedentary
We would like to use EMGs signal to monitor our muscles (real time) and judge whether the sitting posture is standard or not.

1.1.2. Exercise
We also want to use EMGs signal to explore the muscle power during exercise and find in which gesture can our muscles get the most effective exercise.

2. Instrument Introduction
The instrument used in the experiment is a six-lead muscle electrical sensor [5] which has six channels. That means it can measure EMG signals in six different positions at the same time.

The six-lead myoelectric module includes the front-end analog circuit acquisition and the back-end digital signal filtering processing.

The front-end acquisition circuit collects the muscle electrical signals of the human arm or leg through 1 to 6 channels. Then the signal will undergo a series of amplifications and filters.

The mid-end can switch between Envelope Mode and RAW Mode output signals through the SPDT switch; Envelope Mod adopts envelope detection processing to obtain EMG dynamic detection signal; RAW Mode can output EMG original signal,

The back end uses the Arduino UNO development kit that is matched with the six-lead muscle module to collect the output of the mid-end signals.

Fig.1. shows the circuit of one channel and the principle of other channels are basically the same.

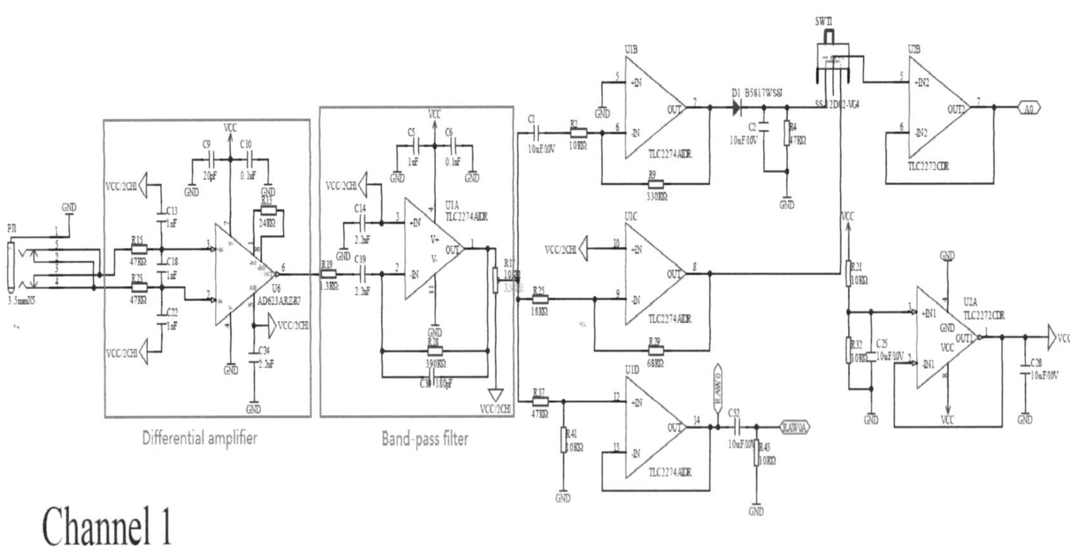

Fig. 1. Amplification and filter of signals

After signal is received by the electro cardio-electrode, it will go through the differential amplifier first. The differential amplifier is used to Integrate dual-channel signals into single-channel signal and amplify the input signal. After that, the signal will be further filtered by the band-pass filter. Because this band-pass filter consists of an amplifier, it also amplifies the signal again.

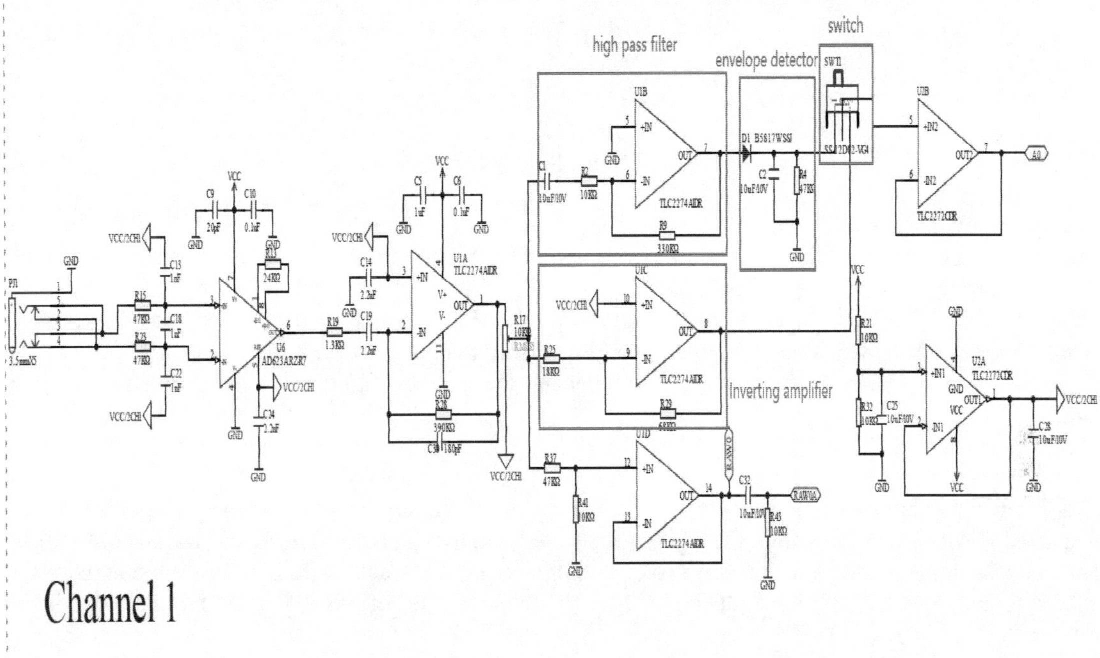

Fig. 2. Integration and further filter of signals

As mentioned above, it can switch between Envelope Mode and RAW Mode output signals. In Envelope Mode [6], the signal goes through the amplifier with high pass filter and envelop detector [7]. In RAW Mode, the signal just goes through an inverting amplifier. Those two modes are controlled by the switch showed in Fig.2. The signal changes of these two modes are shown in the figure below.

3. Experiment and Data Acquisition

3.1 Selection of muscle groups and electric pole position

Medically, the main force bearing muscles in sitting occurs in the rear part of human body while the front muscle is relatively more relaxed in comparison to that. An assumption is made that the rear part of human body is more likely to be hurt through long-term sitting, which could be supported by several clinical evidence. In accordance with the position of muscle groups, the rear part is divided into 3 sections—neck, back and waist, which are highlighted in the figure. The precise muscle groups are termed as cervical muscle, latissimus dorsi and respectively. To reflect the overall process, the position of two detecting electric poles should go across the whole section of muscle group. And the third reference pole should be attached in places where there is apparent bone bulge to avoid the disturbance of muscle signal.

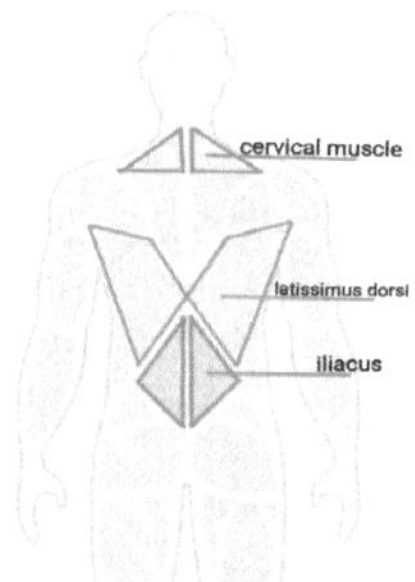

Fig. 3. Focused muscle groups

3.2 Experiment process

The whole experiment will be divided into two main parts: Sitting experiment and exercise experiment. The sitting experiment will detect the sEMG signal of workers who always work in one sitting position for a long time. The exercise experiment will detect the sEMG signal of people who are doing some specific exercises. In these two experiments, a 20-year-old man is selected as the subject. Moreover, all of the experiment will be repeated for three times and the average value will be calculated as the raw data.

3.3 Sitting experiment

This experiment simulates the sedentary behavior of office workers. To ensure the consistency of the measured sEMG signal, the three muscle groups (cervical muscle, latissimus dorsi and iliacus) will be detected simultaneously. Every muscle group will have 6 electric poles, with 2 on the left side, 2 on the right side and 2 on the reference places. Hence, the total number of electric poles on the body of the volunteer should be 18.

In order to better simulate the daily sitting posture of office workers, the experiment will simulate two working scenes. In the experiment, the volunteer is requested to sitting in front of the laptop for 15 minutes in both normal and hunchbacked sitting posture, including browsing web pages for 8 minutes and typing for 7 minutes.

3.4 Exercise experiment

The experiment simulated some of the daily exercises that office workers can do anywhere and anytime. There are several kinds of exercises postures chose to relax and exercise neck, back and waist. The three muscles groups will be tested respectively in different exercise postures and the sEMG signal will reflect the degree of muscle exercise to some extent.

3.5 Neck exercise movements

Three postures will be included in the neck exercise, with neck stretches, head movements and raise shoulders & neck.

Neck stretches: volunteers will anterior and posterior expansion neck for 15 times in a specific style.

Head movements: volunteer will rotate head clockwise and counterclockwise for 4 times of each, each time will continue for 8s.

Raise shoulders & neck: volunteer will lift shoulders while tuck neck for 4 times, each time will include 10s exercise and 5s rest.

3.6 Back exercise movements

Similar to movements in neck muscle group, the exercises for back muscle could be easily done in most occasions. Four movements were chosen as follows:

Superman fly: candidate lie down to the carpet on the front, suspend his/her arms and legs in the air and stretch them. The force needed to retain the position is taken from the back and it provides a good stimulation to the overall parts of the body. 15s each and rest for 5s, totally 3 times a minute.

Bend I shape stretching: stay in half squat position and then vertically wave two arms back and forth. 5s each, lasts for 1 minute.

Bend T shape stretching: stay in half squat position and then horizontally wave two arms back and forth. 5s each, lasts for 1 minute.

Kneeling back stretching: candidate should lie down to the carpet on their back, raising his/her 2 legs and hold them still with 2 arms. The whole process lasts for 1 minute.

3.7 Waist exercise movements

Crunch: candidate lies down to the carpet on their back and bend their knees. Place the hands behind the head, slowly contract the abdominals back and forth. 3s each, lasts for 1 minute.

Air-cycling: candidate lies down to the carpet on their back, raise 2 legs and mimic the motion of cycling in the air. Lasts for 1 minute.

Single leg raising: candidate lies down to the carpet on their back, raise one leg and bend the other leg for a certain time and change sides afterwards. 12s each and rest for 3s. Totally 4 groups per minute.

Bend kneels abdominal holding: candidate kneels down and bend his/her upper body forward. The upper muscle should be relaxed completely in movement.

4. Signal Processing Techniques

After raw surface EMG signals of target muscles are acquired, several processing techniques based on time domain analysis are applied to extract specific features from raw signals, in order to compare the force level between different muscles and movements. The relationship of force level and EMG time domain features are defined as positive linear correlation, in particular between constant force and RMS, dynamic force and MAV [8]. Therefore, the RMS values of sitting posture EMG signals are extracted while EMG signals related to movements are featured by MAV.

4.1 Root mean square value (RMS)

The square root average of the EMG signal is represented by root mean square value (RMS) and calculated by equation.1

$$\mathrm{RMS} = \sqrt{\frac{1}{N}\sum_{n=1}^{N}(\mathrm{EMG}_{(n)})^2} \qquad (1)$$

4.2 Mean absolute value (MAV)

The mean absolute value (MAV) refers to the average rectified values of EMG signal and is given by equation.2

$$\mathrm{MAV} = \frac{1}{N}\sum_{n=1}^{N}|\mathrm{EMG}_{(n)}| \qquad (2)$$

Where, $\mathrm{EMG}_{(n)}$ is the nth sample of EMG signals and N is the total number of EMG signals.

The RMS values of EMG signals in four groups of muscles were calculated in terms of both normal sitting posture and hunchbacked sitting posture. The EMG signals from each muscle group were windowed into several sets with each set containing 50 sequential data points, which is equivalent to a period of 250ms, in order to reduce number of data points while with acceptable precision and accuracy. The RMS values of each data set were calculated separately to construct the RMS values data set.

To demonstrate the force level of muscles more apparently, the normalization is implemented before the signals are compared. The most voluntary contraction (MVC) values of each muscle group are measured by asking subject to contract particular muscle as hard as possible and record the highest EMG signal readings. Hence MVC value represents the maximum possible activation of a muscle. The normalization can be obtained by comparing RMS values with MVC value of the same muscle correspondingly.

4.3 Normalization

$$\text{MVC\_Normalization} = \frac{\text{RMS envelope}}{\text{MVC}} \times 100\% \qquad (3)$$

The MVC values of each muscle groups of the subject is shown in TABLE 1 below.

TABLE 1 MVC Values of Each Muscle Group

	Cervical Vertebra	Latissimus dorsi	Waist
Right	970	746	486
Left	972	757	492

By implementing normalization, the muscle activation level can be represented in percentage and hence can be compared more intuitively.

As mentioned previously, it is the MAV of EMG signal which has the highest correlation with dynamic force. Hence the obtained EMG signals during movements are processed by calculating MAV instead of RMS before conducting normalization.

To calculate MAV, the EMG signals were windowed into several data sets with each one containing 100 sequential data points, which is equivalent to 0.5s time period. The MAV values of doing specific movement for one minute were combined together and the average MAV for one second was calculated to quantify the force level of the stimulated muscle group when doing corresponding movement. The normalization was conducted by comparing average MAV with MVC value of the activated muscle group. Accordingly, by comparing normalization ratios of different movements, we are able to figure out the movement that can mostly activate target muscle group, which can be considered as the most effective movement to exercise target muscle group.

5. Results and Discussions

The MVC normalization percentages of muscle groups for normal and hunchbacked sitting posture in 15 minutes are compared in Fig.4. as below.

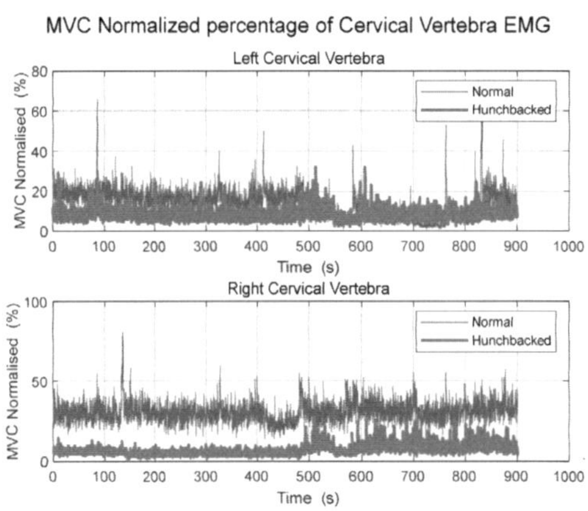

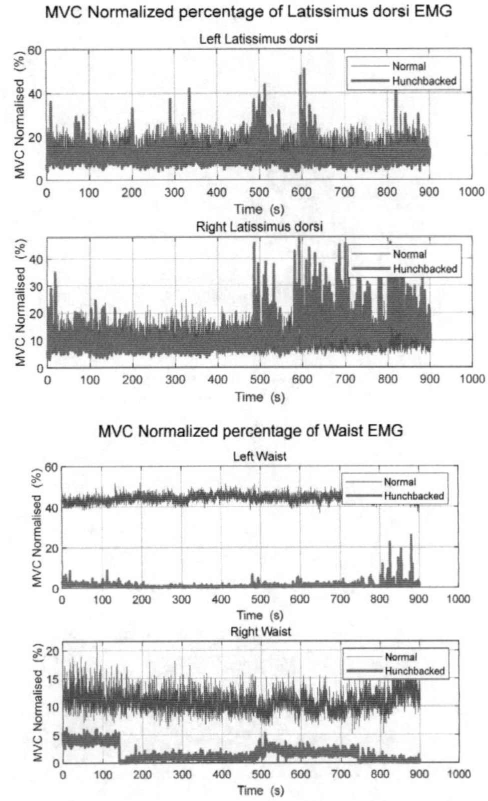

Fig. 4. MVC normalized percentage comparison between normal and hunchbacked sitting posture

As clearly seen, despite some spikes, the data points in red which represent hunchbacked sitting posture normalization percentage are basically below the blue points representing normal sitting posture, which indicates lower force level for these muscle groups when sitting in hunchbacked posture. People are fond of sitting in hunchbacked posture because they are using less force which makes them feel more comfortable.

To determine movement that exercise these muscle groups effectively and safely, the average normalization percentages of different movements activating same muscle groups are compared. TABLE 2 list average normalization percentages correspondingly and Fig.5. shows the comparison outcomes.

TABLE 2. Average Normalization Percentages of Cervical Vertebra Movements

	Neck Stretches	**Head Movements**	**Raise Shoulders and Neck**
Right	14.16%	19.14%	68.66%
Left	10.05%	18.21%	62.36%

Average Normalization Percentages of Latissimus dorsi Movements

	Kneeling with Back Stretched	**Superman**	**Bend "I" Stretched**	**Bend "T" Stretched**
Right	12.84%	50.27%	57.40%	37.90%
Left	13.68%	52.09%	53.77%	40.50%

Average Normalization Percentages of Waist Movements

	Bent Knees Hold abdomen	**Crunches**	**Lying Leg Raises**	**Air Cycling**
Right	0.36%	6.50%	6.84%	55.30%
Left	3.76%	10.11%	7.49%	47.82%

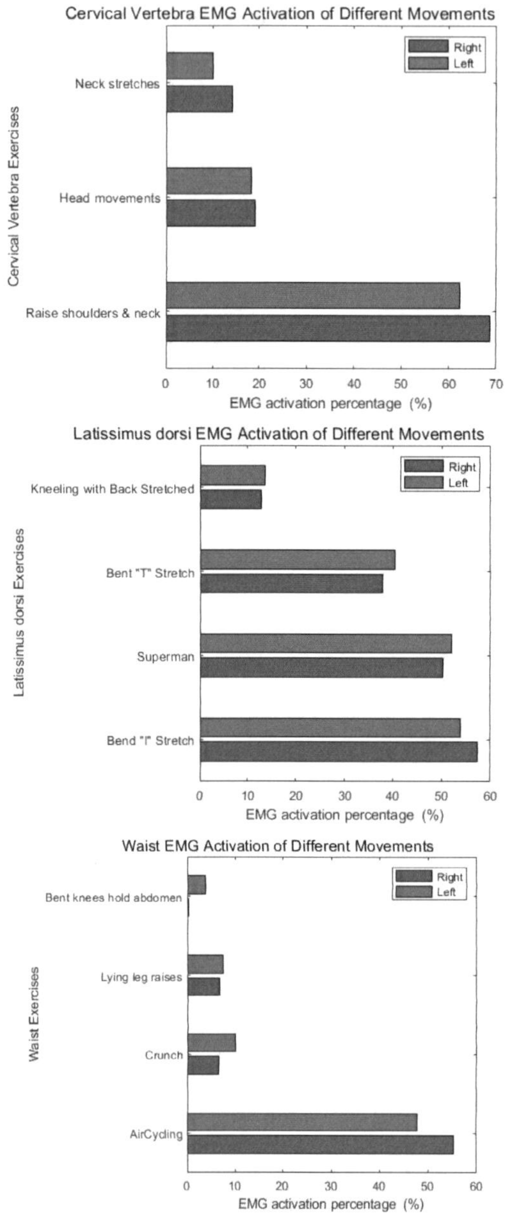

Fig. 5. EMG activation normalized percentages comparison

Hence based on the comparison of EMG activation percentage, the most effective movements to exercise corresponding muscle groups can be chosen accordingly.

6. Applications

6.1. Existing applications:

(1). For barbell bench press, push-ups and other closed chain movements, it is easy to generate force on one side. Therefore, the imbalance between the left and right should be reported immediately to prevent injury.

(2). As for football shooting action, lower limb power order should be the first to send force after the swing leg, mainly for the active contraction of the gastrocnemius muscle, then before the mass force, the thigh to drive the calf, complete the whipping action. The wrong sequence of forces can be analyzed to correct the action.

(3). Compare the degree to which different movements or loads stimulate the same muscle

(4). Training and diagnosis of sports rehabilitation: compare the movements of patients tested with normal movements to better arrange the rehabilitation plan and movements.

6.2. Our goals and applications:

(1) Dynamic application and expansion of experimental objectives

Our goal is to use EMG signals to determine whether muscles are being activated during various movements. With this idea in mind, we can apply this technique to many other areas of technology, such as:

i. In the state of long-distance running, the lower limb skeletal muscle system is prone to the symptoms of muscle fatigue, so the diastolic condition of lower limb muscles in running can be analyzed. Based on the analysis of time domain and frequency domain, muscle fatigue was determined by surface electromyography (EMG), and different quantitative protection models were established [9].

ii. Study on EMG Characteristics of Pull-Up Surface with Different Grip Span and Movement Rhythm [10]: it provides a theoretical basis for scientific training for fitness enthusiasts. By processing the collected EMG signals and using the root mean square amplitude value and muscle contribution rate as experimental indexes, the influence of different pull-up modes on muscles can be analyzed.

(2) Static application and expansion of experimental targets

i. The evaluation method of operating comfort of musculoskeletal load was studied, and weights were set for different joints. In this method, the EMG time domain indexes and frequency domain indexes of the main muscle surfaces in different joint activities are used as the evaluation basis [12]. The joint load data are obtained from the virtual human body model simulation experiment, and weights are set for different joints, so as to evaluate the comfort level during the work.

ii. The comfort and safety of astronauts is very important in aerospace, especially in the narrow working cabin, with high pressure/hypoxia/cold, etc. This requires a combination of ergonomics, biomedicine, artificial intelligence and other disciplines [12]. The fatigue characteristic of human muscle can be detected by electromyography (EMG) signals.

iii. The study on the fatigue characteristics of shoulder and neck muscles based on EMG signals. We chose to measure the muscles in the lower back, and this can also be applied to many other muscles in the body, such as the rotation of the shoulder blades, the flexion and extension of the neck, etc.

7. CONCLUSION

In this paper, a time domain analysis of surface electromyography signal sEMG are performed in order to evaluate force level contraction of muscles. The mean average value (MVC), the root mean square value (RMS), the most voluntary contraction (MVC) and the average activation percentage are extracted and evaluated with the varying of the force level contractions.

In the first part, the results show clearly that lower force level for these muscle groups are required when sitting in hunchbacked posture. Therefore, by comparing certain muscles' value of EMGs, we can judge whether the sitting posture is standard or not.

In the latter part, with the average normalization percentages of different movements activating same muscle groups, we can determine which way of exercise is most effective.

REFERENCES

[1] F. Sadikoglu, C. Kavalcioglu, and B. Dagman, "Electromyogram (EMG) signal detection, classification of EMG signals and diagnosis of neuropathy muscle disease," Procedia Comput. Sci., vol. 120, pp. 422–429, 2017

[2] S. Panahi and A. Tremblay, "Sedentariness and Health: Is Sedentary Behavior More Than Just Physical Inactivity?" Front. Public Heal., vol. 6, no. September, pp. 1–7, 2018

[3] A. O. Andrade and C. I. Andrade, "On the relationship between features extracted from EMG and

force for constant and dynamic protocols," 2012 Annual International Conference of the IEEE Engineering in Medicine and Biology Society, San Diego, CA, USA, 2012, pp. 3392-3395.
[4] A. Kandola and D. Bubnis, "What to know about muscle hypertrophy." M.S. NASM-CPT. NASE Level II-CSs-w, September 28,2020 [Online] Available: https://www.medicalnewstoday.com/articles/muscle-hypertrophy
[5] Polou Robotics & Electronics, ' Advancer Technologies Muscle Sensor v3 ', 2001. [Online]. Available: https://www.pololu.com/product/2726.
[6] A. CHEN, 'Envelope', Mar 23, 2021. [Online]. Available: https://www.investopedia.com/terms/e/envelope.asp. [Accessed: April 10, 2021]
[7] W. Ciciora, J. Farmer, D. Large and M. Adams, "Headend Signal Processing -Response of an Envelope Detector", Modern Cable Television Technology (Second Edition), Pages 333-392, 2004.
[8] A. Abdelouahad, A. Belkhou, A. Jbari and L. Bellarbi, "Time and frequency parameters of sEMG signal — Force relationship," 2018 4th International Conference on Optimization and Applications (ICOA), Mohammedia, Morocco, 2018, pp. 1-5.
[9] X.Zhao. Research and Product Development on the Protection of Lower Limb Musculoskeletal System during Marathon Running. Shanghai University of Engineering Science, 2016.
[10] P.Xing. Study on EMG Characteristics of Pull-Up Surface with Different Grip Span and Movement Rhythm. North University of China.
[11] Q.Shao. Research on Evaluation Method of Operating Comfort Based on Musculoskeletal Load. Shanghai Ocean University.
[12] W.Fan . Study on Fatigue Characteristics and Layout Optimization of Small Working Cabin. Northwestern Polytechnical University, 2015.

Comparison of calculating methods and applications of different concrete maturity

Da-jiang Geng[1,a]; Ning Dai[2,b*]; Xue-sheng Jin[1,c]; En-xin Miao[3,d]

[1] China Construction 4th Engineering Bureau 6th Co., Ltd, Shanghai 201100, China

[2] Key Laboratory of Road and Traffic Engineering of the Ministry of Education, Tongji University, Cao An highway.4800, 201804 Shanghai, China

[3] China Construction 4th Engineering Bureau Co., Ltd, Shanghai 201100, China

[a]email: gdj1410704@alumni.tongji.edu.cn; [b]email: 1710512@tongji.edu.cn;

[c]email: 273959903@qq.com; [d]email: 97658201@qq.com

[*]Corresponding author's e-mail: 1710512@tongji.edu.cn

Abstract. Concrete maturity theory has great potential in practical engineering. In this paper, the calculation and application of the concrete maturity are presented based on the qualitative analysis and quantitative calculations. Firstly, the maturity calculations are summarized by the methods of direct calculation and equivalent age calculation, as well as a theoretical discussion of the advantages and disadvantages of each calculation method. Secondly, the main methods of compressive strength and elastic modulus predicted by the concrete maturity are also summarized, and the calculation accuracy of the main calculation methods is compared with the published test results. Finally, based on actual engineering project, the research on concrete maturity is prospected to clarify the further research direction. The results show that the calculation methods of concrete maturity consist of direct calculation method and equivalent age calculation method, and the equivalent age calculation method is more reasonable in theory. Considering the wide range of construction temperature, the Freiesleben-Hansen equivalent age equation and exponential function prediction equation are recommended to predict the strength and elastic modulus of concrete.

1. Introduction

Generally, the compressive strength and elastic modulus of concrete are easy to obtained under indoor standard curing conditions, while the conditions of engineering site are obviously different with those of indoor standard curing conditions. Hence, there is inevitable error for the concrete strength parameters obtained under indoor standard curing conditions to guide on-site engineering. Considering that the maturity of concrete is a function of time and temperature in the case of sufficient moisture, if the compressive strength and elastic modulus of concrete are reasonably modified with maturity, the accuracy of concrete strength parameters used for on-site engineering can be improved to a certain extent. Therefore, this paper concentrates on literature analysis, qualitative analysis and quantitative comparison methods to study the calculation and application of concrete maturity, so that better predict the strength parameters of concrete and guide engineering practice.

The compressive strength and elastic modulus of concrete are generally used in engineering practice. While in actual calculation, the maturity of concrete is firstly calculated, and then the required concrete indexes are predicted based on the maturity. At present, many researches have been done on the maturity prediction of concrete performance indexes. However, some of the researches are not systematic and objective. For example, most of the calculation methods were verified by their own test results, without verifying by others. In addition, most results were verified by the indoor testing results at constant temperature, which is significantly different from the actual engineering. To this end, three parts of the calculation and application of concrete maturity, that are maturity calculation, maturity application and research prospects are discussed based on qualitative analysis and quantitative calculation methods in this study, so that use the concrete strength parameters predicted by maturity to put forward engineering suggestions.

2. Maturity calculation

The calculation methods of concrete maturity include: direct calculation method and equivalent age indirect calculation method.

2.1. Direct calculation

Basically, the higher the curing temperature and the longer the age of concrete, the greater the maturity should be. The initial maturity calculating formula was based on this understanding, and the corresponding Nurse-Saul maturity equation [1] is

$$M=\sum_{0}^{t}(T-T_0)\Delta t=\int_{t_0}^{t}(T-T_0)\,\mathrm{d}t \tag{1}$$

Where, M is maturity (°C·d), T denotes temperature changes with time, T_0 stands for reference temperature, at which the concrete stops hydrating and generally valued -10°C, while the value of concrete mixed with antifreeze is -15°C [2].

The derivative of time of eq. (1) in both sides is

$$\frac{\mathrm{d}M}{\mathrm{d}t}=(T-T_0) \tag{2}$$

There is an assumption in eq. (1) that the development speed of maturity over time is proportional to temperature. While concrete cured at indoor standard temperature, according to eq. (1), the maturity should linearly change with time, and their relation is curvilinear. Therefore, the assumption in Nurse-Saul maturity (eq. (1)) is unreasonable.

Considering that it is easy to determine the temperature changes with time, the value of the reference temperature T_0 has greater impact on the value of maturity in eq. (1). Regarding the value of reference temperature, researches show that the value of reference temperature is significantly affected by the temperature range of the curing environment, the type of cement, admixture the curing method [3-4].

In order to characterize the influence of temperature and cement type on the hydration rate, based on the Nurse-Saul maturity, De and Tegelaar [5] proposed a modified maturity calculation formula

$$M=\int_{t_0}^{t}T\bullet C^{0.1T-1.245}\mathrm{d}t \tag{3}$$

Where, C is a parameter related to the type of cement, which can be valued according to the literature [6], generally the value ranged in [1.00,1.65]. Similarly, the derivative of time of eq. (3) in both sides is

$$\frac{\mathrm{d}M}{\mathrm{d}t}=T\bullet C^{0.1T-1.245} \tag{4}$$

Obviously, the changing rate of maturity over time is related to temperature, which behaves a non-linear relationship. In other words, the relationship between cement hydration rate and temperature is non-linear. Therefore, compared with eq. (1), the modified maturity calculation formula can simultaneously consider the influence of cement type and temperature on the cement hydration rate, which should be more reasonable in theory. However, this formula is not suitable for negative temperature conditions since it will result in negative maturity, which is obviously unreasonable.

2.2. Equivalent age indirect calculation method

Because the Nurse-Saul maturity was empirically established, the physical meaning of the calculated maturity is not clear. For this reason, Rastrup, E. [7] proposed the concept and calculation method of equivalent age that supposing the concrete age under real temperature history is transformed into an age under a fixed temperature (constant, such as 20°C), the same type of concrete at same mix ratio should share the same strength when the transformed age is equivalent. The equation of equivalent age is

$$t_e = \int_{t_0}^{t} \frac{k(T)}{k_r} \mathrm{d}t = \int_{t_0}^{t} \gamma \, \mathrm{d}t \tag{5}$$

Where, t_e means the equivalent age (d), $k(T)$ represents the hydration reaction rate (1/d) of concrete at a temperature T, k_r denotes the hydration reaction rate (1/d) at a contrast temperature T_r, the equivalent coefficient $\gamma=k(T)/k_r$. Obviously, the key to calculating the equivalent age in eq. (5) is the calculation of the hydration reaction rate $k(T)$ and k_r, as presented in the following equation [8,9]

$$k(T) = A \cdot \exp\left(-\frac{E_a}{RT}\right) \tag{6}$$

Where, A represents the hydration reaction parameter, E_a is the apparent activation energy of cement (kJ/mol), T denotes the Kelvin temperature (K), and R means the gas constant (valued 8.314J/(mol·K)).

Substituting eq. (6) into (5), considering the conversion of Kelvin temperature and Celsius temperature, then

$$t_e = \int_{t_0}^{t} \exp\left(-\frac{E_a}{R}\left(\frac{1}{T+273.15} - \frac{1}{T_r+273.15}\right)\right) \mathrm{d}t \tag{7}$$

When using eq. (7) to calculate the equivalent age, the gas constant R is a constant, and the contrast temperature T_r is a constant (20°C for European standards and 23°C for North American standards). The temperature T is easy to be measured, the apparent activation energy E_a has greater influence on the calculated value of equivalent age. Hence, it can be regarded as a function of the degree of hydration. Theoretically, the degree of hydration or apparent activation energy should be a function of time and temperature. The classical functional expressions of apparent activation energy E_a are shown in Table 1.

Table 1. Summary of apparent activation energy calculation formula.

Author	Expression	Remarks
Freiesleben-Hansen [8]	$E_a = \begin{cases} (33.5+1.47(20-T))\,\mathrm{kJ/mol} & T \le 20^\circ\mathrm{C} \\ 33.5\,\mathrm{kJ/mol} & T \ge 20^\circ\mathrm{C} \end{cases}$	Ignored the time influence
Kim, J. K. [10]	$E_a(t,T) = (42830-43T)e^{(-0.00017Tt)}\,\mathrm{J/mol}, t \ge t_0$	t_0 denotes the time of final set of concrete (d)。 Equation considered the influences of time and temperature

Rilem [11]	$\frac{E_a}{R}=\begin{cases}4000 & T\geq 20^{\circ}\mathrm{C}\ \text{Portland cement}\\ 4000+175(20-T) & T<20^{\circ}\mathrm{C}\ \text{Portland cement}\\ 6000 & \text{Slag cement}\end{cases}$	Ignored the time influence
CEB-FIP [12]	$\frac{E_a}{R}=4000$	Suitable for ordinary Portland cement, ignored the time influence
ASTM [13]	$\frac{E_a}{R}=5000$	Suitable for type I cement without any admixtures. Ignored the influences of time and temperature

Considering that the equivalent age calculation method in eq. (7) is complicated, Rajesh simplified it and obtained following expression [14]

$$t_e=\int_{t_0}^{t}\exp\left(\beta\left(T-T_r\right)\right)\mathrm{d}t \tag{8}$$

Where, β is the temperature influence factor (1/°C).

According to the expression of Nurse-Saul maturity equation in eq. (1), when T_0 = -10°C and concrete at a constant curing temperature T=30°C for 1d, the maturity M was 40°C·d. While at a constant temperature T=2°C for 3.3333d, the maturity M reached 40°C·d. It is obvious the both maturities obtained in different cases are the same. Theoretically, both concrete strengths at a same grade and a same type ought to be the same. However, the concrete strength at second curing condition was higher than the first one, which do not agree with the calculation results. Guo, C. [15] also found the similar phenomenon, a same type and grade concrete cured at different temperature history. Even though both maturities obtained by the Nurse-Saul maturity equation were the same, their strengths were different. Which attributes to the assumption of the development speed of maturity over time is proportional to temperature in the Nurse-Saul maturity equation.

As to the same composition of concretes made of the same cement, as long as the degrees of hydration are the same, the strengths should be the same. Obviously, the equivalent age calculation method based on eq. (7) and (8) fairly meet this requirement. According to the qualitative analysis, the equivalent age model may be more reasonable.

3. Maturity application

According to above analysis, many maturity calculation methods have been effectively established, which is generally aimed at engineering application. Generally, indexes used in engineering application mainly involves in elastic modulus, compressive strength, etc. The key to maturity application is to establish a precise relationship between concrete indexes and maturity.

3.1. Calculation of compressive strength

3.1.1. Calculation formula

The formula of predicting strength using concrete maturity in the early time was a logarithmic form [16]

$$S=a+b\log(M) \tag{9}$$

Where, S represents the compressive strength of concrete (MPa), a and b are fitting constants. According to eq. (9), when the maturity M increases, theoretically, the compressive strength also increases. Particularly, when M approaches infinity, the compressive strength also approaches infinity, which is obviously inconsistent with the facts. In order to deal with this question, researchers proposed hyperbolic and exponential concrete maturity prediction strength formulas [17], which are

$$S = \frac{M}{p + qM} \tag{10}$$

$$S = p\exp(-\frac{q}{M}) \tag{11}$$

Where, p and q are fitting parameters, and the corresponding hyperbolic and exponential ultimate strengths are $1/q$ and p, respectively. In addition, researchers have also proposed many calculation formulas of maturity prediction strength, as shown in Table 2. Actually, the application of these formulas in actual engineering is extremely limited.

Table 2. Calculation formula of maturity prediction strength.

Author	Expression	Remarks
Li, Q. F. [18]	$S = \begin{cases} 10.76M^{0.331} & M \le 840^{\circ}\mathrm{C}\bullet\mathrm{d} \\ 27.4M^{0.1923} & M \ge 840^{\circ}\mathrm{C}\bullet\mathrm{d} \end{cases}$	Suitable for ordinary Portland cement. Unable to solve the problem of tending to infinity.
Li, Q. F. [18]	$S = \begin{cases} 2.32M^{0.559} & M \le 840^{\circ}\mathrm{C}\bullet\mathrm{d} \\ 27.4M^{0.1923} & M \ge 840^{\circ}\mathrm{C}\bullet\mathrm{d} \end{cases}$	Suitable for slag Portland cement. Unable to solve the problem of tending to infinity.
Zhao, Y. P. [19]	$\frac{S}{S_{28}} = \begin{cases} 0.26\ln M - 0.65 & M \le 560^{\circ}\mathrm{C}\bullet\mathrm{d} \\ 0.13\ln M + 0.18 & M \ge 560^{\circ}\mathrm{C}\bullet\mathrm{d} \end{cases}$	S_{28} is the strength cured for 28days under the standard condition. Unable to solve the problem of tending to infinity.

Many existing test results showed that the early curing temperature of concrete significantly affects the later strength. While the maturity calculated according to the Nurse-Saul maturity equation (1) and logarithmic strength prediction equation (9) cannot reflect this effect. One suggestion is to predict strength by combining with the equivalent age, which are [20-22]

$$\frac{S}{S_u} = \frac{k(t_e - t_0)}{1 + k(t_e - t_0)} \tag{12}$$

$$\frac{S}{S_u} = \frac{\sqrt{k(t_e - t_0)}}{1 + \sqrt{k(t_e - t_0)}} \tag{13}$$

$$\frac{S}{S_u} = \exp\left(-\left(\frac{\tau}{t_e}\right)^{\alpha}\right) \tag{14}$$

Where, S_u represents the strength when the concrete reaches the limit of hydration degree, k is the goodness of fit constant, τ denotes the time normalization parameter, and α means the goodness of fit constant. Obviously, according to eq. (12) and (13), if the final setting time t_0 is the same for the same type and mix ratio, only when the equivalent age t_e is the same, the concrete strength ought to be the same. When the equivalent age t_e is equal to the final setting time t_0, the concrete strength S becomes zero. If the equivalent age t_e tend to infinity, the concrete strength S tends to S_u, which is theoretically consistent with the actual situation. Equation (14) also satisfies that when the equivalent age t_e tends to infinity, the concrete strength S tends to S_u. While the most obvious difference between eq. (12), (13) and (14) is the form of expression. As eq. (12) assumes that the relationship between strength and equivalent age is a linear hyperbolic function, and the relations in eq. (13) and (14) are linear hyperbolic function and parabolic hyperbolic, respectively. Obviously, according to the qualitative analysis, eq. (12), eq. (13) and (14) are more reasonable than eq. (9).

Journal of Physics: Conference Series **2011** (2021) 012022 doi:10.1088/1742-6596/2011/1/012022

3.1.2. Comparison and analysis of formula calculation accuracy

Since the temperature ranges of concrete in different seasons are different in actual engineering, it is important to analyze the calculation accuracy of different maturity calculation methods and different strength prediction methods at different temperatures, so that facilitate practical engineering applications. Generally, the calculation methods of maturity or equivalent age mainly include formula (1), formula (3) and (7), and the strength prediction method mainly includes formula (9) to formula (14), and the comprehensive prediction method is finally adopted in this paper. More details can be seen in Table 3.

Table 3. Summary of forecast methods.

Prediction method number	Formula of maturity or equivalent age calculation	Formula of strength prediction
1	Eq. (1)	Eq. (9)
2	Eq. (3)	Eq. (9)
3	Eq. (1)	Eq. (10)
4	Eq. (3)	Eq. (10)
5	Eq. (1)	Eq. (11)
6	Eq. (3)	Eq. (11)
7	Eq. (7)	Eq. (12)
8	Eq. (7)	Eq. (13)
9	Eq. (7)	Eq. (14)

In past five years, many experimental studies have been conducted on the strength evolution law of concrete [23-33]. To enhance the objectivity of comparative work, the calculation accuracies of various prediction formulas are compared based on the existing test results.

(1) Negative temperature (-10°C~0°C)

Since the original experimental data provided by Qiu, Y. [32] and Duan, Y. [33] are relatively comprehensive, their test results are used for the calculation of maturity and strength prediction at negative temperature. The seven calculational formulas are used to compare the calculation and experimental results at constant negative temperature and variable negative temperature, which is shown in Fig. 1. Considering that formula (3) is inapplicable for the negative temperature situation, hence the corresponding prediction method 2, method 4 and 6 are not involved in the comparison. Specifically, when the apparent activation energy E_a is required in the calculation, the first formula in Table 1 can be adopted. When eq. (1) is used to calculate the maturity, the reference temperature T_0 is set -10°C. Therefore, when the curing temperature reaches -10°C, the maturity calculated by eq. (1) is zero, which cannot be used to predict the strength. As a result, the results in Fig. 1(d) do not involve the results from method 1, method 3 and 5.

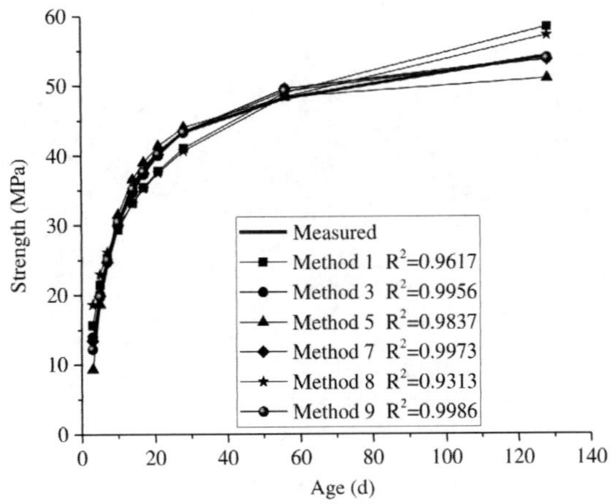

(a) Constant negative temperature -3°C

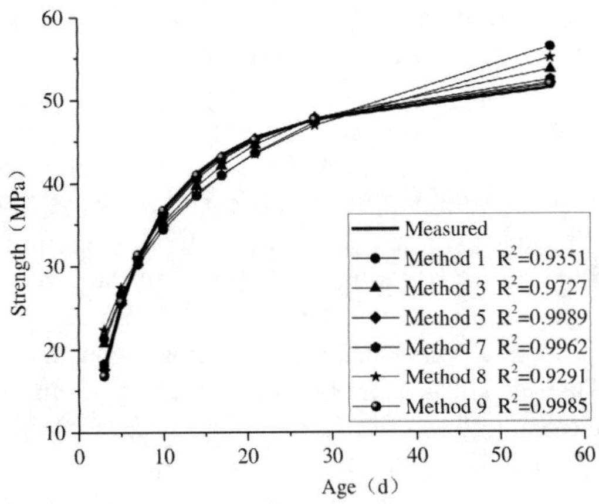

(b) Variable negative temperature at -3°C (5°C, 3°C, and 1°C are cured in sequence for 12h and then -3°C is cured)

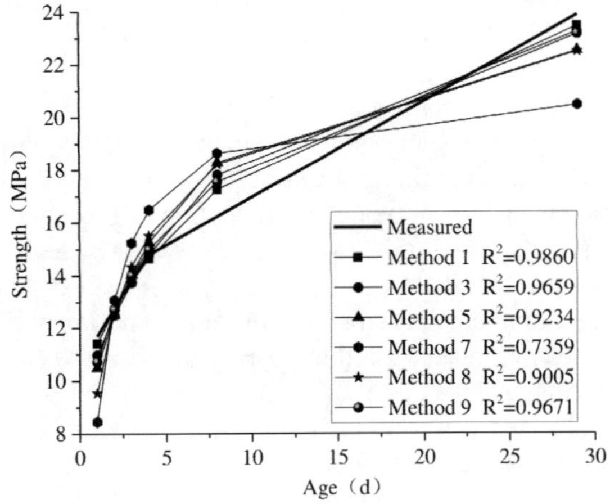

(c) Variable negative temperature at -5°C (10°C curing for 1d and keep -5°C curing)

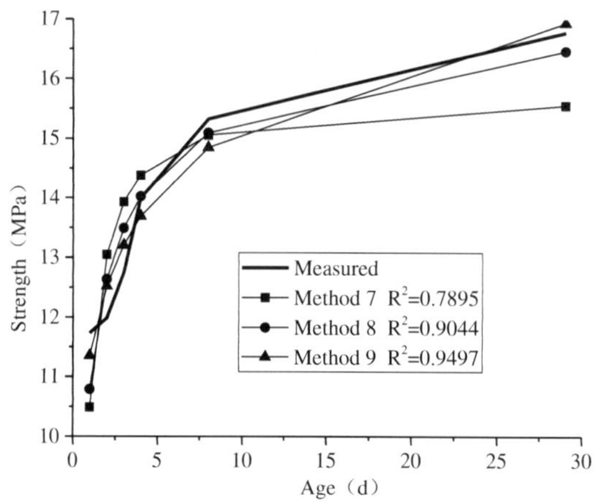

(d) Variable negative temperature at -10°C (10°C curing for 1d and keep -10°C curing)

Fig. 1 Comparison of strength prediction methods under negative temperature

As shown in Fig. 1(a), at a constant negative temperature of -3°C and an age of 128 days, the method 1, method 3 and 5 are the equations corresponding to the Nurse-Saul maturity equation, with the calculation result of method 3 being the best and correspondence to a hyperbolic strength prediction method. While in the case of a constant negative temperature and a long age, the combination method of the Nurse-Saul maturity equation and the hyperbolic strength (combination I) has high accuracy in predicting concrete strength. The method 7, method 8 and 9 are the equations corresponding to the equivalent age calculating method of Freiesleben-Hansen, with the calculation result of method 9 being the best and correspondence to an exponential function strength prediction method. In addition, in the case of a constant negative temperature and a long age, the combination method of the Freiesleben-Hansen and the exponential function strength (combination II) has high accuracy in predicting concrete strength, which is much higher than that in the combination method of the Nurse-Saul maturity equation and the hyperbolic strength.

According to Fig. 1(b), in the case of middle age with variable negative temperature, both the combination method of the Nurse-Saul maturity equation and the exponential strength (combination III) and combination II have high calculating accuracy in predicting concrete strength, and the accuracy of the former is higher than the latter.

According to Fig. 1(c) and 1(d), in the case of variable negative temperature and short age, both the combination method of the Nurse-Saul maturity equation and the logarithmic strength (combination IV) and combination II have high calculating accuracy in predicting concrete strength, and the accuracy of the latter is higher than the former.

In general, when using the Nurse-Saul maturity equation to predict the concrete strength under negative temperature conditions, the hyperbolic, exponential, and logarithmic strength prediction equation are suggested to be adopted in the condition of long age, middle age and short age, respectively. However, when using the equivalent age equation of Freiesleben-Hansen, the exponential strength prediction equation is suggested to be used regardless of the age.

(2) Low positive temperature (0°C~20°C)

In the case of low positive temperature, since the original experimental data provided by Xu et al (2020) and Qiu, Y. (2016) are comprehensive, their test results are used for the calculation of maturity and strength prediction. The nine calculational methods presented in Table. 3 are used to compare the calculation and experimental results at constant low positive temperature and variable low positive temperature, which is shown in Fig. 4. When the curing temperature keeps 0°C, the maturity calculated by eq. (3) is zero. As a result, the predicting method 2, method 4 and 6 eventually get invalidated.

Table 4. Comparison of strength prediction methods at low positive temperature.

Temperature	Fitting goodness R^2								
	Method 1	Method 2	Method 3	Method 4	Method 5	Method 6	Method 7	Method 8	Method 9
Constant 0°C	0.9967	/	0.9964	/	0.9817	/	0.9946	0.9922	0.9968
Constant 10°C	0.9148	0.9148	0.9328	0.9328	0.9770	0.9770	0.9677	0.8973	0.9950
Constant 20°C	0.9613	0.9613	0.9997	0.9997	0.9776	0.9776	0.9891	0.9803	0.9988
Changes around 20°C [23]	0.9804	0.9863	0.9979	0.9977	0.9742	0.9743	0.9905	0.9786	0.9972

Seeing from Table 4, at a constant temperature of 0°C, both the combination I and IV have high accuracy in predicting concrete strength. While at a constant temperature of 10°C, the combination III has high accuracy in predicting concrete strength, which is still lower than that of the combination II. Both the combination I and II have high accuracy in predicting concrete strength at a constant temperature of 20 °C, with the accuracy of combination I higher than combination II. Concrete cured at variable temperature of 20 °C, the method 3 and 4 have the highest calculating accuracy, followed by method 9, which indicates concrete at variable low positive temperature. Both combinations of I and II are the better prediction method.

(3) Positive temperature (20°C~60°C)

In the case of positive temperature, the experimental results by Xu, J. [23] and Qiu, Y. [32] are used to verify the accuracy of strength predicting method. Since there is no test result of variable positive temperature exceeding 20°C at present, the concrete strength cured at conditions of constant 35°C, constant 50°C and 60°C are verified in this paper. As shown in Table 5.

Table 5. Comparison of strength prediction methods at low positive temperature.

Temperature	Fitting goodness R^2								
	Method 1	Method 2	Method 3	Method 4	Method 5	Method 6	Method 7	Method 8	Method 9
Constant 35°C	0.9444	0.9444	0.8913	0.8913	0.9669	0.9669	0.9763	0.9098	0.9892
Constant 50°C	0.9047	0.9047	0.9232	0.9232	0.9649	0.9649	0.9935	0.9457	0.9987
Constant 60°C	0.8934	0.8934	0.9505	0.9505	0.9955	0.9955	0.9787	0.8672	0.9990

Obviously, seeing from Table 5, both the combination I and combination II have the highest accuracy in predicting concrete strength regardless of the curing temperature of constant 35°C, 50°C or 60°C.

Based on the results of negative temperature (-10°C~0°C), low positive temperature (0°C~20°C) and positive temperature (20°C~60°C), it is applicable for combination II to predict concrete strength at all temperature conditions.

3.2. Calculation of elastic modulus

3.2.1. Calculation formula

In addition to strength, the common concrete performance index is the elastic modulus. Compared with the strength predicting experiment, there are fewer experimental results related to elastic modulus prediction [34,35]. Similar to the method of strength prediction, the prediction method of elastic modulus mainly includes the establishment of the relation between elastic modulus and maturity or relation between elastic modulus and equivalent age. Correspondingly, the calculating method is the same as the nine methods in Table 3, where the strength is replaced by the elastic modulus

3.2.2. Comparison and analysis of accuracy of formula calculation

As shown in Table 6, the fitting goodness of all methods are compared when the elastic modulus is predicted according to the nine methods in Table 3.

Table 6. Comparison of strength prediction methods at low positive temperature.

Temperature	Fitting goodness R^2								
	Method 1	Method 2	Method 3	Method 4	Method 5	Method 6	Method 7	Method 8	Method 9
Constant 20°C	0.9756	0.9756	0.9271	0.9271	0.8567	0.8567	0.8828	0.9690	0.9778
Variable 50°C	0.9835	0.9839	0.9284	0.9288	0.8716	0.8720	0.8810	0.9734	0.9843
Constant 60°C	0.9197	0.9197	0.9766	0.9766	0.9544	0.9544	0.9617	0.9512	0.9716

According to Table 6, the combination method of equivalent age equation of Freiesleben-Hansen and exponential function-type elastic modulus method has the highest accuracy in predicting the elastic modulus of concrete at constant 20°C and variable 20°C. While in the case of constant high temperature (60°C), the combination method of Nurse-Saul maturity equation and hyperbolic elastic modulus has higher accuracy in predicting the elastic modulus of concrete, followed by the combination method of equivalent age equation of Freiesleben-Hansen and exponential functional elastic modulus, but the difference in fitting goodness between the two combinations is merely 0.005.

4. Research Outlook

Much achievement has been made on the theory and application of concrete maturity by researchers, but the application in actual engineering is relatively limited. Some recommendations are made for the following research.

(1) Most concrete maturity calculation methods developed based on the assumption of 100% of curing humidity. However, the change in curing humidity significantly affects the hydration reaction rate of cement, and hence the maturity is changed. It is of great importance to develop the calculation formula of maturity based on the concrete tests at various curing humidity [36].

(2) With the continuous development of new types of concrete such as lightweight concrete, recycled concrete and high-performance concrete, whether the existing maturity calculation methods applies to these new concretes or not, it remains needing to be tested and verified [37]. Correspondingly, how to use maturity to judge the performance indexes of the new concrete is also an urgently problem for research.

(3) Since conducting indoor tests is costly whatever in time or expense, it is recommended to conduct numerical simulation to verify the maturity after experimental verification [38].

(4) In order to meet the requirements for the real estate market, rapid construction (e.g., five-day one-story) has become rather common. At present, the concrete of the calculations of connection node between formwork and concrete structure, and the compression deformation of each floor were assumed to be completely hardened and reached strength limit. Nevertheless, concrete in actual engineering generally takes a long time to reach the ultimate strength. Hence, it is unsafe to use the ultimate strength to verify the concrete strength when the actual strength does not reach the ultimate value. As a result, it is meaningful to establish a constitutive model of concrete considering the maturity and apply the constitutive model to finite element calculations [39,40].

5. Concluding remarks

In this study, the importance of concrete maturity in engineering is considered. The maturity calculation methods of concrete are summarized by literature induction, theoretical qualitative analysis and quantitative calculation. The calculating accuracy of maturity in predicting compressive strength and elastic modulus are compared, and some future research suggestions are pointed out. The conclusions are as follows:

(1) The calculation methods of concrete maturity include direct calculation method and equivalent age calculation method. Theoretically, the equivalent age calculation method should be more reasonable.

(2) Since the wide range of on-site construction temperature, it is recommended to use the combination of equivalent age equation of Freiesleben-Hansen and exponential function strength to predict the concrete strength, and the combination of equivalent age equation of Freiesleben-Hansen and exponential function elastic modulus to predict the elastic modulus of concrete.

(3) It is suggested to conduct experiments and numerical simulations to establish a concrete maturity calculation method considering the curing humidity. The existing research results can be used to infer the performance indexes of the new type of concrete. The constitutive model of concrete considering maturity is also suggested to be considered so that the maturity theory a play greater role in the finite element calculation of engineering.

Acknowledgments
The Program of China Construction (No. CSCEC4B-2020-KT-25) and High-Tech Program of China Construction 4th Engineering Bureau (No. CSCEC4B-2020-KT-25) are greatly appreciated for providing financial support for this research.

References

[1] Saul, A. (1951). Principles underlying the steam curing of concrete at atmospheric pressure. Magazine of Concrete Research, 2(6),127-140. doi: 10.1680/macr.1951.2.6.127
[2] JGJ/T104-2011, (2011). Winter construction regulations for construction engineering. Beijing, China Construction Industry Press, 000(011), 140-141.
[3] Ge, Z. (2005). Predicting temperature and strength development of the field concrete. Ames: Iowa State University.
[4] Kim T. (2004). Concrete maturity: A quantitative understanding of how early-age temperature affects the maturity concept. Boulder: University of Colorado at Boulder. doi: 10.1016/S0008-8846(00)00481-6
[5] De. V. R., Tegelaar, R. (1998). Gewichtete Reife des Betons. Beton, 48(11), 674-678.
[6] CUR-RECOMMENDATION. (1986). Determining the strength of young concrete on the bases of weighted maturity. Gouda: The Dutch Centre for Civil Engineering, Codes and Specifications,
[7] Rastrup, E. (1954). Heat of hydration in concrete. Cement Concrete Reseach, 6(17), 127-140.
[8] Freiesleben, H. P., Pedersen, J. (1977). Maturity computer for controlled curing and hardening of concrete. Nordisk Betong, (1), 19-34.
[9] Brown, T. L. Lemay, H., Bursten, B. E. (2009). Chemistry: The Central Science. Prentice Hall. doi: doi:10.1016/S0140-6736(02)79752-3
[10] Kim, J. K., Han, S. H., Lee, K. M. (2001). Estimation of compressive strength by a new apparent activation energy function. Cement and Concrete Reasearch, (31), 217-225. doi: 0.1016/S0008-8846(00)00481-6
[11] Anonymous. TCE1. (1997). A diabatic and semi-adiabatic calorimetry to determine the temperature increase in concrete due to hydration heat of the cement. Materials & Structures, 30(8), 451-464. doi: 10.1007/BF02524773
[12] CEB -FIP MODEL CODE 1990. (1990) Comite Euro-international du Beton. London: Thomas Telford.
[13] ASTM. (2011). Standard practice for estimating concrete strength by the maturity method: ASTM C1074—2011. Pennsylvania: West Conshohocken.
[14] Rajesh, C. T., Nicolas, C. (1991). Rate constant functions for strength development of concrete. Materials Journal, 88(1), 74-83.

[15] Guo, C. (1989). Maturity of concrete: method for predicting early-stage strength. Aci Materials Journal, 86(4), 341-353.
[16] Plowman, J. M. (1956). Maturity and the strength of concrete. Magazine of Concrete Research, 8(22),13-22. doi: 10.1680/macr.1956.8.22.13
[17] Kee, C. F. (1977). Relationship Between Strength and Maturity of Concrete. ACIJ, 54(12), 196-203.
[18] Li, Q. F. (2002). New Theory on Concrete. Concrete, (10), 12-15.
[19] Zhao, Y. P., Sun, Y. L., Yu, T., Yu, J. H. (2009). Test Study on Concrete Strength Growing Rules in Winter at Cold Area. Bulletin of the Chinese Ceramic Society, 28(4), 854-858. doi: 10.16552/j.cnki. issn1001-1625.2009.04.039
[20] Carino, N. J. (1984). The Maturity Method: Theory and Application. Astm Journal of Cement Concrete and Aggregates, 6(2), 61-73. doi: 10.1520/CCA10358J
[21] Knudsen, T. (1980). On particle size distribution in cement hydration. Paris: Editions Septima.
[22] Hansen, P. F, Pedersen, E. (1985). Curing of concrete structures. Switzerland: Draft DEB-Guide to Durable Concrete Structures.
[23] Xu, J., Zhu, R., Zhan, Y. J., Zhu, M. T., Li, X. K. (2020). Experimental Study on Strength Evolution of Concrete Structures Under Different Strength Grades and Curing Conditions. Building Construction, 42(12), 2314-2317. doi: 10.14144/j.cnki.jzsg.2020.12.034
[24] Dai, J. P., Wang, Q. C., Qu, S., Duan, Y., Xie, Z. G. (2018). Strength Prediction Model of Low Temperature Curing Concrete Based on Maturity Theory. Journal of Materials Science and Engineering, 36(02), 263-267.
[25] Ma, Y. Ma, L., Pan, B. L. (2019). Research on Prediction of Early Concrete Strength Based on Maturity Method. Highway Transportation Technology (Application Technology Edition), 15(05), 1-4.
[26] Yang, W. J., Deng, C. L., Yang, H. C., Wang, S. L. (2019). Comparison of prediction models at home and abroad for mechanical properties of concrete. Building Structure, 49(S1), 697-700. doi: 10.19701/j.jzjg.2019.S1.144
[27] Feng, S. X. (2018). The influence of setting agent and curing temperature on the development of concrete strength. Highway Transportation Technology (Application Technology Edition)), 14(09), 141-143.
[28] Song, Y. T., Hu, J., Hu, J. B., Qin, M. Q. (2017). Use of maturity method to estimate compressive strength of marine concrete. New Building Materials, 44(09), 15-18.
[29] Zhu, X. L., Ding, J. T., Cai, Y. B. (2016). Prediction of mechanical properties of anti-abrasion concrete based on equivalent age method. Yangtze River, 47(23), 97-101. doi: 10.16232/j.cnki. 1001-4179. 2016.23.020
[30] Li, J., Cao, D. W., Han, S. (2016). Application Analysis of Maturity Integral Calculation Method for Cement Concrete. Transport Research, 2(06), 60-66. doi: 10.16503/j.cnki.2095-9931.2016.06. 010
[31] Liu, G. (2016). Study on the equivalent method of concrete age based on electrical resistivity. China Three Gorges University.
[32] Qiu, Y. (2016). Maturity method to predict the strength of concrete in winter construction. Harbin Institute of Technology.
[33] Duan, Y. (2016). Study of concrete strength under constant minus-temperature and varied minus-temperature and calculation formula optimization of early strength of concrete. Lanzhou Jiaotong University.
[34] Hou, D. W., Zhang, J., Chen, H. Y., Liu, W. (2012). Development of strength and elastic modulus of concrete under moisture and drying curing conditions. Journal of Hydraulic Engineering, 43(02), 198-208.
[35] Zhu, R., Xu, J., Zhan, Y. J. (2020). A New Model for Predicting Elastic Modulus of Lightweight Concrete with Singly Mixed Fly Ash Based on Maturity Theory. Building Construction, 42(11), 2108-2110+2116. doi: 10.14144/j.cnki.jzsg.2020.11.026

[36] Wang, T., Wang, X. L. (2017). Maturity of Early Strength of Concrete Slab. Bulletin of the Chinese Ceramic Society, 36(05), 1783-1789. doi: 10.16552/j.cnki.issn1001-1625.2017.05.056
[37] Peng, C. Y. Zhang, X. P., Yu, B., Peng, J, Y., Jiao, C. J. (2020). Study on maturity of recycled-fine-powder concrete based on equivalent age. Concrete, (01), 22-26.
[38] Lv, F. Y., Meng, X. L., Li, B., Jia. J., Wang, B., Liu. C. Y. (2019). Simulation of the Maturity of New Cast Concrete After Overwintering. Journal of Water Resources and Architectural Engineering, 17(05), 123-127.
[39] Geng, D. J., Guo, P. J., Zhou, S. H. (2018). Implicit numerical integration of an elasto-plastic model for structured clays. Chinese Journal of Theoretical and Applied Mechanics, 50(01), 78-86.
[40] Geng, D. J., Dai, N., Guo, P. J., Zhou, S. H., Di, H. G. (2021). Implicit numerical integration of highly nonlinear plasticity models. Computers and Geotechnics, 132, 103961. doi: 132(2), 103961. 10.1016/ j.compgeo.2020.103961

Analysis on Concrete Construction Technology in Civil Engineering Construction

Yuzhen Guo

Jiangxi Vocational Technical College of Industry & Trade, Nanchang, Jiangxi, 330038, China

*Corresponding author's e-mail:739960103@qq.com

Abstract. An important structural material in civil engineering construction is concrete. Concrete is widely used in various building structures. Concrete construction technology directly determines the construction quality of the concrete structure itself, which plays a very important role in the stability of the building structure. This paper mainly summarizes the current construction situation of concrete, and deeply analyzes some problems and technical points in the construction process, which is of great significance to improving the construction quality of concrete and optimizing the construction management of concrete. On this basis, this paper expounds the maintenance ideas of concrete pouring from the aspects of concrete mold removal, later repair and concrete maintenance operation, so as to provide a reference for the technical improvement of concrete construction.

1. Introduction

At this stage, concrete is the most widely used building materials in many civil engineering construction, because the concrete itself has the irreplaceable advantages of other building materials. With the rapid expansion of the infrastructure scale of civil engineering in China, the quality problem of construction engineering has increasingly attracted wide attention from the society. The safety of concrete technology is related to the interests of the broad masses and the economic benefits of construction enterprises. Although concrete materials have many properties and advantages, in the actual construction process, the performance of concrete is affected by many factors. In the construction process, if there is an improper control of each construction elements, it will lead to a great negative impact on the quality of concrete. This negative effect is also very similar during the construction process. In the actual construction process, the main construction problems are mainly the following categories.

2. Current status of concrete construction

In the current construction materials, concrete is one of the most widely used construction materials in many civil engineering construction. Because the concrete itself has the irreplaceable advantages of other building materials, which makes the concrete construction technology level also determines the construction level to a certain extent. With the rapid expansion of the infrastructure scale of civil engineering in China, the quality problem of construction engineering has increasingly attracted wide attention from the society. The safety of concrete technology is related to the interests of the broad masses and the economic benefits of construction enterprises. Although concrete materials have many properties and advantages, in the actual construction process, the performance of concrete is affected by many factors. In the construction process, if there is an improper control of each construction elements, it will lead to a great negative impact on the quality of concrete. This negative effect is also very similar during the construction process. In the actual construction process, the main construction problems are mainly the following categories.

Table 1. Main problems of concrete construction.

sequence	**Rprincipal problem**
1	Raw materials used for concrete do not meet the relevant technical requirements
2	Various additives added to the concrete do not conform to the actual site conditions
3	Lack of effective management during the concrete pouring process
4	Concrete lacks effective post-stage maintenance
5	Do not strictly observe the relevant technical requirements during the concrete matching process

There are two main reasons for this situation. The first reason is that China does not implement the corresponding industry standards and norms for concrete construction, and many construction personnel can only carry out the construction based on their own basic experience. Another reason is the lack of relevant training among most of the concrete construction technicians. The professional and technical level is low, and it cannot take reasonable and effective technical measures according to the actual situation of the site to ensure the quality of concrete construction[1].

3. Key points of the concrete construction technology

In the actual construction process, the main points of concrete construction technology are mainly divided into the following major categories.

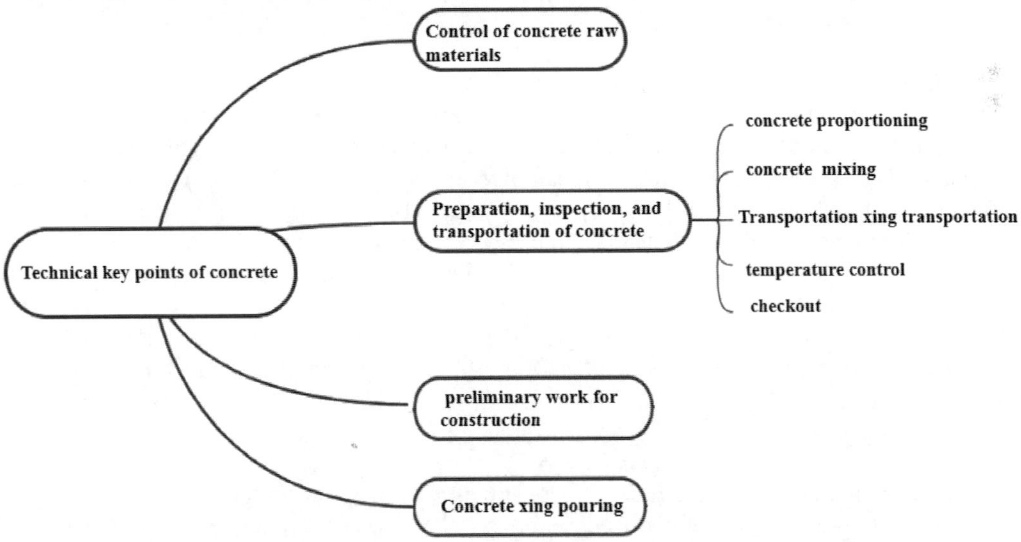

Figure 1. Technical key points of concrete.

3.1. Control of the concrete construction materials

The raw materials for making concrete are mainly cement, water, stone materials and various additives. The choice of concrete production materials will greatly affect the physical and chemical properties and mechanics of concrete, in the configuration of concrete materials, the first step to do is to carry out a series of investigation of raw materials, to ensure that the raw materials comply with the relevant technical regulations. Relevant qualified enterprises can also carry out a series of comprehensive tests on the rational performance and geometric parameters of the raw materials through on-site testing, so as to ensure that the raw materials we choose meet the cost technical standards of the enterprise. The choice of cement is a very important part in a selection of material. When the conditions and environment of the enterprise allow, the preparation of concrete should choose the same time, the same batch of cement, avoid the mixing of cement products of different manufacturers as far as possible, avoid the technical requirements of different manufacturers, and have a certain impact on the quality of concrete[2].

3.2. Preparation, inspection and transportation of mixed concrete

3.2.1. Making of the concrete. In the process of raw materials, concrete needs to ratio different types of raw materials according to some relevant theories. In the process of concrete ratio, we should accurately understand some basic requirements of buildings for concrete performance, and determine the proportion of concrete materials on this basis. In the actual construction process, the ratio of concrete should also be appropriately adjusted according to the changes of the site construction environment, so that the performance of the concrete itself can adapt to the on-site environment. When the choice of additives, different buildings have different use requirements, and different buildings, different environments and different areas will also put forward different requirements for some main frame materials-concrete. In different buildings, we in order to better meet the use requirements of buildings, we need to add various additives to the concrete, which can make the concrete have a certain stability to qualitative, after the addition of additives. A series of experiments should also be carried out in the corresponding actual environment, and sufficient data analysis is obtained through the experiments, so as to ensure that the produced concrete can effectively meet the use requirements of the buildings. In addition to the strict control of these situations, we also need to control the construction process in the proportional pairing of concrete. Avoid the construction personnel from failing to prepare the concrete according to the specific requirements, causing a series of adverse effects to the project progress.

3.2.2. concrete mixing. The pairing ratio of concrete has a great impact on the use degree of the project. In addition to this effect, the mixing concrete also has a great impact on the preparation of concrete. In the proportional pairing of the concrete. We should effectively control the mixing process and mixing equipment, so that the concrete mixing work can meet the construction requirements of the concrete. Ready-mixed materials should also be ready-mixed materials before the actual situation begins, and then arrange the basic steps of the concrete mixing operation, so that the relevant process parameters can be obtained in the actual concrete mixing. Reprepare the concrete for buildings, effectively manage the various raw materials in the mixing process, and the effective control of the concrete mixing process can be realized with a series of modern mixing equipment. The concrete mixer shall also meet the relevant technical requirements during its use. The material in the mixer shall not exceed its maximum capacity[3].

3.3. preparations for construction
Concrete is just a link in the construction. The construction quality of concrete depends not only on its own quality, but also on a series of construction processes. Before the concrete pouring, there are two main steps, the first is to build the formwork, the second is to lay the reinforcement. In the early stage of concrete construction, we should comprehensively check a series of quality problems to ensure that the size of the formwork meets the needs of the relevant building design. The parts of the formwork are effectively supported, so that we are able to build a solid building core, which can avoid a series of effects caused from the insufficient strength of the template structure. Reinforcement plays a very big role in the building, the effective combination of concrete and reinforcement can play their carrying

function, before the concrete construction, a series of comprehensive inspection to ensure that the reinforcement we use meet the relevant technical requirements, and the spacing and quantity of reinforcement should also meet the building design standards, on this basis, the concrete constructors should according to the erection of formwork and concrete laying to develop the characteristics of a reasonable construction plan. Concrete It is a very professional and technical work, and each different building will have different requirements for concrete construction. The construction of concrete should also formulate the corresponding construction technical measures according to the actual situation of the corresponding construction project. Concrete construction personnel should carry out a series of comprehensive understanding and learning of such measures, so as to ensure that the construction personnel can understand the key points of concrete construction. In the construction process, the concrete should also fully consider the external influence factors, and try to choose the period with small temperature change for construction. This is beneficial to exert the maximum performance of the concrete[5].

3.4. Casting of concrete
In the pouring construction of concrete, the construction environment should be clean up on the site, and eliminate all kinds of waste and debris such as formwork, such as domestic waste and some abandoned construction materials. Avoid mixing various impurities and have some adverse effects on the quality of concrete. Secondly, in the pouring of concrete, it should also be necessary to carry out a series of cleaning of the site, control the water consumption, to avoid because of the more formwork water, causing a series of impact on the concrete ratio. The aim is to remove a series of dust inside the formwork, and improve the water content and crystallization degree of the formwork, which can effectively improve the stability of the concrete. During the process of pouring, we should also control the distance between the pouring surface and the feeding pipe to avoid the excessive distance between the pouring surface and the feeding pipe and causing the concrete to fall to the pouring surface. During the construction process of the concrete pouring. We should also choose the appropriate pouring method according to the site situation, and conduct the concrete pouring in strict accordance according to the relevant technical requirements. According to the characteristics of the watering structure, choose scientific and effective sales methods. The main point of pouring is vibration guide, the purpose is to ensure that the concrete material can effectively fill the gap in the formwork, rather than the gap inside the concrete, in the process of vibration leakage and vibration, this reason is the construction of a series of professional level are not enough, in this case, this pouring work should select some experienced construction personnel to vibration work, so as to meet the needs of concrete pouring operation, in order to ensure the continuity of concrete pouring operation. The construction process can not be interrupted, so as to effectively reduce the possibility of concrete producing various cracks.

4. Maintenance work of the concrete pouring

4.1. Concrete removal removal
After the concrete pouring, after a certain time, the concrete was solidified, and the comprehensive mechanical performance of the concrete has reached a certain standard. In this case, the mold removal of the concrete can be done. In the concrete formwork removal work, we should strictly abide by the relevant technical specifications and requirements, in general circumstances, the concrete formwork removal work is mainly to follow the principle of first dismantling, after first dismantling. When the demolition of concrete formwork, the part of lower importance should be removed first, and then the part of higher importance should be removed. The demolition time will directly affect the comprehensive performance of the concrete. Only the corresponding formwork can be removed when the properties of the concrete meet the corresponding requirements. In order to ensure the quality and construction progress during construction, the properties of concrete in real time. When the relevant technical parameters meet the relevant technical requirements, the formwork should be immediately

removed, and dismantled scientifically and reasonable, to avoid the blind and violent construction of causing a certain impact on the construction materials. In this process, we should remove the formwork in time, so that it can create favorable conditions for the maintenance and construction of concrete.

4.2. Later-repair process

After the concrete formwork removal work is completed, relevant technical personnel to conduct a comprehensive inspection of the structure of the concrete. Effective measures are taken to repair the defects and problems existing in the soil structure to enable the structure of the concrete to meet the technical requirements of the specification. The two main common problems of concrete after the formwork removal are explicit and honeycomb problems. For surface honeycomb problems, polish the cover with a wire ball, and then according to the matching ratio of the concrete. With suitable grout, in order to solve the problem of exposed reinforcement and exposed bone, it is necessary to remove the concrete layer at the exposed tendon and exposed position, making the original concrete layer have better comprehensive performance. In doing such work, we may need to re-erect the formwork and vibrwork, threatening the safety of the safety of the building. In this case, it is necessary to timely organize the construction and design, formulate effective solutions, and eliminate a series of problems existing in the concrete in time, to ensure the safety of the buildings. In the construction environment of concrete, due to the limitation of the site construction environment, the poured concrete may have some problems, so that the late repair work can solve this problem very well. In this restoration work in the later stage, the relevant enterprises should pay more attention to actively carrying out a series of inspection work in the later stage, so as to eliminate the safety hidden dangers in the construction process and improve the stability of the building[6].

5. Conclusion

In the actual construction process, the concrete has many design problems and technical problems. In order to ensure the construction quality, it is necessary to avoid a series of engineering quality problems. Relevant enterprises construction before construction should consider good technical and material problems. The quality of concrete will determine the quality of the engineering foundation, it is both a safety problem, but also a technical problem. In the construction, the relevant personnel should strictly control the matching proportion of concrete, and control a series of sudden problems in the construction process of concrete. At the same time, the construction unit should also actively take a series of effective measures to improve the relevant technical level of the construction personnel, to ensure the construction safety from the source, and ultimately can also promote the benign development of concrete construction technology.

References

[1] Zhao,D.Y.(2019)On the Construction Technology of Mass Concrete Structure in Civil Engineering.Residential Property & Real Estate.28:211.
[2] Xu,J.Y.(2019)Analysis on the Construction Technology of Mass Concrete Structure in Civil Engineering Construction.Residential Property & Real Estate.27:165.
[3] Sun,G.F.(2019)Discussion on the Construction Technology of Mass Concrete Structure in Civil Engineering Buildings. doors and windows .16:89-91.
[4] Sun,J.(2019)Discussion on the Construction Technology of Mass Concrete Structure in Civil Engineering Buildings.Residential Property & Real Estate.22:186.
[5] Cao,Q.(2019)Application of Concrete Construction Technology in Civil Engineering.Building materials and decoration.26:16-17.

Journal of Physics: Conference Series **2011** (2021) 012024 doi:10.1088/1742-6596/2011/1/012024

Key technology of overlapping tunnel construction with earth pressure balance shield method in rheological strata

Qingru Yu[1,2], Shihong Zhai[1,4], Wen Liu[1,4,5], Tianrui Liu[6]*, Hu Song[1,3]

[1]CCCC Second Harbor Engineering Co., Ltd, Wuhan, Hubei, China

[2]Key Laboratory of Large-span Bridge Construction Technology for Transportation Industry, Wuhan, Hubei, China

[3]Research and Development Center of Transportation Industry of Intelligent Manufacturing Technologies of Transport Infrastructure, Wuhan, Hubei, China

[4]CCCC Highway Bridge National Engineering Research Centre Co., Ltd, Wuhan, Hubei, China

[5]School of Civil Engineering & Mechanics, HuaZhong University of Science and Technology, Wuhan, Hubei, China

[6]School of Economics & Management, China University of Geosciences, Wuhan, Hubei, China

*LIU Tianrui's e-mail:yaoxiangchuan@foxmail.com

Abstract: Based on the existing bridge structure of Foshan Metro Line 2, this paper summarizes the engineering geological conditions of the overlapping tunnels, uses three-dimensional finite element analysis, establishes an engineering geological model, and analyzes the parameters of the upper tunnel in the disturbed surrounding rock, shield attitude control and its influence on the tunnels already built on the lower floors. Finally, the monitoring and control results during the construction of the shield tunneling of the overlapping tunnels are analyzed to verify that the construction of the overlapping tunnels in the rheological strata has a greater impact on the ground settlement of about 20m, and the soft foundation reinforcement of the lower tunnels in advance can make the soil settlement greatly reduced, and the stress concentration area is reduced. After the double-tube jet-grouting piles are reinforced, the shield tunneling of the upper overlap tunnel has little effect on the existing tunnel below, and the reinforcement effect is better, which can make the construction go successfully. The analysis results of this paper can provide decision support for the construction of overlapping tunnels in rheological strata.

1. Introduction

As the urban rail transit soars in China and gradually develops into a cluster or network in Cities in recent years, the underground tunnel construction is inevitably restrained by space, resulting in more and more overlapping tunnel cases. [1] The overlapping tunnel is a construction method for small spacing tunnels, which often involves above crossing, undercrossing, or oblique crossing. In China, the shield method is barely applied for overlapping tunnel construction in soft rheological strata, let alone the construction of small spacing tunnels. It is essential to establish a complete technology system for overlapping tunnel construction in rheological strata with the shield method.

Rao et al. (2018) [2], aiming at the short distance overlapped tunnel in silt reclamation stratum, explored the effects of the construction of the new shield tunnel on the surrounding environment and proposed a specific reinforcement scheme. Wei et al. (2020) [3] studied the impact of the upper shield excavation at the overlapping section of the tunnel on the vertical displacement of the existing tunnel underneath, and deduced the equation for calculating the vertical deformation of the existing tunnel. Wang et al. (2017) [4] took the overlapping section of Zaoshanlu-Licun Station Interval of Qingdao Metro as the research object to analyze the surrounding rock stability and construction economy under different excavation spacings. Nematollahi et al. (2019) [5] researched the impact of parallel crossing of tunnels at the two-tube section upon the pile foundation of the ring road overpass through the three-dimensional digital simulation based on the Shiraz Metro Project in Iran. Li (2020) [6] adopted the numerical simulation to study the deformation settlement of an old bridge and internal force variation of overlapping tunnels, and verified the effectiveness of the numerical simulation and reinforcement measures by monitoring data. Jiang et al. (2018) [7], based on the special construction conditions of two-tube shield tunnel of Shenzhen metro line 9 overlapped over the existing two-tube tunnel of existing metro line, studied the laws of horizontal and vertical deformation of existing tunnel tubes induced by new shield construction, and analyzed the influence of earth pressure on the deformation of existing tunnel tubes. Fang et al. (2020) [8] investigated the characteristics of vertical soil deformation induced by fully overlapping tunnels constructed using a tunnel boring machine (TBM). Li et al. (2017) [9] used the three-dimensional finite element method to simulate and analyze the tunnel with a horizontal tunnel crossing under the existing tunnel. Ngoc-Anh et al. (2014) [10] applied the full three-dimensional numerical simulation method to study the impact of the two-tube tunnel construction in soft soil on the excavated tunnel or the ground.

The above analysis reveals that although scholars both at home and abroad have fully studied the overlapping tunnel shield construction technology, there are few researches on the overlapping tunnel construction with the shield method in rheological strata. By taking Foshan Metro Line 2 as an example, this paper analyzed the settlement parameters of the ground in the shield construction area induced by the rheological strata, studied the soft strata reinforcement technologies, and summarized the mutual effect between overlapping tunnels during construction.

2. Project Overview

Liantang~Zhangcha station interval runs west to east, parallel to the planned Line 4, laid along Jihua 2nd Road. The minimum clear distance between the outer contours of the two tunnels is 5 m. After passing through Foshan-Kaiping Expressway and Huabao Road under-crossing tunnel, the right track of line 2 crosses the left track at YDK29 + 447, as shown in Figure 1.

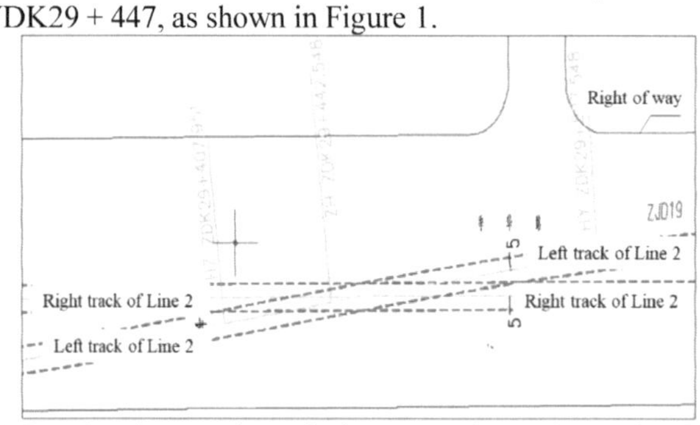

Fig. 1 Overlapping section of Liantang~Zhangcha station interval

The overlapping section of Liantang~Zhangcha station interval mainly crosses the strata from top to bottom: <1-1> plain fill, <1-2> miscellaneous fill, <2-1A> silt, <2-4-2> soft plastic cohesive soil, <3-1> silt and fine sand, <6> completely weathered mudstone and sandstone, <7-1> strongly weathered mudstone, and <8-1> strongly weathered sandstone composite stratum. As shown in Figure 2, these strata feature large thickness variation, poor uniformity, and heterogeneous hardness of rock, greatly affecting the shield construction.

In addition, within the station interval, the surface water is not developed. Only a river flows 35 m away from the left side of the YDK29+100～+490 m Section. This river flows all year round with abundant water.

Groundwater is mainly divided into Quaternary pore water and bedrock fissure water. The Quaternary pore water occurs in the upper Quaternary loose accumulated, residual, and fully weathered rock stratum. Because the aquifers are slight~weak permeable and only the 1.1-3.4 m thick silt and fine sand layers lenticularly distributed in the YDK29+300~+450 Section are medium permeable, the Quaternary pore water content is low. According to the survey and borehole water level observation, the long-term stable water table depth is generally 1~2 m (elevation 1.0~3.0 m) in the station interval. The bedrock fissure water mainly occurs in the joints and fissures of strongly and moderately weathered rock strata, featuring non-uniformity. Because joints and fissures develop irregularly and unevenly, in the areas where joints and fissures of the rock mass develop, the groundwater is rich and highly permeable, but not vice versa. As the overlaying Quaternary soil layer is less permeable, forming a relative aquiclude, the bedrock fissure water has a certain bearing capacity. Generally, in the station interval, the joints and fissures in the argillaceous cemented sand and mudstone are poorly interconnected, so the bedrock fissure water is not rich.

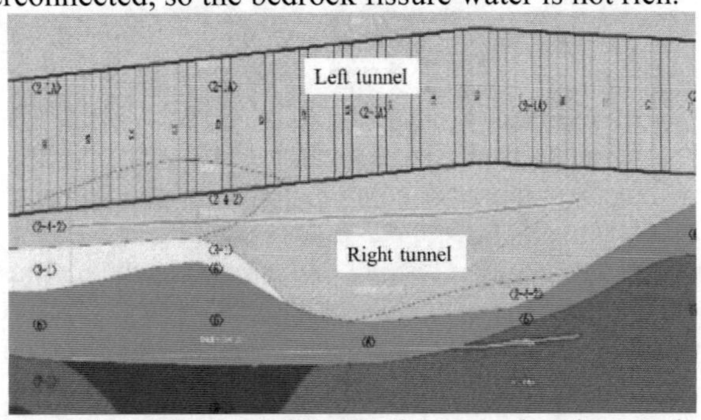

Fig. 2 Geological conditions of the overlapping section

3. Shield Construction Plan for Overlapping Tunnels

The current shield construction method for overlapping tunnels is mainly from the lower tunnel to the upper tunnel. For the lower tunnel, the shield excavation is a conventional earth pressure shield construction, and the high-strength segments and bolts are used at the short-distance overlapping section, to build a permanent structure which acts as a stable support for the construction of the upper tunnel. In order to alleviate the impact of the upper tunnel construction on the completed tunnel structure, a temporary support system matching with the lower tunnel structure should be erected within a certain distance in front and rear of the TBM working plane of the upper tunnel. At the same time, the shield excavation parameters and attitude in the upper tunnel should be controlled, and the impact of the construction on the completed tunnel should be monitored in real time.

At the overlapping section of Liantang-Zhangcha station interval, the left line is mainly in the <2-1A> silt stratum, and the right line is in the <6> completely weathered mudstone and sandstone strata. During the left line tunneling, the natural water content of the silt stratum is higher than the liquid limit, the void ratio is large, the compressibility is particularly high, and the strength is extremely low, so the shield construction may disturb the silt stratum, causing uneven settlement easily. Due to the high risk, the shield construction must be optimized and controlled in the process. The thrust is controlled between 1000 KN ~ 1500 KN; the cutter head rotates at 2.0 rmp; the excavated soil volume per ring is within 56 m³; the synchronous grouting pressure is between 0.2 ~ 0.3 MPa; and the pressure at the top of the soil bin is within 0.3 MPa.

In this paper, the relevant data models were established, together with the GTS NX2018 software and three-dimensional finite element method, to analyze the deformation of the existing right tunnel and the ground on the left line at YDK29 453.182 during the shield construction. The calculation assumes that: (1) the ground surface and soil horizon are distributed in homogeneous horizontal layers;

(2) lining segments are regarded as a whole; and (3) the thrust of the cutter head is simulated by applying a force on the shield excavation surface. The established finite element model is a three-dimensional network with a length of 60 m, a width of 30 m, and a height of 30 m, consisting of totally 38,025 elements and 52,914 nodes. The lower is the existing tunnel on the right line, and the upper is the excavation surface of the left line, as shown in Figure 3.

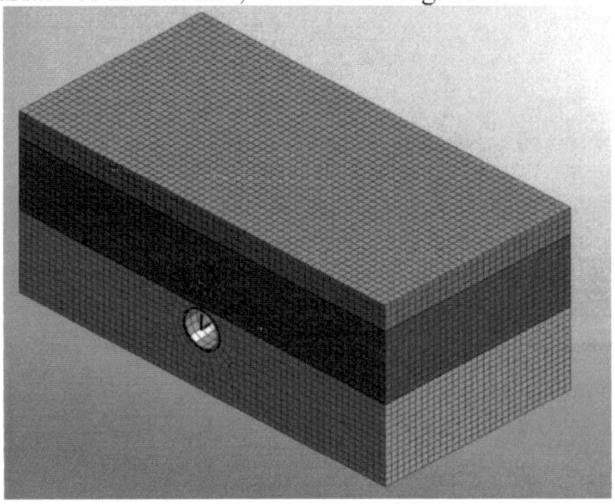

Fig. 3 Finite element grid layout

This manifests the impact of the upper tunnel shield construction on the ground and the right tunnel segments. As shown in Figure 4, the soil displacements concentrate on the upper left part of the shield excavation. This is because the left tunnel is turning to the left, and the maximum ground displacement reaches 14 cm. There will be a safety risk, so the strata must be reinforced before the left tunneling. From Figure 5, to reinforce the overlapping section of the tunnel, double-tube jet grouting piles should be arranged 3 m outside the tunnel on the plane and vertically 0.5 m from the ground to the stable stratum at the bottom of the tunnel. As shown in Figure 6, double-tube jet grouting piles are Φ600 @450mm engaged and staggered.

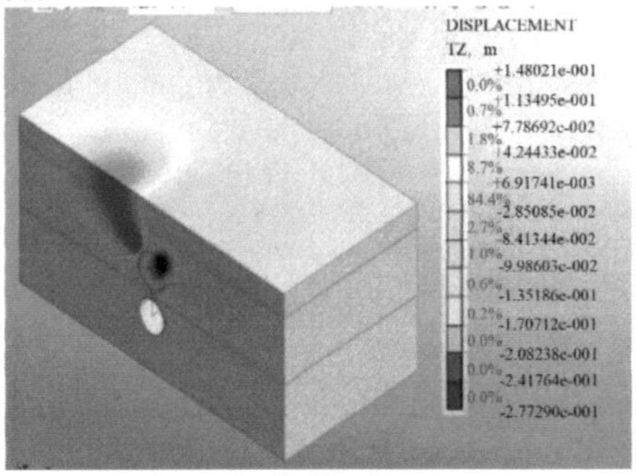

Fig. 4 Soil displacement

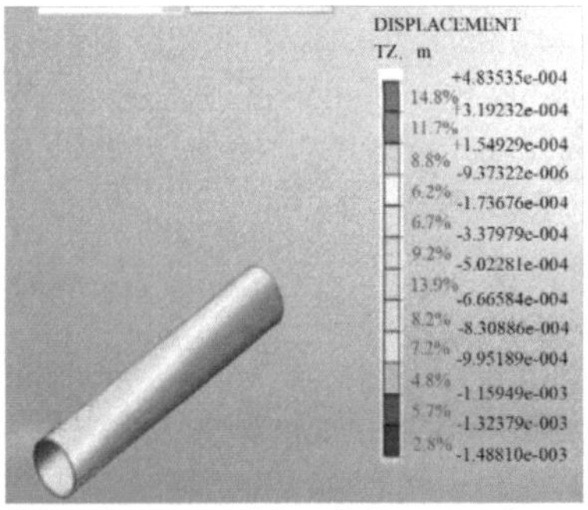

Fig. 5 Settlement of the right tunnel segment

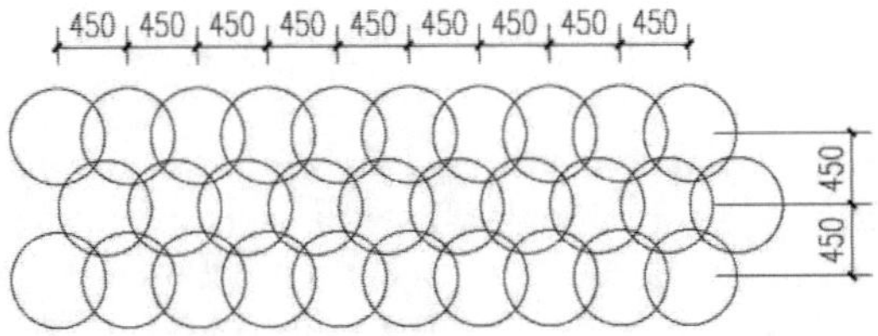

Fig. 6 Layout drawing of double-pipe jet grouting pile

To re-analyze the tunneling on the reinforced left line at YDK29 453.182, the deformation of the right tunnel and ground should be considered. The reinforced finite element model is a three-dimensional network with a length of 60 m, a width of 30 m, and a height of 30 m, consisting of totally 54,709 elements and 61,451 nodes, as shown in Figure 7.

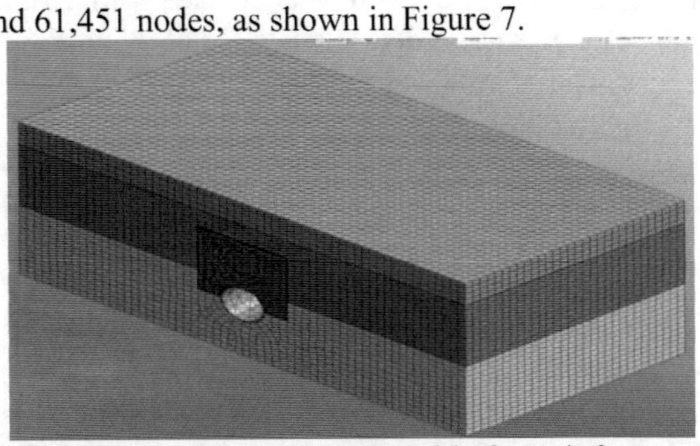

Fig. 7 Layout of finite element grid after reinforcement

As shown in Figure 8, after the double-pipe jet grouting pile is reinforced, the stress concentration area is reduced, and the soil settlement is decreased to 6.9 mm, which avoids the risk of uneven settlement induced by the disturbance to the underlying silt stratum. As shown in Figure 9, the maximum displacement of the existing tunnel segments on the right line is 0.9 mm, indicating a good reinforcement effect; therefore, the construction can be carried out smoothly.

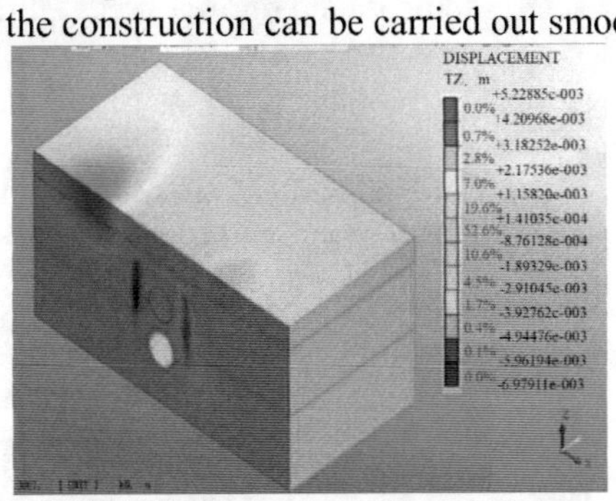

Fig. 8 Soil displacement after reinforcement

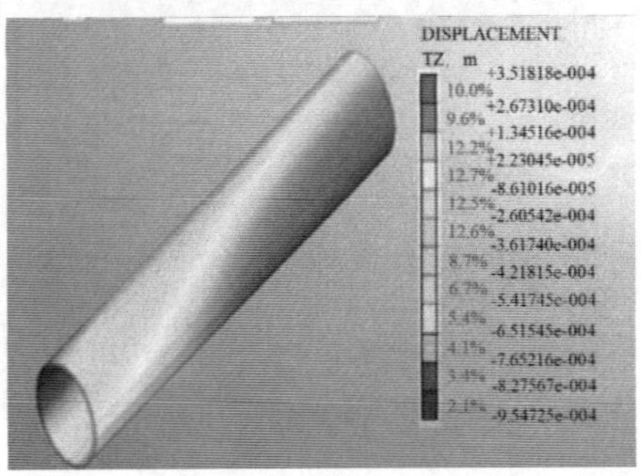

Fig. 9 Settlement of tunnel segments after reinforcement

To sum up, using the three-dimensional finite element method for numerical simulation on the construction method of the lower right tunnel first and the upper left tunnel later, the following conclusions can be drawn:

(1) Before reinforcement, when the lower right tunnel is holed through, the maximum surface settlement will be 14 cm after the upper left tunnel is excavated, the stresses will concentrate along the shield construction direction, and the uneven settlement will be induced by the disturbance to the silt stratum; therefore, in the subsequent double-tube jet-grouting pile reinforcement, in addition to the soil mass around the two tunnels, the soil mass in the middle of the tunnels should also be reinforced. The simulation shows that this method can obtain good results.

(2) The upper left tunnel is holed to a single side groove width of about 20 m, and within about 20 m in front of the shield construction, the soil mass is disturbed by the construction.

(3) It is expected that the surface settlement will be controlled within 10 mm during the lower right tunneling. After the upper tunnel is excavated, it is expected to be controlled within 30 mm.

(4) Before reinforcement, when the existing right tunnel is excavated on the upper left line, the maximum settlement of the segments reaches 1.4 mm. After the double-tube jet grouting piles are reinforced, the maximum settlement is 0.9 mm, greatly reducing the segment deformation.

4. Research on Monitoring and Control during Overlapping Tunnel Construction

The overlapping section of Liantang~Zhangcha section interval is a major risk point, and requires intensive monitoring. For this reason, when the left line is crossing the overlapping section, the monitoring focuses are: ①surface settlement; ②pipeline settlement; ③convergence of the right line segment; and ④ right line pipeline settlement. By observing the surface settlement and the segment convergence and settlement in the right tunnel, the deformation of the ground and segments in the right tunnel can be ascertained in time, for decision-making of the best on-site construction method.

4.1 Surface settlement

As shown in Figure 10, the surface survey points of the overlapped section concentrate on Monitoring Sections 16-18. The specific surface settlement data during the left line tunneling are shown in Tables 1, 2, and 3.

It can be seen from Tables 1, 2, and 3 that during the left line tunneling, the ground surface was slightly uplifted to the maximum value of 24.78 mm, corresponding to the survey point DBC-17-17, and the grouting was stable after the TBM passed through the point. As of July 4, the cumulative settlement value was 21.83 mm at the survey point which was stable. During the shield construction, there were no obvious abnormal conditions such as slurry eruption and collapse on the surface.

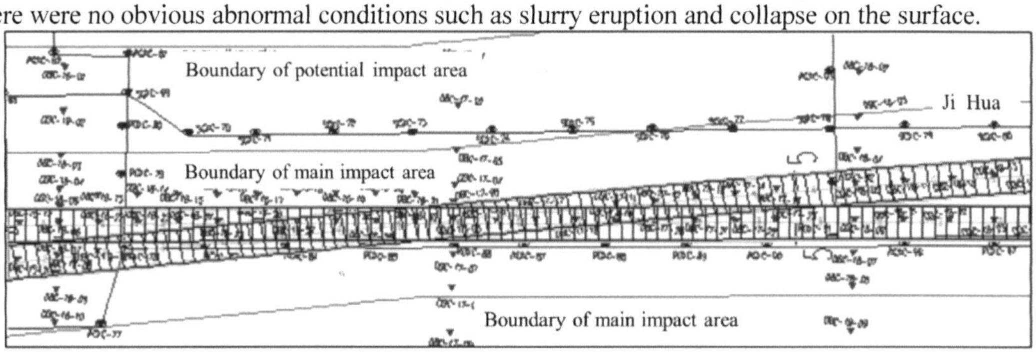

Fig.10 Layout of survey points on the surface of the overlapping section

Table 1 List of Settlement of Monitoring Section 16

Survey Point No.	2018/5/28	2018/6/1	2018/6/5	2018/6/9	2018/6/13	2018/6/17	2018/6/21	Average Speed (mm/d)
DBC-16-08	0.00	4.21	9.32	9.71	7.67	7.22	7.35	0.31
DBC-16-20	0.00	5.22	4.76	7.62	6.37	3.98	3.69	0.15
DBC-16-21	0.00	1.47	1.84	2.18	3.85	0.89	0.53	0.02
DBC-16-22	0.00	0.73	1.26	1.49	5.04	1.54	1.36	0.06
DBC-16-24	0.00	0.79	1.26	1.00	9.60	5.31	5.09	0.21
DBC-16-25	0.00	0.92	1.76	1.22	5.64	0.16	0.13	0.01
DBC-16-26	0.00	0.85	1.68	0.68	6.55	0.37	0.03	0.00
DBC-16-27	0.00	1.56	1.58	1.28	9.10	4.88	4.58	0.19
DBC-16-28	0.00	0.84	-0.21	1.04	8.95	9.62	9.28	0.39

Table 2 List of Settlement of Monitoring Section 17

Survey Point No	2018/6/9	2018/6/12	2018/6/15	2018/6/18	2018/6/21	2018/6/27	2018/7/4	Average Speed (mm/d)
DBC-17-06	0.00	0.73	15.86	11.49	11.46	8.94	9.14	0.37
DBC-17-11	0.00	1.97	11.26	23.05	21.54	20.69	20.70	0.83
DBC-17-12	0.00	0.73	3.38	13.47	12.11	11.66	11.19	0.45
DBC-17-13	0.00	-0.49	0.99	3.33	3.06	2.98	2.94	0.12
DBC-17-14	0.00	-0.57	0.60	4.74	4.06	3.65	2.38	0.10
DBC-17-15	0.00	-1.65	-2.39	-0.66	0.08	-0.60	-2.18	-0.09
DBC-17-16	0.00	0.88	20.24	16.96	16.11	13.85	13.55	0.54
DBC-17-17	0.00	0.24	23.60	24.78	21.56	21.82	21.83	0.87
DBC-17-18	0.00	0.02	13.18	21.57	18.32	18.64	18.23	0.73

Table 3 List of Settlement of Monitoring Section 18

Survey Point No.	2018/6/15	2018/6/19	2018/6/24	2018/6/28	2018/7/3	2018/7/8	2018/7/15	Average Speed (mm/d)
DBC-18-05	0.00	5.97	11.90	9.85	9.78	9.56	9.58	0.32
DBC-18-11	0.00	1.94	10.96	9.90	10.10	10.33	10.45	0.35
DBC-18-12	0.00	0.55	7.88	4.35	3.89	3.96	3.96	0.13
DBC-18-13	0.00	-3.47	-0.35	-3.28	-4.03	-4.26	-4.34	-0.14

4.2 Pipeline settlement

The pipelines in the affected area of the overlapping section of Liantang~Zhangcha station interval are mainly water supply pipes and gas pipes, all laid along the direction of Jihua Road. The gas pipes have been temporarily relocated to the secondary affected area. The corresponding gas pipeline settlement survey points are RGXC-77 ~ RGXC-90. Table 4 shows the settlement changes at these points when the TBM on the left line was crossing above the overlapping section.

Table 4 List of Gas Pipeline Settlement

Survey Point No.	2018/5/28	2018/5/31	2018/6/3	2018/6/6	2018/6/9	2018/6/12	2018/6/17	Average Speed (mm/d)
RGXC-77	0.00	4.06	6.12	8.28	11.49	11.32	10.25	0.51
RGXC-78	0.00	0.52	3.51	7.03	7.83	7.94	7.07	0.35
RGXC-79	0.00	1.24	2.02	0.75	0.39	-0.90	-0.11	-0.01
RGXC-80	0.00	-0.58	-0.37	1.34	1.14	0.61	1.39	0.07
RGXC-81	0.00	-1.54	-2.24	-1.43	-1.10	-1.43	0.57	0.03
RGXC-82	0.00	-0.66	-1.15	0.59	0.86	1.42	0.41	0.02
RGXC-83	0.00	0.26	2.30	4.31	4.59	6.51	7.91	0.40
RGXC-84	0.00	-0.24	1.29	2.91	2.51	3.69	3.99	0.20
RGXC-85	0.00	-0.69	0.39	1.23	0.96	1.50	1.82	0.09
RGXC-86	0.00	0.82	0.62	1.05	-0.65	-1.26	-2.48	-0.12

It can be seen from Table 4 that during the left line tunneling, the gas pipeline settlement survey points were uplifted to the maximum value of 11.49 mm, corresponding to the survey point RGXC-77, and then became stable gradually after the TBM passed through the point. As of June 17, the cumulative settlement at the survey point was 10.25 mm, and the survey point was stable, with an average speed of 0.51 mm/d. During the Shield construction, there was no obvious abnormality such as slurry eruption or collapse around the pipelines.

4.3 Convergence of right line segment

Before the left line crossed above the overlapping section of Liantang~Zhangcha section interval, a full-hall scaffolding was erected in the existing right tunnel for reinforcement. In the reinforced area of the right line, monitoring sections were arranged every 10 rings to set out the convergence and vault settlement survey points. Table 5 shows the monitoring data changes at each survey point for convergence observation of the right line segments when the TBM on the left line was crossing above the overlapping section.

It can be seen from Table 5 that during the left line tunneling, the maximum convergence value in the right line was 3.51 mm, and the corresponding survey point was GGS-Y-23 with an average speed of 0.13 mm/d. During the shield construction, there was no obvious abnormality such as cracking, dislocation, or water seepage at the arch waist of the right line segment.

4.4 Settlement of right line segment

The settlement survey points of the right line segment were set out at the interval same as that of the convergence survey points. They were arranged on the same monitoring section, and the survey points were right above the vault. Table 6 shows the changes of monitoring data at each survey point for settlement observation of the right line segment when the TBM on the left line was crossing above the overlapping section.

It can be seen from Table 6 that when the left line crossed above the overlapping section, the inner vault of the right tunnel was ascending to the maximum value of 6.72 mm, and the corresponding survey point was GGC-Y-26 with an average speed of 0.25 mm/d. During the shield construction, there was no obvious abnormality such as cracks, dislocation, or water seepage on the top of the right tunnel.

To sum up, when the left line crossed above the overlapping section, the maximum convergence value of the right tunnel was 3.51 mm, the maximum rise value of the vault on the right tunnel was 6.72 mm, and there was no obvious abnormality such as cracks, dislocation, or water seepage on the top of the segment. This indicates that after the double-tube jet-grouting piles are reinforced, the shield construction on the left line at the overlapping section has little impact on the existing right tunnel, and that the reinforcement has good effect.

Table 5 List of right line segment convergence monitoring data

Survey Point No.	2018/6/2	2018/6/6	2018/6/10	2018/6/14	2018/6/18	2018/6/25	2018/6/29	Average Speed (mm/d)
GGS-Y-23	0.00	1.75	2.2	2.46	3.28	3.55	3.51	0.13
GGS-Y-24	0.00	0.51	1.17	1.42	2.24	2.3	2.9	0.11
GGS-Y-25	0.00	0.18	0.76	0.97	2.19	2.1	2.52	0.09
GGS-Y-26	0.00	0.84	1.37	2.21	2.35	2.56	3.01	0.11
GGS-Y-27	0.00	0.61	0.94	1.99	2.58	2.84	2.95	0.11
GGS-Y-28	0.00	0.77	1.2	2.14	2.31	2.66	2.95	0.11
GGS-Y-29	0.00	1.2	1.68	2.06	2.33	2.74	2.68	0.10

Table 6 List of Monitoring Data for Settlement of Right Line Vault

Survey Point No.	2018/6/2	2018/6/6	2018/6/10	2018/6/14	2018/6/18	2018/6/25	2018/6/29	Average Speed (mm/d)
GGC-Y-23	0.00	1.67	2.17	3.16	3.02	3.75	3.64	0.13
GGC-Y-24	0.00	2.47	4.06	4.36	5.17	5.37	5.69	0.21
GGC-Y-25	0.00	2.95	3.16	5.75	6.28	6.54	6.15	0.23
GGC-Y-26	0.00	3.12	3.53	5.25	6.39	6.63	6.72	0.25
GGC-Y-27	0.00	0.61	2.95	3.93	4.35	5.63	5.54	0.21
GGC-Y-28	0.00	0.33	2.99	3.89	4.57	5.15	5.24	0.19

5. Conclusion

By analyzing the overlapping tunnel in rheological strata, the following conclusions can be made:

(1) The current shield construction method for overlapping tunnels is mainly from the lower tunnel to the upper tunnel.

(2) After the soft foundation of the lower tunnel is treated in advance, the stress concentration area is reduced, and the soil settlement is decreased to 6.9 mm, which avoids the risk of uneven settlement induced by the disturbance to underlying silt stratum, indicating a good reinforcement effect; therefore, the construction can be carried out smoothly.

(3) The overlapping tunnel construction in rheological strata has an impact on surface settlement of about 20 m, so protection measures are generally not required for surrounding structures.

(4) During the construction of the upper tunnel, the lower existing tunnel must be monitored. After the double-tube jet grouting piles are reinforced, the shield construction on the upper overlapping tunnel has little impact on the existing tunnel, indicating the reinforcement effect is good.

References

[1] GAO, S.T. (2020) Numerical Analysis of Appraching Construction of Overlapping and Parallel Subway Shield Tunnels. J. Modern Tunnelling Technology, 57(S1): 480-484.

[2] RAO, Z.H., LIU, J.G. (2018) Construction Influence and Control Analysis of Short Distance Overlapped Tunnel in Silt Reclamation Stratum. J. Journal of East China Jiaotong University, 35(02): 16-23.

[3] WEI, G., QI, Y.J., SUN, Q., WU, H.J. (2020) Calculation method of existing tunnel displacement under action of support trolley in overlapping tunnel. J. Rock and Soil Mechanics, S2: 1-10.

[4] WANG, W.M., WANG, C., QIN, Z.B., CAO, Z.L., SUN, J.C. (2017) Optimization for Excavation

Spacing of Two Lower Tunnel Tubes at Overlapped Section with a Minimum Clear Distance. J. Modern Tunnelling Technology, 54(05): 200-208.
[5] NEMATOLLAHI M, DIAS D. (2019) Three-dimensional numerical simulation of pile-twin tunnels interaction - Case of the Shiraz subway line. J. TUNNELLING AND UNDERGROUND SPACE TECHNOLOGY, 86: 75-88.
[6] LI, H.Y. (2020) Analys is of Settlement Control of Bridge Induced by Undercrossing of Four-line Overlapping Small-spacing Shield Tunnels. J. Tunnel Construction, 40(S1): 343-349.
[7] JIANG, H., YIN, M.L., JIANG, Y.S., SUN, Z.Y., WU, Q.L. (2018) Structure Deformation Caused by Shield Tunnel Excavation Above Existing Tunnels of Shenzhen Metro. J. Modern Tunnelling Technology, 55(01): 194-202.
[8] FANG, K., YANG, Z., JIANG, Y., et al. (2020) Surface subsidence characteristics of fully overlapping tunnels constructed using tunnel boring machine in a clay stratum. J. COMPUTERS AND GEOTECHNICS, 125(103679).
[9] Li, M., Wang, Y.H., Liu, F., Wang, Y. (2017) Research on the Interaction of Upper and Lower Tunnels. J.Chinese Journal of Underground Space and Engineering, 13(02): 486-492.
[10] NGOC-ANH, D., DANIEL, D., PIERPAOLO, O. (2014) Three-dimensional numerical simulation of mechanized twin stacked tunnels in soft ground. J. Journal of Zhejiang University SCIENCE A, 15(11).

Application of Ecological Technology of Concrete Retaining Wall in Urban Dyke

Yimei Yuan[1], Kun Liu[2*]

[1]Guangdong Polytechnic of Water Resources and Electric Engineering, Guangzhou, Guangdong, 510635, China

[2]CCCC-FHDI Engineering Co. Ltd., Guangzhou, Guangdong, 510230, China

*Corresponding author's e-mail: liuk@fhdigz.com

Abstract. Traditional concrete and mortar-masonry retaining walls are mature, safe and reliable, widely used in urban river embankment, and play a great role in disaster prevention and mitigation in towns. However, such retaining walls lack of ecological characteristics. In order to give the traditional concrete and mortar-masonry retaining walls ecological characteristics and provide conditions for the unfortunate people who fell into the water to rescue themselves and go ashore, two traditional retaining wall ecological technologies were invented, which are suitable for new retaining walls and existing retaining walls. When the new retaining wall is built, the slope ratio of the wall should be appropriately slowed down and several rows of planting grooves should be arranged. For the retaining wall that has been built, the planting groove is anchored on the wall. Landscape plants are selected and planted in the planting groove, and automatic irrigation system is arranged according to the needs. The engineering application shows that applying these two technologies to river improvement projects at the same time can form a full coverage of ecological beauty on the water-side wall of the traditional retaining wall. People who fall into the water can climb on the planting trough and wait for rescue, or climb along the planting trough to reach the goal of harmonious coexistence between people and water.

1. Introduction

Due to the limitation of land for urban river embankments, traditional retaining walls such as vertical concrete and mortar and masonry are widely used. Although this type of traditional retaining wall saves land, has stable structure, mature technology, safety and reliability, and strong resistance to water erosion, it weakens the ecological function of the river and the ecological coordination between the river and the surrounding environment, and is not conducive to the harmonious coexistence of humans and nature. For example, concrete and mortar-masonry retaining walls cut off the connection between water and land, preventing frogs, snakes and other creatures from freely moving between land and water. In addition, the smooth wall reduces the roughness of the river course and increases the water flow velocity of the river course, which is not conducive to the growth of aquatic organisms. For those who fall into the water, it is difficult to climb on a smooth and steep wall. Therefore, the traditional retaining wall is not conducive to the self-rescue of those who fall into the water.

In order to solve the above problems of traditional retaining wall, two kinds of ecological technology for traditional retaining wall were put forward. The first ecological technology is for the new retaining wall, the slope of the wall is appropriately slowed down, several rows of planting grooves are set on the wall, and then landscape greening is carried out. The second ecological technology is for the existing retaining wall, where the planting grooves are directly anchored to the wall and then landscaped. These two ecological technologies have been patented in China, and the patent numbers are ZL201610076592.4 and ZL201720747501.5 respectively.

The two ecological technologies were written into DB44/T2283-2021 "Guidelines for Ecological Design of Water Conservancy Projects", participated in the China Innovation and Entrepreneurship Fair in 2018 and 2019, and selected in the "Guangdong Province 2020 Water Conservancy Science and Technology Promotion Guide".

2. Ecological technology of traditional retaining walls

Ecological technologies are proposed for the newly built retaining wall and the existing retaining wall respectively.

2.1 Ecological technology for new retaining walls

When building a new traditional concrete or mortar-masonry retaining wall, the steep wall should be appropriately slowed down, and several rows of planting troughs should be set on the wall. The distance, depth and width of the trough meet the needs of plant growth and meet the climbing or climbing requirements of the falling water personnel[1]. When building a new retaining wall, according to the sequence of construction of the groove wall and the retaining wall, there are two construction methods: one is the integral type, and another is the stepped first and then the groove wall type.

Integral ecological technology refers to the ecological technology in which the planting trough is used as a part of the wall and the concrete or mortar masonry of the wall is constructed at the same time[2](Figure 1). The eco-technique of building the trough wall after the steps is to construct the retaining wall concrete or mortar masonry in advance, and form a ladder on the wall surface, and then construct the trough wall on the edge of the ladder to form the ecological technology of the planting trough (Figure 2).

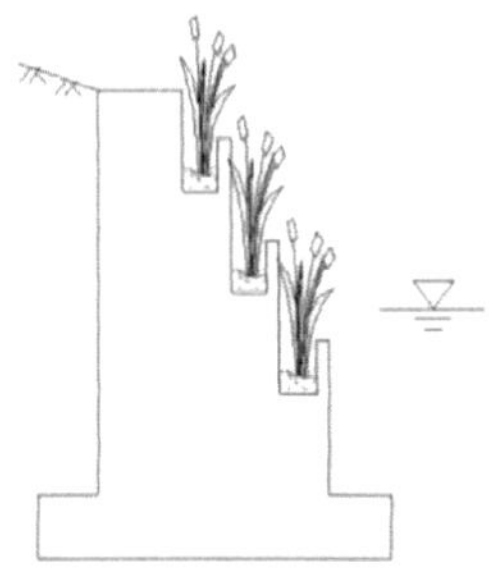

Figure 1. Sectional view of integrated ecological technology

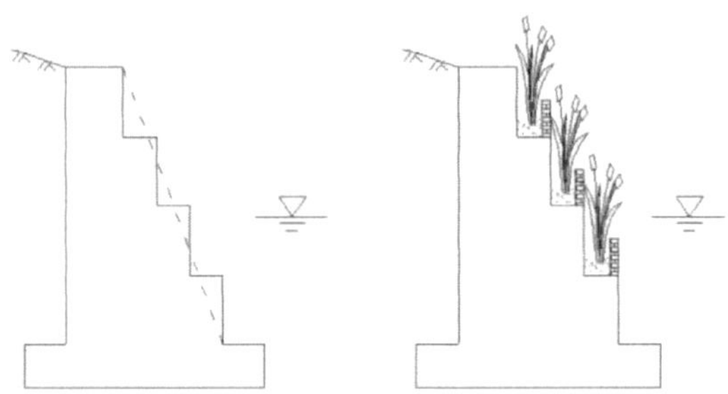

Figure 2. Sectional view of ecological technology of building trough wall after steps

In order to meet the requirements of the ecological characteristics of the retaining wall and the convenience of climbing and self-rescue functions for the falling into the water, Table 1 gives a set of economical and feasible key parameters[3].

Table 1. Key parameters of traditional retaining wall ecological technology

Construction method of planting trough	**Net width of planting trough （cm）**	**Net height of the planting trough （cm）**	**Thickness of planting trough （cm）**	**Distance between adjacent planting troughs（cm）**	**Comprehensive slope ratio of retaining wall**
integrated ecological technology	≥40	40~50	8~12	80~100	1:0.4~1:0.6
ecological technology of building trough wall after steps	≥40	40~50	12~15	90~100	1:0.4~1:0.6
prefabricated ecological technology	≥30	40~50	3~8	90~120	1:0.4~1:0.6

2.2 Ecological technology of built retaining walls

The existing retaining walls in cities and towns have invested huge investment and made great contributions to urban flood control. If they are still safe and stable, they should not be demolished. Planting troughs can be anchored on the waterfront wall to achieve ecological purposes. The prefabricated ecological transformation technology refers to the ecological transformation technology of retaining wall anchored to the built wall in the later stage of the planting trough[4].

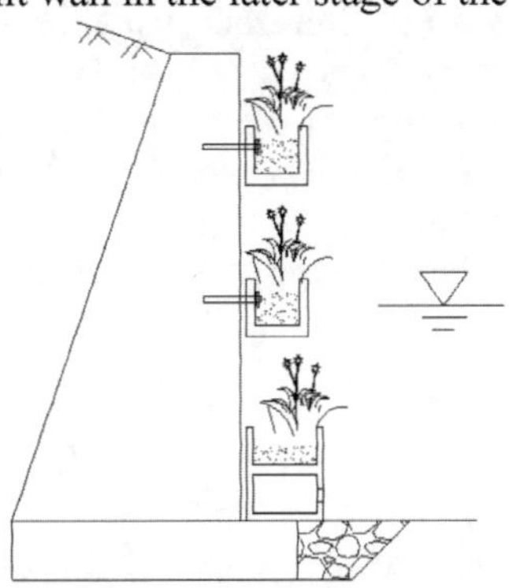

Figure 3. Sectional view of prefabricated ecological technology

2.3 Key points for implementing ecological technology of traditional retaining walls

The engineering level, design standards, engineering layout, load, stability calculation, structural calculation, and foundation treatment of traditional retaining walls should implement relevant technical regulations[5]. The ecological technology of traditional retaining walls should follow the principles of adapting measures to local conditions, ecology, applicability, economy, beauty, safety, and energy saving. According to environmental conditions and viewing needs, appropriate plant materials and appropriate construction techniques should be selected. Before the ecological technical design of the traditional retaining wall, technicians should survey the project site to investigate the geological conditions and hydrology of the river section where the retaining wall is located, the orientation of the planting trough arrangement, the illumination, the utilization of rainwater, and the facade conditions of the retaining wall.

The prefabricated ecological transformation technology shall not affect the safety performance and use function of the existing retaining wall. The planting trough can be directly fixed to the retaining wall by using anchor bolts (planting bars), or can be fixed on the supporting frame anchored to the retaining wall[6]. The anchoring design should meet the requirements of self-weight, wind load resistance, water erosion resistance, resistance to floating objects impact, and climbing of people falling into the water. The anchoring form should be selected reasonably and the anchoring safety should be ensured in accordance with relevant regulations[7][8].

2.4 Implementation points of planting trough plant configuration

It needs to be configured according to factors such as climate conditions, light conditions, water and fertilizer conditions, functional requirements, ornamental effects, maintenance and management. Plants should be mainly native plants, and backbone plants with strong stress resistance should be selected, and the planting gap, layered configuration, and color matching should be coordinated. At the same time, the corresponding plant watering and maintenance methods should be determined[9].

In addition, when selecting suitable plants, factors such as water level, water depth, flood duration, distance between planting troughs, and retaining wall height should also be considered. Submerged plants with strong water purification performance should be selected for underwater areas; shallow-rooted plants that are resistant to flooding and drought, extensive management, well-developed root systems, and resistant to water erosion should be selected for drawdown areas; herbaceous and woody mixed configurations should be selected for water areas. Plants that combine flower-viewing species and foliage-viewing species. The purpose of these practices is to ensure the effect of the landscape.

3. Engineering application example

3.1 Project overview

A river improvement project is located in Guangzhou City. The main task is to prevent floods and drain waterlogging as well as improve the water environment, water ecology, and beautify the city. There are villages and schools on both sides of the river. The total length of the river is 4.5km, the current river width is 8m-20m, and the drainage area is 6.04km^2. The annual average rainfall is 1670mm, and the annual average temperature is 21.8°C.

The current river line is tortuous, the floodplain is chaotic, the comprehensive average slope is 28‰, and the designed flood flow in 20 years is 21.98m^3/s. The regional geological structure is relatively stable. The strata on both sides of the river channel are mainly artificial fill, silty silty clay, silty clay, medium-coarse sand, etc.

3.2 Section design of river embankment

The population on both sides of the river of this project is densely populated, and the land space for the river improvement project is limited. After comparison and selection of plans, the dikes on both sides of the bank adopt the traditional ecological technology of retaining walls, which is safe and reliable, and has ecological characteristics, and it also provides climbing conditions for the unfortunate people who fell into the water, as shown in Figure 4.

The dike is a compound section, divided into upper and lower parts. The lower part of the dike adopts a steep concrete retaining wall with a wall height of 4.0m and a slope ratio of 1:0.15. The anchored planting trough on the lower retaining wall of the dike has a net width of 0.45m and a net depth of 0.48m, made of engineering plastics. First, the bracket is anchored to the lower retaining wall, and then the planting trough is fixed in the bracket. The upper part of the dike adopts a slope-type retaining wall with a wall height of 1.5m and a slope ratio of 1:1.7. 3 rows of planting troughs are arranged on the wall. The gabion net is equipped with a gabion as the wall of the planting trough, with a thickness of 0.3m and a height of 0.45m.

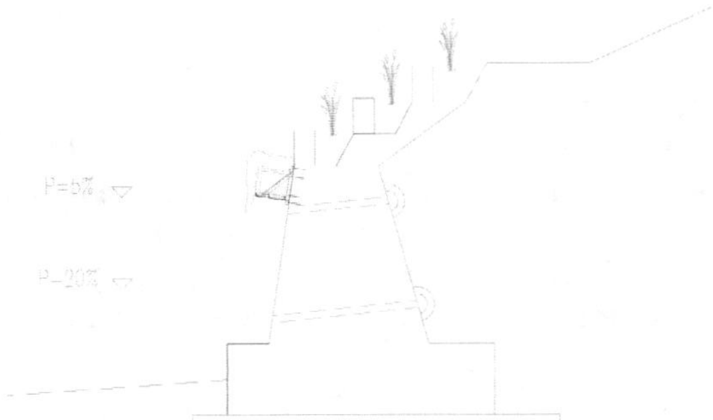

Figure 4. Diagram of river embankment section design

3.3 Implementation effect

According to the local natural geography, climate, and river hydrological conditions, suitable landscape plants have been selected and an automatic irrigation system has been arranged. After the completion of the treatment project, an ecological retaining wall with multiple functions, such as beautifying the ecological environment, facilitating falling ashore, and providing living space for animals, has been formed, which promotes the harmonious coexistence of man and water (Figure 5 and Figure 6).

Figure 5. Ecological effect at the completion of the river improvement project

Figure 6. Ecological effect of the river improvement project 5 months after the completion

4. Conclusion

Ecological retaining wall technology is based on traditional concrete and mortar-masonry retaining walls. By setting rows of planting troughs on the wall, the traditional concrete and mortar-masonry retaining walls are endowed with ecological characteristics. At the same time, the proper slot distance can facilitate the unfortunate overboard to climb ashore to save themselves. The planting trough can be implemented in an integral type, stepped first and then trough wall type, or assembled ecological technology. The main design parameters of the ecological retaining wall technology are: the slope ratio of the retaining wall, the distance between adjacent planting troughs, the thickness of the planting trough, the depth of the trough, and the width of the trough. These parameters directly affect the project cost and ecological performance.

The engineering application shows that the two kinds of ecological retaining wall technology are safe and reliable, and can form an ecological beauty where humans and water coexist harmoniously.

Acknowledgments

This article was funded by the Characteristic Innovation Project of Guangdong Province (2019GKTSCX048).

References

[1]Yuan Y.M. (2019) Review on research and application of ecological retaining wall for river course. Guangdong Water Resources and Hydropower, (11):67-70+75.

[2]Yuan, Y.M., Ye, H.X., Luo, R.H. (2020) Research on planting ecological retaining wall and its automatic irrigation system. China Rural Water and Hydropower, (4):177-180+185.

[3]Yuan Y.M., Ye, H.X., Chen, J.S. (2018) Design and application of a new type of retaining wall from the perspective of ecological management. Pearl river, 39(9):43-46.

[4]Yuan Y.M., Chen, J.S. (2019) Method for determining ecological groove parameters of new retaining wall and its cost analysis. Pearl river, 40(5):8-11+17.

[5]Ministry of Water Resources of the People's Republic of China. (2007) SL379-2007 Design specification for hydraulic retaining wall. China Water Conservancy and Hydropower Press, Beijing.

[6]Ministry of Housing and Urban-Rural Development of the People's Republic of China. (2013) JGJ 145-2013 Technical specification for post-installed fastenings in concrete structures. China Architecture & Building Press, Beijing.

[7]Karki, K.S., Malhotra, R.S. (1972) Supercritical flow over sills at incipient jump conditions. Journal of the Hydraulics Division, ASCE. 98(10):1753-1764.

[8]Ohtsu, I., Yasuda, Y., Awazu, S. (1991) Drag on vertical sill of forced jump. Journal of Hydraulic Reseach, 29(1):29-47.

[9]Ministry of Housing and Urban-Rural Development of the People's Republic of China. (2012) CJJ82-2012 Code for construction and acceptance of landscaping engineering. China Architecture & Building Press, Beijing.

Structural analysis of a school building with viscous dampers applied

Xiangxiang Wu[1*], Dawei Zhou[2], Tianrong, Huang[1]

[1] Shanghai Urban Construction Vocational College, Shanghai, 200438, China

[2]Shanghai Yingliang Construction Technology Co. Ltd., Shanghai, 200092, China

[*]Corresponding author's e-mail: wuxiangxiang@succ.edu.cn

Abstract. Viscous damper, as a type of energy-dissipating device, is applied in the 3-story school building. To investigate the function of viscous dampers, seismic nonlinear analysis is performed on the FE model with viscous dampers simulated in. The analysis is done for both minor earthquake and severe earthquake. The analysis results show that the viscous dampers start to dissipate energy in minor earthquake input and dissipate more energy in severe earthquake. Therefore, application of viscous dampers significantly reduce the seismic response of the building and effectively protect the structure building from severe damage or collapse.

1.Introduction

In recent years, with the developing investigations, the energy-dissipating and seismic isolation technics are widely applied in the high severity seismic zones [1]. Some engineering structures with such technics have shown better seismic performance in the earthquake events, such as Wenchuan earthquake or Lushan earthquake. As illustrated by the facts, the energy-dissipating and seismic isolation technics can effectively reduce the seismic response of the buildings and therefore improve their earthquake resistance abilities[2]. Among these technics, viscous damper, as a type of energy-dissipating device, is widely applied in the engineering structures due to its stable technical features and easy applications[3] [4]. To investigate the function of viscous dampers in the low-rise buildings, the FE model of a 3-story school building with viscous dampers is simulated and seismic analysis is performed on it.

2. Viscous Dampers

2.1 Principal of energy dissipation of viscous dampers

As shown in figure 1, viscous damper is generally made up of the cylinder, piston and viscous liquid. The viscous liquid inside the cylinder is silicone oil or other viscous liquid; and there is a small hole in the piston. When the piston moves in the cylinder, the liquid will go through the small holes in the piston and therefore prevent the relative movement between the piston and cylinder. In this way, the kinetic energy is dissipated to some extent and therefore the seismic response of structures is reduced.

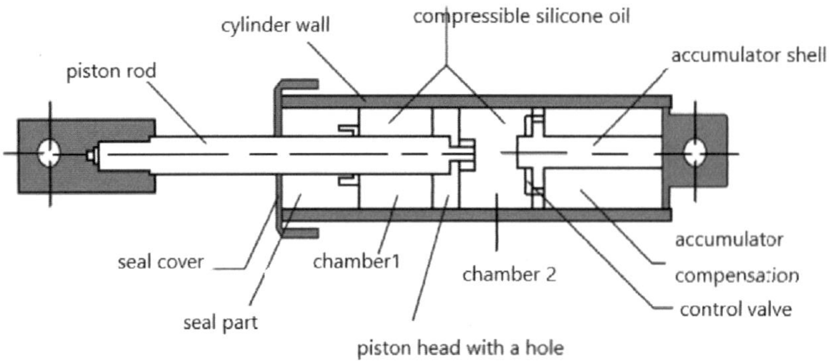

Figure 1. The composition of viscous damper

2.2 Restoring force characteristics of viscous damper

As shown in figure 2, the viscous damper will develop a stable spindle shape loop when dissipating the kinetic energy, and it is clearly seen that the viscous damper is of good energy absorption capacity.

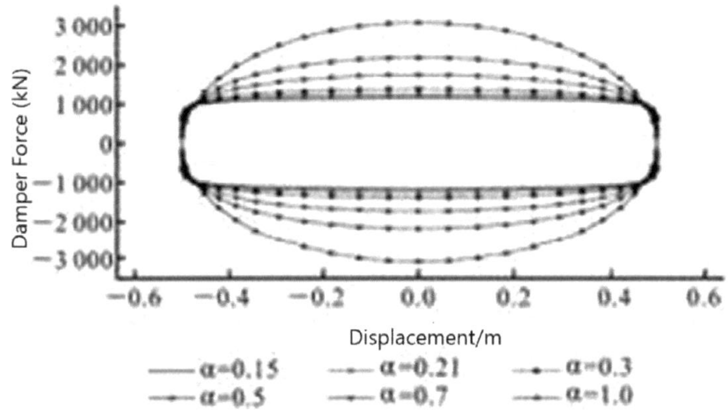

Figure 2. Hysteresis curve of viscous damper

2.3 The advantages of viscous damper

The viscous damper has following advantages:

(1)Viscous dampers in other fields already have a long history of application and have been proven to be of good feasibility and stability. So it is reliable to extend its application to civil structural buildings.

(2)For ordinary dampers, maintenance such as check or repair is often required. But for viscous dampers, such maintenance is rarely needed. The viscous dampers aren't damaged even after the low to medium seismic event and they are easily replaced while after severe seismic event.

(3)As a type of energy dissipating device, the energy dissipating process will cause the temperature to rise or decline. But the temperature change doesn't have much impact on its energy dissipating function.

(4)Viscous damper is a type of velocity - dependent damper. It doesn't add stiffness to the structure, therefore it only increases the structural damping ratio without reducing the vibrating periods, which effectively reduces the structural seismic response (displacement, velocity, and accelerations).

2.4 The arrangement pattern of viscous dampers

The viscous dampers are usually arranged as follows: Wall-embedded, Lateral-braced, K-braced.

3. Application of viscous dampers in this project

3.1 Project Introduction

This project, a three-story school building, is a reinforced concrete frame structure. The seismic fortification intensity of this area is 8 degrees (0.3g). According to the requirement of the province, where the project is in, the technics of earthquake isolation or energy dissipating shall be applied in this project. By weighing the pros and cons of different technologies, the viscous dampers are finally selected as the energy dissipation parts for this project.

The seismic design goals here are: 1) under the minor earthquake, the viscous dampers will start dissipating energy and increase the damping ratio of the structure, so that the structure itself will be in the elastic stage; 2) under the medium or severe earthquake, the viscous dampers will dissipate energy to a large extent so as to reduce the seismic response of the structure.

3.2 Arrangement of viscous dampers

The arrangement of viscous dampers in plan view is shown in Figure 3; their arrangement in sectional view is shown in Figure 4.

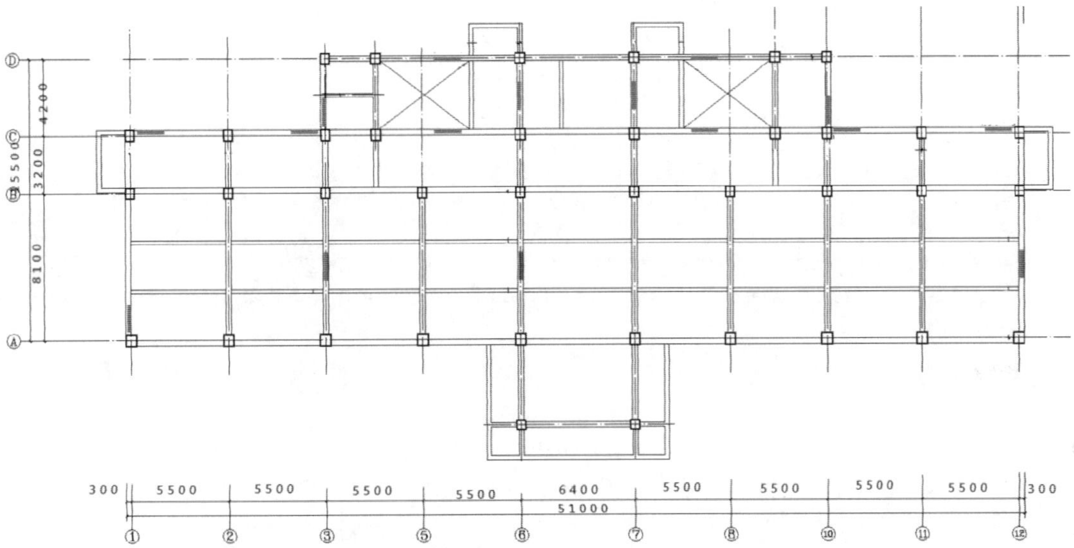

Figure 3. Arrangement of dampers in plan view (red part is damper, same for each story)

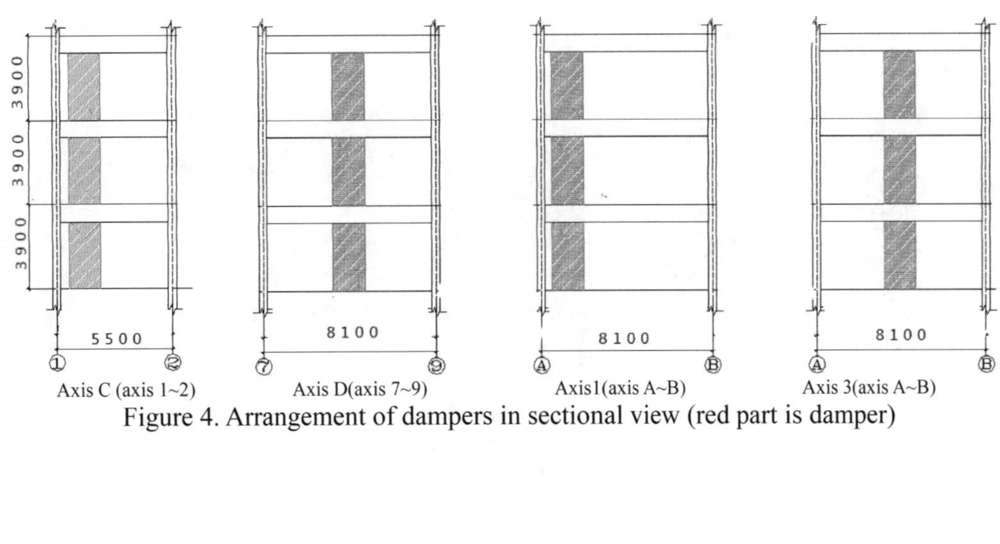

Axis C (axis 1~2) Axis D(axis 7~9) Axis1(axis A~B) Axis 3(axis A~B)

Figure 4. Arrangement of dampers in sectional view (red part is damper)

3.3 Connection between the viscous damper and surrounding structural members

The connection between the viscous damper and its surrounding structural members is shown in Figure 5.

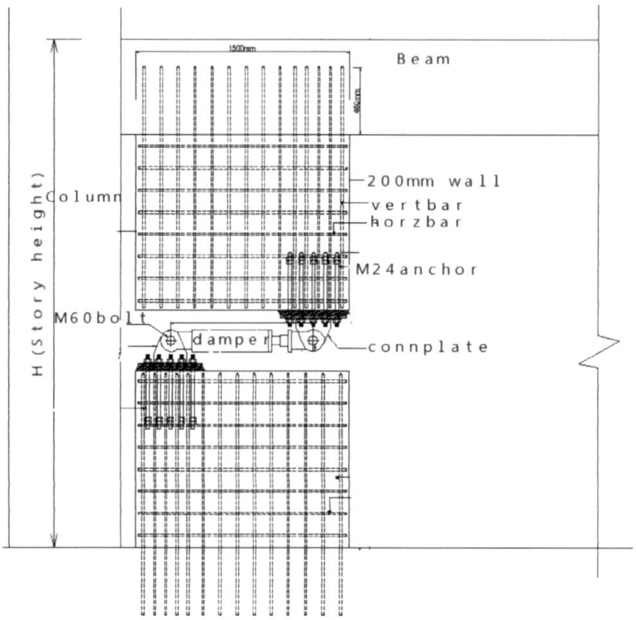

Figure 5. Connection Detail of Dampers

4. Parameters and earthquake wave input in nonlinear seismic analysis

The structural analysis software MIDAs/gen is applied to perform the nonlinear seismic response analysis in this project.

4.1 selection of seismic wave input

An artificial earthquake wave (RH3TG035) and two natural earthquake waves (TH079TG035 and TH082TG035) are selected in the structural analysis. The time history curves of three waves are shown in Figure 6. The characteristics of response spectra of three waves as shown in figure 7, with the design spectra drawn in the same figure.

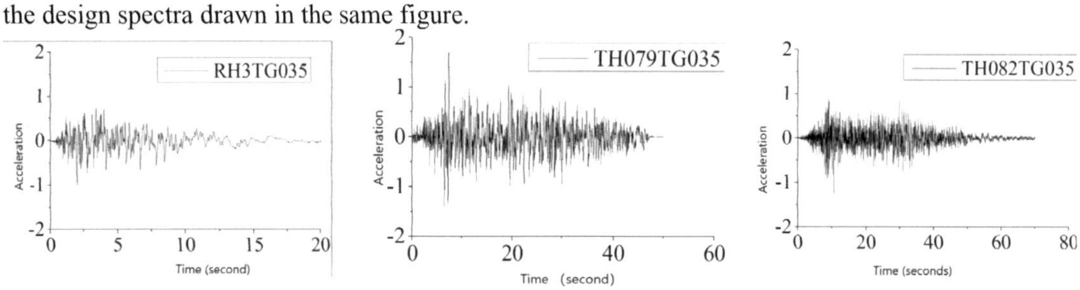

(a) RH3TG035 (b) TH079TG035 (c) TH082TG035

Figure 6. Three earthquake waves applied in the seismic analysis

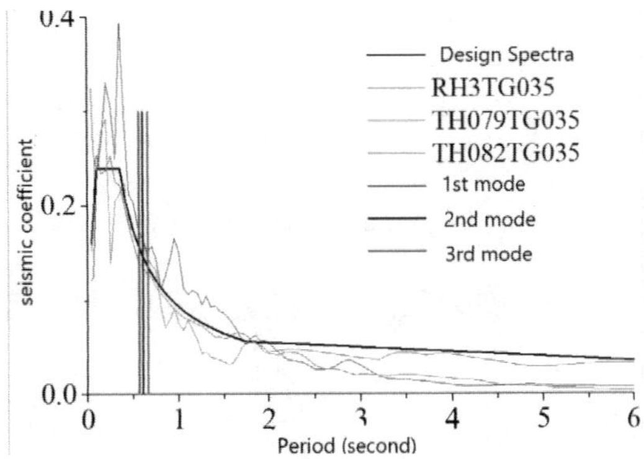

Figure 7 Seismic coefficient of input waves and design spectra

Table 1. Seismic coefficient on structure periodic points for different waves

Input	Seismic coefficient at 3rd mode（α）	difference (%)	Seismic coefficient at 2nd mode（α）	difference (%)	Seismic coefficient at 1st mode（α）	difference (%)
Design Spectra	0.135		0.147		0.156	
RH3TG035	0.127	5.93%	0.135	8.16%	0.142	8.97%
TH079TG035	0.152	12.59%	0.164	11.56%	0.174	11.54%
TH082TG035	0.144	6.67%	0.165	12.24%	0.164	5.13%

As shown in Figure 7 and Table 1, the difference between the response spectra of three seismic input waves and the design spectra is within 20% at the structural main vibration periods. Therefore, the selected three earthquake waves are acceptable according to the requirements in <Code for seismic design of building> GB50010-2010（2016 Edition）[5].

4.2 Constitutive model for component settings

(1)frame beam : M-φ (Moment-rotation angle) hinge model

The max deformation of beams occurs on both ends of beams under the given loads. The plastic deformation can be simulated as plastic hinge and the moment-rotation relationship at the beam end is characterized as M-φ (Moment-rotation angle) curve of plastic hinge. Here, Modified Takeda Trilinear model, as shown in Figure 8, is applied to simulate the M-φ relationship at beam ends. In this trilinear model, stiffness degeneration rather than strength degeneration is considered; the first turning point represents the concrete crack and the second turning point represents the member yielding.

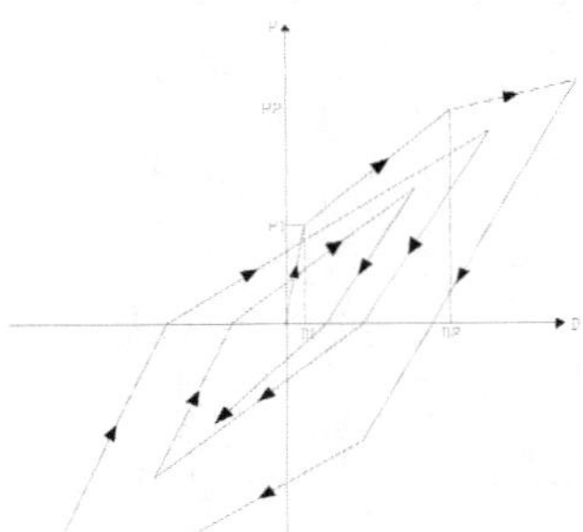

Figure 8. M-φ plastic hinge for beam

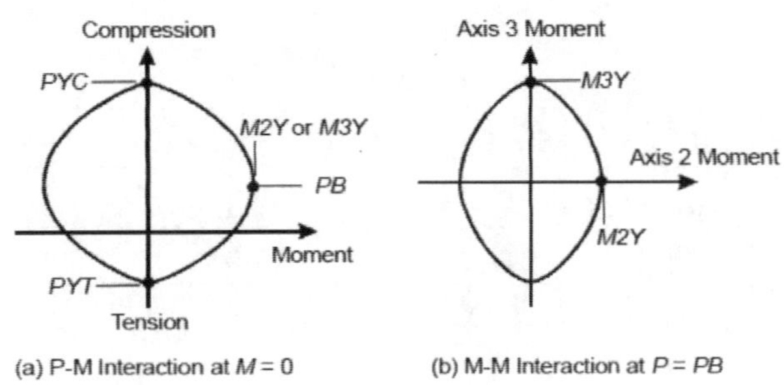

Figure 9. P-M1-M2 plastic hinge for column

(2)columns : P-M1-M2 axial force - bi - directional bending moment hinge model

The max deformation of columns occurs on both ends of columns under the given loads. The plastic deformation can be simulated as P-M1-M2 plastic hinge, where P represents axial loads, M1 and M2 represent moments in two directions. The model applied to simulate the PMM relationship is shown in Figure 9.

4.3 Analysis model of viscous damper
The viscous damper is simulated as a connection element (damper) in software MIDAs/gen, as shown in Figure 10.

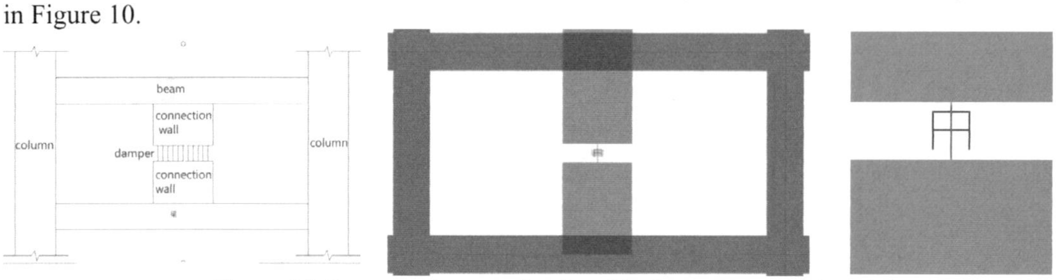

Figure 10. Damper element simulated in the structure FE model

4.4 Verification and validation of MIDAs Model
To verify and validate the structure model with MIDAs/Gen, another widely used software PKPM is applied to simulate the structure. The FE models of two software are shown in Figure 11.

To run the two FE models by inputting the minor earthquake response spectrum, the results are listed in Table 2. It can be seen that the differences of results from two models are insignificant. Therefore, the MIDAs/Gen model is deemed acceptable and is applied in the subsequent analysis.

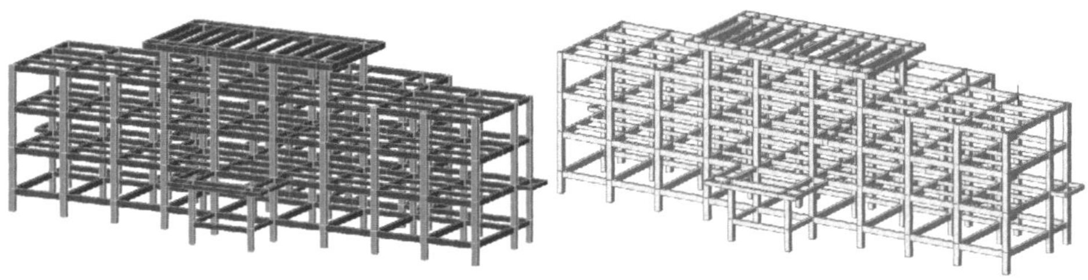

（1） PKPM model （2） MIDAs model
Figure 11. Structural FE model in PKPM and MIDAs

Table 2. Analysis results out of PKPM model and MIDAs model

Item		PKPM	MIDAs/GEN	Difference
Gross mass		3299.244	3326.156	0.82%
Period (s)	Mode 1	0.6672	0.6618	0.81%
	Mode 2	0.6074	0.6032	0.69%
	Mode 3	0.5679	0.5613	1.16%
Max story drift angle	X dir	1/ 637.	1/649	1.88%
	Y dir	1/ 597.	1/618	3.52%
Base Shear (kN)	X dir	3954.86	3970.5	0.40%
	Y dir	3601.09	3632.3	0.87%

5. Seismic analysis of structure
Here, software MIDAs/gen is applied to perform the dynamic time history analysis in two steps:

1. time history analysis in minor earthquake
2. time history analysis in severe earthquake

The structure model with viscous dampers is shown in Figure 12.

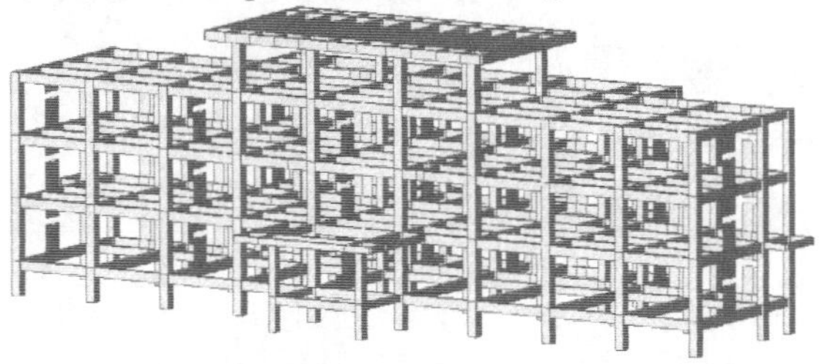

Figure 12. Structure FE model with Viscous Damper simulated in MIDAs

5.1 Dynamic time history analysis under minor earthquakes

The maximum base shear of structure from three earthquake waves are listed in Table 3. In the same table, the base shear from the response spectra analysis is also listed for X and Y directions. As compared in Table 3, the base shear from each wave is larger than 65% of what from response spectra analysis; and the average base shear from three waves is larger than 80% of what from response spectra analysis. Therefore, the results out of time history analysis can meet the minimum safety requirement according to the specifications in <Code for seismic design of building> GB50010-2010 （2016 Edition）[5].

Table 3. Base shear under minor earthquake (kN)

Item		RS input	RH3TG035	TH079TG035	TH082TG035	Average
X dir	Base shear	3971	3163	3876	3413	3484
	Ratio		79.66%	97.61%	85.96%	87.74%
Y dir	Base shear	3632	3095	3584	3122	3267
	Ratio		85.21%	98.67%	85.95%	89.94%

The maximum story drift angle in three earthquake waves are listed in Table 4. It can be seen that the max story drift angle from all three waves is 1/810, which is less than 1/550. 1/550 is the maximum requirement of story drift angle specified in <Code for seismic design of building> GB50010-2010（2016 Edition）[5].

Table 4. Maximum story drift angle (1/rad)

	X dir	Ydir
RH3TG035 wave	991	1056
TH079TG035 wave	810	847
TH082TG035 wave	929	1013
Enveloped	810	847

The sum of forces in the viscous dampers in each story are listed in Table 5. The total shear of each story is also listed in the same table. By comparison, it can be seen that the sum of forces in the dampers are about 1/3 of the total shear of the story. The energy dissipating curve of the viscous damper in the minor earthquake wave is shown in Figure 13.

Table 5. Force sum of dampers and the story shear under minor earthquake (kN)

Earthquake wave	story	X dir		Y dir	
		Shear force	damper forces	Shear force	damper forces
RH3TG035	3	1354.7	562.23	1297.4	712.53

	2	2279.8	804.62	2121.2	964.94
	1	2994.7	1087.84	2928.4	1154.82
TH079TG035	3	1842.6	752.57	1858.8	924.10
	2	2932	987.62	2893.3	1100.33
	1	3747.2	1198.48	3519.4	1245.59
TH082TG035	3	1487.6	623.27	1415.3	755.33
	2	2494	878.55	2277.7	980.05
	1	3263.2	1111.17	3024.2	1142.00

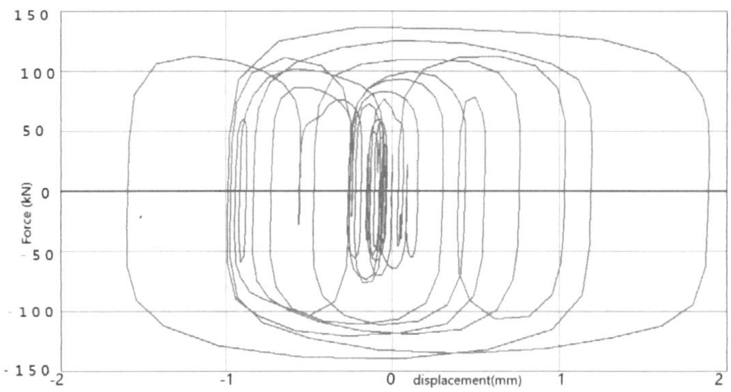

Figure 13. Energy dissipation curves under minor earthquake input

5.2 Time history analysis under severe earthquake

Load combination of 1.0DL+0.5LL+1.0E is applied for time history analysis under severe earthquake input. Here, DL stands for dead load, LL stands for live load and E stands for earthquake load.

5.2.1 Story drift angle and the story shear force

The maximum story drift angle and base shear force are listed in Table 6.

Table 6. Maximum story drift angle and the base shear force

Item		RH3TG035	TH079TG035	TH082TG035	Enveloped value
Max drift angle	X dir	1/184	1/131	1/178	1/131
	Ydir	1/192	1/144	1/186	1/144
Base shear （kN）	X dir	9210.4	10645	9119.8	10645
	Ydir	8094	8788.3	7631.5	8788.3

From the data in Table 6, it can be seen that:

(1)The story drift angle under wave TH079TG03 is the greatest, followed by wave RH3TG035 and then by wave TH082TG035.

(2) The greatest story drift angle under each wave is 1/131 and is less than 1/50, which is the required maximum story drift angle in <Code for seismic design of building> GB50010-2010（2016 Edition）[5].

(3) The structure stands stably even when the large deformation effect and the 2nd order gravity effect are both considered in the analysis. That means the goal of "no collapse in severe earthquake" is obtained.

5.2.2 Damage of plastic hinges on the member ends

Under the seismic input in both X direction and Y direction, the damages of the members are as follows:

Plastic hinges occur at the ends of most of beams and at the ends of a small number of columns. The failure mode appears to be strong column and weak beams, which is the preferred by the design requirements as specified in <Code for seismic design of building> GB50010-2010（2016 Edition）[5].

The sum of forces in the viscous dampers in each story are listed in Table 7. The total shear of each story is also listed in the same table. By comparison, it can be seen that the sum of forces in the dampers are about 1/3 of the total shear of the story. The typical energy dissipating curve of the viscous damper in severe earthquake wave is shown in Figure 14. The viscous dampers show good capability of energy dissipating.

Table 7. Force sum of dampers and the story shear under severe earthquake (kN)

Earthquake wave	story	X dir		Y dir	
		Shear force	damper forces	Shear force	damper forces
RH3TG035	3	3891.3	1407.23	3471.3	1610.54
	2	6291.6	1555.76	5751.9	1655.25
	1	8711.9	1710.75	7777.7	1685.14
TH079TG035	3	5626.5	1599.28	5416.4	1802.49
	2	8968.1	1783.85	8031.8	1955.96
	1	10795	1964.96	9082.2	1980.56
TH082TG035	3	4401.6	1435.18	3790.2	1589.15
	2	7310.2	1669.20	6007.1	1721.58
	1	8899.5	1814.84	7139.4	1778.08

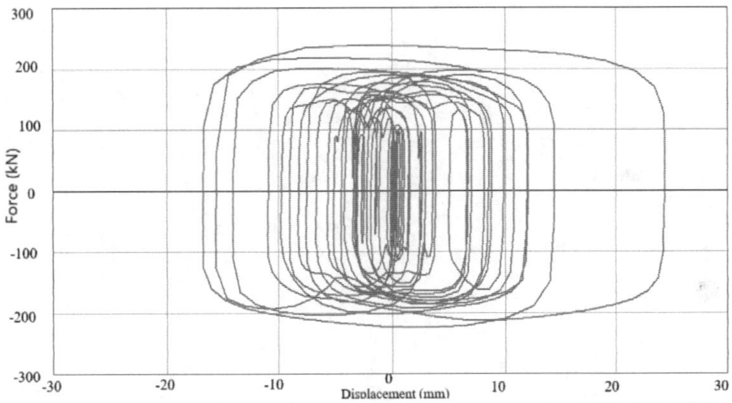

Figure 14. Hysteretic energy dissipation curve of damper under TH079TG035 wave

6. Conclusion

From the structure analysis results, the following conclusions could be drawn :

1) Under the minor earthquake input, the viscous dampers start to dissipate energy and the structure is in the elastic stage.

2) Under the severe earthquake input, the viscous dampers continue to dissipate energy to a greater extent, which reduces the seismic response of structures. Most beams and a small number of columns gradually come to yield; the failure mode of structure is strong column and weak beam.

3) The viscous dampers are a type of feasible device to reduce the seismic response in the minor or severe seismic event.

To better understand the function of viscous dampers, more investigation is expected on their application on high-rise buildings or more complicated structured buildings such as steel-concrete Composite Structures.

Acknowledgments

This project is financially supported by Shanghai Urban Construction College Project cjky202110.

Reference

[1] Feng,H.,Zhou,F.,Ge,H,Zhu,H.,Zhou,L.(2021) Energy dissipation enhancement through multi-toggle brace damper systems for mitigating dynamic responses of structures. J. Structures, Volume 33: 2487-2499

[2] Zhou,Y., Liu,X.,Wang,M.(2021) Experimental Study on Mechanical Properties of Two Types of Viscoelastic Dampers, J. Engineering Mechanics, Vol.38 Suppl:167-177

[3] Wang,C., Shen,T., Wei,J., Pan,Q.(2021), Energy-dissipated design of an office building with viscous dampers, J. Shanxi Architecture, Vol 47(12):43-45

[4] Ye, Z., Li,A., Chen,W.(2001), Study on vibration energy dissipation design of structures with fluid viscous dampers. J. Journal of Building Structures, Vol.22(4):61-66

[5] Code for seismic design of buildings［S］ (GB 50011—2010) . Beijing: China, Architecture & Building Press，2010． (In Chinese)

Journal of Physics: Conference Series **2011** (2021) 012027 doi:10.1088/1742-6596/2011/1/012027

Flexural Behaviours of CFRP Sheets Reinforced Steel-Concrete Composite Beams

Lei Tang[1,*], Dong Wei[2], Zhao Gao[2]

[1]School of Civil Engineering, Qingdao University of Technology, Qingdao 266033, China

[2]Shandong Hi-Speed Engineering Construction Croup Co. Ltd., Jinan 250014, China

[*]Corresponding author's e-mail: tang.lei2018@outlook.com

Abstract. The finite element simulation was performed for 8 carbon fiber reinforced plastic (CFRP) sheets reinforced steel-concrete composite beams via ABAQUS software, in an effort to further investigate their mechanical properties. Next, the influences of shear connection degree and elasticity modulus, width and constitutive relation of CFRP sheets on the mechanical properties of steel-concrete composite beams were studied. Subsequently, their load-dependent change laws of midspan deflection were reported in detail. The simulation results show that the reinforcement effect on the negative moment region of each composite beam is improved with the increase in the elasticity modulus, width of CFRP sheets and the shear connection degree.

1. Introduction

The form of continuous multi-span beams is usually adopted to enhance the integrity of building structures and bridge structures [1]. However, in the negative moment region of multi-span continuous composite beam, the concrete slabs are susceptible to cracking because the tensioning force and steel beams will easily experience local buckling under compressive force, which leads to the degradation of structural bearing capacity and durability [2-3]. In comparison to the replacement of existing composite beams, a method featured by lower cost and shorter time consumption is to reinforce the damaged composite beams. Therefore, carbon fiber reinforced plastic (CFRP) has been added as an external reinforcement material onto the surface of concrete slabs in the negative moment region of composite beams, thus improving the structural bearing capacity. Many experimental studies show that CFRP sheets integrate the merits of light weight, high tensile strength, corrosion resistance, convenient construction, etc. [4].

It has been proven by an experiment on full-size steel-concrete composite beams [5] that the ultimate bearing capacity of composite beams can be strengthened by bonding the CFRP sheets with epoxy resin glue. Al-Saidy et al. [6] used CFRP sheets to reinforce the web and lower flange of steel beams in composite beams, and the results showed that the ultimate bearing capacity of CFRP sheets-reinforced specimens was improved by 45%, and the deflection upon the failure was much lower than that of unreinforced composite beam specimens. El-Zohairy et al. [7] conducted CFRP sheet reinforcement test in negative moment regions of 6 composite beams, performed a finite element analysis, and explored the influence of shear connection degree on the ultimate bearing capacity of composite beams. The results manifested that the optimal performance of CFRP sheet reinforcement could be achieved when the shear connection degree between steel beam and concrete slab reached

80%. In addition, both theoretical analysis and experimental results of composite beams indicate that the interface is probe to horizonal slippage if partial shear connection is adopted, which will result in the degradation of ultimate bearing capacity and rigidity of composite beams and the increase of their deformations. Therefore, the mechanical properties of negative moment region in CFRP sheets reinforced steel-concrete composite beams were probed in this study.

2. Finite Element Model

2.1 Modelling

The composite beam specimens were designed into the same size, as shown in Figure 1. The total length and height of each specimen were 3,000 mm and 344 mm, respectively. The width and height of concrete slab were 500 mm and 100 mm, respectively. Two-layer (upper layer and lower layer) longitudinal tensile reinforcements with diameter of 10 mm were arranged in the concrete slab to resist against the tensile stress in the concrete. H-shaped steel sections (dimensions: 244×175×7×11 mm) were selected for the steel beams. In order to simulate the influence of CFRP width on the mechanical properties, the spacings of two-row studs with the diameter of 13 mm and height of 80 mm were set as 140, 300 and 350 mm, respectively. The pasting length of CFRP sheets was 2,400 mm. The CFRP sheets with width of 175 and 500 mm were selected to analyse the influence of CFRP width on the mechanical properties. The composite beams were reinforced using the CFRP sheets with elasticity modulus of 245 and 400 GPa, aiming to analyse the influence of their elasticity modulus on the mechanical properties of composite beams.

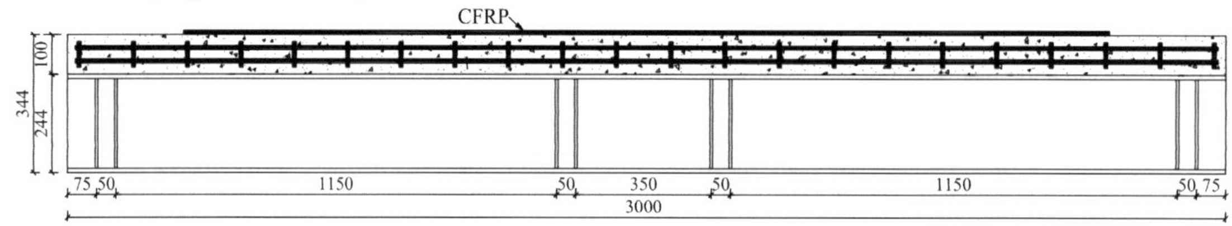

Figure 1: Elevation of Model (unit: mm)

2.2 Constitutive models

The Concrete Damaged Plasticity model was used as the constitutive model of C50 concrete. The stress-strain curves of the concrete under tensile state and compressive state are as shown in Figure 2. The three-broken-line model was taken as the constitutive model of Q355 H-shaped steel, as shown in Figure 3 (a), while double-broken-line model was used for HRB400 ribbed bars and studs, as shown in Figure 3 (b). The data provided by Liu [8] were input into the constitutive models of steel and concrete. The constitutive models of CFRP sheets were elastic material models, namely, Orthotropic model [9-10] and Lamina model [11].

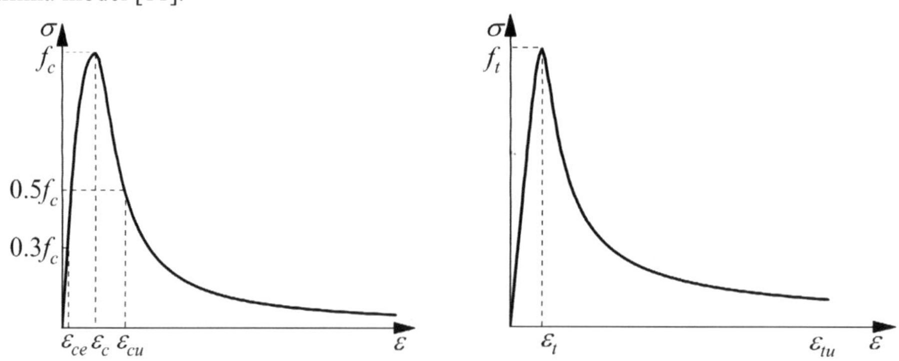

(a) Compressive State (b) Tensile State

Figure 2: Stress-Strain Curves of Concrete

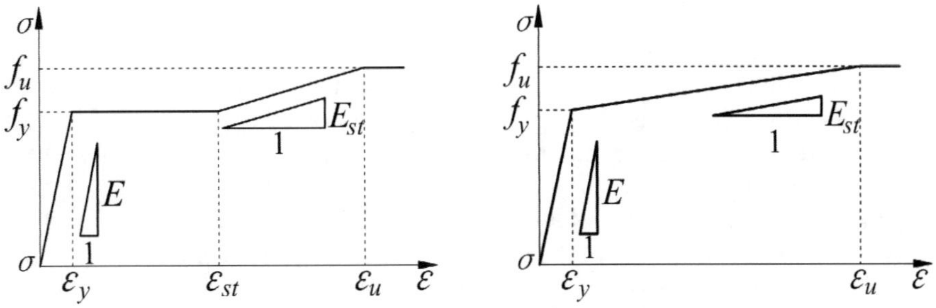

(a) Three-Broken-Line Model (b) Double-Broken-Line Model

Figure 3: Stress-Strain Curves of Steel

2.3 Interaction

The top flange of steel beam served as the master plane and lower surface of concrete slab as the slave plane, and their interactive relationship was defined by the designated contact pairs in general contact. Penalty was used as the tangential behaviour, and the multinomial coefficient of friction was taken as 0.2. Hard contact was used as the normal behaviour to prevent the mutual penetration of contact surface in case of deformation.

Tie constraints were used between steel beam and backing plate and between concrete and backing plate in the finite element model of steel-concrete composite beam, in order to ensure their synchronous deformation. Kinematic Coupling constraint was applied to the steel backing plates at loading points and steel beam parts, and the motion at coupling node was restricted to rigid body motion of loading point. Reinforcement layer and stud were respectively embedded into the concrete slab. Meanwhile, the under surface of studs was tied together with the top face of steel beam.

2.4 Element types and boundary conditions

S4R shell elements were used to simulate the web and flange of steel beam and CFRP sheets. The S4R contained 4 nodes, being insensitive to the shear self-locking effect. The concrete, studs and steel backing plates were simulated using C3D8R solid elements, which included 8 nodes with 1 integral point in the element centre. Under complicated stress state, C3D8R elements could not be so easily affected as to be locked. The adverse effect on calculation accuracy could be mitigated only by dividing the C3D8R elements into at least four layers along the thickness direction. The T3D2 bar elements that did not bear shear force or bending moment were selected to simulate the tensile reinforcements. The T3D2 elements contained two nodes and one integral point, with the plastic and large strain functions. The boundary conditions of simply supported beam were applied to the steel backing plate tied to the left and right ends of concrete slab, as shown in Figure 4. The displacement boundary conditions were applied to two reference points.

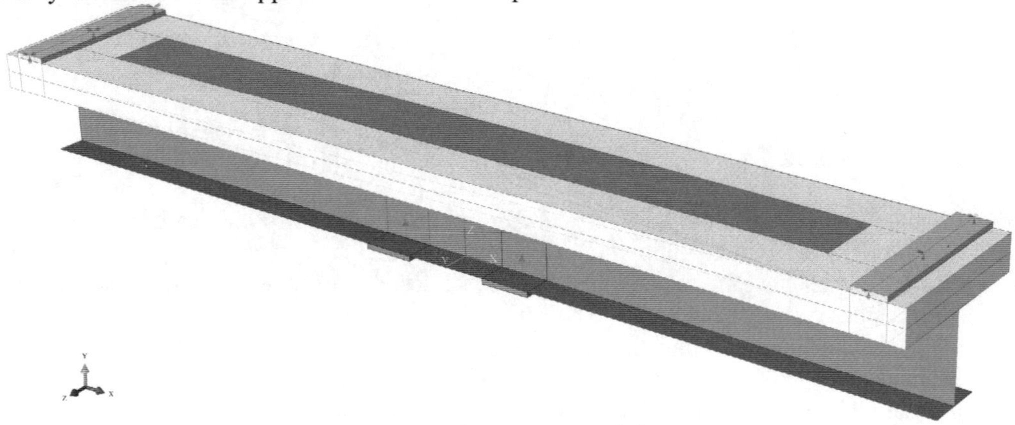

Figure 4: Boundary Conditions

3. Analysis results and discussion

The load-deflection curves of different steel-concrete composite beams are as shown in Figure 5. The finite element simulation results of three reinforced composite beam specimens with different constitutive models of CFRP sheets showed that the ultimate bearing capacity of each steel-concrete composite beam was strengthened thanks to the use of CFRP sheets, and the bearing capacity was affected a little by the constitutive model of CFRP sheets. According to the finite element simulation results, the CFRP sheets started exerting their effect only after the composite beams yielded, while bearing a very small tensile stress before the yielding of composite beams.

By analyzing the load-deflection curves of three composite beams with different elasticity moduli of CFRP sheets, it could be known that the ultimate bearing capacity of each composite beam was enhanced with the increase in the elasticity modulus of CFRP sheets. According to the finite element models of three composite beams with different widths of CFRP sheets, the width of CFRP sheets had a great impact on the ultimate bearing capacity of composite beams, and this influence was gradually enhanced with the increase in the width of CFRP sheets, but the width of CFRP sheets had a minor influence on the rigidity of composite beams.

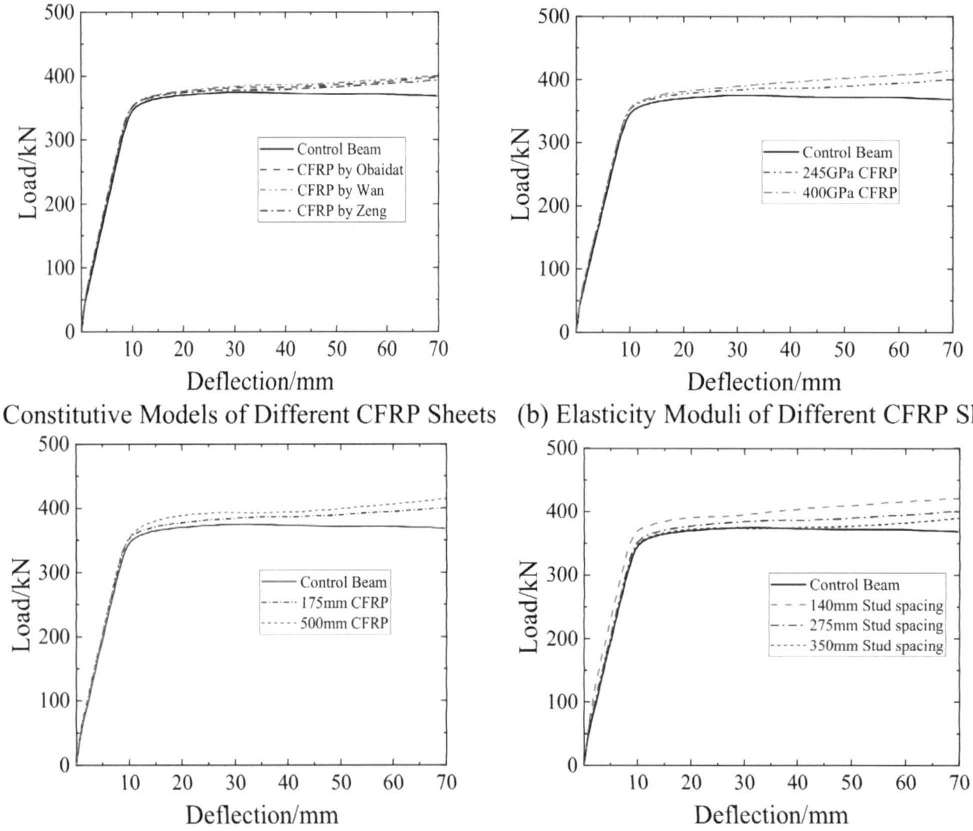

(a) Constitutive Models of Different CFRP Sheets (b) Elasticity Moduli of Different CFRP Sheets

(c) Widths of Different CFRP Sheets (d) Different Stud Spacings

Figure 5: Load-Deflection Curves

It was found through analyzing the finite element simulation results of 4 composite beams that the shear connection degree was an important factor influencing their flexural rigidity and ultimate bearing capacity. Under the same number of CFRP sheet pasting layers, the ultimate bearing capacity of composite beams was enhanced with the reduction of stud spacing, so was the flexural rigidity. By comparing the composite beam and the control beam with stud spacing of 350 mm, it could be discovered that the CFRP sheets started functioning only after the composite beam reached the yield state for a certain while. When the stud spacing was small, the CFRP sheets started fuctioning as early as the initial loading phase, and they significantly enhanced the flexural rigidity of composite beams.

Therefore, the tensile characteristics of CFRP sheets can be exerted to the greatest extent if they are applied to steel-concrete composite beams with a high shear connection degree in practical engineering.

4. Conclusions

The short-term structural mechanical properties in the negative moment regions of 8 CFRP sheets reinforced steel-concrete composite beams were explored in this study. The following conclusions were drawn according to the finite element simulation results:

(1) The ultimate flexural bearing capacity of midspan section can be enhanced by using the CFRP sheets to reinforce the negative movement region of steel-concrete composite beams. The CFRP sheets start functioning only after the composite beams yield, while bearing a small tensile stress before the yielding.

(2) The reinforcement effect on the negative moment region of steel-concrete composite beams is associated with the elasticity modulus and width of CFRP sheets as well as the shear connection degree. The ultimate bearing capacity is elevated with the increase in the elasticity modulus and width of CFRP sheets and the shear connection degree.

(3) The mechanical properties of negative moment region can be evidently improved by using two-layer CFRP sheets to reinforce the composite beams with a high shear connection degree. The higher the shear connection degree, the more the CFRP sheets exert their strength, and the more obvious the improvement effect on the bearing capacity and rigidity of composite beams.

References

[1] Nie, J.G., Fan, J.S., Cai, C.S. (2004) Stiffness and Deflection of Steel–Concrete Composite Beams under Negative Bending. J. Struct. Eng., 130: 1842-1851.

[2] Ryu, H.K., Chang, S.P., Kim, Y.J., Kim, B.S. (2005) Crack Control of A Steel and Concrete Composite Plate Girder With Prefabricated Slabs Under Hogging Moments. Eng. Struct., 27: 1613-1624.

[3] Su, Q.T., Yang, G.T., Wu, C. (2012) Experimental Investigation On Inelastic Behavior of Composite Box Girder Under Negative Moment. Int. J. Steel Struct., 12: 71-84.

[4] Afefy, H.M., Sennah, K., Akhlagh-Neja, H. (2016) Experimental and Analytical Investigations On The Flexural Behavior of CFRP Strengthened Composite Girders. J. Constr. Steel. Res., 120: 94-105.

[5] Lin, W.W., Yoda, T., Taniguchi, N. (2013) Fatigue Tests On Straight Steel–Concrete Composite Beams Subjected To Hogging Moment. J. Constr. Steel. Res., 80: 42-56.

[6] Al-Saidy, A.H., Klaiber, F.W., Wipf, T.J. (2007) Strengthening of Steel–Concrete Composite Girders Using Carbon Fiber Reinforced Polymer Plates. Constr. Build. Mater., 21: 295-302.

[7] El-Zohairy, A., Salim, H., Shaaban, H., Mustafa, S., El-Shihy, A. (2017) Experimental and FE Parametric Study On Continuous Steel-Concrete Composite Beams Strengthened With CFRP Laminates Constr. Build. Mater., 127: 885-898.

[8] Liu, Y., Tong, L.W., Sun, B., et al. (2014) FEA and Bending Capacity Calculation for Mechanical Behavior of Steel-Concrete Composite Beams Under Negative Bending. J. Build. Struct., 35: 10-20.

[9] Obaidat, Y.T., Heyden, S., Dahlblom, O., (2010) The Effect of CFRP and CFRP/Concrete Interface Models when Modelling Retrofitted RC Beams with FEM. Compos. Struct., 92: 1391-1398.

[10] Zeng, J.J., Gao, W.Y., Liu, F. (2018) Interfacial Behavior and Debonding Failures of Full-Scale CFRP-Strengthened H-Section Steel Beams. Compos. Struct., 201: 540-552.

[11] Wan, S.C., Huang, Q., Guan, J., Guo, Z.Y. (2018) Strengthening of Continuous Steel-Concrete Composite Beams in Negative Moment Region Using Prestressed Carbon Fiber-Reinforced Polymer Plates. J Jilin Univ., 48: 1114-1123.

Stability Analysis of Unsaturated Red Clay Slope under Different Heavy Precipitation Patterns

Dengkai Chen[1], Jiuqiang Zhong[2], Bitang Zhu[2] and Yalong Jiang[2,*]

[1] Nanchang Rail Transit Group Co.Ltd., 330038 Nanchang Jiangxi Province, China

[2] School of Civil Engineering and Architecture, East China Jiaotong University, 330013 Nanchang Jiangxi Province, China

*Corresponding author: yalongjiang@whu.edu.cn

Abstract. A fluid-solid coupling numerical model was established for unsaturated red clay, which is prone to causing clustered landslides under three typical heavy precipitations in southwest China, to investigate the influence of the precipitation patterns on the infiltration mode, pore water pressure and stability of unsaturated red clay slopes. The investigation showed that (1) during heavy precipitation, much rainwater tends to form a transient saturated zone in the shallow layers of the slopes in a short time, thereby reducing the matric suction of the soil matrix; (2) under short-term continuous rainstorm, the pore water pressure dropped to the greatest extent; and (3) the safety factor experienced the largest decrease under persistent heavy precipitation following 24-hour rainfall, while slopes displaced the most under sudden rainstorm.

1. Introduction

In China, where geological disasters occur frequently, extreme precipitation is subject to increasing intensity and dominates in the total precipitation. In 2019, the country witnessed a total of 6,181 geological disasters, of which 4,220 were landslides, accounting for 68.3% of the total. Southwest China is the most severely afflicted area of landslide disasters due to its complex terrain, mountainous and hilly land, abundant precipitation and concentrated distribution of red clay [1]. Rainstorm is among the key factors behind landslide disasters which are prevalent under short-term extreme precipitation. According to the geological disaster investigation in 290 Chinese counties and cities, rainfall-induced landslides account for 90% of the total landslides [2]. Precipitation during the flood season, which lasts from May to September under the monsoon climate and the El Nino-Southern Oscillation, makes up more than 70% of the total rainfall, including 20% in 1-2 weeks [3]. Such short-term heavy precipitation infiltrates rainwater into side slopes, and leads to a rapid decline in matric suction, the disappearance of negative pore water pressure in the unsaturated zone and a decrease in shear strength, hence a landslide. Thus, given the same amount of precipitation, the precipitation patterns play a critical role in precipitation infiltration [4], and the occurrence of landslide is closely associated with the precipitation pattern and intensity.

Red clay in Southwest China, generally characterized by high swelling, high shrinkage, high engineering sensitivity and poor engineering properties, often cause landslide clusters. In dry condition, red clay slopes present slope fracture development, high soil strength, and great slope stability due to the effects of drying and wetting cycles; In water-rich condition, the slopes are prone to closed fractures, markedly decreased soil strength, deformation and instability [5].

Studies have been conducted on slope stability under different precipitation patterns. Tsai [6] modeled one-dimensional infiltration of water in soils to study the influence of precipitation patterns on shallow landslide and slope failure; Kim et al. [7] strongly linked slope failure to transient seepage of unsaturated soil slopes induced under different precipitation patterns; Jaesung et al. [8] evaluated the stability of an agricultural embankment under five precipitation patterns; And Ng et al. [9] categorized the precipitation profiles into four groups: uniform rainfall, post-peak rainfall, pre-peak rainfall and median rainfall.

These studies, however, focus mostly on stability analysis of homogeneous soil slopes under common rainfall types, leaving aside the seepage and stability of unsaturated red clay slopes in Southwest China under different heavy precipitation patterns. In such case, this paper was designed to summarize three heavy precipitation patterns causing large-scale landslides in Southwest China, fit several measured soil-water characteristic curves of red clay using MATLAB plotting, and establish a two-dimensional soil slope model by ABAQUS. Then, the infiltration and stability of unsaturated red clay slopes were explored under the heavy precipitation patterns, considering the water-force coupling.

2. Theory and numerical calculation

2.1. Seepage equations of unsaturated soils

Seepage equations of unsaturated soils include the soil-water characteristic curve (SWCC) and permeability coefficient equations, which represent the changes in volume water content-matrix suction and permeability coefficient-matrix suction respectively. To characterize unsaturated red clay soils in Southwest China, the measured data of the soils [10] were put into the Variance Gamma (Vg) Model showing high applicability of unsaturated clay, and the five parameters were worked out by MATLAB nonlinear fitting function Lsqcurvefit to fit the SWCC. Subsequently, the obtained parameters were substituted into Equations (2) and (3) to obtain the solution of the equation between the moisture absorption curve and the permeability curve. Finally, the solution was input into the material definition module of ABAQUS to assign seepage properties to the slope for fluid-solid coupling analysis.

SWCC in VG model was expressed as:

$$\theta=\theta_r+\frac{\theta_s-\theta_r}{[1+(\alpha\varphi)^n]^m} \tag{1}$$

Relationship between saturation and matric suction is:

$$S=\left[\frac{1}{1+(\alpha\varphi)^n}\right]^m \tag{2}$$

Relationship between permeability coefficient and saturation is:

$$K_{vg}=K_s S^{\frac{1}{2}}\left[1-(1-S^{1-m})^m\right]^2 \tag{3}$$

Where θ is volume water content of soil;

φ *is* matrix suction;

θ_s is soil water content in saturated volume;

θ_r is soil water content in residual volume;

S is saturation;

K_{vg} is permeability coefficient of unsaturated soil;

K_s is saturated permeability coefficient; and

α, n and m are fitting parameters, among which $m = 1-1/n$.

The parameters of the VG model were shown in Table 1. The moisture absorption curve (a) and permeability curve (b) were illustrated in Figure 1.

Table 1. VG Model Calculation Parameters

θ_s	θ_r	Ks(cm/s)	α	n
0.5385	0.2628	6.77×10^{-5}	0.1640	1.5444

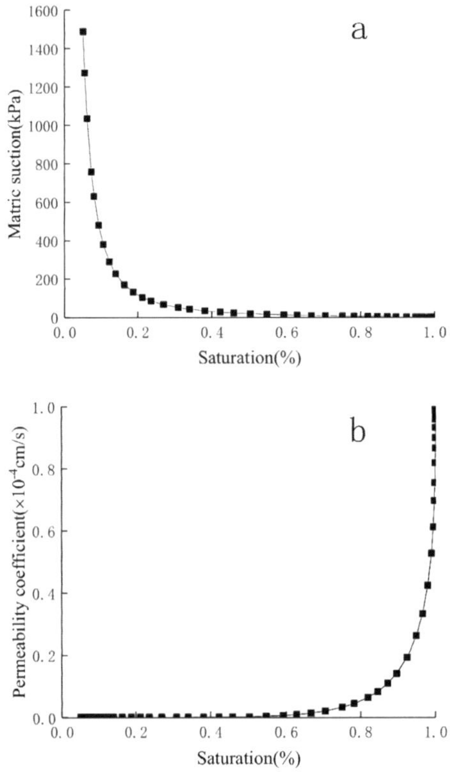

Fig. 1. Unsaturated Red Clay Seepage Control Equation
a. moisture absorption curve b. permeability curve

2.2 Finite element strength reduction technique

The finite element strength reduction method is in essence an approach to reduce shear strength indexes, i.e. cohesion and friction angle, by the Shear Strength Reduction Factor (SSRF), so as to lower the shear strength of soil. By definition, the SSRF is the ratio of the maximum soil shear strength of a slope to the actual shear stress when an external load applied to the slope remains constant.

In case all parts of the slope share the shear strength, the shear stress from the external load is equal to the actual shear strength after reduction when the slope is in limit equilibrium, and the value of SSRF (*Fr*) is taken as the safety factor of the slope (*Fs*).

By defining *Fr* in the Property module, the cohesion and friction angle reduced with the change of field variables were obtained. Then *Fr* is increased until the Property module did not converge, and the *Fr* value was the value of *Fs*.

2.3 Finite element model and parameters

A two-dimensional slope model was constructed in ABAQUS to study the influence of heavy precipitation patterns on pore water pressure distribution of unsaturated red clay slopes and the change of slope stability during precipitation infiltration.

In the finite element model (Figure 2), AB, BC and CD were precipitation boundaries with the boundary function expressed by precipitation intensity q, and the slope angle is β. Here, rainfall fell vertically to the soil surface on AB and CD, the infiltration intensity on the slopes was 0.025 m/h, equal to the precipitation intensity, and the infiltration rate I of Slope BC shares the figure (0.018 m/h) with the normal component of precipitation intensity ($I = q \cos\beta$).

AG and DE fix their horizontal displacements to 0, while GE fix both its horizontal and vertical displacement to 0. With initial groundwater level at the slope toe, the hydrostatic pore pressure of DE increased linearly with depth, i.e. 10×(10-y), the pore pressure of CD was 0, and the remaining boundaries were undrained. Physical and mechanical parameters were shown in Table 2.

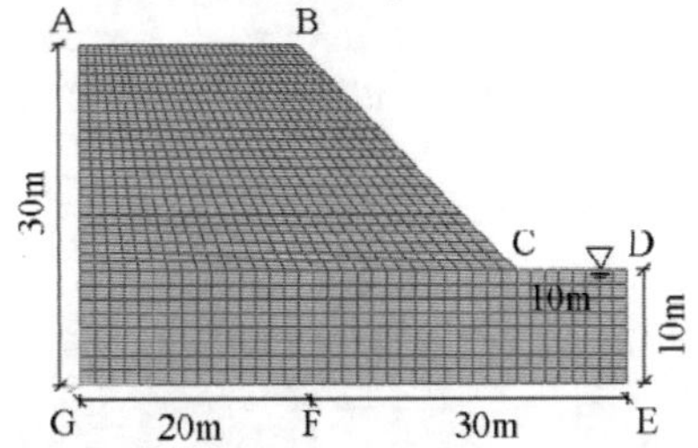

Fig. 2. Unsaturated Red Clay Slope Finite Element Model

Table 2. Physical and mechanical parameters of red clay [10]

Dry density ρ_0/ (g·m^{-3})	Cohesion c/kPa	Friction angle φ/(°)	Initial void ratio e_0	Elastic modulus E/Mpa	Poisson's ratio μ
1.30	30	15	0.84	10	0.3

3. Parameter analysis

3.1. Influence of heavy precipitation patterns on pore water pressure distribution on slope

Based on the characteristics of rainfall intensity in the heavy precipitation-concentration period from May to September in Bazhong, Sichuan Province, three typical heavy precipitation patterns: persistent heavy precipitation, sudden rainstorm and continuous rainstorm were selected for subsequent analysis [11]. Combined with the features of precipitation time-history classification in China, the three patterns had their time-history changes briefed in Figure 4, with the same total precipitation of 480 mm. Afterwards, an analysis was performed on how these patterns affected the distribution of pore water pressure along the depth direction of the slope within 24 hours.

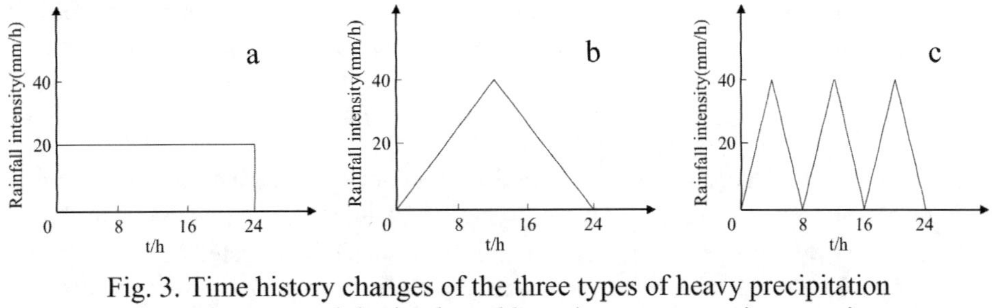

Fig. 3. Time history changes of the three types of heavy precipitation
a. persistent heavy precipitation b. sudden rainstorm c. continuous rainstorm

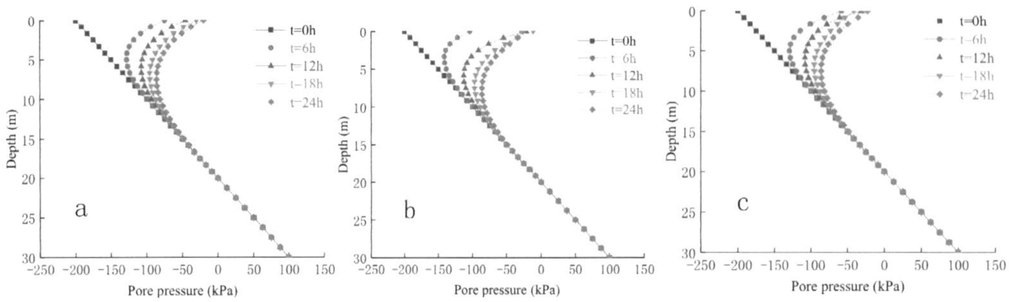

Fig. 4. Pore-water pressure profiles
a. persistent heavy precipitation b. sudden rainstorm c. continuous rainstorm

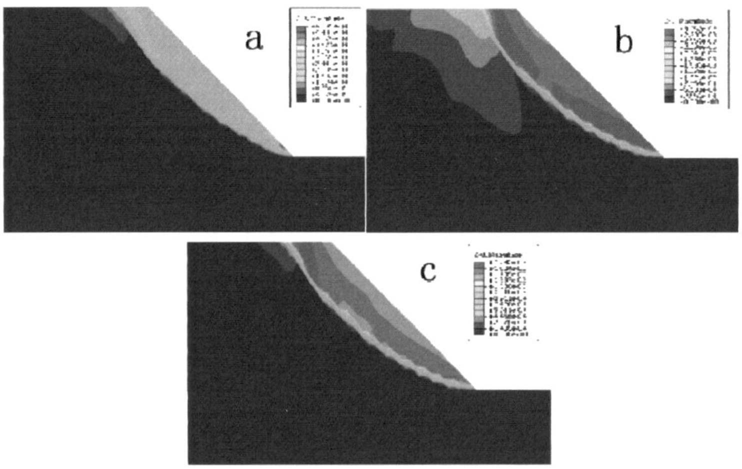

Fig. 5. Displacement increment diagram of soil slope under three kinds of heavy precipitation
a. persistent heavy precipitation b. sudden rainstorm c. continuous rainstorm

Figure 5 illustrated that under the three patterns, the pore water pressure resumed its linear change at a depth of 20 m, which verified the condition that groundwater is below the slope toe. Without precipitation, the pore water pressure presented a linear distribution under hydrostatic pressure. In the early stage of precipitation, when t = 6 h, the pore water pressure at the slope crest rose the fastest under continuous rainstorm, from -200kPa to -58kPa, similar to that under persistent heavy precipitation but much higher than that under sudden rainstorm. As time went on, the pore water pressure under persistent heavy precipitation and continuous rainstorm changed less, while that under sudden rainstorm varied greatly.

As for main reasons, persistent heavy precipitation and continuous rainstorm brought large amount of precipitation when t < 6h, at which time the precipitation intensity was larger than the infiltration rate of red clay with poor soil permeability. As a result, a transient saturated zone was formed in the shallow layer of the slope in a short time, the negative pore water pressure gradually disappeared, and the matrix suction approximated zero. The early stage of the sudden rainstorm, on the other hand, was characterized by low rainfall and less rainfall intensity than the infiltration rate of red clay, and the wetting front moved downward through the soil.

3.2. Influence of heavy precipitation patterns on displacement increment

The location of a potential slip surface was determined using the displacement increment nephograms of unsaturated red clay slopes under the three heavy precipitation patterns (Figure 5). The potential slip surface was arc-shaped, and it was the least deep under persistent heavy precipitation, while the largest displacement increment with the largest change range under sudden rainstorm, mainly in the middle and upper part of the slope. This revealed that persistent heavy rainfall is most likely to cause shallow landslides, whilst sudden rainstorm is prone to triggering deep large-scale landslides.

This is mainly because the instantaneous rainfall of the early persistent heavy precipitation far topped the infiltration capacity of the soil. At this juncture, the permeability coefficient changed little, but the matric suction of the shallow layer decreased rapidly, so the rain could induce a shallow landslide although it failed to penetrate deep into the slope. Under sudden rainstorm, despite low rainfall in the early stage, the rainwater was infiltrated into a wide range of soil, which was prone to triggering plastic strain in the depth of the slope and thus deep landslide.

3.3. Influence of heavy precipitation patterns on slope stability

ABAQUS was used to process Combine Function provided by patterns. The evaluation standard was that the safety factor of slope crest did not converge with horizontal displacement. The curve of the safety factor is obtained and shown in Figure 6.

As can be seen from Figure 6, the safety factor of the slope decreased with precipitation infiltration, with the greatest change under persistent heavy precipitation, followed by continuous rainstorm and sudden rainstorm. At the point of 6 hours, the safety factor of sudden rainstorm with less accumulative rainfall decreases most rapidly and then slows down. During the first 6 hours, the safety factor curve of continuous rainstorm is flatter. After the 6-hour point, the safety reserve decreases more quickly and its overall decreasing trend is close to the continuous rainstorm, but the amplitude of its variation is close to the sudden rainstorm within 24 hours.

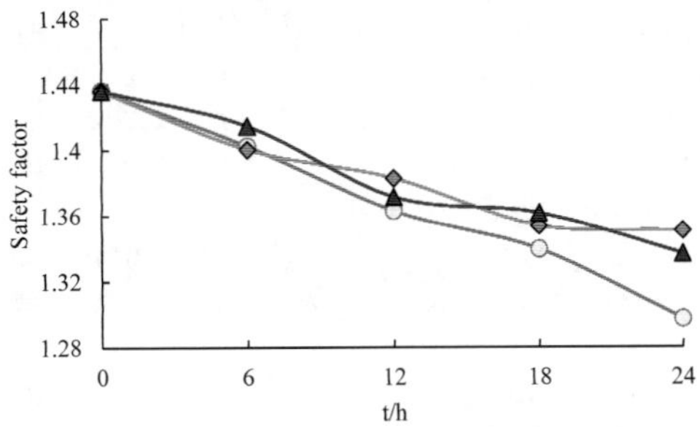

Fig.6. Time history diagram of safety factor

4. Conclusion

By ABAQUS, a two-dimensional soil slope model was established on the measured soil-water characteristic curves of unsaturated red clay fitted in MATLAB in this paper. Considering the fluid-solid coupling, the precipitation infiltration and stability change were investigated in widely-distributed red clay slopes under three heavy precipitation patterns in Southwest China and concluded that:

(1) A transient saturated zone formed in the shallow layer of red clay slope under all the three patterns, reducing the matrix suction in a very short time and thus quickly lowering the shear strength of soil.

(2) Heavy precipitation patterns have a significant impact on precipitation infiltration in red clay slopes, of which continuous rainstorm contributed the most to pore water pressure of unsaturated red clay slope, especially that at the slope crest within the first six hours of precipitation, which was likely to form a transient saturated zone in shallow layer.

(3) The three heavy precipitation patterns moved significantly different towards the time-history changes of the safety factor, which changed the fastest, decreased the most, and induced the greatest risk of slope instability under persistent heavy precipitation.

(4) Persistent heavy precipitation induced the least deep slip surface in the shallow layer, while sudden rainstorm was subject to the deepest slip surface with the largest displacement. It indicated that

sudden rainstorm is more likely to infiltrate rainwater into the deep layer of soil slope, causing deep slip and large-scale landslide disasters.

References

[1] Huang, R.Q. (2007) Large-scale landslides and their sliding mechanisms in China since the 20th century. J. Chinese Journal of Rock Mechanics and Engineering, 26(3):433–454.

[2] Wu, L., Shan, J.S., Zhang, Y.Z., et al. (2005) The formative mechanism and the research idea of prediction method about coast disaster of rainstorm model. J. Meteorology and Disaster Reduction Research, 28(3):17–22.

[3] Wang, W.P., Zhang, Z.Q., Xu, X.Z., et al. (2015) Character of the precipitation period in China. J. Acta Meteorologica Sinica, 73(06):1052 – 1065.

[4] Xu, Q., Tan, X.H., Sheng, M.F. (2012) Stability analysis of soil slopes under rainfall infiltration. J. Chinese Journal of Geotechnical Engineering, (34)11: 254 – 259.

[5] Cheng, Y.G., Fan, A.J., Wang, M.Q. et al. (2019) Treatment of Typical Landslides in a Highway with Red Clay. J. Journal of Water Resources and Architectural Engineering, 17(01):183–189.

[6] Tsai, T.L. (2011) Influences of soil water characteristic curve on rainfall-induced shallow landslides. J. Environmental Earth ences, 64(2):449–459.

[7] Kim, J.M. (2000) A fully coupled finite element analysis of water-table fluctuation and land deformation in partially saturated soils due to surface loading. J. Int J Numer Meth Eng 49(9):1101–1119.

[8] Park, J., Vilayvong, K., Son, Y., et al. (2019) Slope stability evaluation of an agricultural embankment by statistically derived rainfall patterns. J. Paddy and Water Environment, 17(3):303–313.

[9] Ng, C.W., Wang, B., Tung Y.K. (2001) Three-dimensional numerical investigations of groundwater responses in an unsaturated slope subjected to various rainfall patterns. J. Canadian Geotechnical Journal, 38(5):1049–1062.

[10] Tao G.L., Wu X.K., Gan S.C., et al. (2019) Experimental study and model prediction of permeability coefficient of unsaturated clay with different initial void ratios. J. Rock and Soil Mechanics, 40(05):1761–1770+1777.

[11] Fang Y.J., Xiao T.G., Xu Q. (2015) Different types of heavy rainfall which triggered the landslide in sichuan bazhong area and typical case analysis. J. Advances in Geosciences, 5(03):171–185.

Study on mechanical properties of basalt fiber reinforced coral concrete under loading

Hao Zhou[1], Qing Wu[1,*]

[1]College of Civil Engineering and Architecture, Jiangsu University of Science and Technology, Zhenjiang, Jiangsu, 212100, China;

[*]Corresponding author's e-mail: 199100011@stu.just.edu.cn

Absrtact. Basalt fiber and cement are easy to disperse concrete and mortar, so basalt fiber reinforced concrete plays the role of reinforcement, crack prevention and crack resistance. The basalt fiber has good ductility and corrosion resistance, which can replace the traditional concrete and is widely used in construction engineering.In order to improve the strength of coral concrete in marine engineering, the compressive strength and creep properties of coral concrete mixed with basalt fiber were tested, and the creep prediction model suitable for coral concrete was established. The experimental results show that the effect of basalt fiber on the compressive strength of reinforced concrete is not significant; The creep behavior of basalt fiber reinforced coral concrete is higher than that of ordinary concrete; By modifying the existing creep model, a creep prediction model suitable for coral concrete is established. These experimental results can provide a reference for the design of coral concrete structures in ocean engineering.

1. Introduction

Concrete is a kind of brittle material, which has the defects of low tensile strength and low tensile strain limit. The existing literature shows that adding fiber to concrete is a common means to improve its tensile strength and brittleness. Fiber can inhibit the generation and development of internal cracks in concrete, improve the strength and ductility of concrete. Therefore, adding fiber to concrete is an effective method to improve its basic mechanical properties[1-4].

Basalt fiber, as a kind of inorganic fiber, has good aging resistance compared with organic polypropylene fiber and polyacrylonitrile fiber. Its application in concrete can improve the cohesion and stability of concrete; It improves the impact resistance of concrete, reduces its brittleness, improves the impermeability, freeze-thaw cycle resistance and shrinkage resistance of concrete[5-7]. Therefore, this paper studies the changes of compressive strength and compressive creep properties of coral concrete mixed with basalt fiber, and establishes a creep prediction model suitable for coral concrete.

2. Materials and Methods

The specimens are made of C30 concrete with P·O 42.5 grade produced by Helin cement plant; Coarse aggregate: 5 ~ 16mm continuous graded coral from the South China Sea; Fine aggregate: coral sand with fineness modulus of 2.4 ~ 2.8, belonging to medium sand; According to the existing relevant research and experimental results, the compressive strength and compressive creep performance of coral concrete with 1 kg / m^3, 2 kg / m^3 and 3 kg / m^3 Basalt Fiber (BF) were determined, and compared with ordinary coral concrete without fiber (blank group). The basic parameters and specific mix proportion of basalt fiber are as follows Table 1:

Table 1 Basic parameters of basalt fiber

Type	Elongation at break /%	Tensile strength /MPa	Elastic modulus /GPa	Diameter /μm
Basalt Fiber	3.1	2063	56	13

Table 2 Mix proportion of coral concrete with basalt fiber

Number	Cement (kg/m^3)	Coral reef (kg/m^3)	Coral sand (kg/m^3)	Seawater (kg/m^3)	Fiber content (kg/m^3)	Water reducing agent (kg/m^3)
CK	557	749	749	195	0	4
BF1	557	749	749	195	1	4
BF2	557	749	749	195	2	4
BF3	557	749	749	195	3	4

The compressive strength test specimen is 150 mm × 150mm × 150 mm cube, according to the mix proportion in Table 2, the concrete test block is poured and put into the standard curing room for curing. According to the standard GB / T 50081-2009, the compressive strength of the test block of each group is measured and the average value is calculated.

The compressive creep test of coral concrete is carried out by using 100 mm × 100mm × 300 mm prism. Compression creep test: the creep measuring device is embedded in advance. After demoulding the specimen, the compression creep value is tested after the specimen is cured to the age under standard curing conditions. The test method refers to GB / T 50082-2009 standard for test methods of long term performance and durability of ordinary concrete.

3. Results & Discussion

3.1. compressive strength test of coral concrete with basalt fiber

The results of compressive strength of coral concrete with basalt fiber at different times are shown in Table 3. The compressive strength of basalt fiber coral concrete specimens with different contents and ages of 3, 7, 14 and 28 days are analyzed, as shown in Figure 1. The compressive strength of coral concrete with basalt fiber increases with age, and its early strength develops rapidly. It can reach 80% of its 28d compressive strength in 3 days, but the growth rate of strength gradually becomes gentle after 7 days. Basalt fiber has good chemical stability and thermal stability, and does not participate in or change the hydration reaction of cement. Therefore, the strength development law of fiber coral concrete is similar to that of ordinary coral concrete. The early strength growth rate is fast, and the later growth rate is slow. The effect of basalt fiber on the compressive strength of concrete is not significant.

Table 3 Compressive strength of coral concrete with basalt fiber

Number	Compressive strength /MPa			
	3d	7d	14d	28d
CK	21.5	24.8	29.2	31.4
BF1	22.8	26.7	28.6	29.1
BF2	20.9	25.2	27.1	28.2
BF3	21.9	24.0	26.3	28.7

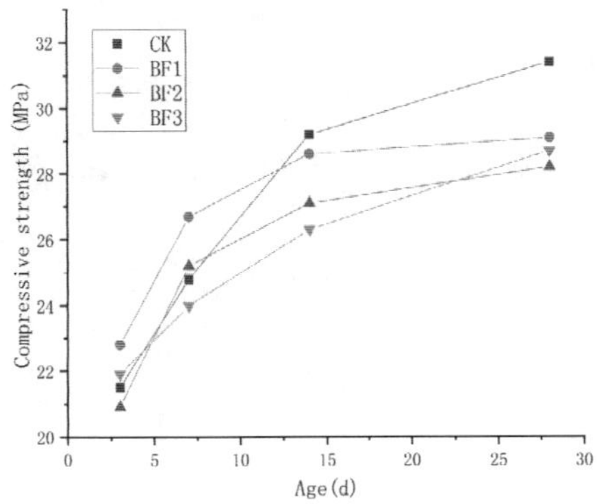

Figure 1. Compressive strength of coral concrete with basalt fiber at different ages

3.2. Compressive creep behavior model of coral concrete with basalt fiber

Because the ability of coral coarse aggregate to restrain the shrinkage deformation of hydrated cement paste is obviously weaker than that of ordinary gravel, and the influence of its low elastic modulus, the instantaneous deformation and final creep of coral concrete are greater than that of ordinary concrete during creep test.

The ACI-209 model proposed by ACI-209 Committee of American Concrete Association in 1982 can better simulate the creep of ordinary concrete. There is a large deviation in the simulation of compressive creep of coral concrete mixed with basalt fiber by using this model directly. It is found that R^2 ranges from 0.34 to 0.87. The model is a hyperbolic power function of the product of creep coefficient and time function, as shown in equation (1).

$$\varphi(t,\tau)=2.35\gamma_{\alpha}\gamma_{s}\gamma_{h}\gamma_{\phi}\gamma_{la}\gamma_{RH}\frac{(t-\tau)^{0.6}}{10+(t-\tau)^{0.6}} \quad (1)$$

Where: γ_{α} - Air content correction factor, γ_{s} - Slump correction factor, γ_{h} - Correction factor of specimen size, γ_{ϕ} - Fine aggregate content correction factor, γ_{la} - Correction factor of concrete loading age, γ_{RH} - Correction factor of environmental relative humidity, t - Calculated age, τ - Loading age.

Compared with other models, the simulation value of ACI-209 model has higher fitting degree with the actual creep data of basalt fiber reinforced coral concrete in this paper, and the two parameters of coral aggregate and fiber content are the influencing factors of creep of basalt fiber reinforced coral concrete, This model can better reflect the compression creep state of coral concrete with basalt fiber. The specific expression is shown in equation (2).

$$\varphi(t,\tau)=K\gamma_{ag}\gamma_{fb}2.35\gamma_{\alpha}\gamma_{s}\gamma_{h}\gamma_{\phi}\gamma_{la}\gamma_{RH}\frac{(t-\tau)^{0.6}}{10+(t-\tau)^{0.6}} \quad (2)$$

Where: K - Basic correction coefficient of coral concrete, with the value of 0.9068, γ_{ag} - correction coefficient of coral aggregate substitution rate, γ_{fb} - correction coefficient of basalt fiber content

Comparing the test results of ordinary concrete and coral concrete with the model calculation results, taking the replacement rate of coral aggregate as a parameter, through multivariate nonlinear fitting, the expression of the correction coefficient γ_{ag} is obtained.

$$\gamma_{ag}=0.501\,8e^{0.506\,a} \quad (3)$$

Where: a - replacement rate of coral aggregate

Comparing the test results and model calculation results of each group of coral concrete mixed with basalt fiber, taking the fiber content as a parameter, the expression of correction coefficient γ_{fb} is obtained.

$$\gamma_{fb} = -0.0024\, b + 1.0010 \quad (4)$$

Where: b - volume content of fiber.

Using the modified model, it is found that the R2 of each group of coral concrete with basalt fiber has been improved to a certain extent. Finally, the creep prediction model of coral concrete with different basalt fiber content is obtained. Through calculation, it is found that the fitting effect of the model fitting value and the test result is good, and the linear regression judgment coefficient R2 is above 0.95.

4. Conclusion

Fiber reinforced concrete can make up for the shortcomings of traditional concrete, improve the performance of traditional concrete, so it will be more and more widely used in the field of construction engineering, and the mechanical properties of basalt fiber reinforced concrete under load has become a hot issue. It is found that the early strength of coral concrete with basalt fiber increases rapidly, reaching 80% of its 28d compressive strength in 7d, and the strength growth rate tends to be flat after 7d; ACI-209 model can simulate the creep of coral concrete mixed with basalt fiber. The ACI-209 model modified by coral aggregate replacement rate and basalt fiber content has high fitting degree and good prediction effect for creep prediction of coral concrete mixed with basalt fiber. But in practical engineering, in addition to load, coral concrete often faces the problem of chloride ion erosion. Therefore, the future research work should focus on the mechanical properties of coral concrete structure with multi factor coupling, which is closer to the engineering practice and more valuable.

Acknowledgments

This work was supported by the Postgraduate Research & Practice InnovationProgram of Jiangsu Province (SJCX20_1479).

References

[1] Li W, Xu J. (2008). Strengthening and toughening effect of basalt fiber on concrete. Acta silicate Sinica (04), 476-481 + 486.

[2] Rong H, Wang H, Chu S, Liang Y, Zhi Z. Mechanical properties of basalt fiber reinforced concrete with different content under high temperature [J]. Comprehensive utilization of fly ash, 2020,34 (01): 56-60.

[3] Xu Z. (2019). Effect of basalt fiber content on concrete performance and strengthening mechanism. Comprehensive utilization of fly ash (06), 23-26.

[4] Cui Y, Zheng Y, Rao L. (2020). Experimental study on mechanical properties of basalt fiber coral concrete. Concrete (12), 74-76.

[5] Xie G, Shen X, Liu J, Zhang B. (2021). Frost resistance and damage deterioration model of basalt fiber reinforced recycled concrete. Composite materials science and Engineering (04), 55-60.

[6] Chen F, Xu B, Jiao H, Chu H, Wang Y, Liu J. (2021). Fiber distribution and pore structure characterization of basalt fiber reinforced concrete. Journal of China University of mining and Technology (02), 273-280.

[7] Hou M, Tao Y, Tao Z, Chai D. (2020). Study on strengthening mechanism of basalt fiber reinforced concrete. Concrete (02), 67-71 + 75.

[8] B Han, H Xie, L Zhu, P Jiang.(2017).Nonlinear model for early age creep of concrete under compression strains. Construction and Building Materials(147).

[9] X Yao, Y Wei.(2014).Design and verification of a testing system for strength, modulus, and creep of concrete subject to tension under controlled temperature and humidity conditions. Construction and Building Materials(53).
[10] N Ranaivomanana, S Multon, A Turatsinze.(2013).Basic creep of concrete under compression, tension and bending. Construction and Building Materials(38).
[11] L. Vandewalle.(2000).Concrete creep and shrinkage at cyclic ambient conditions. Cement and Concrete Composites(3).
[12] S Jin,S Cha, H Jung.(2018).Improvement of concrete creep prediction with probabilistic forecasting method under model uncertainty. Construction and Building Materials(184).

Shear properties of studs in alkali-activated ultra-high performance concrete

Yanpeng Gao[1*], Dong Wei[2], Zhao Gao[2]

[1] School of Civil Engineering, Qingdao University of Technology, Qingdao 266033, China

[2] Shandong Hi-Speed Engineering Construction Croup Co. Ltd., Jinan 250014, China

*Corresponding author: gao.yanpeng@outlook.com

Abstract: With the wide application of ultra-high performance concrete in strengthening orthotropic steel bridge deck, the fatigue damage of steel bridge deck and the cracking of pavement layer are effectively solved. More green and environmentally friendly alkali-activated ultra-high performance concrete is also proposed. The finite element analysis method was used to study the shear performance of different diameter and length of the stud in alkali-activated UHPC. The results show that the diameter of the stud has the greatest influence on the shear bearing capacity of the stud, and the length-diameter ratio has a certain influence on the slip.

1. Introduction

Ultra-high performance concrete (UHPC) has excellent performance in strengthening orthotropic steel bridge deck [1], but the consumption of a large number of silicate cement increases CO2 emissions and brings a burden to the environment, which runs counter to the mode of green development, environmental protection, energy conservation and emission reduction [2]. To solve this problem, some researchers proposed to optimize the design of UHPC mixture to make it more environmentally friendly, but none of them completely abandoned the cement component [3, 4]. Aydin et al. [5] made alkali-activated UHPC by completely replacing cement with slag. The compressive strength of alkali-activated UHPC was similar to that of UHPC, but the cost of alkali-activated UHPC was slightly lower, and it was more environmentally friendly. Compared with ordinary silicate concrete, alkali-activated concrete has the advantages of low carbon emission, fast setting and hardening, and saving aggregate resources.

The connector plays a very important role in strengthening orthotropic steel bridge deck by alkali-activated UHPC. If the shear performance of the connector is insufficient, the concrete layer and the steel bridge deck cannot be cooperatively stressed, and the stiffness of the steel bridge deck cannot be improved. Therefore, it is very important to study the shear performance of the connector in alkali-activated UHPC. Kruszewski et al. [6] selected different diameter, arrangement and different concrete compressive strength as parameters to study the performance of studs in UHPC. Cao et al. [7] studied the stud embedded in thin UHPC by finite element simulation, established a finite element model with the diameter and height of the stud and the compressive strength of concrete as parameters, and carried out experimental verification. Liu et al. [8] studied the static and fatigue properties of studs in high performance fiber reinforced concrete. A large number of studies have shown that different concrete types and strength have an impact on the shear performance of studs. The shear bearing capacity of studs in high performance concrete is higher but the ductility is poor. However, the existing research on the shear performance of alkali-activated UHPC studs is rarely involved. In this chapter, the shear performance of alkali-activated UHPC studs by referring to the research experience of some researchers on the shear performance of studs in high performance concrete.

2. Finite element analysis

ABAQUS was used to simulate the shear test of studs in alkali-activated UHPC. Compared with the finite element analysis of experimental analysis, it has the advantage of low cost. The influence of the length and diameter of studs on the shear bearing capacity of studs was discussed in ABAQUS.

Table 1. Finite Element Simulation Parameters

Specimen	Diameter (mm)	Length (mm)	L/D
NS13-35	13	35	2.7
NS13-55	13	55	4.2
NS13-75	13	75	5.8
NS13-95	13	95	7.3
NS16-55	16	55	3.4
NS16-75	16	75	4.7

2.1 Finite element modeling

The steel plate, alkali-activated UHPC plate and stud are simulated by C3D8R solid element, and the steel bar is simulated by T3D2 two-node three-dimensional truss element and embedded into the alkali-activated UHPC plate. In order to save the calculation cost, a quarter of the specimen is selected for modeling. The constitutive relationship of steel is a two-line model, and alkali-activated UHPC is simulated by concrete damage plasticity model. The loading mode of the vertical displacement simulation test is selected in ABAQUS. Because the maximum slip of the stud in the test is about 2 mm, the vertical displacement is 3 mm. The top surface of the H-beam is coupled to the reference point, and the vertical displacement is applied at the reference point. The root of the stud is bound to the steel plate, and the alkali-activated UHPC plate drills out the hole of the stud size. The contact relationship between the surface of the stud and the corresponding alkali-activated UHPC surface is defined. The contact is divided into hard contact and penalty function contact, and the friction coefficient is defined as 0.2. The contact between steel plate and alkali-activated UHPC is also divided into hard contact and penalty function contact friction coefficient defined as 0. When meshing in ABAQUS, the calculation cost and accuracy are considered at the same time. The global fine grid of 2 mm is adopted for the stud, and the grid of 2 mm is distributed in the contact area between the steel plate and the stud, and then the global grid of 15 mm is used. The contact part of the alkali-activated UHPC plate and the stud is distributed in 2 mm fine grid, and the rest of the less relevant parts are distributed in 17 mm global grid, as shown in Fig. 1.

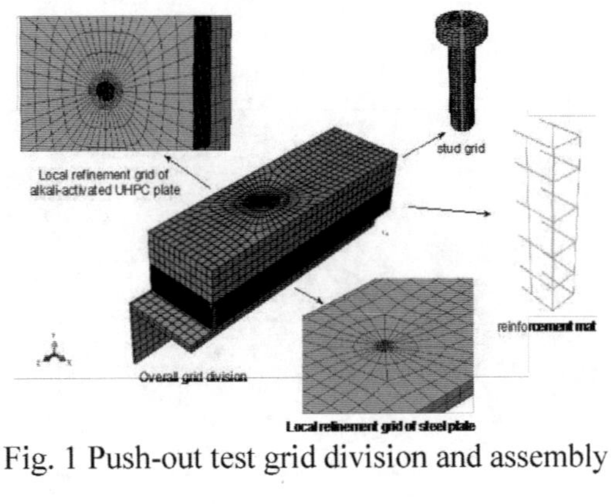

Fig. 1 Push-out test grid division and assembly

3. Finite element results and analysis

3.1 Finite element results

The ultimate bearing capacity and slip value of each specimen in finite element simulation are arranged and shown in Table 2.

Table 2. Finite element results

Specimen	Load-carrying capacity per rivet (kN)	Slippage (mm)	L/D
NS13-35	46.4	0.55	2.7
NS13-55	49.3	0.56	4.2
NS13-75	47.93	0.76	5.8
NS13-95	47	0.74	7.3
NS16-55	79	1.1	3.4
NS16-75	76.5	1	4.7

As shown in the table, diameter has the greatest influence on the bearing capacity and slip of single nail, while length has less influence on the bearing capacity and slip of single nail.

3.2 Stress and strain

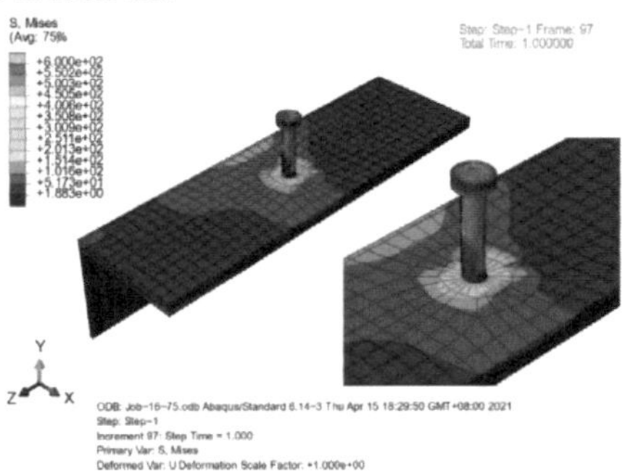

Fig. 2 Stress of stud and steel plate

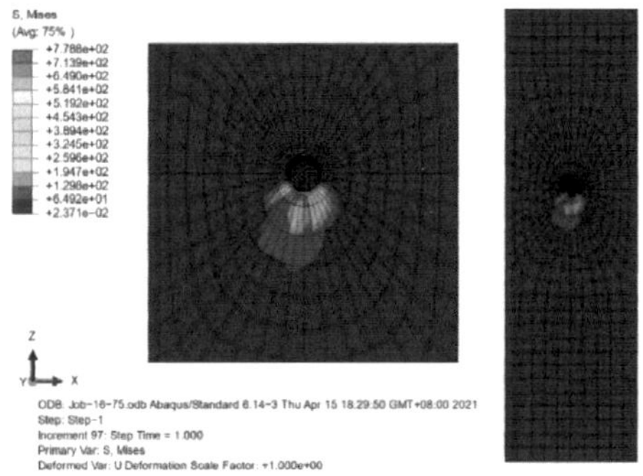

Fig. 3 Equivalent plastic strain of the alkali-activated UHPC plate

The stress of the stud and the equivalent plastic strain diagram of the alkali-activated UHPC plate are output in ABAQUS. As shown in Fig. 2, the maximum stress of the stud appears in the root of the stud and the binding part of the steel plate and the bottom of the stud rod indirectly under the load, so the failure mode of the stud is that the bottom of the stud is cut, and the farther away from the root of the stud, the smaller the contribution to the shear bearing capacity. For the concrete part, as shown in Fig. 3, the maximum plastic strain is located in a small part of the lower part of the stud rod (away from the loading position), which is consistent with the failure mode of the stud in high performance concrete.

3.3 Load slip curve

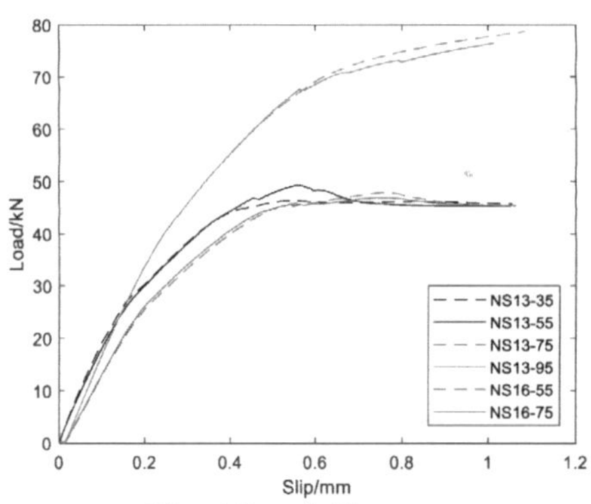

Fig. 4 Load slip curve

As shown in Fig. 4, the bearing capacity of the stud with diameter of 16 mm was significantly higher than that of the stud with diameter of 13 mm. The bearing capacity of the single nail was about 80 kN, and the initial elastic stiffness was also greater than that of the stud with diameter of 13 mm. The slip of the stud in alkali-activated UHPC did not reach 6 mm. The ultimate bearing capacity of each model appeared when the slip reached about 0.6 mm. It may be that the ultra-high compressive strength of alkali-activated UHPC hindered the continuous development of the ductility of the stud, and the accumulated energy led to the sudden shear of the stud. The bearing capacity of the single stud with diameter of 13 mm is about 50 kN. It can be seen from the figure that when the aspect ratio is less than 4, the shear capacity of the stud increases slightly but not obviously with the increase of the length of the stud. The length of the stud has no effect on the slip. When the ratio of length to diameter is greater than 4, the slip of the stud increases, and the shear bearing capacity of the stud decreases with the increase of the length of the stud. It can be seen from the figure that the initial elastic stiffness of the stud with the same diameter is basically the same, and the difference begins at the yield stage. The figure shows that increasing the diameter of the stud is the most direct way to improve the bearing capacity of the stud. The diameter increases by 3mm from 13mm to 16mm and the bearing capacity of the single nail increases by about 50%.

4. Conclusions

Alkali-activated UHPC is a more environmentally friendly alternative cementitious material for UHPC. At present, there are few studies on its structural mechanical properties. In order to better guide engineering practice, it is necessary to study the mechanical properties of bolts in alkali-activated UHPC. In this paper, six groups of bolts in alkali-activated UHPC were simulated by finite element method, and the effects of length and diameter of bolts on shear properties of bolts were studied. According to the results, the following conclusions can be drawn:

(1) The diameter of the stud has the greatest impact on the shear capacity of the stud. In the finite element simulation, the bearing capacity of the two groups of specimens increases by about 50% from 13 mm to 16 mm. The length of the stud has a certain influence on the bearing capacity of the stud, but it has little effect.

(2) Length-diameter ratio has a certain influence on slip and bearing capacity, and there is a critical value of about 4. The increase of length-diameter ratio has the opposite effect on slip and bearing capacity when it is greater than or less than the critical value.

(3) The slip value of studs in alkali-activated UHPC is generally small, and the alkali-activated plate is only damaged in a small range near the studs. The farther the stud is away from the root, the smaller the contribution of shear capacity is.

(4) Limited to personal time and energy, only several aspect ratios of studs are discussed, and a large number of experimental verification and finite element analysis are needed to further explore the shear properties of studs in alkali-activated UHPC.

References

[1] Zhang Y., Cai S.K., Zhu Y.P., Fan L., Shao X.D. (2020) Flexural responses of steel-UHPC composite beams under hogging moment. Eng. Struct., 206: 110134.
[2] Murthi P., Awoyera P., Selvaraj P., Dharsana D., Gobinath R. (2018) Using silica mineral waste as aggregate in a green high strength concrete: workability, strength, failure mode, and morphology assessment. Aust. J. Civ. Eng., 16(2): 122-8.
[3] Abdulkareem O.M., Ben F.A., Bouasker M., Khelidj A. (2018) Mixture design and early age investigations of more sustainable UHPC. Constr. Build. Mater., 163: 235-46.
[4] Ghafari E., Costa H., Júlio E. (2015) Statistical mixture design approach for eco-efficient UHPC. Cem. Concr. Compos., 55: 17-25.
[5] Aydin S., Baradan B. (2013) Engineering properties of reactive powder concrete without Portland cement. ACI. Mater. J., 110(6): 619.
[6] Kruszewski D., Wille K., Zaghi A.E. (2018) Push-out behavior of headed shear studs welded on

thin plates and embedded in UHPC. Eng. Struct., 173: 429-41.

[7] Cao J.H., Shao X.D. (2019) Finite element analysis of headed studs embedded in thin UHPC. J. Constr. Steel. Res., 161: 355-68.

[8] Liu Y.M., Zhang Q.H., Bao Y., Bu Y.Z. (2019) Static and fatigue push-out tests of short headed shear studs embedded in Engineered Cementitious Composites (ECC). Eng. Struct., 182: 29-38.

Research on structural design and anti-penetration performance of ceramic composite armor

Youchun Zou[1], Chao Xiong*[1], Junhui Yin[1], Huiyong Deng[1], Kaibo Cui[1]

[1] Department of Artillery Engineering, Army Engineering University of PLA, Shijiazhuang 050003, China

email: ljgcdxxiongchao@plaust.edu.cn

Abstract. Two types of composite armor were designed (Composite armor I alumina ceramic/Kevlar/616 armored steel; Composite armor II alumina ceramic/616 armored steel/Kevlar). Through penetration tests and numerical simulations, the anti-penetration performance and anti-penetration mechanism of the composite armors were studied, and the influences of material arrangement on its anti-penetration performance were discussed. The results show that the established finite element models can reliably simulate the penetration process of composite armors. The failure modes of Kevlar and 616 armored steel in composite armor I are tensile failure and spalling damage, respectively. The failure mode of Kevlar and 616 armored steel in composite armor II is shear failure. The shear failure of 616 armored steel and Kevlar is a low energy consumption damage mode. Therefore, the structure form of composite armor II can give full play to the anti-penetration performance of the material, and the anti-penetration performance is better than that of composite armor I.

1. Introduction

Modern war puts forward higher requirements for the mobility and protection of equipment. Ceramic composite armor has become one of the research hotspots in the field of protection due to its relatively low quality and good protection performance.

The materials used in composite armor mainly include ceramics, fiber composite materials and metals. The more widely used structural forms include ceramic/metal, ceramic/fiber composite materials and metal/fiber composite materials. An [1] et al. studied the anti-penetration mechanism of ceramic/metal composite armor. The results show that the ceramic/metal structure has good anti-penetration performance and effectively reduces the quality and thickness of the composite structure. Through the penetration tests of SiC/UHMWPE composite armor, Hu [2] et al. found that the structural form of the hard face plate and the flexible support plate has guiding significance for the lightweight design. Cai [3] et al. studied the failure mechanism of foamed aluminum/ultra-high molecular weight polyethylene composite armor under explosion and fragment loading. It is found that foamed aluminum/ultra-high molecular weight polyethylene has good comprehensive protection capabilities.

At present, there are few researches on the anti-penetration performance and anti-penetration mechanism of ceramic composite armor composed of ceramic, metal and fiber composite materials. Two composite armors were designed using ceramic, Kevlar and armored steel. Through penetration tests and numerical simulations, the anti-penetration performance and anti-penetration mechanism of composite armors were studied, which provides a theoretical basis for the design of multilayer composite armor.

2. Materials and methods

The materials that make up the composite armors include alumina ceramic, Kevlar and 616 armored steel, with thicknesses of 10 mm, 10 mm and 6 mm, respectively. As shown in Table 1, based on three materials, two composite structures were designed. The cross-sectional dimension of the composite armor is 150 mm × 150 mm. Projectiles made of T12 steel were used in the tests. The size of the projectile and the test device are shown in Figure 1. In order to reduce the error, three samples of each structure were prepared for three tests and the average of the results was taken. As shown in Figure 2, the DOP method was used to evaluate the protection performance, and the protection coefficient N was calculated from the reference penetration depth and the remaining penetration depth of 603 steel [1, 4]. The formula is as follows.

$$N = \frac{\rho_s (T_0 - T_r)}{\rho_c \delta_c} \tag{1}$$

T_0 is the reference penetration depth; T_r is the remaining penetration depth of the 603 steel after the projectile penetrates the composite armor to be tested at the same speed; ρ_s is the density of 603 steel; ρ_c and δ_c are the average density and thickness of the composite armor to be tested.

Table 1. Designed composite armor

Composite armor number	Structure form
Composite armor I	10 mm alumina ceramic/10 mm Kevlar/6 mm 616 armored steel
Composite armor II	10 mm alumina ceramic/6 mm 616 armored steel/10 mm Kevlar

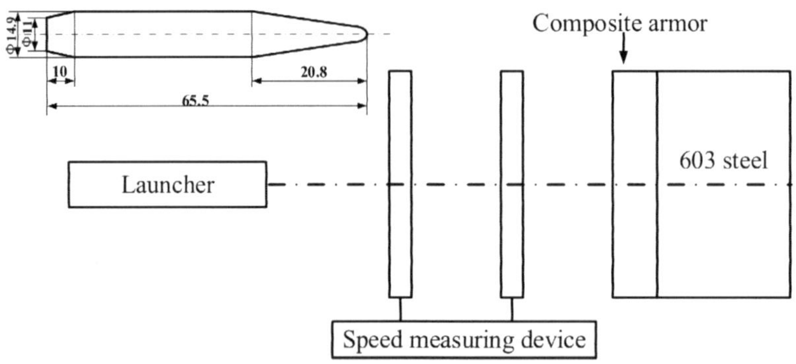

Fig.1 Projectile and test device

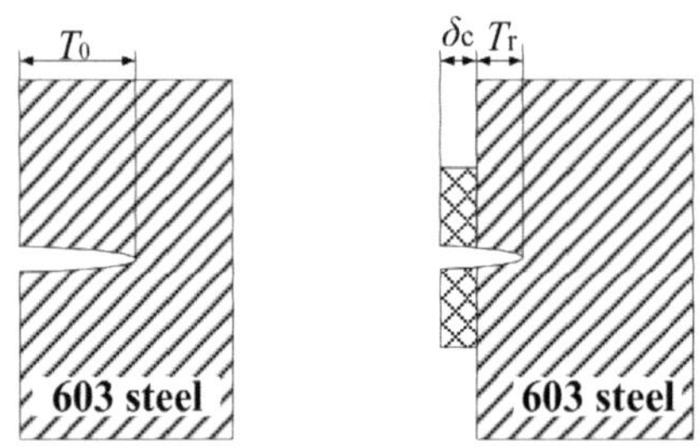

Fig.2 Schematic diagram of DOP method

LSDYNA finite element software was used to simulate the penetration process. In order to improve the calculation efficiency, a quarter model of the structure was established. The model and mesh division are shown in Figure 3. The fixed boundary condition was set at the boundary of the composite structure, and the displacement constraint was imposed on the symmetry plane. CONTACT_ERODING_SURFACE_TO_SURFACE was used to define the interfaces between the projectile and each layer of composite armor, and CONTACT_SURFACE_TO_SURFACE was used to define the interfaces between composite armor. The material parameters are shown in Table 2-4.

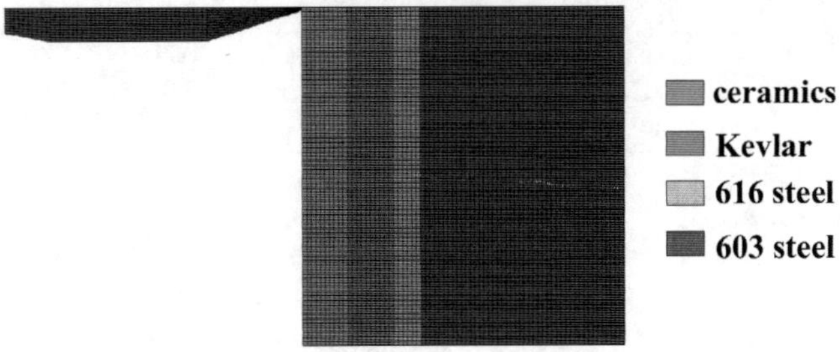

Fig.3 The model and mesh division

Table 2. Material parameters of ceramic[5]

ρ(g/cm^3)	G(GPa)	$\dot{\varepsilon}$ (s^{-1})	T(GPa)	P_{HEL}(GPa)	HEL(GPa)	A	B	C
3.7	90.16	1	0.2	5	10	0.93	0.31	0
M	N	β	D_1	D_2	K_1(GPa)	K_2(GPa)	K_3(GPa)	
0.6	0.6	1	0.005	1	220	191	112	

Table 3. Material parameters of Kevlar

ρ(g/cm^3)	E_a(GPa)	E_b(GPa)	E_c(GPa)	v_{ba}	V_{ca}	V_{cb}	G_{ab}(GPa)
1.35	40.6	40.6	2.6	0.008	0.044	0.044	1.75
G_{bc}(GPa)	G_{ca}(GPa)	$MACF$	KF(GPa)	S_c(GPa)	X_t(GPa)	Y_t(GPa)	Y_c(GPa)
1.6	1.6	3	2.2	0.35	0.725	0.725	0.69

Table 4. Material parameters of bullet, 616 steel and 603 steel [6]

	ρ(g/cm^3)	G(GPa)	A(GPa)	B(GPa)	n	C	m	T_m(K)	$\dot{\varepsilon}$ (s^{-1})
bullet	7.8	81.8	1.7	0.2	0.12	0.016	1.27	1793	1
616 steel	7.83	77	1.47	0.51	0.26	0.014	0.9	1723	1
603 steel	7.86	77	0.95	0.6	0.26	0.016	1.03	1760	1
	C_v(J/kg/K)	γ_0	c(m/s)	S_1	D_1	D_2	D_3	D_4	D_5
bullet	477	2.17	5054	1.49	0	0.33	-1.5	0	0
616 steel	455	2.17	5328	1.49	0.3	0	0	0	0
603 steel	383	1.67	4570	1.33	0.07	1.7	0.07	-0.012	0.61

3. Results and discussion

3.1. Test results and simulation results

The test results and simulation results are shown in Table 5. The penetration resistance of composite armor I is better than that of composite armor II. The test results and the simulation results have relatively small errors, and the established finite element model can accurately simulate the penetration process of the composite armor.

Table 5. Test results and simulation results

Armor number	Projectile velocity m/s	ρ_c (g/cm^3)	T_r /mm	N	Simulation error /%
I-1	1016.5	3.74	0.34	3.45	
I-1	1017.8	3.77	0.46	3.41	0.60
I-1	1019.4	3.76	0.21	3.44	
II-1	1016.3	3.69	2.02	3.36	
II-1	1012.1	3.74	2.19	3.30	3.05
II-1	1009.6	3.77	2.21	3.27	

3.2. Failure Mechanism Analysis

As shown in Figure 4-5, the typical state and material damage morphology of the composite armor I in the penetration process are combined to analyze the damage mechanism. In the process of projectile penetrating ceramic, pressure is generated at the interface between projectile and ceramic. The pressure reaches the yield limit of the projectile, the head of the projectile is plastically deformed, and the ceramic is broken. With further penetration, the ceramic begin to act on the Kevlar. The strength and rigidity of 616 armored steel are greater than that of Kevlar, which leads to tensile failure of Kevlar on the front surface. The 603 armored steel provides support for the 616 armored steel, which leads to spalling damage of 616 armored steel.

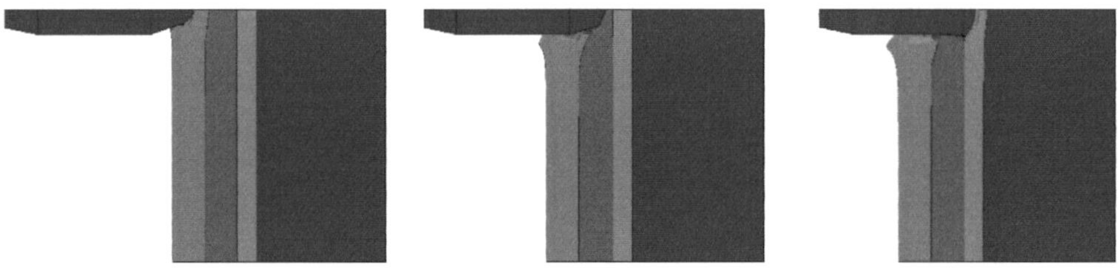

(a)t=0.012ms (b)t=0.021ms (c)t=0.034ms

Fig.4 Penetration process of composite armor I

(a) ceramics (b)Kevlar (c) 616 armored steel

Fig.5 Damage morphology of composite armor I

Figure 6-7 shows the typical state and material damage morphology of composite armor II. The strength and rigidity of Kevlar are much smaller than that of 616 armored steel, so it is not enough to provide support for the deformation of 616 armored steel, which leads to shear failure of 616 armored steel. The Kevlar suffered shear failure due to the restraining effect of 603 armored steel and 616 armored steel. The shear failure of armored steel and Kevlar is a low energy dissipation damage mode, so the penetration resistance of composite armor II is less than that of composite armor I.

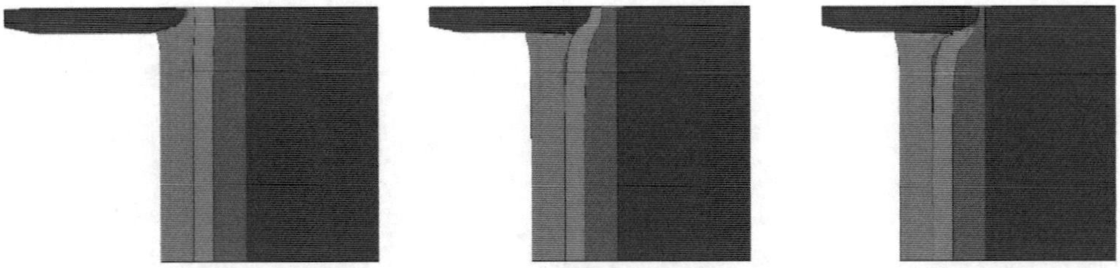

(a)t=0.012ms (b)t=0.026ms (c)t=0.038ms

Fig.6 Penetration process of composite armor II

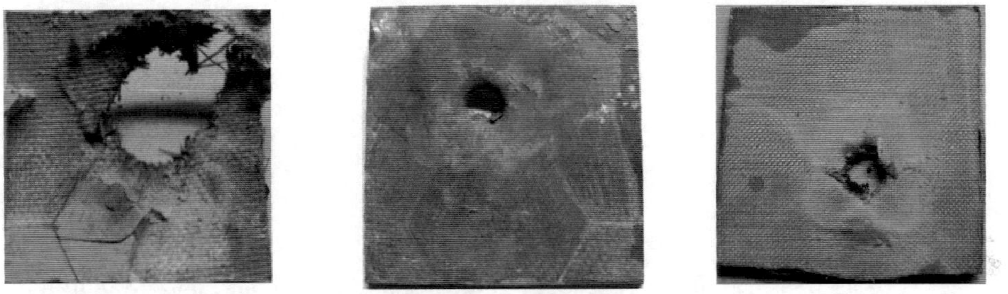

(a) ceramics (b) 616 armored steel (c) Kevlar

Fig.7 Damage morphology of composite armor II

4. Conclusion

(1) The failure modes of Kevlar and 616 armored steel in composite armor I are tensile failure and spalling damage respectively.

(2) The failure mode of Kevlar and 616 armored steel in composite armor II is shear failure.

(3) The shear failure of 616 armored steel and Kevlar is a low energy consumption damage mode, so the penetration resistance of composite armor II is less than that of composite armor I. Under the same material conditions, the structure form of composite armor II can give full play to the anti-penetration performance of composite armor.

References

[1] An X, Tian C, Sun Q and Dong Y 2020 Effects of material of metallic frame on the penetration resistances of ceramic-metal hybrid structures *Defence Technology* **16** 77-87

[2] Hu D, Zhang Y, Shen Z and Cai Q 2017 Investigation on the ballistic behavior of mosaic SiC/UHMWPE composite armor systems *Ceramics International* **43** 10368-76

[3] Cai S, Liu J, Zhang P, Li C, Cheng Y and Chen C 2021 Experimental study on failure mechanisms of sandwich panels with multi-layered aluminum foam/UHMWPE laminate core under combined blast and fragments loading *Thin-Walled Structures* **159** 107227

[4] Anderson C E and Royal-Timmons S A 1997 Ballistic performance of confined 99.5%-Al203 ceramic tiles *International Journal of Impact Engineering* **19** 703-13

[5] Li R, Sun Y, Zhou L, Sun Q, Zhao Y and Feng J 2017 Influence of heat transfer on long-rod projectiles penetrating into ceramic targets *Baozha Yu Chongji/Explosion and Shock Waves* **37** 332-8

[6] Yan K B, Huang Z x and Liu R Z 2014 Numerical and experimental research on ceramic/rubber/steel composite armor penetrated by jet *Gaoya Wuli Xuebao/Chinese Journal of High Pressure Physics* **28** 467-72

Experimental study on rock mechanics characteristics of clastic reservoir with different stress paths

Jiageng Liu[1], Chongzhi Wang[2], Xiaoqian Li[3], Zhaoyang Chen[2], Jing Li[2*], Feng Yang[3] and Hongshan Wang[2]

[1]School of Geosciences in China University of Petroleum, Qingdao, Shandong, 266580, China

[2]College of Pipeline and Civil Engineering in China University of Petroleum, Qingdao, Shandong, 266580, China

[3]Research Institute of Petroleum Engineering, Shengli Oil Field. SINOPEC, Dongying, Shandong, 257200, China

[*]Corresponding author's e-mail: jingli@upc.edu.cn

Abstract. At present, the development of clastic reservoir has become an important way to increase oil and gas productivity in China, and the exploitation of oil and gas will inevitably affect the mechanical properties of rocks. Therefore, it is of great significance to study the rock mechanical properties of clastic reservoir rock. In this paper, the path tests of axial pressure rising-confining pressure rising and axial pressure constant-confining pressure unloading are carried out on the clastic rock samples. The results show that the clastic rock exhibits obvious elastoplastic characteristics under the path of rising confining pressure, but has certain characteristics of elasticity and brittleness under the path of unloading confining pressure; The increase in confining pressure can effectively inhibit the expansion of the rock and increase its bearing capacity; At the same initial confining pressure, the loading failure is less severe than the unloading failure; In the unloading path, the damage of the rock sample during the deformation and failure process is higher; As the confining pressure increases, the degree of rock damage also increases.

1. Introduction

With the continuous improvement of oil and gas exploration and development, China's high-quality oil and gas fields are gradually in the stage of high water cut development, and their reserves and production are decreasing year by year. Oil and gas exploration is bound to move into a new field[1]. At present, the utilization of clastic reservoir to develop resources has gradually become an important way to improve oil and gas exploration in China[2-3]. Oil and gas production is a very complicated stress loading and unloading process. For this reason, it is necessary to study the mechanical properties of reservoir rocks under different stress paths. Many scholars have carried out related researches on the mechanical properties of reservoir rocks. Jiang Yongdong[4] conducted uniaxial and triaxial compression tests on sandstone, and studied the influence of different initial confining pressures and increments on rock mechanical properties; Wang Lehua[5] studied the mechanical properties of deep-buried soft rock under the influence of different confining pressure levels, different unloading stress levels, and different unloading rates; Li Xinping[6] carried out pre-peak and post-peak unloading tests on marble to explore the influence of different stress paths on rock strength; Li Diyuan[7] proposed that the main controlling factor of granite's resistance to failure in the constant axial pressure unloading confining pressure test is the friction force, and the cohesive force in the axial unloading confining pressure test; Li Xiang[8] conducted triaxial compression tests with various confining pressures in order to study the compression failure mechanism of breccia, and found that the peak strength and residual strength of the rock increased with the confining pressure. Although the above-mentioned scholars have studied the mechanical properties of various rocks under different stress paths, there are few reports on the mechanical properties of clastic rocks. Therefore, this paper has carried out different stresses on the rocks of the Hudson deep clastic reservoir Path test research to explore its mechanical properties.

2. Experimental research

2.1. Test rock sample

The test samples were taken from Hudson HD4-H100, HD4-22H and HD4-9-2H wells in the 5056-5078.5m interval, a total of 8 groups. According to the standard size recommended by the International Society of Rock Mechanics (ISRM), They were processed into cylindrical specimens of Φ25mm×50mm. The test rock sample is shown in figure 1.

Figure 1. Core photographs required for testing; (a) Sample A1~A4; (b) Sample B1~B4.

2.2. Test scheme

This paper adopts two test schemes:

Scheme I: Axial pressure rising-confining pressure rising path test. Test procedure: (1) According to the hydrostatic pressure, gradually apply $\sigma_2=\sigma_3$ to the predetermined value at a rate of 0.1MPa/s; (2) Keep σ_3 unchanged, and slowly increase by σ_1 until failure.

Scheme II: Axial pressure constant-confining pressure unloading path test. Test procedure: (1) Apply $\sigma_2=\sigma_3$ gradually to the predetermined value at a rate of 0.1MPa/s according to the hydrostatic pressure; (2) Keep σ_3 unchanged, and increase to the set value $\sigma_1=0.7\sigma_{A1}$ at a rate of 0.1MPa/s, σ_{A1} is the peak stress of the rock sample under the path of axial pressure rising-confining pressure rising; (3) Keep σ_1 unchanged, and slowly decrease σ_3 until it fails.

Table 1. Test scheme.

Test scheme	Sample number	Initial axial Pressure (MPa)	Initial confining pressure (MPa)	Loading rate ($MPa\cdot s^{-1}$)
Scheme I	A1		0	0.1
	A2	-	25	0.1
	A3	-	40	0.1
	A4	-	55	0.1
Scheme II	B1	85	10	0.1
	B2	140	25	0.1
	B3	192	40	0.1
	B4	245	55	0.1

3. Analysis of test results

3.1. Deformation characteristics analysis

The axial and radial stress-strain curves of clastic rocks under two stress paths are shown in figure 2. It can be seen from figure 2 that the clastic rock exhibits obvious elastoplasticity in the axial pressure rising-confining pressure rising path test. The stress-strain curve can be divided into five stages: crack compaction stage, linear change stage, pre-peak plastic yield stage, post fracture stage and residual deformation stage. In the residual deformation stage, the load-bearing capacity of the specimen dropped rapidly, but did not drop to zero, indicating that the ruptured rock still has a certain load-bearing capacity, which is maintained by the friction between the fracture surfaces of the rock. The clastic rock exhibits a certain degree of elasticity and brittleness in the axial pressure constant-confining pressure unloading path test, and the pre-peak stress-strain curve shows a nearly linear relationship, when the peak strength is reached, the expansion of the rock is gradually significant. In the post-fracture stage, the accumulated energy inside the rock is released, the slope of the stress-strain curve becomes negative, the deformation of the rock increases drastically, and the bearing capacity is lost.

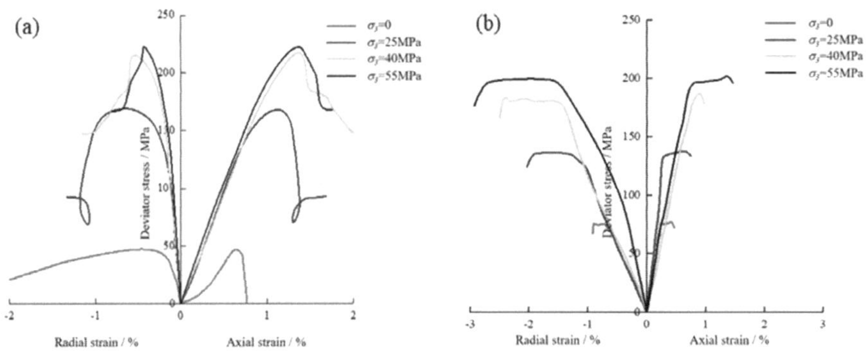

Figure 2. Stress-strain curves of clastic rock; (a) Axial pressure rising-confining pressure rising path test; (b) Axial pressure constant-confining pressure unloading path test.

In order to analyse the influence of different stress paths on rock deformation characteristics, figure 3 shows the comparison of stress-strain curves of rock samples under the two stress paths when the confining pressure is 40 and 55 MPa. It can be seen that the stress-strain curves of the rock samples with different stress paths are significantly different. As the test progresses, the radial strain of the rock sample in the confining pressure unloading test is always smaller than the confining pressure rising test, indicating that the confining pressure has increased to a great extent suppresses the expansion of rocks. In addition, the axial stress and axial strain during rock failure in the rising confining pressure test are greater than those in the unloading confining pressure test, while the radial strain during rock failure in the unloading confining pressure test is larger, indicating that under the same initial stress level, the unloading confining pressure test results in more obvious brittleness characteristics and stronger volume expansion of the rock sample. This is because the rock has accumulated a large amount of energy before unloading. During the unloading process, the rock still absorbs energy from the outside, which leads to its failure. Violent volume expansion occurs at the time, and there is an obvious stress drop phenomenon.

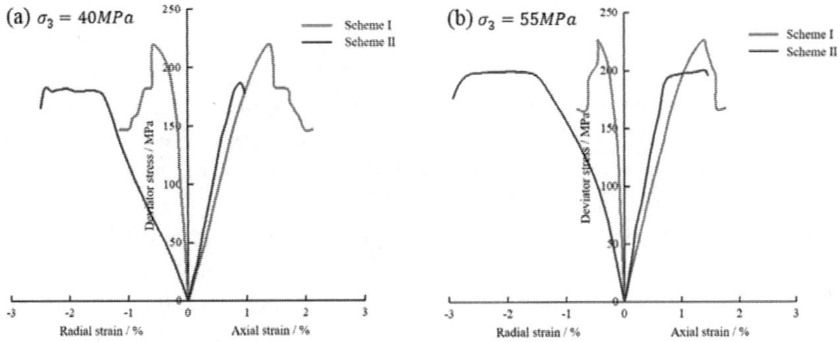

Figure 3. Comparison of stress-strain curves of clastic rocks with different stress paths.

3.2. Strength characteristic analysis

Table 2. The test results of clastic rocks under different stress paths.

Sample number	Initial axial pressure (MPa)	Initial confining pressure (MPa)	Strength σ_1-σ_3(MPa)	Elastic Modulus (GPa)	Poisson's ratio
A1	-	0	48.02	8.32	0.283
A2	-	25	170.50	20.40	0.232
A3	-	40	219.65	19.46	0.205
A4	-	55	226.58	20.19	0.182
B1	85	10	79.60	9.73	0.279
B2	140	25	123.11	19.52	0.235
B3	192	40	159.14	19.78	0.215
B4	245	55	204.22	20.87	0.191

It can be seen from table 2 that the peak strength and elastic modulus of the clastic rock under the two stress paths will increase with the increase of the initial confining pressure, because the confining pressure has a certain restrictive effect on the hoop strain of the rock. It limits the crack development and hoop deformation of the rock, and at the same time improves the compressive strength of the rock, thereby increasing the axial ultimate strain of the rock.

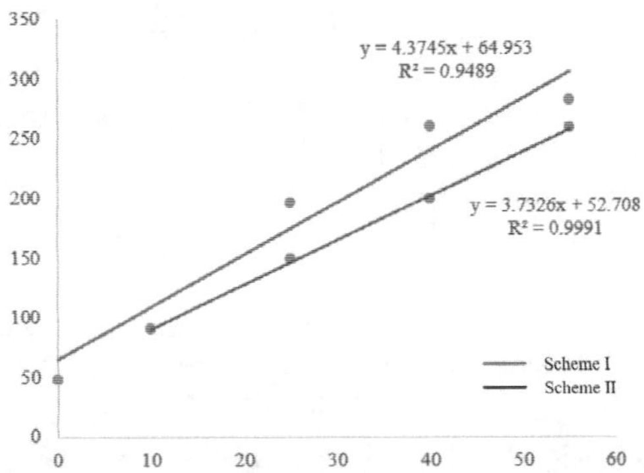

Figure 4. Stress state of specimen failure point under different stress path conditions.

It can be seen from figure 4 that the location of the failure point under the path of constant axial pressure unloading confining pressure is below the path of rising confining pressure as a whole, and the difference of the axial stress at the failure point under the two stress paths increases with the increase of the confining pressure, This shows that in the path of unloading confining pressure under constant axial pressure, the damage degree of the rock sample deformation and failure process is higher and the bearing capacity of the sample is lower. It can be seen that under the path of confining pressure relief under constant axial pressure, with the increase of confining pressure, the degree of rock damage tends to increase. Under the condition of high confining pressure, the difference of the stress path causes the difference of the strength of the specimen to be more significant.

4. Conclusion

In order to study the mechanical properties and complex stress loading and unloading changes of clastic rock during the exploitation of clastic rock reservoir, this paper has carried out the path tests of axial pressure rising-confining pressure rising and axial pressure constant-confining pressure unloading. The main conclusions are as follows:

(1) Clastic rocks show obvious elastoplastic characteristics in the path of rising confining pressure, but have certain elastic and brittle characteristics in the path of unloading confining pressure, The radial deformation of the rock sample is not obvious under the path of rising confining pressure, while the deformation is significant under the path of unloading confining pressure, and the rock failure shows obvious volume expansion phenomenon.

(2) Confining pressure has a certain restraining effect on the hoop strain of the rock, which greatly inhibits the expansion of the rock, improves the plasticity of the rock, and improves the bearing capacity of the rock.

(3) The severity of unloading damage is obviously greater than loading damage. The rock absorbs more energy before failure under the path test of the unloading confining pressure. When it reaches failure, a dramatic expansion phenomenon will cause the rock sample to fail, and there will be a significant stress drop phenomenon.

(4) In the path of unloading the confining pressure, the damage of the clastic rock is higher during the deformation and failure process. As the confining pressure increases, the damage degree also increases.

Based on the research content of this paper, the author thinks that temperature is an important factor affecting the mechanical properties of deep reservoir rocks, and the influence of temperature should be further considered in the follow-up research.

Acknowledgments

This work was financially supported by National Natural Science Fund of China (41972138) and National Great Science Fund of China (ZD2019-183-007).

References

[1] Chen, D. (2017) Innovation of geological exploration technology in the process of petroleum development. J. China Petroleum and Chemical Standard and Quality, 16: 177–178.
[2] Zhang, J.L., Zhang P.H., Xie J. (2013) Diagenesis of clastic reservoirs: Advances and prospects. J. Advances in Earth Science, 28: 957–967.
[3] Li, J., Li, X.R., Zhan, H.B., Song, M.S., Liu, C., Kong, X.C., Sun, L.N. (2020) Modified method for fracability evaluation of tight sandstones based on interval transit time. J. Petroleum Science, 17: 477–486.
[4] Jiang, Y.D., Xian, X.F., Xu, J., Xiong, D.G. (2004) A research on sandstone uniaxial and triaxial compression tests. J. China Mining Magazine, 2004: 68–71.
[5] Wang, L.H., Niu, C.Y. (2019) Experimental study on mechanical properties of deep-buried soft rock under different stress paths. J. Chinese Journal of Rock Mechanics and Engineering, 38: 973–981.
[6] Li, X.P., Xiao, T.L. (2012) Experimental study of jinping II hydropower station marble under loading and unloading stress paths. J. Chinese Journal of Rock Mechanics and Engineering, 31: 882–889.
[7] Li, D.Y., Sun, Z., Li, X.B., Xie, T. (2016) Mechanical response and failure characteristics of granite under different stress paths in triaxial loading and unloading conditions. J. Chinese Journal of Rock Mechanics and Engineering, 35: 3449–3457.
[8] Li, X., Wang, R.B. (2017) The experimental study and simulation of conventional triaxial compression of the breccia[J]. Henan Science, 35: 269–275.

Study on Bending Strength of Cementitious Composites Based on Fiber Alignment

Ping Zhang[1,a], and Xiao-jun Lu[2,b]

[1]Wuhan Zhenghua Architectural Design Co., Ltd, Wuhan, China

[2]Hubei Urban Construction Vocational And Technological College, Wuhan, China

[a]email: zhangping@zhdi.net, [b]email: 458324140@qq.com,

Corresponding author: [a]email: zhangping@zhdi.net,

Abstract: The traditional cement-based composites can't meet the needs of actual 3D printing, people are committed to get breakthroughs in printing materials and extrusion process. This paper introduces the preparation of fiber-reinforced concrete materials, determining the optimum content of glass fiber in composite cement-based materials, and studying the fiber alignment during extrusion. From the bending strength, the performance of 3D printed fiber concrete materials are comprehensively analyzed. As a result, the bending strength of the sample does increase to the fiber volume ratio. The composite cement-based material with 2 vol.-% aligned fiber can have an excellent flexural strength of 12.05MPa, and the samples along the extrusion direction are more excellent than that along the transverse direction.

1. Introduction

3D printing concrete technology is a new kind of concrete moldless forming technology, through the printer material layer by layer is accumulated to realize the free design of the building. At present, though there is a major breakthrough made in the 3D printing equipment, the printing materials are still limited to the traditional ordinary concrete which only show good compressive strength, and it is slightly insufficient in flexural properties respects.

In the 1970s, R.F.Zollo et al [1] reported that a mortar containing high-performance synthetic fibers (glass fiber or carbon fiber) was added to the concrete, and the obtained cementitious composites exhibited a flexural strength of 50MPa. Le T T et al [2] produced a high-performance fiber-reinforced fine aggregate concrete getting the 28d compressive strength of cast samples up to 107MPa and the flexural strength up to 11MPa, while the 28d compressive strength of the samples printed with the composite concrete was from 75 to 102MPa and had a flexural strength from 6 to 17MPa in 2012. Effects on different kinds of fibers of the mechanical properties of 3D printed cement-based materials were studied by Hambach M [3]. It was found that cement-based materials admixed 1 vol.-% of carbon fiber showed a tremendous increase of its flexural strength upon 30MPa and the compressive strength up to 82.3MPa. Admixing 1% (by vol.) of basalt fibers, the flexural strength could reach 13.8MPa and the compressive strength advanced 85MPa. 1 vol.-% percent glass fibers had the flexural strength of 12.4MPa and the compressive strength of 84.5MPa. Although adding the fibers to the cement-based materials would directly improve the ductility of the printing construction, the fact that fibers randomly dispersed in fiber-reinforced cementitious materials fabricated by traditional methods(e.g. casting) was wasting much of their potential on flexural strength. During the 1990s, in order to increase the density of the cement slurry and affect the fiber orientation, people had made lots of attempts to extrude fiber-reinforced cement mortar by using a simple extrusion process [4]. B.Mu [5] introduced fly ash in a glass fiber extruded composite as the replacement of 70% volume of the cement increased the strength by 10% and the toughness by 50% in 2002. To 2006, experiments had shown that fibers could be aligned as a reinforcing agent for cement substrates by extruding, and the toughness of aligned fiber composites was higher than the equal volumes of randomly distributed fiber cement mortar [6]. Manuel Hambach et al [7] had put forward the aligned reinforcing cement slurry having the alignment of carbon fibers guided by the nozzle got a high flexural strength up to 100MPa in 2016. Jong-Han Lee [8] developed a nozzle with blades called the B-nozzle improving the fiber orientation and distribution in fiber-reinforced cement-based materials. Differ from conventional circular nozzles, the B-nozzles increased the mean fiber orientation coefficients by approximately 31–39% and the fiber distribution coefficients by 3–23% in 2019.

The essence of fiber alignment process forced the fibers aligned in the concrete material along the stress direction by the high shear force which was generated from the extrusion process to improve the tensile properties and thereby take full advantage of the toughness of the fibers [9]. Based on the research about extruded fiber-reinforced composites, the mechanical properties of 3D printing fiber reinforcement specimen by controlling fiber volume and alignment were studied in this paper.

2. Experiment

2.1. Raw materials and mix design

The basic materials used in this investigation were: Portland cement, water reducing agent, silica fume, glass fiber (see Table 1), water. Among them, the rapid hardening Portland cement P.O.42.5 were obtained from Lanke Environmental Water Purification Material Factory(Zhengzhou, China) . The polycarboxylate superplasticizer was purchased from Suzhou Xingbang Enterprise, as cement dispersing agent to improve the dispersion of cement particles and reduce the slump loss. The high-performance silicon powder could increase the fiber-cement bond strength in cement matrix [10,11], which was got from Sichuan Langtian Company.

In addition, Glass fibers were pre-chopped having an average length of 12 mm and they were obtained from Yongxing Glass Fiber Factory. Water was ordinary drinking water.

Table 1 Properties of the short glass fibers

Density (g/cm^3)	Tensile strength (MPa)	Elastic modulus (MPa)	Length (mm)	Elongation at break (%)	Melting point (°C)	Color
0.91	346	4286	12	36.4	169	White

There were four groups prepared in the experiment: Group 1 was plain cement composite, fabricated by casting. Group 2 was 1 vol.-% of glass fiber cementitious composites fabricated by conventional mold-casting. Group 3 and Group 4 were fiber-reinforced cement extrudates with fiber volume ratio of 1%, 2%, respectively. The mix proportions were listed in Table 2. By controlling the fiber volume content and alignment, the mixing ratio of Group 2, Group 3 and Group 4 were obtained on the basis of the G1. It was used to test the bending strength of ordinary concrete materials as the compared group to the other groups. While comparing the results of specimens of G1 and G2, the flexural properties of randomly distributed fibers in concrete materials could be analyzed. Furthermore, comparing G2 and G3, the mechanical properties of different fiber alignment in the cementitious composites could be found. And then,we could investigate the influence of different fiber volume which the fibers were aligned along the stress direction between the G3 and G4.

Table 2 Mix proportions

Mix ID	Cement	Water reducing agent	Silica fume	water	Fiber (%)	result
G1	0.53	0.17	0.26	0.02	0	Flow state
G2	0.53	0.17	0.26	0.02	0.01	Viscous mortar
G3	0.53	0.17	0.26	0.02	0.01	Squeeze easily
G4	0.53	0.17	0.26	0.02	0.02	Squeeze difficultly and block the nozzle

2.2. Specimen preparation

After all solid materials, apart from glass fibers, were mixed in dry state, 20 g water were added every time and mixed until a homogeneous cement mixture was obtained, the glass fiber was added and the mixture was stirred for 150s to get the fiber-reinforced cement paste. The prepared mixed mortar was directly poured into a mold with the dimensions of 40×40×160 mm for 3-point-bending test and 50×50×50 mm for compressive strength test, and the surface of the mortar composites became smooth using the spatula. The mold was placed on a vibrating table when the filling was completed. After shaken to 30S, it was taken down and sealed with plastic wrap. The group was placed in the standard curing room as a G2 group. The preparation of the G3 and G4 groups required a simple injection technique that injected the cement paste containing admixed chopped fibers in a specific mold and extruded mixed cement paste through a smaller nozzle. Since the nozzle diameter was smaller than the average length of fibers, when the specimens were fabricated by a sample extrusion technique through high compression and high shear, the fibers within the mixed mortar in the extrusion process would be aligned along the stress direction [12]. The mold taken in this study was a disposable syringe having a nozzle diameter of 10 mm along with 12 mm fibers, as illustrated in Fig 1. And in the same way, the samples were placed on a vibrating table, shaken for 30 s, and sealed with plastic wrap, which were placed in a standard curing room as the G3 and G4 groups.

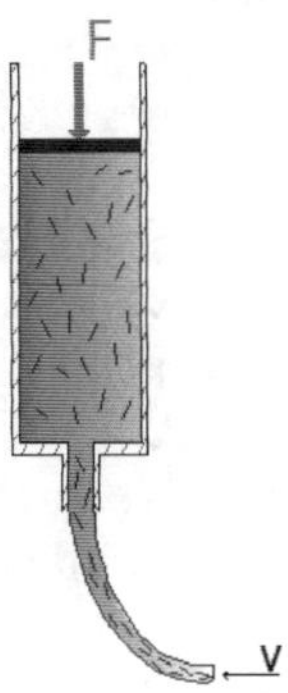

Figure 1 Fibers are aligned by extrusion to produce solid samples containing oriented fibers.

2.3. Flexural strength measurements

The three-point bending test was carried out, according to GB/T 17671-1999, to determine the flexural tensile strength in a 3-point bending test setup. The testing machine was a Micro servo hydraulic universal testing machine with a 100kN load cell attached. The size of prism specimens was 40×40×160mm with 3 specimens for each testing series. Both the mid-bearing part and the bottom support were made of round steel with a radius of 1 cm as sketched in Fig. 2b. Besides, the free span l between supports was equal to 100 mm. In the bending test, the self-made bracket was placed in the center of the lower plate, adjusting the position of the test block and then placing the pressure-bearing round steel on the mid-span of the test block. Additionally, the pressure-bearing round steel was compacted by the upper dial compact before formal measurement. The samples were loaded from 0kN and the speed was set as 1mm/min until to the failure. By measuring the maximum force F the flexural strength fs could be determined through:

$$fs = \frac{3}{2} * \frac{F * l}{b * h^2} \tag{1}$$

Where l represents the distance between the supports (100 ± 0.1 mm for specimens 160 mm in length), b is the specimen's width, here equal to 40 mm, and h is specimen's height, here equal to 40 mm. Before the measurement, the specimen's dimensions were determined within an accuracy of about 0.1 mm. The fiber orientation of samples containing aligned glass fibers and the testing setup can be seen in Fig. 2a.

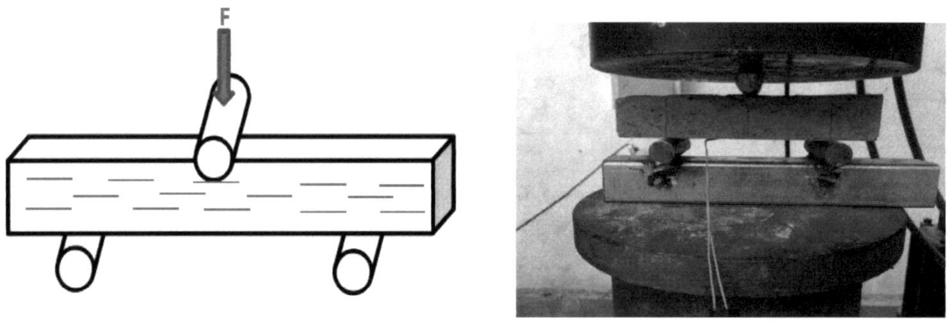

a b

Figure 2 (a)Schematic illustration of aligned fiber-reinforced cement test specimens used in 3-point bending tests.(b)Photograph of three-point bending test in actual experiment.

3. Results and discussion

For the comparison purpose, the group of plain cement composite fabricated by casting and 1% randomly distributed glass fibers composites were placed together as the fiber content control group. The group of 1% randomly distributed glass fibers composites and the glass fiber-reinforced cement extrudates with fiber volume ratio of 1%, 2% are placed together as a control group for fiber alignment.

Table 3 showed the average strength of different fiber volume content and fiber alignment samples. As the measured results, when the fibers of the test samples were aligned, the bending strength was increased from 1 to 12MPa. In addition, samples prepared by conventional mold-casting achieving excellent compressive strength up to 24.63MPa were higher than extrusion processing of fiber reinforced cementitious composites samples G4 in the test direction II achieving 17.7MPa. It is obvious that compressive strength values were slightly decreasing, if fibers were admixed to the prepared cement paste samples, this was attributed to fiber induced damage effects leading to a composite with higher porosity. At least 3 specimens were tested for each three-point bending test and compression test.

Table 3 The strength of different fiber content and alignment samples

Sample	Flexural strength (MPa)	Compressive strength (MPa)		
		Test direction I	casting	Test direction II
G1 (without fibers)	1.61±0.3	-	24.63±0.1	-
G2 (1 vol.-% glass fibers)	5.28±0.2	-	20.32±0.2	-
G3 (1 vol.-% aligned glass fibers)	9.38±0.1	17.34±0.2		19.32±0.1
G4 (2 vol.-% aligned glass fibers)	12.05±0.2	16.30±0.1		17.70±0.3

Typical concrete exhibited the high compressive property, the poor flexural and tensile strength. Generally speaking, in order to improve the overall bending performance of the buildings, the corresponding steel bars were arranged at the specific position to exert the flexural properties in practical engineering. However, for 3D printed buildings, the technical craft of adding steel directly while houses were printed had been realized difficultly. Therefore, it was acknowledged that fibers had taken the place of bars to play a tremendous role in the printing material. This was likely a result of the fiber reducing the occurrence of micro cracks on the concrete surface, decreasing the brittleness of concrete, improving the ductility and having an active influence on the fiber bridging strength to affect the mechanical properties of samples. In order to determine the effect of fiber volume fraction on the mechanical properties and the gain of strength, 3-point bending and compressive strength tests were performed. The bending strength-deflection curve was obtained, as shown in Fig. 3.

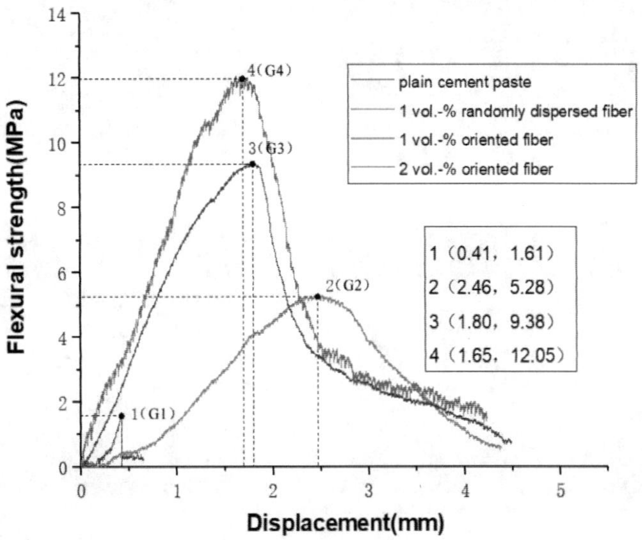

Figure 3 Stress-displacement plots of 3-point bending tests for plain cement paste, mold casted and nozzle-injected glass fiber-reinforced cement paste for 1 and 2 vol.-% glass fibers.

What we could get from the bending strength-deflection plot was that the flexural performance was improved by increasing the fiber volume content (Fig. 3). Comparison of the spots 1 and 2, it was obvious that flexural properties of specimens exhibiting a remarkable improvement due to the increase of fiber volume content. The failure pattern of ordinary concrete composite without fiber can be seen in Fig. 4a. Plain cement composite reached the maximum strength of 1.61MPa and it exhibited obvious brittleness when broken. It can be seen from the experimental process that as the amount of load increased, several small cracks were generated on the surface of the test samples. With the load further increased, the cracks were continuously developed and extended. When the load was close to the ultimate bearing capacity, the specimens were disconnected from the middle part directly and there was no obvious sign before the failure. Then the curve fell in a straight leading to the bending strength dropped suddenly. However, when the fiber content was about 1%, the curve became very smooth and the bending strength was increased to 5.28MPa (relative enhancement of 340%). It was because the mechanical bonding between the matrix and the fibers would improve performance of fiber reinforcement composite showing a strong reinforcement mechanism. In order to restrict the deformation and cracks, a great number of energies were consumed in the fracture and pullout processes resulting in the mechanical capability thus being improved. Hence, due to the great anchor force generated by the fiber end hook bundled with the cement matrix, the ductility was enhanced in bending breakage. There were not collapse and directly break for samples after reaching the peak (Fig. 4b). G4, which fiber content was increased based on G3. In the loading process, the curve went up bit by bit before the bending strength up to 10.5MPa. There were no obvious deformation and small cracks generated on the surface. If the flexural performance achieved 10.5MPa, the mid-span portion displayed some deflection cracks which slowly expanded to the supports. When the maximum bending strength 12.05MPa was achieved, the specimen was broken and the curve began to drop, and then the bending strength declined. After the samples losing effectiveness, it was observed that no large area of soil was falling off. Except for the main crack in the span, there wasn't other crack generating (Fig. 4d). Therefore, the addition of fiber improved the ductility and the plastic deformation property of concrete composites.

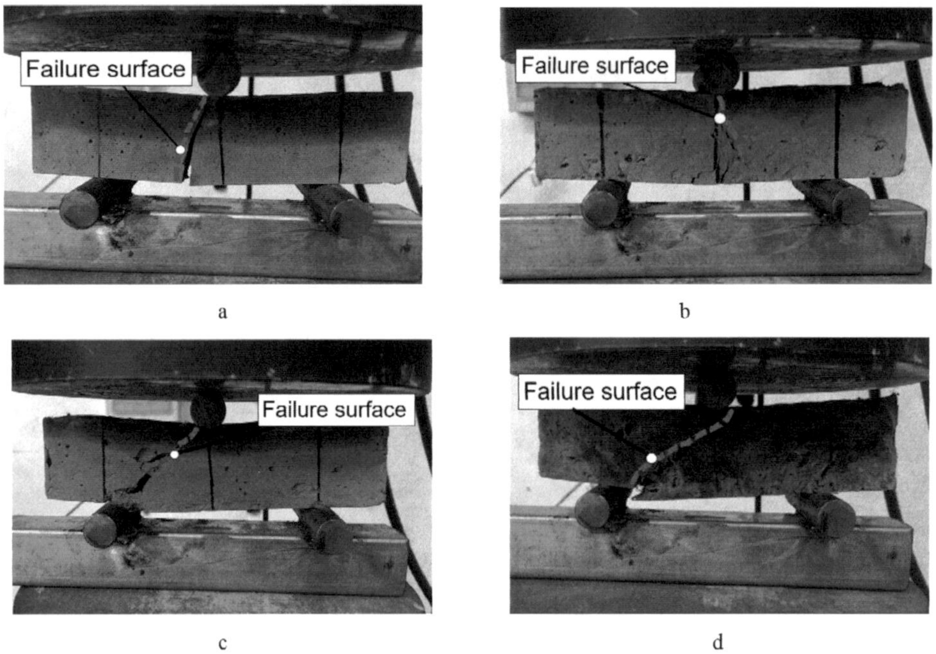

Figure 4 Failure pattern of (a) plain cement sample, (b) 1% randomly distributed glass fibers sample, (c) Aligned fiber-reinforced cement sample (1 vol.-% percent fibers), (d) 2 vol.-% aligned glass fibers sample.

4. Conclusion

In this paper, the effects of fiber alignment on the mechanical properties of concrete are analyzed by comparing the samples produced by extrusion process and conventional mold-casting. The following conclusions have been derived from this study:

(1) The extrusion process plays an important part in aligning the fibers along the stress direction in the concrete composites, especially for 1% volume of glass fibers. Moreover, when the fiber content exceeds 2%, the fiber reinforced mortar will become very viscous, resulting in nozzle blocked typically during extrusion, which will affect the performance of printed products.

(2) Compared with plain concrete paste, the flexural performance of samples with fiber alignment has been greatly improved. A net increase of flexural strength by 600% with the fiber content of 1 vol.-% percent fibers and an ultimate flexural strength of admixing 2% (by vol) of aligned fiber cement of about 12.05MPa can be reached.

(3) There are still different challenges to the current extruding fiber reinforced concrete technology. Because the fiber length is larger than that of the nozzle, it is easy to cause the conveying pipeline and the printing nozzle clogging in the printing process, which results in the condensation of the printed lower layer material, and the upper layer material has not yet been ejected, thus the bond between the layers is not tight, which affects the safety performance of the printing building. Additionally, it will be the key of 3D printing high-rise buildings to further study the method of reinforcing steel bar introduction combined with the concept of assembly building or to make breakthroughs in material mix ratio.

References

[1] J S Lawler, T Wilhelm, D Zampini, et al. Fracture processes of hybrid fiber-reinforced mortar [J]. Materials and Structures, 2003, 36(3):197-208.
[2] T T Le, S A Austin, S Lim, et al. Hardened properties of high-performance printing concrete [J]. Cement and Concrete Research, 2012, 42(3):558-566.
[3] M Hambach, D Volkmer. Properties of 3D-printed fiber-reinforced Portland cement paste [J]. Cement and concrete Composites, 2017, 79(3):62-70.
[4] B Panda, S C Paul, M J Tan. Anisotropic mechanical performance of 3D printed fiber reinforced sustainable construction material [J]. Materials Letters, 2017, 209(12):146-149.
[5] Y Akkaya, A Peled, S P Shah. Parameters related to fiber length and processing in cementitious composites [J]. Materials & Structures, 2000, 33(8):515-524.
[6] B J Mohr, H Nanko, K E Kurtis, Aligned kraft pulp fiber sheets for reinforcing mortar [J], Cement and Concrete Composites, 2006, 28(2):161-172.
[7] M Hambach, H Möller, T Neumann, et al. Portland cement paste with aligned carbon fibers exhibiting exceptionally high flexural strength (> 100 MPa) [J]. Cement and Concrete Research, 2016, 89(11):80-86.
[8] J H Lee, J W Hu, J W Kang, Effects of blades inside a nozzle on the fiber orientation and distribution in fiber-reinforced cement-based materials [J], Composite Structures, 2019, 221(8):1-12.
[9] Longbang Qing, Kelai Yu, et al, Uniaxial tensile behavior of aligned steel fiber reinforced cementitious composites [J]. Materials & Structures, 2019, 52(4):70-82.
[10] Guowei Ma, Zhijian Li, Li Wang, et al, Mechanical anisotropy of aligned fiber reinforced composite for extrusion-based 3D printing [J], Construction and Building Materials, 2019, 202(1):770-783.
[11] E L Kreiger, M A Kreiger, M P Case, Development of the construction processes for reinforced additively constructed concrete [J], Additive Manufacturing,2019, 28(8):39–49.
[12] L Sun, Q Hao, J Zhao, et al. Stress strain behavior of hybrid steel-PVA fiber reinforced cementitious composites under uniaxial compression [J]. Construction & Building Materials, 2018, 188(10):349-360.

Study on Frictional Force between Lining Concrete and Rubber Blocks

Zhiyong Zhou[1,2,*], Sen Gao[3], Xiaolu Chen[3] and Chengying Guan[4]

[1] Wuda Jucheng Structure Co., Ltd. Research Institute, Wuhan, Hubei,430223, China

[2]Engineering Research Center of Building Inspection and Reinforcement, Ministry of Education, Wuhan University, Wuhan, Hubei 430072, China

[3]Central Route Construction Management Bureau of South to North Water Transfer Project, Beijing, 100053, China

[4]School of Civil Engineering, Wuhan University, Wuhan, Hubei, 430072, China

[*]Corresponding author's e-mail: zp@wudajucheng.com

Abstract. This paper mainly studies the friction coefficient between lined concrete and rubber blocks, designs the Friction resistance test device, puts forward two different Shaw hardness rubber blocks in different environments, and carries out the experimental research, obtains the corresponding coefficient of friction results, and provides the parameters and basis for the development of the special enclosure for the research of the subject.

1. Introduction

The first phase of the South-North Water Transfer Central Line is 1432 km long and has a wide variety of building types, including 1100 km of aqueducts. The operation of the project is facing safety issues, especially the soil channel slope instability, lining structure damage and other dangerous situations are difficult to avoid.

Since the main canal of the central line is put into operation, it is responsible for the task of supplying water to the residents and industrial production of large and medium-sized cities such as Beijing and Tianjin. Once the water delivery is interrupted, the impact will be wide and the social impact will be large. It is necessary to repair the main canal lining board without interrupting the channel water delivery.

According to the characteristics of the South-North Water Transfer Project, the National Key Research and Development Plan puts forward the project of "Research on Emergency Rescue and Rapid Repair of Key Technologies and Equipment of the South-North Water Transfer Project", and this paper and one of the research contents of the special topic of the siege in the project under the topic 2017YFC0405001. When the enclosure system is working, it is subject to the impact of upstream water flow and the pressure of the side static water, and it is necessary to verify whether the friction between the perimeter equipment and the lining plate under this operating condition can offset the force of the water flow on the surrounding equipment, so that the enclosure does not slip.

Therefore, the friction coefficient between the rubber pressure block at the bottom of the enclosure and the lining plate is studied experimentally. The friction coefficient between the two is obtained, which provides calculation parameters for the development of special cofferdams.

2. Pre-research

The research on the frictional properties between hard rubber and concrete has carried out literature review and research. The literature "The effect of rubber pad on the contact characteristics of tunnel lining tube sheet"[1] is analyzed by straight shear test, and the influence of rubber pad on the contact characteristics of tunnel lining pipe sheet can be used as a qualitative reference in determining the coefficient of friction between concrete and rubber. The paper "Rubber resistance skateboard in China's water transportation engineering for the first time"[2] in the paper introduced the concrete structure bottom surface set rubber resistance skateboard and stone-throwing machine tool friction coefficient test and related results, analysis of the friction coefficient of each test process - block displacement relationship curve, the determination of the machine machine under different pressures(40kPa,60kPa and 80kPa)coefficient of friction - horizontal slide relationship curve, the study of this paper (pressure in 90kPa) is of reference value. The paper "Experimental study of many friction interfaces and coefficients of friction under water"[3] gives the conclusion that the mean coefficient of friction between concrete and rubber plate in underwater environment is 0.56. JTG 3362-2018"Road reinforced concrete and prestressed concrete bridge culvert design specification"[4] shows that the friction coefficient between rubber bearing and concrete is 0.3. "Mechanical Design Manual"[5] gives the conclusion that the friction coefficient of concrete and rubber is 0.4 ~ 0.75 under wet conditions. It is shown that the coefficient of friction between concrete and rubber is greatly affected by wetting conditions, and this research project is also in the underwater environment, while the channel lining plate may actually have some silt accumulation, so this study should design different test conditions for these influencing factors when conducting a test study on the frictional force between concrete and rubber plate.

3. Friction test

3.1. Pilot scheme

Based on the research situation, the Shore hardness of rubber suitable for underwater concrete is generally 60±5. Therefore, two rubbers are selected in this test, and the Shore hardness is 65 and 55 respectively. Corresponding test conditions are designed for different needle environments, as shown in Table 1.

Working conditions 1 and 2 are friction tests in a conventional dry environment.

Working conditions 3 and 4 are to soak the concrete slab in water for 15 days, then take it out and conduct a friction test on the surface of the concrete without clear water.

Working conditions 5 and 6 are to soak the concrete slab in water for 15 days, then take it out and place it in a test tank for friction test, which is to simulate a clean water environment.

Working conditions 7 and 8 are to soak the concrete slab in water for 15 days, then smear the silt from the bottom of the pond and place it in a test tank for friction test, which is a simulated silt environment.

Table 1. Friction test conditions

Conditions	Concrete Slab Environment	Rubber Shore Hardness
One	Regular dry environment	65
Two		55
Three	Remove after 15 days of soaking in water. There is no clear water on the surface	65
Four		55
Five	Remove after 15 days of immersion,	65
Six	Water environment resistance test	55
Seven	After 15 days of soaking in water, the silt at the bottom of the pond is applied and the water environment is tested	65
Eight		55

The coefficient of friction can usually be determined by the method of friction angle. The method is generally to place one of the two objects to be measured as a slope, and the other is placed on the slope and slides down along it, gradually reducing the slope angle θ, It can be found that when θ reaches a certain value θ_0, the object will slide down at a constant speed.As shown in Figure 1. then from $mg\sin\theta_0=\mu mg\cos\theta_0$, we get: $\mu=\tan\theta_0$, where θ_0 is the angle of friction, that is, the value of μ between two objects is obtained by measuring the angle of friction.

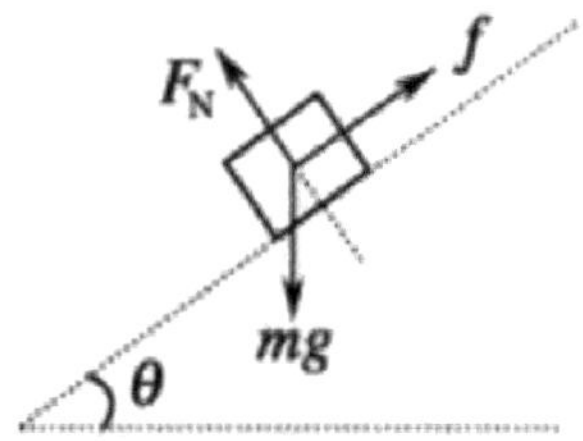

Figure 1. Uses friction angles to determine the coefficient diagram

Because the above-mentioned method of measuring friction angle is difficult to control the friction angle θ_0 accurately in practice, the device as shown in Figure 2 is designed for Friction testing. The basic principle of the device is: the rubber block and concrete block are pressurized vertically by the universal material tester, the rubber block is placed on one piece of concrete each, and the concrete horizontal direction is limited, and the rubber block is placed in the carrier. Connect the carrier to the pull load structure with a traction rope and install the pull sensor. The vertically loaded part of the structure is placed in the sink, and the rubber and concrete blocks can be added to the underwater environment.

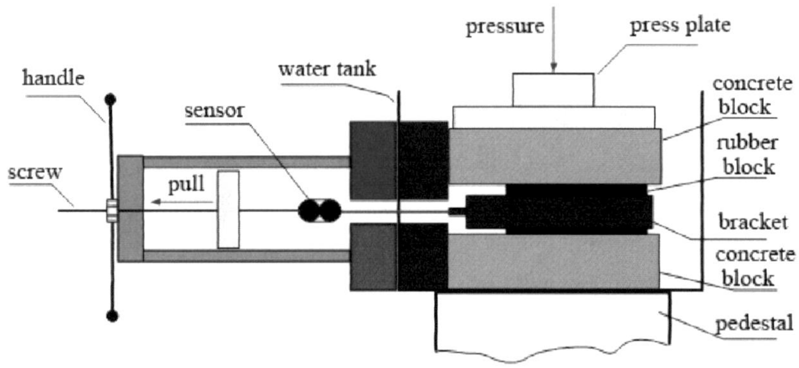

Figure 2. Frictional force test device

3.2. Test conclusions

Tests are conducted according to design test conditions, and the test analysis data are shown in Table 2. Figure 3 is the test state of the test device under the universal material test machine, and Figure 4 is a photo of the mud at the bottom of the reservoir being smeared after 15 days of concrete immersion and placed in a water environment for a drag test.

Table 2. Results table for the coefficient of friction test

Serial Number	Test Conditions		Coefficient of Friction			
	Environment	Shore hardness	The Measured Value			Average Value
1	Regular dry environment	65	1.88	1.73	1.56	1.72
2		55	1.55	1.43	1.62	1.53

3	Remove after 15 days of soaking in water, There is no clear water on the surface	65	1.12	1.24	1.21	1.19
4		55	1.09	1.13	0.88	1.03
5	Remove after 15 days of immersion,Water environment resistance test	65	0.88	0.83	0.84	0.85
6		55	0.84	0.76	0.74	0.78
7	After 15 days of soaking in water, the silt at the bottom of the pond is applied and the water environment is tested	65	0.57	0.59	0.6	0.59
8		55	0.42	0.43	0.47	0.44

The test found that the test value of the number of Friction systems of rubber and concrete slabs with a hardness of 65 under different operating conditions was between 0.57 and 1.88, of which the underwater Friction coefficient was between 0.57 and 0.88 The test value of the molybdo coefficient of rubber and concrete slab with a hardness of 55 is between 0.42 and 1.62 under different operating conditions, of which the coefficient of underwater Friction is between 0.42 and 0.84.

In the dry environment, the Friction coefficient is the largest, the water environment with pond mud situation is the smallest, for this project, you can refer to the water environment with pond mud Friction coefficient test value. The rubber with a Shore hardness of 65 has a greater resistance factor to concrete slabs than the rubber with a Shaw Hardness of 55, which is more conducive to anti-slip during operation of the perimeter system.

Due to the relative softness of the rubber of Shaw's hardness 55, it can be more adapted to the uneven condition of the bottom of the lining plate, and can better disperse the pressure and avoid the disadvantage of excessive local stress to the lining plate, but the rubber block is applied to the pond mud after the average Friction coefficient in the water environment 0.44, the measured minimum value is 0.42,the hardness rubber is used as a pressure block, under the conditions allow, the lining plate can be properly desilted first, so as to improve the anti-slip capacity of the enclosure.

4. Conclusion

This paper introduces the study of frictional force tests of two types of Shore hardness (65 and 55) rubber and concrete slabs under different environmental conditions. The friction coefficient between the rubber and the concrete slab under different working conditions is obtained. The friction coefficient of the two materials is the largest in a dry environment, and the humid environment has a greater impact on the friction between the two materials. In addition, when there is silt, the two are not completely in direct contact. Under this working condition, the friction coefficient is the smallest, which is the most unfavorable for the cofferdam to resist overturning. Through the experimental research, the resistance parameters between the rubber pressure block and the lining plate, which are used with the special fence of the south-north water-north midline channel lining plate, are obtained, which provides the calculation parameters for the development and application of the special fence.

Acknowledgments

This work was financially supported by the National Key R&D Program of China (project No. 2017YFC0405001).

References

[1] Zhang ZY, Jia CH , Li HJ , Feng JL. Experimental study on the influence of rubber pads on the contact characteristics of tunnel lining segments. J. Railway Construction, 2017.07.18:1003-1995.

[2] Xie SW, Li YB. The first application of rubber anti-sliding plate in my country's water transportation project. J. Water Transport Engineering, 2006, B12:53-57.
[3] Zhang JX , Zang B , Liu JK. Experimental Research on Underwater Multi-friction Interface and Friction Coefficient. J. China Harbour Construction, 2017,2(37):65-67.
[4] JTG 3362. (2018) Road reinforced concrete and pre-stressed concrete bridge culvert design specifications. People's Transportation Press. Beijing.
[5] Wen BJ. (2018) Mechanical Design Manual. Machinery Industry Press. Beijing.

Study on axial compression behavior of rectangular concrete filled steel tubular columns with Perfobond Leister Ribs

Hanjian Cheng[1], Zheyuan Lu[2] and Xinke Cao[3]

[1]School of Highway, Chang 'an University, Xi 'an, 710064, China

[2]School of Highway, Chang 'an University, Xi 'an, 710064, China

Email: 2019121060@chd.edu.cn

Abstract: The built-in open-hole type stiffened rectangular steel tube concrete refers to a new combined structure formed by forming an open-hole steel plate in the longitudinal direction of the inner wall of the steel pipe and then filling the concrete. The PBL stiffener has the double functions of longitudinal stiffener and shear connector. In order to study the influence of stiffeners on the ultimate bearing capacity of rectangular CFST columns, the finite element model was established by ABAQUS and the parameters of the stiffeners were analyzed, including the aperture of the stiffeners, the spacing of the stiffeners and the width-thickness ratio. The results show that the finite element model established in this paper is in good agreement with the existing test data. The bearing capacity of rectangular CFST columns increases first and then decreases with the increase of the opening aperture of stiffeners, increases with the increase of the opening spacing, and increases with the decrease of the width to thickness ratio of stiffeners. The conclusions of this paper can provide some suggestions for the design of PBL stiffened rectangular CFST columns in practical engineering.

1. Introduction

The inner wall of the steel pipe is welded with a long stiffening rib, and holes are cut along the length direction, and then filled with concrete to form a PBL stiffened steel pipe concrete structure. This kind of stiffening rib has the function of both stiffening rib and shear connectors, which can not only strengthen the bond between steel panel and concrete, but also improve the local buckling resistance of steel tube. Based on the above advantages, PBL reinforced steel tube concrete structure has been used more and more in the belly bar of truss bridge, the tower of cable-stayed bridge and the pier of continuous rigid frame bridge.

Huang Hong et al. [1] carried out 14 axial compression tests of concrete-filled square steel tubular columns with ribs, taking the width thickness ratio of square steel tube and the height thickness ratio of stiffener as the main parameters. The stress-strain relationship, the longitudinal stress distribution of core concrete and steel tube and their interaction are compared no rib, single rib and double rib square concrete filled steel tubular axial compression mechanical properties of short column. The results show that the stiffener not only increases the longitudinal stress of the core concrete, but also significantly reduces the tensile stress zone of the pipe wall and improves the stability of the pipe wall. Liu Yongjian et al. [2] conducted an experimental study on a group of concrete-filled steel tubular short columns with high steel content and formed by welding, and found that after longitudinal ribs are opened, the stability of the panels can be guaranteed, the interaction with concrete can be strengthened, and the deformation ability is enhanced. By analyzing the force transfer performance of PBL stiffener at the interface of steel concrete, the mechanical characteristics of PBL stiffened concrete filled square steel tube are analyzed.

Cheng Gao et al. [3] conducted tests on 4 groups of 23 PBL stiffened square concrete filled steel tubular short columns, studied the influence of parameters such as width to thickness ratio of steel tube, stiffening rib type and aperture on their mechanical properties, and gave a formula for calculating the minimum stiffening stiffness coefficient of PBL.

The above research mainly studies the mechanical properties of PBL stiffened steel tube concrete column through the experimental method, and does not give the specific influence of stiffening rib parameters on the concrete filled steel tube column.

In this paper, an axial compression finite element model of PBL reinforced rectangular concrete-filled steel tubular column was established. The influence of initial defects and welding residual stress was considered, and the influence of stiffening rib parameters on the bearing capacity of PBL reinforced steel tubular concrete column was studied.

2. Finite element study

The finite element model of PBL reinforced rectangular concrete-filled steel tubular column under axial compression was established by the universal finite element software ABAQUS. The model was compared with the existing test results to verify the accuracy of the model, and the parameters of the stiffeners of the PBL reinforced steel tubular concrete column were analyzed.

2.1 Establishment of finite element model

The explicit method in ABAQUS can overcome the numerical convergence problem caused by the large deformation and contact of the implicit method. The research of Thai et al. shows that the explicit method can accurately predict the quasi-static behavior when the loading rate is controlled. At the same time, it gives the specific method of display analysis in ABAQUS/ Explicit. Based on this method, the finite element simulation of axial compression process of PBL stiffened rectangular concrete-filled steel tubular column is carried out in this paper.

The finite element model of PBL stiffened rectangular concrete-filled steel tubular column is shown in Fig. 1. In this model, the steel plate and PBL stiffener are simulated by 4-node reduction integral shell element (S4R), and the concrete filled by 8-node reduction integral solid element (C3D8R). It is found that when the mesh size of steel panel is about 5% of the width of the specimen, the calculation efficiency can be improved effectively while the calculation accuracy is guaranteed. Use the Surface to Surface (Explicit) contact option for contact interactions between steel plate and concrete. A "hard" contact is specified in the normal direction, which allows the surface to separate during stretching and does not penetrate during compression; In the tangential direction, the friction coefficient between steel plate and concrete is 0.25. The steel plate and concrete surfaces at the end are coupled to the reference points 1 and 2 at the center of the end section by using the "Coupling" constraint to facilitate the setting of boundary conditions and the application of loads. The steel section and concrete section at the bottom of the column are constrained by fixed connection. The steel section at the top of the column is only allowed to have axial displacement, and the translational and rotational degrees of freedom in other directions are restricted. The load acts on the steel plate and concrete at the same time. Displacement loading is adopted at the reference point 1 on the top surface of the column, and the average loading speed is 1mm/s, about 5~10 times of the actual test loading speed, which can shorten the calculation time on the premise of ensuring the calculation accuracy.

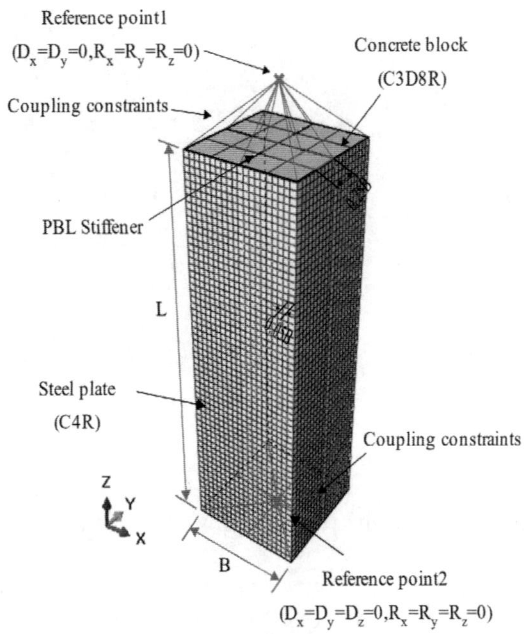

Fig. 1 Finite element model

The five-fold elastic-plastic constitutive model proposed by Liu Wei, which is adopted for the stress-strain curve of steel, as shown in Fig. 2.

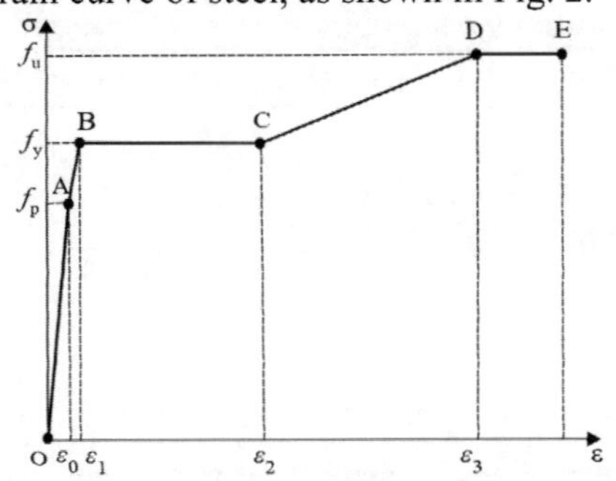

Fig. 2 Steel constitutive model

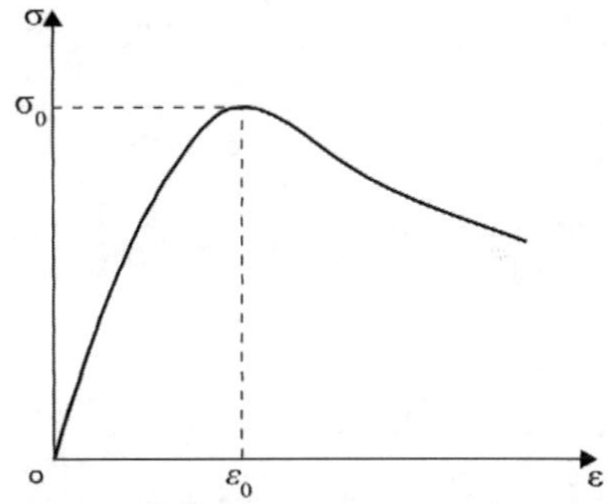

Fig. 3 Concrete constitutive model

The stress-strain curve of concrete adopts the constitutive relation of core concrete considering constraint effect proposed by Han [4], as shown in Fig. 3.

The Technical Specification for Concrete-filled Steel Tube Structure (CECS159:2004) [5] stipulates that the defect function of rectangular concrete-filled steel tube is as follows:

$$\omega = \frac{\omega_0}{4}(1 - cos\frac{2m\pi x}{B})(1 - cos\frac{2n\pi y}{L}) \quad (1)$$

Where, ω_0 is the maximum allowable deviation outside the plane of the plate, which is 0.01B according to the Technical Code for Cold-formed Thin-walled Steel Structures GB50018-2002[6]. The initial defect shape of ribbed rectangular CFST members is shown in Fig. 4.

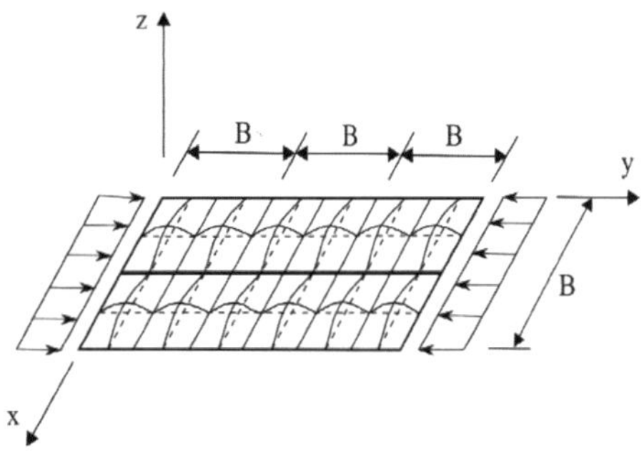

Fig. 4 Defect shape

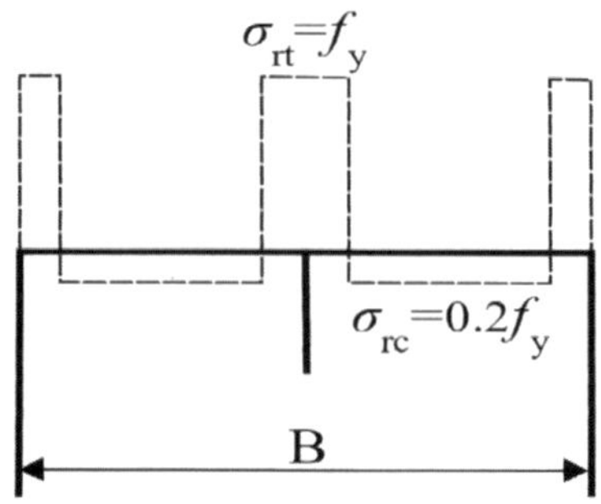

Fig. 5 Residual stress model

Predefined residual stresses are applied in the model. The residual stress model adopts the simplified residual stress distribution model of welded rectangular steel pipe section proposed by [7], as shown in Fig. 5.

2.2 model validation

The axial load-displacement curve and ultimate strength predicted by the finite element model in this paper were compared with the experimental results in the literature, and It was shown in Table 1 and Figure 6.As can be seen from Fig. 6, this model can accurately predict the load-axial displacement behavior of PBL stiffened rectangular concrete-filled steel tubular columns, and the average and standard deviation of the ratio of finite element results to test results are 0.96 and 0.08 respectively (Table 1). In conclusion, the finite element model established in this paper is reliable. It can be used to study the mechanical properties of PBL stiffened rectangular CFST columns.

Table 1 Comparison of ultimate strength

The data source	Specimen number	H	B	t	bs	ts	d	y	F/T
Literature 3	SCB20-1	600	200	4	60	4	-	-	0.92
	SCC20-1	600	200	4	60	4	30	50	0.98
	SCB30-1	900	300	4	90	4	-	-	0.99
	SCC30-1	900	300	4	90	4	30	100	0.95
The average									0.96
The standard deviation									0.08

Note: F/T is the ratio of the finite element of the specimen to the axial compressive bearing capacity of the test.

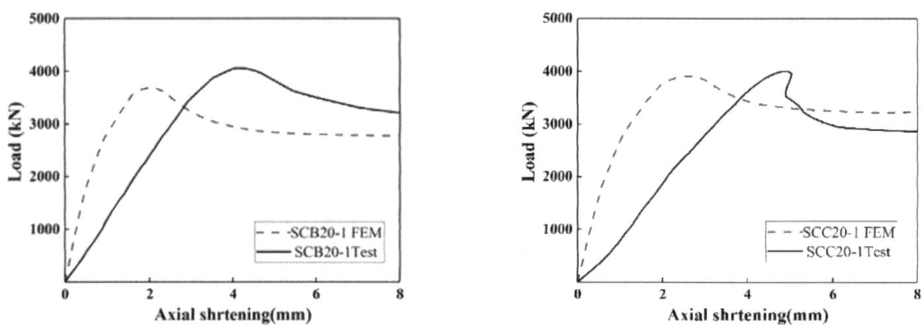

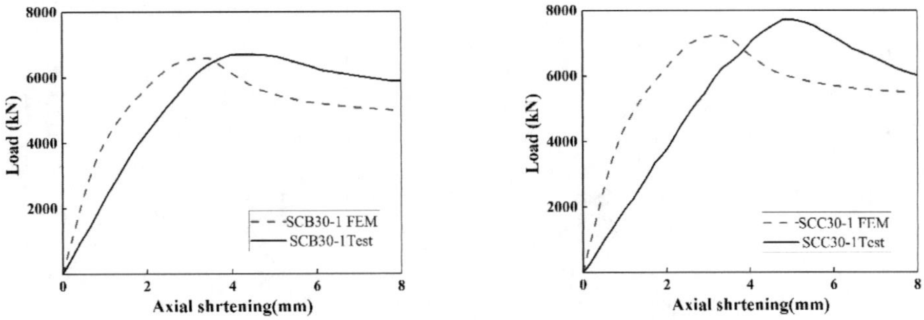

Fig. 6 Load-axial displacement curve comparison

3. Parameter analysis

3.1 parameter design

In order to analyze the influence of opening aperture, opening spacing and stiffening rib width-thickness ratio on the mechanical properties of PBL stiffened rectangular CFST columns, three groups of specimens were designed using finite element method to study the change of bearing capacity. LSC~PSC5 changed the opening diameter (0~2/3b_s), PSC6~PSC10 changed the opening spacing (1.5d~2.2d), and PSC11~PSC15 changed the width to thickness ratio of stiffening rib (10-27).

Table 2 Finite element parameter design

Specimen number	L	B	t	b_s	t_s	d	y
LSC	900	300	4	90	4	-	-
PSC1	900	300	4	90	4	20	100
PSC2	900	300	4	90	4	30	100
PSC3	900	300	4	90	4	40	100
PSC4	900	300	4	90	4	50	100
PSC5	900	300	4	90	4	60	100
PSC6	900	300	4	90	4	45	70
PSC7	900	300	4	90	4	45	80
PSC8	900	300	4	90	4	45	90
PSC9	900	300	4	90	4	45	100
PSC10	900	300	4	90	4	45	110
PSC11	900	300	4	80	3	40	80
PSC12	900	300	4	80	4	40	80
PSC13	900	300	4	80	5	40	80
PSC14	900	300	4	80	6	40	80
PSC15	900	300	4	80	8	40	80

Note: L- column height, B- column width, t - plate thickness, b_s - rib width, t_s - rib thickness, d - rib opening aperture, y- rib opening spacing.

3.2 Influence of hole diameter

The figure 7 shows that compared with concrete filled steel tubular column of stiffener is not opening, the opening of stiffener can largely improve the bearing capacity of concrete filled steel tube column, PBL stiffening the bearing capacity of steel tube concrete column first increases with the increase of hole diameter of stiffening rib, after opening aperture is about 1/3 of the stiffener width, the bearing capacity is the largest.

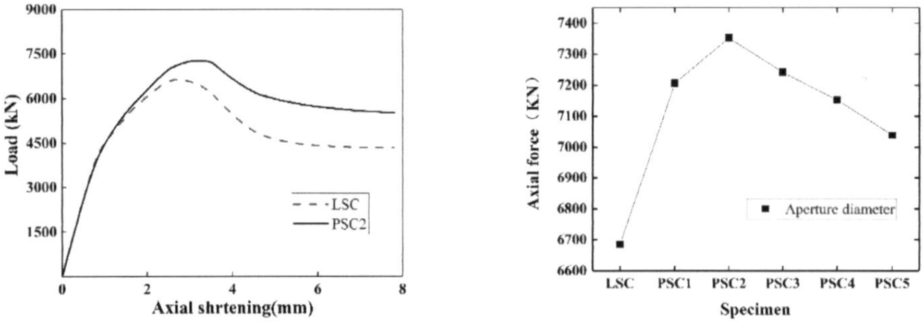

Fig. 7 Bearing capacity varies with opening size

3.3 Influence of opening spacing

The figure 8 shows that PBL type stiffening the bearing capacity of rectangular concrete-filled steel tube column with the change of hole spacing is relatively small, the influence of the overall bearing capacity increases with the increase of hole spacing, hole when spacing greater than 2 times of hole diameter, bearing capacity increases slowly, suggested that the design, PBL hole spacing of stiffening rib 2 times of hole diameter.

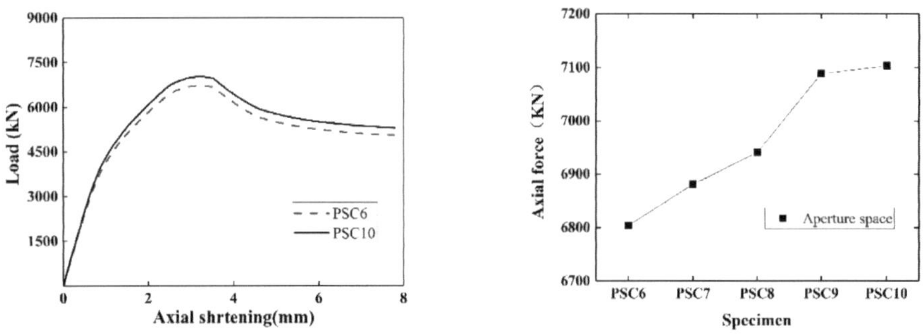

Fig. 8 Variation of bearing capacity with opening spacing

3.4 Influence of stiffening rib width to thickness ratio

It can be seen from Fig. 9 that the bearing capacity of PBL reinforced rectangular concrete-filled steel tubular columns has a relatively large influence with the change of stiffener width-thickness ratio, and the bearing capacity increases with the increase of stiffener width-thickness ratio. When the minimum stiffness of the stiffeners is satisfied, the ratio of the stiffeners to the width of the stiffeners can be appropriately reduced to improve the bearing capacity of PBL stiffened rectangular CFST columns.

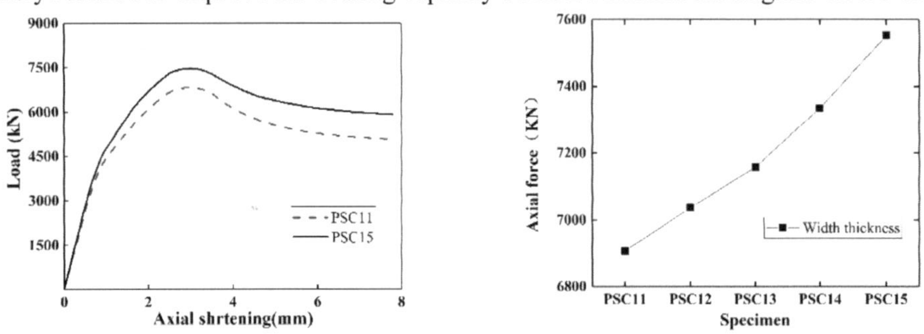

Fig. 9 Variation of bearing capacity with stiffening rib width-thickness ratio

4. Conclusion

The bearing capacity of PBL reinforced rectangular concrete-filled steel tubular columns is less affected by the spacing of stiffeners, but is relatively affected by the aperture of stiffeners and the width-thickness ratio of stiffeners. Some suggestions are given for the design of web members of truss bridges, pylons of cable-stayed bridges and piers of continuous rigid frame bridges. However, there is no specific analysis on the interface interaction between the opening stiffener and the concrete in this paper, which needs further research.

Reference

[1] Huang Z, Zhang A G, Li Y, Chen M C. (2011) Experimental study and finite element analysis of reinforced concrete-filled square steel tubular short columns under axial compression [J]. Journal of Building Structures ,32(02):75-82.
[2] Liu Y J, Zhang J G, Xu K L, Yao W Q, Sun C. (2011) Experimental study on concrete short columns with longitudinal ribs under axial compression[J]. Journal of Building Structures,32(10):159-165.
[3] Liu Y J, Cheng G, Zhang N, Zhang J G. (2014) Experimental Study on Axial Compressed Short Columned Square Concrete-filled Steel Tubular with Open-hole Steel Plate Stiffened [J]. Journal of Building Structures,35(10):39-46.
[4] HAN L H. (2004) Concrete-filled Steel Tubular Structure: Theory and Practice [M], Beijing: Science Press.
[5] CECS159:2004 technical specification for structures with concrete filled rectangular steel tube members [S]. China Planning press, Beijing.
[6] GB50018:2002 Technical Code for Cold-formed Thin-walled Steel Structures [S]. China Planning press, Beijing.
[7] Zhong T, Uy B, Han L H. (2009) Analysis and design of concrete-filled stiffened thin-walled steel tubular columns under axial compression[J]. Thin -Walled Structures, 47(12):1544-1556.

Experimental study on particle breakage and fractal characteristics of coarse-grained soil

Du Jun*, **Shen Xing-gang**

College of Architecture and Civil Engineering, Kunming University, Kunming, Yunnan, 650214, China

*Corresponding author's e-mail: dujun0605@kmu.edu.cn

Abstract. Particle breakage is an important factor affecting the strength and deformation of coarse-grained soil. In order to study the compaction characteristics and particle breakage effect of coarse-grained soil, the compaction tests of coarse-grained soil with different stone content and water content were designed. The results show that the dry density of coarse-grained soil first increases and then decreases with the increase of water content. With the increase of stone content, the maximum dry density and optimal moisture content of coarse-grained soil increase. After compaction test, the coarse-grained soil has obvious particle breakage phenomenon, and the particle breakage index increases with the increase of stone content and water content. The fractal dimension *D* can characterize the coarse-grained content of coarse-grained soil. With the increase of coarse-grained content, the fractal dimension *D* decreases gradually.

1. Introduction

Coarse-grained soil is a kind of granular accumulation, which is widely used in fill engineering such as railway, earth rock dam, expressway and so on [1]. Under the action of load, coarse-grained soil usually produces particle breakage, which changes the grading composition of soil, affects the mechanical properties of soil, and has a significant impact on the safety and stability of high fill slope. Through the large-scale compaction test of earth rock mixture, Wang [2] shows that compaction work and stone content are two important factors affecting the maximum dry density of soil. Du [3] studied the factors influencing the shear properties of rockfill materials, and the results showed that there was an exponential relationship between the internal friction angle and the coarse grain content. Liang [4] has carried out compaction test and shear test of coarse-grained soil with different stone content. The research shows that the compaction characteristics of coarse-grained soil is significantly affected by stone content, and the shear strength of soil sample is the largest under the condition of maximum dry density. Zhang [5] studied the deformation of rockfill materials under the action of cyclic stress history, which shows that the rockfill materials will slip and rotate relative to each other, resulting in particle breakage. Liu [6] carried out large-scale shear tests on four kinds of soil with different water content, which showed that the water content affected its macro mechanical properties by affecting its micro particle breakage. In order to describe and reflect the pore structure characteristics of granular materials, fractal theory is applied to the study of mechanical properties of coarse-grained soil [7-8]. Although scholars at home and abroad have carried out a variety of tests to study the mechanical properties of coarse-grained soil, it still needs to be further studied to consider the water content and stone content of coarse-grained soil, and to analyse the particle breakage characteristics of soil using fractal theory. In order to provide theoretical support for coarse-grained soil filling engineering, the compaction tests of different coarse-grained content of coarse-grained soil were carried out, and the fractal theory was used to analyse the breakage effect of soil particles after compaction.

2. Test process and method

2.1. Test materials and instruments

The materials used in the test are selected from the filling materials of an earth rock dam. In order to meet the loading requirements of the test, the on-site gradation of soil is scaled according to the equivalent substitution method. The results are shown in table 1. DDJ30-6 electric compaction instrument is used for compaction test. The sample size of the instrument is Φ 300x288mm, hammer drop height is 600mm ±1, hammer speed is 8 blows/min, hammer mass is 15.5kg, and compaction work is 591.9kJ/m^3.

Table 1. Content of each granule group before and after scaling

gradation	percentage content of each particle size /%						
	<2mm	2~5mm	5~10mm	10~20mm	20~40mm	40~60mm	>60mm
site grading	42.23	20.75	3.89	6.52	10.23	8.02	8.36
scale grading	42.23	20.75	5.03	8.42	13.21	10.36	

2.2. Test plan

Stone content refers to the percentage of block stone in the total mass of earth rock mixture. Generally, the particle size of 5mm is used as the dividing line between block stone and soil, which is expressed by P_5. The compaction test is carried out according to different stone content and water content. The stone content is designed as 15%, 30%, 45% and 60%, and the water content is designed as 4.7%, 6.5%, 8.2%, 9.8% and 11.4%. The cumulative curve of soil gradation under different stone content is shown in figure 1.

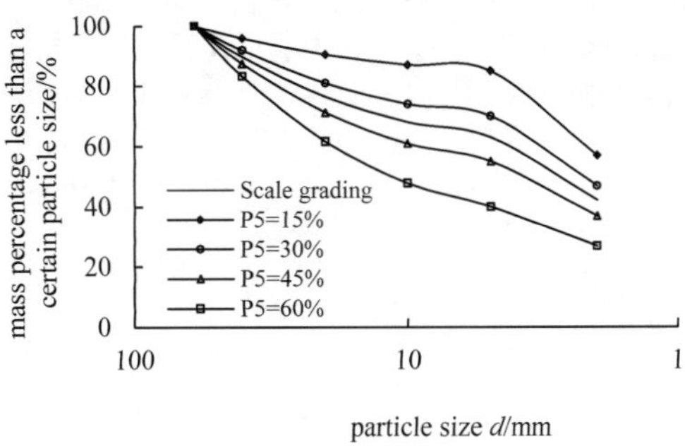

Figure 1. Cumulative curve of particle size distribution

2.3. Test process

The soil samples were prepared according to the stone content and water content of the designed soil samples. The prepared soil samples were loaded into the compaction machine in three layers, and each layer was compacted 44 times. After the test, 2kg representative soil sample was taken from the middle of the soil sample to determine the moisture content. The soil samples after compaction test were dried and screened, and the change of particle size was counted.

3. Test results and analysis

3.1. Compaction characteristics of coarse-grained soil

The compaction curve of coarse-grained soil with different stone content is shown in figure 2. It can be seen from figure 2 that the dry density of coarse-grained soil first increases and then decreases with the increase of water content. The reason is that when the moisture content is small, the whole soil particle, including the gap between soil particles, is surrounded by film water. At this time, due to the lubrication of water, the shear resistance is large, the soil is not easy to be compacted, and the dry density is small. The water content of soil will be increased continuously, and the above thin film layer will thicken, which will reduce the shear resistance, and the dry density will gradually reach the peak, namely the maximum dry density. As water is incompressible and its density is generally smaller than that of soil particles, the dry density decreases with the increase of water content. When the water content is the same, with the increase of coarse content, the maximum dry density shows an upward trend. This is because when the stone content is different, the coarse and fine content are also different, resulting in different dry density.

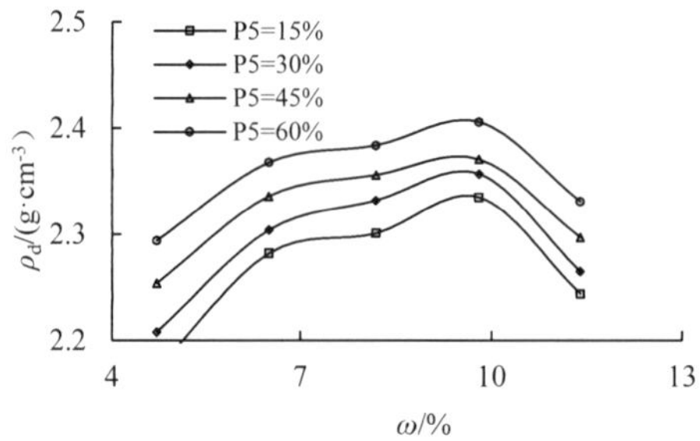

Figure 2. Compaction curve of soil samples with different stone content

The changes of maximum dry density and optimal water content under different stone content conditions are shown in figure 3. It can be seen from Figure 3 that the optimal moisture content increases approximately linearly with the increase of coarse grain content. The reason is that there are more fine particles in coarse-grained soil and the water holding capacity of soil is strong. The maximum dry density increases with the increase of stone content, and the increasing rate is from slow to steep. This is due to the different content of coarse particles, resulting in the different proportion of soil and block stone. When there are less coarse particles, the soil is the skeleton, the block stone is the filling material, and the block stone is suspended in the soil. At this time, the pore of the soil is large, and the soil is not easy to be compacted. With the increase of the content of rock, the soil gradually becomes the filling material, and the block stone becomes the skeleton. Under the hammer load, the large particles have obvious particle breakage, which makes the block rock in the soil become more and more overturned, occluded and filled with each other, and the pore volume of the soil decreases and the dry density increases.

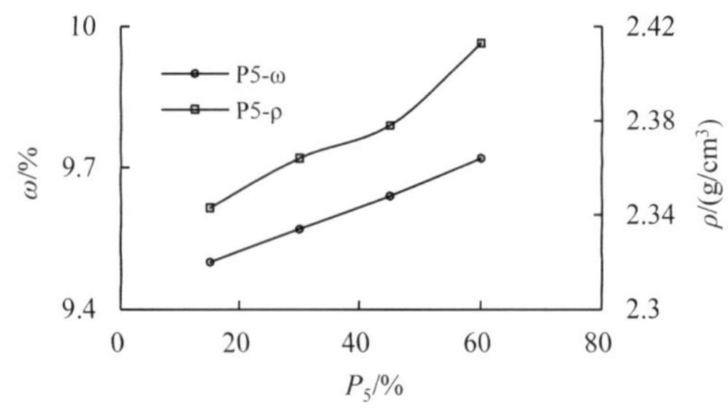

Figure 3. Variation of optimum water content and maximum dry density with stone content

3.2. Particle breakage characteristics of coarse grained soil

The degree of particle breakage is measured by multi particle size index Bg. Marsal [9] reflects the particle breakage by summing the absolute value of the percentage difference of particle composition before and after the test. It can be expressed as:

$$B_g = \sum |w_{ki} - w_{kf}| \tag{1}$$

Where w_{ki} is the content of a certain fraction on the grading curve before compaction, w_{kf} is the content of a certain fraction on the grading curve after compaction.

The gradation changes of coarse-grained soil before and after test under different stone content and water content are shown in table 2. According to table 2, the mass percentage of 60-40mm and 40-20mm coarse-grained soil decreased, the mass percentage of 10-5mm and less than 2mm increased. The test results show that the coarse-grained soil has obvious particle breakage during compaction.

Table 2. statistical results of soil gradation after compaction test

P_5/%	w/%	mass percentage content of different particle sizes /%					
		<2mm	2~5mm	5~10mm	10~20mm	20~40mm	40~60mm
15%	4.7	58.31	27.48	2.09	3.78	4.87	3.47
	6.5	58.49	27.37	2.00	3.98	4.76	3.4
	8.2	58.74	27.14	2.13	4.10	4.64	3.25
	9.8	58.89	27.04	2.26	4.05	4.59	3.17
	11.4	59.08	26.88	2.39	4.08	4.51	3.06
30%	4.7	48.45	22.65	4.56	6.81	10.17	7.36
	6.5	48.81	22.48	4.73	6.79	10.04	7.15
	8.2	49.11	22.41	4.89	6.75	9.77	7.07
	9.8	49.43	22.32	5.03	6.69	9.54	6.99
	11.4	49.73	22.24	5.18	6.64	9.28	6.93
45%	4.7	39.56	18.89	6.42	9.93	14.46	10.74
	6.5	40.02	19.12	6.65	9.7	14.17	10.34
	8.2	40.49	19.29	6.86	9.51	13.99	9.86
	9.8	40.98	19.46	7.06	9.27	13.81	9.42
	11.4	41.45	19.62	7.26	9.06	13.64	8.97
60%	4.7	30.16	14.42	7.78	12.63	20.06	14.95
	6.5	31.13	14.72	7.98	12.24	19.4	14.53
	8.2	31.38	15.02	8.36	12.06	19.02	14.16
	9.8	31.59	15.34	8.64	11.89	18.68	13.86
	11.4	31.82	15.64	8.96	11.72	18.33	13.53

According to formula (1), the particle breakage index B_g of coarse-grained soil after compaction test is calculated, as shown in table 3. It can be seen from table 3 that under the same water content condition, with the increase of stone content, the particle breakage index B_g increases. Under the same stone content, the particle breakage index B_g increases with the increase of water content.

Table 3. Calculation results of particle breakage index *Bg*

P_5/%	different moisture content *w*/%				
	4.7	6.5	8.2	9.8	11.4
15	3.64	4.10	5.22	5.70	6.38
30	4.20	5.10	6.18	6.86	8.00
45	8.16	9.74	11.70	13.02	15.08
60	9.28	12.02	14.08	15.50	17.40

3.3. Fractal characteristics of coarse grained soil

The fractal dimension can be used to characterize the grading characteristics of soil [10]. The calculation method of fractal dimension of coarse-grained soil is as follows:

$$\lg[M_1(d_i)/M_t] = (3-D)\lg(d_i/d_{\max}) \tag{2}$$

Where M_t is total mass, $M_1(d_i)$ is total mass of particles larger than a certain particle size, $M_1(d_i)/M_t$ can be obtained by particle screening statistics, d_{max} is maximum particle size, D is fractal dimension.

According to formula (2), the fractal dimension D of coarse-grained soil before and after the test is calculated as shown in Table 4.

Table 4. Fractal dimension *D* value of coarse grained soil before and after test

P_5 /%	initial	different moisture content *w* / %				
		4.7	6.5	8.2	9.8	11.4
15	2.8791	2.8855	2.8861	2.8871	2.8878	2.8886
30	2.8088	2.8184	2.8204	2.8224	2.8243	2.8261
45	2.7225	2.7479	2.7526	2.7571	2.7617	2.7661
60	2.6020	2.6514	2.6625	2.6670	2.6710	2.6751

It can be seen from the analysis in table 4 that the fractal dimension *D* of coarse-grained soil can characterize the grading characteristics of coarse-grained soil. The decrease of fractal dimension *D* results in the increase of coarse-grained content in coarse-grained soil, that is, the fractal dimension of coarse-grained soil is inversely proportional to the coarse-grained content. After compaction test, the fractal dimension of coarse-grained soil decreases with the increase of stone content under the same water content. Under the same stone content, the fractal dimension of coarse-grained soil increases with the increase of water content.

4. Conclusions

(1) With the increase of coarse-grained content, the optimal moisture content and maximum dry density of coarse-grained soil increase.

(2) The larger the content of coarse particles, the more significant the particle breakage. Particle breakage is directly proportional to water content and coarse particle content of coarse-grained soil.

(3) The fractal dimension *D* value can be used to characterize the grading characteristics of coarse-grained soil, and the fractal dimension decreases when the coarse-grained soil particles are broken.

(4) The research of coarse-grained soil particle breakage has a good theoretical support to guide the practice of high fill engineering. The micromechanical mechanism of coarse-grained soil particle breakage needs to be further studied by combining with the discrete element numerical simulation method.

Acknowledgments

This research is supported by the Joint special fund project of basic research in local universities of Yunnan Province(202001BA070001-066).

References

[1] Shi Zhen-ming., Zhao Xiaowei., Peng Ming. (2014) Review of studies on large scale triaxial tests of coarse-grained soils. Journal of Engineering Geology., 22(5):792-796.
[2] Wang Jiang., Gong Hui., Guo Na-na., et al. (2021) Large scale compaction test of earth rock mixture. Highway., 66(01):297-302.
[3] Du Jun., Hou Ke-peng. (2017) Effect of percentage of coarse-grained aggregate on shear deformation and strength of rockfill materials in waste dump. Mining and Metallurgical Engineering., 37(01):14-17.
[4] Liang Wei., Zhao Wei-sheng., Deng Dai-qiang., et al. (2020) Shear characteristics of rockfill material under maximum dry density condition. Journal of Henan Polytechnic University (natural science)., 39(06):152-158.

[5] Zhang Zhen-dong., Li Guo-ying. (2017) Experimental study on particle breakage behaviors of rockfill under cyclic loadings. Chinese Journal of Geotechnical Engineering., 39(08):1510-1516.
[6] Liu Xin-rong., Tu Yi-liang., Wang Peng., et al. (2017) Particle breakage of soil-rock aggregate based on large-scale direct shear tests. Chinese Journal of Geotechnical Engineering., 39(08):1425-1434.
[7] Shi Xiu-song., Cheng Zhan-lin. (2010) Fractal behavior in breakage of rockfill material. Chinese Journal of Rock Mechanics and Engineering., 29(S2): 3852-3857.
[8] Du Jun., Hou Ke-peng., Liang Wei., et al. (2013) Experimental study of compaction characteristics and fractal feature in breakage of coarse-grained soils. Rock and Soil Mechanics., 34(S1):155-161.
[9] Marsal R.J. (1967) Large-scale testing of rockfill material. Journal of the Soil Mechanics & Foundation Division, 93(2): 27—43.
[10] Tyler S.W., Wheatcraft S.W. (1992) Fractal scaling of soil particle size distributions: analysis and limitations. Soil Science Society of America Journal, 56(2): 362-369.

Journal of Physics: Conference Series **2011** (2021) 012037 doi:10.1088/1742-6596/2011/1/012037

Method for generating mesostructure of soil-rock mixture based on Minkowski sum

Peng Zhang[1*], Junqi Chen[1] and Qingjun Xian[1]

[1] College of Civil Engineering, Henan University of Technology, Zhengzhou, Henan, 450001, China

[*]Corresponding author's e-mail: zhangpeng@haut.edu.cn

Abstract. The soil-rock mixture is widely distributed in nature, and the mechanical properties of it is a key concern in geotechnical engineering field. The mechanical response of soil-rock mixture can be obtained using numerical approaches. As the precondition for numeric simulation, the mesostructure of soil-rock mixture is indispensable. The random rock packing is the main step in the construction of the mesostructure. The classical trial packing method are inefficient when the rock area fraction is high. Permissible positions to pack new rock can be predicted in the algorithm using entrance block and background mesh, resulting in dramatic improvement in efficiency and robustness. Entrance block and Minkowski sum can be transformed into each other, and there are more algorithms for Minkowski sum. Therefore, a random rock packing algorithm based on Minkowski sum is constructed, each rock can be packed successfully at one time, avoiding multiple trials in the traditional trial packing method and improving the efficiency of packing.

1. Introduction

The soil-rock mixture consists of rock and soil, which is not only widely distributed in nature but also used as engineering materials. The mechanical response of soil-rock mixture can be obtained using numerical approaches. The mesostructure of soil-rock mixture, as the precondition for numeric simulation, is indispensable. The random rock packing is the main step in the construction of the mesostructure model of soil-rock mixture, in which the rock packed later should not penetrate with the previously packed rock. The trial packing method[1-2] first randomly determines a location to try to pack rock, and then judges whether the rock overlaps with the previously packed rock. If there is a penetration, reselect a new location and try again. Repeated trials are inefficient when the rock area fraction is high. The layering disposition method[3] tends to produce more blank areas, thus reducing the maximum rock area fraction.

An algorithm using entrance block[4] and background mesh is proposed by Chen et al[5-6], in which permissible positions to pack new rock can be predicted, resulting in dramatic improvement in efficiency and robustness. Entrance block and Minkowski sum can be transformed into each other[7]. The concept of entrance block is relatively new and there are fewer algorithms for it, while there are more algorithms for Minkowski sum. Therefore, this paper introduces Minkowski sum into the random rock packing and uses the existing algorithm for Minkowski sum. A random rock packing algorithm based on Minkowski sum is constructed.

2. Polygon penetration judgment

2.1. Polygon penetration judgment based on entrance block

In order to accurately judge the contact between two blocks, Shi proposed the concept of the entrance block and formed the complete contact theory[4]. Therefore, the position relationship between two blocks is reduced to the position relationship between a reference point and the entrance block, the contact judgment of two blocks is converted into the inclusion judgment of a reference point and the entrance block.

Select any point a_0 in block A as the reference point, the entrance block $E(A,B)$ of block A and block B is defined as

$$E(A,B) = \mathrm{B} - \mathrm{A} + a_0 = \{b - a + a_0 | a \in A, b \in B\}. \quad (1)$$

where $a \pm b$ represents the addition and subtraction between the position vectors corresponding to point a and point b. The entrance block $E(A,B)$ of block A and block B is illustrated in figure 1. In this paper, convex polygon is used to represent blocks (rock) in two-dimension.

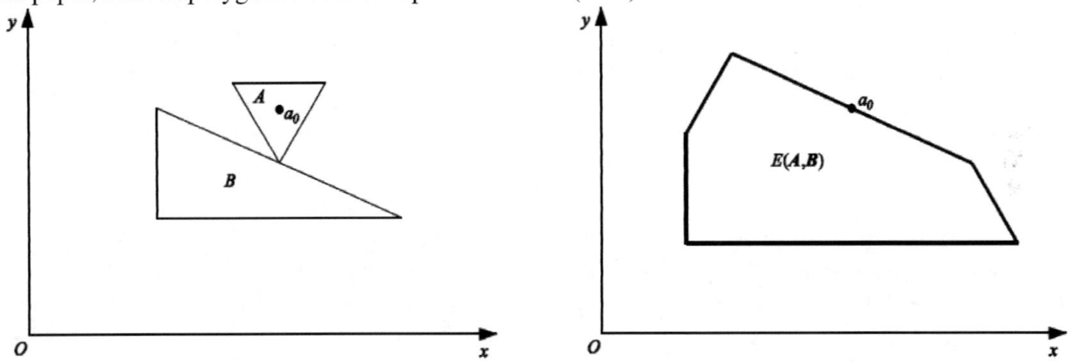

Figure 1. The entrance block $E(A,B)$ of block A and block B.

When the reference point a_0 is on the boundary $\partial E(A,B)$ of the entrance block $E(A,B)$, block A and block B contact with each other (Figure 1 shows this state); when the reference point a_0 is located outside the entrance block $E(A,B)$, block A and block B are separated with each other (Figure 2(a) shows this state); when the reference point a_0 is located inside the entrance block $E(A,B)$, block A and block B penetrate each other. The positional relationship between the reference point a_0 and the entrance block $E(A,B)$ is the direct and clear expression of the positional relationship between block A and block B.

2.2. Polygon penetration judgment based on Minkowski sum

The Minkowski sum of two objects A and B in an Euclidian space was defined by the German mathematician Hermann Minkowski (1864-1909) as the position vector addition of each point a in A and each point b in B

$$A \oplus B = \{a + b | a \in A, b \in B\}. \quad (2)$$

By comparing equations (1) and (2), we can immediately see that the entrance block $E(A,B)$ is actually the Minkowski sum $B \oplus \bar{A}$, $E(A,B) = B \oplus \bar{A} = \bar{A} \oplus B$, where $\bar{A}$ is obtained through translating the centrally symmetric counterpart of A by vector a_0, $\bar{A} = -A + a_0$, which is illustrated in figure 2.

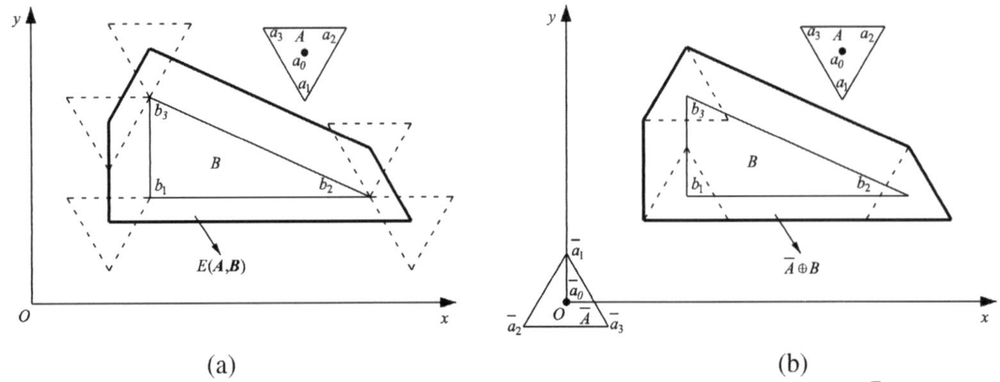

Figure 2. (a)The entrance block $E(A,B)$, (b) the Minkowski sum $B \oplus \bar{A}$.

Therefore, $\bar{A}$ is firstly obtained by changing A, and then the positional relationship between block A and block B can be determined by using the position relationship between the reference point a_0 and $B \oplus \bar{A}$. When the reference point a_0 is on the boundary $\partial(B \oplus \bar{A})$ of $B \oplus \bar{A}$, block A and block B contact with each other; when the reference point a_0 is located outside $B \oplus \bar{A}$, block A and block B are separated with each other (Figure 2(b) shows this state); when the reference point a_0 is located inside $B \oplus \bar{A}$, block A and block B penetrate each other.

Before the position relationship judgment, Minkowski sum $B \oplus \bar{A}$ should be formed firstly. Minkowski sum algorithms include slope diagram algorithm[8], contributing vertices method[9] and so on. The concept in contributing vertices method is intuitive and simple, and it will be utilized to form Minkowski sum in this paper.

3. Random rock packing

3.1. Target region discretizing and boundary expanding

In order to quantitatively represent the position of rock in the target region, selecting a reference point in the rock, and the position of the rock is represented by the coordinate of the reference point. If the reference point is located at a certain point in the target region, the position of the rock is determined. A rock to be packed can be anywhere within the target region, and the reference point can be anywhere on the rock, so the reference point can be located at any point in the target region. There are an infinite number of such potential points. In order to facilitate the packing operation, the number of potential points needs to be limited. A background grid covering the target region is arranged to discretize it, and the background grid nodes within it are taken as the potential points. The red points in figure 3(b) are the potential points after the discretization of the target region in figure 3(a), that is, the initial potential points for the first rock.

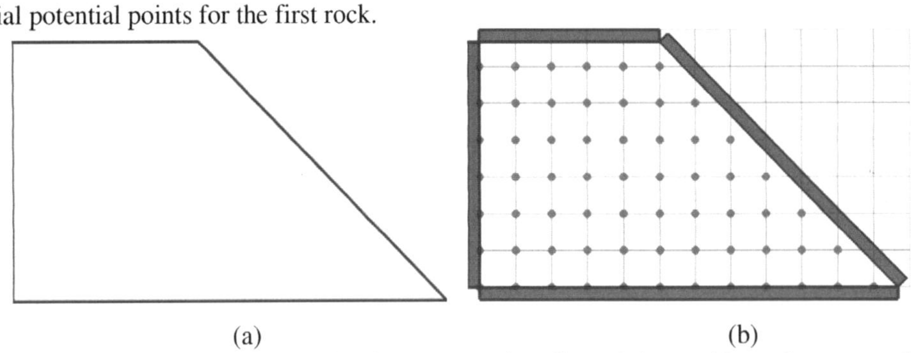

Figure 3. (a)The target region, (b) The target region discretizing and boundary expanding.

The Minkowski sum algorithm based on contributed vertices is applicable to two convex polygons, so it is necessary to expand each boundary line of the target region into a convex polygon outside the

Journal of Physics: Conference Series **2011** (2021) 012037 doi:10.1088/1742-6596/2011/1/012037

target region. In this paper, we chose to expand it into a rectangle. The four blue rectangles in figure 3(b) are the expansion of the four boundary lines of the target region in figure 3(a) to the outside of the region.

3.2. The first rock packing

According to the particle size requirement, the first rock polygon A_1 is generated randomly, and any point in it is selected as the reference point a_{01}. The yellow polygon on the right of figure 4 is A_1, and the blue point in it is a_{01}. Calculate $\overline{A}_1$ from A_1 and a_{01}, and then calculate the Minkowski sum of $\overline{A}_1$ and each boundary polygon. The four light blue convex polygons in figure 4 are the Minkowski sum polygons formed by $\overline{A}_1$ and the four boundary rectangles. According to section 2.2, if the selected point is located inside Minkowski sum, the packed rock and boundary block penetrate each other, so the potential points inside Minkowski sum must be deleted, so that the potential points are located outside Minkowski sum. The red points in figure 4 are the potential points for the first rock after the boundary Minkowski sum modification.

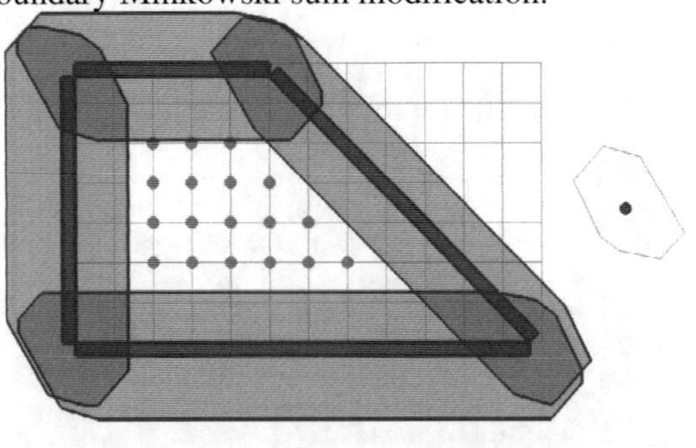

Figure 4. The potential points for the first rock.

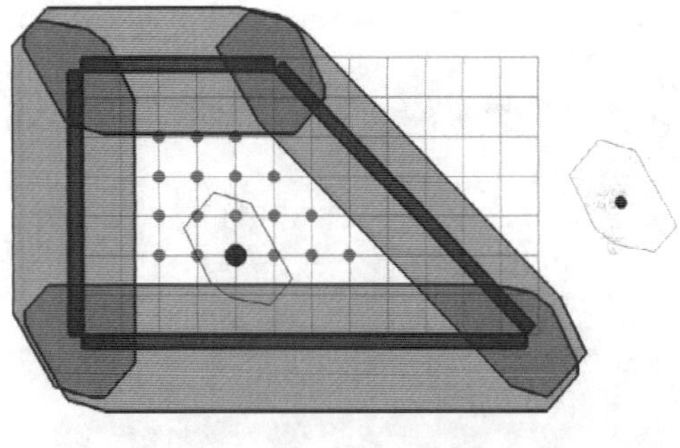

Figure 5. The result of the first packed rock.

Randomly select a point among the potential points for the first rock to locate the reference point a_{01}, and then the first rock is successfully packed. The big blue point in figure 5 is the selected point, and the yellow polygon in the target region is the first packed rock. It can be seen from the figure that it is completely within the target region and does not overlap with the boundary.

3.3. The orther rock packing

Randomly generate the i-th rock polygon A_i $(i > 1)$, and select any point in it as the reference point a_{0i}, and calculate $\overline{A}_i$ from A_i and a_{0i}. The yellow polygon on the right of figure 6(b) is A_2, and the blue point in it is a_{02}. The reference point of the i-th rock can be placed anywhere in the target region except for the previously packed i-1 rock, so the initial potential points of the i-th rock needs to be reset. The potential points within the previously packed i-1 rock should be deducted at this time. The red points in figure 6(a) are the potential points for the second rock in theory.

But the Minkowski sum of $\overline{A}_i$ and the previously packed i-1 rock should be calculated later, and the potential points within the Minkowski sum will be deducted. The k-th rock A_k is located in the Minkowski sum of $\overline{A}_i$ and A_k $(0< k < i)$, so the subsequent deduction of the potential points within the Minkowski sum can ensure that the potential points in previously packed i-1 rock are deducted. In order to facilitate the operation, the potential points in previously packed i-1 rock are no longer deducted at this time. The red points in figure 6 are the potential points for the second rock in the algorithm.

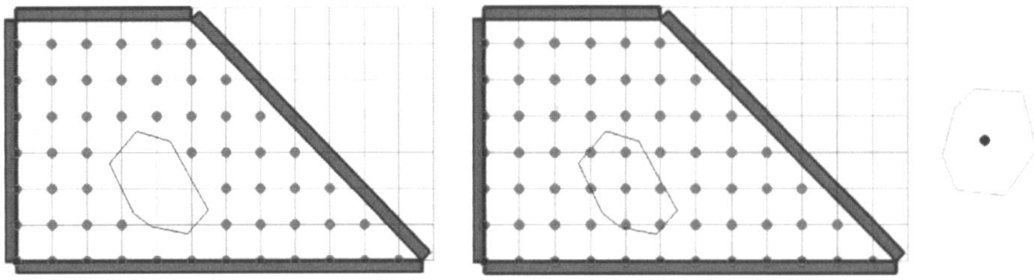

(a) (b)

Figure 6. The initial potential points for the second rock (a) in theory (b) in the algorithm.

Calculate the Minkowski sum of A_i and each boundary polygon and the previously packed rock, delete the potential points in the Minkowski sum, and the potential points for the i-th rock can be obtained. The five light blue convex polygons in figure 7 are the Minkowski sum polygons formed by $\bar{A}_2$ and the four boundary rectangles and the packed first rock, and the red points are the potential points for the second rock after the Minkowski sum modification.

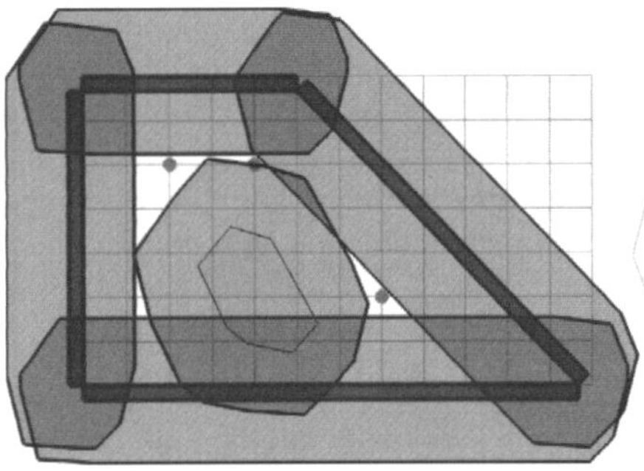

Figure 7. The potential points for the second rock.

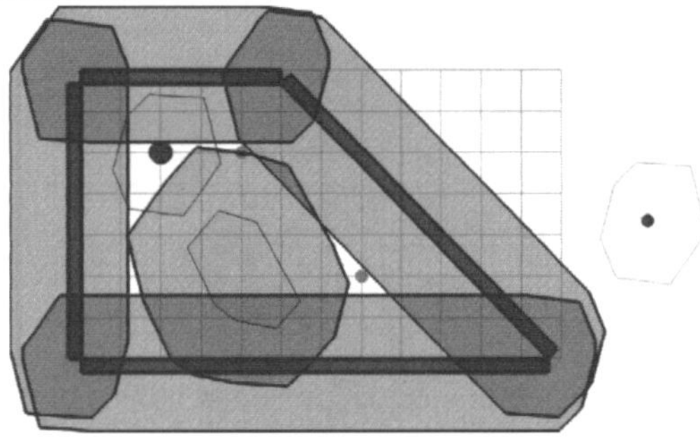

Figure 8. The result of the second packed rock.

Randomly select a point among the potential points of the i-th rock to locate the reference point a_{0i}, and then the i-th rock is successfully packed. The big blue point in figure 8 is the selected point, and the yellow polygon containing the big blue point is the second packed rock. It can be seen from the figure that it is completely within the target region and does not overlap with the boundary and the first packed rock.

Pack the other rock according to the method, and stop packing when the packing conditions are met (such as reaching the specified rock area fraction or there is nowhere to pack).

4. Implementation

Based on the algorithm described above, using C++, a system for generating a mesostructure of soil-rock mixture based on Minkowski sum was compiled. Using the system, a two-dimensional soil-rock mixture slope model and a two-dimensional soil-rock mixture dam model were generated respectively, as shown in figure 9 and figure 10 respectively.

The total length of the two-dimensional soil-rock mixture slope is 105m, the height is 40m, the step length is 30m, the step height is 20m, the slope gradient is 45°, the diameter of the circumscribed circle of the rock is 10m, the background grid size is 0.25m, and the rock area fraction is 0.20. Figure 9 shows three packed results. It can be seen from the figure that the size and shape of the generated rock and the distribution of rock are random.

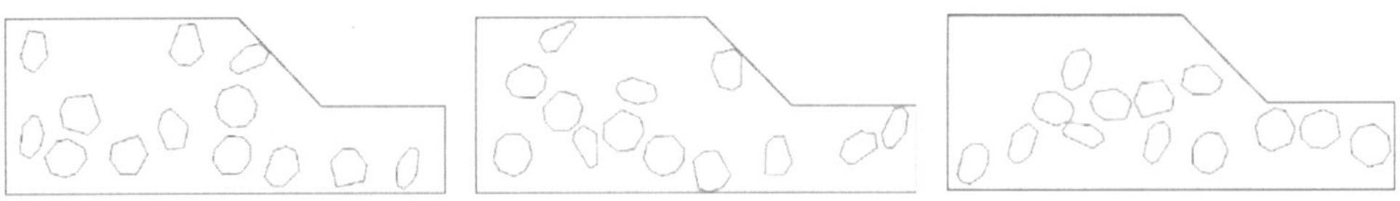

Figure 9. Three packed results of a two-dimensional soil-rock mixture slope.

The bottom length of the two-dimensional soil-rock mixture dam is 5m, the top length is 2m, and the height is 2m. The slope on the left is 60°, and the slope on the right is 45°. The diameter of the circumscribed circle of the rock is 0.25m, and the background grid size is 0.0125m. Figure 10(a-b) shows the packed results of three kinds of rock area fraction of 0.10, 0.30. Figure 10(c) shows the packed result of the maximum rock area fraction, which is 0.56.

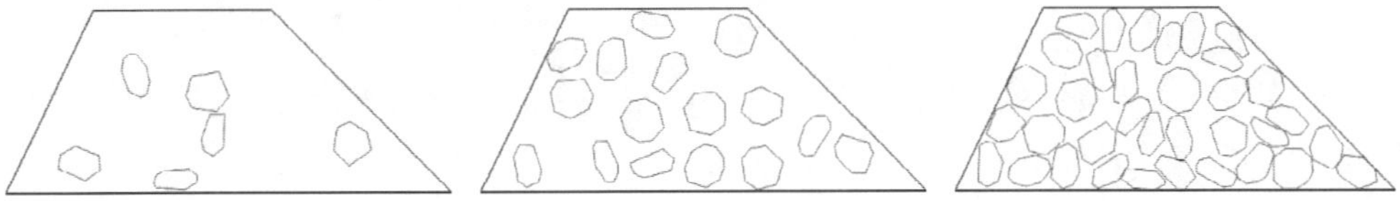

(a) (b) (c)

Figure 10. Three packed results of a two-dimensional soil-rock mixture dam of rock area fraction of (a) 0.10, (b) 0.30, (c) 0.56.

5. Conclusion

The relationship between the entrance block and the Minkowski sum is described in this paper. The Minkowski sum is introduced into the random packing algorithm of the soil-rock mixture. The existing algorithm of Minkowski sum can be fully utilized to promote the study of soil-rock mixture. In this paper, a random packing algorithm based on Minkowski sum is proposed for soil-rock mixture, which can obtain the potential points for each rock, so that each rock can be packed successfully at one time, avoiding multiple trials in the traditional trial packing method and improving the efficiency of packing.

In addition, Minkowski sum is not only suitable for two-dimensional rock penetration judgment, but also for three-dimensional rock penetration judgment. The random packing algorithm for three-dimensional soil-rock mixture based on Minkowski sum will be studied in the future.

References

[1] Xu, W., Hu R., Yue Z. (2009) Development of random mesostructure generating system of soil-rock mixture and study of its mesostructural mechanics based on numerical test. Chinese Journal of Rock Mechanics and Engineering, 28: 1652- 1665.

[2] Qin, C., Guo C., Zhang n. (2011) A pre-processing scheme based on background grid approach for meso-concrete mechanics. Journal of hydraulic engineering, 42: 941- 948.

[3] Tang, X, Zhang, C. (2008) Layering disposition and FE coordinate generation for random aggregate arrangements. Journal of Tsinghua University (Science and Technology), 48: 2048- 2052.

[4] Shi, G. (2015) Contact theory. Science China Technological Sciences, 58: 1450-1496.

[5] Chen, L., Zhang, P., Zheng, H. (2017) Mesostructure modeling of soil -rock mixtures and study of its mesostructural mechanics based on numerical manifold method. Chinese Journal of Rock and Soil Mechanics, 38: 2402-2410+2433.

[6] Chen, L., Yang, Y., Zheng, H. (2018) Numerical study of soil-rock mixture: Generation of random aggregate structure. Science China Technological Sciences, 61: 359–369.

[7] Zhang, P., Zheng, H. (2015) Generalized Contributing Vertices-Based Method for Minkowski Sum Outer-Face of Two Polygons. In: the 8th International Conference on Image and Graphics. Tianjin. pp. 333-346.

[8] Ghosh, P.K. (1993) A unified computational framework for Minkowski operations. Computers & Graphics, 17: 357–378.

[9] Barki, H., Denis, F., Dupont, F. (2009) Contributing vertices-based minkowski sum computation of convex polyhedra. Computer-Aided Design, 41: 525-538.

Study on constraint degree calculation of frozen shaft inner wall mass concrete

Li Fang-zheng[2,3], Zhang Ji-wei[1] *, Wu Wei-dong[4], Yao jin[4], Jin Xiao-gang[4]

[1] University of Science and Technology Beijing, Beijing, 100083, China

[2] Shaft Branch, China Coal Research Institute, Beijing, 100013, China

[3] Beijing China Coal Mine Engineering Co., Ltd., Beijing, 100013, China

[4] Xinjiang Bingtuan Water Conservancy and Hydropower Construction Engineering Group Co., Ltd., Xinjiang,830000, China

*Corresponding author's e-mail: zhangjiwei@ustb.edu.cn

Abstract. Based on the circle-shaped air shaft of Hong Qing-He coal mine located in Inner Mongolia China, the thermal-mechanical numerical model of frozen shaft inner wall mass concrete was established under complex thermal-mechanical boundary, including internal heating, external cold source, friction constraints, above shaft load and lateral pressure. The evolution characteristics of temperature-displacement field of inner wall were analysed. According to numerical simulation results, the constraint degree at the middle section of inner wall concrete is calculated. In the early stage (0-10d), the constraint degree of inner wall fluctuates within 60-99% as curing time goes. Three protection measures to delete confinement degree of mass concrete shaft wall were proposed, including (1) Reducing the friction between inner and outer walls; (2) Reducing hydration heat release of mass concrete; (3) Reducing the temperature difference between inner and outer walls during curing.

1. Introduction

Coal production areas of China has been transferred from the eastern region to the western region after 2000. The main stratum in the western region of China is Cretaceous and Jurassic, and its characteristics are low compressive strength, soften by water and easy to weather. Therefore, it is hard to build shaft by conventional shaft sinking method. Artificial ground freezing (AGF) is an effective method to prevent water burst and large deformation in above stratum during shaft sinking [1]-[4].

However, the thickness of inner wall of freezing shaft increases gradually with the depth reach to 1000m, and it will exceed 2m according to current standards. High strength concrete often be used as a main method to reduce inner wall thickness, but early temperature-constrained crack often emerges after curing under low-temperature environment because this kind concrete is higher hydration exothermic than ordinary concrete [5]. Early temperature microcracks would expansion, penetration and cracking gradually with the shaft load and hydraulic pressure increases after frozen wall thawing. The shaft structure will unstable and damage if most of temperature microcracks propagation because inner shaft structure is the last sealing water and anti-pressure structure.

Several previous studies have investigated mechanical properties of shaft concrete under high hydraulic pressure, freezing shaft wall structure of deep bedrock and the mechanism of disruption of shaft in the epipedon [6]-[9]. Most of their results seldom consider the impact of early temperature stress on confinement degree of shaft wall under complex thermal-mechanical boundary, such as internal heating, external cold source, friction constraints, above shaft load and lateral pressure. Above problem is one of the most important issue of freezing shaft stability study.

Here, we report a novel numerical simulation method that considers complex thermal-mechanical boundary to grasp the change of early temperature stress and calculate relating confinement degree of inner wall. This enables us to determine how to delete confinement degree of mass concrete shaft wall and propose relevant protection measures.

2. Project overview

The circle-shaped air shaft of Hong Qing-He coal mine located in Inner Mongolia, which diameter and depth are 7.6m and 689m respectively. The main stratum is Quaternary, Cretaceous (534.23m) and Jurassic (390.32m). A schematic diagram of the frozen shaft is shown in Fig.2, the shaft consists of outer wall, plastic plate and inner wall. Based on the design documents, the thickness of frozen wall is 3.7m, the lowest saline temperature is -28～-32°Cand the average temperature is not higher than -10°C.In this study, the C70 inner wall of shaft at -440~-636m are selected as a prototype, which thickness is 1.2m can be treat as mass concrete.

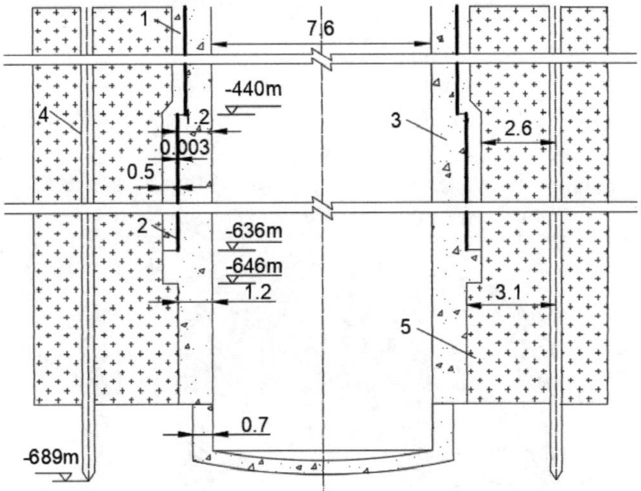

Fig.1 shaft structure and frozen wall of Hong Qing-He coal mine which consist of (1) outer wall, (2) plastic plate, (3) inner wall, (4) freezing hole, (5) frozen wall

3. Model description of numerical simulation

3.1Basic assumptions

Considering the actual physical process, some basic assumptions are introduced as below:(1) outer wall, inner wall and frozen rock is homogeneous and can be treated as linear elastic material;(2) Concrete and rebar of inner wall is compatible deformation;(3) Temperature, stress and strain of cylinder-shaped satisfies spatial axisymmetric finite element model;(4) Frozen wall simplified as constant temperature cold source.

3.2 Governing equations

The solid heat transfer axisymmetric model is expressed as:

$$\frac{\partial T_n}{\partial \tau} = \frac{\lambda_n}{c_n \rho_n}\left(\frac{\partial^2 T_n}{\partial r^2} + \frac{1}{r}\frac{\partial T_n}{\partial r}\right) + \frac{Q}{c_n \rho_n} \left(\tau > 0, 0 < r < \infty\right) \quad (1)$$

where T, τ , c, ρ, λ and Q are temperature, time, specific heat capacity, density and hydration releases heat, respectively. Subscript n can be divided 1,2 and 3 represent inner wall, outer wall and frozen wall. r is the distance between any point and innermost of inner wall.

The thermal stress axisymmetric model has been derived as below [10]:

$$\begin{cases} \sigma_r = -\dfrac{\alpha_f E}{(1-\mu) r^2} \displaystyle\int_a^r T_1 r dr + D_1 + \dfrac{D_2}{r^2} \\ \sigma_\theta = -\dfrac{\alpha_f E}{(1-\mu) r^2} \left(\displaystyle\int_a^r T_1 r dr - T_1 r^2 \right) + D_1 - \dfrac{D_2}{r^2} \\ \sigma_z = \mu(\sigma_r + \sigma_\theta) - \alpha_f E T_1 \\ \tau_{r\theta} = 0 \end{cases} \tag{2}$$

where σ_r, σ_θ, σ_z and $\tau_{r\theta}$ are radial, circular, vertical and tangential stress, respectively. D_1、D_2 is undetermined coefficient. α_f is coefficient of thermal expansion of inner wall. a is radius of inner wall. μ and E are poisson ratio and elasticity modulus, respectively.

The confinement degree of inner wall can be calculated by below equation:

$$k = (1 - L_c / L_f) \times 100 \tag{3}$$

where L_c and L_f are displacement under constraint and free conditions.

3.3 Physical parameters and boundary conditions

The numerical model presented in this study included the inner wall concrete pouring at 620 m and frozen wall. As is shown in Fig.2, A axisymmetric numerical model with a range width 4.3m × height 5 m, consisting of 0.5m outer wall, 1.2m inner wall and 2.6m frozen wall. Using the descriptions above, an axisymmetric numerical model consisting of 8,600 quadrilateral elements was created. Physical parameters and boundary conditions at different stage as follow:

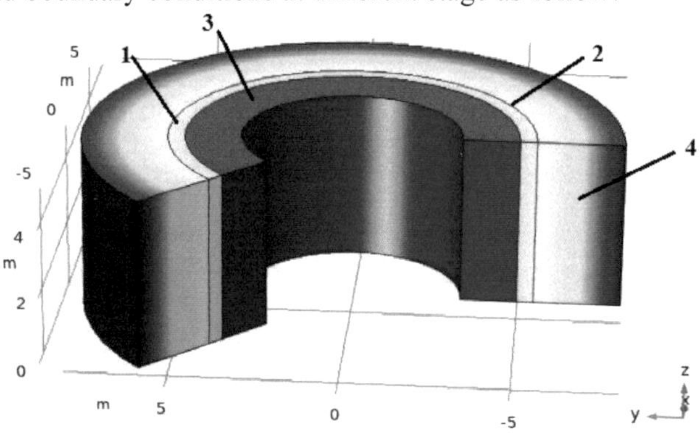

Fig.2 axisymmetric finite element model which consists of (1) outer wall, (2) plastic plate, (3) inner wall, (4) freezing hole.

(1) The pre-pouring stage of the inner wall

The initial temperature of stratum is 26°Cand outer temperature boundary of frozen wall is -20°C. The initial temperature boundary of inner wall can be grasped when the temperature of outer wall reach 0°C according to field monitoring. The physical and thermodynamics parameters for all materials are listed in Tables 1.

Table 1 Thermophysical parameters

Materials	Density / (kg·m^{-3})	Thermal conductivity/ (W(m·K)$^{-1}$)		Specific heat capacity / (kJ (kg·K)$^{-1}$)	
		-20°C	20°C	-20°C	20°C
Stratum	2040	1.82	0.95	0.96	1.20
Outer wall	2350	2.53		0.97	
Inner wall	2525	3.2		0.91	

(2) The 0-10d poured stage of the inner wall

The heat convection between concrete surface of inner wall and air is inevitable during concrete poured stage and its transfer coefficient can be calculated by below equation [11]:

$$N_u = 3.06v + 0.035\Delta t_2 + 9.55 \tag{4}$$

where v is the speed of air usually 2m/d according to field measurement. Δt_2 is the temperature difference between concrete surface and air.

The internal heating of inner wall attribute to the hydration heat of cement in concrete. Three hydration heat of cement model has been proposed by Zhu et al, such as exponential, hyperbolic and compound exponentials. The adiabatic temperature rises of inner wall concrete deduced from hydration heat is expressed as [12]:

$$\theta(\tau_1) = \frac{Q(\tau_1)\cdot(W + kF)}{c_1\rho_1} \tag{5}$$

where $Q_i(\tau_1)$ is cumulative hydration heat at τ_1. W is cementitious material consumption of per cubic meter concrete. F is mixed material consumption of per cubic meter concrete. k is reduction factor.

Using the descriptions above, the thermal boundary condition of inner wall is as follow: 1) the top and bottom boundary was set thermal insulation. 2) the left vertical boundary was set thermal flux and can be calculated by Eq. (2). 3) the right vertical boundary was set thermal contact which simulate heat transfer of plastic plate. 4) The initial temperature was set as 10°C, and the hydration heat was calculated by Eq. (5).

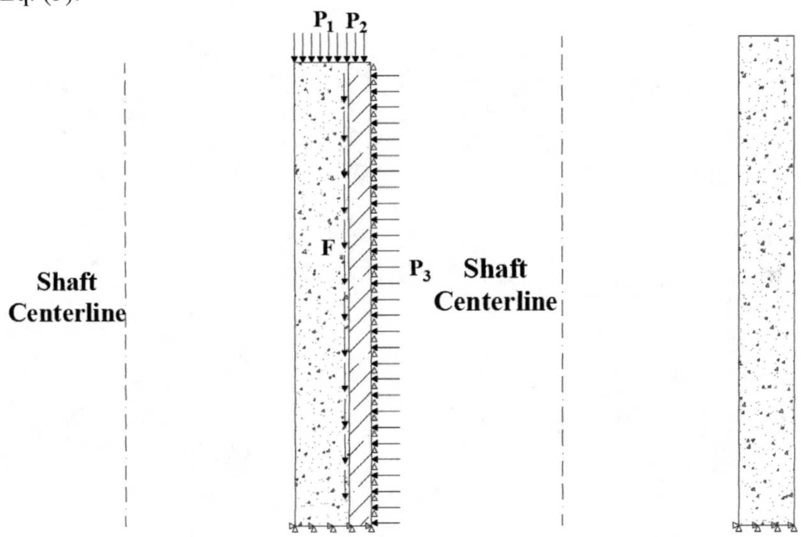

(a) Constraint model (b) Free model

Fig.3 Mechanical model of inner wall of (a) constraint model and (b) free model.

As is shown in Fig.3, the mechanical boundary can be divided constraint and free model. The in situ mechanical boundary is simplified as Fig.3 (a), the top boundary was set boundary load P_1, the left vertical boundary was set free, the right vertical boundary was set friction by plastic plate, the horizontal and vertical displacements of the bottom boundaries were constrained, the sum of surrounding rock pressure and frost heave force P_3 on outer wall was set as 2MPa based on the field monitoring data. The free model only constrains the horizontal and vertical displacements of the bottom boundaries.

4. Verification and analysis

4.1 Temperature field analysis

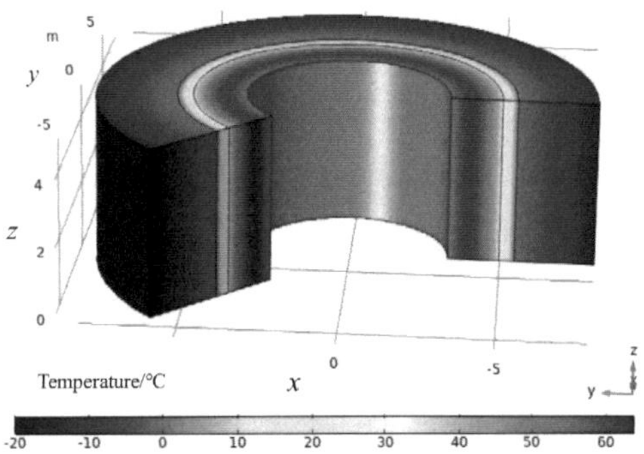

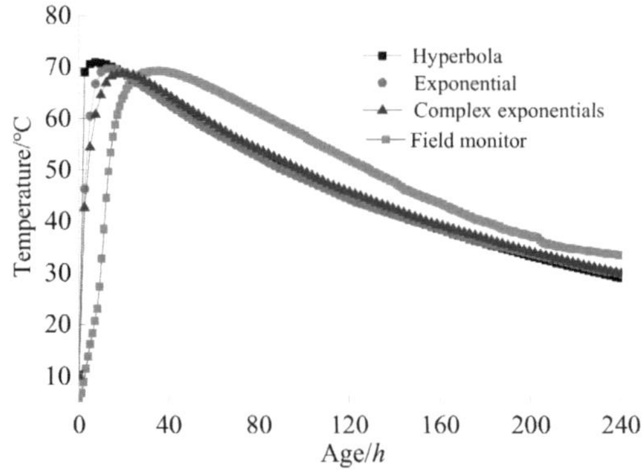

(a) Temperature field of concrete curing 48h (b) Middle temperature history

Fig.4 Temperature field of inner wall concrete

As is shown in Fig.4 (a), the middle temperature is obviously higher than other part in the inner wall. The main reason is that the heat conduction ability of concrete is so poor that the heat transfer in the middle part is hindered, but air and cold source exchange heat with inner and outer edge of inner wall continually in the meanwhile.

Based on the field monitoring data, the middle temperature of inner wall increased sharply to 70°C in 0 to 48h, and then dropped gradually (shown in Fig. 4(b)). The trend of temperature history calculated by three hydration heat of cement model is similar as field monitoring data, and compound exponential error is the minimum. Therefore, compound exponential is more suitable than other models on early hydration heat calculation of freezing shaft concrete.

4.2. Displacement field analysis

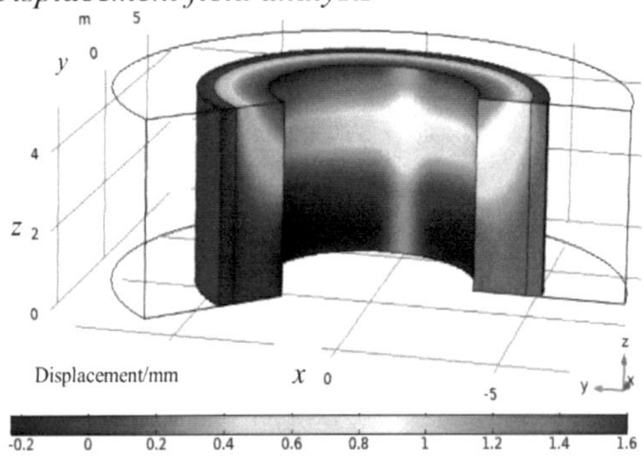

(a) 24 h

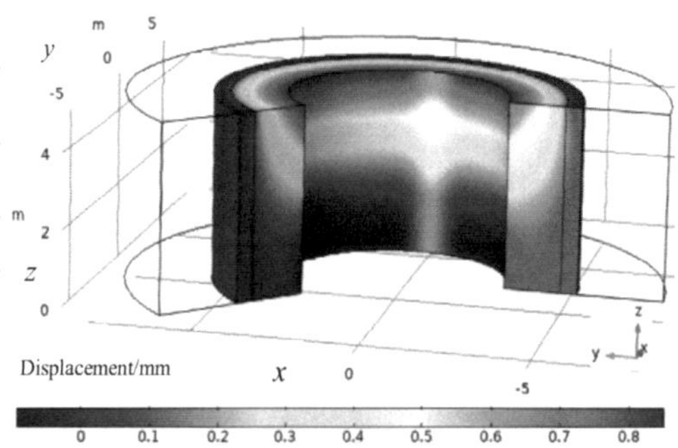

(b) 120 h

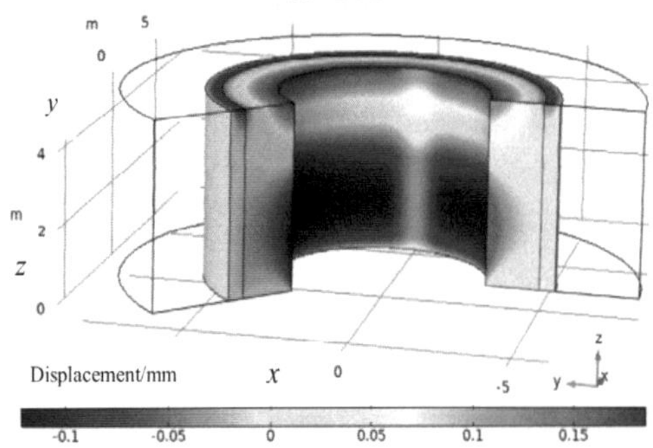

(c) 240 h

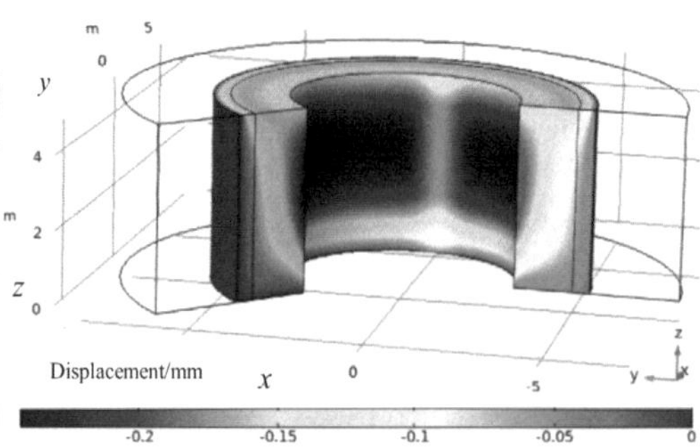

(d) 360 h

Fig.5 Displacement field of inner wall concrete in (a) 24h; (b) 120h;(c)240h;(d)360h

Fig.5 shows that the evolution of displacement field of inner wall concrete in different periods. The concrete expansion at the early period (0-24h) with temperature increases leading to the maximum upward displacement 1.6mm, whereas the displacement decreases gradually as curing time goes. The displacement of inner wall goes down to negative at 360h because of concrete shrink and increased vertical load. Therefore, concrete expansion and shrink of inner wall is obviously during 0-15d.

4.3. Constraint degree analysis

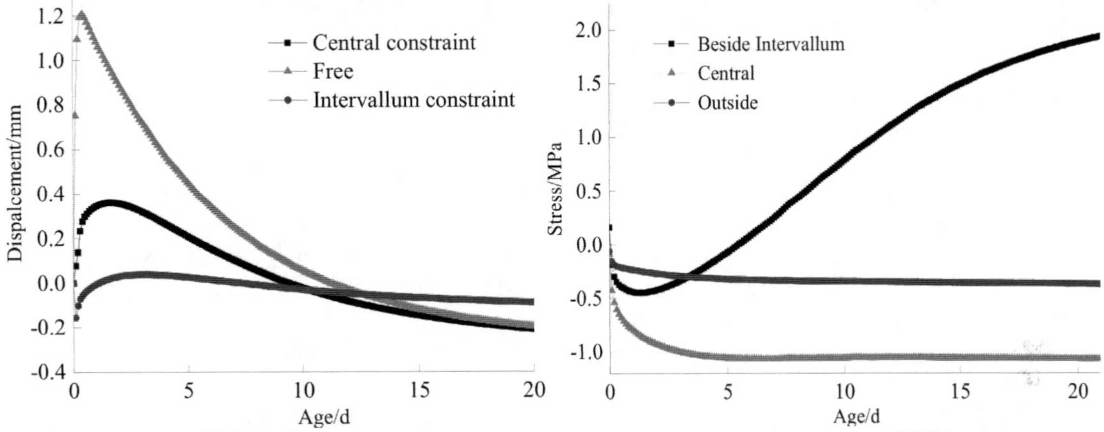

(a) Displacement (b) Stress

Fig.6 Displacement and Stress at the middle section of inner wall concrete in the different curing age

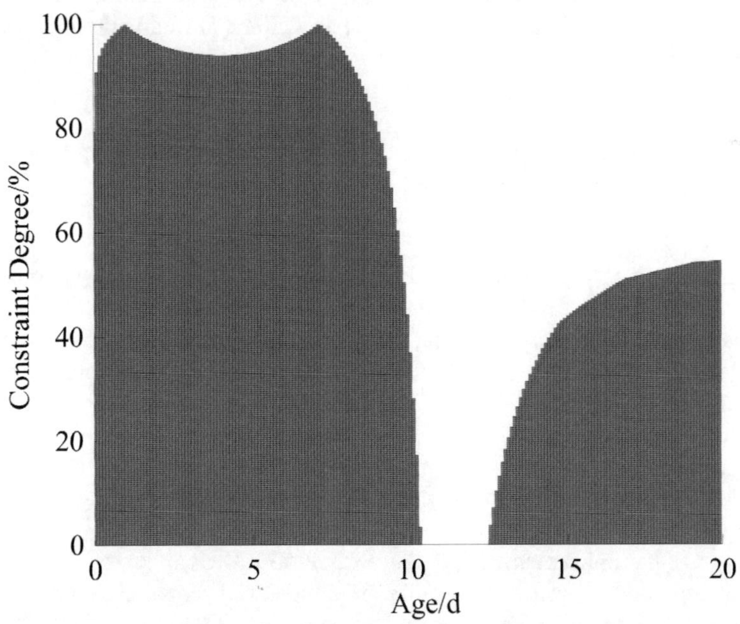

Fig.7 Constraint degree beside intervallum of inner wall concrete in the different curing age

As is shown in Fig.6(a), the maximum displacement at the middle section of inner wall under free condition is 1.2mm, while central and intervallum constrain are 0.4 and 0.038mm respectively. Obviously, the intervallum constrain is higher than central constrain leading to higher stress beside intervallum. The stress beside intervallum sharply increases as curing time goes and reach 2MPa finally (Fig.6(b)). Therefore, the cracking risk beside intervallum is higher than other area of inner wall.

The constraint degree beside intervallum at the middle section of inner wall concrete is calculated by Eq.(3) and shown in Fig.7. In the early stage (0-10d), the constraint degree of inner wall fluctuates within 60-99% as curing time goes, and two peaks in the curve emerge at the time of rapid temperature change. It is indicated that constraint degree is closely associated with the temperature change, and the cracking or damage risk time of inner wall in early-age curing is 0-10d.

In summary, the main relevant protection measures to delete confinement degree of mass concrete shaft wall as follow: (1) Reducing the friction between inner and outer walls; (2) Reducing hydration heat release of mass concrete; (3) Reducing the temperature difference between inner and outer walls during curing.

5. Conclusions

In this report, a numerical method for calculating predicting constraint degree of frozen shaft inner wall mass concrete was proposed. This proposed method is easily applicable in practical construction. Based on the novel method, the evolution characteristics of temperature-displacement field of inner wall were analyzed. The following conclusions can be drawn from the study:

(a) The thermal-mechanical numerical model of frozen shaft inner wall mass concrete is established under complex thermal-mechanical boundary, such as internal heating, external cold source, friction constraints, above shaft load and lateral pressure.

(b) The cracking risk beside intervallum is higher than other area of inner wall and the cracking or damage risk time of inner wall in early-age curing is 0-10d.

(c) Base on numerical simulation results, the constraint degree at the middle section of inner wall concrete is calculated. In the early stage (0-10d), the constraint degree of inner wall fluctuates within 60-99% as curing time goes.

(d) The main protection measures to delete confinement degree of mass concrete shaft wall as follow: (1) Reducing the friction between inner and outer walls; (2) Reducing hydration heat release of mass concrete; (3) Reducing the temperature difference between inner and outer walls during curing.

Acknowledgments

The research was financially supported by the National Natural Science Foundation of China (grant No. 51804157; 51774183; 51604114), which are all gratefully acknowledged.

References

[1] Ji-wei Zhang, Julian Murton, Shu-jie Liu, Li-li Sui, etc (2020). Sensitivity and regression analysis of acoustic parameters for determining physical properties of frozen fine sand with ultrasonic test. Quarterly Journal of Engineering Geology and Hydrogeology, 54,1-12.

[2] Russo, G., Corbo, A., Cavuoto, F., & Autuori, S. (2015). Artificial ground freezing to excavate a tunnel in sandy soil. Measurements and back analysis. Tunnelling and Underground Space Technology, 50, 226-238.

[3] Ji-wei Zhang, Julian Murton, Shu-jie Liu, Li-li Sui, Song Zhang (2021). Detection of the freezing state and frozen section thickness of fine sand by ultrasonic testing. Permafrost Periglacial Process, 32,76-91.

[4] Farazi, A. H., & Quamruzzaman, C. (2013). Structural design of frozen ground works for shaft sinking by practicing artificial ground freezing (AGF) method in Khalashpir coal field. International Journal of Engineering and Science, 2(3), 69-74.

[5] Zhang Ji-wei, Li Fang-zheng, Liu shujie (2020). Influence of different fibers on early-age cracking resistance of shaft wall mass concrete. Key Engineering Materials, 852, 150-159.

[6] Ohlsson, U., Nyström, M., Olofsson, T., & Waagaard, K. (1998). Influence of hydraulic pressure in fracture mechanics modelling of crack propagation in concrete. Materials and Structures, 31(3), 203-208.

[7] Xue, W., Yao, Z., Jing, W., & Song, H. (2020). Mechanical damage and failure behavior of shaft-lining concrete after exposure to high pore-water pressure. Journal of Materials in Civil Engineering, 32(1), 04019339.
[8] Barghamadi, H., Nater, P., & Huwyler, P. (2016). Hydraulic fracturing stress measurements and their implication for the design of the first concrete lined pressure shaft in Sri Lanka. Journal of Hydraulic Engineering, 2(1), 60-66.
[9] Peng, S., Rong, C., Cheng, H., Wang, X., Li, M., Tang, B., & Li, X. (2019). Mechanical properties of high-strength high-performance reinforced concrete shaft lining structures in deep freezing wells. Advances in Civil engineering, 2019.
[10] Zhou Yang, Zhou Guoqing (2010). Axisymmetric deformation analysis for the double layer shaft with a plastic interlayer. Journal of China Coal Society,35(9),1470-1475.
[11] Zhang Jianrong, Liu Zhaoqiu (2006). Experimental study on convective Heat transfer coefficient of concrete in wind tunnel. Journal of Civil Engineering, 39(9),39-42,61.
[12] Bofang, Z. (2013). Thermal stresses and temperature control of mass concrete. Butterworth-Heinemann.

Application of non-destructive testing technology in construction steel structure industry

Han Wei[1]*, Liu Lina[2]

[1]*Shandong Water Conservancy Vocational College, Rizhao, Shandong, 276826, China

[2]Rizhao Polytechnic, Rizhao, Shandong, 276826, China

Email: tg667788@xzcstudio.com

Abstract. High-rise buildings are mostly made of steel structures, and the damage detection technology for the welding process of steel structures still needs to be improved. Especially in the current welding process, there are problems such as the lack of data-based control mechanism and the uneven quality of the staff. The use of non-destructive testing technology can effectively detect defects in the welding process, while ensuring the quality of use of the steel structure. This article will study the application of non-destructive testing technology in steel structures, and provide some opinions and suggestions on the development of the construction steel structure industry by analyzing the problems that occur in the welding process.

1. Introduction

Steel materials have a series of characteristics such as stable properties, strong structure, and strong corrosion resistance. Its low cost makes it widely used in the construction industry, but the steel structure requires a lot of welding processes during use. Once the welding quality is not up to standard, it will affect the overall construction quality. Therefore, management personnel must integrate non-destructive inspection technology to achieve rapid analysis and positioning of steel structures, clarify the types of failures, and propose effective solutions to ensure the stable development of the construction steel structure industry.

2. Development of non-destructive testing technology

The test technology evidence task has different directions and can be divided into three stages. In the early stage, the method of simulation experiment was first used to judge whether the steel structure is firm and safe through 3D model analysis, but in the actual simulation process, the simulation effect cannot be kept completely consistent with the real object. The second stage is to apply destructive experimental methods to spot-check the building steel structure. During the inspection process, the overall sample structure needs to be destroyed. The performance of the material is comprehensively evaluated through complex mathematics and garlic[1]. This method can directly and clearly find that there are connection methods and weld quality in the construction process, but the test results have a certain randomness and cannot represent the quality of the overall structure. With the continuous development of science and technology, the application of non-destructive testing technology emerges. Non-destructive testing technology uses detectors to monitor the internal environment of the square structure and analyze the internal physical parameters of the steel structure without affecting the overall structure of the building. Non-destructive testing technology includes three aspects. The first is non-destructive detection. Non-destructive detection is to integrate ultrasound and magnetic field changes under the premise of ensuring that the building materials are not destroyed and preserved intact to monitor the internal environment to determine whether the building is distributed on average with damage or not. The second is non-destructive testing[2]. Non-destructive testing is developed on the basis of non-destructive testing. It is mainly used to determine the specific location of the scar and the shape of the defect. The last step is non-destructive evaluation. Non-destructive evaluation has a broader meaning. In a narrow sense, it refers to a comprehensive and scientific rating of the tested object when the tested object has a defect, and subsequent solutions are given.

3. Problems in welding of building steel structure

Steel structure is very important in overall building materials, and welding is often required to connect different steel structures. Common types of welds mainly include butt welds, trapezoidal welding and electric welding slag welding. During the welding process, due to the external environment and human factors, a series of problems such as slag inclusion, pores, cracks, and incomplete penetration are likely to occur, resulting in the strength of the overall steel structure not meeting the requirements and causing major safety hazards.

3.1. Backward view of quality supervision

As a personnel-intensive industry, the construction steel structure industry still has the idea of management in the current mechanism. Some managers lack a sense of responsibility, the inspection mechanism is too rigid, and the impact factor of non-destructive testing technology is not considered. The adoption of a single assessment mechanism in a large area has severely discouraged the enthusiasm of welders and affected the construction period of the project. In the actual management process, the focus of supervision is all on the technical and construction schedule, and the lack of quality and safety awareness has affected the development of the construction steel structure industry.

3.2. Welding personnel vary in quality

Welding as a special operation, different building steel structure needs to take different welding process, but the steel structure is constantly reformed. The traditional construction workers disappear, replaced by most of the migrant workers without preparation. The actual operation is strong mobility, and most of the employees have no qualification certificate. The phenomenon of unlicensed post is more serious, and the overall welding technology level is uneven, the lack of technical support. The welding efficiency is poor, and the construction quality is difficult to guarantee[3].

3.3. Lack of data control mechanism

There are a lot of data information and material storage management in steel structure industry. The internal technological process, organizational measures and inspection means are more complex. There are many problems in the process of data analysis. Once the monitoring means are not reasonable, it will lead to the difficult implementation of standards and norms.

4. Application of Non-destructive testing technology in construction steel structure industry

4.1. Comprehensively use a variety of testing methods

The important supporting role of steel structure in the building can effectively improve the stiffness and strength of the whole building and extend the service life of the building. Therefore, it is necessary to clearly analyze the common defects of steel structure welding, strictly implement the inspection standards of welding technology, constantly refine the key points of internal use, and improve the internal processing technology[4]. Several common test methods will be discussed below.

4.1.1 Visual inspection.
Appearance inspection is the basis of nondestructive inspection in the building structure steel structure industry. Appearance inspection is mainly artificial application of some simple mechanical equipment for the exterior surface of the building structure for investigation. It avoids a series of problems such as cracks, sags and crests and uneven color on the outer surface of the steel structure. Appearance inspection mainly uses artificial vision to test. In the actual process of time-consuming and laborious, due to human factors are likely to appear data problems for the comprehensive performance of the building with a certain impact.

4.1.2 X-ray detector.
Radiographic inspection is a commonly used method in appearance inspection. It analyzes the interior of the tested object by emitting some physical rays without damaging the steel structure, and analyzes whether there are defects, cracks and a series of problems in the interior of the tested object with the help of the strength of the change in the direction of rays from different structures.

Figure 1. Ray test launcher

4.1.3 Penetration inspection.
Penetration testing is also a kind of nondestructive testing. Osmosis testing is mainly the use of molecular permeability, fine scar for secondary examination. Penetration inspection is different from other inspection operation process, time-consuming, and it needs to wait a certain time[5]. In the process of operation, it is necessary to apply penetrant on the surface of the measured object to observe its infiltration into the inside of the building, which can clearly show the internal defects and cracks of the building, and the data is more accurate. But in the actual inspection process, the general penetration inspection is used for porous material not widely in the steel structure.

4.1.4 Magnetic particle inspection.
Magnetic particle inspection uses steel magnetism, which is a more extensive inspection method, relatively simple. Magnetic particle inspection uses the magnetic field principle through the analysis of the magnetic circuit phenomenon of the measured object. It can be more intuitive with clear expression of the measured object defect position. There is no complicated instrument in the inspection process, and the operation steps are relatively simple, but it has a certain impact on the surrounding ecological environment. Therefore, before and after inspection, magnetic powder and waste after inspection should be strictly treated.

4.1.5 Ultrasonic examination.
Ultrasonic inspection compared with other test methods have good directness. The detection depth is proportional to the law of the ground, and it can form on the different interface reflection, refraction, and other phenomena. Only inspection personnel through the electronic image on the display can be clear metal internal porosity, cracking, slag inclusion defects, such as rapid locating defects types, so that the subsequent repair measures. The management personnel should make comprehensive use of the

nondestructive inspection method to quickly locate the fault point, judge the fault type, supervise the construction personnel to take effective measures to optimize the internal performance of the steel structure. For example, in the process of ultrasonic inspection, different probes should be selected according to the different thickness of the inspection and the different carbon. Ultrasonic probe can be divided into straight probe, inclined probe, small tube probe, surface probe and so on[6]. In order to improve the accuracy of detection, inspectors need to choose the probe properly and ensure that the probe chip area within the range of 25 square millimeters to 500 square millimeters. Before inspection, pre-treatment should be done, and the weld surface should be cleaned to ensure smooth movement of the probe. Through grinding sound, it can improve the acoustic effect and reasonable control the room temperature. It can ensure that the steel structure in the room temperature or cooling state, as far as possible to reduce the damage to the probe. In the process of using ultrasonic inspection, digital ultrasonic can be comprehensively used to complete the production of DAC curve for the weld. According to the occlusal standard of the weld in different periods, the follow-up quality control points can be clarified, and the inspection surface cleaning and follow-up management can be completed by fully combining the thickness of the workpiece and the probe. At the same time, in the inspection process, the monitoring personnel need to choose the appropriate coupling agent according to the standard of the flaw detector probe, and give priority to the coupling agent with sound permeability and strong mobility to ensure that the ultrasonic wave can be effectively introduced into the surface of the steel structure, so as to improve the inspection efficiency.

4.2. Enhance data detail control
The inspection process of steel structure is relatively complex, so inspectors need to reasonably select the inspection surface according to the construction situation and the convenience of inspection, and complete the drawing of welding seam distance waveform diagram. It can determine the accuracy of flat line, quantitative line and scrap line, and comprehensively manage the use of steel structure to avoid overlapping probes as far as possible. The DAC curves and defects are analyzed quantitatively and qualitatively by direct contact method.

For example, the compensating gain classification should be adjusted properly during the nondestructive testing of welds. Generally between 6 dB and 4 dB in industry, the inclined probe and parallel scanning method should be selected as far as possible to achieve a comprehensive and quick inspection of the weld. It should observe the echo signal displayed in the wave screen in time, and quickly remember the abnormal weld position, so as to provide reference for the subsequent qualitative and quantitative defects. It can distinguish the grade frame rate between the defects, focus on the location of the second and third grade cracks. Various factors affecting the weld are considered comprehensively, and various parameters of echo are judged by K value to fit the horizontal and vertical distance of wave value.

4.3. Reasonable choice of testing methods
The specific parameters of different welds are different, and the management personnel should choose the monitoring method reasonably according to the groove of the welded joint. Welding industry is a highly practical course. During the inspection process, inspectors must interpret the steel structure description and the corresponding construction quality inspection specifications. It needs to clarify the physical and chemical properties of each component, confirm the inspection location, inspection quantity and the level and requirements of the weld quality, and meet the different needs of different buildings[7]. It should clarify the type of steel used internally, welding process, and welding type. Follow-up welding can only be carried out during the preliminary investigation. In the process of welding inspection, it is necessary to carry out a comprehensive quality acceptance based on the "Code for Construction Quality Acceptance of Steel Structure Engineering" in strict accordance with national standards. The general monitoring mainly includes welding seam and welding foot size to ensure that the deviation of actual size and standard size is within a reasonable range. And it should check the appearance to ensure that there are no obvious cracks, pores, arc scratches and other important defects

on the surface.

Test object	Test method
Fillet weld, T-weld	Ultrasonic (UT), Magnetic (MT)
Groove position	Ultrasonic (UT), Magnetic (MT)
Root clearing site	Magnetic (MT)
Butt weld	Ultrasonic (UT), Radiographic (RT)

Figure 2. Main test methods

For example, when inspecting trapezoidal weld defects, an oblique probe is generally used for the first inspection at the position of the trapezoidal weld near the web. At the same time, it is combined with a straight probe to conduct ultrasonic inspection on the outer welding seam. It should adopt a combination of secondary and primary inspection to improve the accuracy. It can quickly locate the defect type according to different waveform characteristics such as defect wave, external wave, internal wave[8]. It can try to choose a 5MHz dual crystal straight probe to cooperate. At the same time, the basic sensitivity value should be clarified during the inspection process. It ensures that the hole equivalent is within a certain range and fits the welding process of this project to the maximum extent. The waveform is compared with the classic defect echo characteristics to improve the accuracy of curve evaluation and qualitative analysis, laying a data foundation for subsequent maintenance methods.

5. Conclusion

Steel structure has been widely used in high-rise buildings and civil buildings due to its light weight and easy construction. However, due to the need for a large number of welding processes during use, a series of problems such as slag inclusion and bubbles often occur, which seriously affects the service life of the steel structure. Therefore, the corresponding management personnel should apply non-destructive inspection techniques, such as radiographic flaw detection, penetrant inspection and ultrasonic inspection, to strictly control the quality of welds and improve internal monitoring points. In the future, non-destructive testing technology will be organically combined with digital image processing technology and computer technology, and it will be faster and more scientific in terms of three-dimensional image acquisition and timeliness to ensure the safe use of building structures.

References

[1] Ding Aixiang. Application of ultrasonic nondestructive testing technology in weld inspection of building steel structures J. Building Materials and Decoration, 2019, (19):63-64.

[2] Liu Xianglun. Application of nondestructive testing technology in building steel structure J. Building Materials and Decoration, 2019, (12):49-50.

[3] Wu Xiaosong. Analysis of the application of ultrasonic non-destructive testing technology in welding seam testing of building steel structures J. China Real Estate Industry, 2019, (17):192.

[4] Zhu Mingzhang. Application of non-destructive testing technology in construction engineering J. Journal of Shandong Vocational and Technical College of Commerce, 2020, 20(03): 103-106.

[5] Wang Bairan, Tong Renyuan, Li Qing, Jin Ying, Jin Weiliang, Lin Zheng. Research on amusement equipment steel strop detection device based on magnetic flux leakage detection technology J. Journal of China Jiliang University, 2020, 31(04): 430- 436.

[6] Li Hanchao, Gu Yang, Yu Yating. Three-dimensional simulation analysis of metal surface defects based on eddy current nondestructive testing technology J. Journal of University of Electronic Science and Technology of China, 2020, 49(06): 942-948.

[7] Dong Ming, Xu Ruyi, Chen Yuan, Zhang Guangming, Wan Xiang. Current status and prospects of ultrasonic testing for shaft parts of coal mine machinery in service[J]. Journal of Xi'an University of Science and Technology, 2020, 40(05): 775-780.
[8] Tao Dongyang, Mao Guoqing. Overview of the application of non-destructive testing technology in bridge testing J. Heilongjiang Transportation Science and Technology, 2020, 43(09): 130-131.

Experimental research on combustion and explosion characteristics of medium density fiberboard on building

Jiameng Wu[1], Shunbing Zhu[1,2], Zizheng Pang[1], and Yi Duan[1]

[1]College of Safety Science and Engineering, Nanjing Tech University, Nanjing, Jiangsu, 211816, China

[2] Key Laboratory of Urban and Industrial Safety , Nanjing Tech University, Nanjing, Jiangsu, 211816, China

Corresponding author's e-mail: cjsyzx@njtech.edu.cn

Abstract. In recent years, the amount of medium density fiberboard(MFD) used in the construction industry is huge. The dust produced in the production has a great risk of dust combustion and explosion in different thermal runaway environment. Therefore, this paper takes the MDF dust as the research object to explore its combustion and explosion characteristics. The minimum ignition temperature of dust layer and 20L explosion vessel were used to study the changes of minimum ignition temperature(MIT) and combustion morphology of dust layer with different particle sizes of wood powder, and the influence of different influencing factors on the maximum explosion pressure was analyzed. The results show that with the decrease of particle size, the MIT decreases from 350°Cto 260°C, the maximum heating rate increases, the carbonization phenomenon becomes more obvious, and the maximum explosion pressure increases gradually; Under 200 mesh dust, with the increase of mass concentration and ignition delay, the maximum explosion pressure first increases and then decreases, and 50 g/m^3 and 60 ms are the best mass concentration and ignition delay.

1. Introduction

With the improvement of the quality of life, people's demand for wood products has gradually increased, and the probability of dust explosions in wood product processing enterprises has also increased significantly[1]. A large amount of wood powder will be produced in the production process of sawing, drilling, grinding, etc. This dust not only threatens human health and pollutes the environment, but also the personal injuries and property losses caused by dust explosions are huge. Further research on dust explosions has become especially critical, it has great theoretical and practical significance for preventing and reducing the occurrence of dust explosion accidents.

Dust explosion is a special form of combustion, and there is the possibility and hazard of dust combustion transforming into dust explosion in dust explosion accidents. The current research on dust combustion and explosion is also the focus of Chinese and foreign scholars. Zhang et al[2] used MIE and MIT test device to conduct experimental research on polyamide fiber dust . The lowest ignition temperature of dust layer was 307°C, and the lowest ignition temperature of dust cloud is 300°C. Zhong et al[3] used a self-developed experimental device to carry out a hot surface ignition experiment with stone pine nut dust and found that the thicker the dust thickness is, the longer the initiation time and duration of dust layer ignition are. Qing et al[4] used a 20L spherical explosion device to conduct a comparative explosion test study on nanometer and micrometer aluminum powder. Tests show that the maximum explosion pressure and explosion index mainly depend on the dust concentration, which increases first and then decreases as the dust concentration increases. Matsuda et al[5] conducted tests on relevant parameters of dust that may explode in the industry. The experimental results show that the characteristics of dust explosion are related to the experimental method, dust category and form, external environment and other factors. Cashdollar et al[6] comprehensively and systematically studied the relationship between the volatile content of coal dust and the lower explosion limit of the mixture in the state of coal dust and gas mixture, and shows that the increase of ignition energy, the decrease of volatile content and the decrease of coal dust particle size will lead to the decrease of ignition lower limit of coal dust to some extent.

Research on coal powder, starch dust, aluminum dust and alloy dust has been relatively common. However, there has not been much research on the associated dust of the MFD that has appeared in the construction or manufacturing industry in recent years. Therefore, studying the minimum ignition temperature and maximum explosion pressure of fiberboard dust is of great significance.

2. Materials and Methods

2.1 Materials
The experimental materials are selected from wood products processing plants, and the experimental dust is obtained through cutting, mechanical polishing and other links. The mechanically polished dust was sieved and dried, and sieved out 20, 40, 60, 80, 100, 150 and 200 mesh. The dust of the target sample is dried in a high-temperature oven at 50°C for 12 hours. After drying and cooling, it is placed in a sealed bag for storage.

2.2 Experimental method of the lowest ignition temperature of the dust layer

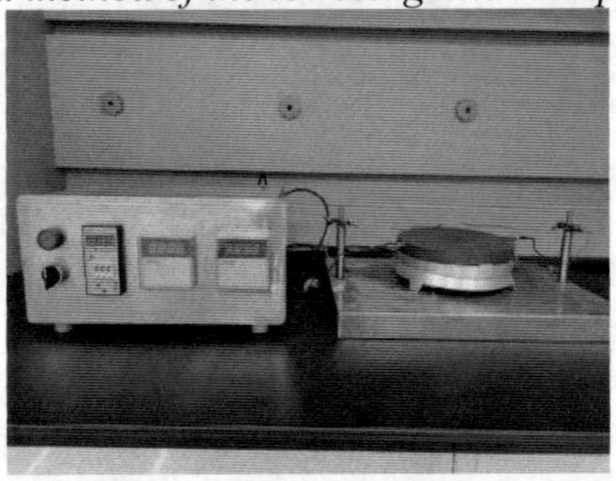

Figure1. Minimum ignition temperature test system for dust layer

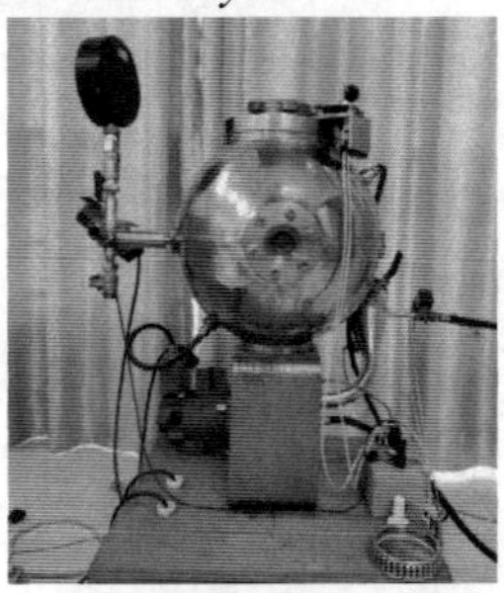

Figure2. 20L spherical explosion vessl

The MIT of dust layer refers to the minimum temperature at which the accumulated dust ignites and burns under the action of external energy. In this experiment, the MIT device of dust layer is used, the diameter of metal ring is 9cm and the height is 1.3cm, as shown in Figure 1. The initial heating temperature of the experiment is 300°C. If the dust catches fire, the temperature will be increased by 20°C, otherwise, the temperature will be lowered by 10°C to determine the MIT of the wood dust layer of different particle sizes. 20, 40, 60, 80, 100, 150 and 200 mesh wood flour were selected and spread on the hot plate with metal ring thickness of 1.3cm to observe the dust combustion morphology at different temperatures.

2.3 Experimental method of maximum explosion pressure of dust cloud
The 20L explosive vessel was used in the experiment. The experimental device is shown in Figure 2. The powder injection pressure is 2.0MPa. The maximum explosion pressure of wood dust with different particle size, concentration and ignition delay was studied, and the influence of different factors on the maximum explosion pressure was analyzed.

3. Results & Discussion

3.1 The effect of particle size on the minimum ignition temperature and combustion form of the dust layer

According to the combustion criteria[7], the MIT of dust layers with different particle sizes are obtained, as shown in Table 1.

Table 1. MIT of dust layer with different particle size

Particle size /Mesh	20	40	60	80	100	150	200
MIT/°C	350	310	290	280	270	260	260

(a) 130°C (b) 200°C (c) 319°C

(d) 608°C (e) 642°C (f) 668°C

Figure 3. Combustion state of wood dust at different stages

The hot plate temperature was set at 300 °C, and the combustion process of 200 mesh wood dust was illustrated. Figure 3 shows the combustion state of wood powder at 130, 200, 319, 608, 642 and 668°C respectively. When the dust was heated to 130°C, a little white smoke appeared on the surface, accompanied by the fragrance of wood powder; At 200°C, the smoke increased, the aroma became stronger, and the surface color of wood flour became darker; At 233°C, black carbonization occurs locally on the surface; At 319°C, the area of surface carbon expanded, and the rate of smoke production and temperature rise accelerated; At 608°C, the rising rate of temperature slowed down, and the white smoke turned to green smoke with a pungent burning smell; At 642°C, the surface carbonization is basically completed, and there are some sparks on the edge; The maximum temperature of wood powder combustion was reached at 668°C, the mars began to disappear, the whitening phenomenon appeared in the outside and center of the sample, the temperature decreased, and the combustion was completed. The phenomenon of dust layer during combustion is: with the decrease of particle size, the MIT decreases, the temperature increases at the end of carbonization, the carbonization phenomenon is more obvious, and the carbonization time is gradually delayed.

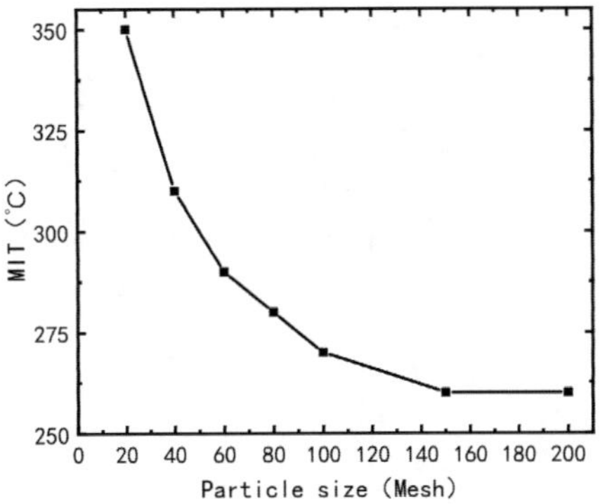

Figure 4. Effect of MIT on particle size in dust layer

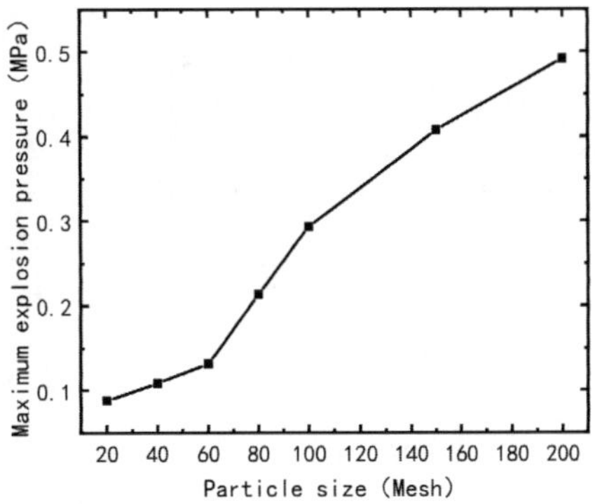

Figure 5. Variation of maximum explosion pressure with particle size

Figure 4 shows the relationship between the MIT and particle size of dust layer. It can be seen from the figure that with the decrease of particle size, the MIT of dust layer decreases from 350°C to 260°C. This is because the smaller the particle size is, the larger the specific surface area is, the more complete the reaction with combustion supporting materials is[8], and the reaction heat increases, which makes it easier to reach the MIT.

3.2 The effect of particle size on maximum explosion pressure

Figure 5 shows the curve of the maximum explosion pressure with the particle size measured at the mass concentration of $50g/m^3$. It can be seen from the figure that with the decrease of particle size, the maximum explosion pressure increases gradually, and the growth rate begins to slow down. This is because the larger the mesh, the smaller the particle size, the larger the specific surface area of dust, the more the reaction with oxygen, the more the heat generated by the increase of combustion rate, and the explosion pressure increases correspondingly. When the specific surface area increases to a certain extent, due to the lack of oxygen, the maximum explosion pressure increases slowly.

3.3 The effect of concentration and ignition delay on the maximum explosion pressure

Figure 6 shows the curve of maximum explosion pressure with mass concentration under ignition delay of 30, 60, 90 and 120 ms. Taking 60ms ignition delay as an example, when the initial dust concentration increases, the maximum explosion pressure increases. When the mass concentration reaches $50g/m^3$, the maximum explosion pressure is 0.4916 Mpa, reaching the maximum value. As the mass concentration continues to increase, the maximum explosion pressure begins to decrease. This is because with the increase of dust concentration, the number of effective molecular collisions increases, the reaction rate accelerates, and the energy generated increases. When the mass concentration continues to increase, the oxygen concentration in the explosion ball is relatively insufficient, resulting in more and more dust not participating in the reaction and absorbing the heat generated around, which reduces the maximum explosion pressure.

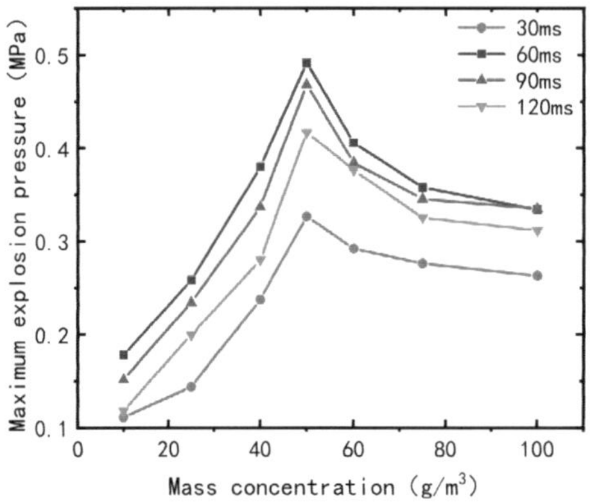

Figure 6. Variation of maximum explosion pressure with mass concentration under different ignition delay

With the increase of ignition delay, the maximum explosion pressure of wood dust increases gradually at the same concentration, and reaches the maximum value in 60ms. When the ignition delay continues to increase, the maximum explosion pressure of wood dust begins to decrease, because with the increase of ignition delay to 60ms, the turbulent motion of gas in the container makes the wood dust disperse more evenly and reach a good explosion state; The ignition delay time will be increased continuously, and the wood powder will enter the settlement process[9]. The dust participating in the reaction will decrease, and the maximum explosion pressure will continue to decline, so the best ignition delay time is 60ms.

4. Conclusion

With the construction and manufacturing industry's huge demand for wood products, especially medium density fiberboard, leading to safety risks are also increasing. In order to reduce the damage caused by dust explosion accidents, this paper studies the characteristics of dust combustion and explosion, and obtains the following conclusions:

(1) Using the MIT of dust layer device, it is found that within a certain range, the MIT of wood dust with a thickness of 1.3cm decreases with the decrease of the dust particle size.

(2) Using 20L explosive vessel, under the condition of powder injection pressure of 2.0MPa, ignition energy of 10kJ and ignition delay of 60ms, with the decrease of particle size, the maximum explosion pressure rises and the growth rate of maximum explosion pressure slows down.

(3) Under the same ignition delay, with the increase of 200 mesh dust concentration, the maximum explosion pressure first increases and then decreases, and 50g/m3 is the best concentration; When the mass concentration is 50g/m3, with the increase of ignition delay, the maximum explosion pressure first increases and then decreases, and 60ms ignition delay is the best ignition delay time.

Acknowledgments

This work is funded by the Postgraduate Research & Practice Innovation Program of Jiangsu Province with account number SJCX21_0448. The authors wish to thank the research group for their kind support in this study.

References

[1] Xie, J., Liu, D.L., Lu, Y.Y. (2019) Situation of dust explosion in wood processing enterprises and analysis for regulation measures. China Wood-Based Panels, 26: 19-22

[2] Zhang, X.L., Zhang, Z.K., Shen, Q., et al. (2016) Explosion characteristics of polyamide dust. China Powder Science and Technology, 22: 22-26.

[3] Zhong, S.J., Zhou, L.G., Wang, J. (2014) Smoldering combustion of dust layer on hot surface. Journal of Combustion Science and Technology, 20: 199-207.
[4] Li, Q.Z., Lin, B.Q., Li W.X., et al. (2011) Explosion characteristics of nano-aluminum powder–air mixtures in 20 L spherical vessels. Powder Technology, 212: 303-309.
[5] Matsuda, T., Yashim, M., Nifuku, M., et al. (2001) Some aspects in testing and assessment of metal dust explosions. Journal of Loss Prevention in the Process Industries, 14: 449-453.
[6] Cashdollar, K.L., Weiss, E.S., Montgomery, T.G., et al. (2007) Post-expiosion observations of experimental mine and laboratory coal dust explosions. Journal of Loss Prevention in the process Industries, 20: 607-615.
[7] GB/T 16430-2018, Determination of the minimun ignition temperature of dust layer[S].
[8] Wan, H.W. (2018) Experimental study on dust explosion characterisitics in wood products processing. Xi'an University of Architecture and Technology, Xi'an.
[9] Cao, W.G. (2016) Experimental and mechanism study on explosion characteristic of lignite coal dust. Nanjing university of science & technology, Nanjing.

Construction Key Techniques of Upper Cross-beam for Main Tower of Super-Kilometer Span Rail-cum-Road Suspension Bridge

FU Wen-xuan[1,2,3,*]

[1] CCCC Second Harbor Engineering Co., Ltd., Wuhan, Hubei 430040, China

[2] Key Laboratory of Large-span Bridge Construction Technology, Wuhan, Hubei 430040, China

[3] CCCC Highway Bridge National Engineering Research Centre Co., Ltd., Wuhan, Hubei 430040, China

[*] E-mail: 280583432@csnec.cn

Abstract: As the span of the suspension bridge increases, the height of the main tower and the force on the tower increase. The beam profile volume and pouring weight become larger, which increases the risk of high-altitude construction of the main tower and the upper cross beam. The main bridge of Wufengshan Changjiang River Bridge of Lianzhen Railway is the first rail-cum-road suspension bridge in China. The main span reaches 1092m and the total height of the tower is 191.5m. Taking the engineering practice of the upper cross beam of the main tower of Wufengshan Changjiang River Bridge of Lianzhen Railway as an example, this paper introduces the construction difficulties of the upper cross beam of the main tower, the selection of the upper cross beam construction support, the structural characteristics of the support and the construction process. The upper cross beam is constructed asynchronously with the tower. From the bottom of the upper cross beam to the top of the lower cross beam is 121.5m. The upper cross beam was poured in six times with a total concrete volume of 1918m^3. The method of small trusses with bracket and floor steel pipe support is used for construction. The finite element analysis software is used to calculate the super-high support of the upper cross beam, and the support strength, rigidity and stability meet the requirements of the specification. At present, the bridge has been partially put into use, which verifies the rationality of the design and construction of the beam support on the main tower. It has strong reference and guiding value.

1. Introduction

With the development of urban construction and expressway and railway construction in China, the span of bridges is constantly breaking through. The long-span cable-stayed bridge or suspension bridge will also become an inevitable trend. When the span of the bridge exceeds 600m and above, the suspension bridge is often considered first. Compared with highway suspension bridges, railway suspension bridges started a little later. At present, a number of rail-cum-road suspension bridge have been built at home and abroad, and the span has exceeded the threshold of 1000 meters [1]. The parameters are shown in Table 1.

Table 1. The large span rail-cum-road suspension bridge at home and abroad

Name of the bridge	Year of completion	**The main span**	**tower height**
The New Bay Bridge, US	1936	704m	160m
The April 25th Bridge, Portugal	1966	1013m	190m
The Great Naruto Bridge, Japan	1985	876m	125.93m
The Seto Bridge, Japan	1988	1100m	194m
The Tsing Ma Bridge, Hong Kong	1997	1377m	206m
The Messina Strait Bridge, Italy	Under construction	3300m	370m
Wufengshan Changjiang River Bridge, Chain	2020	1092m	191.5m

The Wufengshan Changjiang River Bridge is the first rail-cum-road suspension bridge in inland China. It is also the suspension bridge with the fastest running speed and the largest running load in the world. Its calculation theory, design method and construction technology have all faced many technical tests, thus breaking a number of new technical records. The cross beam is not only an important force transmission structure for the tower, but its construction is also a key link in the construction of the tower. Due to the large size, irregular shape, and large height of the upper cross beam structure, the construction is generally divided into multiple pourings and the process is complicated. Therefore, how to design a safe, economic, and convenient super-high support system to complete the construction of the upper cross beam is an important technical problem.

2. Project Overview

The Wufengshan Changjiang River Bridge of Lianzhen Railway is an important part of the railway line in Lianzhen, which adopts a double-tower single-span suspension steel truss girder suspension structure. The main cable span is (350+1092+350) m and the stiffening beam span is (84 + 84 + 1092 + 84 + 84) m. The overall layout of Wufengshan Changjiang River Bridge is shown in Figure 1.

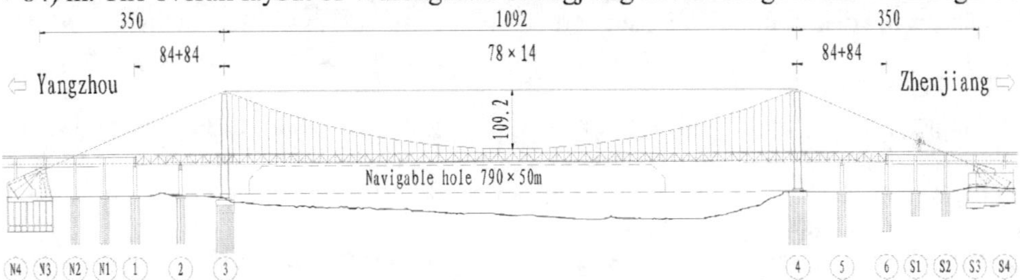

Figure 1. Arrangement of Main Bridge of Wufengshan Changjiang River Bridge (Unit: m)

The Wufengshan Changjiang River Bridge adopts a double deck layout, the upper level is an eight-lane highway and the lower level is a four-line high-speed railway. The bridge tower adopts reinforced concrete frame structure. The total height of the tower is 191.5m (calculated from the top face of the tower to the bottom of the saddle), which is respectively composed of upper, middle and lower tower columns, upper and lower cross beams and saddle housings on the top of the tower. The tower is made of C55 grade concrete and HRB400 steel bar. In order to reduce the wind resistance coefficient, the tower section is chamfered at 100×150cm around. The upper and lower cross beams are prestressed concrete structures.

The upper cross beam of the main tower is a single-box and single-chamber structure. The height gradually changing from 11m in the middle (not counting the height of the beam crown) to 32.8m on both sides. The length of the transverse bridge is about 34.4m, and the width of the bridge is 7.5m. The thickness of the top, bottom and web are 0.8m. There are manholes on the top surface. The structure of main tower and beam is shown in Figure 2.

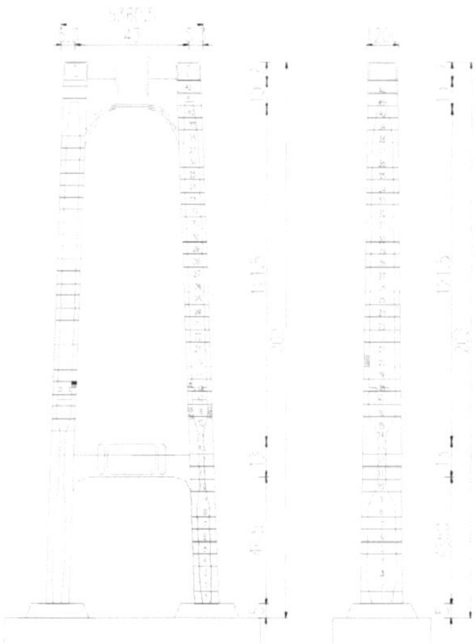

Figure 2. Elevation plan of main tower (Unit: m)

3. Difficulties in construction of upper cross beam

Different from the lower cross beam and tower construction synchronous, the upper cross beam and tower construction asynchronous in order not to affect the progress of the top sealing of the tower. The upper cross beam has a large height and a complex structure. The construction difficulties are as follows:

- The bottom of the upper cross beam is 121.5m away from the lower cross beam. The installation and disassembly of the support is difficult, and the safety risk is high.
- The total casting height of the upper cross beam is as high as 32.7m, with many times of layered casting and long casting cycle.
- The height of the upper cross beam gradually changes from 10.0m in the middle to 23.2m on both sides. The decorative block on both sides is a special-shaped structure and has a large height difference. The support and other formwork system should consider the construction needs of the beam and decorative blocks.

The concrete cubic volume of upper cross beam is 1918m3, which is divided into six layers for pouring. The upper cross beam support structure is considered to bear the total weight load of the first, second, and third layers of concrete pouring. The volume of the concrete is shown in Table 2.

Table 2. The concrete cubic volume of upper cross beam

Layered pouring	Pouring part	Layered thickness (m)	Volume (m^3)	Pouring time (h)
The first layer	Lower decorative block	6.7	194.2	4
The second layer	Top of lower decorative block	6	198.1	4
The third layer	Beam bottom plate and side wall	5.5	660.7	14
The fourth layer	Beam side walls	3	219.3	4.5
The fifth layer	Beam side wall and roof	3	387	8
The sixth layer	Upper decorative block	8.5	258.9	5.5
Total		32.7	1918	

4. Upper cross beam support design

4.1 Determination of upper cross beam support

The construction of the upper cross beam on the main tower is a typical high-altitude operation, which is difficult to construct and has a large safety factor. Generally, the form of large brackets with corbels is more [2]. Considering economic efficiency, construction period, and hazard source control, since the lower cross beam of Wufengshan Changjiang River Bridge uses floor-standing supports and the amount of steel pipes is large, the upper cross beam support adopts the form of small truss with bracket and steel column support. After the lower cross beam component is removed, it is repaired and used for the upper cross beam support. The steel column of the lower cross beam can be recycled, which is convenient for disassembly and transportation. In order to ensure that the force condition at the lower decorative block of the upper cross beam meets the requirements, a tripod and a corbel are provided at the lower ends of the upper cross beam. The bottom mold and the outer mold of the upper cross beam special-shaped decorative block adopt large-scale combined steel template, which is designed by a professional template factory. After being processed, it is transported to the site.

4.2 The structure and characteristics of the upper cross beam support

The upper cross beam supports are arranged in order from top to bottom: transverse distribution beam, bailey beam (type 321), main transverse beam, triangle truss and bracket, steel column, horizontal bracing, buttress and inclined strut. The main component material types are shown in Table 3. The bracket layout is shown in Figure 3.

Table 3. Main component materials of upper cross beam support

component	type specification	materials
Transverse distribution beam	I14a/2I14a	Q235B
Bailey beam	type 321	Q235B
Main transverse beam	2HN900×300	Q235B
Triangle truss	2HN900×300、2HM588×300	Q235B
Bracket	2HN900×300	Q235B
Steel column	Ø1000×16	Q235B
Horizontal bracing	Ø630×8，Ø426×6	Q235B
steel pipe attached to the tower	Ø630×8、Ø426×6	Q235B
Inclined strut	Ø426×6	Q235B

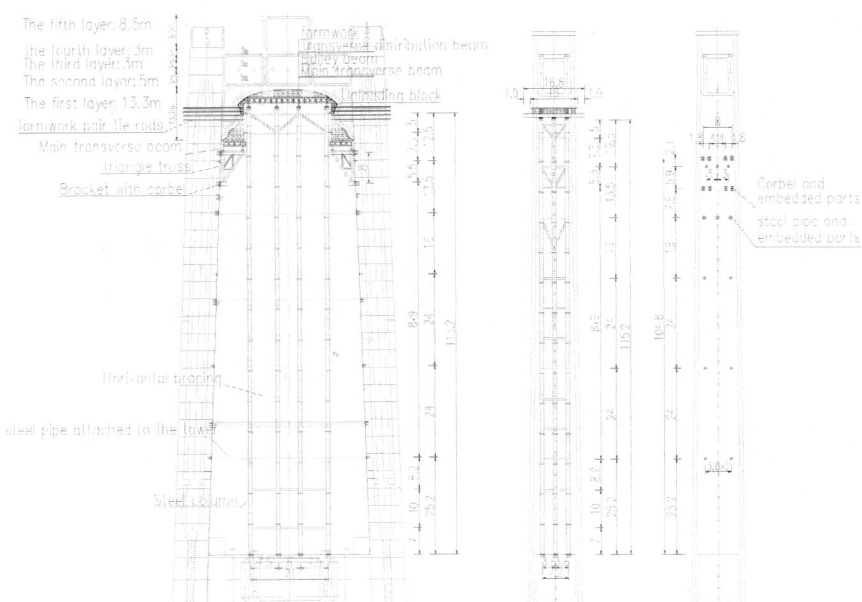

Figure 3. Elevation plan of upper cross beam

Journal of Physics: Conference Series **2011** (2021) 012041 doi:10.1088/1742-6596/2011/1/012041

Features of upper cross beam support:

- The overall height of the upper cross beam bracket is more than 100 meters, and it is a high-rise structure. Wind vibration is more likely to occur under strong wind. The "Building Structure Load Code" should be used to calculate the wind load [3].
- There are 4 groups of triangle truss at both ends of the support, with an interval of (1.8×2+4.4+1.8×2) m. The upper bracket of the triangle truss is under tension and the lower bracket is under pressure.
- The lower part of the upper cross beam and the adjacent tower should be equipped with formwork pair tie rods, and two layers of cushion beams and support beams should be added between the bottom mold of the decorative block and its adjacent Steel columns to increase the horizontal rigidity and reduce the lateral deformation of the bottom mold.
- Affected by wind load, the steel pipe attached to the tower is subject to greater horizontal force. The steel pipe and the tower are connected in a consolidated form. The horizontal bracing and inclined strut are added between adjacent steel pipes attached to the tower to increase the rigidity of the steel pipe.
- According to the horizontal bracing elevation of the upper cross beam support, a reasonable arrangement of the active lateral brace of tower. The support can take into account the active lateral brace of the middle tower, which provides convenience for its installation and removal.
- In order to realize the asynchronous construction of the upper cross beam and the main tower, pre-embedded parts of the triangle truss bracket are embedded in the tower. The upper cross beam is constructed by layered pouring of the floor steel pipe support and triangle truss. The support force is clear and the construction is convenient.

4.3 Finite element calculation of the upper cross beam support

The finite element calculation model of the upper cross beam support and the calculation results of the most unfavorable conditions are shown in Figure 3. The calculation of the support does not consider the beneficial effects of prestressing.

The constraint conditions of the model is as follows:

- The steel column is consolidated at the bottom of the column.
- The steel pipe attached to the tower is consolidated at the connection of the tower.
- The main transverse beam is hinged with the steel column.
- The main beam is elastically connected with bailey and the triangle truss.
- The transverse distribution beam is elastically connected with the Bailey.
- The bracket of triangle truss is consolidated with the tower.

The load of the upper cross beam support includes: the weight of the support and the upper cross beam, the weight of the formwork system, the load of the construction personnel and equipment, the concrete dumping and vibrating load, and the wind load.

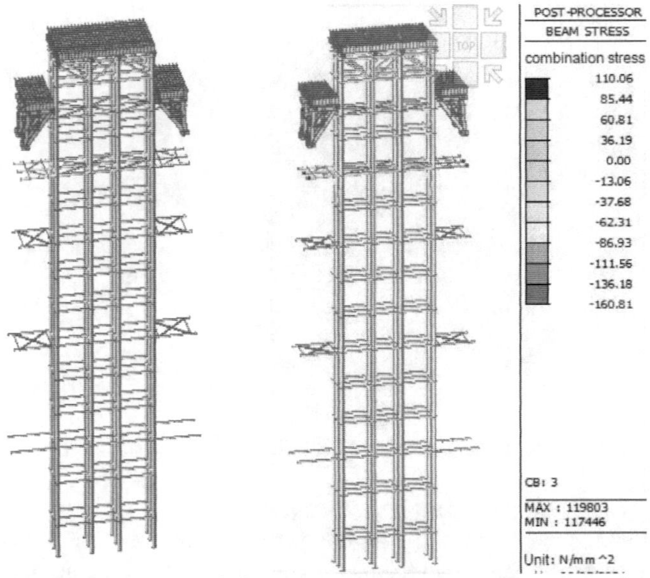

(a) Calculation model (b) Combination stress nephogram (MPa)

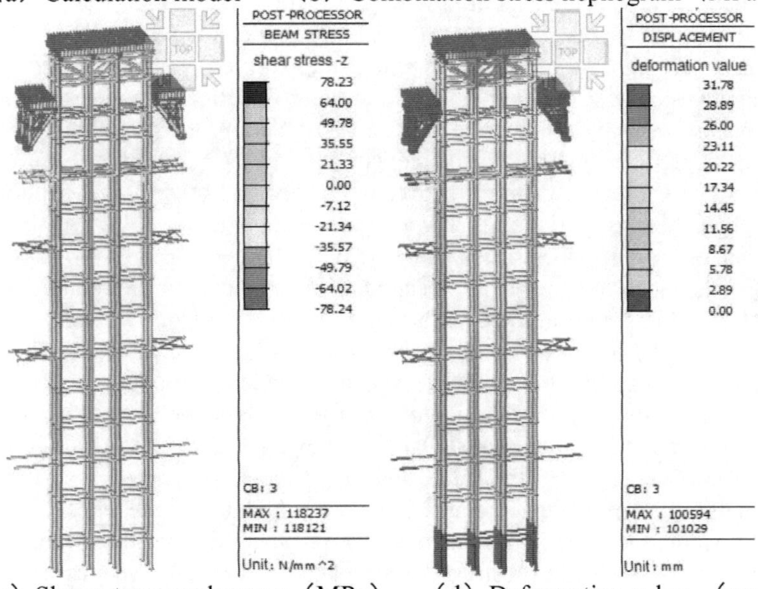

(c) Shear stress nephogram (MPa) (d) Deformation value (mm)

Figure 4. The calculation model of upper cross beam support and its calculation results

5. Checking the overall stability of the upper cross beam support

The finite element software is used to analyze the buckling of the overall beam support structure (Figure 5). The buckling analysis principle is: constant load + buckling coefficient × variable load = buckling load. The calculated eigenvalue coefficient of the buckling mode of the structure is 5.1, which is greater than 4. The overall stability of the support meets the requirements of use [4].

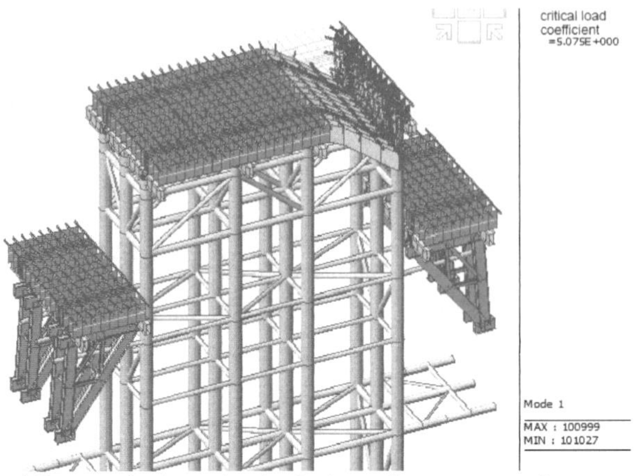

Figure 5. Buckling analysis and calculation of upper cross beam support

6. Construction process of upper cross beam support

Due to the asynchronous construction of the upper cross beam and the tower, the upper cross beam steel bars are pre-embedded during the construction of the tower column. At the same time, the installation of the upper cross beam support and the formwork system is carried out. After the topping of the upper tower column is completed, the inner climbing frame is removed. The bottom formwork of the beam is installed, and the upper cross beam is constructed in layers from bottom to top by conventional techniques. The steel columns are connected by flanges. Each section is about 8m long and the unit weight is about 4t. All components are assembled on the ground as much as possible to reduce the amount of aerial welding. After the construction of the lower cross beam is completed, the dismantled steel columns, beams, distribution beams, bailey and other components are repaired and turned over for use in the construction of the upper cross beam.

The construction process of upper cross beam support installation is as follows:

Install support embedded parts →Repair treatment of lower cross beam components →Install the first section of steel column and the first and second floors of horizontal bracing by tower crane → Installation of the third and fourth floors of steel columns →Active lateral brace installation →Install the remaining steel columns and horizontal bracing →Installation of triangle truss and main transverse beam →Bailey beam installation→Installation of transverse distribution beam →Special-shaped mold of the decorative block and triangle truss installation →Lay and fine-tune the bottom formwork.

The upper and lower concrete of the same floor are poured simultaneously to reduce the uneven deformation of the support. When the strength and elastic modulus of the last layer of concrete reach 90% of the design value, the upper cross beam can be prestressed. The support can be removed only after the construction of the beam prestress is completed. The top-down demolition technology is adopted. The pre-pull force of the active lateral brace is removed in advance when the active cross brace is encountered.

The construction process for the removal of the upper cross beam support is as follows:

Cleaning up debris→ Cutting off the unloading block→ Removing the bottom formwork→ Removing the distribution beams one by one →Removing the Bailey frame and the triangle truss→Removing the main transverse beam of the 10-ton winch →Removing the steel column and horizontal bracing→ Repair the embedded parts of the tower and the lower cross beam.

7. Conclusion

Wufengshan Changjiang River Bridge is the most difficult part of the whole route construction. Its railway bridge has been officially put into use with the opening of Huaiyang Town Railway in December 2020, and the construction of the highway bridge is still in progress and is expected to be officially opened to traffic at the end of June 2021. The main tower is not only the key control project for the construction of Wufengshan Changjiang River Bridge, but also the facade project of the bridge.

The existing recommendations are as follows:

- The beam of the main tower is large in volume and span. Using the existing construction environment conditions, it adopts the construction form of small truss and floor steel pipe support. The floor support design reaches 120 meters and the overall construction of the upper cross beam is smooth with good quality control. The on-site installation of the cross-beam support is shown in Figure 6, which can provide technical support and reference for similar projects.
- The asynchronous construction of the upper cross beam and the tower can reduce the influence of the upper cross beam construction on the construction progress of the tower top sealing, steel truss beam erection, and suspension cable erection [5]. The safety of the tower structure can be ensured and construction requirements can be met with a reasonable arrangement of active cross brace of the tower.
- There are many forms of the beam cross support. The upper beam support of this project uses floor-standing steel pipes, which the amount of steel used is very large. On the condition that it not only meets the construction requirements but also saves costs, other support forms can also be used. At present, the type of bracket with corbel is used more and more in the construction of the upper beam, but the force of corbel is usually not big. The type of bracket with large-tonnage and heavy-duty corbel are suitable for the construction of large-height and large-volume beams. It will be a development trend in the future beam construction process, which is worthy of further study.

(a) The lower cross beam (b) The upper cross beam

Figure 6. On-site cross beam construction of Wufengshan Changjiang River Bridge

References

[1] CHEN Liang. Study on the Characteristics of Long-span Road-Rail Plate-Truss Combine-suspension Bridge[D].Changsha University of Science &Technology,2019.

[2] PENG Zhichuan. Construction technology research for Cross-Beams support of Main Tower of Suspension bridge[J]. Highway,2021,66(02):143-151.

[3] GU Chenglei, HUANG Xiuping. Computational comparative study of Temporary engineering wind load[J].China Water Transport,2019,19(03):228-229.

[4] Li Fengyun. Mechanical Analysis of the Double Limb Lattice Column - Bailey Beam Formwork Supports System in Space [D].Chongqing University, 2014.

[5] YIN Zhen-jun, RUAN Zhong-jun, RUAN Zhong-ge, CHEN Ning-xian. Construction Technique for Pylons of Huanggang Changjiang River Rail-cum-Road Bridge[J]. Bridge Construction,2013,41(06):20-23.

Treatment technology for large deformation of tunnel passing through soft and fragmented zone

Liang Xiaoteng[1,2*], Shi Zhangru[1]

[1]CCCC Second Harbour Engineering Co., Ltd., Wuhan, Hubei, 430040, China

[2]Underground Engineering Catastrophe and Protection Laboratory of Chinese Academy of Sciences,Wuhan, Hubei,430040,China

[*]Corresponding author's e-mail: liangxiaoteng1@ccccltd.cn

Abstract. Large deformation is a major problem for tunnels, threatening engineering safety and increasing construction costs. In order to solve the large deformation problem of deep-buried tunnels passing through the soft and fragmented zone, relying on the Dafalang tunnel project under construction, the mechanism of the large deformation of the fracture zone tunnel is analyzed, and deformation control measures suitable for geological conditions are proposed accordingly. The large deformation of the Dafalang tunnel is the combined effect of many factors. The complex tectonic stress environment is the mechanical condition for the large deformation of the tunnel. The weak and broken surrounding rock is the material basis and internal condition of the large deformation of the tunnel.Optimizing construction parameters, grading deformation control, and combined length and short anchor rods are adopted to effectively reduce tunnel deformation and ensure construction safety and progress.

1. Introduction

The large deformation of the tunnel has always been a major problem for mountain tunnels to pass through the soft and fragmented zone. It often causes problems such as tunnel face collapse, roof collapse, initial support cracking and spalling, and arch distortion[1,2]. The tunnel was forced to undergo arch replacement construction[3], which slowed down the construction progress and threatened the safety of people and property. In Southwest China, due to its complex and changeable geological conditions and abundant groundwater, the problem of large tunnel deformation is particularly prominent.

Scholars have made a lot of useful explorations and obtained many important results for the type and classification, the mechanism and the cause of the large deformation of the surrounding rock of the tunnel in the soft and fragmented zone. Li Guoliang[4] proposed to use relative deformation and rock mass strength-stress ratio as the grading index in the design stage, and divide the large deformation of extrusion into three grades, and predict the large deformation according to the rock mass strength-stress ratio. Relying on the beam tunnel, Dai Yonghao[5] applied model tests and numerical simulation methods to analyze the rheological mechanics of the rock mass, and proposed a joint support scheme based on steel arch + anchor + steel mesh, which solved the construction support problem in weak surrounding rock sections. Yu Tianci[6] relying on the Shiziyuan tunnel project of the Chenglan Railway, comparative analysis of the support effect found: improved excavation methods, strengthened initial support, asymmetrical reserved deformation, radial grouting to strengthen surrounding rock, and long bolts to reinforce surrounding rock and strengthen monitoring and measurement can effectively control the large deformation of soft rock tunnels. Guo Jian[7] monitored the entire construction process of the displacement of the surrounding rock of the carbonaceous slate, the internal force of the initial support and the internal force of the second lining, and explored the characteristics of the lining of the tunnel with large deformation in different construction stages.

In summary, scholars have made certain studies on the large deformation of the tunnel, but the treatment measures for the large deformation of the tunnel are also different in different sections and different lithologies, especially when passing through weak and broken zones and other unfavorable geology. The concept and method of treatment of large deformation needs to be further improved. This article combines the large deformation problems encountered by the Da Franc deep-buried tunnel through the soft and soft fracture zone, and adopts measures such as optimized construction parameters, deformation grading control, and combined length and short bolts to ensure the progress of the tunnel construction and the safety of the structure, providing a reference for similar projects in the future.

2. Project overview

2.1. Engineering design overview

Dafalang Tunnel is located in Wenshan, Yunnan. It is a separated tunnel. The tunnel is spread out in a curve. The axis direction of the entrance section of the tunnel is about 98°, and the axis direction of the exit section is about 88°. The total length of the left tunnel is 3390m, and the maximum buried depth of the tunnel is about 206.5m. The total length of the right-line tunnel is 3375m, and the maximum buried depth of the tunnel is about 195.5. It is a deep-buried super-long tunnel. The large deformation section of the Dafalang Tunnel has a span of 12.78m and a height of 10.29m. The three-step method is adopted for excavation. The excavation sequence is shown in Figure 1.

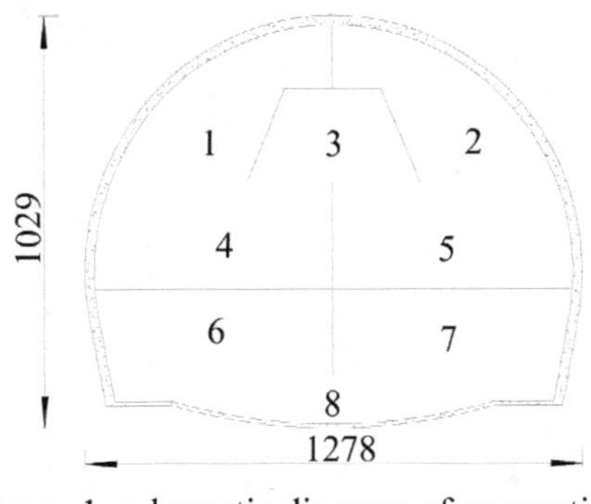

Figure 1. schematic diagram of excavation

Figure 2. tunnel excavation reveals surrounding rock

2.2. Geological profile

The designed elevation of the entrance of the Dafalang Tunnel is 1278.891m, the designed elevation of the tunnel exit is 1269.591m, and the relative elevation difference between the entrance and exit is 9.3m. The tunnel area is located in the southeastern region of Yunnan, which belongs to the low-medium mountain plateau landform area with large undulations.

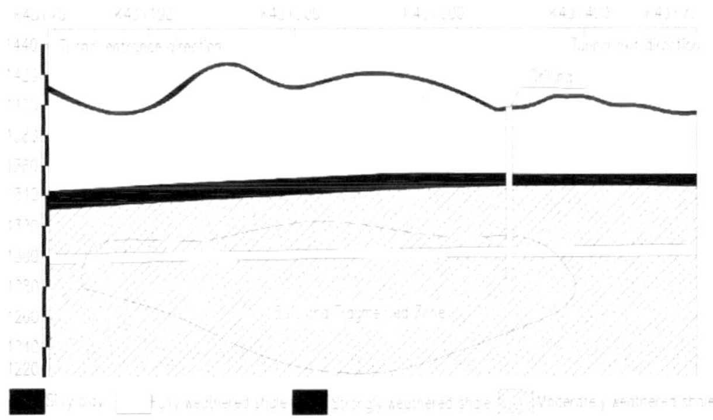

Figure 3. tunnel geological profile

Figure 4. steel support bending

According to the survey and design geological data, the Dafalang tunnel passes through the fracture zone at K43+100-ZK43+400, where the surrounding rock is dominated by shale and belongs to the fracture zone. Its own conditions are weak and severely affected by the structure. There is a high risk of water and mud inrush, collapse, roof fall, and large deformation of the tunnel. The longitudinal section of the tunnel geology is shown in Figure 3.

3. Overview of tunnel large deformation

3.1. Construction overview

When the tunnel is broken during construction, according to the third-party monitoring data, the monitoring sections of the left and right tunnels are at a large deformation rate, and the maximum single-day deformation of the vault is 58.4mm and 49.2mm, respectively, exceeding the warning value (5mm/d) . In addition, there are missing pieces in the cave, the partial arch of the initial support is twisted, and the concrete has many cracks. Take immediate measures to control the large deformation of the tunnel. After the measures were taken, the deformation of the left and right holes of the Dafalang Tunnel was basically stable, and the excavation continued forward.

On October 7, 2020, each monitoring section of the left and right tunnels was at a relatively large deformation rate (the maximum single-day deformation was 53.9mm), which exceeded the warning value. In the cave, there are missing blocks, local arches of the initial support are twisted, and there are many cracks in the concrete. Subsequently, the deformation of the Da Franc tunnel exceeded the warning value continuously.

3.2. Tunnel large deformation mechanism

The large deformation of the soft rock in the Dafalang Tunnel is the combined effect of multiple factors. According to the survey data, a fault fracture zone developed in the tunnel site area within the construction area. The tectonic stress environment in the fracture zone is complex, and the surrounding rock stress redistributes after tunnel excavation, which further changes the stress environment, which is the mechanical condition for the large deformation of the tunnel. The survey lithology of the large deformation section of the tunnel is mainly moderately weathered shale, but the tunnel excavation reveals that it is strongly weathered argillaceous shale. Affected by faults, the surrounding rock joints and fissures are developed, and the rock mass in the fault zone is weak and broken, and the strength is low. The surrounding rock fragmentation zone is the material basis and internal conditions for the large deformation of the tunnel. The tunnel excavation revealed that the strata strata, the structural axis and the line intersect, the strata dip angle is 60°, and the flaky structure. After the tunnel is excavated, the vault to the right arch is deformed seriously, and it is prone to bending and deformation under the complex stress environment.

4. Tunnel large deformation control measures

4.1. Optimize construction parameters
In order to solve the problem of poor geological conditions in the fracture zone of the Dafalang tunnel, which led to the large deformation of the tunnel. The large deformation of the right side of the initial support for the left and right tunnels need to be replaced by the arch, according to the opening of the hole to the face of the tunnel. The direction of the machine is gradually dismantled and replaced, and the net shotcrete is reapplied. The weak and broken zone adopts three steps + temporary inverted arch construction method; strengthen the advance support, the original arch design length is 4mgrouting small pipe shed, and the length is changed to 4.5m. Strengthen the initial support of the weak and broken zone of the left and right tunnels. The original design steel arch I20b is changed to I22b, and the spacing is changed from 60cm to 50cm; the single-layer steel mesh of shotcrete is changed to double-layer steel mesh; the original design of the upper step is unilateral 2 φ42 grouting lock foot anchor pipes were changed to 4; the original design of the middle step was changed to 2 φ42 grouting lock foot anchor pipes on one side to 2 6 meters Φ76mm*6mm lock foot anchor pipes, with a downward tilt The angles are 20° and 45° respectively; the grouting pressure of the lock foot anchor pipe is controlled at 0.5-2.0 MPa. Enhance the mechanical properties of the reinforced concrete for the secondary lining in the weak fracture zone under the eccentric pressure. The distance between the main reinforcement of the secondary lining invert and the arch wall is changed from 20cm to 15cm. Strengthen the advance geological prediction and monitoring measurement of the tunnel section in the weak and fracture zone with the left and right bias.

4.2. Long and short combined anchor rod reinforcement
In the tunnel, a multi-point displacement meter is used to measure the deep displacement of the surrounding rock. The measuring rods are divided into three lengths: 0.5m, 1m, and 1.5m. The arrangement position is shown in Figure 5, and each measuring point is numbered one by one. The test of the surrounding rock loose zone shows that the displacement change trend of each measuring point of 1m, 3m, and 5m is relatively consistent, indicating that the surrounding rock loose zone is larger than the range of 5m. The original anchor rod did not extend into the elastic area of the surrounding rock for anchoring, resulting in the failure of the anchor shotcrete support of the soft rock tunnel, so the anchoring length of the anchor rod needs to be increased.

The construction technology of combining long and short bolts to control the deformation of the surrounding rock of the tunnel is a combined support method of long and short bolts. The short anchor rod in the early stage is used to limit the deformation of the shallow loose and broken surrounding rock, and form a reinforced composite arch structure with the shallow surrounding rock to jointly assume the surrounding rock deformation in the plastic zone. On the one hand, the long anchor rod controls the deformation of the deep surrounding rock; on the other hand, the reinforced composite arch structure is suspended from the deep stable rock mass. The shallow and deep surrounding rocks work together to coordinate the deformation.

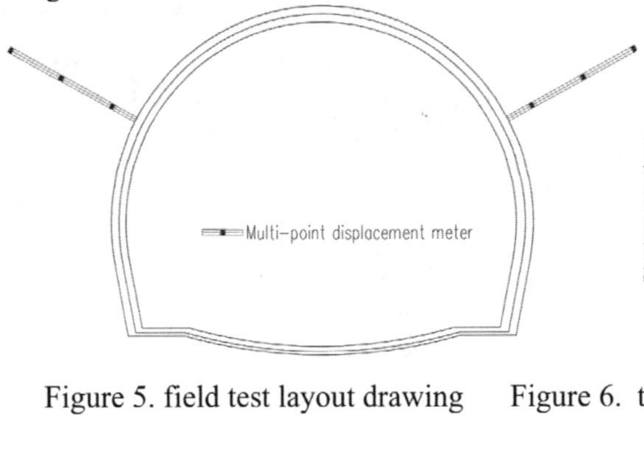

Figure 5. field test layout drawing

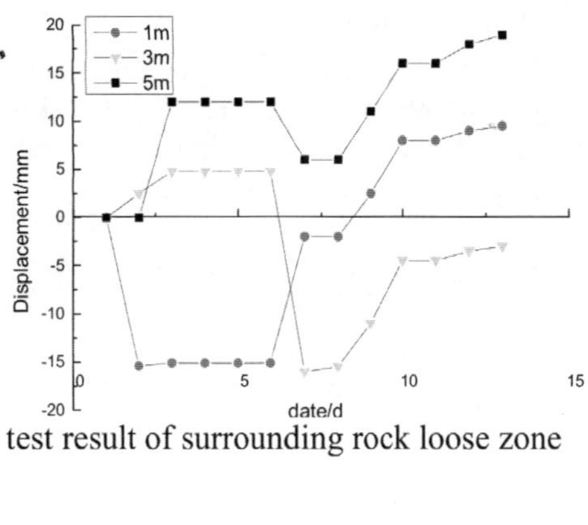

Figure 6. test result of surrounding rock loose zone

The combination of long and short anchor rods uses 3-4m short anchor rods and 6-8m long anchor rods . The short anchor rods can be resin (medicine-packed) anchor rods; the anchor rods longer than 6m should be hollow anchor rods. The severely deformed section where the deformation develops rapidly, the short bolt (<4m) is used for quick and convenient construction, which is used for the initial deformation control, and creates the construction time for the long bolt.

4.3. Hierarchical control of large tunnel deformation

In order to better control the deformation of the surrounding rock, quantitative analysis of the deformation of the surrounding rock in each construction stage can be carried out for hierarchical control and management of the reserved deformation, which is conducive to the connection of various construction procedures and coordinated deformation control. Propose the following classification control standards for surrounding rock deformation in each construction stage, as shown in the following table. If the surrounding rock deformation control exceeds the limit, corresponding strengthening measures should be taken in time.

Table 1. rock deformation control standard

Deformation grade	Reserved deformation	Characteristic description of surrounding rock
Slightly deformation	20cm	The strength of the surrounding rock is average, the joints and fissures are generally developed, and there is no water or wetting.
Moderate deformation	40cm	The surrounding rock is weak and broken, with well-developed joints and fissures, without water or wetting.
Severely deformation	60cm	The surrounding rock is weak and broken, joints and fissures are developed, and there is water or seepage or running water.

5. Conclusion

Affected by geological conditions, large deformations often occur during the construction of tunnels in soft and fragmented zones, threatening the safety of the project. By analyzing the large deformation mechanism of the Dafalang Tunnel and putting forward corresponding treatment measures accordingly, the tunnel deformation can be effectively controlled, and the following conclusions are drawn:

(1) Relying on the Dafalang tunnel project under construction, analyzing the large deformation mechanism of the fracture zone tunnel, the results show that the large deformation of the soft rock of the Dafalang tunnel is the combined effect of multiple factors, and the complex tectonic stress environment is the mechanical condition for the large deformation of the tunnel; The weak and broken surrounding rock is the material basis and internal condition of the large deformation of the tunnel.

(2) The test of the tunnel surrounding rock loose circle shows that the displacement change trend of each measuring point of 1m, 3m, and 5m is relatively consistent, indicating that the loose zone of the surrounding rock is larger than the range of 5m. Long and short combined anchor rods are used to control the deformation of the tunnel. The length of the short anchor rod is 3-4m, and the length of the long anchor rod is 6-8m.

(3) Aiming at the large deformation of Dafalang Tunnel, the deformation control measures adapted to the geological conditions are proposed. Optimizing construction parameters, deformation grading control, combined length and short anchor rods are used to effectively reduce tunnel deformation and ensure construction safety and progress.

Acknowledgments

This research was financially supported by the Applied Basic Research Project (main subject) of the CCCC Second Harbour Engineering Co.,Ltd.(No. HT-2017-A-07-007). The author thanks scholars who have made outstanding achievements in this field

References

[1]Li S C, Xu F, Li L P, Wang W M, Zhang W, Zhang Q Q, Shi S S. State of the art ： challenge and methods on large deformation in tunnel engineering and introduction of a new type supporting system. 2016, Journal of Architecture and Civil Engineering, 35(07):1366-1376.

[2]Sun T J, Wang W, Luo M R，et al. Influences of different lining structure defects on overall safety of tunnel structure. 2017, Journal of Architecture and Civil Engineering, 34(3):82-90.

[3]Liu G, Zhang F Y, Li X Z. Research on large deformation andits mechanism of Muzhailing tunnel, 2005, Chinese Journal of RockMechanics and Engineering, 24(Supp.2):5521-5526.

[4]Li G L, Liu Z C, Zhu Y Q. On the Large Squeezing Deformation Law and Classification Criteria for the Lanzhou-Chongqing Railway Tunnels in Soft and High Geostress Rocks. 2015, modern tunnelling technology,52(01):62-68.

[5]Dai Y H, Chen WZ, Tian H M. Study on large deformation and support scheme of soft rock in Daliang tunnel. 2015, Chinese Journal of Rock Mechanics and Engineering,34(S2):4149-4156.

[6]Yu T C. Study on large deformation control technology of tunnel in soft rock, 2017, China Civil Engineering Journal, 50(S2):112-117.

[7]Guo J, Yang J S, Chen W. Research on large deformation of surrounding rock and mechanical characteristics of lining of carbonaceous slate tunnel based on field measurement. 2019, Chinese Journal of Rock Mechanics and Engineering, 38(4):832-841.

Research on Application Effect Evaluation of Steel Box Girder Bridge Based on Combination Weighting

Lin-jun Zhang[1], Hongwen Zhou[1*], Heng Li[1], Baoyue Yuan[1]

[1] School of Urban Construction and Safety Engineering, Shanghai Institute of Technology, Shanghai, 201418, China

[*]Corresponding author's e-mail: hngfzhou@sit.edu.cn

Abstract. At present, the research on the application effect of steel box girder bridges is still only in the qualitative analysis link, and the calculation of index weights is greatly affected by personal subjective factors, and the quantitative evaluation of the application effect of steel box girder bridges has not been achieved. In this paper, we proposes a combination of subjective and objective combined weighting fuzzy evaluation model, which avoids the unreasonable problem of a single weighting method in determining the weight coefficients, and makes the weight distribution results more accurate and effective. Among them, the coefficient of variation method and the AHP method are used to calculate the subjective and objective weights of the indicators, and the linear combination calculation method is used to determine the comprehensive weights of the indicators. Then, combined with the fuzzy evaluation model, a comprehensive analysis of the application effect is carried out. Finally, the effectiveness of the combined weighting evaluation model is verified with a practical project case.

1. Introduction

In recent years, with the continuous extension of my country's highway network, problems such as long construction period and many potential safety hazards of traditional cast-in-place concrete bridges have become increasingly prominent. With the continuous development of steel structure technology and large-span bridge construction, road bridge steel box girder structures with short construction period, low environmental impact, and simple maintenance have been widely used.

However, the current evaluation of the application effect of steel box girder technology is still only at the qualitative level, lacking quantitative analysis and the evaluation results are general and unobjective. Therefore, it is particularly important to comprehensively evaluate the application effects of steel box girder bridges, find out the shortcomings in project construction, and improve the evaluation level.

Domestic and foreign scholars have carried out a lot of research on the evaluation of the application effect of engineering projects, and have achieved certain results in related theories and models. Established a comprehensive evaluation model for steel box girder construction safety risks by combining triangular fuzzy number method, Bayesian theory and fuzzy comprehensive evaluation method[1]; based on the combination of BP neural network technology and fuzzy theory, the welding quality of steel box girder is evaluated effectively[2]; used the analytic hierarchy process and improved value approximation ideal solution sequencing model to evaluate the risk of the highway construction plan, and made good progress in the highway construction[3]; the coefficient of variation method and the approach to ideal point method, combined with the grey relational analysis theory to construct the safety facility evaluation model of the expressway[4]; used game theory to optimize the fuzzy analytic hierarchy process and the critic method to establish a finite cloud model, so as to determine the risk level of each tailing pond[5].

At present, there are relatively few comprehensive evaluation studies on the application effects of steel box girder technology. The commonly used evaluation methods are mainly Delphi method and Monte Carlo simulation method. However, in view of the many influencing factors in the construction process of engineering projects, the single weighting method is more one-sided in determining the weight coefficient, and it is difficult to guarantee the accuracy and reliability of the evaluation results.

Therefore, in accordance with the idea of combining subjective and objective, this paper proposes a combined weighting model to determine the index weights based on the coefficient of variation method and the AHP method. Through the comprehensive evaluation of each section of the steel box girder bridge by using the fuzzy comprehensive evaluation method, the comprehensive evaluation grade of steel box girder bridge is obtained and the shortcomings in the project construction are pointed out, which provides an effective reference for the construction and management of steel box girder bridges. And use this model to comprehensively evaluate and analyze the application effect of a steel box girder bridge on a certain expressway in Henan, so as to verify the feasibility of this method.

2. Fuzzy Evaluation of Coefficient of Variation-AHP Combination Weighting

The single weighting method may be too subjective when determining the index weight, or only rely on data, so it cannot accurately and comprehensively reflect the importance of the index. AHP (Analytic Hierarchy Process) method of subjective weighting, coefficient of variation method objective weighting and linear combination calculation methods are used to determine the index weight coefficients. At the same time, the fuzzy comprehensive evaluation method is used to achieve quantitative analysis of fuzzy things in the evaluation process, to a certain extent, to avoid the shortcomings of a single empowerment method.

2.1. AHP (Analytic Hierarchy Process) method subjective empowerment

(1) Establish a judgment matrix

A judgment matrix is constructed by comparing the importance of each level of evaluation indicators. The importance of the indicators is determined by the scale of 1-9 proposed by Saaty[6]. The values and meanings of the judgment matrix are shown in table 1.

Table 1. Scale and implication of judgment matrix

Scale a_{ij}	**Meaning** (i,j)
1	i,j indicators are equally important
3	i index is slightly more important than j index
5	i index is more important than j index
7	i index is much more important than the j index
9	i index is more important than absolute j index
2、4、6、8	Between the above adjacent judgment scale values
Reciprocal	If the judgment value of index i is compared with index j is a_{ij}, then the judgment value of index j is $1/a_{ij}$ product.

(2) Calculate the weight of the evaluation index

This paper uses the square root method to calculate the weight of each indicator $B=(b_{ij})_{n\times n}$ in the criterion layer to the target layer A.

$$W_i=\sqrt[n]{\prod_{j=1}^{n} b_{ij}}\ ,\quad W_i=\frac{\overline{W_i}}{\sum_{j=1}^{n}\overline{W_j}} \tag{1}$$

That is, the weight of index $B=(B_1, B_2, B_3, \cdots B_n)^T$ to index A is: $W=(W_1, W_2, W_3, \cdots W_n)^T$.

(3) Consistency inspection

Calculate the maximum eigenvalue of the judgment matrix and judge the consistency

$$\lambda_{\max}=\frac{1}{n}\sum_{i=1}^{n}\frac{(BW)_i}{W_i}, \quad CI=\frac{\lambda_{\max}-n}{n-1} \tag{2}$$

Among them, for the average random consistency index RI of the judgment matrix, whether the judgment matrix has satisfactory consistency, the RI values under different orders are shown in table 2.

Table 2. RI value table

n	1	2	3	4	5	6	7	8	9
RI	0.00	0.00	0.58	0.90	1.12	1.24	1.32	1.41	1.45

When the order is greater than 2, the ratio of the consistency index Cl to the average random consistency index RI is called the random consistency ratio, which is recorded as CR; at the same time, when CR<0.1, it can be judged that the judgment matrix has passed the consistency test and the evaluation result is feasible; otherwise, the judgment matrix should be adjusted appropriately.

2.2. Steps to determine the objective weight by the coefficient of variation method

In view of the one-sidedness of the single subjective weight method in determining the weight, the objective weighting of the coefficient of variation method is to further determine the coefficient of variation by calculating the ratio of the average of each evaluation index to the standard deviation. If the degree of variation of the actual calculation data is greater, the weight assigned is positively correlated; on the contrary, the weight of the indicator is smaller, which can objectively and effectively reflect the degree of differentiation of various indicators, which is an objective method of calculating weights, the main calculation steps are as follows[7]:

(1) Calculate the coefficient of variation of each index:

$$S_j=\sqrt{\frac{1}{n-1}\sum_{j=1}^{n}(x_{ij}-\overline{x_j})^2}, \quad V_j=\frac{S_j}{\overline{x_j}} \tag{3}$$

(2) Normalize each evaluation index to calculate the weight of each index:

$$W_j=\frac{V_j}{\sum_{j=1}^{n}V_j} \tag{4}$$

In the formula, $\overline{x_j}$ is the average value of the j-th evaluation index; V_j is the coefficient of variation; W_j is the evaluation index weight.

2.3. Coefficient of Variation-AHP Method Combination Weighting

In order to further improve the accuracy of the evaluation index weights, a linear combination calculation method based on subjective and objective weighting is proposed for the weights obtained by the above-mentioned coefficient of variation method and AHP method to determine the combined weights of the indicators[8]. In order to reduce the subjective and objective weight deviation and improve the reliability of the weight value, the main calculation steps are as follows:

$$W_{ij}=\alpha W_i+(1-\alpha)W_j \tag{5}$$

In the formula, W_i is the weight calculated by the AHP method, W_j is the weight calculated by the coefficient of variation method, and α is the proportion of the subjective weight in the combination.

The objective function is used to find the optimal value of a, so as to obtain the optimal combination weighting value.

$$\min y=\sum_{i、j=1}^{n}\left[\left(W_{ij}-W_{i}\right)^{2}-\left(W_{ij}-W_{j}\right)^{2}\right] \quad (6)$$

Through the objective function calculation, the α optimal solution is 0.5.

2.4. Fuzzy comprehensive evaluation

As a quantitative evaluation method for fuzzy things, the fuzzy comprehensive evaluation method is conducive to the realization of quantitative analysis of fuzzy things on the basis of qualitative analysis.

For the evaluation index system, establish multiple evaluation index factor sets $U=[u_1,u_2,u_3,\cdots,u_n]$, and based on the single factor evaluation results, establish the corresponding evaluation set $V=[v_1,v_2,v_3,\cdots,v_m]$ and the degree of membership r_{ij}, this paper adopts the triangular distribution function and establishes the corresponding membership function, and constructs the membership function formula according to the corresponding index.

$$B=W\bullet R=\left(r_1,r_2,\cdots,r_n\right)\times\begin{bmatrix} r_{11} & r_{12} & \cdots & r_{1n} \\ r_{21} & r_{22} & \cdots & r_{2n} \\ \vdots & \vdots & \ddots & \vdots \\ r_{m1} & r_{m2} & \cdots & r_{mn} \end{bmatrix}=\left(b_1,b_2,\cdots,b_n\right) \quad (7)$$

In the formula, r_{ij} represents the degree of membership of the index factor u_n and the evaluation set v_i; W is the importance weight set; R is the evaluation matrix; B is the corresponding fuzzy comprehensive evaluation set.

3. Example application

This article uses the construction of a steel box girder project of an expressway in Henan as the research background. Divide it into 4 sections and use the actual test data as the analysis sample to verify the coefficient of variation-AHP method in expressway steel box girder. Effectiveness in the comprehensive evaluation of the application effect of beam technology.

The comprehensive evaluation of the application effect of steel box girder technology is mainly conducted from five aspects that affect the quality, schedule, safety, cost and environmental impact factors of project management. The evaluation system is divided into target level: comprehensive evaluation index for the application effect of steel box girder technology; criterion level: engineering quality, project progress, engineering safety, engineering cost and engineering environment regarding the application effect of steel box girder technology; index level: evaluation of specific indicators for relevant elements in the application process of steel box girder technology in the criterion layer, so as to construct a comprehensive evaluation system for the application effect of steel box girder bridges and the details are shown in table 3.

At the same time, according to the relevant specifications for the application of expressway steel box girder technology, this article divides the index levels into four application effect evaluation levels, namely first level (excellent), second level (good), third level (medium), and fourth level (poor), the critical value of each index is determined according to relevant specifications.

Table 3. Comprehensive evaluation index system of application effect of steel box girder bridge

Target level	criterion level	index level
	Engineering quality B1	Steel quality C1
		Geometry size of beam C2
		Weld quality C3
		Anti-rust coating C4
		Linear error C5
	Project progress B2	Interference in the production process C6
		Interference during transportation C7

Comprehensive evaluation index of application effect of steel box girder bridge A		Interference of welding assembly C8 Interference of hoisting erection C9
	Engineering safety B3	Human unsafe behavior C10 Integrity rate of machinery and equipment C11 Construction environment interference C12 Traffic Guarantee Plan C13
	Engineering cost B4	Additional material fee C14 Additional labor cost C15 Additional station fee C16
	Engineering environment B5	Wastewater pollution C17 Solid waste pollution C18 Civilized Construction C19

3.1. *Calculation of combined weight of steel box girder*

The subjective weight is obtained by the AHP method through formula (1)~(2); at the same time, according to the measured value data of the indicators, from formula (3)~(4) uses the coefficient of variation method to obtain the objective weight W; finally, Matlab is used to obtain the combination weight W of the indicators according to the equations (5) ~ (6); the relevant indicator combination weight values are shown in table 4.

Table 4. Weight values of each index

Evaluation index	Subjective weight	Objective weight	Comprehensive weight
C1	0.023	0.001	0.012
C2	0.009	0.010	0.010
C3	0.006	0.002	0.004
C4	0.012	0.004	0.008
C5	0.003	0.027	0.015
C6	0.015	0.095	0.055
C7	0.076	0.142	0.109
C8	0.045	0.142	0.094
C9	0.026	0.177	0.102
C10	0.015	0.113	0.064
C11	0.053	0.071	0.062
C12	0.01	0.026	0.018
C13	0.033	0.018	0.026
C14	0.075	0.006	0.041
C15	0.03	0.013	0.022
C16	0.16	0.009	0.085
C17	0.07	0.046	0.058
C18	0.213	0.080	0.147
C19	0.124	0.017	0.071

It can be seen from table 4 that the combination weighting method is used to calculate the index weight value, and the coefficient of variation method is combined with the traditional AHP method, which can avoid the shortcomings of a single method in calculating weights, and can more accurately reflect each the degree of influence of the evaluation index in the project evaluation.

Among the combined weights, the weight value of solid waste pollution (C18) is 0.147, which is the highest weight, indicating that strengthening the control of solid waste pollution during the construction of steel box girder bridges plays an important role in the overall evaluation effect; the weight values of interference (C7) and interference (C9) of hoisting erection are 0.109 and 0.102, respectively, which account for a higher weight, indicating that the control of the transportation and hoisting progress of steel box girder in the construction of steel box girder bridges is the result of comprehensive evaluation of impact Important factor.

According to the characteristics of steel box girder technology application, a three-level index system was constructed. In the evaluation study based on the combination of coefficient of variation-AHP method, it was found that the index weights of the three calculation methods are significantly different. As a subjective weighting method, the AHP method is affected by subjective experience factors, which makes the gap between the indicator weights larger, and it is difficult to reflect the true evaluation results. The weighting calculation of the comparative analysis method of the coefficient of variation is more objective, but when the average value of the indicator is close to Small disturbances at 0 will cause deviations in the value of the coefficient of variation, thereby reducing the reliability of the calculation and analysis results. Therefore, this paper adopts the combined weighting method to further highlight the important influencing factors on the basis of maintaining the objectivity of the evaluation, so that the weight value of the evaluation index is more scientific and effective.

3.2. Evaluation grade of steel box girder technology application

By bringing the quantified value of the project evaluation index into the membership function, the membership level of the steel box girder technology application evaluation index is further obtained. Here, the index layer of the "engineering quality" element is taken as an example for comprehensive calculation.

By constructing the evaluation index set and evaluation matrix of the "engineering quality" elements, according to the above-mentioned coefficient of variation-AHP method, the comprehensive weight is $W_1 = (0.012, 0.01, 0.004, 0.008, 0.015)$, and the comprehensive evaluation result is $B_1 = (0.004, 0.010, 0.036, 0.011)$. Therefore, according to the principle of maximum degree of membership, it can be known that the quality of steel box girder bridge project quality is "medium".

In the same way, the fuzzy comprehensive evaluation results of other index levels can obtained respectively: $B_2 = (0.034, 0.254, 0.072, 0)$; $B_3 = (0, 0.092, 0.078, 0)$; $B_4 = (0.081, 0.067, 0, 0)$; $B_5 = (0.069, 0.133, 0.074, 0)$.

According to the principle of maximum degree of membership, the comprehensive evaluation results of the application effect of steel box girder bridges are determined, as shown in table 5.

Table 5. Application evaluation results of steel box girder bridge

Comprehensive Evaluation	Evaluation grade				Evaluation result
	Level 1 (Excellent)	Level 2 (Good)	Level 3 (Medium)	Level 4 (Poor)	
steel box girder bridge construction	0.055	0.124	0.052	0.001	Level 2 (Good)

3.3. Evaluation results

In summary, through the evaluation model of coefficient of variation and combined weight of AHP method, the comprehensive evaluation result of the application of steel box girder bridge in an expressway in Henan is level 2 "good", and the best evaluation result of cost control effect is level 1 "excellent". The evaluation result of engineering quality control, engineering progress management and engineering safety is level 2 "good", while the evaluation result of engineering quality is generally the third level "medium".

Therefore, it is necessary to further strengthen the inspection and management of steel quality in the subsequent construction of related projects, and further improve the evaluation level of project quality. The evaluation result is consistent with the actual situation of the project, which verifies the effectiveness of the method in the comprehensive evaluation of the application effect of steel box girder bridges.

4. Conclusion

In this study, by combining the whole application process of steel box girder bridges, 19 main factors influencing the application effect of steel box girder bridges in 5 categories were found. For the first time, a comprehensive analysis of the application effect of steel box girder bridge is carried out by combining the fuzzy evaluation model of composite weighting, and the main control links and critical control points in the construction process are further defined, which enriches the index system and evaluation model of the application evaluation of steel box girder bridge.

The fuzzy comprehensive evaluation model of the coefficient of variation-AHP method effectively avoids the one-sidedness of a single method in determining the index weight, reduces the influence of subjective factors, and further improves the accuracy of the evaluation results. However, due to the current steel box girder technology just a few researches on the appraisal of the application, the index system established in this article still has certain shortcomings, and there is a certain degree of subjectivity for the expert scoring in the AHP method. Therefore, in the follow-up research, it is necessary to further improve the selection of indicators and the optimization of evaluation methods.

Acknowledgments

This article is one of the phase results of the Henan Transportation Science and Technology Fund Project (2018T05). Thanks to the technical staff and the organization for their financial support.

References

[1] Hongyun Xue, Fengkun Cui, Xu Dong, Baoqun Wang, Wei Lu, Kai Wang. (2020) Safety Risk Analysis and Control for Construction of Two-way Jacking Steel Box Girder Based on Hybrid Algorithm [J]. Highway, 65 (11): 124-129.

[2] Hongfei Xu, Yunfei Xian, Baisheng Wu. (2020) Welding Quality Evaluation Method of Steel Box Girder Based on Fuzzy BP Neural Network [J]. Science Technology and Industry, 20(12): 233-236.

[3] Haopeng Jiang, Jiancheng Sun, Wenwei Yang, et al. (2020) Risk Evaluation of Highway Construction Scheme Based on AHP and Improved TOPSIS Weight Algorithm [J]. Journal of Wuhan University (Engineering Science), 53(08): 698-703.

[4] He Gao, Huicong Zhou. (2014) Evaluation of Highway Safety Facilities Based on Grey Approximation Ideal Point [J]. Forest Engineering, 30(06): 124-128.

[5] Yixuan Dong, Hongwen Zhou. (2020) Risk assessment of tailings pond dam break based on game theory-finite cloud model [J]. Hydropower and Energy Science, 38(12): 75-78+168.

[6] SAATY T L. (1977) A scaling method for priorities in hierarchical structures [J]. Journal of Mathematical Psychology, 15(3): 234-281.

[7] Peng Huiwu, Angui Li, et al. (2020) Evaluation system of distributed energy system based on IAHP- coefficient of variation method [J]. Journal of Xi 'an University of Architecture and Technology (Natural Science Edition), 52(04): 572-578.

[8] Ping Zhuang, Yanxi Li. (2011) An Empirical Study on the Risk Assessment Model of Enterprise Investment Based on G1-Coefficient of Variation [J]. Soft science, 25(10): 107-112+120.

Comparison for Solid Concrete-filled Circular Steel Tubes Load-carrying Capacities in Frame Structures Basing on Chinese codes

Duan Yu-tong[1*], Lei Hong-gang[2]

[1]College of Civil Engineering, Taiyuan University of Technology, Taiyuan 030024, China

[2]College of Civil Engineering, Taiyuan University of Technology, Taiyuan 030024, China

[*] Corresponding author's e-mail: lhgang168@126.com

Abstract: Solid steel tubular concrete members have excellent axial compression capacity. In terms of the calculation method of their axial compression bearing capacity, the current ten domestic standards all provide specific provisions. On the basis of main structure provisions such as diameter-thickness ratio and confinement factor, the formula differences in axial-force compression capability of domestic and international standards are compared and analyzed. First, the difference value of bearing capacities reach up to 62.79% according to current standards when diameter -thickness ratio range from 4 to 30. Second, concrete-filled steel tube members with diameter -thickness ratio greater than 20 and diameter less than 200, the current specifications cannot calculate the axial compressive bearing capacity.

1.Introduction

Major in axial compression and small eccentric compression, concrete-filled steel tubular members are widely used in civil construction, industrial construction, bridge construction and other fields[1-4]. In order to establish a design method for the bearing capacity of concrete-filled steel tube members, many well-known scholars from Harbin Institute of Technology, the Chinese Academy of Building Sciences, Tsinghua University and other units have carried out a large number of theoretical and experimental studies over the years, which have provided an important basis and contribution for the compilation of a series of codes and standards in China. There are 10 copies of solid circular concrete-filled steel tube axial compression members in Chinese current codes and standards[5~14]。

In the above standards, there are two main theoretical systems for the design of concrete-filled steel tubes: (1)Unified Theory of Concrete Filled Steel Tubes(hereinafter called "Unified theory"); (2)Experiment-based limit equilibrium theory(hereinafter called "Limit equilibrium theory"). As the comparative study of design methods in the norms and standards, a new method with a simple form and a clear concept is proposed by analyzing the design methods for the bearing capacity of concrete-filled steel tubes in the 7 domestic and foreign codes[15]. Comparing the calculation formulas of concrete-filled steel tube bearing capacity in different regulations in China, it is pointed out that although the formulas of the various regulations have certain differences in form, their essential laws are consistent[16, 17]. Basing on the domestic and foreign design regulations for concrete-filled steel tube columns, a new unified algorithm is proposed on the basis of the Chinese,

American and European standards[18, 19]. Based on the Chinese and British specifications and combined with actual projects, the two specifications on the structure and design requirements of concrete filled steel tube are compared, and it is shows that the requirements of Chinese specifications are more stringent[20]. Introduces the calculation methods of the six codes of China, the United States, Europe and Japan, and compares and analyzes the existing test results, and reveals that the calculation results of the national codes are generally safe[21]. Listing three domestic and five overseas codes and calculated and compared the design formulas for the short column axial compression bearing capacity. The results indicate that the Chinese code is more consistent with the test results[22]. This paper will take the bearing capacity calculation method of the concrete-filled circular steel tube axial compression member as the research object, comparing and analyzing the differences in the 10 norms.

2. Comparative analysis of structural requirements

2.1 regulations of steel and concrete

As shown in Fig.1, regarding the selection of steel and concrete, these codes give a clear selection range of grades and labels, and make clear requirements for its performance: the steel should meet the relevant regulations of the current national standard " Standard for design of steel structures" GB 50017, and the concrete should comply with the relevant regulations of the current national standard "Code for design of concrete structures" GB 50010.

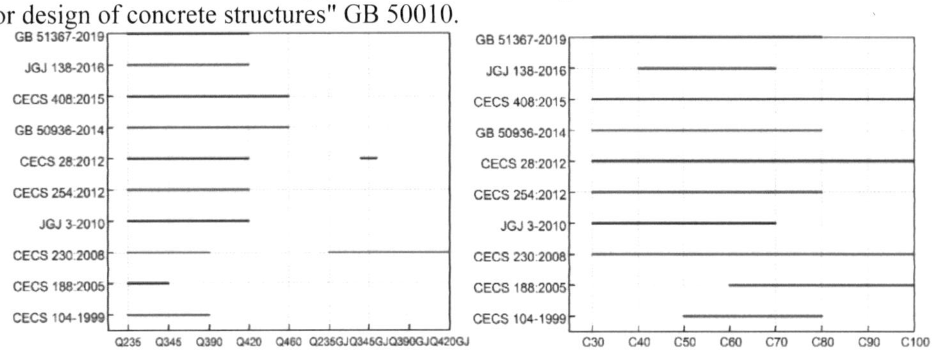

a. Steel selection b. Concrete selection

Fig.1 Selection of steel and concrete specified by the codes

Fig.1 shows that most steels use Q235, Q345, Q390, Q420, but high-strength and high-performance steels of Q460 and above are not in the scope of encouraging applications. At the same time, it is recommended that the strength of concrete is not less than C30, but the high-strength and super-high-strength concrete of C80~C100 is not in the encouragement range of wide application.

2.2 Regulations of steel pipe diameter -thickness ratio and size

Regarding the diameter -thickness ratio and dimensions of steel pipes, these codes give clear regulations, see Table 1 for details.

Table 1 Diameter-thickness ratio D/t and steel pipe size limit requirements

Code name	D/t	L/D or l/i	$D_{min}(mm)$	$t_{min}(mm)$
GB 51367-2019	-	-	200mm	4
JGJ 138-2016	$D/t \leq 135(235/f_{ak})$;		400	8
CECS 408:2015	$D/t \leq 135(235/f_y)$;		168	3
GB 50936-2014	$D/t \leq 135(235/f_y)$;		168	3
CECS 28:2012	$(20{\sim}135)\ 235/f_y$		200	4
CECS 254:2012	20~100	≤20 or≤80	100	3
JGJ 3-2010	$(20{\sim}100)\sqrt{235/f_y}$		400	8
CECS 230:2008	$(20{\sim}90)\ 235/f_y$		300	6
CECS 188:2005	≤90(Q235) or≤75(Q345)		200	6
CECS 104-1999	$(20{\sim}135)\ 235/f_y$		-	8

Firstly, the specification only stipulates the minimum value of steel pipe size, and does not mention the maximum value of steel pipe diameter and wall thickness. And then, the length to diameter ratio of the component is limited to 20, which is equivalent to slenderness ratio no more than 80, which is the same as the building frame column in GB50017. Also, the limit range of steel pipe diameter-thickness ratio is 20~135, and the minimum wall thickness is 3~8mm. Finally, with the needs of engineering, the diameter and wall thickness of steel pipes are increasing, when the outer diameter, wall thickness, and diameter-to-thickness ratio of the steel pipe exceed the specification limits, the scope of application cannot be expanded at will. In-depth theoretical and experimental research should be carried out before supplementary regulations.

2.3 Regulations of confinement factor

The confinement factor is defined as $\theta = A_s f / A_c f_c$, where f and f_c are the design value of steel strength and concrete strength respectively; A_s、A_c are the steel pipe area and the concrete area respectively. In Chinese current standards, the selection range of the confinement factor of concrete-filled steel tube members is basically limited to between 0.5 and 2.5.

3. Comparison of solid CFCST bearing capacity under axial compression

The design method of solid concrete-filled circular steel tube axially compressed components in China can be summarized into two theoretical systems: one is the unified theory, and the other is the limit equilibrium theory. Three of them adopt a unified theory, namely: CECS 408:2015、GB 50936-2014、CECS 254:2012; eight of them use limit equilibrium theory, which are: GB 51367-2019、JGJ 138-2016、GB 50936-2014、CECS 28:2012、JGJ 3-2010、CECS 230:2008、CECS 188:2005、CECS 104-1999。

3.1 Unified theory

The unified theory requires that the axial compression bearing capacity of single-limb concrete-filled steel tube members under a single force state should satisfy equations (1)~(9):

$$N \le N_u \quad (1)$$

$$N_u = \varphi N_0 \quad (2)$$

$$N_0 = A_{sc} f_{sc} \quad (3)$$

$$f_{sc} = (1.22 + \mathrm{B}\theta + \mathrm{C}\theta^2) f_c \quad (4)$$

$$A_{sc} = A_s + A_c \quad (5)$$

$$\alpha_{sc} = \frac{A_s}{A_c} \quad (6)$$

$$\theta = \alpha_{sc} \frac{f}{f_c} \quad (7)$$

$$\varphi = \frac{1}{2\bar{\lambda}_{sc}^2}\left[\bar{\lambda}_{sc}^2 + (1 + 0.25\bar{\lambda}_{sc}) - \sqrt{\left(\bar{\lambda}_{sc}^2 + (1 + 0.25\bar{\lambda}_{sc})\right)^2 - 4\bar{\lambda}_{sc}^2}\right] \quad (8)$$

$$\bar{\lambda}_{sc} = \frac{\lambda_{sc}}{\pi}\sqrt{\frac{f_{sc}}{E_{sc}}} \approx 0.01\lambda_{sc}(0.001 f_y + 0.781) \quad (9)$$

Where N, N_u and N_0 are the design value of axial pressure acting on the component, the design value of the component axial compression stability bearing capacity, design value of the component axial compression strength bearing capacity. A_{sc} and f_{sc} are combined material cross-sectional area, f and f_c are the design values of steel compressive strength and concrete compressive strength respectively. Also, α_{sc} and θ are the steel content and the confinement factor related to the section of the member, B and C are the influence coefficients of the section shape on the confinement factor. What else, φ is the stability coefficient of the axial compression member, λ_{sc} and $\bar{\lambda}_{sc}$ are the slenderness ratio of the member and the regular slenderness ratio of the member respectively.

In the relevant three of these standards, the calculation requirements of CECS 408:2015 basically refer to GB 50936-2014. The difference between CECS 254:2012 and the former is the cross-sectional strength of the combined material and the value of the stability coefficient. And the design value of the axial compression bearing capacity of the 3 is obtained by multiplying the axial compression strength bearing capacity and the stability factor. The calculation formula for the bearing capacity of the component is shown in Table 2.

Table 2 Formulas for axial compressive bearing capacity of unified theoretical

Number	Code name	$A_{sc}(mm^2)$	$f_{sc}(N/mm^2)$	$N_0(KN)$	$N_u(KN)$
1	GB 50936-2014	$A_s + A_c$	$(1.212 + B\theta + C\theta^2)f_c$	$A_{sc}f_{sc}$	φN_0
2	CECS 408:2015			$A_{sc}f_{sc} + N_b$	
3	CECS 254:2012			$k_c k_2 A_{sc} f_{sc}$	φN_0

It can be seen from the table2 that CECS 408:2015 is the same as GB 50936-2014's calculation formula for strength bearing capacity. Also, CECS 254:2012 calculation formula takes into account the effect of concrete creep and the cross-sectional strength of the reliability correction as the member's strength bearing capacity, while GB 50936-2014 stipulates that when the permanent load of a solid axial compression member accounts for more than 50% of the total axial pressure, the stability bearing capacity of the member should consider the influence of concrete creep, and the influence coefficient is taken as 0.9. in conclusion, the difference of them is that one considers the influence of concrete creep in the calculation of the stability bearing capacity N_u, and the other considers the influence of reliability correction and concrete creep in the calculation of the strength bearing capacity of members N_0. In addition, the values of the stability coefficients in the two specifications are also different, which leads to different calculation results. At the same time, GB 50936-2014 composite material section strength f_{sc} gives two solutions. The results show a linear increase with the rise of the steel content, and the calculated value is slightly higher than the look-up table value with an error of 2% to 6%.

3.2 Limit equilibrium theory

The limit equilibrium theory requires that the axial compressive bearing capacity of single-limb concrete-filled steel tube members should satisfy formulas (10)~(15):

$$N \le N_u \quad (10)$$

$$N_u = \varphi_l \varphi_e N_0 \quad (11)$$

$$N_0 = 0.9A_c f_c(1 + \alpha\theta)(\text{when } \theta \le [\theta]) \quad (12)$$

$$N_0 = 0.9A_c f_c(1 + \sqrt{\theta} + \theta)(\text{when } \theta > [\theta]) \quad (13)$$

And in any case: $\varphi_l \varphi_e \le \varphi_0$ (14)

$$\theta = \frac{A_a f_a}{A_c f_c} \quad (15)$$

Where N,N_u and N_0 are the design value of the component axial pressure, the design value of the component axial compression bearing capacity, the design value of the component axial compression short column strength bearing capacity. A_c and f_c are the design values of the cross-sectional area and compressive strength of the core concrete in the steel pipe respectively. A_a and f_a are the design values of the cross-sectional area and compressive strength of the steel pipe respectively. α is the coefficient related to the concrete strength grade, θ is the confinement factor of the concrete-filled steel tube member. φ_l and φ_e are the bearing capacity reduction coefficients considering the influence of slenderness ratio and eccentricity respectively, and φ_0 is the value of φ_l that considered as the axial compression member.

In the relevant 8 codes, the calculation methods of the coefficients related to the concrete grade, such as the section area of the member, the selection of material strength, and the confinement factor are the same. What differs is the calculation methods of the two influence coefficients. And the design value of the axial compression bearing capacity of the member is defined as the product of the short column strength bearing capacity design value and the two influence coefficient. The calculation formula for the bearing capacity of the member is shown in Table 3. Except for JGJ 138-2016 and CECS 188:2005, the calculation formulas for the strength and bearing capacity of short columns under axial compression members are basically the same. The difference of the calculation results is mainly manifested in α, φ_l and φ_e .

Table 3 Formulas for axial compressive bearing capacity of equilibrium theory

Number	Code name	$N_0(KN)$	φ_l	φ_e	$N_u(KN)$
1	JGJ 138-2016	When$\theta \leq [\theta]$: $N \leq 0.9\varphi_l\varphi_e f_c A_c(1+\alpha\theta)$ When$\theta > [\theta]$: $N \leq 0.9\varphi_l\varphi_e f_c A_c(1+\sqrt{\theta}+\theta)$	φ_{l1}	φ_{e1}	—
2	GB 50936-2014	When$\theta \leq 1/(\alpha-1)^2$: $N = 0.9f_c A_c(1+\alpha\theta)$ When$\theta > 1/(\alpha-1)^2$: $N = 0.9f_c A_c(1+\sqrt{\theta}+\theta)$	φ_{l2}	φ_{e1}	$\varphi_l\varphi_e N_0$
3	CECS 28:2012		φ_{l1}	φ_{e2}	
4	JGJ 3-2010		φ_{l1}	φ_{e3}	
5	CECS230:2008		φ_{l3}	φ_{e2}	
6	CECS 104-1999		φ_{l1}	φ_{e4}	
7	GB 51367-2019	When$\theta \leq 1/(\alpha-1)^2$: $N = 0.9\eta_c f_c A_c(1+\alpha\theta)$ When$\theta > 1/(\alpha-1)^2$: $N = 0.9\eta_c f_c A_c(1+\sqrt{\theta}+\theta)$	φ_{l2}	φ_{e1}	
8	CECS188:2005	-	φ_{l1}	—	$\varphi_l f_{cc}A_{cc}(1+1.8\theta)$

It is shown in Table 3 that there are three calculation methods for φ_l. The difference is that φ_{l2} is more detailed than φ_{l1} in calculation formula, but neither formula gives the upper limit of the aspect ratio, and φ_{l3} is given by the look-up table and the maximum aspect ratio is 20. In addition, there are four formula mode for the eccentricity influence coefficient φ_e, and the difference lies in the large eccentricity calculation formula only.

3.3 comparison of stability coefficient

The calculation in this paper is based on the axial compression member with hinged ends, and the eccentricity influence coefficient in the limit equilibrium theory is taken as 1. The bearing capacity of all the codes differs in the value of the stability coefficient and the slenderness ratio influence coefficient mainly. the results of the calculation formulas given in GB 50936-2014, CECS 254:2012, JGJ 138-2016 and CECS230:2008 are shown in Figure 2.

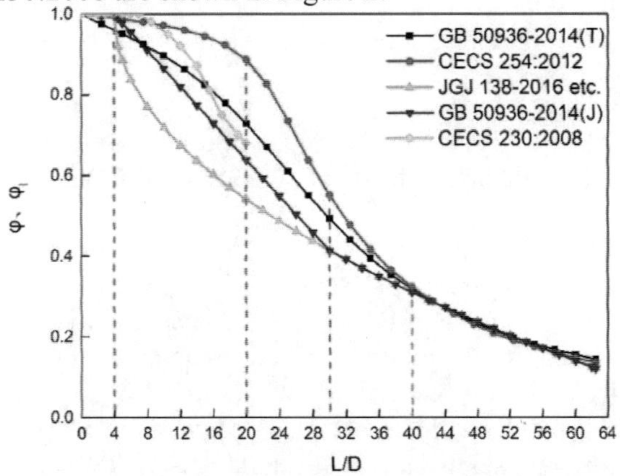

Fig.2 Comparison of φ and φ_l

It can be seen from Fig.2: (1) The main difference in the values of the four codes is the aspect ratio in the range of 4-40, and the stability coefficient of the unified theory is generally greater than the slenderness ratio influence coefficient of the limit equilibrium theory. (2) The length-to-diameter ratio of the member is restricted to less than 20, which is equivalent to $l/i \leq 80$,but the slenderness ratio of the two influence coefficients can be up to 250. (3) The calculation results of the difference in the range of length-to-diameter ratio 0~4, 4~20, 20~30, 30~40, and >40 for each specification are shown in Table 4.

Table 4 φ and φ_l difference percentage

Range of L/D	0~4	4~20	20~30	30~40	> 40
Mean	0.99	0.83	0.58	0.40	0.21
variance	0.01	0.13	0.13	0.07	0.06
Maximum difference percentage	4.10%	64.07%	64.07%	33.37%	18.75%

4 Conclusion

(1) Due to its superior axial compression and low eccentric compression performance, the solid CFCST component exhibits a wide range of application prospects in axially stressed structural systems represented by "spatial grid structure, tower mast structure, and truss structure"

(2)Regarding the material and structure regulations of solid circular steel tube concrete, it is indicated that high-strength steel above Q460 and ultra-high-strength concrete above C100 are not in the scope of encouraged application. As the diameter of the steel tube and the material strength increases, the calculated difference in the strength and bearing capacity of the above-mentioned standard members becomes larger, if there is no basis to rely on, the scope of application must not be expanded at will and in-depth theoretical and experimental research is necessary.

(3)The current regulations in China have different provisions for calculating the bearing capacity of solid circular concrete filled steel tube axial compression members. The calculation difference mainly lies in the value of the stability coefficient and the slenderness ratio influence coefficient. There is a big difference in the length-to-diameter ratio range of 4-30, the maximum difference between the two coefficients is 64.07%, and the difference in the bearing capacity of the components is 62.79%.

(4)In the design of concrete-filled steel tube members, the current codes mostly stipulate that the diameter of the steel tube is not less than 200mm, and at the same time restrict its aspect ratio to not greater than 20, which is equivalent to restricting its slenderness ratio to less than 80. However, in the space latticed structure, the limit of the slenderness ratio of the compression bar is 180, and the diameter of the steel pipe is not clearly specified, and the calculation of the component bearing capacity is difficult to meet the current requirements of the concrete-filled steel tube specification. For the reinforcement method of infilled concrete in the reticulated shell structure, it is recommended that when the diameter of the steel pipe is 60-219, the concrete should be C50 or more, and when the diameter of the steel pipe is more than 219, the concrete can be C40 or more to meet the hoop rate of 0.5-2.5.

References

[1] Han L, Li W, Bjorhovde R.(2014). Developments and advanced applications of concrete-filled steel tubular (CFST) structures: Members[J]. Journal of Constructional Steel Research. 100, 211-228.

[2] Shanmugam N E, Lakshmi B.(2001). State of the art report on steel–concrete composite columns[J]. Journal of Constructional Steel Research, 57(10), 1041-1080.

[3] Han L. L., Mu Y. M., Wang F. C., Fan B. K., L W., Liang J.,...Li C. J. (2020). Design theory of CFST (concrete-filled steel tubular) mixed structures and its applications in bridge engineering. CHINA CIVIL ENGINEERING JOURNAL, 53(05): 1-24.

[4] Chen B. C., Wei J. G., Zhou J., &Liu J. P. (2017) Application of concrete-filled steel tube arch bridges in China: current status and prospects. CHINA CIVIL ENGINEERING JOURNAL, 50(6), 50-61.

[5] 2019). standard for design of strengthening steel structure GB 51367-2019. Beijing: China Architecture & Building Press. (Reprinted.

[6] 2016). Code for design of composite structures JGJ 138-2016. Beijing: China Architecture & Building Press. (Reprinted.

[7] 2015). Design specification for specified concrete filled steel tubular members CECS 408:2015. Beijing: China Architecture & Building Press. (Reprinted.

[8] 2014). Technical code for concrete filled steel tubular structures GB 50936-2014. Beijing: China Architecture & Building Press. (Reprinted.
[9] 2012). Technical specification for concrete-filled steel tubular structures CECS 28:2012. Beijing: China planning Press. (Reprinted.
[10] 2012). Technical specification for solid and hollow concrete-filled steel tubular structures. In China Association for Engineering Construction Standardization (Ed.) CECS 254:2012. Beijing: China Architecture & Building Press. (Reprinted.
[11] 2010). Technical specification for concrete structures of tall building JGJ 3-2010. Beijing: China Architecture & Building Press. (Reprinted.
[12] 2008). Specification for design of steel-concrete mixed structure of tall buildings CECS 230:2008. Beijing: China planning Press. (Reprinted.
[13] 2005). Technical specification for steel tube-reinforced concrete column structure. In China Association for Engineering Construction Standardization (Ed.) CECS 188:2005. Beijing: China Architecture & Building Press. (Reprinted.
[14] 1999). Technical Specification for High-Strength Concrete Structures In China Association for Engineering Construction Standardization (Ed.) CECS 104:1999. Beijing. (Reprinted.
[15] Ma X.B.(2005). Comparison of Design Methods for Bearing Capacity of Concrete-Filled Steel Tubes and Preliminary Exploration of New Methods. Master thesis, Harbin Institute of Technology, china.
[16] Zha X.X., Yu M., &Zhong S.T.(2013). Consistency study on bearing capacity formulae in the Chinese 'Code for design of concrete-filled steel tubular structures'. Journal of Building Structure, A1, 11-20.
[17] Zha X.X., Yu M., Li Y.T., &Zhong S.T.(2011). Unified Theory and Formulation for Concrete Filled Steel Tube under Axial Load. Progress in Steel Building Structures, 13(01), 1-7.
[18] Ou Z.J., &Chen B.C.(2012). Study of unified calculation method for the stability factor of concrete filled steel tubular laced columns. CHINA CIVIL ENGINEERING JOURNAL, (4), 43-47.
[19] Ou Z.J., &Chen B.C.(2018). Study on unified calculation method for the ultimate load carrying capacity of concrete filled steel tubular columns. CHINA CIVIL ENGINEERING JOURNAL, (7): 80-85.
[20] Li Q.(2018). Comparative study on concrete filled steel tubular column design method of Chinese and British codes. Building Structure, 48(22),56-61.
[21] Han L.H., Zhong T., Yao G.H., &Yang Y.F. (2002). Design and calculation of bearing capacity of concrete-filled steel tube basic members--Comparison of national regulations (I). Progress in Steel Building Structures, 4(3).
[22] Zheng L.L., Chen J.W., Yan W.S., &Wang N. (2016). Comparison of Axial Compressive Bearing Capacity of Concrete Filled Steel Tubular Stub Columns at Home and Abroad. Journal of North China University of Science and Technology(Natural Science Edition), 38(4), 133-138.

Study on fault sealing mechanism based on rock physics simulation experiment

Lisha Qu[1], Jiageng Liu[2], Guangqing Zhou[3], Jiandong Wang[3], Jing Li[1*], Chongzhi Wang[1] and Rui Zhu[1]

[1]College of Pipeline and Civil Engineering in China University of Petroleum, Qingdao, Shandong, 266580, China

[2]School of Geosciences in China University of Petroleum, Qingdao, Shandong, 266580, China

[3]Research Institute of Petroleum Engineering, Shengli Oil Field. SINOPEC, Dongying, Shandong 257200, China

[*]Corresponding author's e-mail: jingli@upc.edu.cn

Abstract. At present, the development of fault oil and gas reservoirs has become an important way for to increase oil and gas productivity in China. However, due to the complex internal stress state of rock faults, the phenomenon of oil and gas escape is prone to occur during the mining process, so it is of great significance to study the sealing mechanism of the rock fault. For this reason, based on the theory of fault sealing, rock physical simulation experiments of fault sealing were carried out, and the mechanism of fault sealing and its main controlling factors were deeply studied. The results show that the rock fault dip and the argillaceous content of the fault filling are the main controlling factors for fault sealing; The essence of fault lateral sealing and vertical sealing is the sealing of the displacement pressure difference; As the dip angle of the fault increases, the lateral sealing of the fault becomes better, while the vertical sealing becomes worse; with the increase of the argillaceous content of the fault filling, the lateral and vertical sealability of the fault becomes better.

1. Introduction

Faulted oil and gas reservoirs are widely distributed in China. However, the internal stress of the fault zone is complex, and it is easy to cause oil and gas to escape during the mining process. Therefore, the study of fault sealing mechanism is of great significance to the exploration and development of faulted oil and gas reservoirs. Many scholars have carried out relevant research on this. Yielding[1] analyzed the reasons for fault sealing from the perspective of lithology, and proposed the concepts of fault gouge ratio SGR and smear factor; Clarke[2] used three-dimensional analysis technology for the first time to obtain the SGR distribution map of the fault gouge ratio; Vivek[3] proposed a method of artificially injecting CO2 into the fault to seal the fault through precipitation. Zhou Xingui[4] found that the greater the angle of intersection between the current horizontal principal compressive stress direction and the fault strike, the greater the possibility of the fault blocking oil and gas; Li Qiang[5] combined fault sealability with oil and gas accumulation, and used the orthogonal test design method to determine the factors affecting the displacement pressure of the fault rock. The higher the SGR value, the better the lateral sealing of the fault. The above-mentioned research is mainly aimed at theoretical research on fault sealing, and physical simulation experiments are rarely used to study fault sealing mechanism. Therefore, this paper takes the 4th block of Zhunzhong as the research area, analyzes the mechanism of fault sealing through experiments, and deeply researches the influencing factors of fault sealing.

2. Physical simulation experiment of fault sealing

The essence of fault sealing is the difference in permeability. Through physical simulation experiments to create an ideal experimental environment, restore the true situation of the fault.

2.1. Experimental sample

The rock samples of the fault closure physical model experiment are taken from the Qigu Formation in a well area of Block 4 in Zhunzhong, the depth range is 4454.42~4468.3m, and the core size is Φ25×65mm.

2.2. Experimental program

Set up the two variables of fault dip and argillaceous content of the fault filling to carry out the experimental study of the fault sealing, ignoring other secondary influencing factors.

(1) Experiment on the influence of fault dip on fault lateral sealing

Set up fault physical models at different inclination angles. The argillaceous content of the fracture filling is 50%. It is assumed that the confining pressure corresponding to the conventional permeability during the pressurization process is the critical closing pressure. Taking the fracture surface as a single fracture as an example, the core is cross-sectioned at different angles α, and five groups of models are set to α_1=30°, α_2=45°, α_3=60°, α_4=75°, α_5=90° respectively. The cutting direction is shown in figure 1.

(2) Experiment on the influence of fault dip on fault vertical sealing

As shown in figure 2, the full-diameter core is cut along the vertical direction, which is similar to the lateral closed model. The cut angles are set to β_1=0°, β_2=5°, β_3=10°, β_4=15°, β_5=20°. (Corresponding to the fault dip in the actual formation is 90°, 85°, 80°, 75°, 70°)

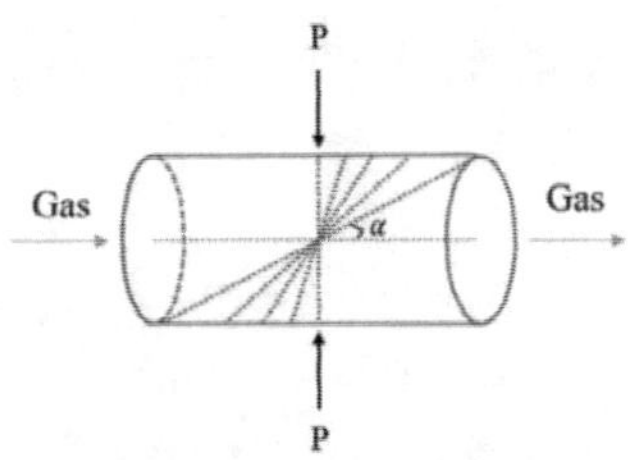

Figure 1. Schematic diagram of lateral closed physical model.

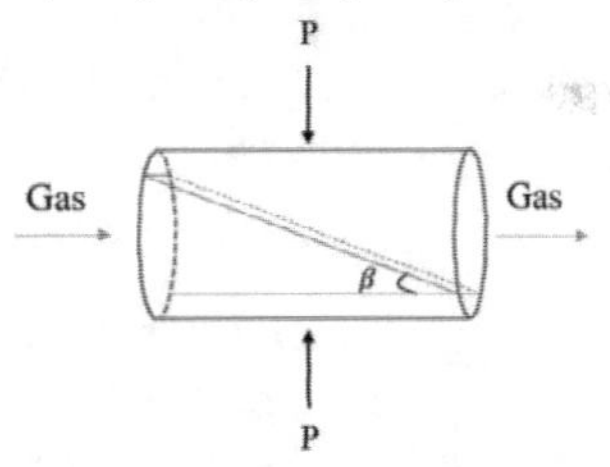

Figure 2. Schematic diagram of vertical closed physical model.

(3) Experiment on the influence of argillaceous content of the filling material in the fault zone on fault lateral sealing

Considering actual reservoir conditions, physical models of 45° dip angle faults with argillaceous content of 30%, 40%, 50%, 60%, and 70% are set.

(4) Experiment on the influence of argillaceous content of the filling material in the fault zone on fault vertical sealing

Considering the actual geological conditions, physical models of 80° dip angle faults with argillaceous content of 30%, 40%, 50%, 60%, and 70% are set.

3. Analysis of results

3.1. Experimental analysis of the effect of fault dip angle on fault lateral sealing

According to the experimental results, the relationship curve between the confining pressure and the permeability of the fault lateral sealing model is established as shown in the figure below.

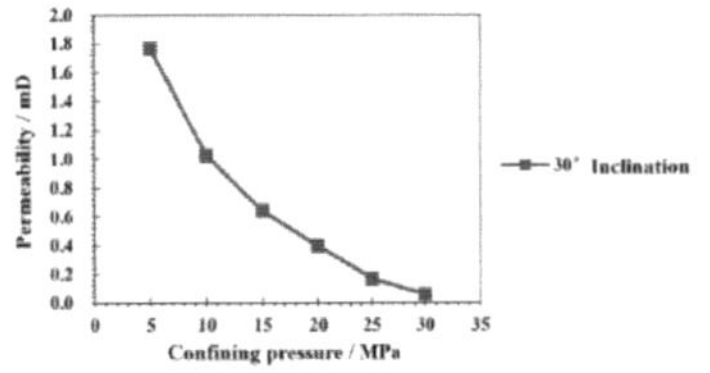

Figure 3. Lateral 30°

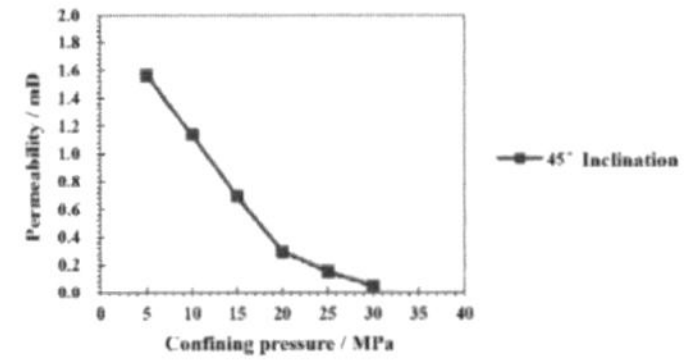

Figure 4. Lateral 45°

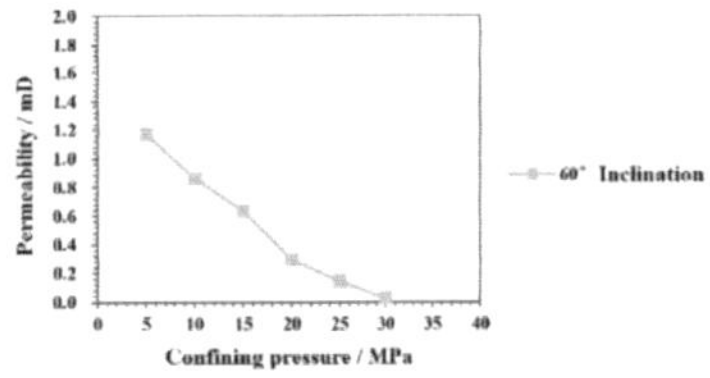

Figure 5. Lateral 60°

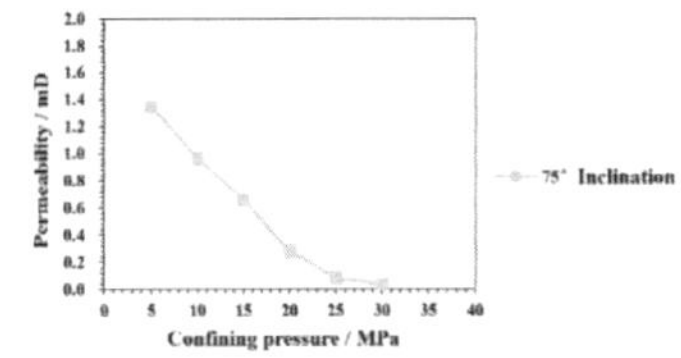

Figure 6. Lateral 75°

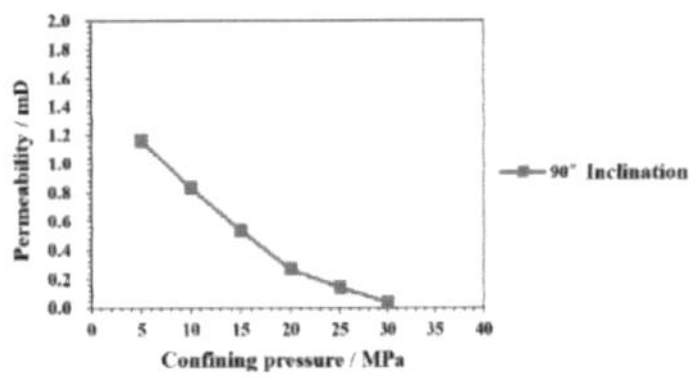

Figure 7. Lateral 90°

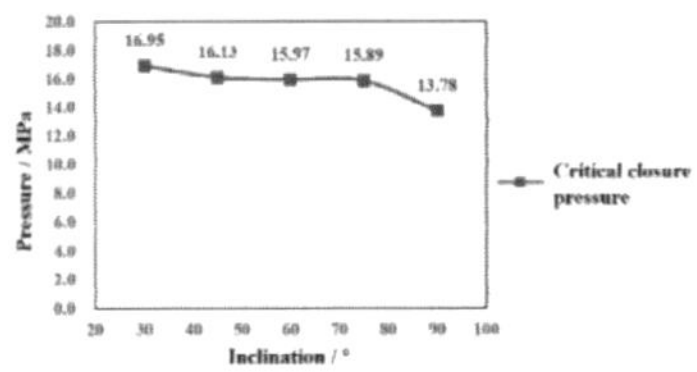

Figure 8. Lateral closed critical closure pressure curve.

It can be seen from figure 3 to figure 7 that the permeability of the fault lateral sealing model is much higher than that of the conventional core. Under the same geological conditions, the ability of the fault to transport oil and gas increases.

It can be seen from figure 8 that as the fault dip increases, the critical closure pressure decreases. In the actual formation, the greater the dip angle of the fault, the better the lateral sealing ability of the fault.

3.2. Analysis of the Experimental Results of the Influence of Fault Dip Angle on fault vertical sealing

According to the experimental results, the relationship curve between the confining pressure and the permeability of the fault vertical sealing model is established as shown in the figure below.

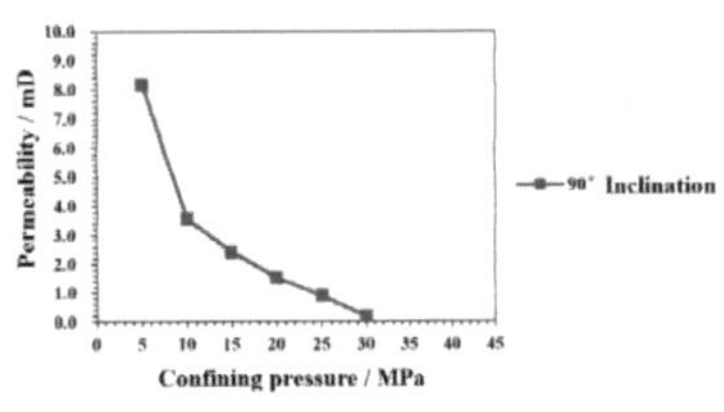

Figure 9. Vertical 90°

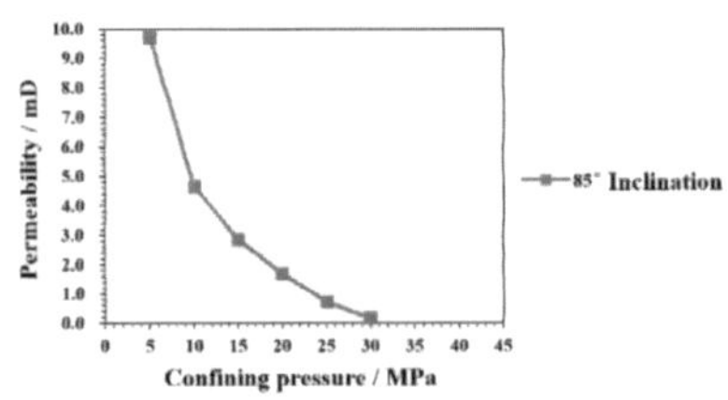

Figure 10. Vertical 85°

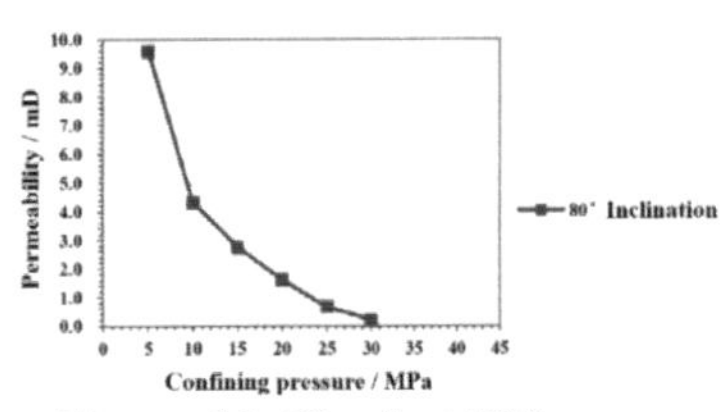

Figure 11. Vertical 80°

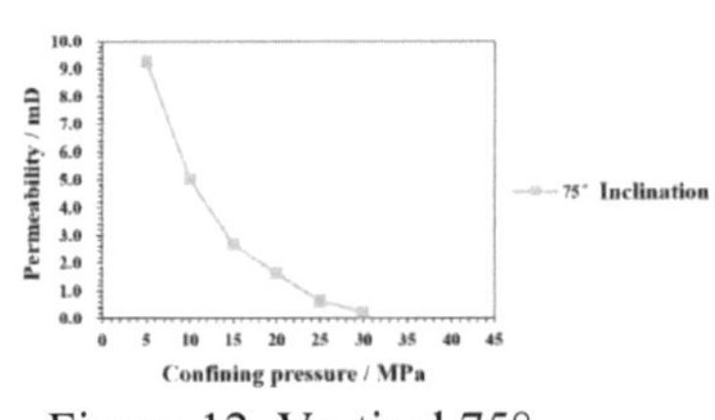

Figure 12. Vertical 75°

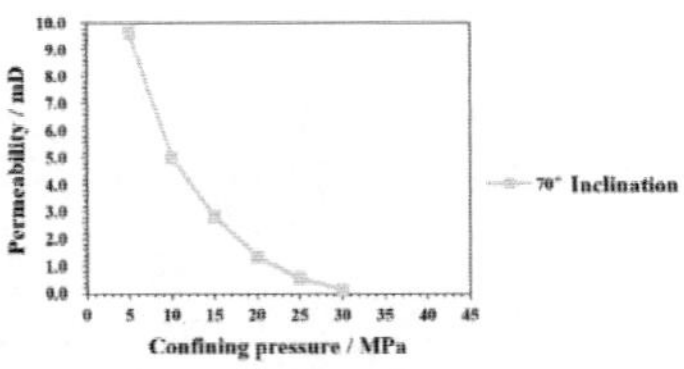

Figure 13. Vertical 70°

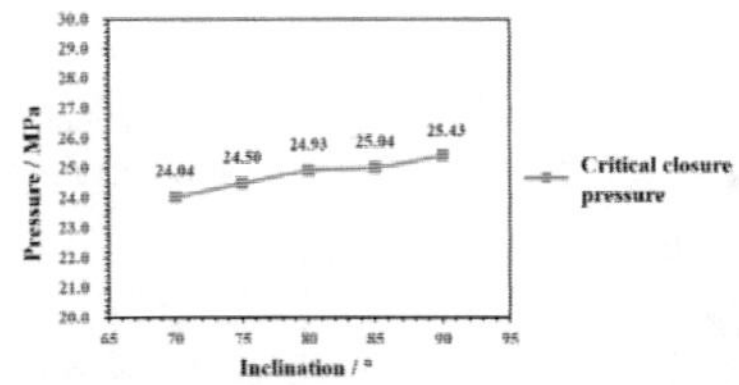

Figure 14. Vertical closed critical closure pressure curve.

It can be seen from figure 9 to figure 13 that the vertical rock permeability of the fault is much greater than the conventional permeability of the rock sample, and the permeability of the fault vertical closure model is approximately 12~15 times that of the conventional permeability.

It can be seen from figure 14 that as the dip angle increases, the critical closure pressure of the vertical sealing of the fault gradually increases, its sealing performance becomes worse, and the effective stress preventing the vertical migration of oil and gas becomes larger.

3.3. The experimental results of the influence of argillaceous content on fault lateral sealing

According to the experimental results, the relationship curve between the confining pressure and the permeability of the fault lateral sealing model is established as shown in the figure below.

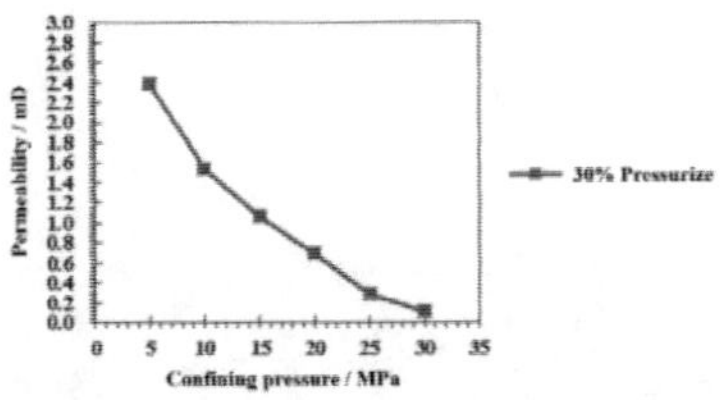

Figure 15. Lateral 30%

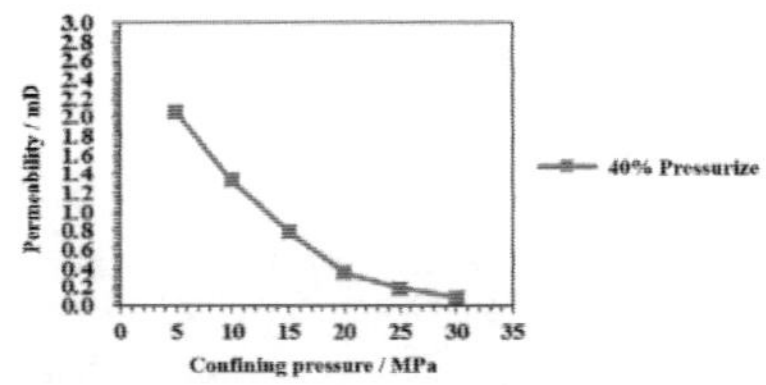

Figure 16. Lateral 40%

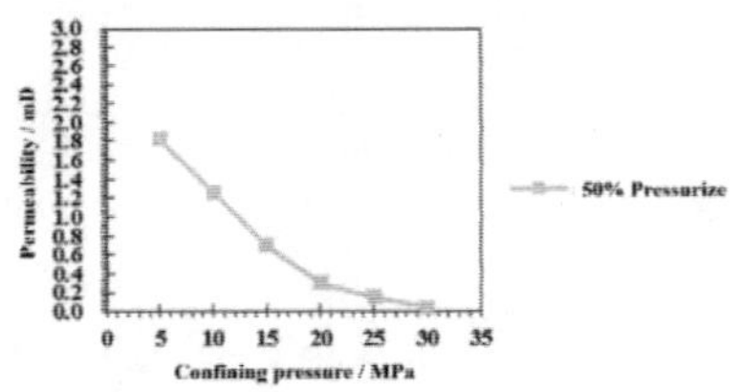

Figure 17. Lateral 50%

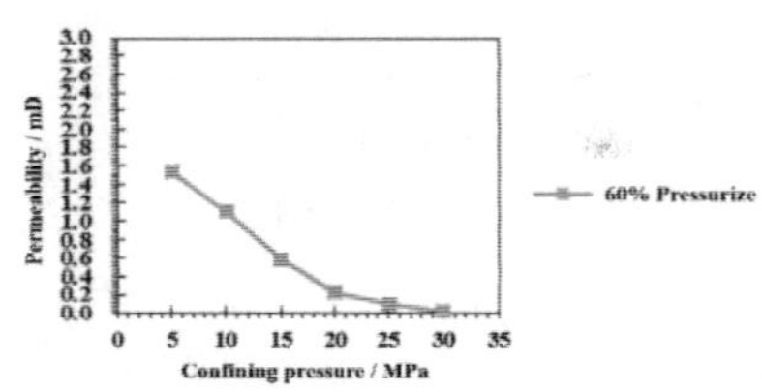

Figure 18. Lateral 60%

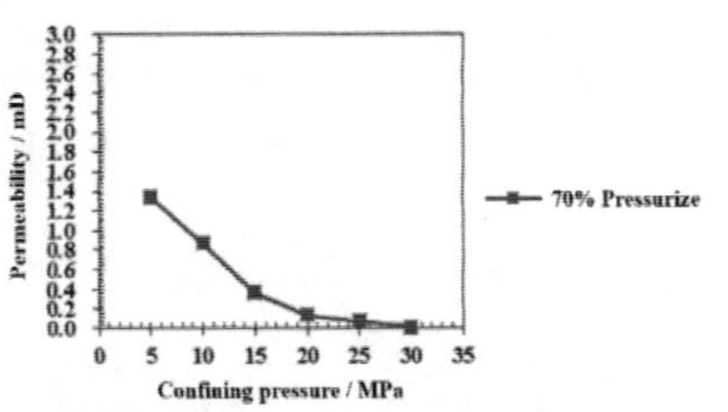

Figure 19. Lateral 70%

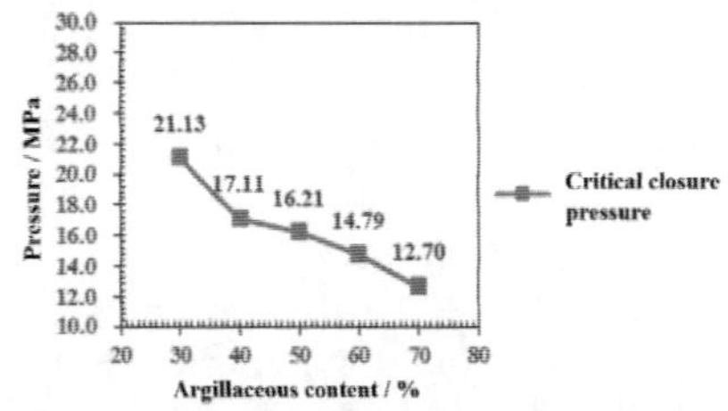

Figure 20. Critical closure pressure curve of lateral argillaceous content.

It can be seen from figure 15 to figure 19 that with the increase of the argillaceous content of the fracture filling, the rock permeability gradually decreases under the same confining pressure. The higher the argillaceous content of the fracture filling, the better the lateral sealing of the fault, and vice versa, the worse the lateral sealing of the fault.

It can be seen from figure 20 that as the argillaceous content increases, the critical closure pressure of the fault lateral closure model gradually decreases, and the critical closure pressure ranges from 12.86 to 20.58 MPa.

3.4. The experimental results of the influence of argillaceous content on fault vertical sealing

According to the experimental results, the relationship curve between the confining pressure and the permeability of the fault vertical sealing model is established as shown in the figure below.

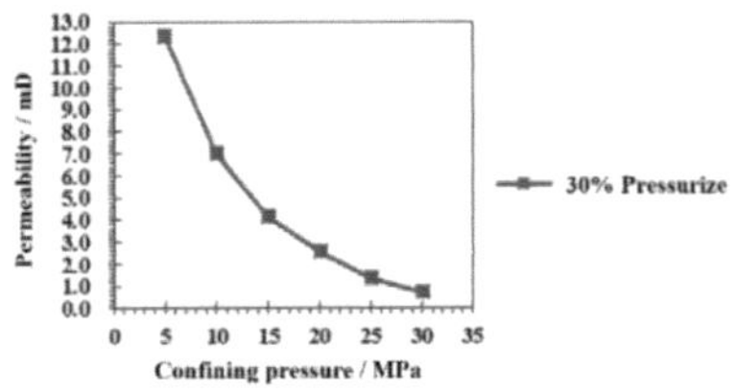

Figure 21. Vertical 30%

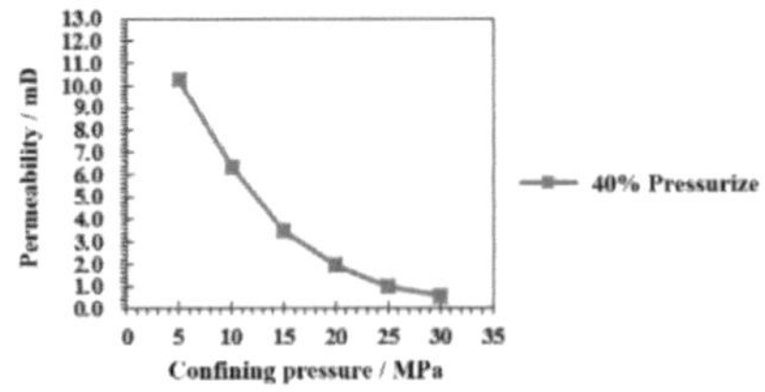

Figure 22. Vertical 40%

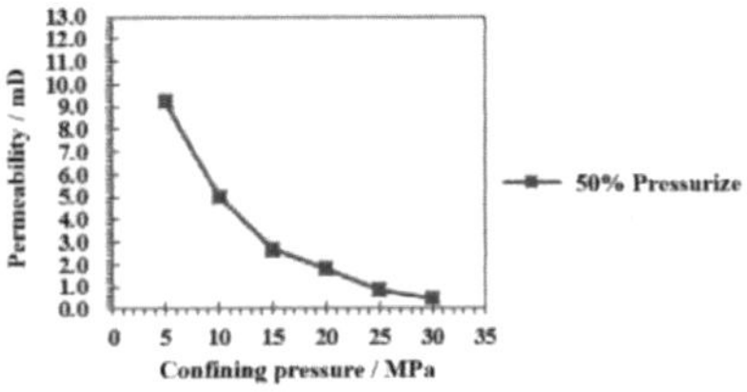

Figure 23. Vertical 50%

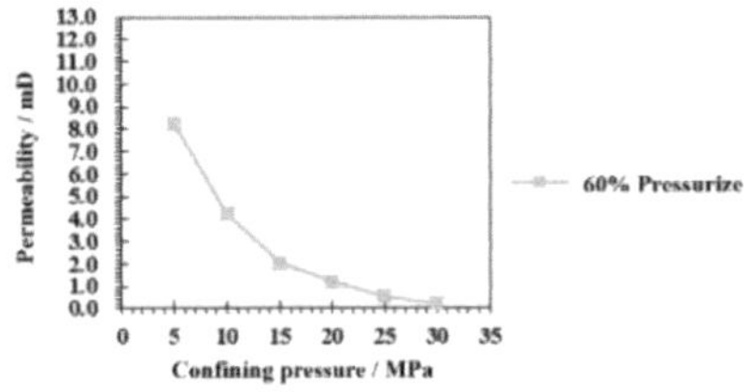

Figure 24. Vertical 60%

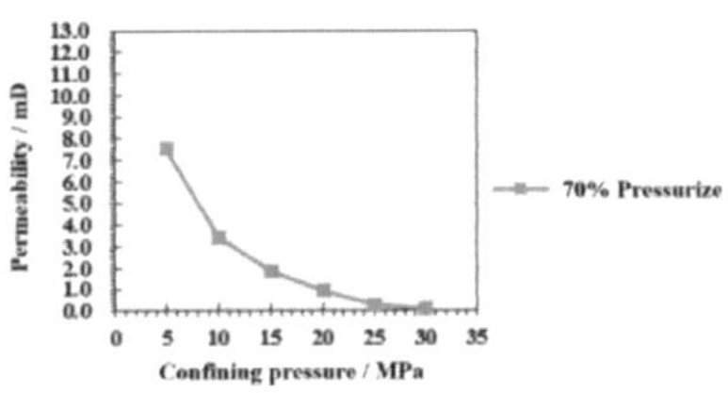

Figure 25. Vertical 30%

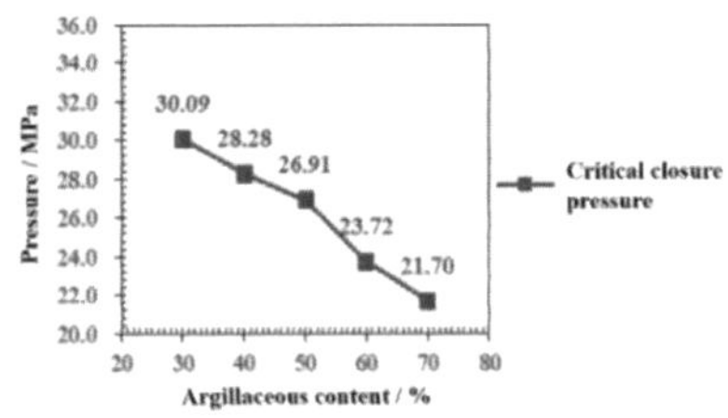

Figure 26. Critical closure pressure curve of vertical argillaceous content.

It can be seen from figure 21 to figure 25 that with the increase of the model argillaceous content, the permeability of the rock decreases under the same confining pressure. In the actual formation, the higher the argillaceous content of the fault filling, the better the vertical sealing of the fault.

It can be seen from figure 26 that as the argillaceous content increases, the critical closure pressure gradually decreases. In fractures with high argillaceous content, only a small confining pressure is needed to achieve the vertical sealing requirements and prevent the migration of oil and gas.

4. Fault Sealing Mechanism

4.1. Fault lateral sealing mechanism

The experimental results of the lateral sealing of the fault are summarized as follows:

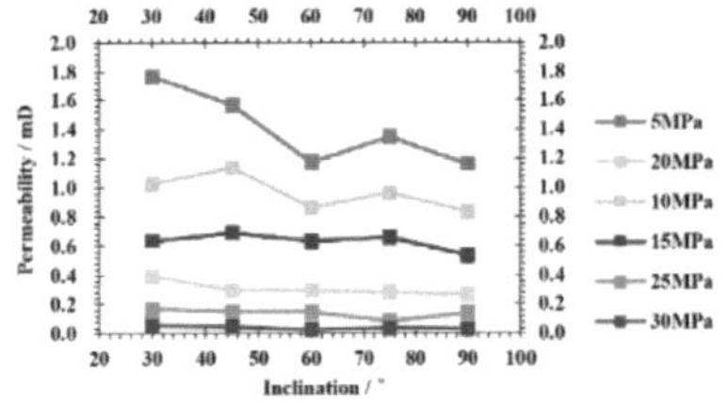

Figure 27. Lateral closed dip-permeability curve.

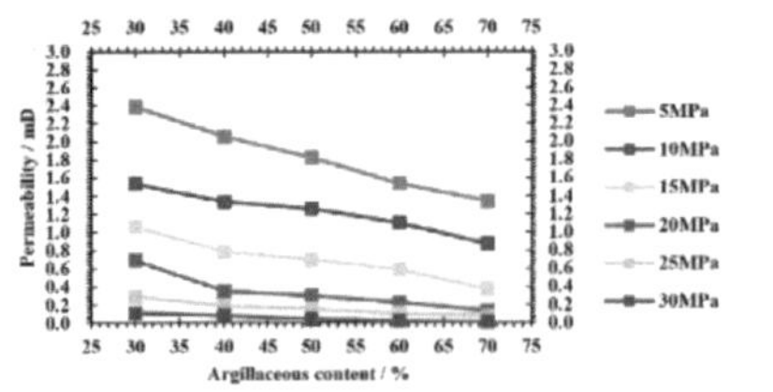

Figure 28. Lateral closed argillaceous content-permeability curve.

It can be seen from figure 27 and figure 28 that as the fracture filling argillaceous content increases, the rock permeability gradually decreases. The higher the argillaceous content, the lower the fracture porosity and the smaller the pore radius. Therefore, the higher the drainage pressure of the fault, the higher the oil and gas sealing height; The essence of the lateral sealing effect of the fault is the sealing of the displacement pressure difference.

4.2. Fault vertical sealing mechanism

The experimental results of the vertical sealing of the fault are summarized as follows:

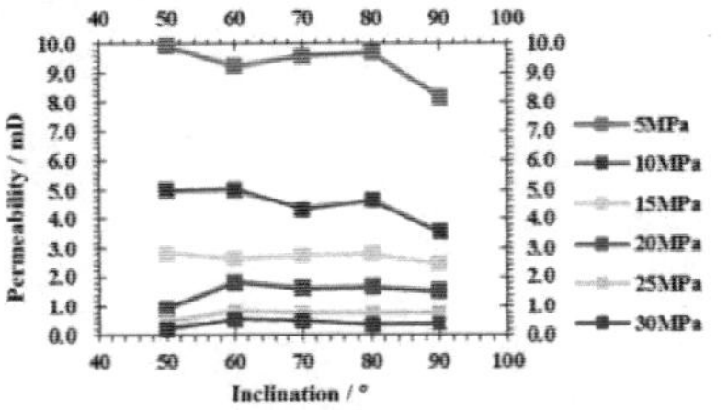

Figure 29. Vertical closed dip-permeability curve.

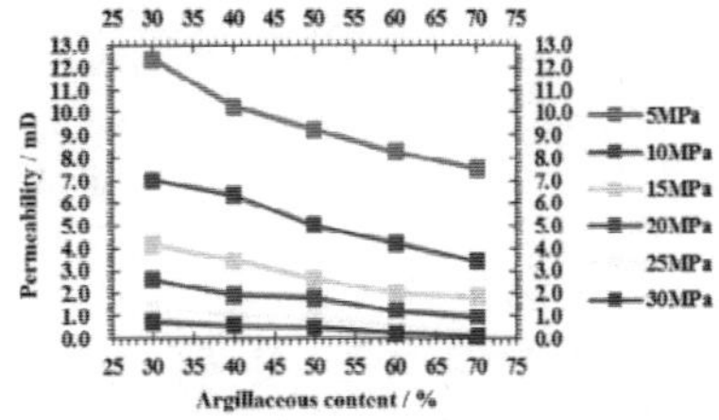

Figure 30. Vertical closed argillaceous content-permeability curve.

It can be seen from figure 29 and figure 30 that under the same confining pressure, as the argillaceous content increases, the rock permeability gradually decreases, the essence of fault sealing is the sealing of the displacement pressure difference. However, the argillaceous content required for vertical sealing is higher than for lateral sealing.

5. Conclusion

The conclusions of this research are as follows:

(1) As the dip angle of the fault increases, the lateral sealing of the fault becomes better; the vertical sealing becomes worse.

(2) As the argillaceous content of the fault filling increases, the better the lateral sealing of the fault, and the better the vertical sealing of the fault; In the actual formation, the higher the argillaceous content of the fracture filling, the better the sealing of the fault.

(3) The dip angle of the fault and the argillaceous content of the fault filling are the main controlling factors for the sealability of the fault. The essence of the fault seal is the seal of the displacement pressure difference.

The results of the research have guiding significance for the exploration and development of faulted oil and gas reservoirs, aiming to improve the efficiency of oil and gas exploration and development.

Acknowledgments

This work was financially supported by the National Natural Science Foundation of China (41972138) and the National Great Science Fund of China (ZD2019-183-007 and 2016ZX05002-002).

References

[1] Yielding, G., Freeman, B., Needham, D.T. (1997) Quantitative fault seal prediction. J. AAPG Bulletin, 81: 897-917.
[2] Clarke, S.M., Burley, S.D., Williams, G.D. (2005) Dynamic fault seal analysis and flow pathway modelling in three-dimensional basin models. In: Dore, A.G., Vining, B.A. (Eds), Petroleum Geology: North-West Europe and Global Perspectives – Proceedings of the 6th Petroleum Geology Conference. Geological Society of London., London. pp. 1275-1288.
[3] Patil, V.V., Mcpherson, B.J. (2020) Identifying Hydrogeochemical Conditions for Fault Self-Sealing in Geological CO2 Storage. J. Water Resources Research, 56: 1-18.
[4] Zhou, X.G., Sun, B.S., Tan C.X. (2000) Current geostress and fault sealing effect. J. Petroleum

Exploration and Development, 27: 127-131.
[5] Li, Q., Tian, X.P., He, J. (2018) Quantitative characterization of fault sealing and its control on hydrocarbon accumulation. J. Journal of Southwest Petroleum University (Science & Technology Edition), 40: 40-50.

Journal of Physics: Conference Series **2011** (2021) 012046 doi:10.1088/1742-6596/2011/1/012046

Calculation on Short-term Stiffness of Simply-supported Beams with Gradient Concrete and Glass Fiber Reinforced Plastic Bars

Lirong Yao, Deyin Zhao, Rongqiang Zhong, jing Guo* and Zhiming Li

Sinopec Northwest Oil Field Company, Wulumuqi, Xinjiang Uygur Autonomous Region, 830000, China

*Corresponding author's e-mail: jijing@nepu.edu.cn

Abstract: It is of the advantages of high bearing capacity and economy for the gradient concrete beam with glass fiber reinforced plastic(GFRP) bars, so it will be applied widely in practical engineering. In order to acquire the short-term stiffness of simply-supported beams with gradient concrete and GFRP bars, 14 simply-supported beams were designed. Based on material constitutive model, finite element models of the beams are established by ABAQUS software. The stiffness variation law of the specimens is acquired. The simulation results are in good agreement with the experimental results, which proves the correctness of finite element modeling. Based on numerical simulation the stiffness calculation formula of the beams is provided, which can provide reference for the application of the beams in the actual engineering.

1. Introduction

In order to make fully use of the advantages of high compressive strength of high strength concrete (HSC) and economy of normal strength concrete (NSC)[1-2], a kind of flexural members with gradient concrete is proposed, which the HSC is used in compression zone and the NSC is used in tension zone[3].

Masjedi et al. proposed the calculation method of composite beams with variable stiffness, and the differential equations of four degrees of freedom were established according to Euler Bernoulli theory. Later the beams with variable stiffness were tested, and the static response of composite beams with non-uniform stiffness was proposed according to Chebyshev collocation method[4-5]. Wang et al. studied the stiffness degradation law of steel-concrete beams under fatigue load, and it was found that with the increase of fatigue cycles, the stiffness degradation law of beams showed an obvious S-type monotonic decreasing trend[6-7]. Zhou et al. carried out seven bending tests of reinforced concrete short beams strengthened with CFRP and based on reasonable calculation assumption, a calculation method for the whole process curve of bending stiffness of reinforced concrete short beams strengthened with CFRP was proposed[8-9]. Four reinforced concrete (RC) beams with non-dismantling ultra-high performance concrete formwork were tested by Liang et al. Based on the assumption of plane section, the effective moment of inertia method was used to establish the calculation formula for bending stiffness of this kind of beams[10-12].

At present, although the stiffness of GFRP concrete beams has already been studied, the research on stiffness of simply-supported beams with gradient concrete and GFRP bars has not been reported and the relative theory is immature. Based on ABAQUS, the short-term stiffness of simply-supported beams with gradient concrete and GFRP bars was analyzed. The calculation formula for short-term stiffness of this type of beams was put forward through statistical regression, which could provide theoretical support for its popularization and application in practical engineering.

2. Specimen design

In order to study the influence of different parameters on the beam stiffness, 14 simply-supported beams with gradient concrete and GFRP bars were designed as shown in table 1. The main parameters included the strength of HSC in compression zone (f_{cu}^{k1}), the reinforcement ratio of GFRP bars (ρ_{GFRP}) and the section heights of HSC in compression zone (h_1). The calculated span of all specimens was 2100mm with a sectional size of 150mm×200 mm(b×h). Two steel bars with a diameter of 12mm were adopted in the compression zone. Stirrups with a diameter of 8mm and a spacing of 80mm were adopted. The compressive strength of NSC (f_{cu}^{k}) was 30MPa, and the thickness of concrete cover was 20mm.

Table 1. The main parameters of specimens

Specimens	f_{cu}^{k} (MPa)	f_{cu}^{k1} (MPa)	GFRP bars	ρ_{GFRP}	h_1 (mm)
SBGCG-1	30	60	2φ14	1.14%	90
SBGCG-2	30	60	2φ16	1.49%	90
SBGCG-3	30	60	2φ18	1.88%	90
SBGCG-4	30	60	2φ20	2.33%	90
SBGCG-5	30	60	2φ14	1.14%	120
SBGCG-6	30	60	2φ16	1.49%	120
SBGCG-7	30	60	2φ18	1.88%	120
SBGCG-8	30	60	2φ20	2.33%	120
SBGCG-9	30	70	2φ18	1.88%	90
SBGCG-10	30	80	2φ18	1.88%	90
SBGCG-11	30	90	2φ18	1.88%	90
SBGCG-12	30	70	2φ20	2.33%	120
SBGCG-13	30	80	2φ20	2.33%	120
SBGCG-14	30	90	2φ20	2.33%	120

3. ABAQUS finite element analysis and verification

3.1 Constitutive model

The linear elastic model was adopted as the constitutive model of GFRP bars, which the ultimate tensile strength was 606 Mpa and the elastic modulus was 40 Gpa. The ideal double-broken line model was adopted as the elastic-plastic constitutive model of steel bars and stirrups, and the constitutive model proposed in the Code for Design of Concrete Structures was adopted as that of NSC, while HSC adopted Ding's constitutive model.

3.2 Finite element model

The HSC, NSC and gaskets were established by C3D8R solid element, while steel bars and GFRP bars were established by T3D2 three-dimensional truss element. Steel bars were embedded into concrete by embedded command, and HSC and NSC were connected by binding. Besides, the displacement of gasket at the left end was constrained in X, Y and Z directions (U1=U2=U3=0), and the displacement of gasket at the right end was constrained in Y and Z directions (U2=U3=0). Hexahedral element was used for the whole mesh generation. The finite element model was shown in figure 1.

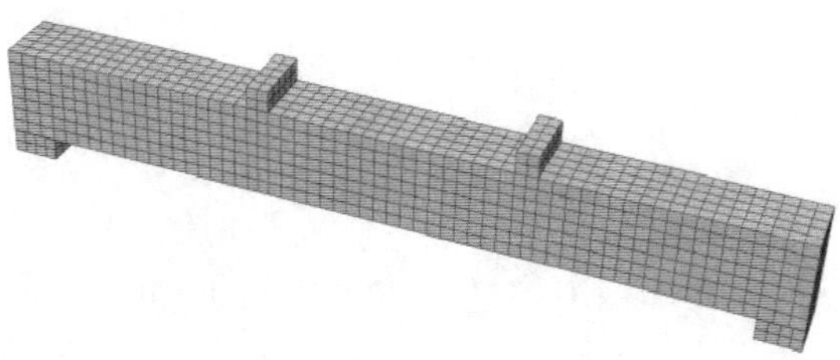

Figure 1. Finite element model of simply-supported beams with gratitude concrete and GFRP bars

3.3 Experiment verification of numerical analysis

Two simply-supported reinforced gradient concrete test beams by W. Zhou and one simply-supported concrete test beam with FRP bars by Y. F. Zheng[13] were selected to carry out finite element simulation. The comparative results of the ultimate load (N_m and N_u) and the deflection (f_m and f_u) obtained by simulation and experiment were shown in table 2, and the results were in good agreement.

Table2. Comparison between simulation and experiment

Specimens	f_m (mm)	f_u (mm)	N_m (kN)	N_u (kN)
B1	42.81	42.62	71.73	69.61
B2	38.66	38.63	143.18	138.34
BL2-1	20.71	20.81	88.91	93.65

4. Stiffness analysis of simply-supported beams

4.1 Formula of the short-term Stiffness

Bending rigidity of section:

$$B_s = \frac{M_k}{\phi} \tag{1}$$

Relationship between the strain of GFRP bars and concrete:

$$\phi = \frac{\overline{\varepsilon_f} + \overline{\varepsilon_c}}{h_0} \tag{2}$$

Substituting formula (2) into formula (1):

$$B_s = \frac{M_k h_0}{\overline{\varepsilon_f} + \overline{\varepsilon_c}} \tag{3}$$

Where ϕ is the curvature of the section of simply-supported beam, $\overline{\varepsilon_f}$ and $\overline{\varepsilon_c}$ are the average strain of GFRP bars and the average strain of high-strength concrete, respectively.

According to the cross-sectional balance, formula (4) can be obtained:

$$M_k = \sigma_f A_f \eta h_0 \tag{4}$$

According to the physical relationship, formula (5) can be obtained:

$$\overline{\varepsilon_f} = \psi \varepsilon_f = \psi \frac{\sigma_f}{E_f} \tag{5}$$

According to the equilibrium relationship of cross-sectional stress distribution, formula (6) and formula (7) can be obtained:

$$M_k = C \cdot \eta h_0 = \omega \cdot \sigma_{ck} \xi h_0 \tag{6}$$

$$M_k = T \cdot \eta h_0 = \sigma_{fk} A_f \eta h_0 \tag{7}$$

The compressive stress at the edges of high-strength concrete:

$$\sigma_{\mathrm{ck}} = \frac{M_{\mathrm{k}}}{\omega\xi\eta b h_0^2} \tag{8}$$

$$\sigma_{\mathrm{fk}} = \frac{M_k}{A_{\mathrm{f}}\eta h_0} \tag{9}$$

According to the balance of the physical relationship, formula (10) and formula (11) can be obtained:

$$\overline{\varepsilon_{\mathrm{c}}} = \psi_{\mathrm{c}}\varepsilon_{\mathrm{c}} = \psi_{\mathrm{c}}\frac{\sigma_{\mathrm{c}}}{\nu E_{\mathrm{c}}} = \frac{M_{\mathrm{k}}}{\zeta E_{\mathrm{c}} b h_0^2} \tag{10}$$

$$\overline{\varepsilon_{\mathrm{f}}} = \frac{\sigma_{\mathrm{fk}}}{E_{\mathrm{f}}} = \frac{M_{\mathrm{k}}}{E_{\mathrm{f}}A_{\mathrm{f}}\cdot\eta h_0} \tag{11}$$

Substituting formula (10), (11) into formula (3). Then short-term cross-sectional stiffness of the beam can be obtained by formula (12):

$$B_{\mathrm{s}} = \frac{E_{\mathrm{f}}A_{\mathrm{f}}h_0^2}{\frac{1}{\eta} + \frac{E_{\mathrm{f}}\rho_{GFRP}}{E_{\mathrm{c}}\zeta}} \tag{12}$$

The deflection formula of the beams:

$$f = \alpha\frac{Ml_0^2}{B_{\mathrm{s}}} \tag{13}$$

Where ρ_{GFRP} is the GFRP reinforcement ratio, that is $\rho_{\mathrm{GFRP}}=A_f/bh_0$. η is the internal force arm coefficient of the cracked section. η *and* α *are* equal to 0.86 and 0 .1375,respectively.

4.2 Comparisons between deflection values of calculated and finite element analysis

The deflection of 14 simply-supported beams (f_{cal}) is calculated according to formula (13), which is shown in table 3. f_{s} represents the mid-span deflection of the specimens obtained by the finite element method. The calculated value of the formula agrees well with the simulated result. The mean value of the ratio of the calculated value to the simulated value is 1.06.

Table 3. The comparisons of f_{s} and f_{cal} for 14 specimens

Specimens	$f_{\mathrm{cu}}^{\mathrm{k}}$ (MPa)	$f_{\mathrm{cu}}^{\mathrm{k1}}$ (MPa)	ρ_{GFRP}	f_{s} (mm)	f_{cal} (mm)	$\frac{\lvert f_{\mathrm{s}} - f_{\mathrm{cal}}\rvert}{f_{\mathrm{s}}}$
SBGCG-1	30	60	1.14%	30.14	28.94	9.0%
SBGCG-2	30	60	1.49%	25.22	25.23	0.1%
SBGCG-3	30	60	1.88%	22.11	21.26	3.8%
SBGCG-4	30	60	2.33%	18.11	17.14	5.3%
SBGCG-5	30	60	1.14%	59.2	54.12	8.6%
SBGCG-6	30	60	1.49%	43.13	38.98	9.5%
SBGCG-7	30	60	1.88%	41.72	37.92	9.1%
SBGCG-8	30	60	2.33%	33.14	30.08	9.2%
SBGCG-9	30	70	1.88%	48.95	46.23	5.5%
SBGCG-10	30	80	1.88%	61.73	58.03	5.9%
SBGCG-11	30	90	1.88%	59.42	58.66	1.2%
SBGCG-12	30	70	2.33%	33.7	36.25	7.4%
SBGCG-13	30	80	2.33%	35.34	38.20	8.1%
SBGCG-14	30	90	2.33%	38.61	41.35	7.1%

5. Conclusions

At present, although the stiffness of GFRP concrete beams has already been studied, the research on stiffness of simply-supported beams with gradient concrete and GFRP bars had not been reported and the relative theory was immature. Based on ABAQUS, the short-term stiffness of simply-supported beams with gradient concrete and GFRP bars was analyzed. The conclusions are as follows:

(1) By numerical simulation of two simply-supported Reinforced concrete beams with gradient concrete and one simply-supported beam with NSC and GFRP bars, the simulated results and experimental data agree well.

(2) Based on the principle of minimum stiffness, the deflection calculation formula of simply-supported beams consisting of gradient concrete and GFRP bars is regressed statistically, which is in good agreement with the finite element calculation. It can provide reference for the application of the beams in the actual engineering.

References

[1] Ji, J., Wang, Y. (2017) Analysis of Axial Compression Behavior of Strength-gradient Composite Columns with Built-in High-strength Concrete Filled Steel Tube. Journal of Northeast Petroleum University. 41: 107-116.

[2] Ji, J., Yu, D. Y. (2020) Research on Axial Compression Bearing Capacity of Different-strength Concrete Filled Double Steel Tube Short Columns. Building Structure. 50: 120-129.

[3] Ji, J., Yu, D. Y. (2019) Effect of Post-fire Curing on the Compressive Properties of Fire-Damaged Ultra-high Toughness Cementitious Composites. Journal of Testing and Evaluation. 49: 140-152.

[4] Masjedi, P. K., Weaver, P. M.(2020) Analytical Solution for the Fully Coupled Static Response of Variable Stiffness Composite Beams. Applied Mathematical Modelling. 81: 16-36.

[5] Ji, J., Xu, Z. C. (2018) Nonlinear Buckling Analysis of H-Type Honeycombed Composite Column with Rectangular Concrete-Filled Steel Tube Flanges. International Journal of Steel Structures. 18: 1153-1166.

[6] Wang, B., Huang, Q. (2021) Stiffness Degradation and Its Calculation Model for Composite Beams after Fatigue. Journal of Vibration and Shock. 40: 265-271.

[7] Ji, J., Zhang, S. L. (2018) Feasibility of Developing Engineered Cementitious Composite with High Volumes of Fly Ash Using Cost-Effective PVA Fiber. Journal of Testing and Evaluation. 48:1-8.

[8] Zhou, Y., Wang, Y. Y. (2020) Calculation Method on Flexural Stiffness of Reinforced Concrete Short Beam Strengthened by CFRP. China Civil Engineering Journal. 53: 9-20.

[9] Ji, J., Zhang, W. F. (2019) Finite Element Model Based Test and Analysis on ACHC Short Columns and Hoop Coefficient. Cluster Computing. 22: 5333-5345.

[10] Liang, X. W., W. P. (2020) Mechanical Behavior and Immediate Stiffness of RC Beams with UHPC Permanent Formwork. Journal of Building Structures. 44: 17-22.

[11] Ji, J., Yang. M. M. (2019) Output-Only Parameters Identification of Earthquake-Excited Building Structure with Least Squares and Input Modification Process. Applied Sciences. 09: 1-8.

[12] Ji, J., Song, H. Y. (2021) Tensile Performance of High Ductility Cementitious Composites with Recycled Powder from C&D Waste. Frontiers in Materials. 8:1-8.

[13] Zheng, Y. F. (2008) Experimental Studies and FEM Analysis of Concrete Beams Reinforced. University of Jinan, Jinan.

Influence of Process Parameters of Unidirectional Glass Fiber Fabric on Resin Infiltration Rate

Xia Qianjin[1], Hao MingYang[2*], Zhang HongSheng[3], Zhong Hai[2]

[1.] Department of Aeronautical Engineering, College of Large Aircraft, Ordos Institute of Technology, Ordos 017000, Inner Mongolia, China

[2.] ChongQing Polycomp International Corporation, Dadukou District, Chongqing 40082

[3.] Faculty of Mechanlcal and Meterial Engineering, HuaiYin Institute of Technology, HuaiAn, JiangSu, China, 223003

haomingyang@cpicfiber.com

Abstract. The tensile and compressive strength, torsional strength, and fatigue strength of wind turbine blades are closely related to the infusion defects during the manufacturing process of blade girders. Inadequate understanding of the resin infiltration process of the glass fiber fabric is one of key factors leading to the infusion defects. This article mainly studies the influence of fabric weft TEX, weft angle, stitch length, and stitch tension on the resin infiltration rate through experiments. The results show that the resin infiltration rate in the 90° direction of fabric increases as weft TEX and weft angle increase and that the infiltration rate in the thickness direction increases when stitch length and stitch tension increase.

1. Introduction

Wind power is a type of clean energy with smaller environmental burdens. In 2019, wind power generation accounted for about 4.5% of the total power generation in China. Wind turbine blades, as core components of wind turbines, have a decisive impact on service life and power generation efficiency of wind turbines. The blade girder of wind turbine, made of composite materials of glass fiber resin matrix, is a primary force-taking structure which is usually produced through a vacuum infusion process with unidirectional glass fiber fabric. Wind power blades are developing in the direction of large-scale size [1]. With the increase of the blade size, the width of unidirectional glass fiber fabric is gradually increasing, so the requirements for infusion of the fabric are getting higher and higher.

The resin infiltration rate is a physical property which indicates flow resistance degree of the fabric or reinforcement materials in the resin. The resin infiltration to reinforcement materials has a decisive effect on the quality and efficiency of parts during production.

Now the Vacuum-Assisted Resin Transfer Molding method (VARTM) is primarily adopted to produce wind turbine blades. [3] Its main process is to place fabric on the die and next to put a layer of demolding cloth on the top layer of the fabric. Then cloth with high infiltration rate is put on the top, and the last step is infusion by vacuum pumps before wrapping with vacuum bags and sealing.

The flow speed of the resin has a great influence on the infiltration result and labor costs of wind turbine blade girders. Slower infiltration rate in the 90°direction of the fabric can easily cause unthorough infiltration of fabric which shows white color, while faster infiltration rate in the same direction can lead to the resin going through the glass fiber quickly, resulting in poor infiltration as well. Therefore, fabric manufacturers should adequately adjust their fabric process parameters according to requirements of clients to meet the needs of the infusion of blade girders and reducing labor costs.

At present, the research on fabric infiltration is mainly divided into non-experimental methods and experimental methods[2]. This study mainly explores the influence of fabric process parameters on the resin infiltration rate by experimental tests, and the results can provide references for glass fiber fabric manufacturers.

2. Experiment

2.1. Experimental Design

A type of frequently used unidirectional high-mode glass fiber fabric is selected for the tests. The thickness of the fabric is 0.2mm, and the width in 90°direction is 650mm. The total weight per unit area is $1200 g/m^2$, in which the weight in the longitudinal direction is about $1164 g/m^2$ and about $51 g/m^2$ in the latitudinal direction.

The manufacturing of unidirectional fabric involves process parameters and equipment. Since the weft TEX, weft angle, stitching tension, and stitching length are the main factors that affect the resin infiltration to the fabric, this study will investigate the influence of the above four factors on resin infiltration rate from the experimental perspective[4].

The ranges of the four factors to be measured are listed in Table 1. In the experiment, the single factor design was used to study the effect of a certain factor on the resin infiltration rate, namely conducting experiment with different values of one factor while keeping the other factors unchanged. The first set of values of the four factors in Table 1 was used as reference. When

Table 1 the first set of values of the four fabric factors

	Value 1	Value 2	Value 3
Weft tex	136	200	300
Weft angle	88	90	
stitching length	4.6	4.8	5.0
stitching tension	1~2	2~4	5~6

2.2. Experimental

At present, the vacuum resin infusion process is generally used to produce wind turbine blades. Fig.1 shows directions of the fabric layering. During infusion, the resin flows primarily in the direction of the internal pressure gradient and along the thickness direction of the fabric, and the pressure gradient caused by vacuum pump is consistent with the 90° direction of the fabric when producing the blade girder. Therefore, this paper measures the resin flow rate in both the 90-degree direction and the thickness direction of the fabric to represent resin infiltration rate.

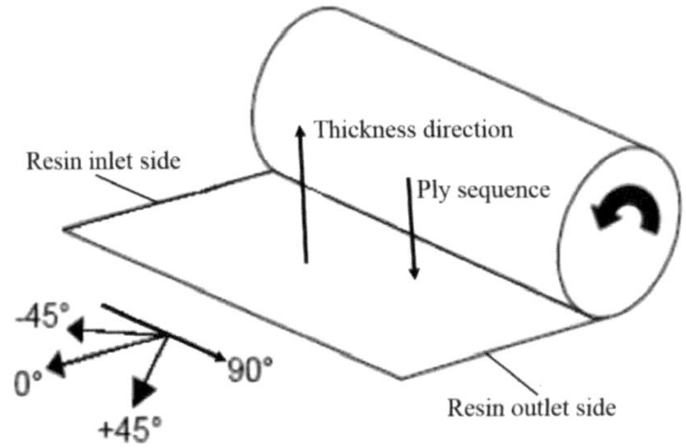

Fig. 1 The directions in fabric

2.2.1. Resin flow rate K_{22} in the 90° direction. In the experiment, 20 layers of fabric were stacked together, and they were infiltrated with the resin along the 90° direction of the fabric. The faster the resin flowed in the fabric, the longer distance of the fabric infiltrated in the 90° direction was. The variable K_{22} in this article denotes the average rate of resin infiltration in the 90° direction of the fabric in five minutes.

2.2.2. Resin flow rate K_{33} in the thickness direction. The resin started infiltrate from the bottom of 20-layer stacked fabric which was kept in vacuum pressure. The time of the infiltration was measured. Since the thickness of the fabric stack is constant, the faster the resin flowed, the shorter time the infiltration took. K_{33} in the article denotes the average rate of resin infiltration in the thickness direction of the fiber.

3. Data Analysis

3.1. The effect of weft tex on infiltration rate

As can be seen from Fig. 2, as the weft TEX increases, the resin flow rate K_{22} in the 90° direction increases, while the resin flow rate K_{33} in the thickness direction has no significant change. The results show that the weft TEX has a significant impact on K_{22} but hardly affects K_{33}.

3.2. The effect of weft angle on infiltration rate

As can be seen in Fig. 3(a), K_{22} in the fabric with 90° weft yarn is larger than that with 90° weft yarn by about 17%. However, K_{33} in the thickness direction with 90° weft yarn is 2.7% smaller than that with 90° weft yarn(see in Fig. 3(b)).

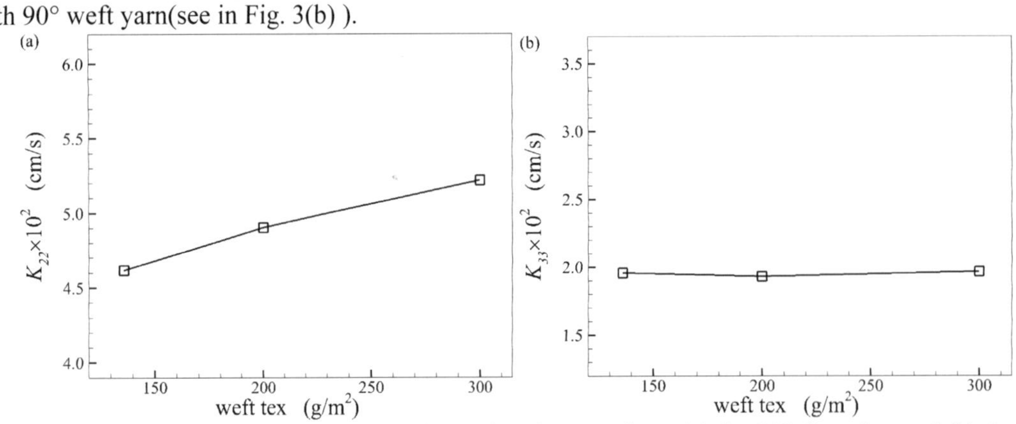

Fig. 2 Resin flow rates at different values of weft TEX along (a) the 90° direction and (b) the thickness direction

Journal of Physics: Conference Series **2011** (2021) 012047 doi:10.1088/1742-6596/2011/1/012047

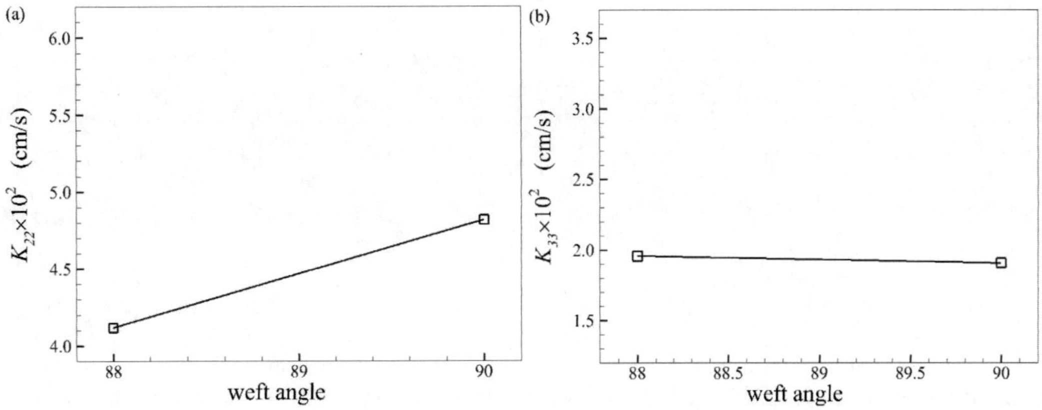

Fig. 3 Resin flow rates at different values of weft angle along (a) the 90° direction and (b) the thickness direction

3.3. *The effect of stitch length on infiltration rate*

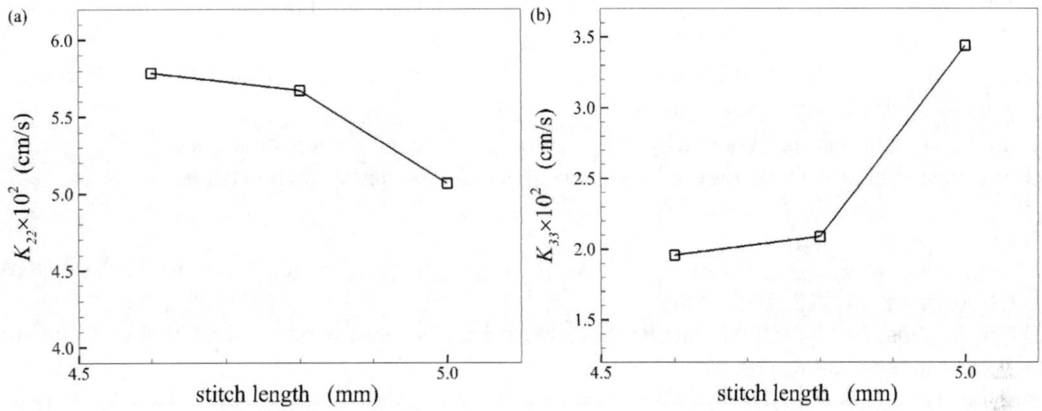

Fig. 4 Resin flow rates at different values of stitch length along (a) the 90° direction and (b) the thickness direction

Fig. 4 shows that K_{22} in the 90° direction is slightly different at stitch length 4.6mm and 4.8mm, but it is significantly reduced at stitch length 5.0mm. The tendency is opposite for K_{33} which decreases when stitch length increases.

3.4. *The effect of stitch tension on infiltration rate*

It can be seen from Fig. 5 that K_{22} has minor change when the stitch tension increases, indicating that the stitch tension has little effect on K_{22}. However, K_{33} increases prominently as the stitch tension increases. This is because the number of slits caused by the stitching threads is the same at certain stitch length, and greater stitch tension cause larger slits. So the resin infiltrates faster along the thickness direction.

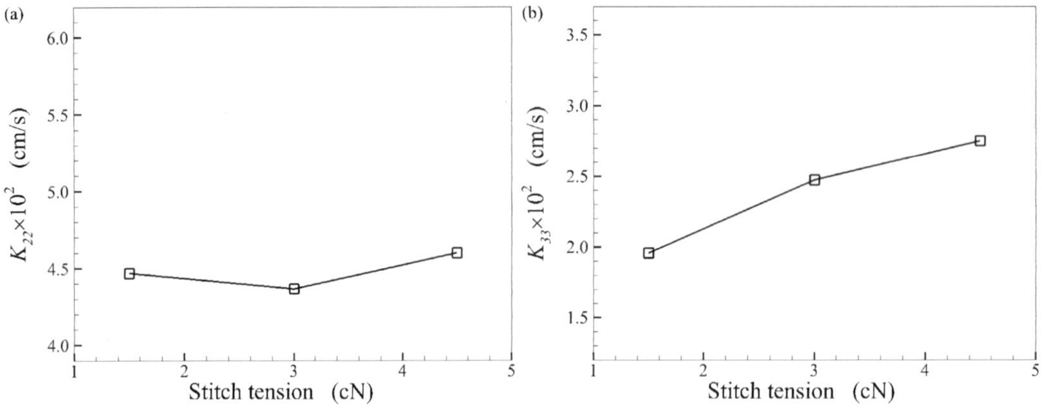

Fig. 5 Resin flow rates at different values of stitch tension along (a) the 90° direction and (b) the thickness direction

4. Conclusion

A type of frequently used unidirectional high-mode glass fiber fabric is selected for the tests.The thickness of the 20-layer fabric is 4mm, and the width in 90°direction is 650mm.From the experimental analysis above, some conclusions can be drawn:

(1) K_{22} and K_{33} should be adjusted together to achieve a better resin infiltration effect during the production of blade girders of wind turbines;

(2) increase of K_{22} can be obtained by augmentation of weft TEX and weft angle;

(3) increase of K_{33} can be achieved by increasing stitch length and stitch tension.

References

[1] H. Zhou, H. Wang, etc., Research process in composite material wind turbine blade, Material Advance A, 2012, 26(2): 65-6.

[2] D. Qi, R. Ning, H. Ning, Study onVRAM technique for blade of wind power, Chinese Cementing Compound, 2008, 17(6):50-52,56.

[3] Ngo Nd,Tarnma K K.Microstructure-based evaluation of the in fluence og woven architecture on permeability by asymptotic homogenization theory.Composites Science and Technology, 2002 (62):1347-1356.

[4] Hayward JS and Harris B. Effect of process variables on the quality of RTM mouldings. SAMPE J 1990; 26(3): 39-46

Mix Design Recycled Aggregate Pervious Concrete and the Influence on Pavement Property

Du Yuxi[1], Tang Feng[*], Zhou Xin[1] and Lv Zhiqian[1]

[1]Hunan Agricultural University, College of water conservancy and civil engineering, Changsha, Hunan 410128

*Corresponding author's e-mail: 21134293@qq.com

Abstract: This paper prepared recycled aggregate pervious concrete by replacing natural aggregate with 100% volume of waste concrete. We carried out experiments to study the influences of strength grade and target porosity of waste recycled aggregate concrete on the recycled one's compressive strength and permeability coefficient. Then, the optimal proportion meeting the requirements of pervious concrete pavement is obtained, which is expected to be applied to the pavement with low strength requirement and water permeability. The test results of the compressive strength and permeability coefficient of 28-day test cubes show that, considering both the compressive strength and permeability, the best proportion is as follows: the strength grade of recycled aggregate is C20 + C30, the mix proportion of aggregate is 50%, the target porosity is 15%, the water-cement ratio is 0.3, the recycled aggregate is composed of 50% recycled aggregate with a particle size of 4.75-9.5 mm and 50% that of 9.5-19 mm, and the optimal mix proportion of recycled aggregate pervious concrete (kg/m3) is: m (recycled aggregate): m (cement): m (water) = 1357.04:564.94:169.88. The compressive strength and the permeability coefficient of recycled aggregate pervious concrete based on this mix proportion are 25.07 Mpa and 6.23 mm/s, respectively, which are in line with the provisions of *Technical Specification for Application of Recycled Aggregate Pervious Concrete* (*CJJ/T 253-2016*) and can be used for pavement. The study results can provide a reference for the promotion and application of recycled aggregate pervious concrete in the pavement.

1. Introduction

With the accelerating urbanization and the gradual increase of construction waste, the disposal and reuse of waste concrete have become a more severe problem. China produces about 1.8 billion tons of construction waste every year. Still, the overall utilization rate is less than 10%, far lower than the 90% in European and American countries and 95% in Japan and South Korea. At present, waste concrete is mainly directly buried or disposed of. At the same time, the long-term excessive exploitation and consumption of natural aggregate result in its severe shortage. According to statistics, China's concrete industry is consuming about 5 billion tons of natural aggregate every year, bringing increasingly severe environmental problems, so it is imperative to conduct this study.

The rainfall in Hunan Province is about 1150 mm-2100 mm with more than 10 heavy rains every year, and the maximum rainfall in one hour of 157 mm, which is relatively high. In recent years, heavy rainfall has caused frequent urban waterlogging disasters and resulted in traffic paralysis, commuting inconvenience, seriously affecting people's production and life, and threatening urban security. Therefore, replacing the natural aggregate with the recycled aggregate of construction waste to prepare recycled aggregate pervious concrete can solve the problems of municipal pavement water, urban waterlogging, and so on. Adhering to the concept of sustainable development can effectively facilitate China's "sponge city" construction, which can bring high economic and commercial benefits [2].

In this study, recycled aggregate pervious concrete with certain compressive strength and permeability is made from 100% volume of waste concrete with different strength grades and particle sizes through crushing, cleaning, and screening, without adding any natural aggregate. Considering the influence of the target porosity of waste on the permeability coefficient, this study carries out combined tests of recycled aggregate with different strength grades to design the recycled aggregate pervious concrete that can be applied to pavements. According to the design requirements of strength and permeability, this study tries to find out the best proportion of waste concrete to help effectively solve the pollution and landfill problems of waste concrete.

2. Raw Materials and Test Methods

2.1. Raw Materials

2.1.1. Recycled Coarse Aggregate

The recycled coarse aggregate used in this test is taken from the construction waste concrete, crushed by jaw crusher and manual labor, and divided into two groups with particle sizes ranging from 4.75-9.5 mm and 9.5-19 mm through square hole sieve. The recycled aggregate pervious concrete is prepared by the aggregate which is cleaned and dried manually to remove surface impurities, as shown in Figure 1. Aggregate gradation is shown in Table 1.

Table 1 Aggregate gradation

Size of Sieve Pore (mm)	19	9.5	4.75
Percent Passing (%)	100	30	0

Figure 1 Test material

2.1.2. Cement and Water

The cement is the South P·O 42.5 grade ordinary portland cement produced by Nanfang Cement Co., Ltd., and the test mixing water is the tap water in the laboratory. The basic property indexes of cement are shown in Table 2[3].

Table 2 The basic properties of cement

Density /kg·m^{-3}	Setting Time / min		Stability	Flexural Strength / MPa		Compressive Strength / MPa	
	Initial Setting Time	Final Setting Time		3d	28d	3d	28d
3100	180	270	Qualified	4.4	7.6	24.4	46.7

2.2. Mix proportion Design

The fundamental property index of recycled aggregate pervious concrete is permeability, and the most significant influencing factor is the target porosity [4]. In the mix proportion design, the target porosity should be designed first as 10%, 15%, 20% and 25%, respectively. The recycled aggregate with different strength grades of C20, C30 and C40 are mixed with each other in the proportion of 50%. There are $3+C_3^{\ 2}=6$ combinations, and each design has three standard test cubes of 150mm×150mm×150mm. According to *Technical Specification for Application of Recycled Aggregate Pervious Concrete (CJJ/T 253-2016)*, the water-cement ratio of recycled aggregate pervious concrete is better between 0.25-0.35, and the effect of slightly adjusting the water-cement ratio on the strength is not obvious. The test adopts the water-cement ratio of 0.3, and the recycled aggregate comprises particles of sizes 4.75-9.5 mm and 9.5-19 mm, with each size accounting for 50% of the total portion. The mix designs of pervious concrete are shown in Table 3.

Table 3 Mix proportions of pervious concrete

Combination	Design Porosity	Water-Cement Ratio	Material Usage (kg/m³)			
			Coarse Aggregate		Cement	Water
			4.75-9.5 mm	9.5-19 mm		
RAC1	10%	0.3	670.62	670.62	661.73	198.52
RAC1	15%	0.3	670.62	670.62	576.79	172.84
RAC1	20%	0.3	670.62	670.62	491.85	148.15
RAC1	25%	0.3	670.62	670.62	406.91	121.48
RAC2	10%	0.3	685.43	685.43	606.42	181.73
RAC2	15%	0.3	685.43	685.43	522.47	157.04
RAC2	20%	0.3	685.43	685.43	437.53	131.36
RAC2	25%	0.3	685.43	685.43	352.59	105.68
RAC3	10%	0.3	709.14	709.14	600.49	179.75
RAC3	15%	0.3	709.14	709.14	516.54	154.07
RAC3	20%	0.3	709.14	709.14	430.62	129.38
RAC3	25%	0.3	709.14	709.14	345.68	103.70
RAC1+RAC2	10%	0.3	678.52	678.52	649.88	195.56
RAC1+RAC2	15%	0.3	678.52	678.52	564.94	169.88
RAC1+RAC2	20%	0.3	678.52	678.52	480.99	144.20
RAC1+RAC2	25%	0.3	678.52	678.52	396.05	118.52
RAC1+RAC3	10%	0.3	670.62	670.62	659.75	198.52
RAC1+RAC3	15%	0.3	670.62	670.62	575.8	172.84
RAC1+RAC3	20%	0.3	670.62	670.62	490.86	147.16
RAC1+RAC3	25%	0.3	670.62	670.62	405.93	121.48
RAC2+RAC3	10%	0.3	709.14	709.14	610.37	182.72
RAC2+RAC3	15%	0.3	709.14	709.14	525.43	158.02
RAC2+RAC3	20%	0.3	709.14	709.14	440.49	132.35
RAC2+RAC3	25%	0.3	709.14	709.14	355.56	106.67

Note: RAC1, RAC2 and RAC3 represent the waste concrete with strength grades of C20, C30 and C40 as recycled coarse aggregate, respectively.

2.3. Production and Maintenance of Pervious Concrete

The concrete is prepared by mixing cement with stones. First, mix 50% water with all coarse aggregate for 30s; and then, mix the remaining mixed water with cement ash for 90s; and finally, fill the mould, and demould and sample it after 1 day. The test cubes of recycled aggregate pervious concrete are placed in the standard curing room, with a temperature of 20 ± 2 °C, a relative humidity of over 95%, and a curing time of 28 days.

Figure 2 Recycled aggregate pervious concrete samples

2.4. Test Method

2.4.1. Mechanical Property Tests
According to *Standard for Testing Methods on Mechanical Properties of Ordinary Concrete (GB/T 50081-2002)*, the compressive strength of the standard test cubes of recycled aggregate pervious concrete aged 28 days under standard curing is tested by a hydraulic press, and the test results are the average value of three 150mm×150mm×150mm standard test cubes[5].

2.4.2. Study on Permeability
According to the *Technical Specification for Pervious Cement Concrete Pavement (CJJ/T 135-2009)*, the permeability coefficient is measured by the head-fixed method. The test device is shown in Figure 3.

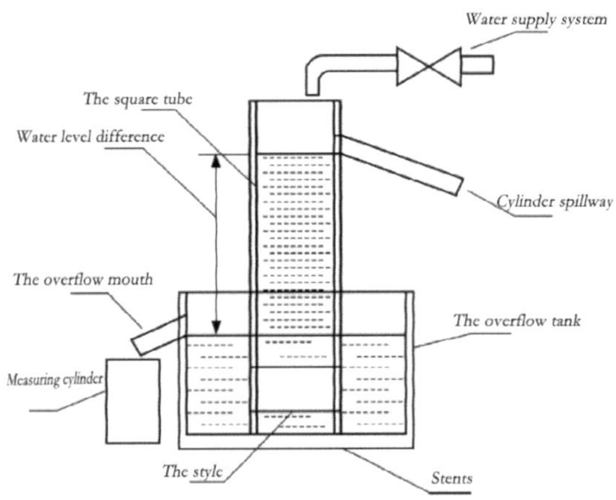

Figure 3 Permeability coefficient test device

Figure 4 Permeability test

As shown in Figure 4, cement is applied on the sides of three 150mm×150mm×150mm standard test cubes formed with different numbers and is put in water with the water level 10 cm higher than the test cubes. Then, they are soaked for 20 minutes. Finally, they are taken out to be connected with the square cylinder of the water permeability test device. The gaps around the test block are sealed with plasticine, and the perimeter of the unit is sealed with clear tape, in which way, the water only penetrates from the surface of the test cubes. An overflow tank is arranged with the valve opened to allow the water flow in. When water flows out of the overflow tank, the inflow is adjusted to keep a certain water level difference of the square cylinder. When the water flows from the overflow tank and the square cylinder become stable, a measuring cylinder is adopted to receive water at the outlet. The water outflow Q is recorded within the corresponding time, and is measured 3 times to obtain the average value. The calculation formula of permeability coefficient is as follows [6]:

$$K_T = \frac{Q \cdot L}{A \cdot h \cdot t} \tag{1}$$

where

K_T --the permeability coefficient when water reaches T °C, mm/s;

Q -- the amount of water flowing through the pervious concrete test cubes within time t, mm^3;

L -- the thickness of pervious concrete test cubes, mm;

A -- cross-sectional area of pervious concrete test cubes, mm^2;

h -- water level difference, mm;

T -- test time, s.

3. Test Results and Analysis

3.1. Influence Analysis on Compressive Strength

The test results are shown in Table 4.

Table 4 Analysis of compressive strength test results

Combination	Design Porosity	Compressive Strength (MPa)			Average Value (MPa)
RAC1	10%	28.20	29.10	31.40	29.57
RAC1	15%	14.00	15.20	16.90	15.37
RAC1	20%	13.40	14.90	16.40	14.90
RAC1	25%	10.30	11.70	12.30	11.43
RAC2	10%	30.80	31.70	34.90	32.47
RAC2	15%	17.50	19.00	20.30	18.93
RAC2	20%	15.10	16.90	18.90	16.97
RAC2	25%	12.80	13.40	14.60	13.60
RAC3	10%	27.90	28.90	31.80	29.53

RAC3	15%	14.60	16.00	18.20	16.27
RAC3	20%	13.70	14.30	15.50	14.50
RAC3	25%	10.40	11.60	12.90	11.63
RAC1+RAC2	10%	24.50	27.70	30.10	27.43
RAC1+RAC2	15%	22.90	25.00	27.30	25.07
RAC1+RAC2	20%	20.70	22.60	23.80	22.37
RAC1+RAC2	25%	15.40	16.30	16.80	16.17
RAC1+RAC3	10%	28.10	29.90	32.20	30.07
RAC1+RAC3	15%	14.40	16.50	17.30	16.07
RAC1+RAC3	20%	11.60	12.50	14.10	12.73
RAC1+RAC3	25%	8.30	8.80	9.20	8.77
RAC2+RAC3	10%	28.20	29.30	31.30	29.60
RAC2+RAC3	15%	16.20	17.40	19.30	17.63
RAC2+RAC3	20%	14.00	14.20	15.00	14.40
RAC2+RAC3	25%	9.40	10.60	11.60	10.53

Note: RAC1, RAC2 and RAC3 respectively represent the waste concrete with strength grade of C20, C30 and C40 as recycled coarse aggregate.

According to the test results in Table 4 and the mix proportion design results, with the increase of target porosity, the amount of cement and water decreases, the cohesive force among aggregates decreases, and the compressive strength of recycled aggregate pervious concrete gradually decreases in the case of the same strength grade combination of recycled aggregate. At the same time, the compressive strength of the recycled aggregate combination with strength grade of C40 is low because a lot of impurities are mixed into the recycled aggregate during crushing. According to *Technical Specification for Application of Recycled Aggregate Pervious Concrete (CJJ/T 253-2016)*, when the compressive strength of pervious concrete reaches 20MPa, it can be applied to the pavement, and there are 8 groups of test combinations meeting the requirements. The test results show that the compressive strength achieves the best performance when the strength grade of recycled aggregate is C30 and the target porosity is 10%. As shown in Figure 5, when the design porosity is from 10% to 15%, the compressive strength of recycled aggregate pervious concrete with strength grade of C20, C30, C40, C20 + C30, C20 + C40, C30 + C40 decreases by 48.0%, 41.7%, 44.9%, 8.6%, 46.6% and 40.4% respectively. When the design porosity is from 15% to 20%, it decreases by 3.0%, 10.4%, 10.9%, 10.8%, 20.8% and 18.3% respectively. When the design porosity is from 20% to 25%, it decreases by 23.3%, 19.9%, 19.8%, 27.7%, 31.1% and 26.9% respectively. As shown in the figure, the compressive strength of recycled aggregate pervious concrete with strength grade of C20 + C30 basically changes linearly, and its decrease is not obvious with the increase of design porosity

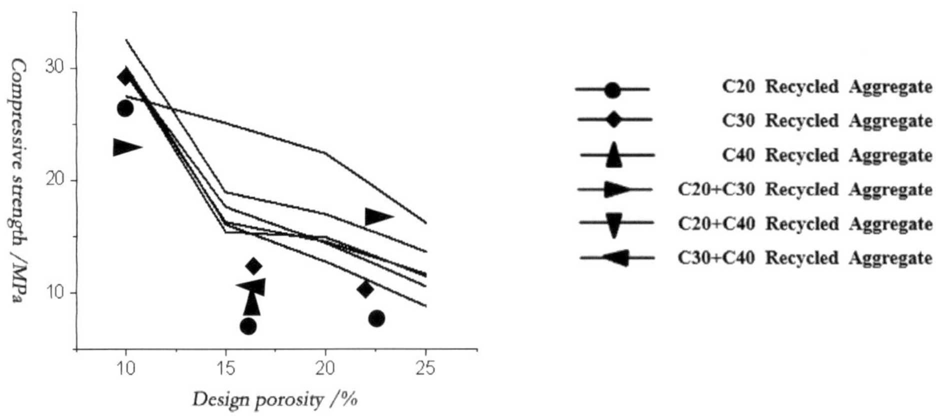

Figure 5 The relationship between design porosity and compressive strength

3.2. Result Analysis of the Permeability Coefficient

The test results are shown in Table 5.

Table 5 Analysis of permeability test results

Combination	Design Porosity	Permeability Coefficient (mm/s)			Average Value (mm/s)
RAC1	10%	3.60	3.91	4.11	3.87
RAC1	15%	7.57	8.05	8.69	8.10
RAC1	20%	9.61	10.12	10.32	10.02
RAC1	25%	11.04	12.00	13.08	12.04
RAC2	10%	2.48	2.64	2.77	2.63
RAC2	15%	6.72	6.86	7.07	6.88
RAC2	20%	9.17	9.65	10.33	9.72
RAC2	25%	15.02	15.48	16.41	15.64
RAC3	10%	2.47	2.63	2.79	2.63
RAC3	15%	7.89	8.30	8.96	8.38
RAC3	20%	12.09	12.21	12.45	12.25
RAC3	25%	18.77	19.35	20.32	19.48
RAC1+RAC2	10%	3.82	3.98	4.10	3.97
RAC1+RAC2	15%	5.85	6.29	6.54	6.23
RAC1+RAC2	20%	6.13	6.71	7.08	6.64
RAC1+RAC2	25%	9.30	10.05	10.83	10.06
RAC1+RAC3	10%	2.22	2.33	2.40	2.32
RAC1+RAC3	15%	5.17	5.37	5.56	5.37
RAC1+RAC3	20%	12.69	13.60	14.73	13.67
RAC1+RAC3	25%	14.51	15.08	15.82	15.14
RAC2+RAC3	10%	2.60	2.75	2.96	2.77
RAC2+RAC3	15%	9.17	9.89	10.79	9.95
RAC2+RAC3	20%	10.58	11.11	11.57	11.09
RAC2+RAC3	25%	19.49	20.22	21.37	20.36

Note: RAC1, RAC2 and RAC3 respectively represent the waste concrete with strength grade of C20, C30 and C40 as recycled coarse aggregate.

The test results show that with the increase of the target porosity, the compactness of pervious concrete decreases and the permeability coefficient increases significantly. At the same time, with the change of the combination of the strength grade of recycled aggregate, the permeability coefficient changes, which may be because different combinations of pervious concrete have different compactness and thus change the water channel among aggregates. According to *Technical Specification for Application of Recycled Aggregate Pervious Concrete (CJJ/T 253-2016)*, the permeability coefficient should not be less than 0.5 mm/s, so they all meet the application requirements of pervious concrete pavement. The test results show that the recycled aggregate combination with strength grade of C30+C40, target porosity of 25%, and aggregate mix proportion of 50%, has the highest permeability coefficient[8]. As shown in Figure 6, when the design porosity is from 10% to 15%, the permeability coefficient of recycled aggregate pervious concrete with strength grade of C20, C30, C40, C20 + C30, C20 + C40 and C30 + C40 increases by 52.2%, 61.8%, 68.6%, 36.3%, 56.8% and 72.2% respectively. When the design porosity is from 15% to 20%, it increases by 19.2%, 29.2%, 31.6%, 6.6%, 60.7%, 10.3% respectively. When the design porosity is from 20% to 25%, it increases by 16.8%, 37.9%, 37.1%, 34.0%, 9.7% and 45.5% respectively. As shown in the figure, when the strength grade of recycled aggregate is C20 and C20 + C30, the permeability coefficient of recycled aggregate pervious concrete does not increase significantly with the increase of design porosity.

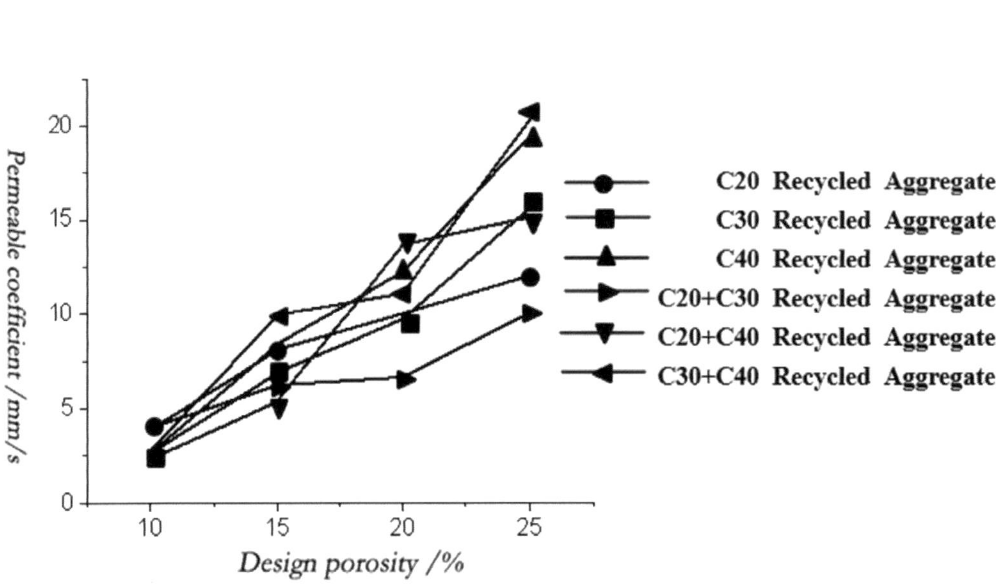

Figure 6 The relationship between design porosity and permeability coefficient

3.3. Comprehensive Analysis

Under the condition of ensuring that the compressive strength and permeability coefficient meet the application specification of pervious concrete, considering the influence of strength grade and target porosity of recycled aggregate on the compressive strength and permeability coefficient of recycled aggregate pervious concrete, the compressive strength of recycled aggregate pervious concrete gradually decreases and the permeability coefficient gradually increases with the increase of target porosity. The strength grade of recycled aggregate has no obvious influence on the permeability coefficient. Considering the compressive strength and permeability, the optimal proportion for pavement is as follows: the strength grade of recycled aggregate is C20 + C30, the mix proportion of aggregate is 50%, the target porosity is 15%, the water-cement ratio is 0.3, the recycled aggregate is composed of 50% recycled aggregate with particle size of 4.75-9.5 mm and that of 9.5-19 mm, and the optimal mix proportion of recycled aggregate pervious concrete (kg/m3) is: m (recycled aggregate): m (cement): m (water) = 1357.04:564.94:169.88. The pervious concrete prepared according to this mix proportion has both good compressive strength and permeability, which are 25.07 Mpa and 6.23 mm/s respectively, and it is in line with the provisions of *Technical Specification for Application of Recycled Aggregate Pervious Concrete (CJJ/T 253-2016)* and can be used for pavement.

4. Conclusion

Recycled aggregate pervious concrete is prepared from 100% volume of waste concrete and is expected to be used in pavements which have low strength requirements and needs to be pervious. This study prepared 24 groups of test samples by mixing cement with stones and explored the effects of strength grade and target porosity of recycled aggregate on compressive strength and permeability coefficient of recycled aggregate pervious concrete. The conclusion is as follows.

1)With the increase of the target porosity, the amount and area of cementation points among recycled aggregate particles decrease, the 28d compressive strength of recycled aggregate pervious concrete gradually decreases, and the permeability coefficient gradually increases, while the strength grade of recycled aggregate has no obvious effect on the permeability, so it is suggested that the porosity of recycled aggregate pervious concrete should be controlled at about 10%.

2)Under the same target porosity, the compressive strength and permeability coefficient of the recycled aggregate pervious concrete prepared with recycled aggregate of mixed strength grade are lower and higher respectively than that made with recycled aggregate of single strength grade.

3)Considering the influence of compressive strength and permeability coefficient, the best proportion of recycled aggregate used in pavement is as follows: the strength grade of recycled aggregate is C20 + C30, the mixing proportion of aggregate is 50%, the design porosity is 15%, the water-cement ratio is 0.3, the recycled aggregate is composed of 50% recycled aggregate with particle size of 4.75-9.5 mm and that of 9.5-19 mm, and the optimal mix proportion (kg / m3) of recycled aggregate pervious concrete is: m (recycled aggregate): m (cement): m (water) = 1357.04:564.94:169.88. The pervious concrete prepared according to this mix proportion has both good compressive strength and water permeability, which are 25.07 Mpa and 6.23 mm/s respectively, and it is in line with the provisions of *Technical Specification for Application of Recycled Aggregate Pervious Concrete (CJJ/T 253-2016)* and can be used for pavement.

***Fund project:**
In 2019, the national undergraduate innovation training program project of Hunan Agricultural University of Hunan province, "Mix Design Recycled Aggregate Pervious Concrete and the Influence on Pavement Property " (Project no. : 201910537074).This paper is the phased results of this project.

About the author:
Du Yuxi (1999—), male, from Shantou, Guangdong province， undergraduate, study and research direction: civil engineering. Email: 15014307509@163.com
*Corresponding author: Tang Feng (1982—), male, from Yueyang, Hunan province, master, lecturer, research direction: structural engineering.

References:
[1] Chen Jialong, Fang Yuanxing. Present Situation and Problems of Concrete Aggregates in China [J]. Architecture Technology, 2005, 36(1):23-24.
[2] Chen Hai, Jiang Yuanhai, Ling Hongjie et al. Experimental Research on Recycled Aggregate in Pervious Concrete [J]. China Concrete and Cement Products, 2018, 10:107-108.
[3] Zhang Dengxiang, Mao ani. Experimental Study on Abrasion Resistance of Pervious Concrete Containing Waste Rubber [J]. Journal of Changsha University of Science and Technology:Natural Science, 2019,16 (2):58-59.
[4] Zhang Ping, Chen Jing, Lan Cong, et al. Study on Performance of Recycled Aggregate Perviouos Concrete [J]. Sichuan Building Materials, 2019,45 (8):1-2.
[5] Huo Hongyuan, Fan Chengcheng, Chen aijiu, et al. Influence of the Recycled Aggregate with Different Strength on Recycled Concrete [J]. Concrete, 2017,2:60-62.
[6] Guo Lei, Liu Siyuan, Chen Shoukai, et al. Research on Mechanical Properties,Permeability and Abrasion Resistance of Fibers Modified Recycled Aggregate Pervious Concrete [J]. Transactions of the Chinese Society of Agricultural Engineering, 2019,35 (2):154-156.
[7] Wang Jiaqin, Xu shanpeng, Yang liansen, et al. Study on Mechanical Properties of Recycled Coarse Aggregate Pervious Concrete [J]. Sichuan Building Materials, 2018,44 (11): 3-4.
[8] Ling Tianqing, Chen Qiaoqiao, Qin Xin, et al. Mix Proportions Design of Pervious Concrete Pavement and Its Performance Influence Factors [J]. Journal of Chongqing Jiaotong University(Natural Science), 2019,38 (3): 39-43.

Applications and Advances in TiO_2 Based Photocatalytic Building Materials

Chongyang Liu [1], Jitao Bai [1, 2, *], Simiao Zhang [3], Zewen Yang [4], Ming Luo [4]

([1.] School of Civil Engineering, Tianjin University, Tianjin 300072, China;

[2.] State Key Laboratory of Hydraulic Engineering Simulation and Safety, Tianjin University, Tianjin 300072, China;

[3.] College of Intelligence and Computing, Tianjin University, Tianjin 300072, China;

[4.] Department of Naval Architecture and Ocean Engineering, Tianjin University, Tianjin 300072, China)

*Corresponding author's e-mail: jitaobai_123@tju.edu.cn

Abstract: TiO_2 is currently the most widely used photocatalytic material. Compared with traditional building materials, TiO_2-based photocatalytic building materials have unique and significant advantages, which endows them with great potential in application. By summarizing the current application and research status of photocatalytic building materials, the existing problems of photocatalytic building materials and corresponding solutions were pointed out, and current research progress of photocatalytic building materials in terms of carbon emission reduction in buildings was introduced. Finally, some prospects were made on photocatalytic building materials, which would provide a reference for the study on the next generation of photocatalytic building materials.

1. Introduction

Among all the photocatalytic materials, TiO_2 is one of the most promising green environmental protection materials [1] and also the most deeply studied photocatalytic material in the field of green building at home and abroad. The potential of TiO_2 as a photocatalyst was first discovered by Fujishima and Honda in 1972[2]. Similar to the photosynthesis of plants, the anode made of TiO_2 can decompose water into hydrogen and oxygen under light conditions [3], which triggers a series of redox reactions. Using this feature and other physical and chemical properties, a variety of building materials have been developed with TiO_2, whose functions mainly involve four aspects: treatment of polluting gases, environmental sterilization, self-cleaning and anti-fogging of building surfaces, and building refrigeration. Most of them exist in the form of photocatalytic coating, self-cleaning glass, functional ceramics, photocatalytic concrete, and other composite building materials [4]. Compared with traditional building materials, photocatalytic materials have special functions such as environment purification and self-cleaning. In addition, photocatalytic materials take natural light as reaction power and hardly need artificial intervention, so they have unique advantages. However, there are some problems in the existing photocatalytic materials. First, the common practice of using photocatalytic materials as admixtures to make composite building materials will cause adverse interactions among different materials [4]. For example, the shading effect of paint or cement on light leads to lower absorption efficiency of the photocatalytic materials; the photocatalyst urges the decomposition of organic molecules in the coating, so that the coating is easy to fall off. Second, under the existing technical conditions, the photovoltage generated by the light on the photocatalytic materials is usually not enough to drive the high-energy reaction. However, the current photocatalysts used to eliminate pollution only rely on light to complete the reaction. As a result, the types of reactions that can occur are very limited and the work efficiency is relatively low. Third, the current photocatalytic building materials are more sensitive to the surrounding environment and their performance is greatly affected by external dynamic light and application sites. These problems have largely limited the promotion of photocatalytic materials.

In addition to the above applications, the role of photocatalytic materials in the control of the greenhouse effect has gradually attracted the attention of the academic community. Richter et al. [5] used photocatalytic technology to remove non-carbon dioxide greenhouse gases from the centralized exhaust of industrial buildings and achieved good results. However, limited by the level of photocatalysis technology, they did not propose an effective and feasible technical route for living buildings that are more closely related to human activities and the main greenhouse gas carbon dioxide. In recent years, with the in-depth study of industrial catalysis, great progress has been made in photocatalytic carbon sequestration based on TiO_2, which makes it possible to control carbon emissions from buildings using photocatalytic materials. In this paper, the application of photocatalytic building materials is reviewed, the existing problems and corresponding solutions are pointed out, and the research progress of photocatalytic building materials is introduced. Finally, the development direction of photocatalytic building materials is prospected, which provides a reference for the research and development of the next generation of photocatalytic building materials.

2. General Application of Photocatalytic Building Materials

2.1. Pollutant Control and Environmental Health

2.1.1. Photocatalytic Mechanism

When the nano-TiO_2 material is illuminated with energy greater than the band gap, the electrons in the low-energy valence state are excited to enter the high-energy conduction band and form electron-hole pairs with the positively charged holes on the low-energy valence band, which then separate and migrate to the surface of TiO_2 particles under electric field. A part of the excited electrons and holes undergo a series of redox reactions with O_2 and H_2O adsorbed on the surface of TiO_2 to form highly active radicals ·OH and O_2^-, which can degrade pollutants such as organic compounds in the buildings. The process is shown in Fig.1.

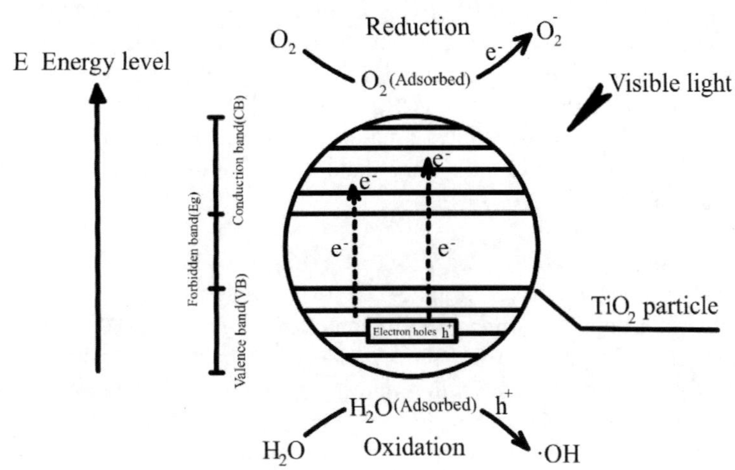

Figure 1 Photocatalytic reaction mechanism of TiO_2 particles (modified from Pei et al., 2011 [6])

2.1.2. Pollutant Treatment

In 1976, Carey et al. [7] first reported the technology of separating biphenyl and chlorobiphenyl through TiO_2/UV photocatalytic reaction. Since then, the use of photocatalysis in the treatment of environmental pollutants has gradually received attention. At present, photocatalytic treatment of pollutants is mainly concentrated in two aspects, one is the treatment of indoor pollutants, such as formaldehyde, and the other is the treatment of automobile exhaust, which is mainly used in road engineering.

For the treatment of indoor pollutants, Guarino et al. [8] coated 70 g/m^2 of TiO_2 on the walls of a 10.95 m × 17.30 m × 2.40 m closed room with a total wall surface of 150 m^2 to treat the mixed gases NH_3, H_2O, CO_2, and CH_4 under the irradiation of 12 fluorescent lamps (36 W). The results showed that the concentration of the mixed gas decreased greatly under the action of photocatalytic coating. Liu et al. [9] studied the purification function of composite diatomite to formaldehyde in the air in a self-made environmental test chamber. The results showed that the degradation rate of formaldehyde could reach 94.08% under natural light by adding the functional coatings of 40% activated alumina and 10% nano TiO_2. Combined with physical adsorption, the coating could completely eliminate formaldehyde. In addition, Wang [10] studied the effect of paint color additives on the absorption efficiency of formaldehyde. The results showed that the degradation effect of pigment on formaldehyde followed an order of green > yellow > red. The reusability of red paint additives was weaker than that of green and yellow. According to this result, both the color aesthetics and the actual effect should be considered when applying indoor air purification to coatings. In fact, due to its large band gap, TiO_2 usually needs UV light to exhibit photocatalytic activity. Therefore, the effect of photocatalytic treatment of organic compounds such as formaldehyde is not satisfactory when the UV band of indoor illumination is less. The problem can generally be solved by doping silver, cerium, and other elements to TiO_2.

In the field of road engineering, nano-TiO_2 materials can be directly mixed with road mixtures or coated evenly on the road surface, so as to achieve effective treatment of automobile exhaust. TiO_2 sol was sprayed on the asphalt pavement on Maoyuan Road in Fengxian District of Shanghai, which could absorb and purify the exhaust emissions of 500 vehicles every day [6]. Each coating could maintain a continuous effect for 12 months. The research of Guan et al. [13] showed that compared with the concrete pavement without TiO_2 coating, the degradation rate of TiO_2 coated pavement to motor vehicle pollutants was about 6%-12%, and the surface still had a certain catalytic effect after it is polished with a thickness of 2 mm.

Whether it is indoor pollutant treatment or application in road engineering, the application forms of TiO_2 can be divided into coating method and mixing method, both of which, however, have some defects. The main problem of the coating method is that the coating is easy to be aged or fall off under the action of external forces, while the disadvantage of mixing method is that the photocatalyst particles inside the material are difficult to contact with the light, leading to low efficiency. However, the mixing method is simple to operate and can give full play to the structural advantages of concrete and other main materials [14]. Hassan et al. [15] tested the durability of concrete mixed with TiO_2. The results showed that the wear resistance of concrete mixed with photocatalytic materials was not affected and the catalytic efficiency did not decrease significantly. Therefore, the photocatalytic building materials obtained by the mixing method have better durability.

2.1.3. Environmental Sterilization

In 1997, Kikuchi et al. coated a thin layer of nano-TiO_2 on glass and found irradiation for 3 hours could completely kill Escherichia coli [16]. Further experiments showed that nano-TiO_2 had good killing ability to many kinds of bacteria and algae, such as Pseudomonas aeruginosa, Escherichia coli, Staphylococcus aureus, Salmonella, Mycobacterium, Aspergillus, and so on. In traditional inorganic antibacterial agents, metal ions such as silver and copper are usually effective ingredients. Compared with photocatalytic sterilization, traditional inorganic antibacterial agents can inactivate the bacteria without light, but the bacteria will release heat-causing or toxic components such as endotoxin when they die. However, the nano-photocatalyst can completely remove endotoxin and other intermediate products [17]. According to this principle, photocatalysts can be used to kill bacteria and algae in cooling towers [18-19].

2.2. *Anti-fogging and Self-cleaning of Building Surfaces*

2.2.1. *Infiltration Characteristics*

Young [20] gave the Young's equation of the static contact angle when a water droplet infiltrated an ideal smooth solid surface (Fig.2(a)). With the introduction of roughness, Wenzel [21] modified the contact angle, introduced the roughness coefficient and proposed the dynamic contact angle. In Wenzel's model (Fig.2(b)), the liquid infiltrated the valley of the entire solid surface. However, under certain conditions, the liquid would not infiltrate the valley, a state which was called the Cassie Baxter state (Fig.2 (c)). When the contact angle was greater than the threshold, the liquid on the solid surface existed in the form of droplets, which refracted the light and made the field of vision white, resulting in fog effect. Therefore, the key to the anti-fogging of the building surface is to construct a super-hydrophilic surface (fig.2(d)), so that the droplets are in the wet state of the Wenzel state [22].

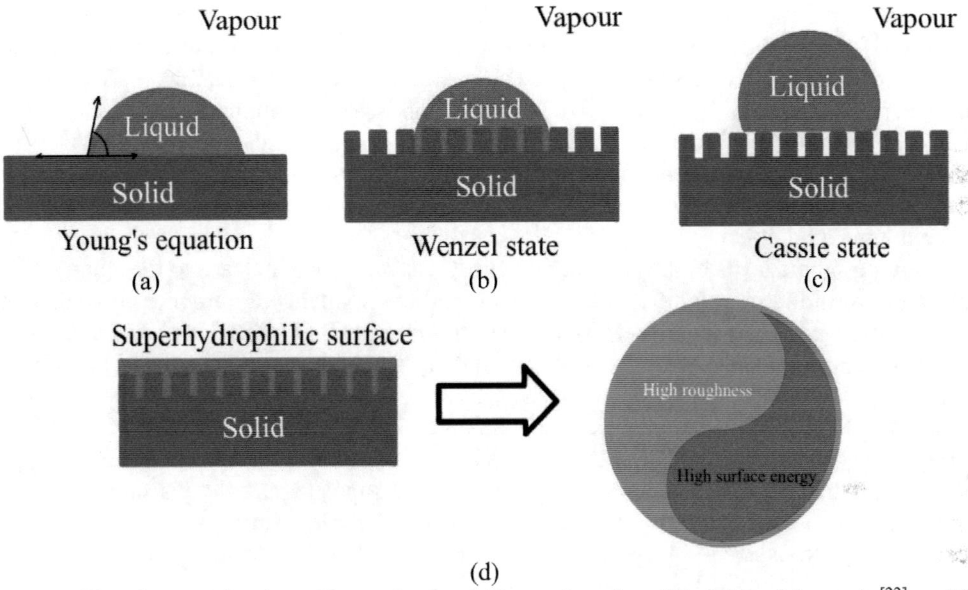

Figure 2 Antifogging mechanism of superhydrophilic surface (modified from Liu et al. [22] and Duran et al. [23])

The methods of constructing superhydrophilic surface include surface physical modification, chemical modification, superhydrophilic coating, and so on. Nano-TiO_2 is an inorganic superhydrophilic material, which was first discovered by Fujishima et al. [24] He coated the glass slide with TiO_2 and dripped it with water droplets. The initial contact angle was tens of degrees. After being irradiated with UV light, the contact angle dropped sharply to almost zero. After the removal of UV light, the superhydrophilicity disappeared and the contact angle increased, which means the process is reversible. Therefore, TiO_2 can be used for anti-fogging of building surfaces. In addition, TiO_2 can degrade pollutants through a photocatalytic reaction under UV light, so that the building has a certain degree of self-cleaning. This is the advantage of TiO_2 compared to other super-hydrophilic materials.

2.2.2. *Progress in Application and Preparation Methods*

There are two types of self-cleaning coatings based on TiO_2: pure inorganic TiO_2 coatings and TiO_2 based nanocomposite coatings [25]. Inorganic TiO_2 coatings include physical vapor deposition (PVD), chemical vapor deposition (CVD), atomic deposition (ALD), and solution-gel method [26]. Because the preparation of pure inorganic TiO_2 coating requires highly of the equipment and vacuum environment, large-area preparation is limited. In addition, the adhesion between TiO_2 and organic coating is insufficient. As a result, the preparation of TiO_2 based nanocomposite coating is introduced.

The composite coating based on organic polymer carrier can effectively improve the catalytic efficiency. Allen et al. [27] showed that the photocatalytic efficiencies of nanometer- and micrometer-sized anatase and rutile are quite different, with higher efficiency of the former. Allen mixed anatase nano-TiO_2 particles of different particle sizes with rutile and then added them into a color paint composed of organic film-forming polymers. In this way, rutile brought alkali stability, while nano anatase provided surface photocatalytic activity. This composite coating could make the surface micronized, and the pollutants attached to the surface could be washed down when being washed by rainwater, so as to achieve a better self-cleaning effect. The micropowder effect was best when the content of nano anatase with particle sizes of 5-10 nm, 20-30 nm and 70 nm was controlled between 10% and 20%. Wang et al. [28] incorporated Fe^{3+} and Ag^{+} into TiO_2-based catalyst and applied the composite nano-TiO_2 to fluorocarbon materials after being modified with a silane coupling agent, so that the coating had antibacterial and self-cleaning effects.

The superhydrophilic coating made of TiO_2 and inorganic SiO_2 can achieve anti-fogging and self-cleaning effects at the same time. The surface of SiO_2 is hydrophilic and can be covered by silanol groups (Si-OH), which can form hydrogen bonds with droplets, leading to film condensation and anti-fog function. The reflectivity of SiO_2 is low, so its transmittance is higher than that of TiO_2 and its light transmission performance is also superior. TiO_2 can decompose some pollutants under light, but the wide band gap of TiO_2 limits the light absorption to the UV region, which only accounts for 4% of the solar spectrum [31] and does not affect the effect of illumination. Therefore, the combination of SiO_2 and TiO_2 can achieve the effects of high light transmission and self-cleaning, and get rid of the influence of anti-fog on the illumination effect.

However, the simple layer-by-layer assembly of SiO_2 and TiO_2 easily leads to the aggregation of the same particles, resulting in a significant decrease in light transmittance. In order to solve this problem, Lee et al. [32] arranged and assembled TiO_2 particles at 7 nm and SiO_2 particles at 22 nm in layers. The results showed that the coating could not only prevent glass from fogging under non-UV light, but also had the effect of self-cleaning and reducing light reflection. Li et al. [33] developed a raspberry-like SiO_2-TiO_2 nano-antireflection film. The particle size of SiO_2 and TiO_2 was modulated by controlling the molar ratio of TEOS/TIPT. After calcination, the amorphous TiO_2 was transformed into dispersed anatase nanoparticles on the SiO_2 core, resulting in raspberry-like composite particles, as shown in Fig.3. This arrangement leads to high light transmittance of the material while ensuring its super-hydrophilic and photocatalytic properties, so it is suitable for applications in solar panels and architectural glass.

Figure 3 Composite particles of SiO_2-TiO_2 (modified from Li et al. [33])

In addition, modern high-rise building glass and coatings put forward higher requirements for durability and strength. Tao et al. [34] used a solution-gel dip coating process to deposit SiO_2 particles, hollow SiO_2 particles, and TiO_2 nanoparticles in order to form a closed-pore structure on the surface of the superhydrophilic film. The light transmittance of this film can reach 96.0%, and it has the characteristics of high temperature resistance and high mechanical strength. It is expected to be used in solar panels and high-rise building glass under severe weather conditions.

At present, the self-cleaning coatings have been applied in some demonstration projects. For example, the roof of the National Centre for the Performing is coated with TiO_2 photocatalytic coating to solve the cleaning problem. However, due to the dry climate in Beijing and the lack of precipitation, this application also has certain limitations because its self-cleaning coating relies on rainwater to wash the surface of the building.

2.3. Building refrigeration
The use of TiO_2 in building refrigeration is mainly due to its high reflectivity. Studies have shown that TiO_2 coated on the building roofs can reflect up to about 85% of solar radiation. Although TiO_2 is highly reflective of most visible and near infrared light, it absorbs both violet and UV light, a property that tends to raise the temperature of the coating and is not conducive to keeping the room cool. The research of Mandal et al. [35] showed that using $BaSO_4$ and polytetrafluoroethylene powder instead of TiO_2 and reducing the concentration of the polymer adhesive with endothermic effect could effectively solve this problem and further improve the reflectivity of the coating. In addition, TiO_2 can also be used for heat transfer enhancement, but this technology is not yet mature and is expected to be used in building air conditioning and refrigeration in the future.

3. Research Progress of Photocatalytic Building Materials

3.1. Photocatalysis and Its Basic Principles
The photovoltage generated by TiO_2 is limited, which limits the types of reactions it can catalyze. For this reason, in recent years, the chemical industry has turned to the research of photoelectrocatalytic technology, that is, the catalytic conversion reaction is carried out with the assistance of external power supply. Take the Cu/Cu_2O-TiO_2 photoelectric catalytic system [36-37] as an example, under the condition of light, the holes generated by the TiO_2 photoanode oxidize H_2O to O_2, the photogenerated electrons conduct to the Cu/Cu_2O dark cathode through an external circuit and reduce CO_2 to CH_3OH in the presence of H^+. In this process, the photoanode reaction needs to be carried out under the condition of light, while the dark cathode reaction does not need light. The microscopic reaction mechanism involved in the dark cathode is shown in Fig.4.

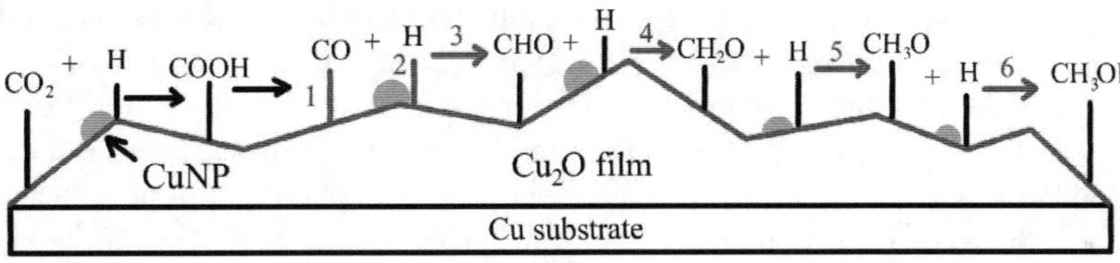

Figure 4 Microscopic reaction mechanism on surface of Cu/Cu_2O (modified from Chang et al. [36])

3.2. Photocatalysis and Building Emission Reduction
The photoelectrocatalytic process described in section 2.1 provides inspiration for artificial carbon fixation and building carbon emission control. Bai et al. [38] from Tianjin University used polymethacrylic acid (2-dimethylamino) ethyl ester hydrogel (PDMAEMA), light absorption agent polyaniline (PANI), silica gel carrier, and photocatalyst TiO_2 to construct a photoelectrocatalytic anode with self-phototropism. On this basis, the concept of reactors on buildings (RB) is proposed. In the RB, the self-phototropic photoelectrocatalytic anode array is assembled on the outer surface of the building, and the Cu/Cu_2O dark cathode array is assembled on the inner surface of the building. The solution system of the two is connected through a pipe passing through the wall with a proton exchange membrane in the middle, forming a Cu/Cu_2O-TiO_2 photoelectrocatalytic system. This system can be simplified to the electrochemical system shown in Fig.5. Under the conditions of sunlight, the RB can reduce the CO_2 enriched by selective membrane to the chemical raw material CH_3OH and release O_2. The total reaction can be expressed as $2\ CO_2 + 4\ H_2O \xrightarrow{RB} 2\ CH_3OH + 3\ O_2$. The electrode reaction that occurs at the autotropic photoelectrocatalytic anode array is $2\ H_2O - 4\ e^- \xrightarrow{TiO_2\ \&\ light} O_2 + 4\ H^+$, the electrode reaction at the Cu/Cu_2O dark cathode array is $CO_2 + 6\ H^+ + 6\ e^- \xrightarrow{Cu/Cu_2O} H_2O + CH_3OH$. RB can effectively eliminate CO_2 in the environment, and its application is expected to achieve ultra-low emissions or even "negative emissions" of buildings. This research provides new ideas for the control of building carbon emission and points out the direction for the development of next-generation photocatalytic building materials.

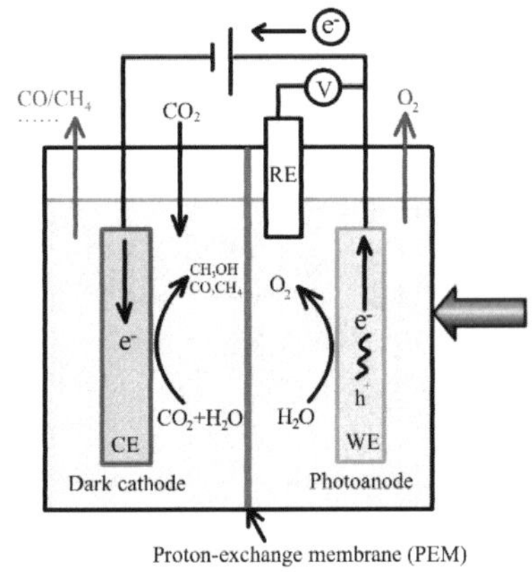

Figure 5 Simplified system of RB

4. Conclusions

At present, the main applications of photocatalytic building materials are pollution gas treatment, environmental sterilization, self-cleaning and anti-fogging of building surfaces, and building refrigeration. In terms of pollutant control and environmental health, the future photocatalytic materials need to focus on solving the problems of working band and using form, that is, developing efficient modification process to make the materials adapt to a wider working band and selecting the appropriate use mode according to the characteristics of coating and mixing methods. In terms of self-cleaning and anti-fogging of building surfaces, photocatalytic materials should move towards stronger light transmission, higher mechanical strength, and less interference from the external environment. For building refrigeration, in the future, photocatalytic materials can be combined with air conditioning in heat transfer enhancement to achieve better refrigeration effect in the future. In addition, the application of photocatalytic technology to carbon emission control in buildings in the form of a dynamic reaction mechanism is also an important direction for the development of photocatalytic materials in the future. For this reason, it is necessary to focus on solving the problems of catalytic efficiency and cost. The above conclusions provide a reference for the improvement and development of photocatalytic materials.

References:

[1] Enbin Shi, Hua Zhu, Xiaolong Chen. Preparation of diatomaceous with modified nano-TiO_2 loaded onto diatomite and its performance study. New Building Materials, 2014, 41(12): 84-88. (in Chinese)
[2] Fujishima A, Honda K. Electrochemical photolysis of water at a semiconductor electrode. Nature, 1972, 238: 37-38.
[3] Fujishima A, Rao TN, Tryk DA. Titanium dioxide photocatalysis. Journal of Photochemistry and Photobiology C: Photochemistry Reviews, 2000, 1: 1-21.
[4] Fei Bi, Shanshan Xiao, Li Zhao, et al. Advancement in research on TiO_2 based photocatalytic green-buildings materials. New Building Materials, 2018, 45(06): 63-66. (in Chinese)
[5] De Richter R, Ming T, Davies P, et al. Removal of non-CO_2 greenhouse gases by large-scale atmospheric solar photocatalysis. Progress in Energy and Combustion Science, 2017, 60: 68-96.
[6] Jianzhong Pei, Yansong Wang, Chundong Zhu, et al. Research progress on automobile exhaust pavement purification materials. China Journal of Highway and Transport, 2019, 32(04): 92-104. (in Chinese)
[7] Carey JH, Lawrence J, Tosine HM. Photodechlorination of PCB's in the presence of titanium dioxide in aqueous suspensions. Bulletin of Environmental Contamination and Toxicology volume, 1976, 16: 697-701.
[8] Marcella Guarino, Annamaria Costa, Marco Porro. Photocatalytic TiO_2 coating to reduce

ammonia and greenhouse gases concentration and emission from animal husbandries. Bioresource Technology, 2008, 99: 2650-2658.
[9] Xiujuan Liu, Yanjun Cui, Yongsheng Xiang, et al. Preparation of reactive alumina/diatomaceous earth/TiO_2 functional coating and their formaldehyde purification. New Chemical Materials, 2018, 46(supp.1): 22-26. (in Chinese)
[10] Ping Wang. Study on the purification of indoor formaldehyde by new photocatalytic coating additives (Master of Science in Engineering). Heilongjiang University, 2019.
[11] Kim C, Lee Y, Han K, et al. Tie redunction effect of VOCs (volatile organic compounds) by TiO_2 coated on the wall in a polluted house. Indoor air 2005: Proceedings of the 10th international conference on indoor air quality and climate, Beijing, 2005, 3228-3232.
[12] Ohno T, Mitsui T, Matsumura M. Photocatalytic activity of S-doped TiO_2 photocatalyst under visible light. Chemistry Letters, 2003, 32(4): 364-367.
[13] Qiang Guan, Meng Chen. Study on decontamination of vehicle emissions by nano-TiO_2 sprayed on highway concrete pavement. Journal of Highway and Transportation Research and Development, 2009, 26(03): 154-158. (in Chinese)
[14] Mingfeng Xu, Huisheng Shi, Kai Wu. Research development of photocatalytic NO_x abatement by TiO_2-cementitious composites. New Building Materials, 2019, 46(05): 46-49. (in Chinese)
[15] Hassan MM, Dylla H, Mohammad LN, et al. Evaluation of the durability of titanium dioxide photocatalyst coating for concrete pavement. Construction and Building Materials, 2010, 24: 1456-1461.
[16] Kikuchi Y, Sunada K, Iyoda T, et al. Photocatalyticbactericidal effect of TiO_2 thin films: dynamic view of the active oxygen species responsible for the effect. Journal of Photochemistry and Photobiology A: Chemistry, 1997, 106: 51-56.
[17] Yanpeng Wu, Chongfang Ma. Status of studies and applications of photocatalytic technology in building environment and facility field. Heating Ventilating and Air Conditioning, 2006, 36(03): 29-36. (in Chinese)
[18] Geichun K, Ding G, Kyoko S, et al. Maintaining the function of cooling tower by using photocatalyst to prevent the algae growth. Proceedings of Air Conditioning and Sanitation Engineering, 2003, 90: 69-77.
[19] Tetsu T, Shuichi T, Shuichi S, et al. Bactericidal effect of an energy storage TiO_2-WO_3 photocatalyst in dark. Electrochemistry Communications, 2003, (05): 793-796.
[20] Young T. An essay on the cohesion of fluids. Philosophical transactions-royal society, biological sciences, 1805, 95(1): 65-87.
[21] Wenzel RN. Resistance of solid surfaces to wetting by water. Industrial and Engineering Chemistry, 1936, 28(8): 988-994.
[22] Jing Liu, Wei Wang, Xinquan Yu, et al. Research progress of superhydrophilic anti-fogging surfaces. Surface Technology, 2020, 49(12): 75-92. (in Chinese)
[23] Duran RI, Laroche G. Current trends, challenges, and perspectives of anti-fogging technology: Surface and materials design, fabrication strategies, and beyond. Progress in Materials Science, 2019, 99 (04): 106-186.
[24] Wang R, Hashimoto K, Fujishima A, et al. Light-induced amphiphilic surfaces. Nature, 1997, 388(6641): 431-432.
[25] Shuxue Zhou, Ling Yang. Research status quo and review of TiO_2-based self-cleaning coatings. Electroplating and Finishing, 2013, 32(01): 57-62. (in Chinese)
[26] Liwu Zhang, Ralf Dillert, Detlef Bahnemann, et al. Photo-induced hydrophilicity and self-cleaning: models and reality. Energy and Environmental Science, 2012, 5, 7491-7507.
[27] Allen NS, Edge M, Sandoval G, et al. Photocatalytic Coatings for Environmental Applications. Photochemistry and Photobiology, 2005, 81: 279-290.
[28] Rongmin Wang, Boyun Wang, Yufeng He, et al. Preparation of composited Nano-TiO_2 and its application on antimicrobial and self-cleaning coatings. Polymers for Advanced Technologies, 2010, 21: 331-336.

[29] Zhao Y, Bai J, Liang J, et al. Analysis of chemical structure and characteristics of building materials based on element chemistry. E3S Web of Conferences, 2021, 248, 01028: 1-9.
[30] Mcmillan PF, Remmele RL. Hydroxyl sites in SiO_2 glass: A note on infrared and Raman spectra. American Mineralogist, 1986, 71: 772-778.
[31] Miyauchi M, Nakajima A, Hashimoto K, et al. A highly hydrophilic thin film under 1 mu W/cm (2) UV illumination. Cheminform, 2000, 32(16): 1923-1927.
[32] Lee D, Rubner MF, Cohen RE. All-nanoparticle thin-film coatings. Nano Letters, 2006, 6(10): 2305-2312.
[33] Li X, He J. Synthesis of raspberry-like SiO_2-TiO_2 nanoparticles toward antireflective and self-cleaning coatings. ACS Applied Materials and Interfaces, 2013, 5(11): 5282-5290.
[34] Tao C, Zhang L. Fabrication of multifunctional closed-surface SiO_2-TiO_2 antireflective thin films. Colloids and Surfaces A: Physicochemical and Engineering Aspects, 2020, 585, 124045: 1-9.
[35] Mandal J, Yang Y, Yu N, et al. Paints as a Scalable and Effective Radiative Cooling Technology for Buildings. Joule, 2020 (4): 1350-1356.
[36] Chang X, Wang T, Zhao Z, et al. Tuning Cu/Cu_2O interfaces for the reduction of carbon dioxide to methanol in aqueous solutions. Angewandte Chemie, 2018, 57, 15415-15419.
[37] Chang X, Wang T, Zhang P, et al. Stable Aqueous Photoelectrochemical CO_2 Reduction by a Cu_2O Dark Cathode with Improved Selectivity for Carbonaceous Products. Angewandte Chemie, 2016, 55, 8840-8845.
[38] Bai J, Jin Z, He X, et al. A design method for green buildings based on the conception of community and conversion of luminous energy. Sichuan Building Science, 2021, 47 (01): 81-89. (in Chinese)

Preparation and Performance of A Novel Type of Concrete Air Entraining Agent

Yongmei Wang[1,2], Serina Ng[1], Tao Ni[1,2], Xin Song[1], Ke Feng[1]

[1] Shijiazhuang Changan Yucai Building Material Co., Ltd., Shijiazhuang 050000, China

[2] Hebei Engineering and Technology Research Center of Functional Materials for Concrete, Shijiazhuang 050000, China

Correspondence should be addressed to Serina Ng; serina.ng@sjzcayc.com

Abstract. This article uses 1,4-dibromobutane, hexadecyl primary amine, sodium 3-chloro-2-hydroxypropionate, formaldehyde and formic acid as raw materials to synthesize a new type of concrete air entraining agent of (1,4)-bis(cetylmethyl hydroxy-propionic acid quaternary ammonium salt) butane through the reaction of ammonia alkylation and quaternization. The article characterizes the molecular architecture of the product, the surface activity of the aqueous solution, the bubble properties, the air entraining and air stabilizing properties of the fresh concrete, and the pore structure of the concrete after hardening. The research results show that the synthesized (1,4)-bis(cetylmethyl hydroxy-propionic acid quaternary ammonium salt) butane has good surface activity in aqueous solution, and its critical micelle concentration (cmc) is 5.3×10^{-4} mmol/L, γ_{cmc} is 28.86 mN/m; (1,4)-bis(cetylmethyl hydroxy-propionic acid quaternary ammonium salt) butane has better air-entraining performance and bubble stability in aqueous solution and concrete than SDS and AOS air entraining agents. The average chord length and spacing coefficient of bubbles in hardened concrete are smaller than that of SDS and AOS air entraining agent, giving rise to a better overall pore structure of the concrete. Compared with SDS and AOS, the concrete mixed with (1,4)-bis(cetylmethyl hydroxy-propionic acid quaternary ammonium salt) butane air entraining agent shows better mechanical property and durability.

1. Introduction

Freeze-thaw damages lead to premature cracking of concrete, a major factor that often seriously damage the service life of concrete and becomes one of the main reasons affects the durability of concrete. Addition of air entraining agent (AEA) to resist freeze-thaw is one of the most widely used technical approaches at present to improve the frost resistance and durability of concrete, where such practices are commonly performed in North America, Northern Europe, Japan and China. The mechanism is through the formation of large number of tiny pores in the hardened concrete, which act as "pressure relief tanks." When excess water freezes, the ice expands and enters the air spaces, avoiding damages to the concrete. In addition, the introduction of air bubbles into fresh concrete can also improve the initial state of fresh concrete, increasing the workability and reducing the risk of concrete segregation and bleeding. The air bubbles in concrete must remain stable in order to remain in fresh concrete for rheological benefits and to maintain the minimum amount of pores in solidified concrete. It has been reported that ensuring the appropriate number, spacing and distribution of micro bubbles in concrete has a significant impact on the mechanical properties of concrete, and is crucial to solve the problem of freeze-thaw cycle failure[1-2]. The essential reason why it is difficult for concrete to entrain air is due to the incompatibility of AEAs in concrete, thus it is particularly important to choose the appropriate AEA.

AEA belongs to the class of surfactant. It is added to concrete and mortar to improve their workability, prevent bleeding, heat insulation, sound resistance and improve frost resistance [3-6]. One of the best classification method is to divide them by their charges, i.e. anionic, cationic and amphoteric type surfactants. The anionic AEA ionizes in the aqueous solution, dissociates the metal cation and forms anionic moiety in solution. In the field of concrete, anionic AEA has been widely used because of its relatively cheap price as this class constitutes largely fatty alcohol sulfonate based and pine soap based AEAs. However, the ionic groups of the anionic surfactant undergo electrostatic attraction with free calcium ions in the concrete, leading to a reduction in the air entraining efficiency. Cationic surfactant is primarily nitrogenous organic amine derivatives. Nitrogen atoms in the molecules contain lone pair of electrons that can undergo H-bonding with the acid, generating positively charged amino groups. In alkaline systems such as concrete with a pH value between 12 to 13, precipitation occurs resulting in a severe loss in surface activity of the surfactant, thus such cationic surfactants are rarely used in concrete as AEAs. Amphoteric AEAs contains both cationic and anionic groups, resulting in good acid and alkaline resistance, good wettability and foaming ability, and is known to produce synergistic effect with other surfactants [7-9]. While amphoteric AEAs can solve the sensitivity problem of anionic engrants, they have little effect on the stability of bubbles in the slurry.

This paper thus target to create a new type of AEA by exploiting the sweet spot. By using 1,4-dibromobutane, hexadecyl primary amine, sodium 3-chloro-2-hydroxypropane sulfonate, formaldehyde and formic acid as raw materials, (1,4)-bis(cetylmethyl hydroxy-propionic acid quaternary ammonium salt) butane gemini type air entraining agent was synthesized. Its surface activity in aqueous solution and in the mortar and the effect of air bubbles in fresh and hardened concrete are investigated. Finally, the novel AEA is compared against different AEAs, and their effect on the mechanical properties and durability of concrete are analyzed.

2. Experimental section

2.1. Materials

1,4-dibromobutane (98%, Aladdin), 1-hexadecylamine (98%, Aladdin), sodium 3-chloro-2-hydroxypropanesulfonate (95%, Aladdin), ethyl alcohol (99.7%, Chron Chemicals), formaldehyde (37%, Chron Chemicals), formic acid (98%, Chron Chemicals), sodium hydroxide (98%, Chron Chemicals), sodium laurylsulfonate (98%, Aladdin) and AOS (Nanjing Qicheng New Material Co., Ltd.) were all purchased and used as supplied. Deionized water was used.

2.2. Synthetic method

(1) Synthesis of alkyl secondary amine: 60 g anhydrous ethanol and 86 g cetylamine are successively added into a 250 mL flask while stirring. Oil bath temperature is maintained at 35°C. 44 g 1,4-dibromobutane dissolved in 40 g anhydrous ethanol was added dropwise into the mixture over 30 min before maintaining reaction for a further 8 h. 30% NaOH solution further added to reach a constant pH value of 10±0.5. The proposed product obtained is as follows:

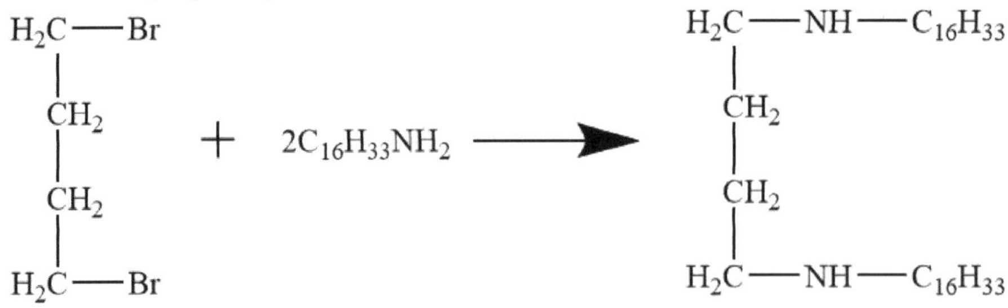

(2) Synthesis of tertiary alkyl amines: To the alkyl secondary amine, formic acid (88wt%) is added dropwise into the flask while rising temperature to 80°C over a period of 1h. Formaldehyde solution (37wt%) is then added dropwise into the flask at a molar ratio of alkyl amine: formic acid: formaldehyde = 1: 5.5 : 5 over 1h. The reaction is then continued for a further 6 h. NaOH (30wt%) is used to adjust the solution to pH 12. The synthesized solution is decanted by oscillating several times and the upper layer of yellowish oil liquid containing unreacted formaldehyde is removed by rotating evaporator. The remaining liquid is washed 2~3 times with 20wt% NaCl solution to achieve the alkyl tertiary amine (the yield is about 90%). The specific reaction formula is as follows:

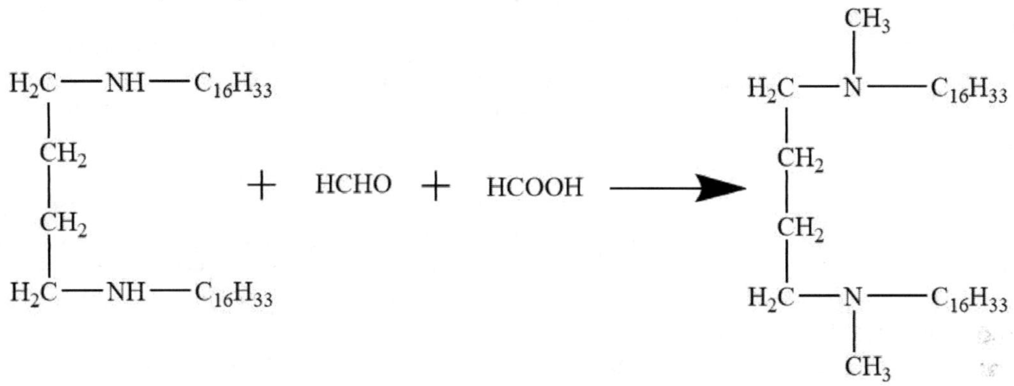

(3) A saturated solution of 3-chloride-2-hydroxyl-propylsulfonates is placed in a three-necked, round-bottom flask and heated to 80°C. The alkyl tertiary amine ethanol solution is dropped into the flask slowly through the constant flow pump, adjusted pH to 8~9 with 30wt% NaOH solution before reacting for 24 h. The unreacted alkyl tertiary amine is removed by the cyclohexane extraction, while solvents by rotating distillation method. The white solid product is achieved by recrystallization of the product with anhydrous ethanol. Yield is about 70%.

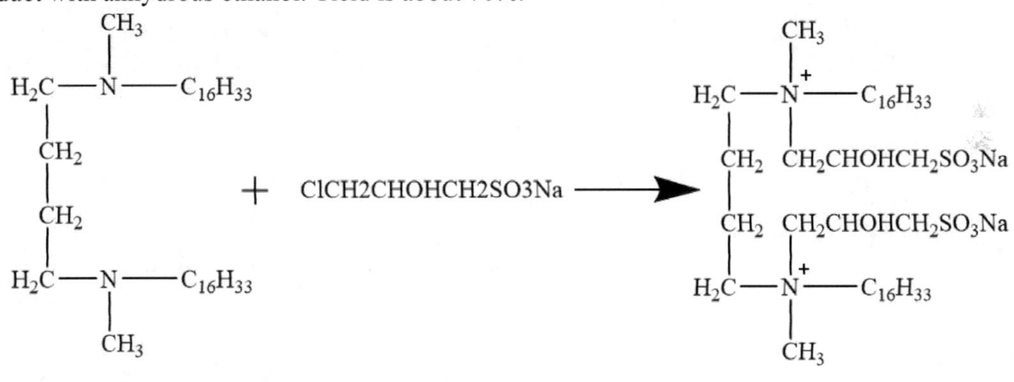

2.3. Infrared analysis of reaction products
The products in (3) are respectively ground with KBr and made into thin slices. Infrared test is carried out, and the structure is analyzed by VECTOR-22 Fourier infrared spectrum from Brock Company in Germany.

2.4. Surface activity test
The surface tension (γ) and critical micelle concentration (CMC) of the product in (3) are determined by hanging drop method using KRUSS DSA30s surface tensiometer at 25°C.

2.5. Bubble performance test
Waring-Blender method is used to test the initial bubble volume V_0 and forming half-life $T_{1/2}$ of the three air entraining agents.

2.6. Concrete test

The fluidity (slump, expansion) and air content of fresh concrete are tested in accordance with the Chinese national standard GB/T50080-2016 Standard "Testing Methods for Properties of Ordinary Concrete Mixes". The dosage of SP employed in all concrete is 1% by weight of binder, while dosage of AEA added varied. In C30 concrete, the dosage of air-entraining agent AEA-1, AEA-2 and AEA-new is 0.02%, 0.01% and 0.008%, respectively. In C50 concrete, the dosage of air-entraining agent AEA-1, AEA-2 and AEA-new is 0.04%, 0.02% and 0.012%, respectively. The raw materials for concrete are shown in Table 1. C30 and C50 are both employed in fresh concrete measurements (Table 2), while only C30 mix proportion is engaged in determining the properties of hardened concrete. All concrete samples are prepared as 150 ×150×150 blocks, vibrated for 10s and stored under standard condition for 28 d before been cut along the horizontal, polished and analyzed. The air content of hardened concrete is controlled within 3.5–4.0%, while the nature of the air voids is measured according to the Chinese water conservancy industry standard SL 352-2006 "Hydraulic concrete test regulations" and ASTM C457 "Standard Test Method for Microscopical Determination of Parameters of the Air-void System in Hardened Concrete". The air content, the number of bubbles, the distribution of bubble aperture, the average chord length of bubbles, the specific surface area of bubbles and the coefficient of bubble spacing of hardened concrete are determined. GB/T50081-2019 "Concrete Physical and Mechanical Properties Test Method Standard" and GB/T50082-2009 "Ordinary Concrete Long Term Performance and Durability Test Method" are employed to determine the mechanical properties of hardened concrete test, and impermeability and frost resistance of hardened concrete. The HP-40 automatic digital display concrete impermeability tester and JW-28 concrete rapid freeze-thaw tester, both from Hebei Jingwei Test Instrument Co., Ltd. were used.

Table 1 Raw materials for concrete experiments

Raw materials	Specifications	Manufacturer
Cement	P.O 42.5 cement	Sichuan Esheng Cement Co. LtdSichuan Esheng Cement Co. Ltd
Fly ash	Class I	Chengdu Bolei Fly Ash Comprehensive Development Co. Ltd
Mechanism of sand	Fineness modulus 2.65	Site self-made
Stone	5-31.5 mm continuous grading	Site self-made
PCE	GK-3000	Shijiazhuang changan yucai building materials co., LTD
AEA-1	SDS	Aladdin
AEA-2	AOS	Aladdin
AEA-new	(1,4) -bis(cetylmethyl hydroxy-propionic acid quaternary ammonium salt) butane air entraining agent	Shijiazhuang changan yucai building materials co., LTD

Table 2 Concrete mix ratio

Concrete grade	Cement /kg	Fly ash/kg	Sand/kg	Stones/kg	Water/kg	Water/cement ratio	Sand ratio / %	PCE dosage /%
C30	300	83	832	1017	165	0.43	45	1.0
C50	390	90	676	1057	158	0.33	39	1.2

3. Results and discussion

3.1. Characterization of AEA-new

FTIR is first employed to characterize the newly synthesized AEA-new. Figure 1 displays the stretching vibration peak at 3488 cm^{-1} is -OH, the stretching vibration peak at 2973 cm^{-1}, 2863 cm^{-1}, and the separation of $-CH_3$ and $-CH_2-$. The absorption peak at 1460 cm^{-1} is the asymmetric stretching vibration of tertiary amine, and the absorption peak at 1123 cm^{-1} is C-N. 1244 cm^{-1} is the absorption peak of $-SO_3^-$ ion, so AEA-new was successfully synthesized.

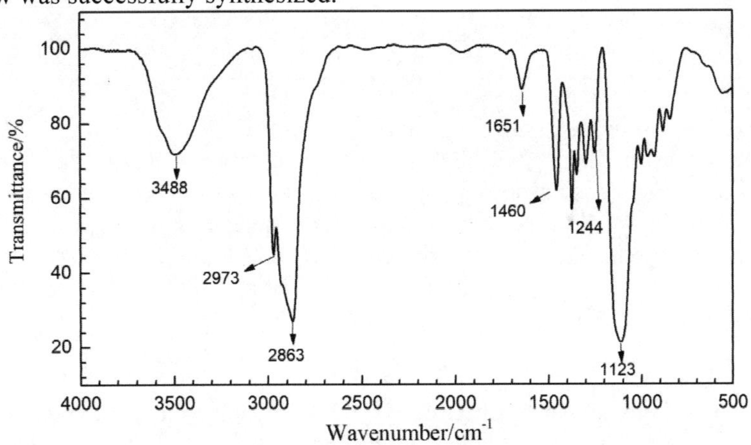

Figure 1 Infrared spectra of products

The surface activity of AEA-new is evaluated based on the critical micelle concentration (CMC) and surface tension (γ_{CMC}), whereby smaller the CMC and γ_{CMC}, higher the surface activity of the AEA in aqueous solution[10-12]. Figure 2 shows the γ-LogC diagram of AEA-new. The lowest turning point of the surface tension curve is its critical micelle concentration (CMC), at which the surface tension corresponding to the surfactant is γ_{CMC}. A comparison of the properties of AEA-new to AEA-1 and AEA-2 are presented in Table 3. AEA-new induce the largest reduction in surface tension of water (28.62 mN/m), while SDS performed the worst (38.0 mN/m). The low surface tension, thus CMC can be attributed to the zwitterionic-amphiphilic distribution of this molecule, causing a weak electrostatic repulsion coupled with closely packed hydrophobic chains, enhancing the stability of the micelle formed[15]. It is of interest to note that the lower the CMC value, the more likely the surfactant is to self-assemble at low concentration [16]. AOS and SDS are both made up of mainly single chain anionic air-entraining agent molecules. AOS is a mixture, usually composed of sodium alpha-olefin sulfonate and sodium hydroxylalkyl sulfonate at a mass ratio of about 7:3. The length of hydrophobic carbon chain in its molecular structure is greater than 12, thus has a more hydrophobic characteristic than SDS. When AEA are more hydrophobic, their arrangements at the gas-liquid interface occurs in a folding and curling modules, decreasing the spatial spacing of the hydrophobic chain of the air entraining agents. In other words, the number of long hydrophobic chain surfactant molecules adsorbed at the gas-liquid interface decreases, thus CMC decreases as well.

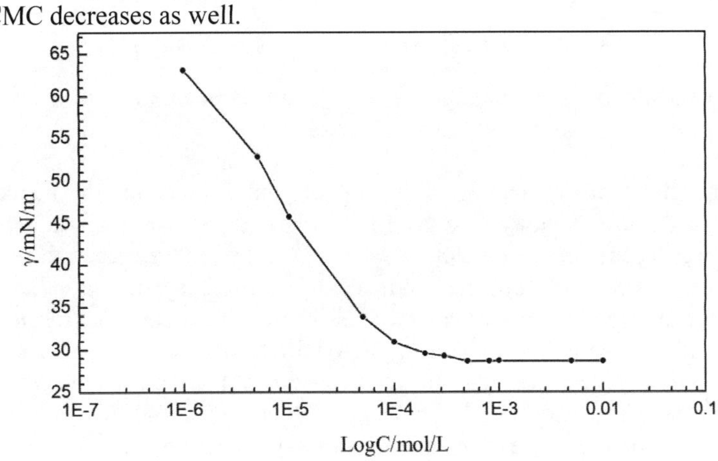

Figure 2 C-γ curve of the product at 25°C

Table 3 CMC and γ_{CMC} of the product.(T=298.15K)

Surfactant	CMC (mol/L)	γ_{CMC} (mN/m)
Product	5.3×10^{-4}	28.62
AOS[13]	6.0×10^{-3}	32.74
SDS[14]	9.8×10^{-3}	38.0

For hydrocarbon based air entraining agents, the value of γ_{CMC} depends on the density of the hydrophobic adsorption layer. The higher the density of the adsorption layer is, the more beneficial it is to reduce the surface tension. The adsorption layer density of (1,4)-bis(cetylmethyl hydroxy-propionic acid quaternary ammonium salt) butane air entraining agent molecules at the gas-liquid interface is the highest and γ_{CMC} is the lowest. The hydrophobic chain of AOS is longer than that of SDS. After the hydrophobic chain folds, the adsorption layer density of the former is higher than that of the latter at the gas-liquid interface. Therefore, SDS has the largest γ_{CMC} value.

3.2. Bubble properties in CPS
Air bubble is in itself a thermodynamic unstable system; transiting from stable bubble to bursting is a spontaneous process of energy reduction. The foaming and air stabilizing ability are indexes to evaluate the effectiveness of surfactants. At room temperature, foaming fluids with different mass percentages (0.1wt%, 0.2wt%, 0.3wt%, 0.4wt% and 0.5wt%) were prepared by the three kinds of air entraining agents. The relationship between the foaming volume V_{max} of the three kinds of air entraining agents and the forming half-life $T_{1/2}$ with the solution concentration was studied respectively. The experimental results are shown in Figure 3 and Figure 4.

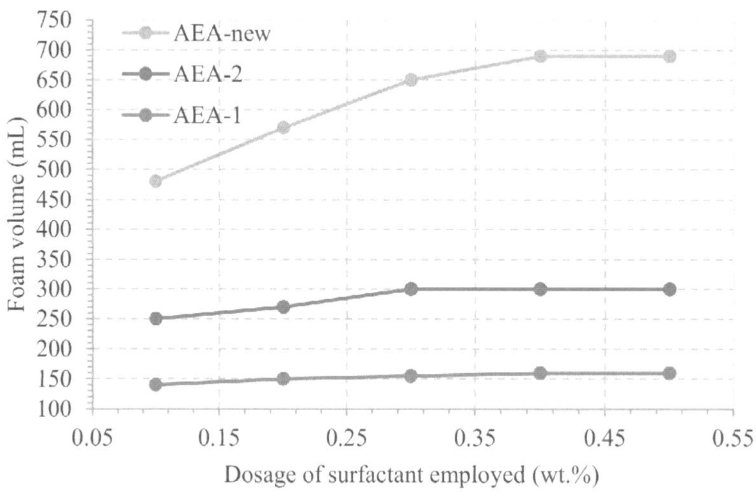

Figure 3 The relationship between the foaming volume and concentration of the three kinds of air entraining agents

Figure 3 demonstrates that the foaming ability of the three air entraining agents greatly differs, whereby foaming is as follow: product > AOS> SDS. When an external force is applied, the inherent energy of the bubbling system, surface tension of the body fluid γ and the area of the gas-liquid interface A ($\gamma\times$A) all increases to maintain equilibrium. When external force is constant, a lower surface tension will then result in the development of larger air bubbles and greater ease in foaming and vice versus. Due to the presence of two hydrophobic chains in AEA-new, a stable directional array on the gas-liquid interface is exerted, thus excellent foaming capacity is observed (Fig 3), whereby AEA-new induces the largest reduction in water surface tension, followed by AOS and SDS respectively. Figure 3 also reveals that when a constant external force is exerted on the foaming liquid, the foam volume first increases and remains constant despite increasing concentration of the air entraining agent. The initial increment is correlated to the decrease in surface tension as concentration of air entraining agent increases. When minimum surface tension is achieved i.e. no further adsorption of air entraining agent at the liquid-gas interface possible, the foam volume will then stay at its maximum.

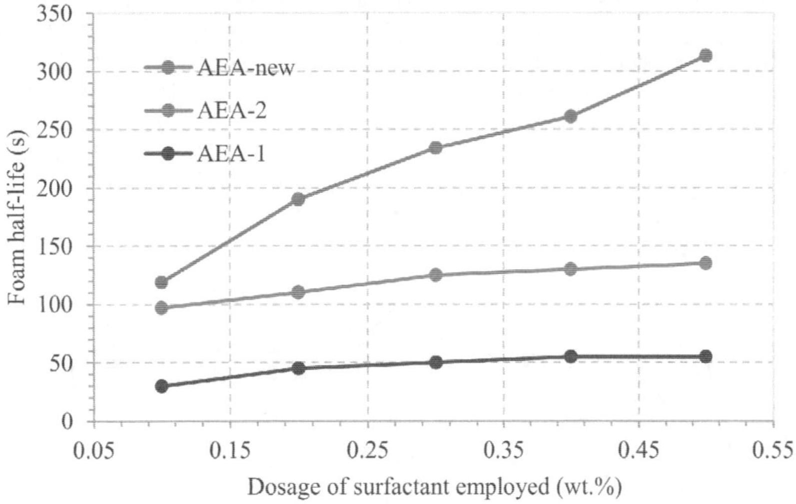

Figure 4 The relationship between the forming half-life of the three air entraining agents and the concentration change

It is interesting to note that the half lives of air bubbles in concrete prepared with AOS and SDS are similar, while that with AEA-new differs greatly (Figure 4). It is speculated here that unlike absolute foaming capability (Figure 3) which is related to the amphiphilicity of the molecules, the similarity in half life is attributed to the molecular architecture of the AEAs. The stability of bubbles formed by SDS is more than a factor lower than AOS, attributing to the decrease in surface tension, whereby energy required to form the film at the gas-liquid interface decreases, increasing bubble thermodynamic stability [17]. When the concentration of AOS and SDS increased to 0.4wt%, the half-life of the air bubbles appeared to be reach a maximum, indicating that the surfactants at the gas-liquid interface may have reached saturation. In comparison, the half life of air bubbles produced in the presence of AEA-new continue to increase with dosage. The phenomenon correlated with the surface tension and CMC values of the air- entraining agents, indicating the importance of these values to achieve a stable porous concrete system.

3.3. *Air entrainment investigation in fresh concrete*

The air entrainment and resulting stability of the air bubbles produced in fresh concrete with the three AEAs in C30 and C50 each respectively is investigated (Table 4). In C30 concrete containing 0.02% AEA-1, the initial air content and at 60 min are 2.5% and 1.8% respectively. When 0.01% AEA-2 is added, the initial air content of C30 is 3.5%, which reduces to 3% at 60 min. In the case of AEA-new, the initial air content of C30 concrete is 4.5%, and maintaine above 4% after 60 minutes when a low dosage of 0.008% is employed. The differences can be attibuted to the surface tension and hydrophilicity of the molecules, where (1) lower surface tension equates greater ease in air entrainment and (2) presence of two different hydrophilic groups in the structure of AEA-new weakens the electrostatic repulsion between them, reducing packing density of the overall molecule, thus allowing a denser arrangement at the gas-liquid interface, causing greater bubble stability.

To achieve similar air entrainment in C50 concrete, the amount of air-entraining agent increases. 0.012% of AEA-new, 50% higher than in C30 is needed to achieve an initial air content of 4.2%. Likewise, increasing AEA-1 and AEA-2 to more than 0.03% failed to increase the air concrete of concrete (Table 4). This may be explained by the high viscosity of concrete; the viscosity of slurry acts as an energy barrier for the formation of bubbles [14]. However, it can be observed that AEA-new is more effective in producing entrained air in fresh concrete than the conventional AEAs.

Table 4 Slump and air contents of fresh concrete

Concrete grade	Type of air entraining agent	Initial slump/mm	Initial workability/ mm	Initial air content/%	60 min air content/%	Admixture/ %
C30	AEA-1	220	550	2.5	1.8	0.02
	AEA-2	220	560	3.5	3.0	0.01
	AEA-new	230	570	4.5	4.2	0.008
C50	AEA-1	225	580	1.8	1.3	0.04
	AEA-2	230	575	2.4	1.8	0.02
	AEA-new	235	590	4.2	4.0	0.012

3.4. *Investigation in hardened concrete*

3.4.1. *Porosity determination in hardened concrete*

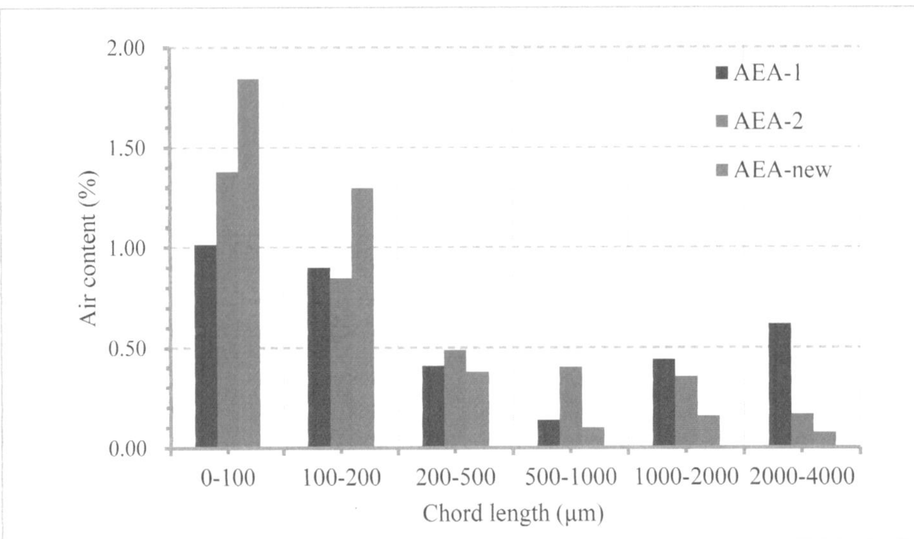

Figure 5 Influence of three kinds of air entraining agents on the air content contributed by bubbles of different sizes in concrete

Figure 5 displays the distribution of air bubbles in the hardened concrete. As observed, air bubbles with chord length of 0-200 μm are the most prevalent, whereby AEA-new produces the highest amount of air voids having chord length between 0–200 μm, similar in the range of 200-500 μm, and greater number of large air voids when AEA-1 and AEA-2 are employed. The variation in air voids can contribute greatly to the strength of concrete, frost resistance and permeability resistance, which will be discussed further.

3.4.2. Corresponding bubble structure parameters in concrete with different air entraining agents

Table 5 Corresponding buble structure parameters in concrete mixed with three kinds of air entraining agents

Type of air entraining agent	Air entraining agent dosage/%	Initial air content of fresh concrete /%	Air content of hardened concrete /%	Average chord length /μm	The bubble number	Bubble specific surface area/(mm²/mm³)	Bubble spacing coefficient /μm
AEA-1	0.035	3.9	3.5	268	3252	26.4	286
AEA-2	0.012	3.8	3.6	204	3835	38.3	187
AEA-new	0.006	3.7	3.8	102	4316	52.7	115

Table 5 presents the air voids in C30 concrete in the presence of different AEAs. By adjusting the amount of air entraining agent, the air content of freshly mixed concrete is close to 3.9%, 3.8% and 3.7%, respectively. The final air voids contents of concrete containing AEA-1, AEA-2 and AEA-new are 10.2%, 5.3% and 0% lower than that in fresh concrete respectively. As observed from Table 5, the overall bubbles in hardened concrete containing AEA-new is more than 10% higher than when the other two AEAs are employed, corresponding to the higher specific surface area: volume ratio and short average chord length of 102 μm. The stability of air bubbles is the lowest when AEA-1 is employed as demonstrated by the high loss in air content from fresh to hardened state (Figure 1 versus Table 5).

3.5. Mechanical properties and durability of hardened concrete

The 28-day compressive strength, impermeability and frost resistance of hardened concrete of the C30 concrete are tested (Table 6). Similar to previous observation, the 28-day compressive strength, impermeability and frost resistance of concrete are superior when AEA-new is employed. The compressive strength of the concrete is 10.7% higher than that of the concrete using the AEA-1, 8.2% higher than that of the concrete using the AEA-2, the impermeability grade and frost resistance grade are three grades higher than that of the AEA-1, two grades higher than that of the AEA-2. All of these are closely related to the concrete bubble structure in Table 5, and are consistent with the data in Table 5. Small bubble size and small bubble spacing coefficient mean that the mechanical properties (28-day compressive strength) and durability (impermeability and frost resistance) of concrete are better in the case of similar air content.

Table 6 Comparison of compressive strength, impermeability and frost resistance of concrete mixed with different air entraining agents

Type of air entraining agent	Mechanical strength	Impermeability		Frost resistance	
	28d compressive strength/MPa	Maximum Anti-seepage pressure value/MPa	Impermeability grade	Maximum number of freeze-thaw cycles	Frost resistance grade
AEA-1	33.5	0.7	P6	100	F100
AEA-2	34.3	1.1	P10	225	F200
AEA-new	37.1	1.3	P12	250	F250

4. Conclusion

In this paper, a new type of concrete air entraining agent (1,4)-bis(cetylmethyl hydroxy-propionic acid quaternary ammonium salt) butane (AEA-new) with CMC and γ_{CMC} values of 5.3×10^{-4} mmol/L and 28.86 mN/m respectively has been successfully synthesized. The effectiveness of this AEA in fresh and hardened concrete are compared with two conventional AEAs–SDS and AOS. It is confirmed that the air entraining performance, ease in producing smaller air bubbles and thus final air void system, and the resulting stability of air bubbles produced by AEA-new are superior than those of SDS and AOS. This results in excellent mechanical properties and durability when AEA-new is used in concrete, whereby 28 d compressive strength is 10.7% and 8.2% higher than when SDS and AOS are employed respectively. The impermeability grade and frost resistance grade are three grades and two grades higher respectively.

References

[1] Ley M T, Folliard K J, Hover K C 2009 Cement. Concrete. Res. **39** 409-16.
[2] Ley M T, Chancey R, Juenger M C G, et al 2009 Cement. Concrete. Res. **39** 417-25.
[3] Du L, Folliard K J 2005 Cement. Concrete. Res. **35** 8 1463-71.
[4] Tehrani-Bagha A R, Holmberg K, Ginkel C V, et al 2015 J. Colloid. Interf. Sci. **449** 72-9,
[5] Chatterji S, 2003 Cement. Concrete. Comp. **25** 759-65,.
[6] Souza M T, Maykot C K, Araujo A C Z, et al 2017 Constr. Build. Mate. **157** 363-71.
[7] Takai M, Hidaka H, Moriya M 1979 J. Am. Oil. Chem. Soc. 56 537-41, .
[8] Soontravanich S, Munoz J A, Scamehorn J F, et al 2008 J. Surfactants. Deterg. **11** 251-61.
[9] Harrison D, Szule R, Fisch M R, J. Phys 1998 Chem. B. **102** 6487-92.
[10] Iyota H, Tomimitsu T, Shimada K, et al, 2006 J. Colloid. Interf. Sci. **299** 428-34, .
[11] Lin Z, Lu B, Zakin J L, et al, 2001 J. Colloid. Interf. Sci. **239** 543-54.
[12] David L D, Erick S G, Garza C, et al, 2010 J. Colloid. Interf. Sci. **348** 152-58, .
[13] Alargova R G, Kochijashky I I, Sierra M L, et al, 2001 J. Colloid. Interf. Sci. **235** 119-29.
[14] Tattersall G H 1976 Mag. Concrete. Res. **28** 143-7.
[15] Heede P, Furniere J, Belie N D 2013 Cement. Concrete. Comp. **37** 293-303.
[16] Menger F M, Keiper J S 2000 Angew. Chem. Int. Edit. **39** 1906-20, .
[17] Razavi S, Shahmardan M M, Nazari M, et al, 2020 Colloid. Surface. A: Physicochemical and Engineering Aspects. **585** 124047.

Effect of Air Entraining Agents on The Air Void Structure of Concrete

Xin Song[1], Serina Ng[1,2*], Tao Ni[1,2], Yong Mei Wang[1,2], Feng Ke[1]

[1] Shijiazhuang Changan Yucai Building Material Co. Ltd., Shijiazhuang 051430, Hebei, China.

[2] Hebei Engineering and Technology Research Center of Functional Materials for Concrete, Shijiazhuang 051430, Hebei, China.

Correspondence should be addressed to Serina Ng; serina.ng@sjzcayc.com

Abstract. Four different air entraining agents are selected. The air entraining effect and their foaming stabilities in cement pore solution; the impact on air entrainment in fresh concrete; and the variations in air voids and their structures in hardened concrete are compared and studied. The results show that the development in porosity of the hardened concrete mixed with different air entraining agents is consistent with the air entrainment in fresh concretes within the first hour, but deviated from the foaming stability of the air entraining agents in cement pore solutions. Carboxylic based air entraining agents generate calcium soap, whereby such moiety can effectively encapsulate air bubbles. While the stability is insignificant in cement pore solution, the impact is obvious in concrete. When employed alone, the air entraining agent failed to improve both air entrainment of fresh concrete and final concrete structure. Instead, a combination of air entraining agents is found to induce a synergetic effect, effectively enabling both the air-entraining performance and the development of an enhanced porous structure of the hardened concrete.

1. Introduction

Air entraining agents (AEA) can reduce the surface tension of fresh concrete to ideally generate uniform, stable and optimally packed micro bubbles during the mixing of concrete, improving the workability, frost resistance, carbonization resistance and other durability parameters[1,2]. As a result, AEAs are widely specified and used in concrete field since it first discovered in the mid-1930s[3,4].Structure and stability of air bubbles depend on the types of AEA. Şahin et al.[5] pointed out that fatty alcohol type AEAs produced small air bubbles, while synthetic and resin type form coarser and mid-sized bubbles. Barfield et al.[6] concluded that different AEAs formed concrete with different air void characteristics, whereby synthetic AEAs produced excellent air-void system. Łaz′niewska-Piekarczyk[7] reported that synthetic AEAs generated the closely spaced and small air voids, followed by AEA made of saponified resin.

Good air void structure distribution are crucial for determining the concrete property[8,9]. Puthipad et. al.[9,10] claimed that fine air bubbles (<0.2 mm) enhance the workability of fresh concrete and improve the frost resistance and final strength of the hardened concrete. On the other hand, large air bubbles are unstable and can escape easily from the concrete, thus reducing the stability and eventual frost resistance of the concrete. Powers[11] reported that the frost resistance of concrete highly depends on the air void size and distribution. Many researchers believe that the frost resistance of concrete, besides mechanical strength, mainly depends on the spacing factor and volume of air content[5,12].

Christodoulou[13] reported that by incorporating AEA, tiny air bubbles form, imparting a ball-bearing effect in fresh concrete that enhances easi-ness for aggregates movement. Bruere[14] found that the floatation energy of air bubbles reduces the settlement rate of cement particles and decreases bleeding and segre-gation. Zhang et. al.[15] observed that small air bubbles (10–600 mm) reduced yield stress and viscosity of mortar, whereby the characteristics of the air bubbles have a greater influence on the rheology than that from the total air content. Despite the dis-tinct correlation, AEA are usually evaluated by the air content of fresh concrete in practice, but not based on the resulting porosity of concrete.

In this paper, four commonly used AEAs in China are investigated for their air entraining abilities in cement pore solution (CPS), their effects on the air entrainment in fresh concrete and final porosity in hardened concrete. The aim of this paper is to present a correlation between the influence of AEAs on fresh and hardened state concrete, and provides a reference for the development and application of AEAs especially in China for the future. To quantify the complex formation and the resulting stability of bubbles in solution, the performance of AEA in CPS is also investigated

2. Materials and Methods

2.1. Materials

Ordinary Portland Cement (OPC) is purchased from Sichuan Esheng Group, Si-chuan, China. Fly ash is purchased from Chengdu Lingyun Fly Ash Comprehensive Utilization and Development Co. Ltd, Sichuan, China. Both are used as per supplied. Their chemical compositions and physical properties are shown in Table 1. Fine aggregates is crushed limestone (fineness modules = 2.7, Table 2). Coarse aggregate is also crushed limestone, with a continuous grading of 5/20 mm. The superplasticizer (SP) and 4 different AEA (Table 3, Figure 1) employed are provided by Shijiazhuang Chang'an Yucai Building Material Co. Ltd, Hebei, China and used as supplied.

Table 1 Chemical and physical compositions of cement and fly ash

	SiO_2 (%)	Al_2O_3 (%)	Fe_2O_3 (%)	CaO (%)	SO_3 (%)	MgO (%)	K_2O (%)	Na_2O (%)	LOI (%)	Specific surface area (cm^2/g)
Cement	22.1	4.8	3.1	64.3	2.1	1.9	1.0	0.2	1.6	3410
Fly ash	54.1	27.4	6.4	2.2	0.1	0.8	1.6	0.4	2.9	3600

Table 2 Sand gradation

Sieve diameter (mm)	Grader retained percentage (%)	Accumulated retained percentage (%)
4.75	1.3	1.3
2.36	19.6	20.9
1.18	18.7	39.6
0.60	17.8	57.4
0.30	14.2	71.6
0.15	10.4	82.0
0.075	9.7	91.7
Residual	8.3	100.0

Table 3 Physical parameters of air-entraining agent

No.	Main chemical type	Color	State of matter
AEA-1	Sodium Abietate	brown	liquid
AEA-2	Lauryl sodium sulfate	white	powder
AEA-3	Sodium Abietate/Lauryl sodium Sulfate compound air-entraining agent	yellow	liquid
AEA-4	Triterpenoid saponins	brown	liquid

Journal of Physics: Conference Series **2011** (2021) 012051 doi:10.1088/1742-6596/2011/1/012051

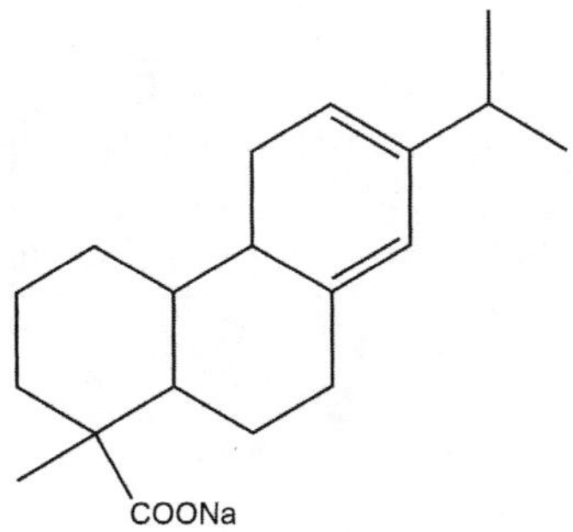

(a)

(b)

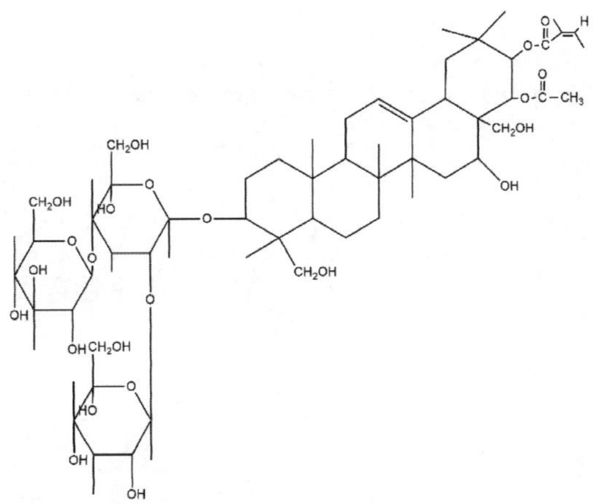

(c)

Figure 1 Molecular structures of AEAs（a: Sodium Abietate; b: Lauryl sodium sulfate; c: Triterpenoid saponins）

2.2. Experiment method

2.2.1. AEA in CPS – Shaking experiment.

Water is added to OPC at a w/c of 2, mixed for 2 min and filtered under pressure to obtain the CPS. CPS is used immediately after production. A 15 mL 1 wt.% AEA–CPS solution is prepared, shake violently for 10 s and let stand. The total volume of the foam at 0 min, 15 min and 30min is observed and recorded[16].

2.2.2. Air entrainment measurement in fresh concrete.
Concrete mixes are prepared with a water to binder ratio (w/b) of 0.46 and 1 wt.% SP to maintain a slump of 180 mm. The mixture proportion is given in Table 4 and ini-tial air content is controlled at 3.5 to 4.0 vol%.

Table 4 Mix design of concrete

Material dosage (kg·m^{-3})					Concrete admixture content (%)
Water	Cement	Fly ash	Sand	Stone	GK-3000
165	290	70	805	1025	1

According to Chinese Standard GB/T 50080-2016, the air contents of fresh concretes are tested with a SANYO direct reading air content tester (SANYO, Japan). The air void structure of fresh concrete was tested by AVA3000 air-void analyzer.

2.2.3. Investigation on the porous structure of hardened concrete.
The porosity, i.e. hardened air void structure of the hardened concretes is obtained with a NELD-BS610 hardened concrete air-void parameters analyzer (Naierde, China) according to Chinese Standard SL 352-2006.

3. Results and Discussion

3.1. Air entraining performance of air entraining agent in cement pore solution

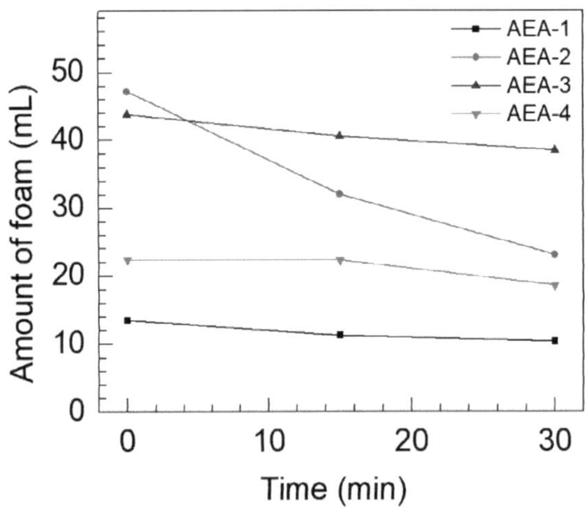

Figure 2 Generated foam volume with different AEAs added in CPS as a function of time

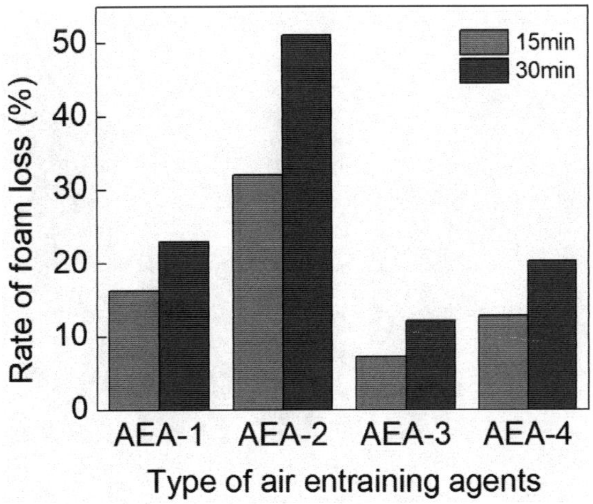

Figure 3 Stability of foam generated by different AEAs in CPS

CPS contains high amount of electrolytes that can impart a viscosifying effect on the AEAs[3,17,18]. Figure 2 displays the amount of foam generated and remained as a function of time in CPS when different AEAs are added. The initial amount of foam from CPS containing AES 1 to 4 are 13.5 mL, 47.2 mL, 43.8 mL and 27.1 mL respectively, i.e. measured foam volumes are: AEA-2 > AEA-1 > AEA-3 > AEA-4. From Figure 3, it can be observed that rate of foam loss of the AEAs at 15 min are 16.3%, 32.0%, 7.3% and 12.9%, respectively, which varied to 23.0%, 51.1%, 12.1% and 20.3% respectively at 30 min. AEA-1 is a rosin based air entraining agent with carboxyl group. In CPS, a large number of electrolytes particularly calcium ions can interact with carboxylic ac-ids to form insoluble calcium soaps. This results in poor foaming capability. However the insoluble calcium soap is adsorbed on the surface of the liquid film to increase the thickness of the bubble film and effectively prevents the aggregation and rupture of bubbles, stabilizing the foam[4]. While having the highest amount of initial foam, CPS containing AEA-2 displayed the highest rate of foam loss as a function of time. This shows that sulphate type AEA may have the best air entraining performance, but when employed alone the introduced bubble type and stability are not ideal. In combination, sulphate and rosin AEA has excellent performance in air entraining ability and air stabilizing ability, as confirmed by the synergetic effect presented when AEA-3 is added. While having similar initial foam amount as AEA-2, the rate of foam loss for AEA-3 is the smallest. The anionicity and low molecular weight of SDS (AEA-2) allow the molecules to be agile and quick in arranging on the air-liquid interface, causing good initial foaming ability. However, as the bubble film viscosity is small (low thick-ness), coalescing of the small air bubbles occur rapidly and can quickly escape and rupture[19,20]. From the observation, the presence of SDS gives good foaming ability (AEA-2) while rosin component (AEA-4) improves the foaming stability of the air-entraining agent. Moreover, SDS as a straight-chain alkane organic acid with a small and soft molecular structure can effectively fill in the void of rosin molecules to form a wedge-shaped structure that can transform into a stable bilayer as the molecules self-assemble[21]. thus further increases the liquid film thickness and improves the stability of the foam.

AEA-4 is a triterpenoid saponin air entraining agent. AEA-4 has the highest molecular weight among the 3 single AEAs, thus is unable to form directional arrangement at the interface causing adsorption at the gas-liquid interface to be lowest. The presence of spatially distributed hydroxyl groups in the saponins glucose unit together with intermolecular van der Waals forces generates thick and elastic bubble films, in-creasing the stability of bubbles formed[19].

3.2. Effect of AEAs on the type of air bubbles and air entrainment in fresh concrete

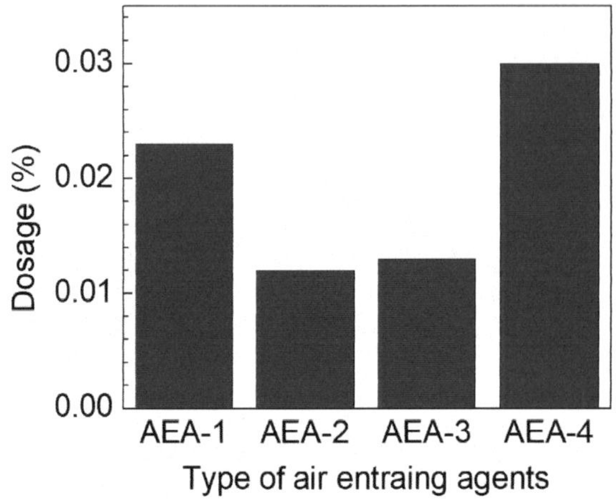

Figure 4 Dosages of AEAs added to achieve the desired air content in fresh concrete (3.5%-4.0%)

Figure 4 shows the dosage of different AEAs when the air content of fresh concrete is controlled at 3.5%-4.0%. As can be seen from Figure 4, when the air content is roughly the same, the amount of AEAs needed are as follow: AEA-2> AEA-3> AEA-1> AEA-4 (0.012%, 0.013%, 0.023% and 0.030% respectively), differing from that in CPS. This difference can be explained by the non-uniformity and complexity of the concrete system. In the process of mixing concrete, the existence of air bubbles is a balance between formation and destruction as particles and aggregates can 'burst' the generated air bubbles. Particularly for AEA-4, whereby air bubbles are stabilized by van der Waals forces, presence of 'bursting agents' can greatly destabilize the air bubbles. In-soluble calcium soap, compatible with the highly electrolyte rich concrete as formed by AEA-1 can effectively prevent the bursting of air bubbles. AEA-2 on the other hand, can migrates quickly to the gas-liquid interfaces to generate air bubbles rapidly, thus air entraining performance remain superior. Similar to in CPS, AEA-3 as a combination AEA is able to display both good air bubble formation and eventual stability.

Table 5 Air contents of fresh concrete containing AEAs, measured by pressure method and AVA method

No.	Air content measures by pressure method (%)	Air content measured by AVA (%)
AEA-1	3.7	2.9
AEA-2	3.5	2.3
AEA-3	3.5	3.1
AEA-4	4.0	3.7

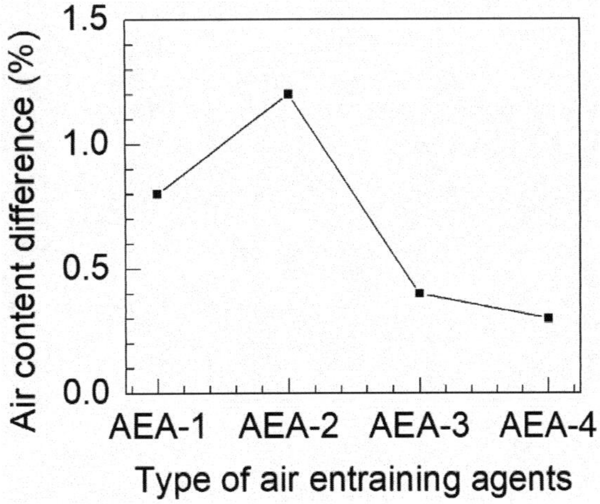

Figure 5 Difference in the percentage of air contents in fresh concrete, measured by pressure method and AVA method

Air contents of fresh concrete containing AEAs, measured by pressure method and AVA method are shown in Table 5, while Figure 5 presents the differences in measured air contents between these two methods. The air contents of the concretes measured by the AVA method is in all less than that measured by the pressure method as the former method only takes into account air bubbles less than 3mm, while pressure method includes all air bubbles present. The difference in air bubble content measured by the two methods is AEA-4> AEA-3> AEA-1> AEA-2, indicating that the percentage of small air bubbles in concrete prepared with AEA-4 is the highest, while concrete with AEA-2 contain mostly larger air bubbles. As the presence of smaller air bubbles indicate better concrete performance, the air void structure of the concrete mixed with AEA-4 and AEA-3 may be the best, followed by AEA-1, and AEA-2. It is of interest to note that the amount of air bubbles below 3 mm is also the highest in AEA-4, indicating that saponin based AEAs are ideal for maintaining fresh concrete properties including workability.

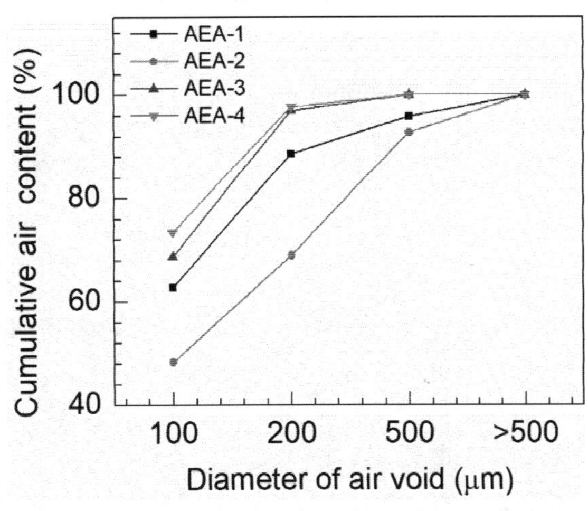

Figure 6 Cumulative amount of air in fresh concrete measured at 5 min by AVA

Figure 6 shows the cumulative air content of the fresh concretes. For concrete containing air bubbles below 100mm, the order is as follow: AEA-4 > AEA-3 > AEA-1 > AEA-2. As expected, AEA-2 produced the largest percentage of large air bubbles (> 0.2 mm) and AEA-4 appeared to have the best stomatal structure and air stabilizing ability, indicating that the air entraining property in fresh concrete is similar to that in CPS. The compound AEA-3 has the advantages of two AEAs, and the synergistic effect between the components can effectively improve the AEA's air entraining performance and improve the air void structure of fresh concrete. There is little difference in the cumulative air voids in fresh concrete containing AEA-3 and AEA-4.

3.3. Influence of air entraining agent on air void structure of hardened concrete

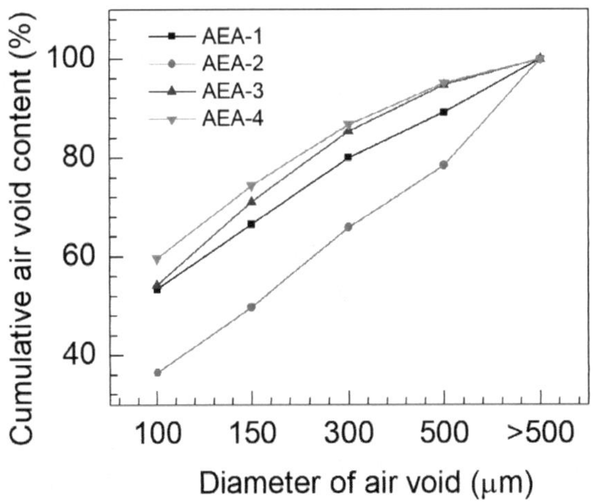

Figure 7 Cumulative air void content of hardened concrete at 28 days

Figure7 shows the cumulative air void content of hardened concrete mixed with different air entraining agents at 28 days. When comparing Figure 6 and 7, the amount of air escaped from fresh concrete to its hardened state is about 10%, indicating that the air loss after first hour is consistent for all concretes, regardless of AEA types. Therefore, the main impact of AEAs on the final hardened concrete property may be determined from the initial air entraining effect. AEA-4 introduces the largest number of small air voids in hardened concrete, with AEA-3 coming close, followed by AEA-1 and finally AEA-2.

4. Conclusions

Here, we showed the performance of 3 commonly employed AEAs on their per-formance in concrete, both fresh and hardened state. As a reference, CPS was employed and it was found that air bubbles development in CPS or in concrete differs slightly, but can provide valuable information. It was found that AEAs can either generate air bubbles or stabilize them, rarely displaying both properties. Sulphate based AEAs may be the most ideal AEA types to generate in situ air bubbles, whereas stabilization depends greatly on the hydrophobicity and anionicity of the molecules or the formation of calcium soap. However, it is possible to formulate AEAs with both good foaming and air bubbles stabilizing properties when AEAs with different properties (e.g. sulphate + rosin based) are formulated together.

References

[1] You Q J, Ding X X, Li J Z 2003 *Concr.* **6** 28-29.
[2] Luo F J, Wang S T, Fu Y W 2009 *Concr.* **9** 14-17.
[3] Chatterji S 2003 *Concr. Compos.* **25** 759-65.
[4] Du L, Folliard X K J 2005 *Cem. Con. Res.* **35** 1463-71.
[5] Şahin Y, Akkaya Y, Boylu F, Taşdemir M 2017 *Cem. Concr. Compos.* **82** 95–104.
[6] Barfield M, Ghafoori N. Constr. 2012 Build. Mater. **26** 490–96.
[7] Łaźniewska-Piekarczyk B. 2013 *Constr. Build. Mater.* **40** 659–71.
[8] Attachaiyawuth A, Rath S, Tanaka K, Ouchi M. 2016 *J Adv. Concr. Technol.* **14** 55–69.
[9] Puthipad N, Ouchi M, Rath S, Attachaiyawuth A. 2017 *Constr. Build. Mater.* **144** 1–12.

[10] Rath S, Puthipad N, Attachaiyawuth A, Ouchi M. 2017 *J Adv. Concr. Technol.* **15** 29–37.
[11] Powers T C. 1954 *J. Proc.* **50** 741–760.
[12] Pleau R, Pigeon M. 1996 *Cem. Concr. Aggreg.* **18** 19–29.
[13] Christodoulou G. 2000 *Soc. Chem. Ind.*
[14] Bruere G. 1955 *J. Proc.* **51** 905–19.
[15] Zhang X, Zhang H, Gao H, He Jiang Y M. Constr. 2019 *Build. Mater.* **196** 245-55.
[16] Atahan H N, Carlos C J, Chae S, Monteiro P J M, Bastacky J. 2008 *Cem. Concr. Compos.* **30** 566-75.
[17] Diao S. Study on high temperature and high salt foam system and its properties. M.E.Sc., Southwest Petroleum University, Sichuan, China, 2006.
[18] Carey E, Stubenrauch C, 2010 *J. Colloid. Interf. Sci.* **346** 414-23.
[19] Ding B, Liu J P, Liu J Z. 2006 *Concr.* **11** 34-5.
[20] Carlson J. Advancement on the application of a flat-bed scanner for hardened portland Cement Concrete Air-Void Analysis. M.C.E., Michigan Technological University, Michigan, USA, 2005.
[21] Chen Y Z, Qiu F, Zhao X J. 2009 *Chem. J. Chinese. U.* **30** 1337-41.

Study on the effect of ion invasion on compressive strength, resistivity and polarizability of concrete

Si Tong Lu, Jun Ge Huang*, Yu Zhang

College of Urban Construction and Safety Engineering, Shanghai Institute of Technology, Shanghai, 201418, P.R. China

*Corresponding author's e-mail: hjg@sit.edu.cn

Abstract. The structural safety of concrete has been attached great importance, especially in saline-alkali area. In order to better study the changes of physical parameters of concrete after corrosion in saline-alkali area, the soil quality in saline-alkali area was investigated and sampled to analyze the main ionic components of concrete corrosion environment. According to the corrosion ionic components, the concrete model was made by mixing carbonate and sulfate solutions with different concentrations. Resistivity method and induced polarization method were used to measure the resistivity and polarizability characterization of the model within the curing period (28d), and the compressive strength of the concrete model during the curing period was tested to study the correlation between the resistivity, polarizability and compressive strength of the concrete model. The results show that the resistivity, polarizability and compressive strength of concrete are closely related to the type and concentration of ions. There is a corresponding relationship between the polarizability and the bearing capacity of concrete test blocks mixed with the same solution. The larger the polarizability, the weaker the bearing capacity. There is no strong correlation between resistivity and bearing capacity. Polarizability data can reflect both ion invasion and strength change of concrete.

1. Introduction

Concrete building science is important composition of the material, the stability of the concrete due to the strong plasticity, are widely used in engineering construction, in recent years, more and more concrete because salt corrosion by coagulation micro pore structure and surface to the interior structure, produce certain physical and chemical reaction, the bearing capacity of structures, The bearing capacity is not only an important feature of rock and soil mechanics for structures, but also an important reflection in engineering technology. Due to the distribution of salt and alkali in Xinjiang and some areas in northern China and coastal areas, concrete is easily eroded in saline-alkali environment [1]. The erosion methods mainly include the following aspects: ① Direct chemical erosion [2]: It mainly refers to the chemical action of acidic components in the environment on the concrete protective layer. ② Alkali-base reaction. ③ concrete carbonization. ④ freeze-thaw damage [3]. ⑤ Sulfate erosion [4]. Land salinization in China covers a wide area, including the western, central and coastal areas, and the environment of ions in the ocean and saline-alkali areas are also similar. The results of salt-alkali sampling in Xinjiang show that there are a lot of SO_4^{2-}, CO_3^{2-} and Cl^- ions in the anions, and the cations are mainly Na^+, K^+ and Ca^{2+}.

Intrusion of ions corrodes concrete, causing corrosion of steel bars inside concrete and affecting concrete strength [5]. The main methods for detecting concrete corrosion and steel corrosion generally fall into the following categories: one is damage detection, which aims to observe the corrosion status of concrete by destroying the concrete surface [6]. One is nondestructive testing, mainly using analytical method, electrochemical method and physical analysis method [7].

Damage detection has been gradually replaced by nondestructive detection because of the damage to the concrete members and the damage to the active members. The analysis method is based on the comprehensive analysis of the environment where the concrete object is located, which needs the support of reasonable and reliable model. Electrochemical methods need to consider many factors, accuracy cannot be guaranteed, and data processing time is long [8]. Physical analysis methods, such as resistance probe method and infrared thermal imaging method, mainly reflect the corrosion conditions inside concrete through the changes of physical parameters, such as resistance value, magnetic permeability, resistivity and polarizability [9].

Induced polarization method is mainly used in geological survey [10], which is characterized by the observation of the ratio of the secondary field to the primary field, and can eliminate some interference factors. If used to observe concrete members, the effect of rebar in concrete can be eliminated [11]. The core of this study is whether the polarization can reflect the corrosion of concrete.

The concrete mixed with sodium sulfate and sodium bicarbonate solutions of different concentrations is used to simulate the state of concrete after erosion. In the eroded environment, the resistivity, polarizability and the change of compressive strength of the model test during the curing period are measured. Study the correspondence between parameters.

2. Concrete model

2.1. Matching ratio and raw materials

The proportion is calculated by 394 kg of cement, 185 kg of water, 571 kg of sand and 1250 kg of pebble to a cubic meter. The strength of concrete is C30, the particle size of stones is 5-20mm, the cement is P.O 42.5, and the sand is standard sand. Tap water is used, and the specific ratio is shown in Table 1.

Table 1. List of various materials.

Group	Solute	Solute /Kg	Water /Kg	Solution weight /Kg
1	3%Na_2SO_4	0.0915	2.9585	3.05
2	6%Na_2SO_4	0.183	2.867	3.05
3	Purified water	0	3.05	3.05
4	3%$NaHCO_3$	0.0915	2.9585	3.05
5	6%$NaHCO_3$	0.183	2.867	3.05

2.2. Forming and curing of specimen

Pour cement, stone, sand into the mixing pot and stir it dry for 40s (cover of the mixer), then add 80% solution, stir it three times for 40s each time, and slowly pour the remaining solvent into the mixing pot while stirring. Remove the cover of the blender when the solvent is poured in. Close the lid again after pouring. Stir well. After the mixing is completed, the strength parts are formed and the mixture is built into the 100mm×100mm×100mm triple test die. Place the triplet test mold on the shaking table, turn on the magnetic suction, and vibrate for three times in numerical control, 60 seconds each. After forming, the specimens were put into the curing room for natural conservation, covered with wet towels, and moistened with water spray regularly. After 24H, the mold is removed and put into the room for maintenance to be measured.

3. Determination of concrete physical property parameters

3.1. resistivity and polarizability test

The four-electrode method was used, the electrodes were uniformly arranged, and the pole distance was set at 2cm. The size of the model made is 100mm×100mm×100mm, which cannot be placed on the rock specimen measurement rack. Four electrodes are placed manually. The power supply electrode AB at both ends is the iron clamp electrode, and the measuring electrode Mn is the calomel electrode. None of the four ABMN electrodes is connected to the fixed device circuit. Five groups were tested according to each concentration test block, and the average value was taken. This is shown in Figure 1.

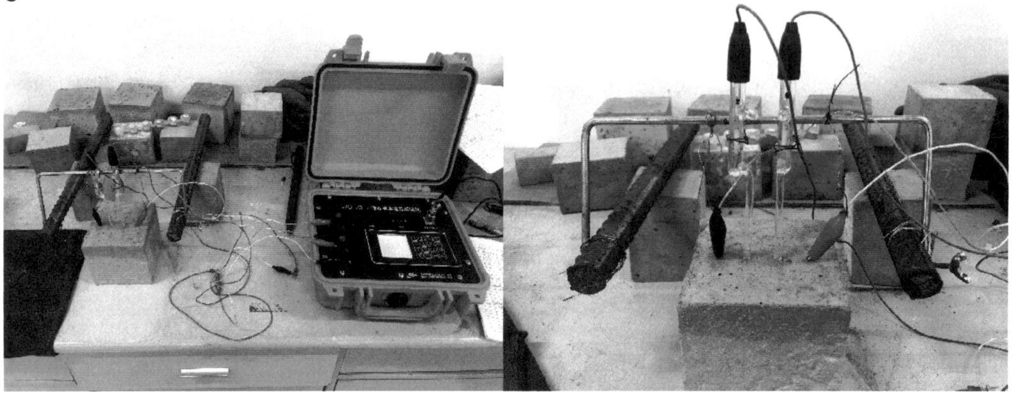

Figure 1. Four electrode measurements of resistivity and polarizability

3.2. Concrete pressure test

Three good test blocks will be set according to each concentration to carry out pressure test, and the pressure value will be averaged for three times. For each group of test blocks with different concentrations, according to the production completion time, the curing time interval from the first group to the third group is 3, 7, 14, 21 and 28 days. The pressure test was carried out in groups 4 to 5 with curing intervals of 3, 6, 13, 20 and 28 days respectively.

4. Correlation analysis of physical parameters

4.1 Concrete test block data collection

The concrete sample block was tested by using the four-electrode method, and the direct contact between the MN calomel electrode and the contact surface was measured. The calomel electrode contact surface is a circular surface with a certain area. In order to ensure good contact, drops of water are used during measurement to enhance the contact effect between concrete and calomel electrode. Try to ensure that the wetted area formed by each drop of water on the contact surface is consistent, but there is still a certain error due to human operation. In addition, the concrete test block is non-uniform body, and the conduction ability of electric signals to the current in the cement part and the sand part is different. Each time the measurement is selected for a fixed point, when placed, there will still be a small difference. Therefore, the measurement will cause the instability of the data, some of the data will be more fluctuation. Therefore, the experimental method of multiple measurements to take a stable value was adopted to reduce the error caused by this.

4.2 Resistivity comparison

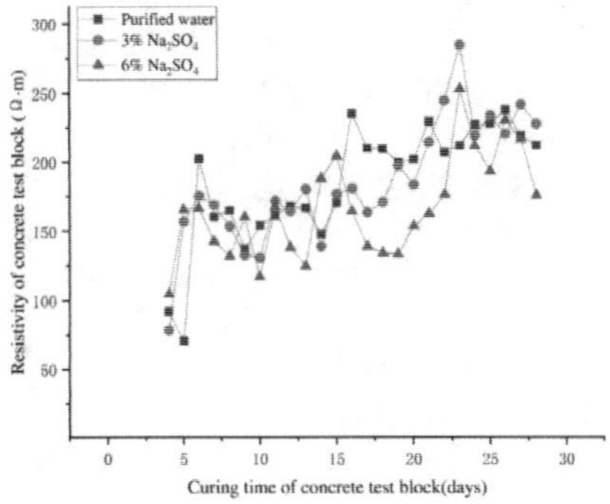

Figure 2. Resistivity contrast diagram of sodium sulfate concrete and plain concrete.

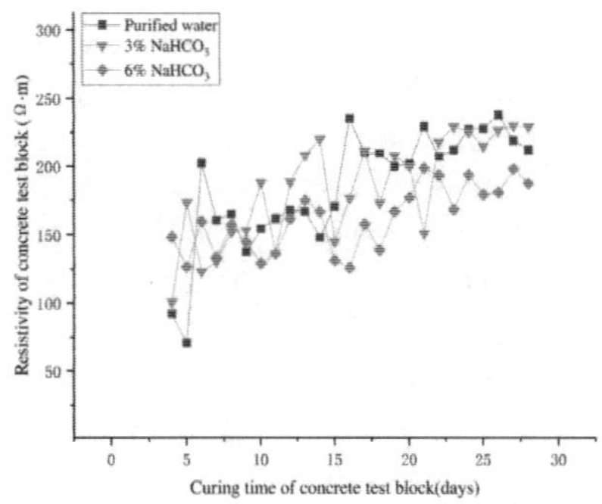

Figure 3. Resistivity contrast diagram of sodium bicarbonate concrete and plain concrete.

As can be seen from Fig. 2 and 3, the resistivity of concrete increases significantly with the increase of curing time. The average resistivity of concrete test blocks mixed with clean water is 185.09 Ω·m, and the maximum value is 235.28Ω·m after curing for 16 days. The average resistivity of concrete test blocks mixed with 6% sodium sulfate is 166.26 Ω·m, and the maximum value is 253.23Ω·m on the 23rd day of curing. The average resistivity of the concrete test block mixed with 6% sodium bicarbonate is 161.4Ω·m, and the maximum value is 198.65 Ω·m at the 21st day. The average resistivity of the concrete test block mixed with 3% sodium sulfate is 184.52Ω·m, and the maximum value is the value of the 23rd day, and the resistivity is 284.77Ω·m. The average resistivity of concrete test blocks mixed with 3% sodium bicarbonate is 184.45Ω·m, and the maximum value is 230.14Ω·m at the 25th day. From the average and overall trend, the resistivity of concrete test blocks mixed with purified water is higher than that of concrete test blocks mixed with 6% sodium sulfate and 6% sodium bicarbonate, the relative difference is about 10%. There is little difference between the resistivity of concrete samples mixed with purified water and that of concrete mixed with 3% sodium sulfate and 3% sodium bicarbonate, the relative difference is only about 0.5%. The resistivity of concrete test blocks is different with different ion concentrations in concrete, and the higher the concentration, the greater the influence on the resistivity is. For the existing observation means, more than 5% of the relative difference can be observed, from this point we know that the concentration of concrete internal ions need to reach a certain limit can be observed.

4.3. Polarizability comparison

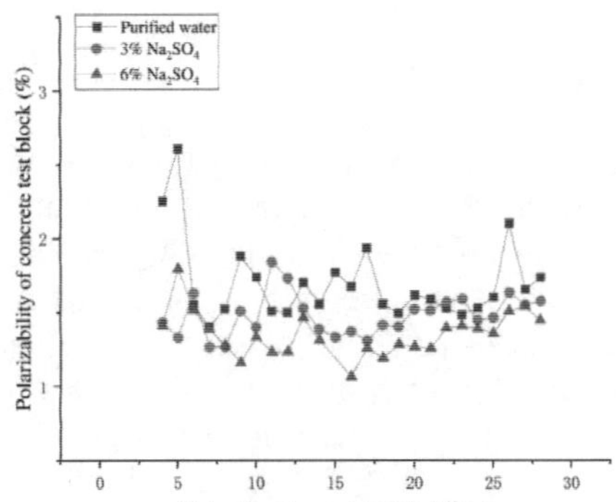

Figure 4. Comparison chart of changes in polarizability of sodium sulfate concrete and plain concrete.

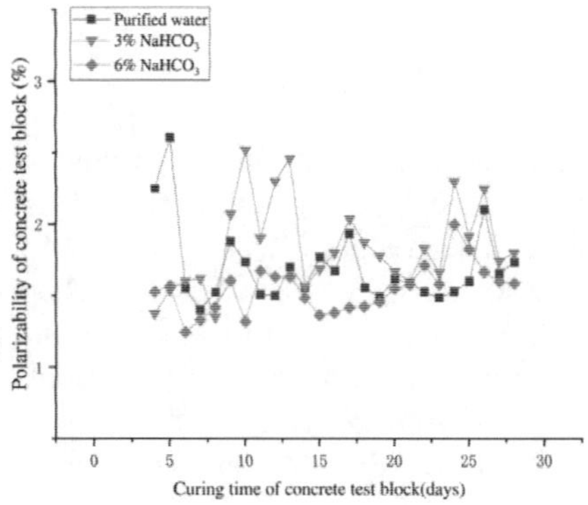

Figure 5. Comparison chart of changes in polarizability of sodium bicarbonate concrete and plain concrete.

It can be seen from Fig. 4 and Fig. 5 that the change of polarizability with curing time is not obvious. Comparing the polarizability of concrete mixed with sodium sulfate solution with that of purified water in the same period, the order of polarizability value from high to low is as follows: purified water, 3% sodium sulfate, 6% sodium sulfate; Compared with purified water test blocks, the polarizability of concrete mixed with sodium bicarbonate solution is in the order of 3% sodium bicarbonate, purified water and 6% sodium bicarbonate. The average polarizability of the concrete test block mixed with water is 1.70%, and the maximum value is the value of the fifth day, and the polarizability is 2.61%. The average polarizability of the concrete test block mixed with 6% sodium sulfate is 1.35%, and the maximum value is the value of the fifth day, and the polarizability is 1.80%. The average polarizability of the concrete test block mixed with 6% sodium bicarbonate is 1.54%, and the maximum value is the value on the 24th day, and the polarizability is 2.00%. The average polarizability of concrete test blocks mixed with 3% sodium sulfate is 1.48%, and the maximum value is 1.84% on the 11th day. The average polarizability of concrete test blocks mixed with 3% sodium bicarbonate is 1.85%, and the maximum value is the value on the 10th day, and the polarizability is 2.52%. From the relative difference of the mean polarizability, the difference between purified water and 3% sodium sulfate and 6% sodium sulfate is 12.9% and 20%, respectively. The relative difference of the mean polarizability between purified water and 3% sodium bicarbonate and 6% sodium bicarbonate was 8.4% and 9.8%, respectively. There are also observable differences in the values of polarizability with different ion types and concentrations.

4.4. Compressive strength comparison

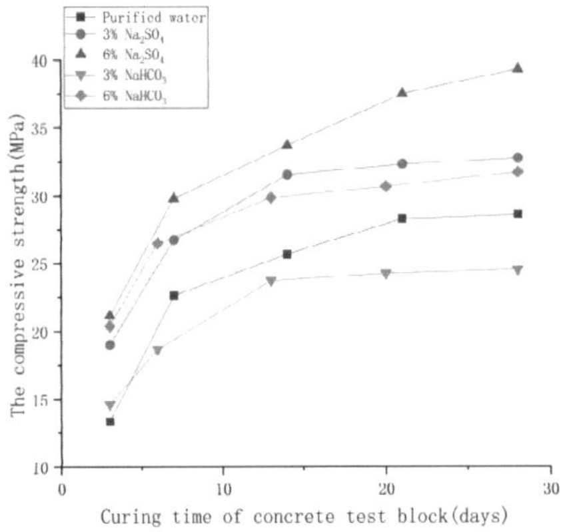

Figure 6. Comparison diagram of compressive strength of concrete test blocks with different ions and concentrations of ions.

As can be seen from Fig. 6, there are significant differences in the compressive strength of concrete test blocks mixed with purified water, 3% and 6% sodium sulfate solution, and 3% and 6% sodium bicarbonate at the age. The strength of concrete test blocks mixed with sodium sulfate solution is higher than that mixed with purified water, and the strength gradually increases with the increase of concentration. This phenomenon is related to the effect of concrete expansion caused by sulfate [12]. The compressive strength of concrete mixed with 6% sodium bicarbonate solution is higher than that of concrete mixed with 3% sodium bicarbonate solution, and the compressive strength of concrete mixed with 3% sodium bicarbonate solution is higher than that of concrete mixed with 3% sodium bicarbonate solution. After curing for 28 days, the compressive strength of concrete samples mixed with 6% sodium sulfate solution is 1.37 times that of concrete samples mixed with 3% sodium sulfate solution, 1.2 times that of concrete samples mixed with 3% sodium sulfate solution, and the compressive strength of concrete samples mixed with 6% sodium bicarbonate solution is 1.1 times that of concrete samples mixed with 6% sodium bicarbonate solution. As a whole, the strength of concrete mixed with sodium sulfate solution is greater.

4.5. Correlation between resistivity and compressive strength of test block

In terms of resistivity, the resistivity increases with the increase of curing time, and there is a certain difference in resistivity due to different ion concentrations and types, but there are also some ion concentrations with little differentiation. For example, compared with the concrete test blocks mixed with purified water solution and 3% sodium sulfate solution, the overall data maintained an increasing trend with the curing time. However, in the whole curing 28d, the resistivity of one solution was not significantly higher than that of the other solution for a long time. Another example is the comparison between the concrete test blocks mixed with purified water solution and those mixed with 6% sodium bicarbonate solution. After curing for 15 days, there is obvious differentiation. The resistivity of the concrete test blocks mixed with purified water solution is higher than that of the concrete test blocks mixed with 6% sodium bicarbonate solution. In several groups of data with large degree of differentiation, the resistivity of the test block mixed with purified water is higher than that of the test block mixed with 6% sodium sulfate solution and 6 sodium bicarbonate solution. The compressive strength of concrete test blocks mixed with three kinds of solutions varies from large to small: 6% sodium sulfate, 6% sodium bicarbonate and purified water. Such data cannot fully show the relationship between concrete resistivity and compressive strength. Different ion types and different resistivity characterization make the corresponding relationship between resistivity and compressive capacity weak.

4.6. Correlation between the test block's polarizability and compressive strength

From the overall data, the polarization rate changes relatively smoothly with the curing time, and the polarization rate has comparable characteristics. The order of the polarization rate from small to large is: 6% sodium sulfate, 3% sodium sulfate, 6% sodium bicarbonate, clean water, 3% sodium bicarbonate. Consistent with the order of compressive strength from large to small. In this experiment, it can be seen that the polarization rate has an anti-correspondence relationship with the degree of compression resistance, and the high polarization rate has a weak compression strength ability.

4.7. Correlation between resistivity and polarizability of test block

According to the comparison of resistivity and polarizability data, the resistivity of the concrete test block mixed with 6% sodium sulfate solution and the resistivity of concrete mixed with 6% sodium bicarbonate solution is lower than that of the concrete test block mixed with clear aqueous solution. , The polarizability is also low; the resistivity of the concrete mixed with 3% sodium sulfate and sodium bicarbonate solution is not much different from that of the concrete test block mixed with water, but the polarizability is high or low. For the comparison between different ions and the same concentration, the resistivity cannot be distinguished from each other, but the polarizability has a more obvious degree of discrimination. There is no obvious correlation between the resistivity and polarizability of the test block.

5. conclusion

In the context of the diversified development of building science, the main body of buildings is still mostly constructed by concrete materials. The importance of concrete safety evaluation is well known. In this paper, some research rules can be found by selecting the resistivity, polarizability and compressive strength data of concrete models mixed with different ions and concentrations of solutions. In the above research results, there are still some deficiencies, such as the selection of invading ion species is less, the concentration gradient accuracy is not enough, it is suggested that the future scientific research can increase the species of invading ion concrete, and the ion concentration is fine, in order to find the correlation in the relevant engineering technology; In addition, the erosion of salt ions on concrete in engineering is the result of a variety of ions mixed together. With the conclusion of the study on the law of single ions, the influence of mixed ions on the compressive strength, resistivity and polarizability of concrete can be expanded and its correlation research can be carried out.

Acknowledgments
Foundation item: National natural science foundation of China (41564005).

References

[1] Wang J B, Niu D T, He H, Song Z P.(2019) Study on durability deterioration law and mechanism of shotcrete in salt lake erosion environment. China Civil Engineering Journal,52 (06):67-80.
[2] Yu H Y.(2016) Analysis of early deformation and freezing damage mechanism of frozen concrete. Harbin Institute of Technology.
[3] Liu Y P, Peng C J, Liu X S.(2009) Chemical mechanism analysis of concrete dam erosion caused by mine acid wastewater. Mining Research and Development, 29 (03):73-74+92.
[4] Ba H J, Liu Z G, Chen W S, Zhao Y D, Yang Y Z, Deng H W.(2005) Effect of sodium sulfate content on early shrinkage cracking of concrete. Journal of the Chinese Ceramic Society , 2005(01):36-41.
[5] Huang R.(2017) Study on Durability of Structural Concrete in Western Environment. Southeast University, in china.
[6] Ma Z C.(2012) Electrochemical Study on Corrosion of Concrete Reinforcement. Changsha University of Science and Technology, in china.
[7] He S Q.(2004) Experimental Study on Durability of Reinforced Concrete Structures in Chloride Ion Environment. Dalian University of Technology, in china.
[8] Lu M, Wang P X, Lu J Y. (2000) Research Status of Reinforcement Rust Corrosion in Concrete. Concrete,2000(02):37-41+27.
[9] Zhang Y Z. (2003) Durability Testing, Evaluation and Optimal Design Method of Concrete Structures. Zhejiang University,2003.
[10] Sun E F, Guo F W. (2019) Application of induced polarization method in urban geological survey. Energy and Environmental Protection, 2019,41 (07):124-128.
[11] Tong M S, Li L, Wang W N, Fan Q H, Jiang Y Z, Zhang J J, Ding Z , Cui J. (2005) Relationship between induced polarization relaxation time spectrum and pore structure and permeability of rocks. Chinese Journal of Geophysics,2005(03):710-716.
[12] Qi Y M, Hung J G, Lu S T, Zhang Y, Liu Y. (2020) Difference of electrical parameters of concrete in saline-alkali environment. Journal of Applied Technology,2020,20(03):270-275.

Molecular Dynamics Simulation of the Mechanical Behavior of Mixed-Layer Clay upon Hydration

Yu Yang[1*] and Guoyuan Xu[1]

[1] School of Civil Engineering and Transportation, South China University of Technology, Guangzhou, Guangdong, 510641, China

[*]Corresponding author's e-mail: 201810101692@mail.scut.edu.cn

Abstract. A proper understanding of the the mechanical behavior of clay minerals is of crucial importance in earth sciences, soil mechanics, and geotechnical engineering. Although more than seventy percent of all clays are of mixed layer types, the vast majority of the previous studies are devoted to pure clays, which make the hydraulic and mechanical properties of the widespread mixed-layer clay (MLC) and its role in soils are little understood, especially the most common illite-montmorillonite (I-M) mixed-layer clay (MLC). This research reports on a molecular dynamics (MD) study of the differences in mechanical behavior between I-M MLC containing mixed K^+ and Na^+ and Na-montmorillonite (Na-MMT). The variations of key components in stiffness matrix with hydration have been analyzed. It is found that the stiffness tensor of two types of clay could be split into two groups: the in-plane coefficients (C11, C12, C22, and C66) and the out-of-plane coefficients (C13, C23, C33, C44, and C55). Due to the different forms of the support force of the clay layered structure in the two directions, they exhibit significantly different hydration mechanisms. Moreover, the I-M MLC is more difficult to hydrate and swell due to the asymmetric I-M interlayer and the existence of K^+, and its ability to resist deformation is stronger than Na-MMT. This work offers insight into the molecular mechanism for initial swelling and mechanical behavior in widespread MLC. This will help to decipher its specific role in soils and controlling clay swelling.

1. Introduction

A proper understanding of the the mechanical behavior of clay minerals is of crucial importance in earth sciences, soil mechanics, and geotechnical and geoenvironmental engineering, etc[1]. It is required not only for rock physics modeling, but also for geomechanics applications such as excavation stability analysis, hydraulic fracture design, and geophysical monitoring, etc. These layered clay minerals are rich in nature and are the main components of clay stone and many soils. Even a tiny amount in the soil could cause significant swelling. The main reason is that these expansive clays have the ability to adsorb water and expand upon hydration. This process significantly impacts the structural and mechanical properties of the clay. Many studies have been attempted to understand the swelling process and identify the evolution of hydraulic and mechanical properties.

However, most of the past studies, whether experiment or computation, have focused on pure clays of one type or another. In reality, there is a wide variety of mixed-layer (ML) clays that contain an intermixed stacking sequence of two or more types of different layers within a single crystal[2]. For instance, a study indicated that more than 70% of sedimentary rocks in America include some form of mixed-layer clays (MLCs). Hence, they are more likely to be found than pure clays in geological formations, and the most interesting and abundant mixed layer type is illite-montmorillonite (I-M) clay

[3]. They may exist in soils more widely, but the relevant studies are very limited. Actually, MLCs have different layer types, layer charge positions, and cation types, so the mechanical properties of the clay and their interactions with water are also distinct.

Traditional researches on clay mechanical behavior mainly include theoretical and experimental methods. Although they can obtain some macroscopic mechanical properties, the available data are also very limited. It is very challenging to measure the essential mechanical properties of clays because of their small grain size, high affinity for water, and layer geometry. In addition, although these properties are very sensitive to water content at the scale of the clay layer, in these measurements, the water content is not controlled. Whereas regarding the assemblage of anisotropic clay minerals and the mixed layer clays, we know very little. This is mainly because it is difficult to prepare the samples of these types of clays, which have the complications such as stacking disorder, poor crystallinity, and mixed charges. However, MD method offers the possibility of solving these problems.

MD simulation based upon classical statistical mechanics is the computational technique used to understand the molecular-scale behavior. Its application in some micro-scale geotechnical problems has become a hotspot[4]. As the main component of soils, the clay minerals have been widely studied, including the micro-mechanisms of surface wettability, matter adsorption, clay swelling and shrinking , and, most importantly, mechanical properties. However, in spite of the numerous studies on clays, few of them were focused on the hydration mechanical properties, especially the widespread MLCs.

In this work, the molecular modeling of I-M MLCs during swelling processes has been carried out to understand the hydration mechanical behavior and the same phenomenon in Na-MMT was reported to grasp their differences. The incremental insertion of water molecules into the clay interlayer has been conducted by writing the script, which simulates the initial phase of water sorption to the greatest extent. Sufficient MD simulations have been run to obtain the hydrated clay model after swelling. Next, a script based on static method has been used to compute the key coefficients of stiffness tensor for each swollen model. The evolutions of key elastic constants have been analyzed, and their different compression and shear mechanisms have been identified. And the mechanical behavior of the two types of clays has been compared and explained from the MD perspective.

2. Theoretical Calculations

One major factor for the abundance of MLCs is that, under a solid-state conversion, pure clay is transformed into MLC under the influence of earth temperature and pressure. Fig. 1 shows the most typical MLC model with 50% Na^+ and 50% K^+ in the form of I-M-M-I, where the bottom and top layers are illite, and MMT builds the two central ones. The details have been shown in the figure.

Among them, the structure units of illite and MMT are both obtained from the latest American Mineralogist Crystal Structure Database, of which the type of illite is RM30, and the MMT type is Arizona-MMT. They are organized as 4 × 2 unit cells and the three dimensions of the entire MLC model are 20.85 Å × 18.01 Å × 39 Å. All the initial structures are established on Materials Studio 8.0 software, for the subsequent adsorption of H2O molecules, which are flexible simple point charge (SPC) water models. The recently developed CLAYFF[5] force field is employed to describe the interatomic interactions, and charges are assigned to each atom.

In the simulation process, the constructed clay structures were first geometrically optimized to obtain a low-energy configuration. Next, the script was executed to realize the incremental insertion of H_2O molecules into the clay interlayer, followed by an equilibrium step before the subsequent insertion. In the equilibrium process, the MD simulation in the NPT ensemble (constant number of particles, pressure, and temperature) was conducted to perform a full swelling using the LAMMPS package. Because the swelling entered a stable state after a certain time, and the MD simulations of 1 ns and 5 ns were proved to yield very similar results; hence, the time of MD equilibrium and MD production were both 1 ns, and the time step was 1fs (1 ns = 10^3 ps = 10^6 fs). After the swelling simulation, a script based on static method has been used to compute the key coefficients of stiffness tensor for each swollen clay model, which is achieved by minimizing the energy of the system and allowing the atomic coordinates and cell parameters to relax.

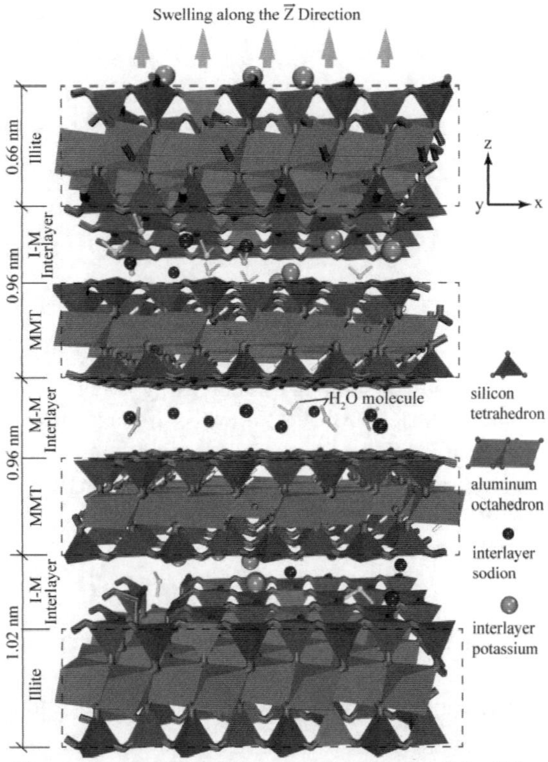

Figure 1. Snapshot of the molecular model of the hydrated mixed layer clay with 50% Na^+ and 50% K^+.

3. Results and discussions

3.1. In-plane elastic constant

The evolution of hydration mechanics of different types of clay can be further obtained by analyzing the key components in the stiffness matrix. In this work, we reported all relevant stiffness tensor values for the two types of clays we studied. The results indicated that these stiffness tensor coefficients could be split into two groups: the in-plane coefficients (C11, C12, C22, and C66) and the out-of-plane coefficients (C13, C23, C33, C44, and C55). Due to the different forms of the support force of the structure in the two directions, they exhibited significantly different hydration mechanisms.

Figure 2 shows the in-plane elastic constants of two types of clay as a function of water content, obtained through the MD calculations, including C11, C12, C22, and C66. It can be found that these in-plane elastic constants have the same trend as a whole: They obviously decrease more or less stepwise with increasing hydration. Among them, the in-plane elastic constant C22 is the largest under the same water content, followed by C11, C12, and finally C66. Compared with other in-plane elastic constants, C11 and C22 have a larger range of variation, which is closely related to the Young's modulus in the X and Y directions. What is interesting is that they both have three approximately flattened segments when the water content is about 40, 120, and 240, which correspond to the one-layer, two-layer, and three-layer hydration state (the 1W, 2W, and 3W states).

However, it can be found that the in-plane elastic constants of MLC are higher than that of MMT under the same water content. And when about two hydration layers were formed, the in-plane elastic constants of MMT began to decrease significantly at the water content of 120, while MLC was postponed to a higher water content of about 150. In addition, the mechanical properties of MLC at this time also declined more slowly than that of MMT. This was because MLC had stronger

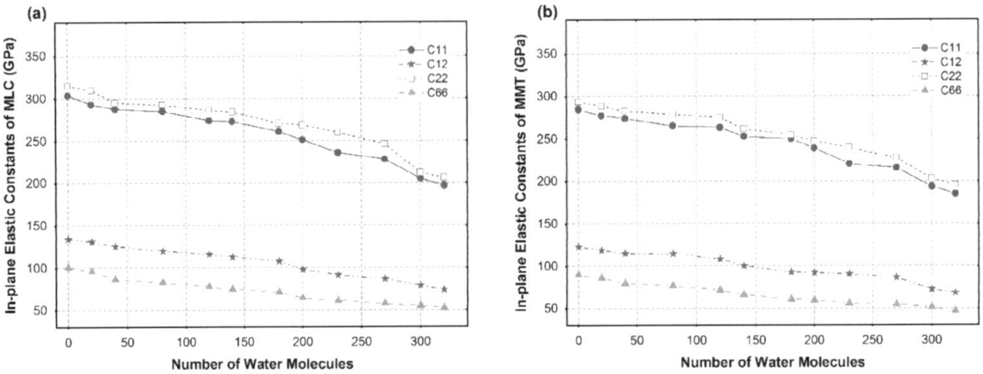

Figure 2. Evolution of in-plane elastic constants with increasing water content for two types of clay; (a) I-M MLC with 50% Na^+ and 50% K^+; (b) MMT-MMT with Na^+ only. The results for C11 are represented by blue, C12 by gray, C22 by red, and C66 by yellow; negative values of interaction energies indicate attractive interactions.

intermolecular interactions, so it was more difficult to swell into the next hydration stage. The trend of in-plane elastic constants for pure MMT was consistent with the study by Ebrahimi et al. [6], which further increased the credibility of these simulated data.

By observing the structure of the hydrated clay and its interlayer, it can be concluded that the difference of the C11, C22, and C12 elastic constants derived from the ion-covalent bonding strength within clay layers, the isomorphic substitution positions, and the change in geometric conformation for clay-water system. It is obvious that the geometric effects of swelling of the interlayer space caused the reduction of the elastic constants. Moreover, these in-plane coefficients did not exhibit strong hydration characteristics and fluctuating changes like the out-of-plane coefficients described later. This was due to the forces needed to apply a given strain in a direction parallel to the clay sheet were mostly supported by the clay crystalline structure and not by the interlayer water. The shear stiffness tensor C66 was mainly controlled by the relative torsional of the bonding atoms. The primary reason for its decrease was that the contribution of interlayer water in the weak shear part increased.

It can also be found that during the hydration process, water molecules were preferentially adsorbed on the clay surface. But with the increase of water content, more outer-sphere surface complexes (OSSCs) were formed around the interlayer Na^+ to promote swelling. Meanwhile, the geometric effects of the interlayer spacing change were also more obvious, and the elastic constants were greatly reduced. Compared with MMT, MLC containing asymmetric I-M interlayer and K+ formed fewer OSSCs. Therefore, its interlayer space was more difficult to swell and its ability to resist deformation was stronger. In addition, the crystalline structure of illite was inherently stiffer than MMT. These factors caused the elastic stiffness of MLC to always be greater than Na-MMT.

3.2. Out-of-plane elastic constant

Figure 3 shows the variations in out-of-plane elastic constants of two types of clay during hydration, including C33, C13, C23, C44, and C55. It can be found the varying trend of these five stiffness parameters fluctuate stepwise determined by the hydrated state. They rapidly decrease in the initial hydration stage and reach a local minimum at the water content of 40. Then they gradually increase and reach a local maximum when about two hydration layers are formed (at the water content of 120). As hydration continues, these out of-plane stiffnesses begin to decline rapidly again, but their descending rate slows down when the water content is 240. On the whole, they stabilize after reaching the 3W state and are significantly lower than that of dry state. Among them, the out-of-plane elastic constant C33 is the largest under the same water content and C44 is the smallest. The variation curves of C13, C23, and C55 are relatively close, and their values are between C33 and C44. They also step into a relatively flat evolution period when different numbers of hydration layers are formed. It can be

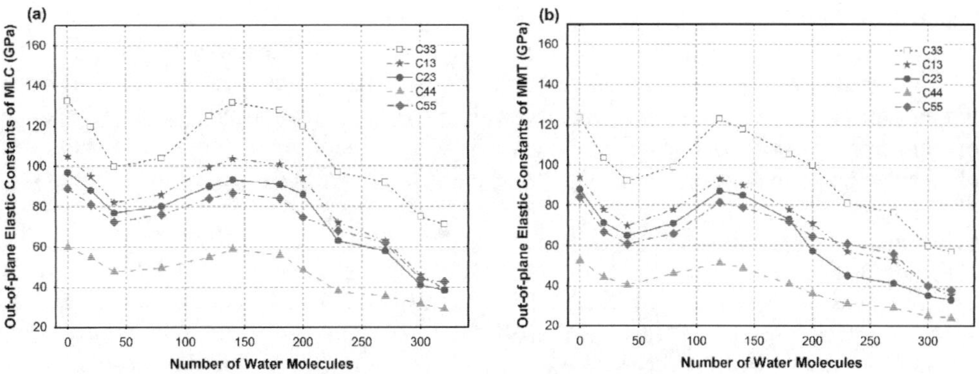

Figure 3. Evolution of out-of-plane elastic constants with increasing water content for two types of clay; (a) I-M MLC with 50% Na^+ and 50% K^+; (b) MMT with Na^+ only. The results for C33 are represented by red, C13 by gray, C23 by blue, C44 by yellow, and C55 by green.

found that these elastic constants are sensitive to water content because water impacts strongly the interaction between the clay layers (i.e., from partially saturated states to fully saturated clay).

By observing the structure of hydrated clay and its interlayer, it can be found that at the beginning of hydration, the interlayer was propped up by water molecules, resulting in the decrease in out-of-plane elastic constants. In the subsequent filling stage, the interlayer spacing was not significantly changed and the water molecules were adsorbed on the clay surface to form the strongly bound-water ordered structure with strong attraction and hydrogen-bond interactions. Therefore, the mechanical stiffness of clay was enhanced at this stage. But when the two hydration layers were formed, the clay began to be excessively stretched and the bound-water film became thicker. The attraction and hydrogen bonding interactions of some water molecules on clay surface were weakened, and its existence as weakly bound-water caused a rapid decline in mechanical properties. Finally, as hydration approached osmotic swelling, hydrogen bonds were continuously formed only between water molecules, and the interlayer atomic pores and these elastic constants were also close to stability.

Among these out-of-plane coefficients of the stiffness tensor, C33 was the most concerned, that was, the ability to resist deformation in the direction perpendicular to the mineral surface. It can be found that the normal stiffness C33 oscillated with interlayer hydration linked to the formation of different numbers of hydration layers. In addition, the in-plane elastic constants C11 and C22 were approximately 3 times that of C33, which indicated that the tensile and compressive strength in the direction perpendicular to the mineral plane was significantly lower than the strength within the clay layers. This anisotropy is due to the difference between iono-covalent bonding inside the crystal structure and screened electrostatic bondings between clay layers.

As for the clay shear, the shear resistance of water in the clay confinement layer was the most concerned, that was, the shear stiffness components C44 and C55 along the mineral plane. They oscillated with hydration like the stiffness tensor C33. This showed that the shear strength of the interlayer water was closely related to the formation of hydrogen bonds and the size of atomic pores in clay interlayer. It can be found that the interlayer ions and water molecules were free to move in the interlayer and the clay layers could move with respect to each other. The elastic constants were very sensitive to water content because there is no covalent bonding in this direction. However, these water molecules did not exhibit the shear-free strength characteristics of free water but increased with hydration at a certain stage. For example, when the interlayer reached a stable 2W hydration state, the shear strength was increased to a local maximum. In this region, the clay surface created strong Coulombic bonding with the adsorbed water, which may create an ordered structure with nonzero shear modulus. This is consistent with the study by Sayers et al. On the whole, the shear resistance between dry clays was much greater than that of the 3W hydration state. Thus, the addition of water had a lubricating effect on shear friction between layers, resulting in a sharp decrease of C44and C55.

It can be concluded that the different hydration mechanisms between the in-plane and the out-of-plane elastic constants are due to the clay layered structure. The mechanical behaviors in the plane of the layer and in the orthogonal direction are governed by different interactions. And mechanisms in the direction of compression and shear are controlled by the formation of hydrogen bonds and the size of atomic pores in clay interlayer. These elastic coefficients of the stiffness tensor can be directly related to the engineering elastic modulus, and related research will be reported in the future.

4. Conclusions

The MD study of clay with increasing hydration was carried out to compare I-M MLC's mechanical behavior containing K^+ and Na^+ with Na-MMT. The variations of key components in the stiffness matrix of two types of clays have been analyzed, and their different compression and shear mechanisms have been identified at the molecular level. It is found that the stiffness tensor of two types of clay could be split into two groups: the in-plane coefficients (C11, C12, C22, and C66) and the out-of-plane coefficients (C13, C23, C33, C44, and C55). Because of the different forms of support force of the clay layered structure in the two directions, they exhibit different hydration mechanisms.

Among them, the in-plane elastic constants are not very sensitive to the hydration state. Their decrease with hydration can be explained by the geometric effect of interlayer swelling. In contrast, the out-of-plane elastic constants are very sensitive to the water content and oscillate significantly with interlayer hydration linked to the formation of different numbers of hydration layers. Due to the asymmetric I-M interlayer and the existence of K^+, the mechanical behavior of MLC is different from that of pure MMT. It is found that the I-M MLC is more difficult to hydrate and swell, and its ability to resist deformation and elastic constants are significantly stronger than Na-MMT. In addition, the most concerned coefficients of the stiffness tensor C33, C44, and C55 in geotechnical engineering have been discussed. The mechanical behavior and micro-mechanism of the two types of clays have been compared and explained from the MD perspective.

This study provides the perception of the molecular mechanism for initial swelling and mechanical behavior in the widespread MLCs. From this study, it will help in disclosing underlying changes in stiffness tensor for MLC and identifying its specific role in practical engineering.

Acknowledgments

This research is financially supported by the National Natural Science Foundation of China under Grant 51508200 and 51078151.

References

[1] Katti, D. R., Schmidt, S. R., Ghosh, P., & Katti, K. S. (2007). Molecular modeling of the mechanical behavior and interactions in dry and slightly hydrated sodium montmorillonite interlayer. Canadian Geotechnical Journal, 44(4), 425-435.
[2] Weaver, C.E. The Distribution and Identification of Mixed-layer Clays in Sedimentary Rocks1. Clays Clay Miner. 1955, 4, 385–386.
[3] Altaner, S.P.; Ylagan, R.F. Comparison of structural models of mixed-layer illite/smectite and reaction mechanisms of smectite illitization. Clays Clay Miner. 1997, 45, 517–533.
[4] Yang, Y., Adhikari, S., & Xu, G. (2020). Molecular simulation to discover rheological properties and soil-binding ability of PHPA polymer on montmorillonite surface. Surface and Interface Analysis, 52(10), 665-675.
[5] Cygan, R. T., Liang, J. J., & Kalinichev, A. G. (2004). Molecular models of hydroxide, oxyhydroxide, and clay phases and the development of a general force field. The Journal of Physical Chemistry B, 108(4), 1255-1266.
[6] Ebrahimi, D., Pellenq, R. J. M., & Whittle, A. J. (2012). Nanoscale elastic properties of montmorillonite upon water adsorption. Langmuir, 28(49), 16855-16863.

Effects of AlN and rare earth fluorides on the thermal conductivity of SiC ceramics with impedance spectroscopy analysis

Jingren Li[1,2], Wenzhong Lu[1,2,*]

[1]School of Optical and Electronic Information, Huazhong University of Science and Technology, Wuhan, China

[2]Key Lab of Functional Materials for Electronic Information (B), MoE, Huazhong University of Science and Technology, Wuhan, China

*Corresponding author. email: lwz@hust.edu.cn Tel./fax: (+86)27 87543134

Abstract: Dense silicon carbide (SiC) ceramics were fabricated through hot-pressing with an innovative additive combination of AlN and rare earth fluorides (ReF_3, Re= La, Nd and Gd). Effects of these additives on the microstructure, phase composition and thermal conductivity of SiC ceramics were evaluated. All samples showed compact microstructures, high densities and clean grain boundaries without visible secondary phases. Sample sintered with 4 wt%AlN and 1 wt%NdF_3 (4A1Nd sample) exhibited the highest thermal conductivity (152.3W/m·K) amongst all. Impedance spectroscopy analysis was employed to associate with the thermal conductivity variations in different samples by analyzing defects and impurities. 4A1Nd sample exhibited the highest fitting grain and grain boundary resistances among all samples, indicated its lowest concentration of V_{Si}'''' vacancies and impurities, which suggested the best impurity purifying ability of AlN and NdF_3 combination. On the contrary, the sample sintered with 4 wt%AlN (4A sample) exhibited a higher concentration of defects than other ReF_3 added samples. Therefore, the AlN-ReF_3 additive combinations are promising in both tailoring and improving the thermal conductivity of SiC ceramics.

1. Introduction

Silicon carbide (SiC) ceramic is a potential substrate and package material that possesses variety of advantages such as high thermal conductivity, promising mechanical properties and outstanding heat resistance [1-3]. However, pure SiC ceramic is difficult to sinter due to its strong covalent bonding wurtzite structure. In pursuit of full densification, the sintering temperature of SiC ceramic is generally over 2200°C [4,5]. The existence of liquid phase during sintering can promote the densification and sinterability, lower the sintering temperature and thus lead to finer microstructure and improved mechanical properties of the SiC ceramics [6]. Therefore, the fabrication of SiC ceramics through liquid phase sintering to achieve a higher densification at lower temperature has becoming a focus in recent decades. Various sinter additives have been utilized to obtain liquid phases during sintering. T.Y. Cho et al. [7] fabricated SiC ceramic with AlN-Y_2O_3-Sc_2O_3 additive combination through pressureless sintering and yielded decent thermal conductivity and bending strength of 110 W/(m·K) and 520 MPa respectively. A.C. Santos and S. Ribeiro [8] observed the mechanical properties of SiC ceramics sintered with AlN and Yb_2O_3, obtained high fracture toughness and bending strength of 3.39 MPa $m^{1/2}$ and 638 MPa respectively.

Unfortunately, ReF_3 additives were seldom employed despite their promising properties of improving the ceramic sinterability. Compared to conventional metal oxide additives, ReF_3 can form liquid phases with SiO_2 at further lower temperature, AlN can form solid solution ($2H_{ss}$ phase) with SiC and promote the sinterability of ceramics [9]. Moreover, the introduction of AlN and ReF_3 can also prevent the extra introduction of oxygen impurity. In addition, in-depth studies referring to the effects of impurity and defect on thermal conductivity of SiC ceramics are still lacking.

In this report, effects of different AlN-ReF_3 (Re= La, Nd, Gd) additive combinations on densification, microstructure development and thermal conductivity of SiC ceramics were investigated. The impedance spectroscopy analysis was utilized to further associate with the variations in thermal conductivity and defects in different samples.

2. Materials and Methods

Commercially available sub-micron α-SiC (H.C. Starck, Germany) with an average particle size of 0.5μm was used as the raw material. ReF_3 (Re= La, Nd, Gd) powders (99.99% purity) and AlN powder (98% purity, Toyo Aluminium, Japan) were added to the SiC raw material as sintering additives.

Mixing ratio of the powder mixtures are 95 wt%SiC-4 wt%AlN-1 wt%ReF_3 (4A1Re samples) and 96 wt%SiC-4 wt%AlN (4A sample) respectively. Firstly, the powder mixtures were milled using a planetary ball mill in ethanol for 24 hours at a speed of 350 rpm. Secondly, the resulting mixtures were dried at 80°C for 12 hours. Thirdly, the mixtures were crushed and sieved (100 mesh) and finally hot-pressed at 1950°C with flowing argon atmosphere. Ambient was kept still at a constant pressure of 40MPa for 2 hours and then cooled to room temperature.

Density of specimens were measured using the Archimedes method with an immersion medium of deionized water and calculated from the theoretical density of raw materials by the rules of mixture (neglecting interaction of mixtures), while the relative density is calculated by comparing the bulk density to theoretical density.

The thermal conductivities of all samples were acquired by measuring their thermal diffusivity through laser flash method at room temperature (Nanoflash Model LFA467, Netzsch, Germany). Specimens used to measure thermal diffusivity were prepared into disk shape (φ12.7×2mm). An average diffusivity value of three samples was taken as the final thermal diffusivity.

Microstructural development and morphology of SiC ceramics were examined by performing scanning electron microscopy (SEM; EM-30 Plus, COXEM CO., LTD, Korea) on their fractured surfaces. The linear intercept method was employed to calculate the average grain sizes of all samples from SEM images. Phase composition of specimens was characterized by X-ray diffraction (XRD; MAXima_X XRD-7000, SHIMADZU, Japan) with Cu-Kα radiation at a step width of 0.02° from 20-80°.

Polished and silver coated SiC ceramic disks (φ12.7×2mm) were prepared for the impedance spectroscopy analysis. Impedance spectroscopy was measured at 550°C using an impedance analyzer (6500B, Wayne Kerr, UK) and a temperature change measurement system (VDMS-2000, Partulab, China).

3. Results and Discussions

Fig.1 shows the XRD patterns of all samples after hot press sintering. The XRD analysis suggest that SiC is the main phase in all samples, among which 4H-SiC, 6H-SiC and $2H_{ss}$ phases are the primary constitutions, none secondary phases were detected from any resultant ceramics.

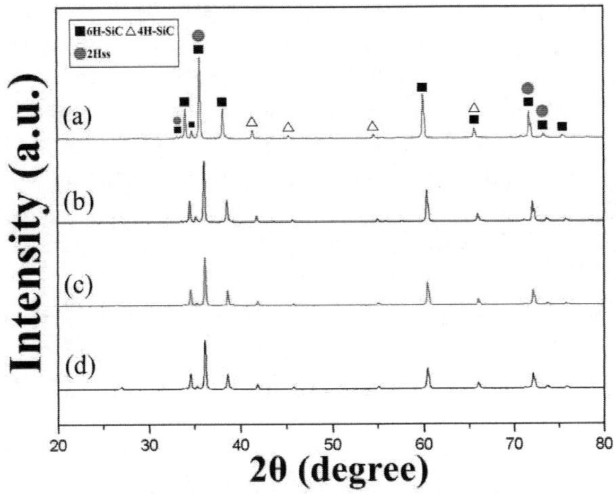

Fig.1 XRD patterns of SiC ceramics hot-pressed with AlN-ReF_3 additive combinations, (a)=4A1La, (b)=4A1Nd, (c)=4A1Gd, (d)=4A

It looks to us that a sintering temperature of 1950°C is relatively high, these secondary phases (from the reactions between ReF_3 and SiO_2 in raw material) tend to volatilize during 2 hours sintering, especially when applied to hot-press method with external pressure.

Fig.2 Typical fractured surfaces of all samples. (a)=4A1La, (b)=4A1Nd, (c)=4A1Gd, (d)=4A

Fig.2 (a)-(d) shows the SEM micrographs of typical fractured surface of SiC ceramics sintered with different additives. All samples exhibit contiguous microstructure, equiaxed grains and clean grain boundaries with no visible secondary phases. Differences in their microstructures are apparent, 4A1La sample shows the largest average grain size amongst all, while 4A sample exhibits the most refined microstructure. The relative density and average grain size of all samples are listed in Table 1.

The relative densities of all samples were over 97%, 4A samples exhibited a lower relative density than 4AlRe samples, which suggested that the ReF_3 additives can evidently improve the densification and sinterability of SiC ceramics.

Such microstructure differences were mainly due to the radius difference among various rare earth cations: La^{3+}(106.1pm) possesses larger cation radius than Nd^{3+}(99.5pm) and Gd^{3+}(93.8pm). The cationic field strength (CFS) of liquid phase is given by:

$$CFS = Z / r^2 \tag{1}$$

Where Z is the valence of corresponding rare earth cation in liquid phase, r is the cation radius.

Accordingly, as the viscosity of resultant liquid phase increases with increasing CFS, it also increases with decreasing cation radius [10]. Thus the viscosity of liquid phases in 4A1La sample was the lowest amongst all samples, which made the sinterability and grain growth of 4A1La sample obviously outstripped others. 4A sample produced very limited liquid phases during sintering, hence yielded the finest grains.

Table 1 Average grain sizes and relative densities of all samples

Samples	Grain size (μm)	Relative density (%)
4A1La	3.411 (±0.092)	98.3
4A1Nd	2.739 (±0.058)	98.7
4A1Gd	2.168 (±0.060)	99.2
4A	1.991 (±0.079)	97.5

Fig.3 shows the thermal conductivity of all samples, 4A1Nd and 4A obtained the highest (152.3 W/m·K) and lowest (122.4 W/m·K) thermal conductivity respectively. It is well known that the mean free path of phonon inside SiC grains is highly responsible for the thermal conductivity value. Larger average grain size leads to further reduced phonon scattering and higher thermal conductivity due to fewer scattering grain boundaries per unit volume [11].

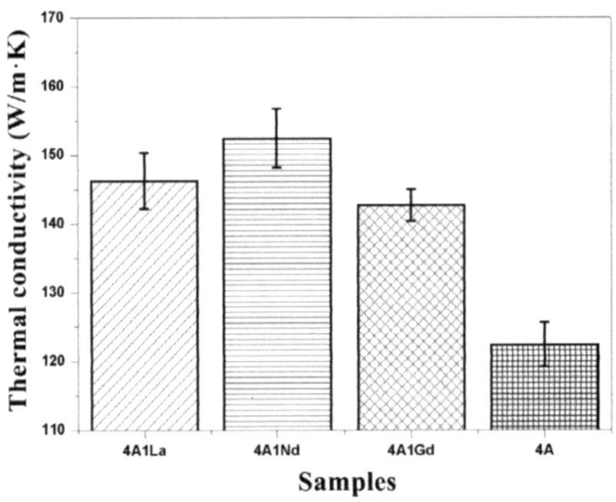

Fig.3 Thermal conductivities of all samples

As listed in Table 1, the average grain size of 4A1La sample is 24% larger than that of 4A1Nd sample, 56% and 71% larger than those of 4A1Gd and 4A sample respectively. Such an advantage of grain size contributed a lot to a high thermal conductivity of 4A1La sample.

However, the 4A1Nd possessed an even higher thermal conductivity despite its distinctively smaller grains than 4A1La sample, which indicated that the average grain size is hardly the sole decisive factor of thermal conductivity in SiC ceramics.

Phonon scattering by the defects inside SiC grains is another dominating factor that greatly affects the thermal conductivity of SiC ceramics. The scattering cross section Γ, which representing the phonon scattering by lattice defects is given by the equation below [12]:

$$\Gamma = X_s(1-X_s)[(\Delta M / M)^2 + \varepsilon(\Delta\delta / \delta)^2] \quad (2)$$

Where X_s is the solute concentration, ε is a dimensionless parameter, ΔM/M and Δδ/δ represent mass and strain misfit (mainly between Si and V_{Si}'''' in this study) respectively. Higher Γ leads to lower thermal conductivity. Since the mass of vacancies are zero, the V_{Si}'''' vacancy possesses a larger cross section value, hence the decrease of V_{Si}'''' concentration will lead to the thermal conductivity improvement in SiC ceramics.

Oxygen is the major impurity in SiC ceramics, which can produce V_{Si}'''' by infiltrating into the SiC lattice, leading to the decline of thermal conductivity by enhancing the phonon scattering effect both from vacancies and oxygen [13]. The formation of liquid phases during sintering can be beneficial to the removal of oxygen impurity from SiC lattice [14]. In order to further investigate the effects of such oxygen extraction mechanism on the thermal conductivity of SiC ceramics sintered with different AlN-ReF$_3$ additives, impedance spectroscopy analyses were performed at a frequency range from 100Hz to 10Mhz at 550°C for all samples, the Cole-Cole circle plot was established to analyze the data.

The summary of all analyses (from 4A1La, 4A1Nd, 4A1Gd and 4A samples) are shown in Fig. 4. Different RC elements were assigned to specific regions of certain sample, a simplified model with two parallel R/CPE and one CPE was adopted to fit the measure data in Fig.4, R/CPE and CPE represent different responses like SiC grains, grain boundaries or electrodes [15], the fitting parameters are shown in Table 2.

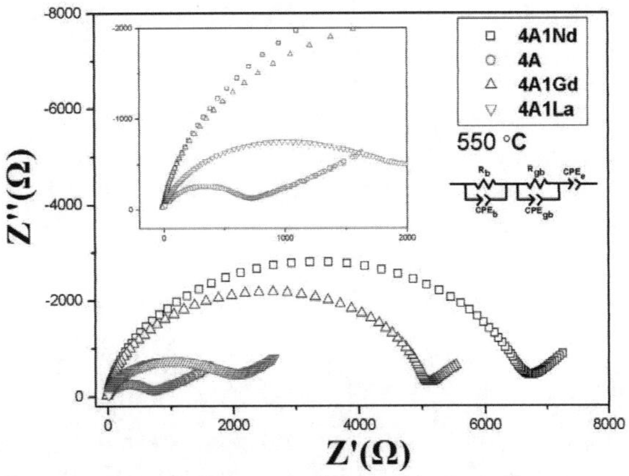

Fig.4 Impedance spectra (Z″-Z′) of all samples with different additives

The impedance of CPE (Z_{CPE}) is introduced to accurately response the behaviors of grain and grain boundary section by fixing the non-ideal behavior of capacitance, which is given by:

$$Z = 1/[(j\omega)^n Y_0] \tag{3}$$

Where n is an empirical exponent, for an ideal capacitor n=1, for an ideal resistor n=0.

From the fitting parameters in Table 2, the grain resistances (R_g) and grain boundary resistances (R_{gb}) of all samples are diversified: the R_{gb} and R_g of 4A sample is comparably low, indicating its high concentration of vacancies and impurities both at grain boundaries and inside grains. Analogically, since the R_{gb} and R_g of 4A1Nd sample are both higher than that of other samples, the vacancies and impurity concentration in 4A1Nd sample are the lowest amongst all, which suggested the best oxygen cleansing ability. In other words, the thermal conductivity improvement from AlN and NdF$_3$ additives is optimal. In addition, the fitting resistances of 4A1Nd and 4A1Gd sample are higher than those of 4A and 4A1La sample, in our explanation, a finer microstructure of ceramic can be more favourable to the extraction process of oxygen impurity due to shorter route from grain to the grain boundary.

Table 2 Fitting parameters of different SiC ceramics measured at 550°C

samples	grain boundary			grain			electrodes	
	R_{gb}/Ω	Y_0/pF	n	R_g/Ω	Y_0/pF	n	Y_0/nF	n
4A1La	1787	18510	0.79	556	1324	0.97	3514	0.76
4A1Nd	6115	3054	0.82	1762	187	0.99	2375	0.64
4A1Gd	4111	3542	0.76	1184	222	0.96	2748	0.63
4A	440	23430	0.83	152	1471	0.92	6823	0.72

Generally speaking, all the AlN-ReF$_3$ (Re= La, Nd, Gd) additive combinations can contribute to the reduce of oxygen impurity content in SiC lattice, decreasing the concentration of $V_{Si}^{''''}$ and improving the thermal conductivity of SiC ceramics.

4.Conclusion

Dense SiC ceramics with high thermal conductivity were fabricated through hot-pressing method with AlN and rare earth fluorides (LaF_3, NdF_3 and GdF_3). Effects of different additive combinations on microstructure, morphology and thermal conductivity were studied. All samples exhibited compact microstructures with high relative densities, indicated that such additive combinations can effectively improve the sinterability of SiC ceramics. Sample sintered with 4 wt%AlN and 1 wt%NdF_3 additives showed a higher thermal conductivity (152.3W/m·K) than other samples. Impedance spectroscopy analysis was utilized to associate with the variations in thermal conductivity and defect concentration in different samples. From the fitting results, 4A1Nd sample showed both the highest grain boundary resistance (R_{gb}) and bulk grain resistance (R_g), which testified its lowest concentration of vacancies after sintering. On the contrary, 4A sample exhibited the lowest R_{gb} and R_g, represented the limited ability of extracting oxygen by AlN as the sole additive. In summary, the present study suggest that AlN-ReF_3 additive combinations are effective in tailoring the thermal conductivity by affecting the morphology and defect concentration of SiC ceramics.

Acknowledgments

The authors are grateful to the financial support from the National Natural Science Foundation of China (No. 51772107) and the Innovation Team Program of Hubei Province, China (No. 2019CFA004).

References

[1] Zhang, C., Yao, X.M., Li, Y.S., Liang, H.Q., Chen, J., Zhang, J.X., Yang, J., Li, X.Y., Qiu, T., Chen, Z.M., Liu, X.J., Huang, Z.R. (2015) Effect of AlN addition on the thermal conductivity of pressureless sintered SiC ceramics. Ceram. Int., 41: 9107–9114.

[2] Šajgalík, P., Sedláček, J., Lenčéš, Z., Dusza, J., Lin, H.T. (2016) Additive-free hot-pressed silicon carbide ceramics—A material with exceptional mechanical properties. J. Eur. Ceram. Soc., 36: 1333–1341.

[3] Besisa, D.H.A., Ewais, E.M.M., Ahmed, Y.M.Z., Elhosiny, F.I., Fend, T., Kuznetsov, D.V. (2018) Thermal shock resistance of pressureless sintered SiC/AlN ceramic composites, Mater. Res. Express., 5: 015506.

[4] Lara, A., Ortiz, A.L., Muñoz, A., Domínguez-Rodríguez, A. (2012) Densification of additive-free polycrystalline β-SiC by spark-plasma sintering. Ceram. Int., 38: 45–53.

[5] Lee, E.J., Lee, D.H.J., Kim, C., Kim, D.J. (2014) Densification behavior of high purity SiC by hot pressing, Ceram. Int., 38: 16389–16392.

[6] Liang, H.Q., Yao, X.M., Zhang, J.X., Liu, X. J., Huang, Z.R. (2014) The effect of rare earth oxides on the pressureless liquid phase sintering of α-SiC, J. Eur. Ceram. Soc., 34: 2865–2874.

[7] Cho, T.Y., Kim, Y.W., Kim, K.J. (2016) Thermal, electrical, and mechanical properties of pressureless sintered silicon carbide ceramics with yttria-scandia-aluminum nitride, J. Eur. Ceram. Soc., 36: 2659–2665.

[8] Santos, A.C., Ribeiro, S. (2018) Liquid phase sintering and characterization of SiC ceramics, Ceram. Int., 44: 11048–11059.

[9] Kim, K.J., Kim, Y.W., Lim, K.Y., Nishimura, T., Narimatsu, E. (2014) Electrical and thermal properties of SiC-AlN ceramics without sintering additives, J. Eur. Ceram. Soc., 35: 2715–2721.

[10] Lofaj, F., Satet, R., Hoffmann, M.J., López, A.R. (2004) Thermal expansion and glass transition temperature of rare-earth doped oxynitride glasses. J. Eur. Ceram Soc., 24: 3377–3385.

[11] Noviyanto, A., Han, S.W., Yu, H.W., Yoon, D.H. (2013) Rare-earth nitrate additives for the sintering of silicon carbide, J. Eur. Ceram.Soc., 33, 2915–2923.

[12] Virkar, A.V., Jackson, T.B., Cutler, R.A., (1989) Thermodynamic and kinetic effects of oxygen removal on the thermal conductivity of aluminum nitride, J. Am. Ceram. Soc., 72: 2031–2042.

[13] Zhou, Y., Hirao, K., Watari, K., Yamauchi, Y., Kanzaki, S. (2004) Thermal conductivity of silicon carbide densified with rare-earth oxide additives, J. Eur. Ceram. Soc., 24: 265–270.

[14] Zhou, Y., Hirao, K., Yamauchi, Y., Kanzaki, S. (2003) Effects of rare-earth oxide and alumina additives on thermal conductivity of liquid-phase-sintered silicon carbide, J. Mater. Res., 18: 1854–1862.

[15] Irvine, J.T.S., Sinclair, D.C., West, A.R. (1990) Electroceramics: characterization by impedance spectroscopy, Adv. Mater., 2: 132–138 .

Preparation of CuO/Ag rough nanostructures as SERS-active substrates

Qiong Yang[1], Jingjing Wang[1], Haoran Wu[1], Chaorong Li[1*]

[1]Department of Physics, Zhejiang Sci-Tech University, Hangzhou, Zhejiang, China

*Corresponding author: email: crli@zstu.edu.cn

Abstract. In this paper, CuO nanostructures with nanometer roughness were firstly prepared by electrochemical anodic oxidation method, and then CuO/Ag composite nanostructures were prepared by magnetron sputtering method on the surface of CuO substrate. The successful deposition of Ag films on CuO was confirmed by XRD, UV-Vis and SEM. In addition, the effects of different Ag sputtering time on substrate morphology and SERS properties were studied. Due to the many gaps between the rough CuO/Ag nanomorphology, the substrate has many "hot spots" and adsorption sites, which can effectively improve the SERS performance of the substrate. We selected R6G dye molecule as the probe molecule to evaluate the SERS performance of the substrate. The final experimental results show that the preparation method of CuO/Ag rough nanostructure is simple, and it has significant SERS performance improvement. The substrate's detection limit for R6G can even reach 10^{-11}M.

1. Introduction

While the development of science and technology brings us a convenient life, it also brings many problems to be discovered and solved in environmental pollution[1], food safety[2], biotechnology[3], etc. With the rapid development of various spectroscopic techniques, these techniques are widely used to detect the presence of harmful substance[4]. Raman spectroscopy is widely used in chemical analysis[5], environmental monitoring[6] and other fields due to its high sensitivity, non-damage to samples and real-time operation. With people's further understanding of SERS enhanced mechanism, nobel metal nanostructures such as gold, silver and copper and their nanocomposite materials were considered to be able to achieve substantial SERS effect[7]. Generally speaking, the surface of high-performance SERS substrate should have enough "hot spots", the generation of these "hot spots" is attributed to the Localized Surface Plasmon Resonance (LSPR) of metal nanoparticles, which brings very strong enhancement of the electromagnetic (EM) field, while the cracks and tips on the surface of nanoscale rough materials can produce a large number of hot spots[8]. Therefore, it is of great significance to prepare nano-sized rough morphology by simple method for SERS performance of reinforced materials.

Three-dimensional SERS structure mainly has two advantages: one is to increase specific surface area, the other is to have excellent light capture effect[9]. Among them, the latter can make the incident light reflect multiple times in the Raman substrate, greatly improving the utilization rate of light. It is worth noting that, in addition to the ultra-sensitive detection ability, high-quality SERS active substrate also needs to have good stability, good repeatability, good durability, simple and economical preparation process and many other factors[10]. Therefore, to design a SERS substrate with practical value, this seemingly simple problem is still facing a big challenge today.

In this paper, a simple anodic oxidation method was used to prepare CuO substrates with nanometer roughness from Cu sheets, and then Ag films were deposited on the surface of CuO/Ag nanocomposite substrates by magnetron sputtering method. The thickness of Ag film on CuO surface was changed by controlling the time of magnetron sputtering, and the Raman enhancement effect of R6G molecule as probe molecule was used to explore more suitable preparation conditions.

2. Experiment

2.1. Materials
Copper foil, sodium hydroxide (AR), acetone (AR), sodium thiosulfate (AR), hydrochloric acid (AR), ethanol (AR) and other chemical reagents were purchased from Gaojing Chemical Reagent Co. Ltd. Rhodamine 6G (R6G) was purchased from Mike Chemical Reagent Co. Ltd. All reagents (analytical grade) can be received without further purification. Deionized water was used throughout the experiment.

2.2. Synthesis of CuO nanostructures
The copper foils were washed with acetone, ethanol and deionized water for 15min. Copper foil was soaked in 0.25M hydrochloric acid for 5min to remove surface oxides, then washed with deionized water and ethanol and dried. A mixture of 50mL 7.5mM $Na_2S_2O_3$ and 0.102M NaOH was used as the electrolyte, heated to 60°C in a water bath, and reacted for a 10 minutes at a constant current of 1.0mA. After the reaction, the sample was quickly taken out, washed with ethanol and deionized water, and dried.

2.3. Synthesis of CuO/Ag nanostructures
Silver nanoparticles were grown on CuO thin films by DC magnetron sputtering. The diameter of Ag target material used is 60mm and the thickness is 5mm. The sputtering vacuum was 6×10^{-4}Pa, the instrument pressure was 1.0Pa, the working gas was Ar gas, the substrate temperature was room temperature, the sputtering time was 10s, 60s, denoted as CuO/Ag10s and CuO/Ag60s, respectively.

2.4. Measurement and instrument
The microstructure of the composite substrate was characterized by scanning electron microscopy (FESEM Hitachi S-4800). The phase composition of the material was analyzed by X-ray diffraction (XRD, Bruker D8 Discover) using CuKα (λ=1.5406A) radiation at 40 kV and 40 mA. UV-Vis absorption spectra were measured by Hitachi U3900 spectrophotometer.

2.5. Raman spectral measurements
The CuO/Ag nanocomposite structure was immersed in 10^{-6} mol/L R6G aqueous solution for 30 min, then washed with deionized water and dried. Raman spectra were recorded by a confocal microreflex (Renishaw, Invia Reflex) with a laser wavelength of 532 nm. The laser power is 0.5mW, the integration time is 5s, and the number of integration is once. Substrate adsorption of other concentrations of R6G solution operation is the same as above.

3. Results and discussions

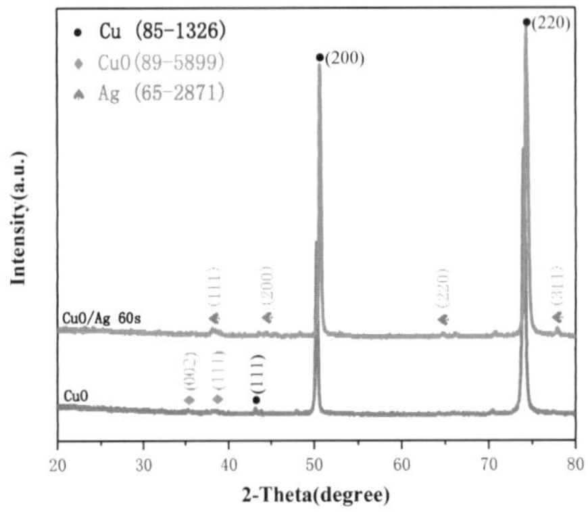

Fig.1. The XRD spectra of CuO and CuO/Ag 60s.

Fig.1 shows the details of crystal directions of CuO and CuO/Ag nanocomposite structures characterized by XRD. It can be seen from the figure that the diffraction peaks of copper correspond to the (111), (200) and (220) crystal planes of copper respectively. After the anodic oxidation reaction, relatively weak diffraction peaks appear around 35.6°and 38.7°. They correspond to the (002) and (111) crystal faces of CuO (JCPDS File No.89-5899), respectively. Four crystal peaks of silver were obtained from the material after 60 s magnetron sputtering, which correspond to the diffraction peaks of (111), (200), (220) and (311) of the metallic silver FCC structure (JCPDS File No.65-2871). The above XRD characterization indicates that the CuO/Ag composite nanostructures have been successfully prepared.

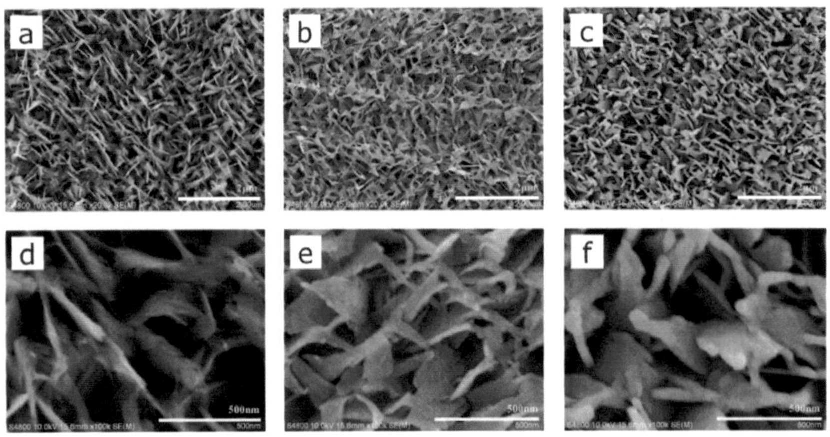

Fig.2. The SEM images of the rough CuO nanomaterials sputter silver at different times. (a, d)CuO;(b,e) CuO/Ag10s; (c,f) CuO/Ag60s.

The composites were characterized by scanning electron microscopy. Fig. 2a shows CuO under low magnification scanning electron microscopy (SEM), indicating that a large number of dense CuO nanoscale structures grow on its surface. According to the high-power scanning electron microscope image (Fig. 2d), the thickness of individual CuO nanosheets is very thin and the edges are flocculent. These nanostructures create many good Raman "hot spots". As can be seen from Fig. 2b and 2e, after 10 seconds of magnetron sputtering, silver adheres to the surface of CuO nanosheets, making the thickness of the lamellar structure significantly increased and the "hot spot" at the tip of the lamellar structure more obvious. As can be seen from Figure 2c and Figure 2f, when the sputtering time increases to 60 seconds, the thickness of the nanosheet further increases and the uniformity is good.

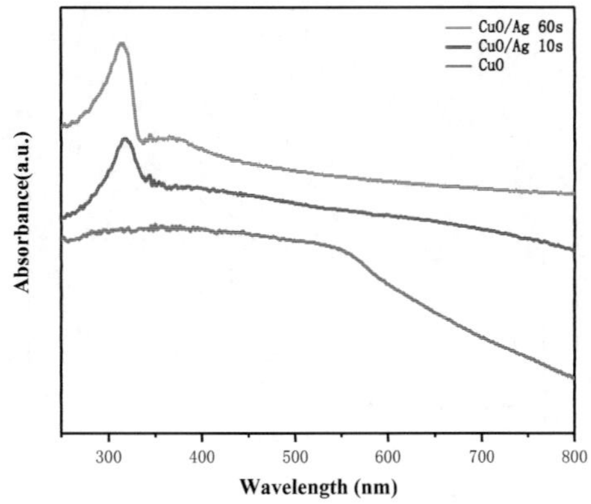

Fig. 3. The UV-Vis absorption spectra of CuO, CuO/Ag10s and CuO/Ag60s.

Fig. 3 shows the UV-Vis absorption spectra of CuO/Ag nanostructures prepared at different sputtering times. CuO has an obvious absorption edge at about 560nm. After loading silver for 10 seconds, the substrate has no significant change in the absorption of visible light bands, and there is an obvious absorption peak at 314nm. After loading silver for 60 seconds, the absorption peak strength at 314nm increases significantly. This indicates that the silver nanofilms are successfully recombined with CuO, and the silver loading capacity increases with the increase of sputtering time.

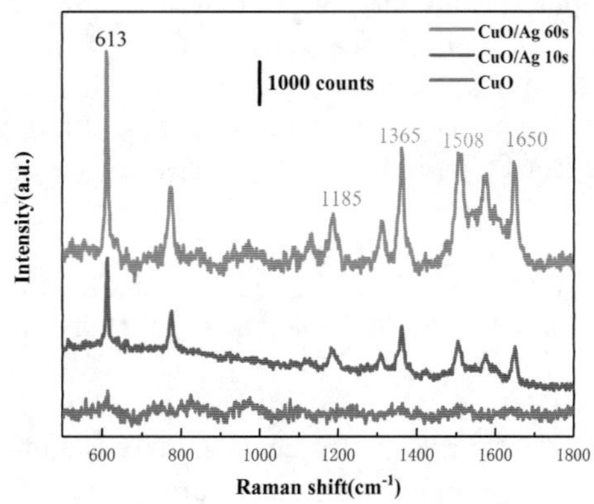

Fig. 4. Raman spectra of 10^{-6} mol/L R6G absorbed on the CuO, CuO/Ag10s and CuO/Ag60s.

In this paper, Rhodamine 6G (R6G) was selected as the probe molecule to study the SERS properties of CuO/Ag composite nanomaterials. SERS spectra of the substrates prepared under different reaction conditions adsorbed R6G are shown in Fig. 4. We can clearly observe from the spectrogram that, compared with CuO, CuO/Ag substrate has a more obvious Raman characteristic peak when the concentration of R6G molecule is 10^{-6}mol/L, indicating that the prepared CuO/Ag nanostructure has a good enhancement effect. Among them, CuO/Ag60S substrate showed the best SERS performance.

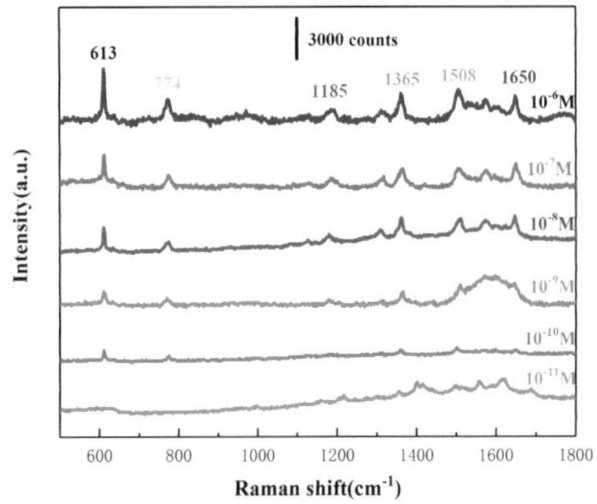

Fig. 5. Raman spectra of R6G with different concentrations absorbed on the CuO/Ag60s.

As shown in Fig. 5, when CuO/Ag60S substrates adsorbed different concentrations of R6G solutions, the spectra and some characteristic peaks of R6G could be clearly identified, even as low as 10^{-11} M, showing good Raman enhancement performance.

4. Conclusions

In conclusion, we propose a simple anodic oxidation strategy to successfully prepare rough CuO nanostructures, and then use magnetron sputtering to plate silver films of different thickness on their surfaces to obtain CuO/Ag composite nanomaterials. According to the Raman spectra of R6G molecules, this rough CuO/Ag nanostructure has a good enhancement effect due to the abundant interludes between CuO/Ag nanosheets. The CuO/Ag60s substrate's detection limit for R6G can even reach 10^{-11}M. Therefore, this rough CuO/Ag nanostructure can be used as a good SERS active substrate for the detection of various dye molecules and pollutants.

Acknowledgments

This work was supported by the Natural Science Foundation of China (Nos. 51802282 and 51672249) and Science Foundation of Zhejiang Sci-Tech University (ZSTU) under Grant No.17062065-Y.

References

[1] Li, S., Wang, Q., Song, X., Bu, Y., (2019) A green and general strategy for the synthesis of hollow Ag/CdS nanocomposites for superior SERS performance. CrystEngComm., 21: 3709-3720.

[2] Song, Y., Huang, H. C., Lu, W., Li, N., Su, J., Cheng, S. B., Lai, Y., Chen, J., Zhan, J., (2021) Ag@WS2 quantum dots for Surface Enhanced Raman Spectroscopy: Enhanced charge transfer induced highly sensitive detection of thiram from honey and beverages. Food. Chem., 344: 128570.

[3] Song, J., Zhou, J., Duan, H., (2012) Self-assembled plasmonic vesicles of SERS-encoded amphiphilic gold nanoparticles for cancer cell targeting and traceable intracellular drug delivery. J. Am. Chem. Soc., 134: 13458-13469.

[4] Ferrari, M., Mottola, L., Quaresima, V., (2004) Principles, Techniques, and Limitations of Near Infrared Spectroscopy. Can. J. Appl. Physiol., 29(4): 463-487.

[5] Zhao, X., Liu, C., Yu, J., Li, Z., Liu, L., Li, C., Xu, S., Li, W., Man, B., Zhang, C., (2020) Hydrophobic multiscale cavities for high-performance and self-cleaning surface-enhanced Raman spectroscopy (SERS) sensing. Nanophotonics., 9: 4761-4773.

[6] Liu, C., Lei, F., Li, J., Wei, Y., Li, Z., Zhang, C., Man, B., Yu, J., (2021) Integrated accurate extraction and fast detection of analyte: Capillarity-Based SERS substrate using in effluent monitoring. Appl. Surf. Sci., 542: 148735.

[7] Liu, Y., Ma, H., Han, X. X., Zhao, B., (2021) Metal–semiconductor heterostructures for surface-enhanced Raman scattering: synergistic contribution of plasmons and charge transfer. Mater. Horiz., 8: 370-382.
[8] Ye, F., Ju, S., Liu, Y., Jiang, Y., Chen, H., Ge, L., Yan, C., Yuan, A., (2019) Ag-CuO Nanocomposites: Surface-Enhanced Raman Scattering Substrate and Photocatalytic Performance. Cryst. Res. Technol., 54: 1800257.
[9] Yu, J., Yang, M., Zhang, C., Yang, S., Sun, Q., Liu, M., Peng, Q., Xu, X., Man, B., Lei, F., (2018) Capillarity-Assistant Assembly: A Fast Preparation of 3D Pomegranate-Like Ag Nanoparticle Clusters on CuO Nanowires and Its Applications in SERS. Adv. Mater. Interfaces., 5: 1800672.
[10] Yang, L., Li, P., Liu, J., (2014) Progress in multifunctional surface-enhanced Raman scattering substrate for detection. RSC Adv., 4: 49635-49646.

Nonequilibrium behavior of polymer nano films cast from different solvents

Xiao Fan, Jinghong Ma and Yu Ma*

State Key Laboratory for Modification of Chemical Fibers and Polymer Materials, College of Materials Science and Engineering, Donghua University, Shanghai 201620, P.R. China

*Corresponding author's e-mail: yma@dhu.edu.cn

Abstract. Properties of materials are largely determined by the processing conditions. This is especially true of polymers, which typically have much longer relaxation times than processing times and therefore possess out-of-equilibrium chain conformations. Here, we demonstrate that the nonequilibrium behavior of polymer nano films can be tuned by controlling the processing pathways. Films cast from different quality of solvents were aged at room temperature for various times, and then carried out the dewetting experiment above the glass transition temperature. The residual stress and relaxation time strongly depend on the quality of the solvents, and the differences may be caused by the boiling point and interaction parameter. This work offered a pathway for preparing a stable nano film, which is crucial for the durability of nanomaterials and devices.

1. Introduction

The processing condition is believed to have a huge impact on the behavior of polymeric samples, especially in the film processing field, which resulting in residual stresses due to fast solvent loss, leading to a nonequilibrium state of the chain conformations[1-3]. This in turn affects macroscopic properties of matters, such as viscoelastic properties. Film thickness, which is controlled by different spin speed or concentration of the polymer solutions, has been considered as one of the most significant aspect for the difference of out-of-equilibrium states[4]. Besides, the characteristic aging time varied strongly with the quality of the solvents, which may be attributed to distorted chain conformations in the as-cast films[5].

Dewetting experiments have been proven to be a simple and high-efficiency method for providing the detailed rheological properties of polymer films. Dewetting, as a prevailing phenomenon for thin films or coatings, occurs when the thin films are heated above its glass transition temperature Tg or placed in the solvent vapor atmosphere. It is generally believed that the dewetting velocity is derived from the balance between the viscoelastic and capillary stresses, and the variations in dewetting velocity are caused by preparation-induced differences.

In this work, we focus on revealing effects of quality of solvents on nonequilibrium behavior of polystyrene (PS) nano films via dewetting experiments[1, 4, 6-8]. To this end, spin-coating method was employed to form PS nano films on a hydrophobic substrate polydimethylsiloxane (PDMS), and different quality of solvents were used. For quantitatively studying the film properties caused by the preparation, we relied on dewetting experiments after aging for different periods to realize the nonequilibrium behavior of polymer films.

2. Physics of dewetting of polymer films

The spreading coefficient, *S*, describes the ability of the drop to spread on or dewet from a surface, and can be obtained by calculation[9]. For a film A, on a substrate B, *S* is given by Young's equation: $S = \gamma_B - \gamma_{AB} - \gamma_A$, where γ_A, γ_B, and γ_{AB} are the energies of the A and B surfaces and the AB interface, respectively. Equation 1 was used to calculate the interface tension, γ_{AB}

$$\gamma_{\mathrm{AB}} = [(\gamma_B^d)^{\frac{1}{2}} - (\gamma_A^d)^{\frac{1}{2}}]^2 + [(\gamma_B^p)^{\frac{1}{2}} - (\gamma_A^p)^{\frac{1}{2}}]^2 \quad (1)$$

Where γ^d, γ^p are the dispersion force and polarity force, respectively. For PS/PDMS system, the value of PS and PDMS were shown in Table 1[10]. Therefore, the calculated *S* was -25.68 mN/m. The negative spreading coefficient indicated the instability of the nanofilms and had the tendency to dewetting.

Table 1. Components of surface energy for various polymers (γ in mN/m)

	γ	γ^d	γ^p
PS	40.7	34.5	6.1
PDMS	19.8	19.0	0.8

3. Experimental section

Materials. PS (M_w = 192000 g mol^{-1}) were purchased from Sigma-Aldrich. Ethyl acetate (EA), Methyl Ethyl Ketone (MEK), 1,3-Dioxolane and 3-Pentanone were used to prepare PS solutions. All reagents were analytical grade and used without further purification. The detailed cleaning pathways of glass substrates and silicon wafers were described in our previous works[9].

PDMS (Sylgard 184, Dow Corning Corporation) kit with base and curing agent was used as received. The PDMS base and curing agent of w/w ratio 10:1 was firstly mixed vigorously and then degassed in vacuum for 15 min. PDMS substrates with smooth surfaces were prepared by coating mixtures onto glass substrates and then cured in the oven at 85 °C for 3 h.

Film preparation. In current work, we focused on the films of thickness h = ca. 100 nm that prepared by spin-coating PS solution on PDMS substrates. Film thicknesses were controlled by changing the initial concentration of the solution, while the spin rate was fixed at ω = 3000 rotations per minute (rpm). Thicknesses of films were measured by the ellipsometer (ELLIP/SR). In the period of spin-coating process, the substrates rotated at high speeds, and the centrifugation force dominates the thinning of the film in the early stage (hydrodynamic thinning), followed by evaporation in the later stage. It should be noted that, there was no significant difference between the air-dried and vacuum-dried samples[4-5]. Therefore, the non-equilibrium effects observed in this work were not due to the possible trapped solvent molecules, and the experiments were performed without further vacuum drying.

Dewetting experiment. Dewetting of PS films was induced by heating samples to 140 °C on a hot stage (Simplex Scientific), i.e., to a temperature above the glass transition temperature. The dewetting process was captured in situ and in real time by optical microscopy (BK-POLR-TR). A series of images were captured with a CCD camera (Figure 1).

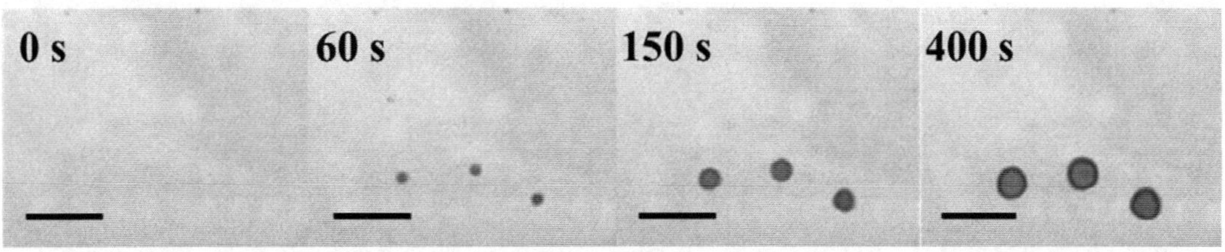

Figure 1. Optical micrographs showing the time evolution of dewetting for the films of thickness h ≈ 100 nm cast from EA. The dewetting experiments were performed at T_{dew} = 140 °C. Scale bars are 100 μm.

4. Results and discussion

In the first series of experiments, films were spin-coated from EA at 25 °C, then stored (aged) at room temperature for different periods, followed by dewetting experiments above Tg. Figure 2A presents typical graphs for the radius (R) of a dewetting hole as a function of dewetting time (t_{dew}) at various values of aging time (t_{aging}). An increase in t_{aging} led to a slowing down of the growth of dewetting holes in 100 nm PS films. At the beginning of dewetting, in Figure 2B, it is observed that the initial dewetting velocity is approximately constant over a period of time, which arise from that a mature rim has not yet formed, a constant dewetting velocity is expected. After the constant region, V_{dew} decreased according to a power-law $V_{dew} \sim t_{dew}^{-\alpha}$ with an exponent $\alpha \approx -1$, which agreed with the value reported by Reiter[8]. To quantify the changes caused by aging, we measured R for a given dewetting time as a function of t_{aging}. As shown in Figure 2C, R showed a decay with aging time that could be fitted by a single exponential. Meanwhile, we present additional curves of the initial dewetting velocity as a function of t_{aging}, which suggested an exponential decay (Figure 2D). Thus, we fitted an exponential function to the measured values of R and V using the following form

$$yt = (y0 - y\infty)\exp\left(-\frac{t}{\tau_{res}}\right) + y\infty \quad (2)$$

with $yt = R(t) = V(t)$, or $y0 = R(0) = V(0)$, or $y\infty = R(\infty) = V(\infty)$. The resulting relaxation times (τ_{res}) were very similar ($\tau_{res} = 21.6$ h from Figure 2C and 23.7 h from Figure 2D). This decrease in dewetting dynamics reflects a reduction of the driving forces for dewetting and strongly suggests a decrease in the amount of residual stress within the film caused by physical aging.

Assuming that all residual stresses (σ_{res}) were released, the ratio of dewetting velocity in as-prepared film and fully aged film $V(0)/V(\infty)$ should be equal to the ratio of total stresses and capillary stress (σ_{cap})[8-9].

$$\frac{V(0)}{V(\infty)} = \frac{\sigma_{res}+\sigma_{cap}}{\sigma_{cap}} \quad (3)$$

σ_{cap} was 256.8 kPa based on $\sigma_{cap} = |S|/h$. Thus, σ_{res} in PS nano film cast from EA was 188.8 kPa.

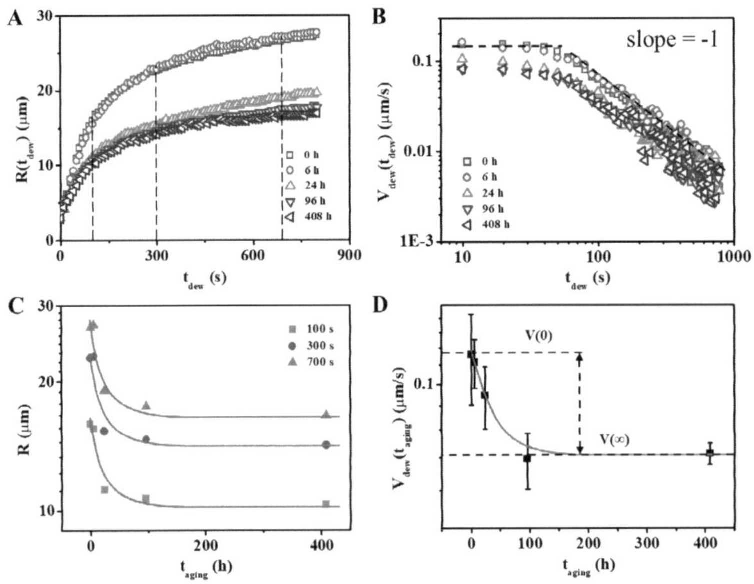

Figure 2. (A) Hole radius (R) as a function of dewetting time (t_{dew}) after being aged at room temperature for different periods (t_{aging}). The PS nano film was cast from EA. The operating temperature of the dewetting experiment is 140 °C. (B) Dewetting velocity V_{dew} (t_{dew}) on logarithmic scales as a function of dewetting time t_{dew} for different aging times. (C) Hole radii taken at three constant times (dashed lines in part A) follow an exponential decay as a function of aging time. (D) Semi-logarithmic representation of the initial dewetting velocities *vs* t_{aging}. Red solid lines represent fits to Equation 2.

After that, the experiment in Figure 2 was repeated, and the characteristic relaxation times for aging at room temperature was determined for all sample series.

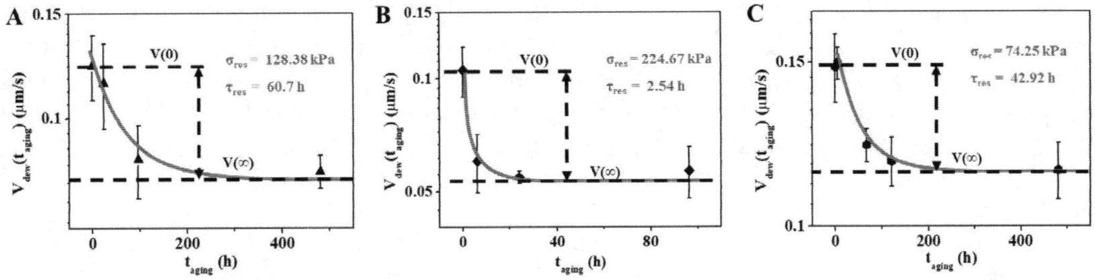

Figure 3. Semi-logarithmic representation of the initial dewetting velocities *vs* t_{aging} prepared by different solvents: (A) MEK, (B) 1,3-dioxolanem and (C) 3-pentanone. Solid lines represent fits to Equation 2.

As shown in Figure 3, all samples follow the exponential decay in the aging experiment at room temperature, and the residual stress and relaxation time calculated by Equation 2 and Equation 3 were shown in the figure, indicating that the PS nano films cast from different solvents led to the difference in the equilibrium state. For PS nano films cast from 1,3-dioxolane, a relatively high residual stress and short relaxation time were found.

For further understand the sources of diverse non-equilibrium states of polymer nano films prepared by different solvents, boiling point of solvent and the Flory-Huggins interaction parameter between PS and solvent were introduced to explain the diversity. Figure 4A shows that PS nano film cast from different quality of solvent possesses dissimilar residual stress, i.e., locates in different equilibrium states. The drying time of the film during spin-coating varied and is governed by the boiling point of the solvent. A higher boiling point determines the relatively low evaporation rate, and a longer film formation time allows more time for the chains to equilibrate during the solvent evaporation process[11]. Besides, the Flory-Huggins parameter χ_{12} has been one of the most widely used quantities, characterizing a variety of polymer-solvent and polymer-polymer interactions, and was calculated by Equation 4. If $\chi_{12} < 0.5$, the polymer should be soluble.

$$\chi_{12} = \frac{V_{mol}[(\delta_{D2}-\delta_{D1})^2+0.25(\delta_{P2}-\delta_{P1})^2+0.25(\delta_{H2}-\delta_{H1})^2]}{RT} \tag{4}$$

where V_{mol}, R and T represent the molar volume of the solvent (mL mol^{-1}), the gas constant and temperature (K), respectively. The detailed solubility parameters come from HSPiP software.

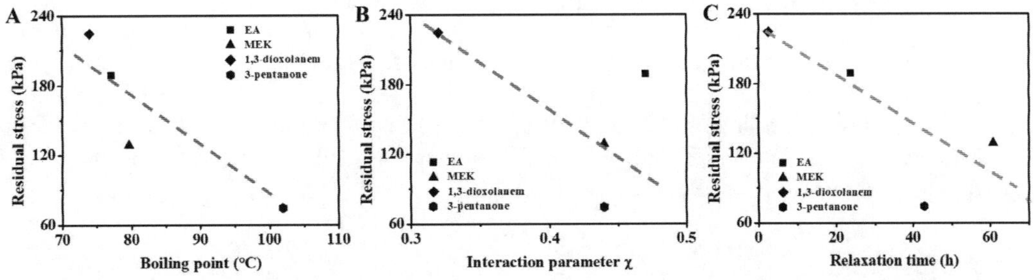

Figure 4. The interrelation between residual stress (σ_{res}) and (A) boiling point, (B) interaction parameter χ and (C) relaxation time τ_{res}. The dashed lines are for eye guide only.

As can be seen in Figure 4B, the smaller value means a more stretchable chain in the solution. The larger the swollen coil size, the stronger the deformation of the coils in the film, the higher the in-plane tensile stress in the film[5]. Figure 4C illustrates an approximate linear relationship between residual stress and relaxation time, revealing the faster relaxation with a higher residual stress.

5. Conclusions

In summary, four kinds of solvents with different boiling point and interaction parameter with PS were adopted to explored the processing induced nonequilibrium behavior in PS nano films by dewetting experiment. Although the same preparation method, the seemingly identical films are in different non-equilibrium states caused by utilizing various quality of solvent, which arise from the difference in boiling point and interaction parameters, causing the difference in the chain conformations. This work demonstrates the role of the quality of the solvent on the nonequilibrium conformation in the film, and provides a strategy for getting a thermodynamically stable thin film.

Acknowledgments

This research was financially supported by the National Natural Science Funds of China (Grant No. 21204011), the Fundamental Research Funds for the Central Universities, and the funds for course development.

References

[1]. Reiter, G.; Hamieh, M.; Damman, P.; Sclavons, S.; Gabriele, S.; Vilmin, T.; Raphael, E. (2005) Residual stresses in thin polymer films cause rupture and dominate early stages of dewetting. Nat. Mater., 4 (10): 754-758.

[2]. Chung, J. Y.; Chastek, T. Q.; Fasolka, M. J.; Ro, H. W.; Stafford, C. M. (2009) Quantifying residual stress in nanoscale thin polymer films via surface wrinkling. ACS Nano**,** 3: 844–852.

[3]. Chandran, S.; Baschnagel, J.; Cangialosi, D.; Fukao, K.; Glynos, E.; Janssen, L. M. C.; Müller, M.; Muthukumar, M.; Steiner, U.; Xu, J.; Napolitano, S.; Reiter, G. (2019) Processing pathways decide polymer properties at the molecular level. Macromolecules, 52 (19): 7146-7156.

[4]. Chandran, S.; Handa, R.; Kchaou, M.; Al Akhrass, S.; Semenov, A. N.; Reiter, G. (2017) Time allowed for equilibration quantifies the preparation induced nonequilibrium behavior of polymer films. ACS Macro Lett., 6 (11): 1296-1300.

[5]. Raegen, A.; Chowdhury, M.; Calers, C.; Schmatulla, A.; Steiner, U.; Reiter, G. (2010) Aging of thin polymer films cast from a near-theta solvent. Phys. Rev. Lett., 105 (22): No. 227801.

[6]. Chowdhury, M.; Freyberg, P.; Ziebert, F.; Yang, A. C.; Steiner, U.; Reiter, G. (2012) Segmental relaxations have macroscopic consequences in glassy polymer films. Phys. Rev. Lett., 109 (13): No. 136102.

[7]. Chandran, S.; Reiter, G. (2019) Segmental rearrangements relax stresses in nonequilibrated polymer films. ACS Macro Lett., 8 (6): 646-650.

[8]. Mulama, A. A.; Chandran, S.; Roumpos, K.; Oduor, A. O.; Reiter, G. (2019) Dewetting rheology for determining viscoelastic properties of nonequilibrated thin polymer films. Macromolecules, 52 (20): 7894-7903.

[9]. Fan, X.; Xu, J.; Chen, L.; Hong, N.; Wang, C.; Ma, J.; Ma, Y. (2020) Processing induced nonequilibrium behavior of polyvinylpyrrolidone nanofilms revealed by dewetting. Langmuir, 36 (50): 15430-15441.

[10].Krevelen, D. W. v.; Nijenhuis, K. t. (2009) Properties of Polymers, 4th Ed, Elsevier, New York. Thomas, K. R.; Chenneviere, A.; Reiter, G.; Steiner, U. (2011) Nonequilibrium behavior of thin polymer films. Phys. Rev. E, 83 (2): 021804.

Catalytic reduction of 4-Nitrophenol using CoPt nanoparticles on the reduced graphene oxide layer nanocomposite catalyst

Xiaolong Zhang[1,2,a], Qian Duan[1,b*]

[1]School of Materials Science and Engineering, Changchun University of Science and Technology, Changchun, 130022, PR China.

[2]Key Laboratory of Functional Materials Physics and Chemistry of the Ministry of Education, Jilin Normal University, Changchun 130103, PR China.

[a]email: zhangxiaolong@jlnu.edu.cn, [b]email:duanqian88@hotmail.com

[*]Corresponding author: email: duanqian88@hotmail.com

Tel: +86 431 85583015; Fax: +86 431 85583015

Abstract: The reduced graphene oxide (rGO) supported CoPt nanoparticles nanocomposite catalyst has been preparation by a facile method. The XRD and TEM results show that the CoPt nanoparticles were successfully loaded onto the rGO layer. The apparent rate constant (k) is about 0.167 min^{-1}. The rGO/CoPt nanocomposite catalyst exhibits excellent catalytic performance for catalytic reduction of 4-nitrophenol. Moreover, the magnetic result shows that the rGO/CoPt nanocomposite is superpara-magnetism.

1. Introduction

In recent years, the environmental pollution is becoming serious. Organic pollutants are one of environmental pollution, which are attracted more attention. 4-nitrophenol (4-NP) is one kind of organic pollutant, which is present in industrial wastewater and may be bring serious health harm to human beings [1]. At present, the catalytic reduction of 4-NP to p-aminophenol (4-AP) with sodium borohydride($NaBH_4$) has been widely concerned because of its low cost and high efficiency. Moreover, 4-AP is an important pharmaceutica intermediate [2]. Binary alloy nanoparticles have attracted much attention due to their outstanding catalytic activity for organic pollutants. As a kind of binary alloy nanoparticles, CoPt nanoparticles have attracted particular interest of researchers, because of their excellent catalytic activity for the 4-NP reduction reaction. As far as we know, the shape, size of nanoparticles play an important roles in the application of catalytic reduction of 4-NP. However, the aggregation process of magnetic CoPt nanoparticles tends to minimize the surface energy, limiting its catalytic performance. As we know, the excellent catalytic materials should have abundant active sites and large surface area [3]. Thus, immobilization of CoPt nanoparticles on a support is an effective way to increase active sites and surface area.

In this work, we reported the synthesis and properties of the nanocomposites catalyst (rGO/CoPt), where reduced grapheme oxide (rGO) was used as a support. The catalytic property of rGO/CoPt nanocomposites catalyst was investigated by catalytic reduction of 4-NP to 4-AP in the presence of $NaBH_4$. The relationships between structures and catalytic performances of the rGO/CoPt nanocomposites catalyst have been investigated in detail. The rGO/CoPt nanocomposites catalyst exhibits excellent catalytic activity for the reduction of 4-NP.

2. Materials and Methods

2.1. Materials

Co(acac)$_2$, Pt(acac)$_2$, 4-nitrophenol, sodium borohydride ($NaBH_4$), OA (AR), OAm (AR), benzyl ether (95%), and hydrazine hydrate were purchased from Sigma Aldrich Ltd. The Graphene oxide (GO) was synthesized according to a modified Hummers' method. All chemicals were used without further purification.

2.2. Preparation of the rGO/CoPt nanocomposites.

Co(acac)$_2$ and Pt(acac)$_2$ were mixed as 1:1 in a solution composed of OA, OAm and benzyl ether in the flask. Ultrasound for 10 min, heated to 300 °C for 30 minutes. Last, CoPt nanoparticles (NPs) were collected after drying at 100 °C for 6 h. 100 mg GO and 10 mg of CoPt were dispersed in the solution of deionized water (30 mL) and ethanol (10 mL). Then, the mixture was heated up to 100 °C and hydrazine hydrate (6 mL) was injected in the mixture. At last, the sample was washed and collected; we got the rGO/CoPt nanocomposite catalyst.

2.3. Catalytic Performance Measurements.

4-NP aqueous solution (1 mL, 0.5 mM/L) and $NaBH_4$ (1 mL, 0.01 M/L) were put in the quartz cuvette. The absorption peak shifts from 317 nm to 400 nm. The reaction started when the rGO/CoPt nanocomposite catalyst (70 μL, 2 mg/mL) was added to the mixture. UV-Vis spectroscopy was employed to monitor the system, and the spectra were collected each 5 min.

3. Results and Discussion

Figure 1 shows the X-ray diffraction (XRD) pattern of rGO/CoPt. The rGO/CoPt peaks located at 2θ = 41.34°, 47.61°, 69.77°, and 84.57°, corresponding to the (111), (200), (220), (311) planes of FCT(face-centered tetragonal) CoPt nanoparticles [4]. The peak located at 23.11° is assigned to the characteristic diffraction peak of the (002) plane, which is confirmed the formation of rGO. The XRD pattern confirms the formation of rGO/CoPt nanocomposite.

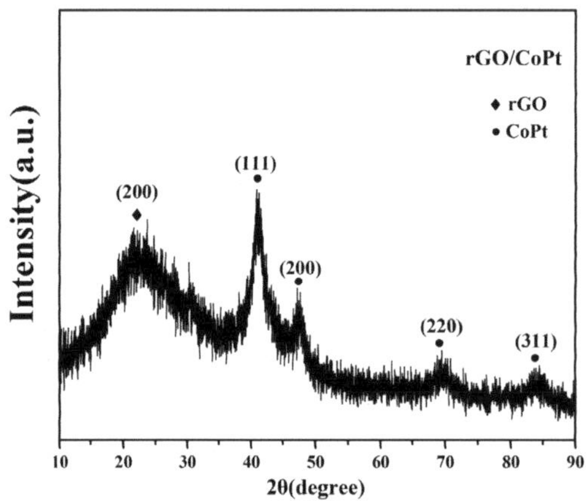

Figure 1. XRD pattern of rGO/CoPt nanocomposite.

X-ray photoelectron spectroscopy (XPS) analysis was carried out to examine the surface composition of the rGO/CoPt nanocomposite in Figure 2. Figure 2a shows the XPS survey spectra of rGO/CoPt. The survey spectrum of rGO/CoPt shows the Pt4f, C1s, N1s, O1s, and Co2p peaks, and there are no other elements. As shown the high-resolution XPS spectra of Co2p and Pt4f, for the Co2p spectrum, the peaks located at about 777.5 eV and 792.4 eV are corresponding to Co2p$_{3/2}$ and Co2p$_{1/2}$, which maybe confirm that there is no oxidization of the alloy NPs in the rGO/CoPt nanocomposite. In the spectrum of Pt4f, as shown in Figure 2c, two peaks located at 71.04 eV and 74.36 eV is correspond to Pt4f$_{7/2}$ and Pt4f$_{5/2}$, respectively [5].

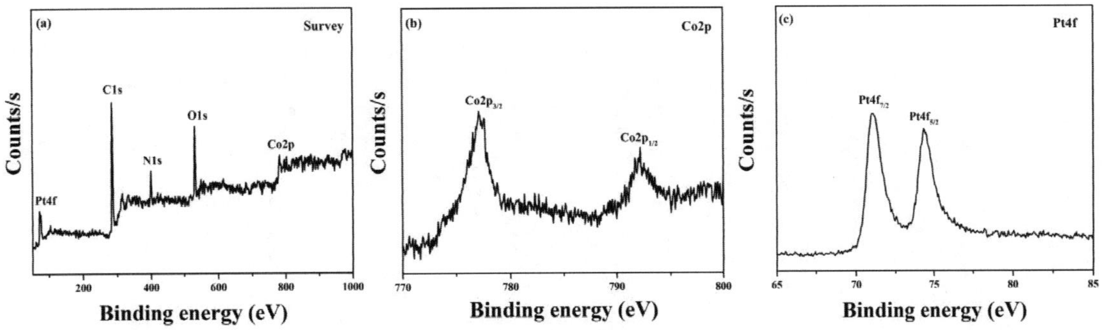

Figure 2. XPS spectrums of survey scan (a), Co2p (b) and Pt4f (c) for rGO/CoPt nanocomposite.

The morphology, size and elements of the rGO/CoPt nanocomposite are shown in Figure 3. As shown in Figure 3a, the nanoparticles are almost spherical in shape with a small size. Moreover, the two dimensional (2D) lamellar structure of rGO is detected in the rGO/CoPt nanocomposite. Figure 3b shows that the lattice distances is 0.213 nm, which is accord to the lattice plane of (111) of FCT CoPt. As shown in Figure 3c, the SAED of the rGO/CoPt nanocomposite is accord with the small size of the particles. Figure 3d shows the EDX of the rGO/CoPt nanocomposite, and there are no other impurity elements.

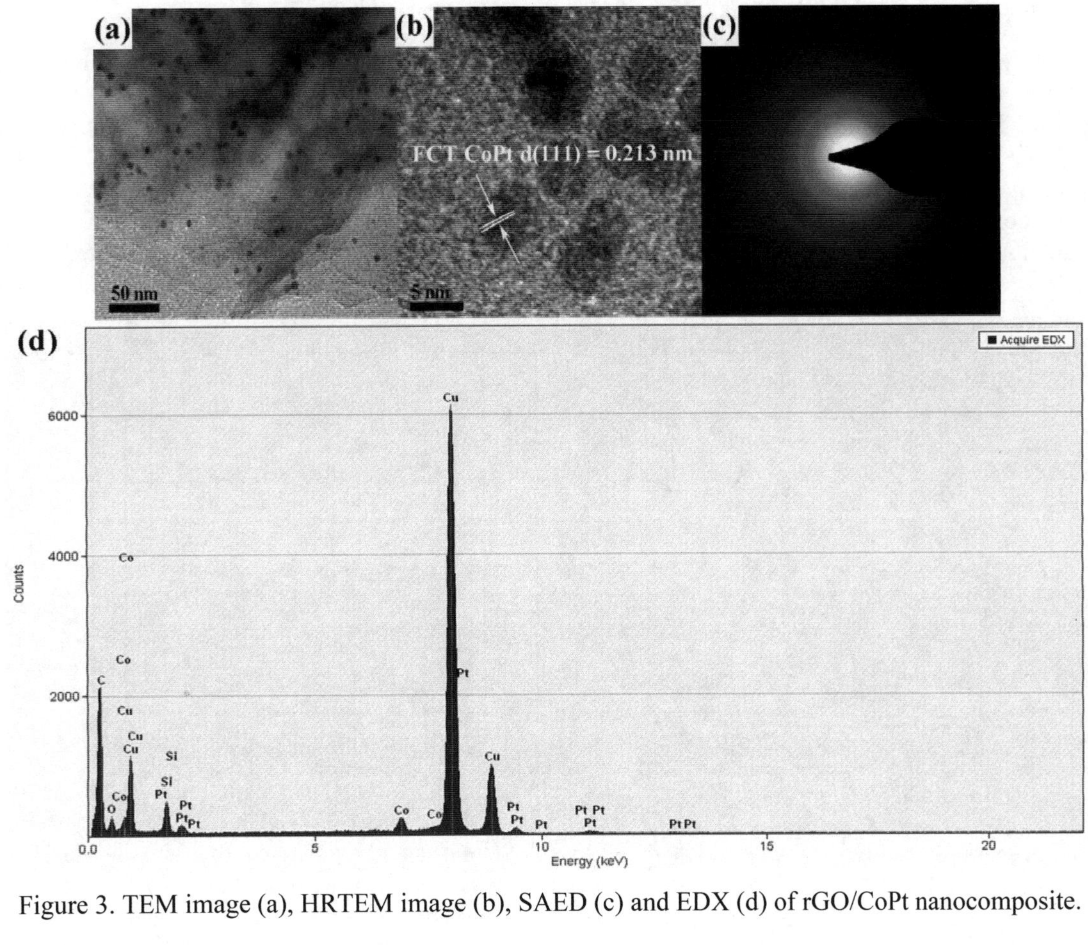

Figure 3. TEM image (a), HRTEM image (b), SAED (c) and EDX (d) of rGO/CoPt nanocomposite.

Figure 4 shows the hysteresis loop of rGO/CoPt nanocomposite at 300 K. As shown in Figure 4, the value of saturation magnetization (Ms) is about 23.2 emu/g, and the coercivity (Hc) is almost close to 0 Oe. These values show that rGO/CoPt nanocomposite owns superpara-magnetism.

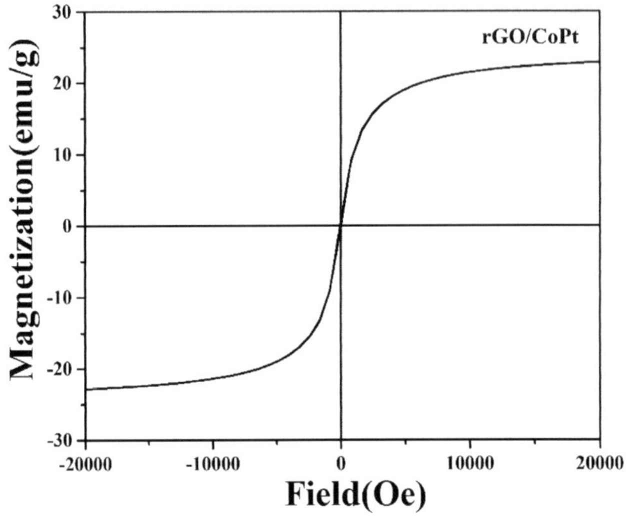

Figure 4. Hysteresis loop of the rGO/CoPt nanocomposite.

The catalytic performance of rGO/CoPt nanocomposite was studied by the catalytic reduction of 4-NP in Figure 5. The UV-Vis was used to monitors the catalytic reduction reaction. Although the absorption peak of 4-NP shifts to 400 nm from 317 nm when $NaBH_4$ is added, the catalytic reduction is limited without a catalyst. After the rGO/CoPt nanocomposite catalyst is added into the solution, the peak (at 400 nm) decreases, and the peak (at 300 nm) increases over time. In addition, no other peaks were observed, which confirms that there are no other products in the process of the catalytic reduction. As shown in Figure 5, rGO/CoPt nanocomposite catalyst completely catalytic reduction 4-NP to 4-AP in about 25 minutes and the value of k is about 0.167 min^{-1}, which confirms the catalyst exhibits the better catalytic performance than noble metal catalyst [6]. Combining with the result, the excellent catalytic performance of rGO/CoPt nanocomposite catalyst may be due to the synergy of rGO support layer and CoPt nanoparticles.

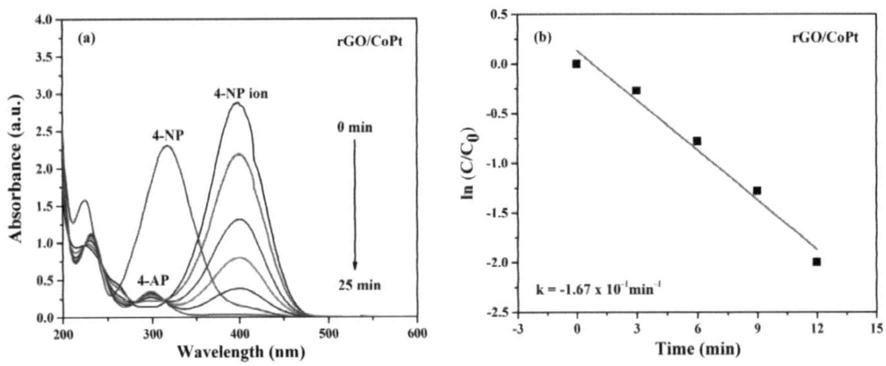

Figure 5. UV-visible spectra and plot of ln(C/C_0) for the catalytic reduction of 4-NP.

The possible catalytic mechanism of 4-NP by rGO/CoPt nanocomposite catalyst within $NaBH_4$ is described as follow: When the rGO/CoPt nanocomposite catalyst is added into the mix of $NaBH_4$ and 4-NP, the BH_4^- and 4-NP ions are quickly adsorbed onto the surface of rGO/CoPt nanocomposite catalyst. Then, the electrons and the hydrogen anion transfer from the rGO/CoPt nanocomposite catalyst to 4-NP. At last, 4-AP is formed and desorbs from rGO/CoPt nanocomposite catalyst.

4. Conclusions

In conclusion, we have prepared rGO/CoPt nanocomposite catalyst via a facile method. The rGO/CoPt nanocomposite catalyst exhibits an excellence catalytic performance for the catalytic reduction of 4-NP. The excellent catalytic performance of rGO/CoPt nanocomposite catalyst may be due to the synergy of rGO support layer and CoPt nanoparticles. This work may be beneficial to the research of catalysis.

Acknowledgements

This work was supported by the National Natural Science Foundation of China (Grant Numbers 21676115 and 61675090), Program for the development of Science and Technology of Jilin province (Grant Numbers 20190303069SF, 20200301043RQ and 20190103002JH) and Program for Science and Technology of Education Department of Jilin Province (Grant Numbers JJKH20200418KJ and JJKH20191018KJ).

References

[1] Nasrollahzadeh, M., Sajjadi, M., Tahsili, M.R. (2020) High efficiency treatment of organic inorganic pollutants using recyclable magnetic N-heterocyclic copper(II) complex and itsantimicrobial applications. Sep. Purif. Technol., 238: 116403.

[2] Pal, J., Mondal, C., Sasmal, A.K., Ganguly, M., Negishi, Y., Pal, T. (2014) Account of Nitroarene Reduction with Size- and Facet-Controlled CuO-MnO_2 Nanocomposites. ACS Appl. Mater. Interfaces, 6: 9173-9184.

[3] Goswami, A., Rathi, A.K., Aparicio, C., Tomanec, O., Petr M., Pocklanova, R., Gawande, M.B., Varma, R.S., and Zboril, R.(2017) In-situ Generation of Pd-Pt Core-shell Nanoparticles on Reduced Graphene Oxide (Pd@Pt/rGO) using Microwaves: Applications in Dehalogenation Reactions and Reduction of Olefins. ACS Appl. Mater. Interfaces, 9(3): 2815-2824.

[4] Martins, A., Fantini, M.C.A., Souza-Neto, N.M., Ramos, A.Y., Santos, A.D. (2006) Alternate monatomic layer sputter deposition of FCT ($L1_0$-type) ordered FePt and CoPt films. J. Magn. Magn. Mater., 305: 152-156.

[5] Motin, A.M., Haunold, T., Bukhtiyarov, A.V., Bera, A., Rameshan, C., Rupprechter, G. (2018) Surface science approach to Pt/carbon model catalysts: XPS, STM and microreactor studies. Appl. Surf. Sci., 440: 680-687.

[6] Arora, N., Mehta, A., Mishra, A., Basu, S. (2018) 4-Nitrophenol reduction catalyzed by Au-Ag bimetallic nanoparticles supported on LDH: Homogeneous vs. heterogeneous catalysis. Appl. Clay Sci., 151: 1-9.

Versatile hydrogel based on polyvinyl alcohol/chitosan/regenerated silk fibroin

Qinyue Sun [1], Jizhu Fu [1], Heng Zhang [1], Huihui Yang [1], Yumin Wang [3], Yi Xiong [2,3,*], Wei Zeng [1,*], Faniu Wang [1], Chao Xu [1]

[1] National Engineering Research Center for Agro-Ecological Big Data Analysis & Application, School of Electronics and Information Engineering, Anhui University, Hefei 230601, Anhui, People's Republic of China

[2] Science and Technology Institute, Key Laboratory of Textile Fiber and Products, Ministry of Education, Laboratory for Electron Microscopy, Wuhan Textile University, Wuhan 430073, Hubei, People's Republic of China

[3] School of Physics and Technology, MOE Key Laboratory of Artificial Micro- and Nano-Structures and Center for Electron Microscopy, Wuhan University, Wuhan 430072, Hubei, People's Republic of China

* Corresponding author's e-mail: Yi Xiong, xiong@wtu.edu.cn; Wei Zeng, zengwei@ahu.edu.cn;

Abstract. Versatile sensors have broad application prospects in human motion detection, health monitoring, wearable electronic devices and flexible electronic skin and other emerging fields. In this work, the enhanced hydrogel was prepared by freezing and thawing process with calcium ion crosslinking. This composite material shows excellent flexibility and elasticity, and after being cut in half, it can automatically heal well in a short time without external force. It shows great potential in flexible and wearable devices.

1. Introduction

With the rapid development of mobile Internet and intelligent terminals, wearable electronic devices have presented a huge market prospect. As one of the core components of wearable electronic devices, flexible wearable sensors have become the focus of attention due to their wide sensitivity range, short response time, portability and comfort [1-3]. Biological hydrogels are three-dimensional network materials formed by cross-linking hydrophilic polymer chains in an aqueous microenvironment [4, 5]. They are chemically or physically crosslinked and have unique structure and mechanical properties, which can be customized according to specific requirements [5]. Compared with other materials, polyvinyl alcohol (PVA) hydrogels have the advantages of good security, high flexibility, good transparency, mild preparation process and extensibility, making them ideal materials for the manufacture of portable wearable sensors [6].

The combination of gelatin, chitosan (CS), cellulose and PVA is a common method to improve the mechanical strength of hydrogels [7-10]. For example, the tensile strength of bacterial cellulose/PVA composite hydrogel prepared by Liu [8] and other gelatin/PVA and Zhao [9] is close to 1 MPa, Wang et al. [10] combined CS and PVA to prepare the hydrogel with sensitivity to the precursor. However, the materials with high tensile strength, self-healing properties, good water retention and biocompatibility are still needed in practical applications [11].

In existing studies, PVA has been combined with silk fibroin (SF) (a fibroin from silk fiber) to achieve biocompatibility and biodegradability [12]. The addition of SF in PVA hydrogel improved the stability of the hydrogel and significantly increased the water absorption of the gel [13]. In addition, because of its unique microstructure (including beta folding), SF can contribute to the water retention of PVA hydrogels [11].

Herein, a self-healing hydrogel was made by polyvinyl alcohol/chitosan/regenerated silk fibroin (PVA/CS/RSF). It was prepared by calcium ion crosslinking freeze-thaw method and exhibits excellent flexibility and elasticity, and it can automatically heal well in a short time without external force after being cut in half.

2. Experimental

PVA/CS/RSF enhanced hydrogel was prepared by calcium ion crosslinking and freeze-thaw method. The specific preparation process of hydrogel is shown in figure 1. First, the modified PVA solution needs to be prepared. 5 g PVA and 2 g CS were put into a beaker, and heated in a water bath at 90 °C for 2.5 h, then cooled to room temperature, and stirred at 1000 rpm to obtain the modified PVA solution.

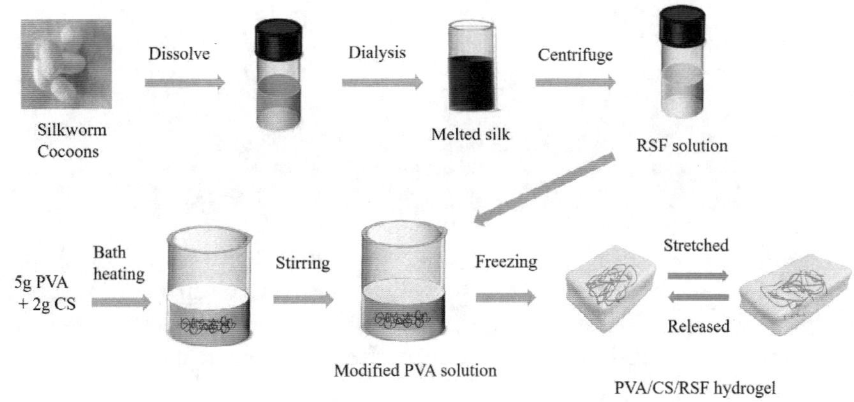

Figure 1. Fabrication process of PVA/CS/RSF hydrogel.

After that, boiled a small amount of cocoons in 0.2 M $NaCO_3$ solution for 30 minutes, washed them with deionized water for several times, and then dried them at 60 °C for 3 days to obtain degummed fibers. Then, the degummed fibers were stirred and dissolved in the mixed solution of $CaCl_2/H_2O/C_2H_5OH$ (1:8:2) at 85 °C, and dialyzed it in deionized water for 3 days. The dialysate was then centrifuged at 6000 rpm for 30 minutes to remove suspended particles. Then, RSF solution was obtained from the supernatant. Finally, 10 ml modified PVA solution and 0.2 ml RSF solution were mixed and placed in a glass container and stirred for 2 hours at 2000 rpm at room temperature. Then put the stirred solution in the refrigerator, froze and thawed at minus 20 °C for 5 times (2 hours each time). After 10 hours, removed it from the refrigerator, cooled at room temperature, and get PVA/CS/RSF hydrogel from the container.

The morphology was analyzed by scanning electron microscopy (SEM, S-4800).

3. Results and Discussion

The macroscopic properties of functional materials are usually closely related to their microstructures [14 – 16]. The structure of PVA/CS/RSF hydrogel was studied by scanning electron microscope (SEM) as shown in figure 2. The surface of PVA/CS/RSF hydrogel is irregular and there are many holes in the internal structure of the hydrogel, as shown in figure 2 (a-b). The optical image of hydrogels is shown in figure 2 (c), they are pure PVA, PVA/CS, RSF and PVA/CS/RSF from left to right, respectively. The original pure PVA showed a white appearance, which turned yellow and thicker after introducing CS, RSF. The loading capacity of hydrogel was investigated by hanging a weight on the hydrogel. It is found that the hydrogel can bear a maximum weight of 500 g, as shown in figure 2 (d). The hydrogel can be highly compressed and then returned to its original shape, showing excellent flexibility, as shown in figure 2 (e). Similarly, as shown in figure 2 (f), the original gel is about 4.5 cm in length, and is about 9.3 cm after stretching. It can be stretched to twice its length and then restored to its original state, as shown in figure 1, the physical crosslinking of the hydrogel was not destroyed after stretching, showing excellent elasticity. Therefore, PVA/CA/RSF hydrogel is an ideal material for flexible devices.

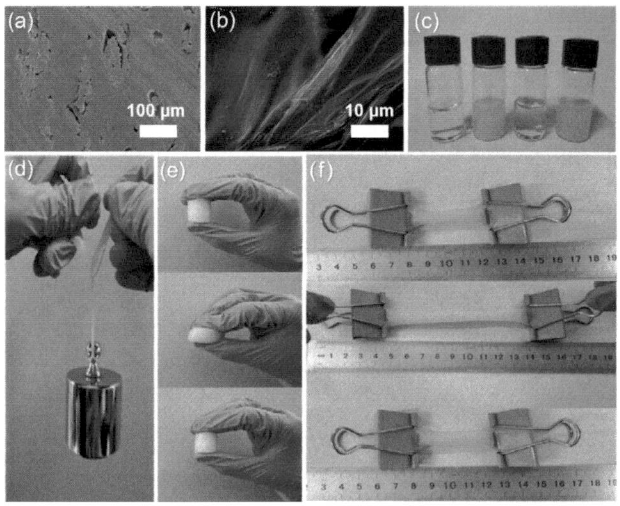

Figure 2. (a), (b) Surface and internal images of PVA/CS/RSF hydrogel. (c) Optical image of hydrogels. (d) Bearing capacity test of PVA/CS/RSF hydrogel. (e) The optical images of cylindrical PVA/CS/RSF hydrogel in its initial state, compressed and restored with fingers. (f) Optical images of PVA/CS/RSF hydrogel before, during and after stretching.

Autonomous self-healing is defined as the property that enables a material to repair damage internally and automatically. It usually occurs in prepared materials without the intervention of external stimulation to promote the occurrence or degree of self-healing [17, 18]. As shown in figure 3 (a), there are two original hydrogels. The transparent one is hydrogel without adding RSF, and the yellow one is hydrogel added with RSF. In order to study the self-healing ability of hydrogels after adding RSF, the PVA/CS/RSF hydrogel was cut into two halves on the table, as shown in figure 3 (b). Without any external interference, two hydrogels can heal autonomously in a short time, as shown in figure 3 (c). In order to detect good self-healing, the self-healing hydrogel was picked up vertically for a period of time and swayed around. PVA/CS/RSF hydrogel can support its weight and shake without breaking, as shown in figure 3 (d).

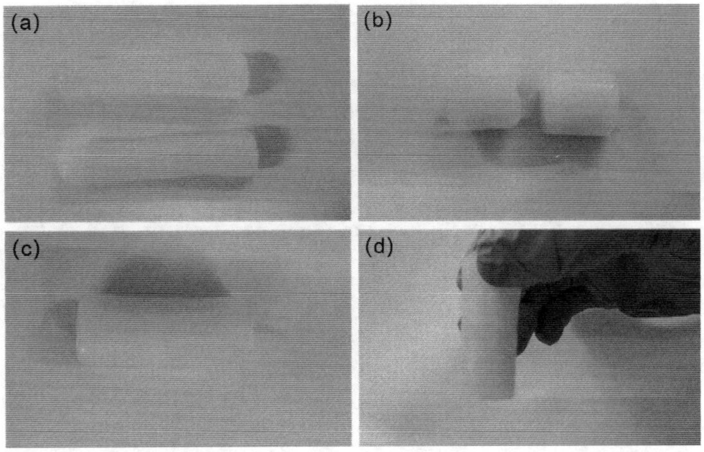

Figure 3. (a) Two original hydrogels with RSF (yellow) and no RSF (transparent). (b) Two cut raw hydrogels. (c) The self-healing condition of the hydrogel after cutting without any external stimulation at room temperature (20 °C). (d) Test whether the PVA/CS/RSF hydrogel can support its own weight without breaking after healing.

4. Conclusions

A self-healing hydrogel was made by polyvinyl alcohol/chitosan/regenerated silk fibroin (PVA/CS/RSF). It has the advantages of simple preparation technology, low material cost, practicability and environmental protection. The composite material shows excellent flexibility and elasticity, and after being cut in half, it can automatically heal in a short time. It shows great potential in wearable devices, human-computer interaction, remote personalized medical data collection and other aspects.

Acknowledgements

This work was supported by the Anhui Collaborative Innovation Project (GXXT-2019-036), Open Research Fund of National Engineering Research Center for Agro-Ecological Big Data Analysis & Application, Anhui University (No. AE201910, AE202001), Open Research Fund of Key Laboratory of Textile Fiber and Products (Wuhan Textile University), Ministry of Education (No. Fzxw2021026).

References

[1] Wei, N., Ruan, L.R., Zeng, W., Liang, D., Xu, C., Huang, L.S., Zhao, J.L. (2018) Compressible Supercapacitor with Residual Stress Effect for Sensitive Elastic-Electrochemical Stress Sensor. ACS Appl. Mater. Interfaces, 10:38057-38065.

[2] Ma, Y., Wei, N., Wang, Q., Wu, C.H., Zeng, W., Gao, Y.F., Xu, C., Huang, L.S., Zhao, J.L. (2019) Ultrathin PEDOT: PSS/rGO Aerogel Providing Tape-Like Self-Healable Electrode for Sensing Space Electric Field with Electrochemical Mechanism. Adv. Electron. Mater, 5:1900637.

[3] Huang, L.S., Zhou, X.W., Xue, R., Xu, P.F., Wang, S.L., Xu, C., Zeng, W., Xiong, Y., Sang, H.Q., Liang, D. (2020) Low-Temperature Growing Anatase TiO2/SnO2 Multi-dimensional Heterojunctions at MXene Conductive Network for High-Efficient Perovskite Solar Cells. Nano-Micro Lett., 12:44.

[4] Mondal, S., Das, S., Nandi, A.K. (2020) A review on recent advances in polymer and peptide hydrogels. Soft Matter, 16(6):1404-1454.

[5] Jayakumar, A., Jose, V.K., Lee, J.M. (2020) Hydrogels for Medical and Environmental Applications. Small Methods, 4(3):1900735.

[6] Jing, X., Li, H., Mi, H.Y., Liu, Y.J., Feng, P.Y., Tan, Y.M., Turng, L.S. (2019) Highly transparent, stretchable, and rapid self-healing polyvinyl alcohol/cellulose nanofibril

hydrogel sensors for sensitive pressure sensing and human motion detection. Sens. Actuators, B., 295: 159-167.

[7] Cipriano, B.H., Banik, S.J., Sharma, R., Rumore, D., Hwang, W., Briber, R.M., Raghavan, S.R. (2014) Superabsorbent hydrogels that are robust and highly stretchable. Macromolecules., 47(13):4445-4452.

[8] Liu, Y.R., Geever, L.M., Kennedy, J.E., Higginbotham, C.L., Cahill, P.A., McGuinness, G.B. (2010) Thermal behavior and mechanical properties of physically crosslinked PVA/Gelatin hydrogels. J. Mech. Behav. Biomed., 3(2):203-209.

[9] Zhao, N.N., Wu, F., Xing, Y., Qu, W.J., Chen, N., Shang, Y.X., Yan, M.X., Li, Y.J., Li, L., Chen, R.J. (2019) Flexible hydrogel electrolyte with superior mechanical properties based on poly (vinyl alcohol) and bacterial cellulose for the solid-state zinc-air batteries[J]. ACS Appl. Mater. Interfaces., 11(17):15537−15542.

[10] Wang, T., Gunasekaran, S. (2006) State of water in chitosan-PVA hydrogel[J]. J. Appl. Polym. Sci., 101(5):3227-3232.

[11] Yang, N.N., Qi, P., Ren, J., Yu, H.P., Liu, S.X., Li, J., Chen, W.S., Kaplan, D.L., Ling, S.J. (2019) Polyvinyl Alcohol/Silk Fibroin/Borax Hydrogel Ionotronics: A Highly Stretchable, Self-Healable, and Biocompatible Sensing Platform. ACS Appl. Mater., 11(26):23632−23638.

[12] Wang, X.Q., Yucel, T, Lu, Q., Hu, X., Kaplan, D. L. (2010) Silk Nanospheres and Microspheres from Silk/pva Blend Films for Drug Delivery. Biomaterials, 31(6):1025−1035.

[13] Kundu, J., Poole-Warren, L. A., Martens, P., Kundu, S. C. (2020) Silk Fibroin/Poly(Vinyl Alcohol) Photocrosslinked Hydrogels for Delivery of Macromolecular Drugs. Acta Biomater, 8(5):1720-1729.

[14] Xue, R., Zhou, X.W., Peng, S., Xu, P.F., Wang, S.L., Xu, C., Zeng, W., Xiong, Y. Liang, D. (2020) Architecturing Lattice-Matched Bismuthene–SnO2 Heterojunction for Effective Perovskite Solar Cells. ACS Sustain. Chem. Eng, 8:10714–10725.

[15] Xiong, Y., Hu, J., Nie, X., Wei, D., Zhang, N.G., Peng, S., Dong, X.W., Li, Y.C., Fang, P.F. (2020) One-step firing of carbon fiber and ceramic precursors for high performance electro-thermal composite: Influence of graphene coating. Mater. Design, 191:108633.

[16] Nie, X., Xiong, Y., Zeng, W. (2020) The interfacial lattice origin of anatase titanium dioxide by hydrothermal architecture. J. Cryst. Growth, 552:125927.

[17] Taylor, D.L., Panhuis, M.I.H. (2016) Self-Healing Hydrogels. Adv. Mater., 28(41):9060-9093.

[18] Hager, M.D., Greil, P., Leyens, C., Zwaag, S.V.D., Schubert, U.S. (2010) Self-Healing Materials. Adv. Mater., 22(47):5424-5430.

A dual-narrow-band near-infrared perfect absorber based on a crossed double nanorods metasurface

Ling Guo [1, 2, a], Mengran Guo[1, b], Yajie Liu [1, c] and Shan Yin[1, d]

[1] School of Electornic Engineering and Automation, Guilin University of Electronic Technology, Guilin, 541004, China

[2] Guangxi Key Laboratory of Automatic Detecting and Instruments, Guilin 541004, China

[a]email: guoling@guet.edu.cn, [b]email: g_mengran@163.com, [c]email: liuyajiejiayou@163.com, [d]email: syin@guet.edu.cn.

Abstract. Perfect absorbing materials have attracted the attention of researchers because of their wide application in solar photovoltaic industry, biosensor and filter. Here, we report on our numerical work concerning a planar metamaterial absorber, which composed of a crossed double nanorods on a dielectric layer with a gold substrate. Based on the resonance in the crossed nanorods and the similar-Bloch-wave resonance, the planar metamaterial structure can achieve multiple peaks with high absorption and narrow band. There are two nearly perfect absorption peaks obtained at 905 nm and 1161 nm, of which absorptivity and the bandwidth are 99.97%, 99.99% and 18 nm, 19 nm respectively. This peculiar feature may envisage the possible application of the structure in solar photovoltaic industry, biosensor, optical switch and filter.

1. Introduction

Metamaterial absorber is a new type of artificial electromagnetic material, which is widely used in solar photovoltaic industry [1, 2], biosensor [3, 4] and filter [5]. In recent years, more and more attention has been paid to the study of perfect absorbing materials. Multichannel absorption is an important aspect of metamaterial absorber. The first double-channel absorber is designed by Landy et al., they used a double-resonant rings to generate two resonances with different frequencies [6]. Later Tao et al. [7] studied an absorber composed of dual band ELC resonator, and the absorptions of the peaks are 85% and 94% respectively. However, the above two structures can only absorb electromagnetic waves with fixed polarization, which will limit the development and application of metamaterial absorber. It is proved that the polarization sensitivity of metamaterial absorber can be effectively eliminated by the electromagnetic resonance ring with symmetrical structure [8]. In 2018, Lei Zhao et al. proposed a dual-band absorber is designed based on the symmetry-dielectric-dielectric-metal (SDDM) structure and the absorptivity above 99% and 96%， respectively [9].

In this study, we propose a planar metamaterial absorber, which composed of a crossed double nanorods on a dielectric layer with a gold substrate. By using the multi-resonances in the crossed nanorods and the similar-Bloch-wave resonance, a high absorption of incident light can be achieve. Using finite-difference time-domain method (FDTD), the simulation is demonstrated that there are absorption peaks at 699nm, 905nm and 1161nm respectively. In particular, the absorption are close to perfect absorption, and the absorption rate can reach 99.97% and 99.99% with bandwidth are 18 nm,

19 nm respectively. The results may be applied peculiar feature may envisage the possible application of the structure in solar photovoltaic industry, biosensor, optical switch and filter.

2. Metamaterial Geometry and Numerical Model

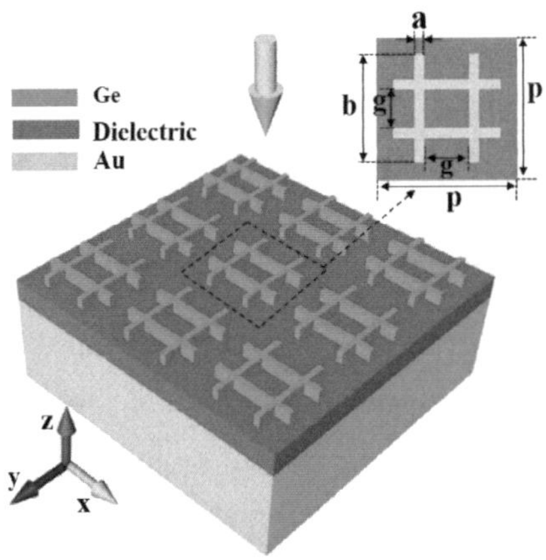

Figure 1. Perspective views of a representative sample of the proposed metamaterial absorber

Here, we designed a dual-narrow-band near-infrared perfect absorber with metamaterial. Figure 1(a) shows the geometrical configuration of a representative sample of the proposed metamaterial absorber composed of two-dimensional germanium grating on a dielectric layer with a gold substrate, in which the period P =850 nm, the nanorods width a=40 nm, the nanorods length b=655 nm, the gap between nanorods g=300nm, the refractive index of the dielectric layer N=1.5, and the height of the nanorods structure is 90 nm, and the dielectric layer thickness is 185 nm.

In the numerical simulations, the method of finite-difference time-domain (FDTD) was used. The propagation direction of the incident light filed is along the z axis, perpendicular to the structure. The permittivity of metal is defined by using Drude dispersion. The periodic boundary conditions were applied in the direction with periodicity (i.e., x-direction), the perfect matching layer conditions were applied in the z-direction boundaries, and the structure is infinite in y-direction in the two-dimensional analysis.

3. Results and Discussion

Then, we studied the proposed metamaterial absorber composed of a crossed double nanorods on a dielectric layer with a gold substrate. The absorption spectrum of the proposed structure is shown in Figure 2(a). There are three absorption peaks at 699nm, 905nm and 1161nm in the spectrum of 650nm~1400nm. In particular, the absorption at 905nm and 1161nm are close to perfect absorption, and the absorption rate can reach 99.97% and 99.99% respectively, which have higher absorptivity compared with the result of Lei Zhao et al. [9]. The absorption peak at 699 nm (Peak1) is caused by the higher-order mode in the crossed double nanorods, the electric filed distribution is shown in Figure 2(b). It is obvious that the field intensity is mainly distributed in the center of vertical bars and corners of the four bars, the corner parts and the center parts of the vertical bars are antiphase which shows two zero nodes. The absorption peak at 905 nm (Peak2) is caused by the lower-order mode in the crossed double nanorods, the electric filed distribution is shown in Figure 2(c). The field intensity is mainly localized in the vertical bars and the corners of the all bars, however the phase signs of the corner parts and the center parts of the vertical bars are identical.

For the absorption peak at 1161 nm (Peak 3), it is mainly produced by the resonance of the similar-Bloch-wave resonance, the electric filed distribution is shown in Figure 2(d). The field is not only localized in the crossed nanorods, but also in the surround medium. It is noted as the similar-Bloch-wave resonance which is formed by the back and forth propagation of the scattered velocity wave.

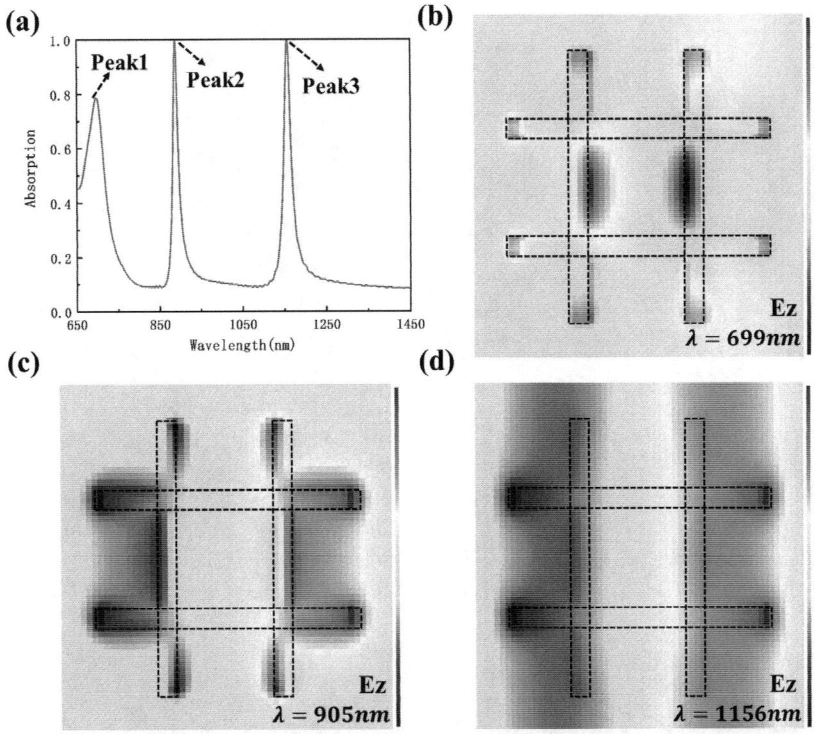

Figure 2. (a) The absorption spectrum of the proposed metamaterial absorber composed of well-shaped germanium structure on a dielectric layer with a gold substrate. The electric field (Ez) distribution of (b) $\lambda = 699nm$, (c) $\lambda = 905nm$, and (d) $\lambda = 1156nm$.

In metamaterial, the geometric parameters determine the position and properties of spectrum. Therefore, in this part, we studied the effects of the period and geometric parameters (a, b and g) of the proposed structure of each resonance. Figure 3(a) shows the absorption spectra of the proposed metamaterial when changing the parameter a. It can be noted that the geometry parameter a shows a little influence on the FWHM while the intensity of the metamaterial and the location of resonance have noticeable red shift with the increase of a. For the geometry parameter b, we can see in Figure 3(b) that all absorption spectra has a slight movement. It is worth noting that the intensity of the third peak decreases greatly when $b = 800\ nm$. In Figure 3(c), the absorption spectra of parameter g vary from 240 to 280 are given. Obviously, the third peak has almost no change, which indicates that the change of g has no effect on the similar-Bloch-wave resonance. Figure 3(d) shows the absorption spectrum of the proposed structure with period p varying from 750 nm to 850 nm. Obviously, with the increase of the period p, the positions of Peak 2 and Peak 3 are red shifted, while the absorption rate is increased but the FWHM full width at half is almost unchanged. By using the sweep and optimization algorithm, we could find the proper geometric parameters to achieve ultra-narrow FWHM of the resonance, which is beneficial to get better performance absorber.

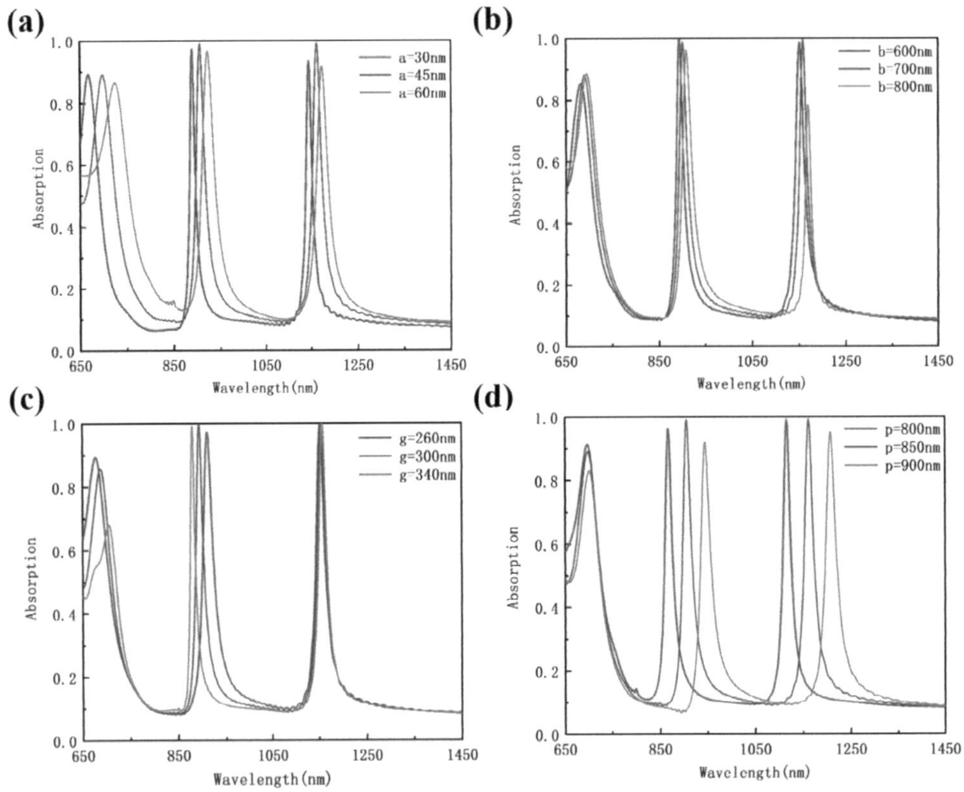

Figure 3. The absorption spectra of the proposed metamaterial absorber with various parameter (a) a while $b = 655nm, g = 300nm, p = 850nm$ (b) b while $a = 45nm, g = 300nm, p = 850nm$ (c) g while $a = 45nm, b = 655nm, p = 850nm$ (d) p while $a = 45nm, b = 655nm, g = 300nm$.

4. Conclusions

In summary, We propose a novel metamaterial absorber composed of two-dimensional grating on a dielectric layer with a gold substrate which is based on higher order resonance, lower order resonance, and similar-Bloch-wave. There are three absorption peaks formed at 698 nm, 905nm and 1161 nm. Especially, the absorption peaks at 905nm and 1161 nm are closed to perfect absorption, and the absorption rate can reach 99.97% and 99.99% with bandwidth are 18 nm,19 nm respectively. The results provides a potential applications in the peculiar feature may envisage the possible application of the structure in solar photovoltaic industry, biosensor, optical switch and filter.

Acknowledgements

This work was partially supported by Natural Science Foundation of Guangxi Province (Grant No. 2018GXNSFBA281175); Science and Technology Projects of Guangxi Province (Grant No.AD18281033); Graduate innovation project of Guilin University of Electronic Science Technology (Grant No.2021YCXS137).

References

[1] Da, Y., Xuan, Y. (2015) Perfect solar absorber based on nanocone structured surface for high-efficiency solar thermoelectric generators. Science China, 58(001):19-28.

[2] Wang, H., Sivan, V., Mitchell, A., et al. (2015) Highly efficient selective metamaterial absorber for high-temperature solar thermal energy harvesting. Solar Energy Materials and Solar Cells, 137:235-242.

[3] Wu, D., Liu, Y., Li, R., et al. (2016) Infrared Perfect Ultra-narrow Band Absorber as Plasmonic Sensor. Nanoscale Research Letters, 11(1):483.
[4] Yin, S., Zhu, J., et al. (2015) High-performance terahertz wave absorbers made of silicon-based metamaterials. Applied Physics Letters, 76(7):26.
[5] Ogawa, S., Okada, K., Fukushima, N., et al. (2012) Wavelength selective uncooled infrared sensor by plasmonics. Applied Physics Letters, 100(2):113-163.
[6] Landy N., Sajuyigbe, S., Mock, J., et al. A Perfect Metamaterial Absorber. (2008) Physical Review Letters, 100(20):207402.
[7] Hu, T., Bingham, C., Pilon, D., et al. (2010) A dual band terahertz metamaterial absorber. Journal of Physics D Applied Physics, 43(22):225102.
[8] Sun, J., Liu, L., Dong, G., et al. (2011) An extremely broad band metamaterial absorber based on destructive interference. Optics Express, 19(22):21155-21162.
[9] Zhao, L., Liu, H., et al. (2018) Design of multi-narrowband metamaterial perfect absorbers in near-infrared band based on resonators asymmetric method and modified resonators stacked method. Optics Communications, 420:95-103.

Research Progress of Effects on Nanomaterials on Pyrotechnic Properties

Qi Qin[1,a], Xiaoping Cui[2,b*]

[1,2]College of Equipment Management and Support, Engineering University of PAP, Xi'an, China

[a]email: 19931693577@163.com,

*Corresponding author: [b]email: liubo@smail.nju.edu.cn.

Abstract. Nanomaterials have been widely concerned by researchersin in recent years, because of its unique physical and chemical properties, which play an irreplaceable role in improving the performance and safety of pyrotechnics. In this paper,in the light of the effects of nanomaterials on pyrotechnic properties such as metal nanoparticles, metal oxide nanoparticles, composite nanoparticles, carbon-based nanomaterials, the latest research progress and current research situation are reviewed from two aspects including thermal decomposition temperature and decomposition heat, and the function mechanism was analysed. The future research directions are expected.

1. Introduction

Pyrotechnics is composed of combustible agent, oxidant, binder and functional additive, which can react quickly and produce many effects such as sound, light, smoke and thermal[1-2]. It is widely used in civil and military fields, but there are some problems, such as waste of resources, unsatisfactory effect, low safety factor during use and storage, which are caused by insufficient reaction of reagents. Therefore, it poses a severe challenge to the development and application of pyrotechnics.

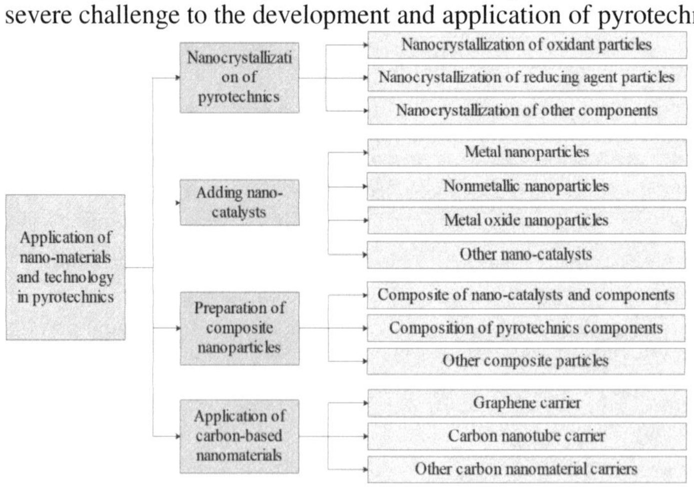

Fig. 1 Application of nano-materials and technology in pyrotechnics

In recent years, the rapid development of research and development and preparation technology of new nano-materials has gradually solved these problems. The specific application in pyrotechnics is shown in Figure 1. In this paper, the effects of nano-materials on thermal decomposition temperature and decomposition thermal of pyrotechnics are reviewed, which provides reference and support for the next pyrotechnics formula optimization.

2. Influence on thermal decomposition temperature of pyrotechnics

2.1. Metal nanoparticles

Aluminum powder is a kind of combustible agent often used in pyrotechnics, and its nanoparticles can greatly reduce the activation energy of oxidant thermal decomposition, thus reducing its thermal decomposition temperature. Zhang et al. [3] found that the initial reaction temperature of pyrotechnic agent system containing nano-aluminum powder (50 nm) decreased by 3.8%, which indicated that the difficulty of thermal reaction of pyrotechnic agent added nano-aluminum powder was effectively improved. Pantoya et al. [4] found that the ignition temperature of thermite containing nano-Al powder (17~202 nm) mixed by mechanical means decreased significantly, which was mainly due to the decrease of nano-material size, which led to a sharp increase in the specific surface area of components, and the contact between combustible agent and oxidant was more sufficient, which effectively reduced the diffusion and reaction distance, thus reducing the reaction temperature.

At the same time, further study[5] found that the initial decomposition temperature of the ultra-refined $KClO_3$ and Al system ($KClO_3$ is submicron, about 300 nm, Al powder is nanometer, 50 nm) is about 14°C lower than that of the samples before refinement. The reason is that the surface active position of particles in ultrafine pyrotechnics increases, and the chemical reaction activity increases. The thermal generated in the mixed system is not easy to dissipate, and the thermal can be absorbed quickly to reach the activation temperature required by the reaction.

In addition, Tan Linghua et al. [6] found that nano-Cu particles also have significant catalytic effect on the thermal decomposition of NH_4ClO_4 (AP). The exothermic peak temperature of AP at low temperature and high temperature decreased by 15℃ and 17.6℃ respectively with 26 μm copper powder, respectively, while that of nano-Cu (20 nm) was 35.1°C and 130.2°C, and the decrease of exothermic peak temperature of AP increased with the decrease of Cu powder particle size.

2.2. Metal oxide nanoparticles

Metal oxide nanoparticles play an important role in catalysis and energy conversion applications of nanotechnology[7]. Chen et al. [8] found that nano-Co_3O_4 particles can reduce the peak temperature of AP at high temperature by 128.5 ℃ ,and the smaller the particle size, the more the thermal decomposition reaction temperature of AP decreases. Ma et al. [9] explained by analysis and comparative study that micron Fe_2O_3 made the first and second exothermic peaks of AP move to low temperature by 1.10℃ and 62.25℃, respectively, while the values of nano-Fe_2O_3 were 61.89℃ and 118.82℃, which indicates that the catalytic activity of nano-Fe_2O_3 was obviously better than that of micron Fe_2O_3. Zhang et al. [10] found that nano-$CoFe_2O_4$,$NiFe_2O_4$ and $ZnFe_2O_4$ reduced the high temperature decomposition peak temperature of AP by 108.99℃, 82.03℃ and 30.91℃, respectively, among which $CoFe_2O_4$ played the most obvious role because the interaction between iron and cobalt was beneficial to the thermal decomposition of AP, but the main reason was that smaller microcrystals and particle sizes contained more lattice defects and active sites, which improved the reactivity and thus reduced the decomposition temperature of AP.

In order to explore the influence of the preparation method of nanomaterials on the thermal decomposition performance of AP, Hao Gazi et al. [11] found through experiments that the catalytic effect of nano-$CuCr_2O_4$ prepared by mechanical crushing method is more obvious than that prepared by coprecipitation method. Figure 2 and 3 are TEM images of nano-$CuCr_2O_4$ directly prepared by coprecipitation method and TEM diagrams of nano-$CuCr_2O_4$ prepared by grinding method, respectively. It can be seen that the particle size of nano-$CuCr_2O_4$ prepared by coprecipitation method is about 30 nm, and its distribution is uniform, but its dispersion is poor. The particle size of $CuCr_2O_4$ prepared by mechanical pulverization is about 60 nm, and the dispersion of $CuCr_2O_4$ is better. Zhu et al. [12] found that the average particle size of spherical nano-CuO samples obtained by adding NaOH at 100℃ is the smallest (about 6 nm), which reduces the decomposition temperature of AP by 102℃, which is better than 96℃ of needle-like nano-CuO, because spherical nano-CuO has larger specific surface area and stronger catalytic activity.

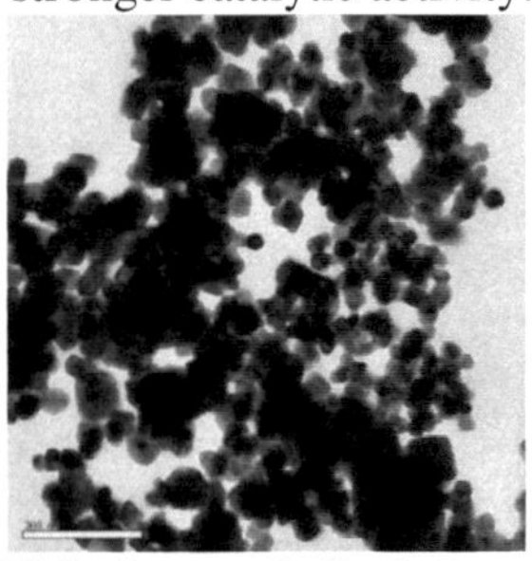

Fig. 2[11] TEM of nano-$CuCr_2O_4$(coprecipitation)

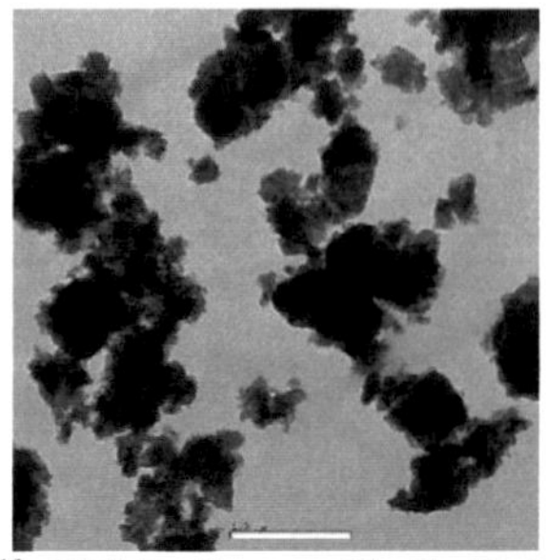

Fig. 3[11] TEM of nano-$CuCr_2O_4$(grinding)

2.3. *Composite nanoparticles*

By preparing nanocomposite materials of nanoparticles and reactants, the reaction distance can be greatly reduced, the contact area can be increased, and the catalytic effect can be improved. Song et al. [13] studied that the high and low temperature decomposition peaks of AP/Fe_2O_3 nano-composite oxidant prepared by sol-gel method were 124.96°C and 113.02°C earlier than those of pure AP, respectively. This is because the low valence oxides on the surface of composite particles play a bridge role, The transfer of electrons from ClO_4^- to NH_4^+ is promoted, and the specific surface area of nano-Fe_2O_3 increases, which increases the proportion of surface atoms. A large number of unsaturated suspended bonds and unsaturated bonds, a large number of active points and lattice defects on the surface would adsorb a large number of gas phase decomposition products of AP, which significantly promotes the high temperature decomposition of AP.

At the same time, composite nano-materials can obviously solve the agglomeration problem of nano-particles. Chen et al. [14] prepared CuO/AP composite particles with nano-CuO as the core and AP as the shell by solvent-non-solvent method, which effectively solved the problem that nano-CuO particles are easy to agglomerate, and at the same time, it advanced the high temperature decomposition peak temperature of AP by 101℃.

2.4. *Carbon-based nanoparticles*

Carbon is the most widely distributed element in nature. Zero-dimensional, one-dimensional, two-dimensional and three-dimensional carbon nanomaterials based on carbon can be widely used as carriers of energetic components by coating or packaging because of their unique physical structure and chemical properties, as well as their green and pollution-free characteristics[15].

The open pore structure of carbon nanotubes (CNTs) is beneficial to electron transfer and thermal conduction. At the same time, loading catalyst particles on the wall of CNTs can effectively prevent the agglomeration of nanoparticles. Compared with simple mixed samples, CNTs and Fe_2O_3 composite particles prepared by Jiang et al. [16] can further reduce the peak temperature of AP pyrolysis and have better catalytic performance. Meng et al. [17] also deposited titanium dioxide coating with a thickness of about 2 nm on CNTs surface by sol-gel method. As seen in Figure 4,the composite can continuously reduce the peak temperature of AP pyrolysis by 52.5℃, which indicates that the composite can further improve the catalytic performance of TiO_2 particles as AP thermal decomposition carriers. Metal Ag/CNTs nanoparticles also catalyzed the pyrotechnics of magnesium/potassium nitrate system, and the exothermic peak of ignition reaction of pyrotechnics decreased by 37℃[18]. It is found that[19] the thermal decomposition temperature would also be reduced and the reaction rate would be improved by combining CNTs with oxidant particles to prepare composite particles. Because it avoids the agglomeration of nano-scale oxidant particles, it would greatly reduce the particle size of oxidant loaded on CNTs, and improve the endothermic and exothermic rates of the reaction.

a. CNTs TEM b. TiO_2/CNTs nanocomposite TEM

Fig. 4[17] TEM images of CNTs and TiO_2/CNTs nanocomposite; the insets are their corresponding higher magnified images

Compared with carbon nanotubes, graphene (GA) has an ultra-thin nanosheet structure, which means that it has a higher specific surface area and can support more catalysts. Li et al. [20] found that nano-Mn_3O_4-GR composite reduced the thermal decomposition temperature of AP by 141.9℃. In AP/GA aerogel nanocomposites (AP average particle size is about 69.41 nm) prepared by Wang Baoxue et al. [21], the peak temperature of AP pyrolysis decreased by 83.7℃. The analysis shows that: On the one hand, graphene has a large specific surface area, which can adsorb NH_3 and $HClO_4$ to the surface when AP is decomposed at low temperature, delay their entry into gas phase and oxidation reaction, and then greatly weaken the decomposition of AP at low temperature; On the other hand, the decomposition product of $HClO_4$ reacts with graphene, which gives off a lot of thermal and causes the surface temperature of AP crystal to increase. As a result, pyrolysis occurs ahead of time.

In conclusion, the application of nano-materials can significantly reduce the thermal decomposition temperature of pyrotechnics. Compared with the nanocrystallization of pyrotechnics components, nano-catalysts and their composite materials, the addition of carbon nano-materials can significantly reduce the thermal decomposition temperature, as shown in Table 1.

Table 1 Effect of nano-materials on thermal decomposition temperature

Nanomaterial Classification	Nanoparticles	Particle Size/nm	Temperature Drop Range/°C	Reduction Range	Researchers
Metal Nanoparticles	Al	50 nm±	4.63	3.8%	Zhang, et al. [3]
	Al sub-micron $KClO_3$	$KClO_3$300 nm± Al50 nm±	14	11.3%	Zhang, et al. [5]
	Cu	20 nm±	130. 2	27.3%	Tan, et al. [6]
Metal Oxide Nanoparticles	Co_3O_4	21.8 nm±	128.5	28.4%	Chen, et al. [8]
	Fe_2O_3	8-25 nm±	118.82	25.3%	Ma, et al. [9]
	$CoFe_2O_4$	220 nm±	108.99	26.9%	Zhang, et al. [10]
	Cu_2O	19-68 nm	103	22.9%	Zhu, et al. [12]
Composite Nanoparticles	AP/Fe_2O_3	AP 19-68 nm Fe_2O_3 20nm±	124.96	28. 7%	Song, et al. [13]
	Nano-CuO/AP	CuO 40nm±	101	22.3%	Chen, et al. [14]
Carbon-based Nanomaterials	Fe_2O_3/CNTs	Fe_2O_3 10-30 nm	122.1	25.5%	Jiang, et al. [16]
	Ag/CNTs	Ag 69.41 nm±	37.1	5.9%	Pouretedal, et al. [18]
	TiO_2/CNTs	TiO_2 2 nm±	52.5	12.3%	Zhang, et al. [19]
	Mn_3O_4-GR	$Mn_3O_4$10 nm±	141.9	32.7%	Li, et al. [20]
	AP/GA	AP 69.41 nm	83.7	20.6%	Wang, et al. [21]

3. Influence on decomposition thermal of pyrotechnics

3.1. Metal nanoparticles

Tan et al. [6] found that when copper powder reaches nanometer level (20 nm), the apparent decomposition thermal of AP pyrolysis can be increased to 1.20 kJ/g, which is 172.7% higher than that of commercial micron copper powder (26 μm) with the same dose. This is due to the large specific surface area of nano-copper powder, which can promote the chemical reaction and thermal release between gases, greatly promote the decomposition of AP, and thus increase the apparent decomposition thermal of high temperature decomposition. Zhang et al. [22] found that in $Zr/KClO_4$, the traditional formula of pyrotechnic laser pump source, with the decrease of Zr particle size, the radiation energy increased obviously, and the peak radiation intensity doubled, which greatly increased the laser output energy.

3.2. Metal oxides and their composite nanoparticles

Ma et al. [9] found that the constant volume combustion thermal of simulated solid propellant increased by 5095.70 J/g and 2350.84 J/g respectively compared with pure AP after adding nano-Fe_2O_3 and micron Fe_2O_3, and the effect was remarkable. The reduction of AP particle size also contributed to the improvement of propellant decomposition thermal[23]. Wang et al. [24] found that the total thermal release of Fe_2O_3/Al nanocomposites prepared by sol-gel method reached 1648 J/g, which was 95.7% higher than that of micron-sized mixtures, and showed better ignition and energy characteristics than traditional materials. Sherif et al. [25] prepared TiO_2 colloidal particles with an average particle size of 10 nm by hydrothermal method. It was found that TiO_2 combined the two main thermal release stages of AP and increased the total thermal release by 18%. Wu et al. [26] successfully prepared a CuO/Al nano-composite energetic material with more uniform microstructure by DNA self-assembled method. Figure 5 is the TEM and SEM images of physically mixed CuO/Al samples and DNA self-assembled. It can be found that most CuO and Al in the physical blend samples (Figure 5a and 5c) are aggregated and only a few are in close contact, and DNA self-assembled samples (Figure 5b and 5d). The distribution of CuO and Al in the samples was uniform, and there was no obvious aggregation of nanoparticles. The experimental results show that the reaction thermal of DNA self-assembled samples is higher than that of physical blend samples with the same ratio, and the maximum increase range can reach 52.15%. The nanocrystallization of Al nanoparticles is also an important factor leading to the obvious improvement of thermal reaction efficiency and rate.

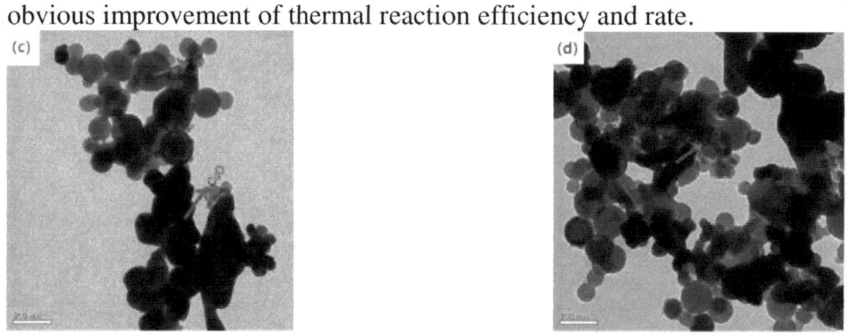

a.physically mixed nanocomposites TEM b.DNA-assembled nanocomposites TEM

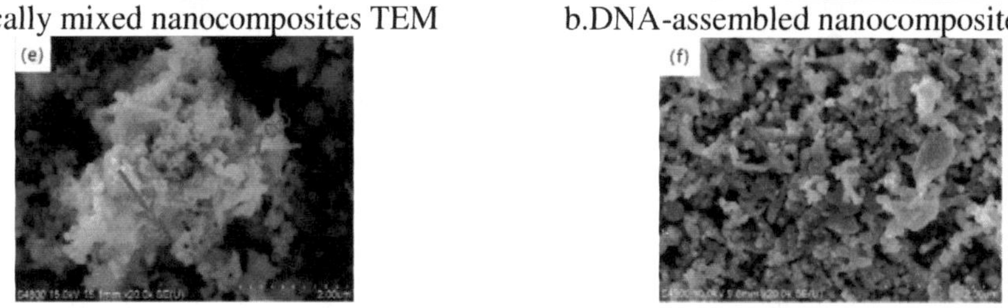

c.physically mixed nanocomposites SEM d.DNA-assembled nanocomposites SEM

Fig.5[26] TEM and SEM images of Al and CuO nanoparticles、physically mixed CuO/Al samples and DNA self-assembled CuO/Al samples(M_{Al}:M_{CuO}=0.80:1.00)

3.3. Carbon-based nanomaterials

Sherif et al. [15] studied the catalytic effect of reduced graphene oxide (RGO) and Fe_2O_3 nanocomposites on AP, and found that the total thermal release of AP increased by 83%. The main reason is that the superior interface surface area of RGO ensures the absorption of gas products on the catalyst surface, and nano-powder can reduce the required endothermic phase transition, and finally make the total thermal released surge. Lan et al. [27] found that compared with the decomposition thermal of pure AN of 1048. 62 J/g, the decomposition thermal of mechanical mixture of GA and AN increased by 251.9 J/g. Graphene aerogel/ammonium nitrate (GA/AN) nanocomposites (AN average particle size 71 nm) were prepared by sol-gel method and supercritical CO_2 drying method. As can be seen from Figure 6, GA has rich pore structure, and its skeleton is wrinkled graphene. The pore structure of GA/AN nanocomposites is obviously less, and most pores are completely covered by AN filling, which makes it increase by 280.88 J/g on the basis of mechanical mixture, and the increase range reaches 50.8%. The reasons are as follows: First, graphene itself has a catalytic effect on the thermal decomposition of KP; Second, graphene is a special two-dimensional structure, which can effectively improve the dispersion of nano-catalysts, prevent agglomeration, and facilitate the adsorption and reaction between products and catalytically active sites during KP thermal decomposition; Third, graphene has excellent conductivity, and nano-oxides are loaded on graphene, which is equivalent to enhancing its conductivity and accelerating the electron transfer process; Fourth, graphene can be used as a thermal conductive medium in the thermal decomposition of KP because of its good thermal conductivity.

a. GA SEM b. GA/AN nanocomposite SEM

Fig. 6[27] SEM images of GA and GA/AN nanocomposite

In conclusion, the application of nano-materials can significantly improve the decomposition thermal of pyrotechnics. Under the same material conditions, different mixing and preparation methods would also affect the decomposition heat to varying degrees, while the addition of carbon nano-materials can obviously improve the agglomeration phenomenon of nano-materials, make each component distribute evenly, and then greatly improve the decomposition heat of pyrotechnics, as shown in Table 2.

Table 2 Effect of nano-materials on thermal decomposition

Nanomaterial Classification	Nanoparticles	Particle Size/nm	Thermal Decomposition/ (kJ/g)	Increase Range	Researchers
Metal Nanoparticles	Cu	20 nm±	0.76	172.7%	Tan, et al. [6]
Metal Oxides and Their Composite Nanoparticles	Fe_2O_3	8-25 nm±	5.095	30.5%	Ma, et al. [9]
	Fe_2O_3/Al	20 nm±	1.648	95.7%	Wang, et al. [24]
	TiO_2/APC	10 nm±	0.089	18%	Sherif, et al. [25]
	CuO/Al	100 nm±	1.52	52.15%	Wu, et al. [26]
Carbon-based Nanomaterials	RGO/Fe_2O_3	30 nm±		83%	Sherif, et al. [15]
	GA/AN	70 nm±	0.533	50.8%	Lan, et al. [27]
	CoO/GA	80 nm±	0.66	200%	Liu, et al. [28]

4. Conclusion

In summary, the application of new nano-materials in pyrotechnics has made great progress, but there are also some problems in the development process, such as uneven dispersion of nano-particles in pyrotechnics, easy agglomeration of nano-particles and carriers, inefficient control of binding degree, and inaccurate quantification of catalytic effect of nano-catalysts. Therefore, the future application of nano-materials in pyrotechnics can be studied from the following directions:

a. Research and innovate the preparation methods and means of nano-materials, and reduce the particle size of nano-materials. At present, the particle size of nano-materials is mostly concentrated in tens to tens of nanometers, which is far from the 0.1 nm level of short-period metal atoms.

b. Innovate and explore the preparation methods of nano-agents and the technical means of combining with nano-carriers, so as to minimize the agglomeration of nano-particles, improve the degree of combination with carriers and make them uniformly distributed in the system.

c. Deepen the research and data analysis of catalytic mechanism of nanoparticles, and explore the lifting function of nanoparticles on reaction rate and effect, so as to achieve accurate and controllable catalytic effect.

d. Looking for green and environmentally friendly alternative nano-materials to reduce the pollution degree after pyrotechnic reaction. In addition, if the manufacturing cost of nano-materials can be reduced and mass production can be realized in the future, it will play an extremely important role in promoting the miniaturization and precision of pyrotechnic products.

Acknowledgements

This work was supported by the Education and Teaching Foundation of Engineering University of PAP (WJX2021141), also funded by the Basic Theory Research Foundation of Engineering University of PAP (WJY202146).

References

[1] Pan G P, Yang S. (1997) Pyrotechnics . Beijing Institute of Technology Press,Beijing.

[2] Pan G P, Yang S. (2005) Advanced Pyrotechnics. Beijing Institute of Technology Press,Beijing.

[3] Zhang F, Wang P L, Wang L Q. (2009) Effect of nano-aluminium powder on thermal safety of pyrotechnics. Journal of Safety Science and Technology,5: 47-50.

[4] Pantoya M. L, Granier J. J. (2005) Combustion Behavior of Highly Energetic Thermites: Nano versus Micron Composites. Propellants,Explosives,Pyrotech-nics,30:53-62.

[5] Zhang F, Wang P L. (2011) Characteristics of potassium chlorate/aluminum powder/sulfur ultrafine pyrotechnics. CIESC Journal, 62: 10-14.

[6] Tan L H, Li F S. (2005) Preparation of nanometer copper powder and its catalytic performance for ammonium perchlorate thermal decomposition. Journal of Nanjing Institute of Technology, 3:6-10.

[7] Chavali M. S,Nikolova M. P. (2019) Metal Oxide Nanoparticles and Their Applications in Nanotechnology. SN Applied Sciences, 1:607.

[8] Chen W F, Li F S, Liu J X, et al. (2005) Preparation of nano-Co_3O_4 and its catalytic performance for the thermal decomposition of ammonium perchlorate. Chinese Journal of Catalysis, 26: 1073-1077.

[9] Ma Z Y, Li F S, Cui P, et al. (2003) Preparation of nano-Fe_2O_3 and its catalytic performance for the thermal decomposition of ammonium perchlorate. Chinese Journal of Catalysis,24: 795-798.

[10] Zhang M,Zhao F,Yang Y,et al. (2020) Catalytic Activity of Ferrates ($NiFe_2O_4$,$ZnFe_2O_4$ and $CoFe_2O_4$) on the Thermal Decomposition of Ammonium Perchlorate. Propellants Explosives. Pyrotechnics. ,45:463-471.

[11] Hao G Z, Liu J, Xiao L. et al. (2015) Preparation of nano-$CuCr_2O_4$ by mechanical crushing and its effect on thermal decomposition properties of ammonium perchlorate. Acta Armamentarii, 36: 1654-1659.

[12] Zhu J W, Zhang W G, Wang H Z, et al. (2004) Morphology-controlled synthesis and properties of nano-CuO. Chinese Journal of Inorganic Chemistry, 07: 863-867.

[13] Song X L Li F S, Zhang J L, et al. (2009) Preparation, mechanical sensitivity and thermal decomposition characteristics of nano-AP/Fe_2O_3 composite oxidant. Journal of Solid Rocket Technology,32: 306-309.

[14] Chen A S, Li F S, Ma Z Y, et al. (2004) Preparation and catalytic performance of nano-CuO/AP composite particles. Journal of Solid Rocket Technology,02:123-125.

[15] Zhang M,Zhao F,Yang Y, et al. (2020) Catalytic Activity of Ferrates ($NiFe_2O_4$,$ZnFe_2O_4$ and $CoFe_2O_4$) on the Thermal Decomposition of Ammonium Perchlorate. Propellants Explosives.

Pyrotechnics,45:463-471.
[16] Jiang W, Liu J X, Liu Y, et al. (2008) Preparation of Fe_2O_3/CNTs composite particles and its effect on the thermal decomposition catalytic performance of AP. Journal of Solid Rocket Technology,31: 65-68.
[17] Meng S H,Liu J F,Kong X T, et al. (2020) Preparation of TiO_2/CNTs Nanocomposite and Its Catalytic Performance on the Thermal Decomposition of Ammonium Perchlorate. Transition Metal Chemistry, 45:545-551.
[18] Pouretedal H. R, Shahmoradi M, Zareh A, et al. (2019) The Catalytic Effect of Ag/CNTs Nanocomposite on the Ignition Reaction of Mg/KNO_3 Pyrotechnic by Determination of the Kinetic Triplet. Journal of Thermal Analysis and Calorimetry, 135:2975-2983.
[19] Zhang B Y, Jiao Q J, Ren H, et al. (2011) Morphology and thermal behavior of CNTs/$KClO_4$ composites. Chinese Journal of Energetic Materials, 19: 401-404.
[20] Li N, Geng Z, Cao M,et al. (2013) Well-dispersed Ultrafine Mn_3O_4 Nanoparticles on Graphene as A Promising Catalyst for the Thermal Decomposition of Ammonium Perchlorate. CARBON, 54:124-132.
[21] Wang X B, Li J Q, Luo Y J. (2012) Preparation and thermal decomposition behavior of ammonium perchlorate/graphene nanocomposites. Chinese Journal of Explosives & Propellants 35: 76-80.
[22] Zhang Q, Kang X L, Luo J S, et al. (2011) Effect of particle size of Zr powder on combustion characteristics of Zr/KClO4 pyrotechnics. High Power Laser and Particle Beams, 23: 1867-1872.
[23] Kohga M. (2011) Burning Characteristics and Thermochemical Behavior of AP/HTPB Composite Propellant Using Coarse and Fine AP Particles. Propellants Explos. Pyrotech. ,36:57-64.
[24] Wang Y, Li F S, (2008) Jiang W, et al. Preparation and reaction characteristics of Fe_2O_3/Al nanocomposite thermite. Initiators & Pyrotechnics, 04: 11-14.
[25] Elbasuney S, Yehia M. (2019) Ammonium Perchlorate Encapsulated with TiO_2 Nanocomposite for Catalyzed Combustion Reactions. Journal of Inorganic and Organometallic Polymers and Materials, 29:1349-1357.
[26] Wu X N, Xian M, Chen F S, et al. (2018) Preparation of CuO/Al nanocomposite energetic materials by DNA self-assembly. Chinese Journal of Energetic Materials, 26:1038-1043.
[27] Lan Y F, Luo Y J. (2015) Preparation and characterization of GA/AN nanocomposite energetic materials. Chinese Journal of Explosives & Propellants, 38:15-18.
[28] Liu P A, Wang M J,et al. (2019) Effect of Nano-metal Oxide and Nano-metal Oxide/graphene Composites on Tthermal Decomposition of Potassium Perchlorate. Chemical Papers,73:1489-1497.

Compressive properties and microstructure evolution of sintered nano-silver

He Gong[1]; Hongcheng Wu[1]; Hongcun Guo[1]; Yao Yao[1,2*]

[1]School of Mechanics, Civil Engineering and Architecture, Northwestern Polytechnical University, Xi'an 710072, China

[2]School of Civil Engineering, Xi'an University of Architecture and Technology, Xi'an 710055, China

*Corresponding author; email address: yaoy@nwpu.edu.cn

Abstract. In this work, the compressive properties of sintered nano-silver applied to the third generation semiconductor at room temperature are tested at a loading rate of $10^{-2}s^{-1}$. The compressive stress-strain curve shows obvious elastic and plastic stage, and with the failure of the pore wall and sintering neck, the damage accumulates gradually until failure. The microstructure is analyzed statistically by scanning electron microscope. A finite element model for porous structure is developed by matching scanning electron microscope analysis. The elastic modulus of pore wall and sintering neck is obtained by combining the macro experimental data with the Ashby model. The local failure process of sintered nano-silver is simulated numerically. It is noted that the failure path is determined by large pore and pore spacing.

1. Introduction

The commonly used lead-free solder (SnAgCu) can hardly match the high temperature working environment of the third generation semiconductor [1, 2]. It is urgent to find alternative packaging materials suitable for high operating temperature. After sintering, solidified nano-silver can withstand temperature up to 961℃[3], and it has excellent electrical and thermal conductivity. A small amount of organic matter in the nano-silver paste can prevent the aggregation of silver nanoparticles during storage. During the sintering process, the volatilization of organic matter leads to the formation of a large number of pores in the solidified nano-silver, which causes discrete of mechanical properties of structural parts welded by nano-silver paste [4]. Therefore, the macro [3-7] and micro mechanical properties [8-10] of sintered nano-silver have attracted wide research interest in semiconductor industry.

Due to the mismatch of thermal expansion coefficient between the chip and substrate, the sintered nano-silver suffers from compressive stress in the actual working environment. At present, the researches of sintered nano-silver of compressive experiments mainly based on nanoindentation technology [8, 9, 11] and using micro-compression specimen [12]. However, the macro and micro mechanical properties of sintered nano-silver are not completely consistent due to existence of internal pores. There are few reports on the experiments and numerical simulations of the mechanical properties of sintered nano-silver under macro pressure. In the current study, rectangular specimens were prepared for compressive experiments at room temperature, and the microstructure of sintered nano-silver was analyzed by scanning electron microscope (SEM). The simulation of local failure of sintered nano-silver was realized by finite element analysis of porous structure using ABAQUS.

2. Experimental method and analysis

The sintered nano-silver adopted in this work is provided by INDIUM company, and the silver nanoparticles are irregular polygons before sintering, as shown in Figure 1 (a). According to statistical analysis, the particle size is approximately 359.26 nm [4]. The nano-silver paste was injected into the pouring tank to prepared the compressive specimen, and heated at 300 ℃ for 50 minutes by QUICK 870 heating table, and cooled to room temperature in the air environment. The size of compressive specimen is 3×3×4.5mm, as shown in Figure 1 (b).

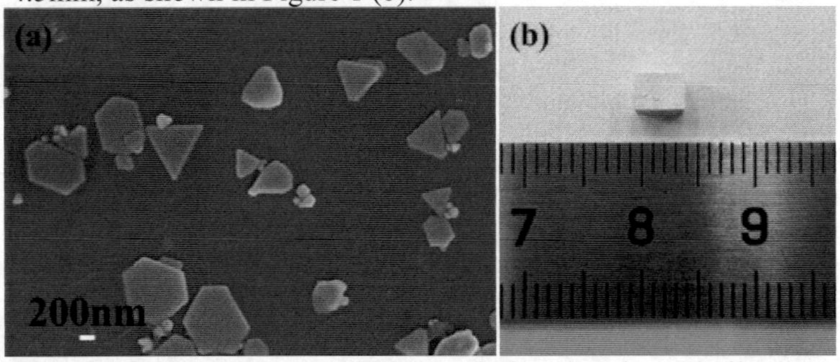

Figure 1 Morphology of nano-silver paste before and after sintering: (a) silver nanoparticles; (b) compressive specimen.

During the sintering process, the silver nanoparticles (About 91% of silver content) in the nano-silver paste are stacked, and the adjacent silver nanoparticles are fused with each other to form the three-dimensional porous solid. The fusion of two adjacent particles forms a sintering neck, as shown in Figure 2 (a). Generally, a plurality of nanoparticles is interconnected to form a supporting structure (pore wall), which is similar to the block in the concrete structure, as shown in Figure 2 (b). Sintering neck and pore wall can conduct force and electricity. Most of the cured structures show ductile failure in the process of tension as observed by SEM. It is noted that the cross section for SEM scanning in Figure 2 was prepared through tension. Therefore, it shows that the curing degree is reasonable at this sintering temperature. In addition, a small amount of interface failure occurs, which is characterized by the smooth fracture surface, as shown in Figure 2 (b). Some of the silver nanoparticles inside the pores have become sphere under the effect of surface tension, as shown in Figure 2 (c). It should be noted that there also exist larger defects, which are not the focus of present study and will not be described in this work.

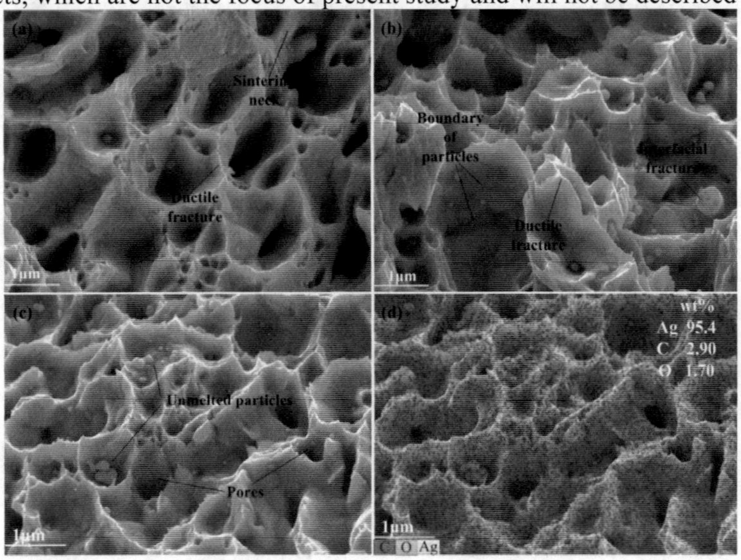

Figure 2 Microstructure and Energy Dispersive Spectrometer (EDS) of sintered nano-silver: (a) sintering neck; (b) boundary of particles; (c) pores and unmelted particles; (d) element distribution.

EDS analysis results of the local cross section is shown in Figure (d). It shows that the internal organic matter of nano-silver paste basically combusts completely after sintering at 300 ℃ for 50 minutes. Thus, the contents of carbon and oxygen are relatively low, as shown in Figure 2 (d). This also confirms that the selected sintering temperature can make the organic matter burn completely, and beneficial to the silver nanoparticles solidify to form an impregnable structure.

The Nano Measurer 1.2 was used to analyze the maximum pore diameter in Figure 2 (a). Here the maximum diameter is defined as the distance between the maximum distance on the edge of the pore. A total of 113 pores are counted, the largest diameter is 2.16 μm and the minimum diameter is 0.03 μm. The average diameter is 0.36 μm, and the porosity is 57.87% as calculated by Image J (Threshold = 114). It should be noted that the porosity in this work represents only meso porosity and does not include macropores.

3. Experimental results and numerical analysis

3.1. Experimental results

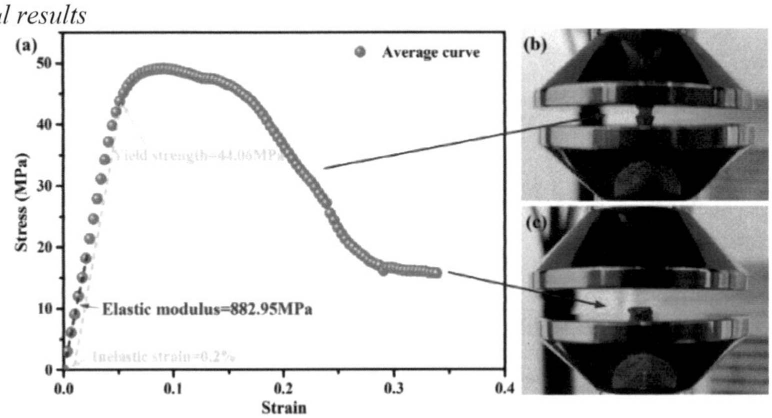

Figure 3 Compressive experiment of sintered nano-silver: (a) stress-strain curve; (b) damage stage; (c) destruction stage

The compressive experiments were conducted for three specimens at a loading rate of $1\times10^{-2}s^{-1}$ at room temperature. The average stress-strain curves of the three specimens are shown in Figure 3. There are obvious elastic and plastic segments in the deformation process, as shown in Figure 3 (a). In the plastic segment, the internal damage occurs due to the failure of sintering neck, and the fracture occurs after approaching the critical strain. In this work, the strength corresponding to 0.2% residual strain is taken as the yield strength, which is 44.06MPa, and the corresponding elastic modulus is 882.95MPa. The maximum strength is 49.07MPa, and the corresponding plastic strain is 0.037. The damage starts to sprout [13] because the sintering neck gradually begins to break as shown in Figure 3 (b), and the failure occurs when the plastic strain reaches the critical plastic strain (About 0.28), as shown in Figure 3 (c).

3.2. Qualitative analysis of failure by finite element method

The elastic modulus of the whole specimen is analyzed in this work, which includes the influence of defects. From the microscopic analysis, it can be observed that there are a lot of pores in the sintered nano-silver. Therefore, it is necessary to make the elastic modulus equivalent when establishing the finite element model.

The elastic modulus E of porous materials can be obtained by determine porosity P and Poisson's ratio υ_0 [13]. Ashby et al. [14] proposed the elastic modulus of porous materials as:

$$E = cE_0(1-P)^n \tag{1}$$

where E_0 is the elastic modulus of solid material; $n \approx 2$ [13]; c can be obtained by [15]:

$$c = \frac{1}{1 + (2 - 3v_0)P} \tag{2}$$

Combined with the results of SEM, a code of random porous model was developed using Python, and the local microstructure of sintered nano-silver was analyzed by finite element model. The model size is consistent with Figure 4 (a), which is 4.5 (height) × 6 (width) μm. In this work, the pore is assumed to be a circle. The radius of pores is generated by random function:

$$r_i = r_{min} + random \times (r_{max} - r_{min}) \tag{3}$$

where r_i is the radius of the *i*th circle; r_{min} and r_{max} are the minimum and the maximum radius of the generated circle, respectively. In this work, $r_{min} = 0.03\mu m$, $r_{min} = 2.16\mu m$, and the porosity is 57.87%; Random is a function.

Equation (4) is used to control that the *i*th and *i+1*th circle do not intersect:

$$D > r_i + r_{i+1} \tag{4}$$

The elastic modulus of the sintering neck with removing porosity is 7558.2MPa from Equation (1) and (2). In this work, the failure mode of microstructure is analyzed qualitatively by finite element method, therefore, the constitutive model of sintering neck is assumed to be an ideal elastic model. The finite element model is shown in Figure 4 (a).

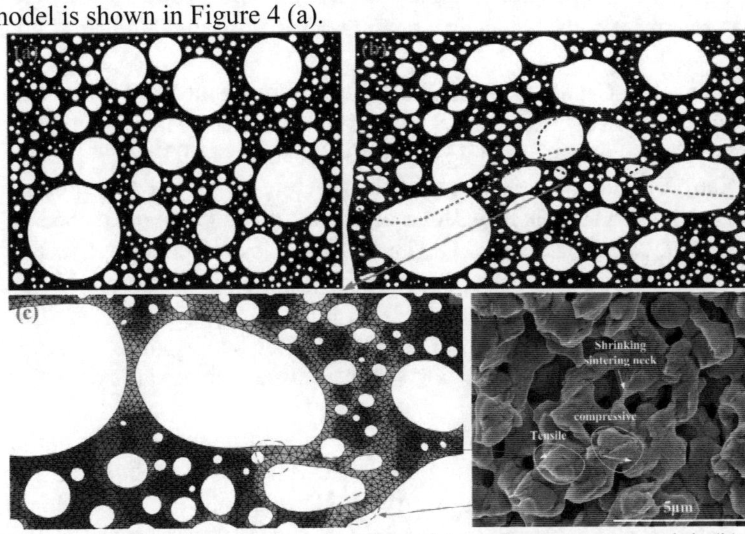

Figure 4 finite element model of sintered nano-silver: (a) random porous model; (b) fracture path; (c) enlarged image of failure region; (d) scanning electron microscopy after failure.

The finite element analysis shows that the failure path of sintered nano-silver is dominated by the pores with larger size in the microstructure, as shown in Figure 4 (b). Smaller minimum distance between pores may cause higher risk of fracture, where the failure of sintering neck and pore wall is the key factor. A large number of internal pores lead to different stress states and failure modes of sintering neck and pore wall in compressive process, which are verified by finite element analysis and experimental results, as shown in Figure 4 (b), (c) and (d).

4. Conclusion

In this work, we found that the internal pores present multi-scale distribution, the pore wall and sintering neck work together to bear the compressive stress, which is confirmed by the microstructure and finite element analysis of the sintered nano-silver. The internal microstructure of sintered nano-silver is established by random porous model according to SEM image. The failure path is determined by the large pores and the interval between the neighboring pores.

Authorship contribution statement

Yao Yao and He Gong designed the experiment, performed the experiments and wrote the manuscript. He Gong and Hongcheng Wu analyzed experimental data. He Gong and Hongcun Guo wrote the numerical code. Yao Yao supervised the research.

Acknowledgment

The authors would like to acknowledge the financial support by the National Natural Science Foundation of China (No. 11772257); Natural Science Foundation of Shaanxi Province (No. 2020JM-103); Innovation Foundation for Doctor Dissertation of Northwestern Polytechnical University (No. CX201948) and Fundamental Research Funds for the Central Universities (No. G2019KY05212).

Conflict of Interest: The authors declare that they have no conflict of interest.

References

[1] D.C. Katsis, J. Wyk, Void induced thermal impedance in power semiconductor modules: some transient temperature effects, Industry Applications Conference.

[2] H.K. Kim, H.K. Liou, K.N. Tu, Three - dimensional morphology of a very rough interface formed in the soldering reaction between eutectic SnPb and Cu, Applied Physics Letters 66(18) (1995) 2337-2339.

[3] J.G. Bai, Z.Z. Zhang, J.N. Calata, G.Q. Lu, Low-Temperature Sintered Nanoscale Silver as a Novel Semiconductor Device-Metallized Substrate Interconnect Material, IEEE Transactions on Components & Packaging Technologies 29(3) (2006) 589-593.

[4] H. Gong, Y. Yao, Y. Yang, Size effect on the fracture of sintered porous nano-silver joints: Experiments and Weibull analysis, Journal of Alloys and Compounds 863(3) (2021) 158611.

[5] G. Chen, X.H. Sun, P. Nie, Y.H. Mei, G.Q. Lu, C. Xu, High-Temperature Creep Behavior of Low-Temperature-Sintered Nano-Silver Paste Films, Journal of Electronic Materials 41(4) (2012) 782-790.

[6] G. Chen, Z.S. Zhang, Y.H. Mei, X. Li, D.J. Yu, L. Wang, X. Chen, Applying viscoplastic constitutive models to predict ratcheting behavior of sintered nanosilver lap-shear joint, Mechanics of Materials 72(may) (2014) 61-71.

[7] H. Gong, Y. Yao, F. Zhao, Corrosion effects on sintered nano-silver joints and the secondary biological hazards, Journal of Materials Science Materials in Electronics (4) (2020).

[8] X. Long, B. Hu, Y. H. Feng, C. Chang, M. Y. Li, Correlation of microstructure and constitutive behaviour of sintered silver particles via nanoindentation, Int J Mech Sci 161 (2019) 105020.

[9] X. Long, Z. Li, X. Z. Lu, H. C. Guo , C. Chang, Q. R. Zhang, A. Zehri, W. Ke, Y. Yao, L. L. Ye, J. Liu, Mechanical behaviour of sintered silver nanoparticles reinforced by SiC microparticles, Mat Sci Eng A-Struct 744 (2019) 406-414.

[10] H. Gong, C. H. Chen, Y. Yao, A void evolution-based damage model for ductile fracture of metallic materials, J Micromech Mol Phys 4 (2020) 1950008.

[11] X. Long, W. B.Tang , Y. H. Feng, C. Chang, L. M. Keer, Y. Yao, Strain rate sensitivity of sintered silver nanoparticles using rate-jump indentation, Int J Mech Sci 140 (2018) 60-67.

[12] C. Chen, K. Suganuma, Microstructure and mechanical properties of sintered Ag particles with flake and spherical shape from nano to micro size, Materials and Design (2018).

[13] J. Carr, X. Milhet, P. Gadaud, S. Boyer, G.E. Thompson, P. Lee, Quantitative characterization of porosity and determination of elastic modulus for sintered micro-silver joints, Journal of Materials Processing Technology 225 (2015) 19-23.

[14] M.F. Ashby, A. Evans, N.A. Fleck, L.J. Gibson, J.W. Hutchinson, H.N.G. Wadley, Metal foams: a design guide, Applied Mechanics Reviews 23(6) (2002) 119-119.

[15] N. Ramakrishnan, V.S. Arunachalam, Effective elastic moduli of porous solids, Journal of Materials Science 25(9) (1990) 3930-3937.

Mechanisms of Sinomenine Nanoparticles in the Treatment of Breast Cancer Based on Network Pharmacology

Ting Chen[2], Xulong Huang[1], Chao Yang[2] and Bin Huang[1*]

[1] Hunan Province Key Laboratory for Antibody-based Drug and Intelligent Delivery System, School of Pharmaceutical Sciences, Hunan University of Medicine, Huaihua, Hunan, 418000, China

[2] Department of Pharmacy, Guizhou Health Vocational College, Tongren, Guizhou, 554301, China

[*]Corresponding author's e-mail: yyan425@aucklanduni.ac.nz

Abstract. This article mainly discusses the mechanism of SINOMENINE nanoparticles in the treatment of breast cancer by using the network pharmacology method. TCMSP and Swiss Target Prediction databases were searched for the target of SINOMENINE. The "breast cancer" target was searched through the GeneCards database. The common targets of both SINOMENINE and breast cancer were considered as the targets of SINOMENINE for treating breast cancer. The software Cytoscape 3.7.1 was used for topological analysis and visualization. The DAVID databases were used for GO enrichment analysis. A total of 57 key targets for the treatment of breast cancer were obtained, of which AKT1, MTOR, EGFR, NOS3, MAPK8, MMP9, etc. The results of pathway enrichment and functional annotation show that SINOMENINE mainly involves protein phosphorylation and positive regulation of nitric oxidebiosynthetic process by regulating Sphingolipid signaling pathway, Pathways in cancer, Measles, Smallcell lung cancer, Focal adhesion, and Insulin resistance. There are 10 major biological processes, including protein phosphorylation, signal transduction, negative regulation of apoptotic process, and protein autophosphorylation. This study not only initially revealed potential molecular mechanism of SINOMENINE, but also provided a reference for the in-depth development of SINOMENINE nanoparticles for the treatment of breast cancer.

1. Introduction

Breast cancer is a highly malignant tumor, endangering the health of women all over the world[1]. According to the statistics, breast cancer is one of the most common cancers in women in the United States, accounting for about 14% of all malignant tumor deaths in women[2]. In northern Europe and other countries, It also shows a high incidence[3]. In China, the incidence of breast cancer ranks first among all cancers in women, and the main cause of death among cancer patients over 45 years of age is breast cancer[4-5]. Therefore, breast cancer is still one of the most difficult cancer diseases in clinical practice. Common treatments for breast cancer include surgery, radiation therapy, chemotherapy, endocrine therapy, biological therapy[6]. The current treatment of breast cancer is mainly a comprehensive treatment based on surgery. In order to prolong the survival rate, reduce recurrence and metastasis, chemotherapy is currently the commonly used treatment option, but some patients cannot tolerate the complications of chemotherapy and receive effective treatment. Therefore, the development of natural medicines are easily tolerated by patients.

SINOMENINE is an alkaloid monomer extracted from the plant of the Caulis Sinomenii[7]. It's effects include immunosuppressive, analgesic, anti-inflammatory, antihypertensive and anti-arrhythmic[8]. With the gradual deepening of research on SINOMENINE, it has been found that SINOMENINE can inhibit a variety of malignant tumor cells. The has obvious inhibitory effects on cervical cancer, lung cancer, breast cancer, liver cancer, gastric cancer and other tumors. However, SINOMENINE has low bioavailability and poor targeting, We prepared it into SINOMENINE chitosan nanoparticles, which greatly improved its anti-tumor effect, thus revealing the mechanism of its anti-breast cancer through the method of network pharmacology.

Network pharmacology is based on the theory of systems biology, starting from the holistic and systemic nature of the interaction between drugs, targets and diseases, and using complex biological network models to reveal the complex overall biological network between drugs-genes-targets-disease. Through such an analysis process, it can predict the pharmacological mechanism of the drug[9-12]. This study, based on the analysis method of network pharmacology, comprehensively explores the mechanism of action of SINOMENINE nanoparticles in the treatment of breast cancer from the perspective of overall and system biology. Therefore, it provide a reference for the pharmacological mechanism of SINOMENINE nanoparticles and clinical application.

2. Materials and Methods

2.1. SINOMENINE-Related Targets

The TCMSP database and SwissTargetPrediction (http://www.swisstargetprediction.ch/) database were used to search the targets of SINOMENINE action and establish the active ingredient target data set.

2.2. Breast Cancer-Related Targets

With "breast cancer" as the key word, our screening criteria are "Gene" and "Homo sapiens". Disease targets are searched through the Genecards database (http://www.genecards.org/).

2.3. Network Construction

The drug and the disease target are intersected. The target is the predicted of SINOMENINE in the treatment of breast cancer. PPI network was constructed by the String database (https://string-db.org/) Cytoscape 3.7.1 software constructed a "drug-disease" intersection network and visually analyzed. Each node satisfied the median of the topological parameters " Degree " , " Betweenness centrality " , and " Closeness centrality " value is screened. Select nodes that simultaneously satisfy more than 3 card values, As the core target of SINOMENINE in the treatment of breast cancer. At the same time, cytoscape 3.7.1 software is used to construct the " ingredient-core target-disease" network.

2.4. Bioinformatic Analysis

We use DAVID (https://david.ncifcrf.gov/tools.jsp) to perform gene ontology (GO) biological process and pathway enrichment analysis based on the Kyoto Encyclopedia of Genes and Genomes (KEGG) on the predicted target.

The biological processes, cell composition, molecular functions and signal pathways with extremely significant differences were screened outwith $P \leq 0.05$ (as the screening criteria) by R language tools.

2.5. Pathway Mapper Construction

Use the KEGG Mapper function in the KEGG (https://www.genome.jp/kegg/) database to map the core targets on the relevant signal pathways.

3. Results

3.1. Target Prediction
All the targets retrieved from the TCMSP and Swiss TargetPrediction databases. The duplicate targets were also deleted. Finally, a total of 62 target corresponding to SINOMENINE was obtained.

In the GeneCards database, with the keyword "Breast Cancer", 15207 breast cancer-related disease targets were retrieved and integrated.

3.2. Screening of Core Targets
Take the intersection of the drug target and the disease target to get 57 intersection targets, and draw the Venn diagram. The results are shown in Figure 1. Enter the intersection target into the String database (https://string-db.org/) to construct the PPI network, and the result is shown in Figure 2. Use cytoscape 3.7.1 software to visually analyze the PPI network, the median values of the three topological parameters of all nodes are 6 (Degree), 0.0040 (Betweenness centrality), 0.4206 (Closeness centrality), and we select to satisfy more than three topologies. the node with the median value of the parameter is used as the core target. The results are shown in Table 1.

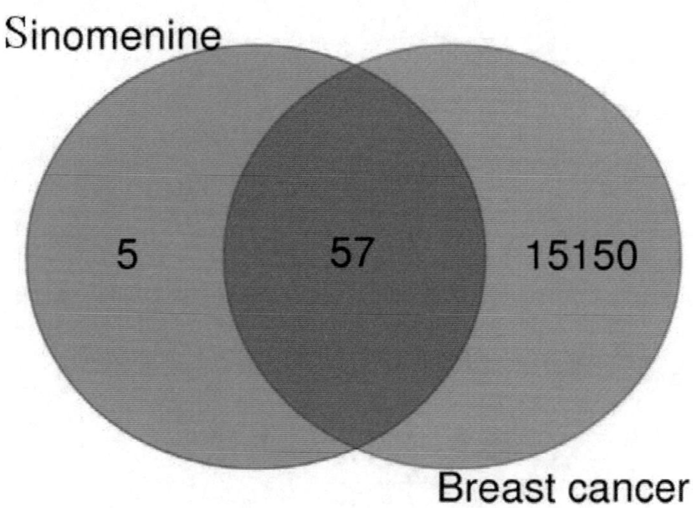

Figure 1. Drug-disease intersection target Venn diagram

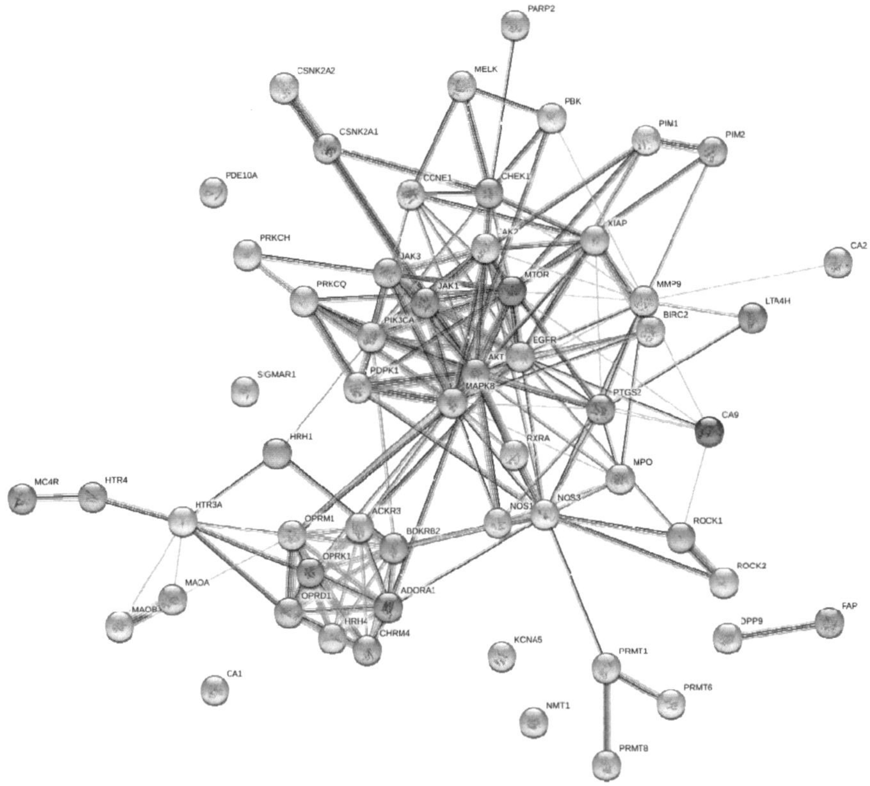

Figure 2. The protein interaction network of drug-disease intersection target

Table 1. Related topological parameters of the core target

UniProt CID	Gene name	Protein name	Degree	Closeness Centrality	Betweenness Centrality
P31749	AKT1	AKT Serine/Threonine Kinase 1	24	0.6125	0.2761
P29474	NOS3	Nitric Oxide Synthase 3	15	0.5158	0.1905
P35372	OPRM1	Opioid Receptor Mu 1	11	0.4757	0.1316
P45983	MAPK8	Mitogen-Activated Protein Kinase 8	18	0.5506	0.1010
P14780	MMP9	Matrix Metallopeptidase 9	16	0.5000	0.0971
P42345	MTOR	Mechanistic Target Of Rapamycin Kinase	19	0.5213	0.0893
O14757	CHEK1	Checkpoint Kinase 1	11	0.4667	0.0771
P00533	EGFR	Epidermal Growth Factor Receptor	18	0.5326	0.0743
P42336	PIK3CA	Phosphatidylinositol-4,5-Bisphosphate 3-Kinase Catalytic Subunit Alpha	13	0.5000	0.0605
P30411	BDKRB2	Bradykinin Receptor B2	11	0.4537	0.0506
P41145	OPRK1	Opioid Receptor Kappa 1	9	0.4579	0.0469
P30542	ADORA1	Adenosine A1 Receptor	9	0.4712	0.0307
P25106	ACKR3	Atypical Chemokine Receptor 3	9	0.4579	0.0278
P35354	PTGS2	Prostaglandin-Endoperoxide Synthase 2	13	0.4804	0.0275
P24864	CCNE1	Cyclin E1	7	0.4375	0.0141
O60674	JAK2	Janus Kinase 2	11	0.4579	0.0115
Q04759	PRKCQ	Protein Kinase C Theta	6	0.4224	0.0091
P98170	XIAP	X-Linked Inhibitor Of Apoptosis	10	0.4495	0.0042

3.3. Drug-Disease-Core Target Network Construction

The "drug-disease-core target" network was constructed by cytoscape 3.7.1 software, and the results are shown in Figure 3

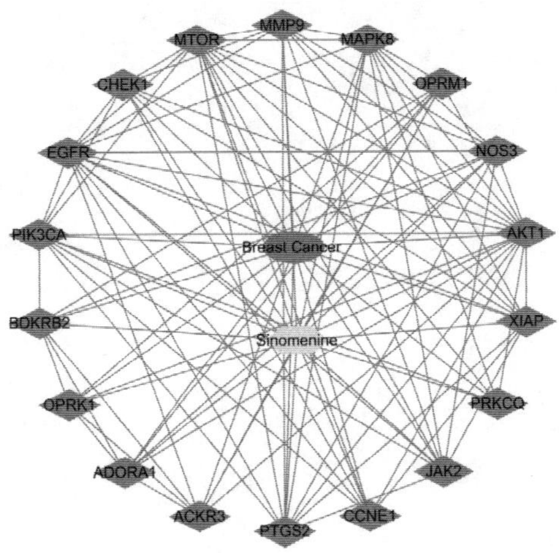

Figure 3. Drugs-Diseases-Core Targets Network

3.4. GO Biological Process Enrichment Analysis

These targets are subjected to GO enrichment analysis and Visualization by DAVID. Retain the results of P≤0.05, and perform visual analysis by R language tools. According to the smaller the Term P-value and the higher the reliability of Term, 134 biological processes were screened. The top 10 biological processes, cell composition, and molecular functions are listed. The results are shown in Figure 4. The results Indicate that the anti-breast cancer effect of SINOMENINE may be produced through protein phosphorylation, positive regulation of nitric oxide biosynthetic process, protein autophosphorylation, negative regulation of apoptotic process, peptidyl-serine phosphorylation and other biological processes.

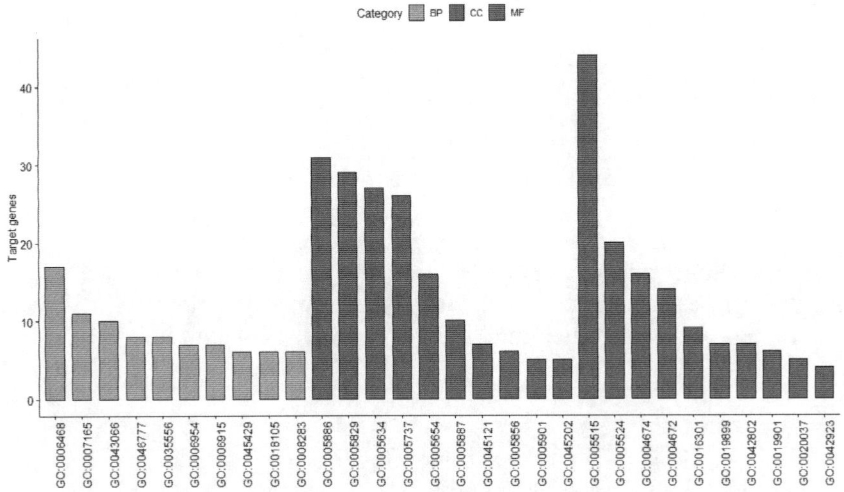

Figure 4. Top 10 biological process enrichment results

3.5. KEGG Pathway Enrichment Analysis

DAVID (https://david.ncifcrf.gov/tools.jsp) was used to perform KEGG pathway enrichment analysis, and the results of P≤0.05 were retained by R language tools. A total of 60 signal pathways were enriched. The results are shown in Figure 5. The main signaling pathways include Sphingolipid signaling pathway, Pathways in cancer, Measles, Small cell lung cancer, Focal adhesion, Insulin resistance, etc. It is suggesting that the anti-breast cancer effect of SINOMENINE may be the result of the participation of multiple signaling pathways.

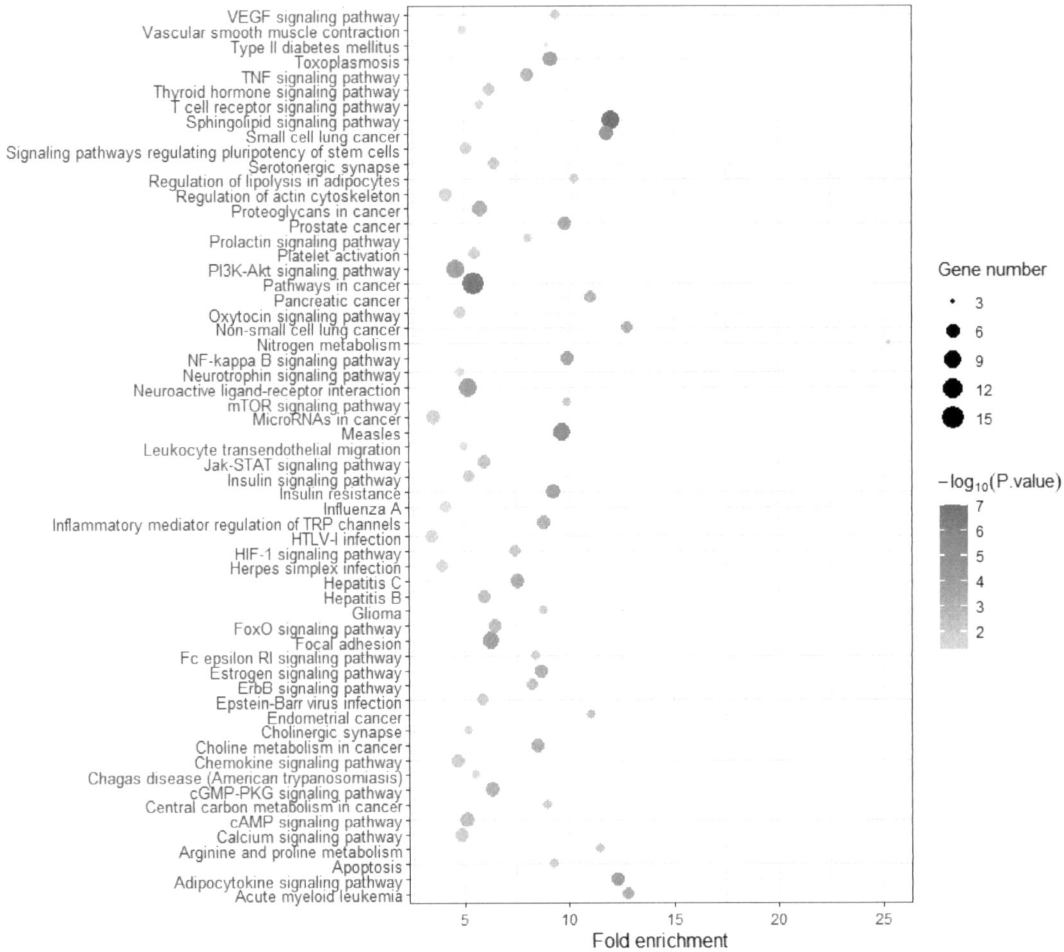

Figure 5. The 60 pathways enriched by major hubs

3.6. Pathway Annotation Diagram of SINOMENINE's Anti-Breast Cancer Effect

Enter the intersection target into the KEGG Mapper function of the KEGG database, and mark the number of the target in each relevant signal pathway. 15 target proteins are involved in regulating Pathways in cancer, revealing the multi-target and multi-pathway anti-cancer effects of SINOMENINE.

4. Discussion

SINOMENINE has a wide range of pharmacological effects. It is mainly used clinically for anti-inflammatory, immunosuppressive, antihypertensive, antiarrhythmic, analgesic, and sedation[13]. In recent years, the anti-tumor activity of SINOMENINE has also been confirmed in many ways, and a large number of results indicate that SINOMENINE has a good inhibitory effect on breast cancer, lung cancer, liver cancer, and other cancers. The network pharmacological analysis results show that AKT1, MTOR, EGFR, NOS3, MAPK8, and MMP9 may be the potential of SINOMENINE in breast cancer treatment. AKT1 is a serine/threonine protein kinase that participates in a variety of biological processes, including metabolism, proliferation, cell survival, growth and angiogenesis. They are all mediated through a series of downstream substrates of serine/threonine phosphate.

GO biological process enrichment analysis shows that the anti-breast cancer effect of SINOMENINE involves biological processes such as apoptosis, signal transduction, biosynthesis, and inflammation. Among them, protein phosphorylation, signal transduction, negative regulation of apoptotic process, protein autophosphorylation, etc. Intracellular signal transduction may be the main process of SINOMENINE's anti-breast cancer effect. Enrichment analysis of KEGG pathway showed that SINOMENINE mainly regulates Pathways in cancer, PI3K-Akt signaling pathway, Sphingolipid signaling pathway, Neuroactive ligand-receptor interaction, Measles, Focal adhesion, Proteoglycans in cancer, Small cell lung cancer, and Insulin resistance, Toxoplasmosis and other pathways. These targets play a very important role. In the PI3K/AKT signal pathway, phosphorylated PI3K and Akt will become important carcinogenic mediators, and mTOR downstream of Akt signal transduction is involved in the regulation of cell autophagy[14]. MDA-MB-231 breast cancer cells of production of ROS and the high expression of COX2 will increase the migration and invasion ability of cells[15]. SINOMENINE may inhibit the production of ROS and inflammatory signal transduction in breast cancer, thereby reducing metastasis.

5. Conclusions

A total of 57 potential targets were obtained through the online database, of which 18 are the core targets for SINOMENINE to inhibit breast cancer. GO biological process enrichment analysis and KEGG pathway enrichment analysis revealed 10 biological processes, cell composition, molecular functions and 60 pathways are closely related to the occurrence and development of inhibiting breast cancer, mainly involving Pathways in cancer, PI3K-Akt signaling pathway, Sphingolipid signaling pathway, Neuroactive ligand-receptor interaction, Measles, Focal adhesion, Proteoglycans in cancer and other molecular events. This study explored the mechanism of SINOMENINE nanoparticles on breast cancer, provided new ways and ideas for elucidating the mechanism of SINOMENINE nanoparticles, and also provided a theoretical reference for its application.

Acknowledgments

The authors would like to thank the financial support from the Scientific Research Foundation of Hunan Provincial Education Department (20B415), the Technology Project Foundation of Tongren ([2018]48), the Technology Project Foundation of Tongren ([2019]55).

References

[1] Chi, L., Zou, Y., Qin, L., Ma, W., Hao, Y., Tang, Y., Luo, R., Wu, Z. (2017) TIMELESS contributes to the progression of breast cancer through activation of MYC. J. Breast Cancer Res., 19: 53.

[2] Siegel, R.L., Miller, K.D., Jemal, A. (2018) Cancer statistics, 2018. J. CA:CancerJ Clin., 68: 7-30.

[3] Belkić, K., Cohen, M., Wilczek, B., Andersson, S., Berman, A.H., Márquez, M., Vukojević, V., Mints, M. (2015) Imaging surveillance programs for women at high breast cancer risk in Europe: Are women from ethnic minority groups adequately included? (Review). J. Int J Oncol., 47: 817-39.

[4] Chen ,W., Zheng, R., Baade, P.D., Zhang, S., Zeng, H., Bray, F., Jemal, A., Yu, X.Q., He, J. (2016) Cancer statistics in China, 2015. J. CA Cancer J Clin., 66: 115-32.

[5] Fan, L., Strasser-Weippl, K., Li, J.J., St Louis, J., Finkelstein, D.M., Yu, K.D., Chen, W.Q., Shao, Z.M., Goss, P.E. (2014) Breast cancer in China. J. Lancet Oncol., 15: e279-89.

[6] Tang, H.B., Cai, J.W., Liu, W.G., Jin, H.Z., Shao, X.Y., Mao, W.M., Wang, X.J. (2014) Effects of Paeonol on Growth Inhibition of Human BreastCancer Cells. J. Journal of Chinese Oncology., 20: 276-281.

[7] Xu, W.J., Yang, W.F., Yan, H., Chu, W.H., Chen, T.X. (2019) Recent advances in anti-inflammatory mechanisms of sinomenine. J. Clin J Tradit Chin Med., 31: 1223-1225.
[8] Qin, F., Cai,H. (2016) Research progress on pharmacological effects of sinomenine. J. ResPract Chin Med., 30: 81-86.
[9] Barabási, A.L., Gulbahce, N., Loscalzo, J. (2011) Network medicine: a network-based approach to human disease. J. Nat Rev Genet., 12: 56-68.
[10] Song, K.K., Bi, T., Zhan, X.R., Li, Z.Y., Wang, J.W., He,W., Tong, Y.Y., Li, Y.W. (2014) Strategy of Discovering Active Ingredient in Traditional Chinese Medicine under Guidance of Network Pharmacology. J. Modernization of Traditional Chinese Medicine and Materia Medica-World Science and Technology., 16: 27-31.
[11] Sheng, S., Wang, J., Wang, L., Liu, H., Li, P., Liu, M., Long, C., Xie, C., Xie, X., Su, W. (2014) Network pharmacology analyses of the antithrombotic pharmacological mechanism of FufangXueshuantong Capsule with experimental support using disseminated intravascular coagulation rats. J. Ethnopharmacol., 154: 735-44.
[12] Li, H., Zhao, L., Zhang, B., Jiang, Y.,Wang, X.,Guo, Y., Liu, H., Li, S., Tong, X. (2014) A network pharmacology approach to determine active compounds and action mechanisms of ge-gen-qin-lian decoction for treatment of type 2 diabetes. J. Evid Based Complement Alternat Med., 2014: 495840.
[13] Tang, L., Lin, Y., Liu, L.P., Lu, X.S., Hu, S.T., Zhang, E.B., Zhang, F., Dai, Z.S., Song, H.P., Huang, Y.M., Cai, X., Liu, L. (2021) Advances in Basic Studies on Antitumor Effect and Underlying Molecular Mechanisms of sinomenine. J .Chin J Exp Tradit Med Form., 27: 175-185.
[14] Saiki, S., Sasazawa, Y., Imamichi, Y., Kawajiri, S., Fujimaki, T., Tanida, I.,Kobayashi, H., Sato, F., Sato, S., Ishikawa, K., Imoto, M., Hattori, N. (2011) Caffeine induces apoptosis by enhancement of autophagy via PI3K/Akt/mTOR/p70S6K inhibition. J. Autophagy., 7: 176-87.
[15] Schexnayder, C., Broussard, K., Onuaguluchi, D., Poché, A., Ismail, M., McAtee, L., Llopis, S., Keizerweerd, A., McFerrin,H., Williams, C. (2018) Metformin Inhibits Migration and Invasion by Suppressing ROS Production and COX2 Expression in MDA-MB-231 Breast Cancer Cells. J. Int J Mol Sci., 19: 3692.

Controllable plasmonic nanostructures induced by ultraviolet laser

Dongfang Li[1,2], Yanping Yuan[1,2*], Chunlian Wang[1,2], Jimin Chen[1,2]

[1] Key Laboratory of Trans-scale Laser Manufacturing Technology (Beijing University of Technology), Ministry of Education, Beijing 100124 China

[2] Institute of Laser Engineering, Faculty of Materials and Manufacturing, Beijing University of Technology, Beijing 100124 China

* Corresponding author: ypyuan@bjut.edu.cn

Abstract. The plasmonic resonance express as strong extinction band and great enhancement of the electromagnetic field in the vicinity of the nanoparticles, which have been exploited in various applications such as data storage, super-resolution imaging, light collimation, solar cells, biosensors and absorbers. However, a simple and effective route to obtain uniform metal nanoparticles is still a challenge. Here, we have demonstrated a cost-efficient, lithography-free, and facile methods to obtain controllable gold nanoparticles (GNPs) for structure color by direct ultraviolet (UV) laser writing. The thermals dewetting process realized by UV laser direct writing, greatly simplifies the manufacturing process of plasmonic nanoparticles, lowers the bar for equipment, and realizes the structure color in selected areas. The direct laser writing process is performed on a 4 nm-thin island-like Au films atop dielectric-metal sandwiches. By local laser heating, the semi-continue gold film merges and reshapes into approximately spherical structure, generating brand colors. The structure color changes as a function of the laser energy density, low laser energy density to green and high laser energy density to red. Especially for lager laser energy density, the processed area shows a high saturation Numerical simulations confirm that the reflectance of the processed area is strongly affected by the absorption of the bottom metal layer and the local surface plasmon resonance of the top GNPs. In addition, the top GNPs play a major role in tunning the peak of reflection spectrum.

1. Introduction

In recent years, plasmonic generated by metallic nanoparticles has drawn enormous interest for routing and manipulating of light at nanoscales. The plasmonic resonance expresses as strong extinction band and a great enhancement of the electromagnetic field in the vicinity of nanoparticles, which has been exploited in various applications, such as data storage, super-resolution imaging, light collimation, solar cells, biosensors and absorbers. [1,2] To tune the plasmonic resonance properties, the nanoparticles have to be controlled over their geometry and separation as well as dielectric environment. [3] To date, ion-beam milling, multistage e-beam and nanoimprint lithography provide an accuracy and high resolution design and fabrication of plasmonic nanostructures for mass production. These processes generally exist the inevitable shortcomings of high cost, small area and time-consuming. [1] The dewetting process has been widely regarded as a cheap and simple method for large-area fabrication of nanoscale plasmonic structures. [4,5] Usually, the dewetting effect carried out on a hot plate or in annealing furnace, leads the entire area dewetting which limits the wide application. Therefore, a way to achieve nanostructures in selected areas with high efficiency and low cost is more desirable.

Here a cost-efficient, lithography-free, and facile method is proposed for fabricating plasmonic nanostructures with controllable arrays. The thermals dewetting process with UV laser direct writing extremely simplifies the fabrication process of gold nanostructures, and the GNPs can be easily obtained in desired area. The sizes and density of the GNPs can be tuned by changing the density of the laser energy. Thereby, producing structure colors span from green to red. Meanwhile, the metal-insulator-metal (MIM) configuration is adopted for enhance the color saturation.

2. Materials and Methods

2.1. Sample Preparation
The 4 nm-thick semicontinuous Au films on the top layer is deposited by ion beam sputtering system (Quorum Q150R Plus) at current of 8 mA. The 100 nm thick Au layer and 40 nm thick SiO2 layer are prepared using a commercial magnetron sputtering equipment DENTON Discovery 635.

2.2. Laser Writing
The sample is mounted on a computer-driven Nano-positioning platform. The UV laser with wavelength of 355 nm, pulse duration 5 ns and frequency 100 kHz is focused onto the sample surface by a NA = 0.43 objective lens (Nikon Engineering NUV 50×). The laser power is adjusted by the continuous attenuator in the optical path.

2.3. Morphology and optical Characterization
The morphology of laser-induced area is investigated by a field emission SEM (SU-8020, Hitachi). The optical image and reflection spectroscopy are conducted on a Ti-U microscope (Nikon) equipped with a halogen light source and spectrometer system (Andor SR-303i-B). The reflected light is collected using an objective with ×20 magnification (NA = 0.4).

2.4. Numerical Simulations
The optical response is calculated using a commercial FDTD solver available from Lumerical Solutions. The refractive indices of Au and silicon are taken from material library included in the software. The periodic boundary conditions are defined along the x-axis and y-axis and perfectly matched layer is applied for top and bottom boundaries. The plane waves are set half a wavelength away from the nanostructure and incident along the -z direction.

3. Results and discussion
Generally, the semi-continuous Au films can be melt and aggregated, forming larger GNPs under annealing process. The Rayleigh instability is identified as the main mechanism for the dewetting process. [6] The size and shape of GNPs can be adjusted by controlling the annealing temperature and time, leading to the modulating of the optical resonance. Figure 1a schematically depicts the processing principle of GNPs based on direct laser writing. A UV laser beam is tightly focused on semi-continuous Au films by a UV objective (NA=0.43) with a diameter of ~1 μm. The sample has a MIM configuration composed by a 4 nm-thick semicontinuous Au films (as shown in Figure 1b), 30 nm-thick silica spacer and a 100 nm-thick gold reflective layer. The Au back-reflector is set to be 100 nm thick to prevent light transmission in the entire visible spectrum. The focused laser has a high energy density, giving a possibility of locally realizing the dewetting process in selected areas. As shown in Figure 1b-c, after the laser irradiation, the island-like gold films transforms into regular Au nanoparticles with an average diameter of 40 nm, which can greatly change the reflectance of the processed area. The new formed GNPs and bottom Au film constitute a gap-surface plasmon (GSP) resonators, which is pivotal for improving the color saturation. [7]

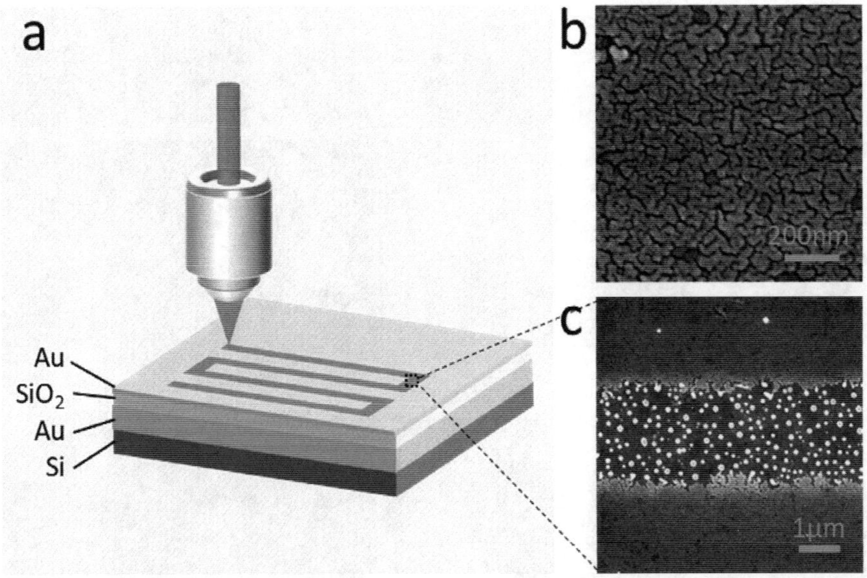

Figure 1. a Schematic illustration of plasmon structure fabricated by UV laser direct writing on the surface of the MIM structure. b SEM image of the semi-continuous Au films on the top layer. c SEM image of the area exposed by the laser beam.

Under irradiation of various laser powers, the semi-continuous gold films form nanostructures with different particle densities and particle sizes. Firstly, the scanning speed of 10 μm/s and an incident power of 40-70 μW are used for direct laser writing, and the corresponding reflection images are shown in Figure 2a. According to previous research7 for the processed area, low laser power corresponds to green, and high power exposure corresponds to red. However, in Figure 2a, the processed area remains to be a red reflection color, as the laser power decreases. To expand the structural color range of the processed area, the different scanning velocity is adopted. The simplified laser energy density E (E=P/V, where P is laser power and V is scanning velocity.) is used to evaluate the energy irradiating on Au film. As shown in Figure 2b, with the increase of E, the color uniformity is improved and the optical image shows the color that span from slightly green to bright red. The SEM image in Figure 2c shows that the GNPs have an average particle size of 150mm, and a particle density of 88/μm2. The GNPs with a diameter of less than 10 nm account for a large proportion. This phenomenon may be related to the merge and explosion of the gold film under laser exposure. As shown in Figure 2d-f, with the increase of E (from 1.65 to 41.6 J/m), the GNPs result an increase of particle density and decrease of particle size. Meanwhile, the particle morphology and color purity shows a tendency to be uniform, especially in high laser energy density.

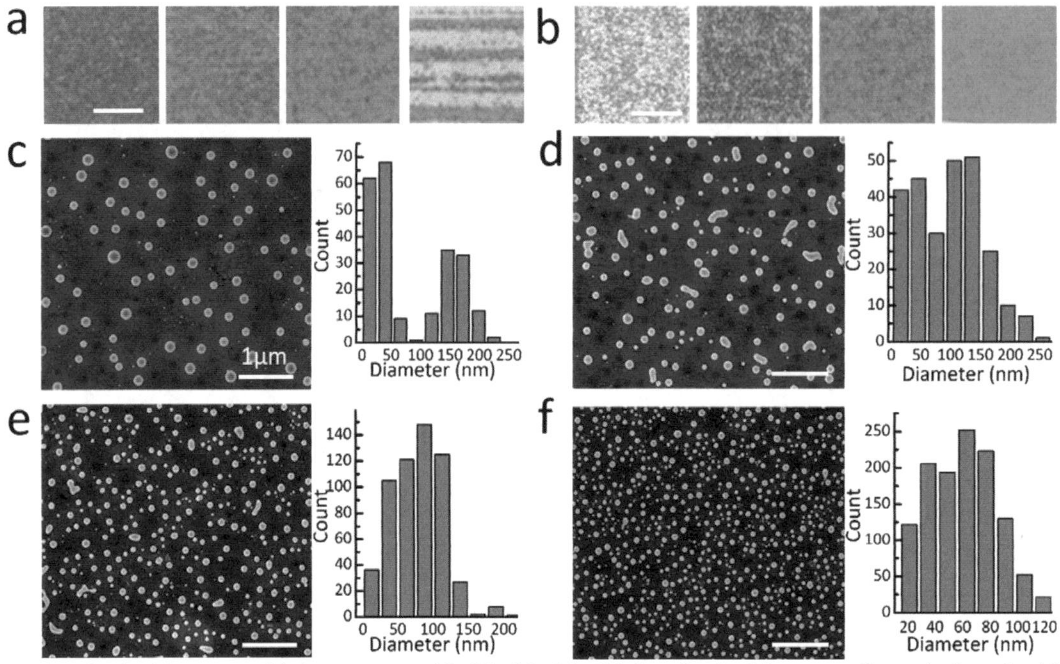

Figure 2 a Optical images with laser power 42, 52, 55, 64 μW. b-c Optical images of area induced with E 1.44, 1.65, 4.8, 41.6 J/m. Scale bar: 20μm(a,b). c-f The SEM images and particlesize distribution for the area in (a).

Figure 3a depicts the reflection spectrums measured on a home-build micro-spectroscopy. The reflectance of the unexposed area is presented by a black curve, showing a reflection peak at 630 nm and reflectance smaller than 40% in the entire spectrum. For the sample exposed at laser energy density of 1.44 and 1.65 J/m, the reflection peaks are blue-shifted to 570nm and 550nm respectively. For samples with laser energy density 4.8 and 41.6 J/m, the corresponding reflection peaks are almost disappeared. While new reflection peaks appear in the 700-800 band, and the reflectance are greatly increase to ~80%. To further understand the physical mechanisms involved in the optical response caused by the morphology of GNPs, numerical simulations are performed with commercially available software (FDTD Solutions) based on the Finite-difference time-domain (FDTD) method. The schematic model is illustrated in Figure 3b I, where the Au nanoparticles are assumed spherical and have plane contact with the SiO2 layer. The geometric parameters of the simulated model are based on the particle distribution and particle size statistics in Figure 2. The simulated reflectance spectra for the sample with brightly saturated red are plotted in Figure 3a by red dash line, which shows a good agreement with the measured reflectance spectra. For detecting the physical origin of the spectral characteristics and corresponding colors of the processed architecture, we explored the optical response of the two additional model II and model III in Figure 3b. These two models are used to calculate the independent absorption spectral of the GNPs and the bottom gold layer, which is set same geometric parameters in model I. As shown in Figure 3c, in the entire wavelength range, the absorption spectrum characteristics of the f-p model are jointly affected by the top GNPs and the bottom gold layer. For the wavelength 500-600nm, the spectrum of the top GNPs is a major contributor for affecting the absorption spectrum of the f-p model. The absorption peak around 520 nm of the top GNPs is associate to the localized surface plasmon resonances. [8] This peak is corresponding well to that of the f-p model, indicating that the top GNPs plays an important role in the entire model. Hence, the optical characteristic of the MIM structure can be effectively controlled by adjusting the laser-induced GNPs.

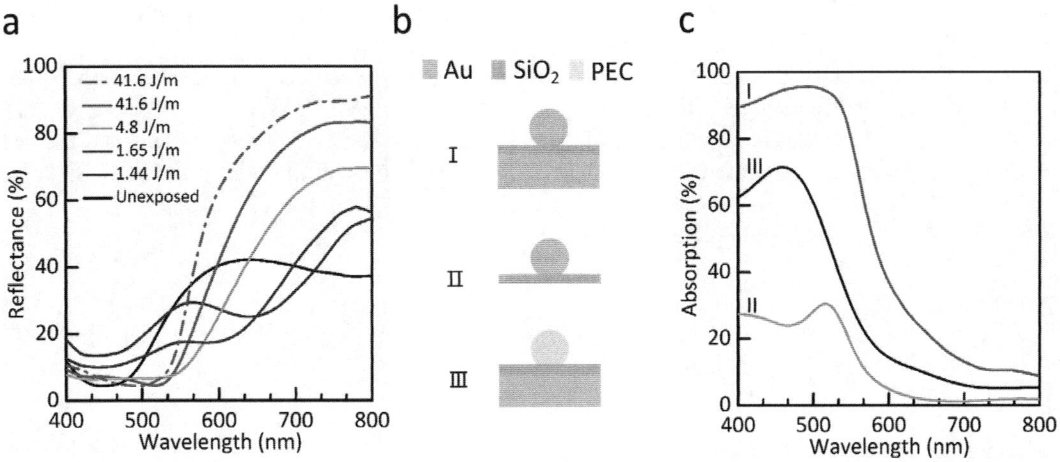

Figure 3. a The experimental and simulated reflectance spectra of the film in Figure 2. b The simulated absorption spectra in for different combination of architecture. The model II is combined by GNP and silica spacer. The model III is composed by perfect electrical conductor (PEC) material, silica spacer and gold reflective layer. c The corresponding simulated absorption spectra in (b).

4. Conclusion

In conclusion, we have demonstrated a simple and effective route to obtain uniform GNPs for structural color by direct UV-laser writing. By laser-heating-induced, the semi-continue gold film transform into the GNPs with different morphologies and particle densities, showing brand structure colors. The structure color varies from green to red, especially showing a high saturation in the red spectrum band. The structure color changes as a function of the laser flounce, low laser energy density to green and lager laser energy density to red color. Numerical simulations confirm that the reflectance of the processed area is main affected by the absorption of the bottom metal layer and the local surface plasmon resonance of the top Au nanoparticles. In addition, the top Au nanoparticles play a major role in tuning the peak of reflection spectrum. This facile direct writing method allows us to prepare GNPs in arbitrary graphics, which is prospective for various applications including data storage, security marking, biosensor and color printers.

Acknowledgments

This research was financially supported by National Natural Science Foundation of China (NSFC) (Grant No. 51805014) National Key R&D Program of China (Grant No. 2018YFB1107401), the Research Foundation from Ministry of Education of China (6141A02033123), Scientific Research Program of Beijing Municipal Education Commission (Grant No. KM201810005012).

References

[1] Yang, K., Yao, X., Liu, B. & Ren, B. (2021) Metallic Plasmonic Array Structures: Principles, Fabrications, Properties, and Applications. Advanced Materials, 2007988.
[2] Lee, S. et al. (2021) Controlled Assembly of Plasmonic Nanoparticles: From Static to Dynamic Nanostructures. Advanced Materials, 2007668.
[3] Goerlitzer, E. S., Puri, A. S., Moses, J. J., Poulikakos, L. V. & Vogel, N. (2021) The Beginner's Guide to Chiral Plasmonics: Mostly Harmless Theory and the Design of Large - Area Substrates. Advanced Optical Materials, 2100378.
[4] Song, J. M. & Lee, J. S. (2016) Self-assembled nanostructured resistive switching memory devices fabricated by templated bottom-up growth. Sci Rep 6, 18967.
[5] Liu, J. et al. (2020) Fabrication of Au network by low-degree solid state dewetting: Continuous plasmon resonance over visible to infrared region. Acta Materialia 188: 599-608.

[6] Ruffino, F. et al. (2012) Formation of nanoparticles from laser irradiated Au thin film on SiO2/Si: Elucidating the Rayleigh-instability role. Materials Letters, 84: 27-30.
[7] Roberts, A. S. et al. (2019) Laser Writing of Bright Colors on Near-Percolation Plasmonic Reflector Arrays. ACS Nano, 13: 71-77.
[8] Tesler, A. B. et al. (2011) Tunable Localized Plasmon Transducers Prepared by Thermal Dewetting of Percolated Evaporated Gold Films. The Journal of Physical Chemistry C, 115: 24642-24652.

An Exploration on Mechanisms of Andrographolide Nanoemulsifying in Treating Gastric Cancer Based on Network Pharmacology

Ting Chen[2], Xulong Huang[1], Kaimin Zheng[2], Jinping Zhong[2] and Bin Huang[1*]

[1]Hunan Province Key Laboratory for Antibody-based Drug and Intelligent Delivery System, School of Pharmaceutical Sciences, Hunan University of Medicine, Huaihua, Hunan, 418000, China

[2]Department of Pharmacy, Guizhou Health Vocational College, Tongren, Guizhou, 554301, China

[*]Corresponding author's e-mail: yyan425@aucklanduni.ac.nz

Abstract. This article preliminarily revealed the mechanism of andrographolide (AD) nanoemulsifying inhibiting gastric cancer(GC). The targets of AD was searched by TCMSP and SwissTargetPrediction database, we entered the keyword "Gastric Cancer" in GeneCards database to find all targets related to GC. The intersection of AD and GC is taken as an important targets for AD to act on GC, Cytoscape 3.7.1 was used for topological analysis of the target, so as to obtain all important targets for AD to inhibit GC. DAVID database was used to perform biological process analysis on all potential targets, mainly through its analysis function. It involves Cell composition, molecular function and signal pathway enrichment analysis. Finally, we obtained a total of 53 potential targets related to GC, among which IL6, MAPK8, CREBBP, CDK2, MAPK14, AR and MAP2K1 may be important targets for inhibiting GC. The effect of AD on GC is mainly through Influenza A, Pathways in cancer, Sphingolipid signaling pathway, Progesterone-mediated oocyte maturation, Herpes simplex infection and other important pathways which are closely related to 10 biological processes such as protein phosphorylation, signal transduction, peptidyl-serine phosphorylation, positive regulation of gene expression, and positive regulation of transcription from RNA polymerase II promoter Sex. This study can provide guidance for the development of anti-GC drugs.

1. Introduction

On a global scale, Gastric Cancer (GC) is malignant gastrointestinal tumor. GC has a high incidence rate in my country[1], and the mortality rate ranks first among many cancers[2]. The clinical treatment of GC is mainly chemotherapy, but the toxic and side effects of chemotherapy drugs have brought much pain to patients, so we need to find a natural drug that is effective and side effects to relieve the pain of the patient.

Andrographolide (AD) is mainly isolated from Andrographis Paniculata plant[3]. Previous studies have shown that andrographolide has antibacterial, anti-inflammatory, antiviral, and immune regulation functions[4]. In recent years, the research of AD in the field of anti-tumor has become a hot spot. AD has a significant inhibitory effect on breast cancer, lung cancer, prostate cancer and other tumor cells[5-7], However, AD has low bioavailability and poor targeting, We prepared it into AD self-nanoemulsifying, which greatly improved its anti-tumor effect, thus revealing the mechanism of its anti-GC through the method of network pharmacology.

Network pharmacology is an emerging discipline that combines pharmacology, biology, statistics, chemistry, network analysis technology and other sciences. Recently, we mainly applied network pharmacology in new drug development[8], drug mechanism research[9], and disease marker discovery[10]. This study used bioinformatics methods to predict the mechanism of AD acting on GC, providing a reference for finding new anti-GC drugs and targets.

2. Materials and Methods

2.1. AD Targets

We employed the TCMSP and SwissTargetPrediction (www.swisstargetprediction.ch) database to search the targets of AD and collect data sets of AD targets.

2.2. GC Targets

In the GeneCards database (www.genecards.org), entered the keyword "gastric cancer", GC-related targets are obtained, and data sets of disease targets are established.

2.3. Network Construction

Drug and disease targets intersect, and the targets in the intersection are the important targets for AD to inhibit GC. String database (https://string-db.org/) analysis function is used to construct PPI network, cytoscape 3.7.1 software is used to visually analyze PPI network, and Degree, Betweenness centrality and Closeness centrality of values is determined. Select nodes that satisfy more than 3 card values at the same time as the core of targets.

2.4. Bioinformatic Analysis

In the DAVID (https://david.ncifcrf.gov/tools.jsp) database, genetic ontology (GO) is used to analyze biological processes(BP), cell composition(CC) and molecular functions(MF). Kyoto Encyclopedia of Genes and Genomes (KEGG) is used for pathway enrichment analysis.

The R language tool performs visual analysis, $P \leqslant 0.05$ as the screening criteria to screen out biological processes, cell compositions, molecular functions and signal pathways with extremely significant differences.

2.5. Pathway Mapper Construction

KEGG Mapper function was used to in the KEGG (https://www.genome.jp/kegg/) database map the core targets on the closely related signal pathways.

3. Results

3.1. Targets Obtained

In the TCMSP and Swiss Target Prediction databases, 61 AD targets were obtained, and in the GeneCards database, 12,202 GC targets were obtained.

3.2. Screening of Core Targets

Cytoscape 3.7.1 software was used for visual analysis, and the core target was selected that satisfies the median value of more than three topological parameters at the same time. The median values of the three topological parameters are 4 (Degree), 0.0065 (Betweenness centrality), 0.4144 (Closeness centrality). The results are shown in Table 1.

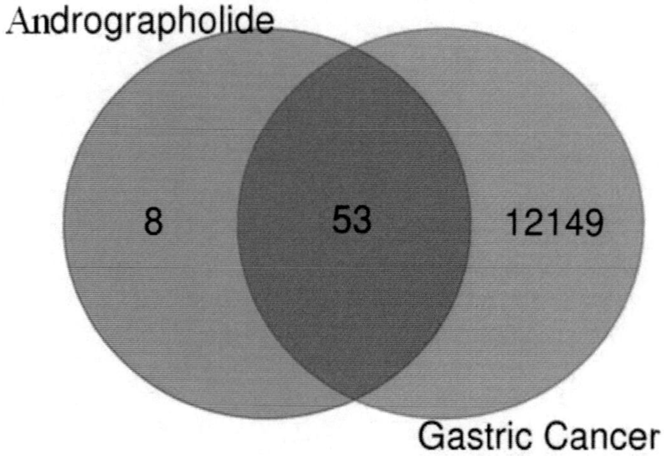

Figure 1. Drug-disease intersection target Venn diagram

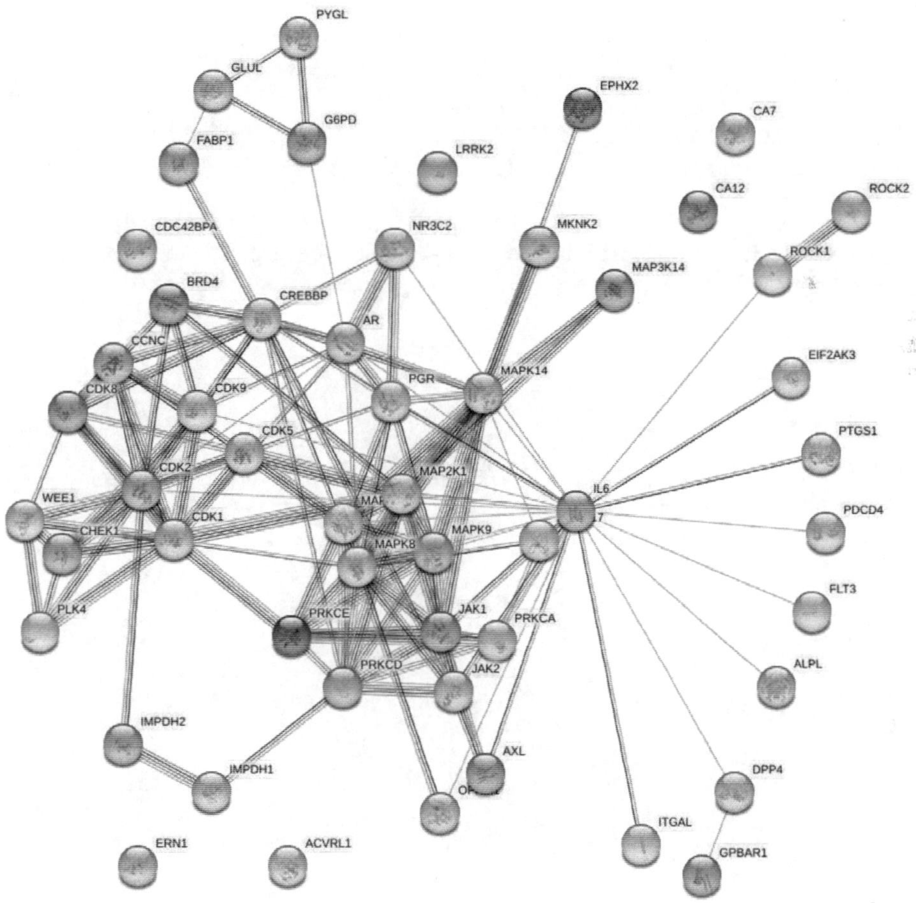

Figure 2. The PPI network diagram

Table 1. Core targets related information

UniProt CID	Gene name	Protein name	Degree	Closeness Centrality	Betweenness Centrality
P05231	IL6	Interleukin 6	25	0.6571	0.4840
P45983	MAPK8	Mitogen-Activated Protein Kinase 8	15	0.5610	0.0672
Q92793	CREBBP	CREB Binding Protein	14	0.5679	0.1243
P24941	CDK2	Cyclin Dependent Kinase 2	14	0.5412	0.1207
Q16539	MAPK14	Mitogen-Activated Protein Kinase 14	13	0.5287	0.1163
P10275	AR	Androgen Receptor	12	0.5543	0.1254
Q02750	MAP2K1	Mitogen-Activated Protein Kinase Kinase 1	12	0.5169	0.0554
P06493	CDK1	Cyclin Dependent Kinase 1	11	0.4510	0.0343
P23458	JAK1	Janus Kinase 1	10	0.4842	0.0168
Q05655	PRKCD	Protein Kinase C Delta	10	0.4510	0.0459
P06401	PGR	Progesterone Receptor	9	0.5227	0.0109
P53779	MAPK10	Mitogen-Activated Protein Kinase 10	9	0.4894	0.0151
P45984	MAPK9	Mitogen-Activated Protein Kinase 9	9	0.4792	0.0082
P50750	CDK9	Cyclin Dependent Kinase 9	8	0.5000	0.0237
Q00535	CDK5	Cyclin Dependent Kinase 5	7	0.4299	0.0082
O60674	JAK2	Janus Kinase 2	7	0.4792	0.0080
Q02156	PRKCE	Protein Kinase C Epsilon	7	0.4220	0.0082

3.3. Interactive network

We used cytoscape 3.7.1 software to draw a "drug-disease-core target" interactive network diagram, as shown in Figure 3.

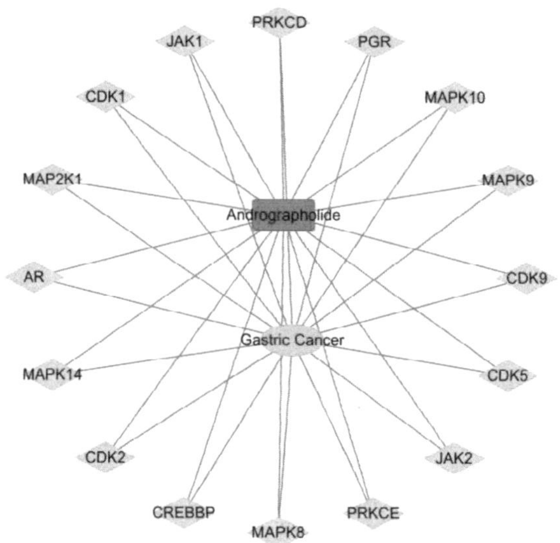

Figure 3. Drugs-Diseases-Core Targets interactive network diagram

3.4 . GO Analysis

A total of 82 biological processes were screened, and the main ten BP, CCand MF were presented, as shown in Figure 4.

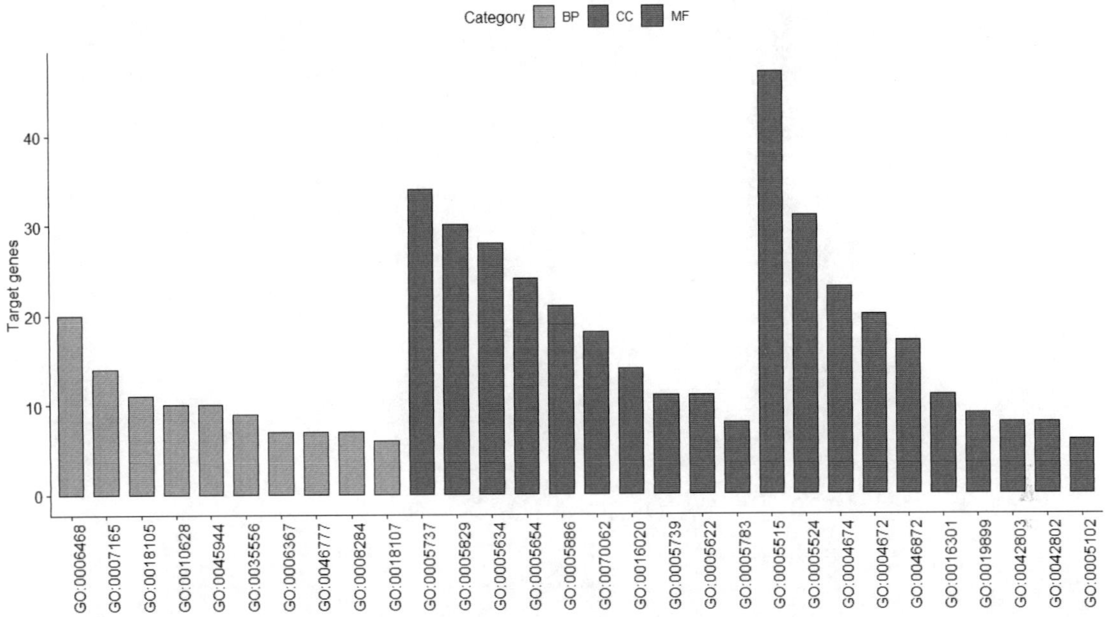

Figure 4. GO Analysis results

3.5. KEGG Analysis

57 pathways were obtained through KEGG pathway enrichment analysis, as shown in Figure 5. The most important pathways are as follows: Influenza A, Pathways in cancer, Sphingolipid signaling pathway, Progesterone-mediated oocyte maturation and Herpes simplex infection. AD may inhibit GC through multiple pathways.

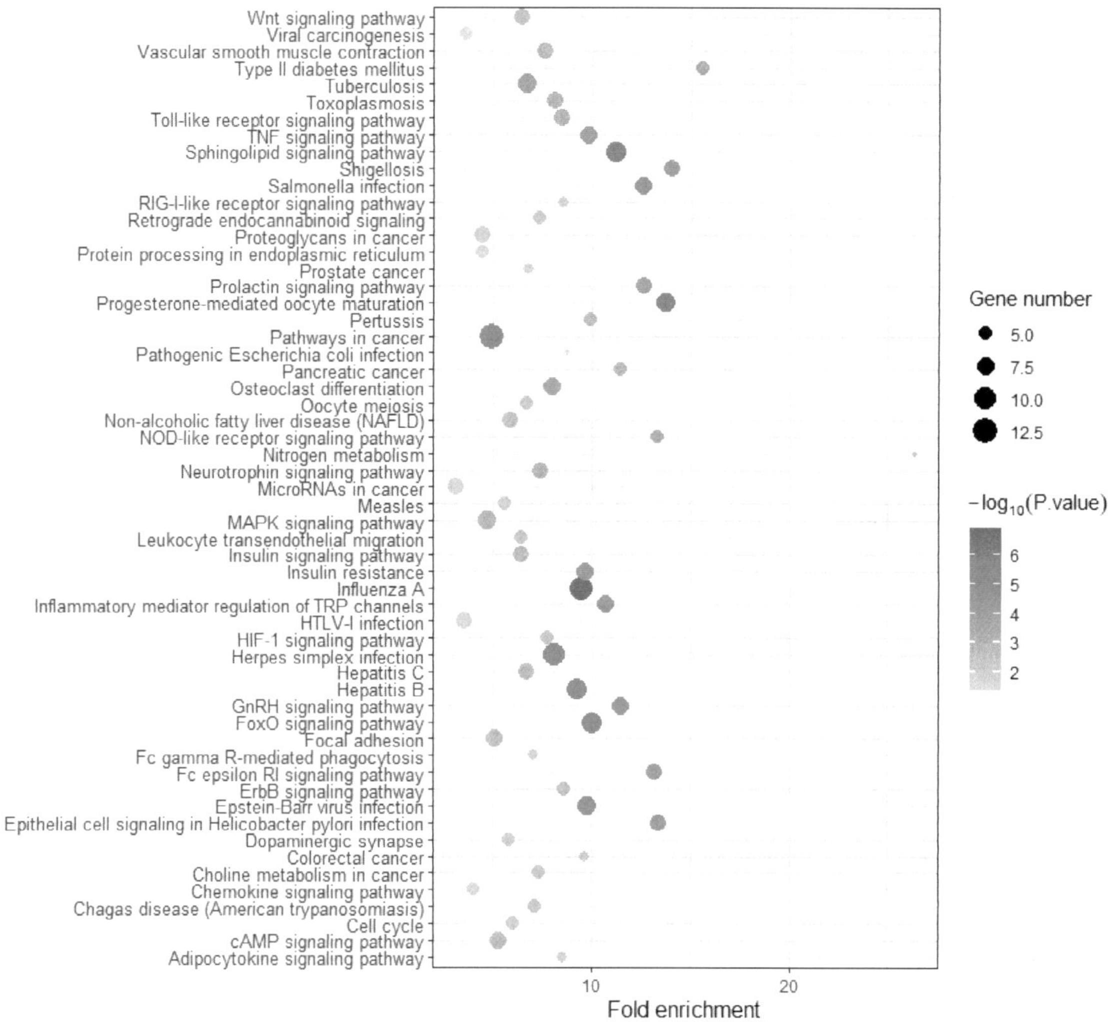

Figure 5. KEGG Analysis results

4. Discussion

Through the research of this subject, the effect of AD on GC may be related to the targets of IL6, MAPK8, CREBBP, CDK2, MAPK14, and AR. IL6 is interleukin-6, which is a lymphokine produced by activated T cells and fibroblasts. It can promote the growth and differentiation of primitive bone marrow-derived cells and enhance the analysis function of natural killer cells. Studies have confirmed that IL-6/STAT3 signaling is related to tumors. Metastasis is closely related. Up-regulation of IL-6 gene expression can induce phosphorylation of STAT3 and promote tumor cell invasion, migration and EMT[11].

GO analysis results show that the biological processes involved in AD acting on GC include apoptosis, signal transduction, biosynthesis, and inflammation, among which protein phosphorylation, signal transduction, peptidyl-serine phosphorylation. The most important biological processes involved in AD acting on GC may include positive regulation of gene expression, Positive regulation of transcription from RNA polymerase II promoter.

KEGG analysis shows that AD mainly acts on GC through the following pathways, such as pathways in cancer, Influenza A, Sphingolipid signaling pathway, Progesterone-mediated oocyte maturation, Herpes simplex infection and other pathways. In the JAK2/STAT3 pathway, phosphorylated JAK2 and STAT3 can promote the proliferation of cancer cells. Studies have confirmed that up-regulation of ROS can block the JAK2/STAT3 signaling pathway and inhibit the growth of a variety of tumor cells[12].

5. Conclusions

53 GC-related targets were obtained through the online database, of which 17 are the core targets for AD to inhibit GC. GO biological process enrichment analysis and KEGG pathway enrichment analysis revealed 10 biological processes, cell composition, molecular functions and 57 pathways are closely related to the occurrence and development of inhibiting breast cancer, mainly involving Pathways in cancer, Influenza A, Sphingolipid signaling pathway, Progesterone-mediated oocyte maturation, Herpes simplex infection. Our research predicted the mechanism of AD nanoemulsifying on GC, and provided guidance for the development of anti-GC drugs. However, the specific mechanism of AD nanoemulsifying in inhibiting GC needs the further study.

Acknowledgments

The authors would like to thank the financial support from the Scientific Research Foundation of Hunan Provincial Education Department (20B415), the Technology Project Foundation of Tongren ([2018]48), the Technology Project Foundation of Tongren ([2019]55).

References

[1] Siegel, R.L., Miller, K.D., Jemal, A. (2020) Cancer statistics, 2020. J. CA:CancerJ Clin., 70: 7-30.
[2] Chen, W.Q., Zheng, R.S., Baade, P.D., Zhang, S., Zeng, H., Bray, F., Jemal, A., Yu, X.Q., He, J. (2015) Cancer statistics in China. J. CA Cancer J Clin, 66: 115.
[3] Mu, Y.C., Xing, X.Y, Yan, Z.J., Chen, Y.Q., Zhang, L.M., Lai, Y., Yuan, X.Y. (2021) Research Progress on Antibacterial Effect of Andrographolide. J. Jounal of Guizhou University of Traditional Chinese Medicine., 43: 85-89.
[4] Lim, J.C., Chan, T.K., Ng, D.S., Sagineedu, S.R., Stanslas, J., Wong, W.S. (2012) Andrographolide and its analogues: versatile bioactive molecules for combating inflammation and cancer. J. Clin Exp Pharmacol Physiol., 39: 300-10.
[5] Peng, Y., Wang, Y., Tang, N., Sun, D., Lan, Y., Yu, Z., Zhao, X., Feng, L., Zhang, B., Jin, L., Yu, F., Ma, X., Lv, C. (2018) Andrographolide inhibits breast cancer through suppressing COX-2 expression and angiogenesis via inactivation of p300 signaling and VEGF pathway. J. J Exp Clin Cancer Res., 37: 248.
[6] Yuan, H., Sun, B., Gao, F., Lan, M. (2016) Synergistic anticancer effects of andrographolide and paclitaxel against A549 NSCLC cells. J. Pharm Biol., 54: 2629-2635.
[7] Forestier-Román, I.S., López-Rivas, A., Sánchez-Vázquez, M.M., Rohena-Rivera, K., Nieves-Burgos, G., Ortiz-Zuazaga, H., Torres-Ramos, C.A, Martínez-Ferrer, M. (2019) Andrographolide induces DNA damage in prostate cancer cells. J. Oncotarget., 10: 1085-1101.
[8] Shi, H.L., Wang, Y.C., Fan, Y.Y., Gong, J.X., Gong, X.R. (2016) High-flux Virtual Screening oI HIV-1 Intergrase Inhibitors from TCMSP Based on Drug Target. J. Chin J Exp Tradit Med Formul., 22: 159-164.
[9] Li, I., Li, R., Gao, S., Zhang, Y., Zang, M., Yu, C.Q. (2013) Explore the research approach for the corresponding relationship of disease-syndrome-prescription base on network pharmacology. J. J Tianjin Univ Tradit Chin Med., 32: 197-199.
[10] Liu, Y.F., Sun, M.Y., Zhao, Y.K., Yao, H.Z., Liu, Y., Gao, R. (2017) Network pharmacology in drug repositioning of C'hinese medicines: application status and thinking. J. Chin J Evidence-Based Med., 17: 1344-1349.
[11] Zhang, H., Song, Y., Yang, H., Liu, Z., Gao, L., Liang, X., Ma, C. (2018) Tumor cell-intrinsic Tim-3 promotes liver cancer via NF-κB/IL-6/STAT3 axis. J. Oncogene., 37: 2456-2468.

[12] Liu, B., Yu, S. (2018) Amentoflavone suppresses hepatocellular carcinoma by repressing hexokinase 2 expression through inhibiting JAK2/STAT3 signaling. J. Biomed Pharmacother., 107: 243–53.

The elastic modulus reduction method for upper bound limit analysis of pressure vessels incorporating material strain hardening effect

Yang Zhang[1], Jingmin Liu[2*], Yanyan Yang[2]

[1] School of Water Conservancy and Ecological Engineering, Nanchang Institute of Technology, Jiangxi, 330099, China

[2]School of Civil Engineering and Architecture, Guangxi University of Science and Technology, Guangxi, 545006, China

[*]Corresponding author's e-mail: 2017994653@nit.edu.cn

Abstract. Pressure vessel is a kind of special equipment which is involved in various industries. Once it is damaged, it will cause great loss to people's life and property. The traditional elastic modulus reduction method for upper bound limit analysis of pressure vessels adopts the elastic-perfectly plastic material model. This method will result in waste of materials without considering the material strain hardening effect. So the material strain hardening effect is introduced into upper bound limit analysis. Firstly, the equivalent yield strength considering material strain hardening effect is established. Then, the formula of limit load multiplier which is expressed by the equivalent yield strength is derived based on the upper bound theorem. Combined with the strategy for elastic modulus adjustment, the elastic modulus reduction method for upper bound limit analysis of pressure vessels incorporating material strain hardening effect is constructed. The limitation of the elastic-perfectly plastic model for the upper bound analysis method is eliminated by the method proposed in this paper. The result calculated by this method is closer to the true limit load of pressure vessels.

1. Introduction

Analytical method and numerical method are the main methods to analyze the ultimate bearing capacity of the structure considering the strain hardening effect of materials. In the analytical method[1], it is assumed that the material is incompressible and satisfies the power hardening constitutive relation; the ultimate bearing capacity of the structure is solved on the basis of the total theory of plasticity. This method is applicable to frame structures with the simple constitutive model. The numerical method mainly includes elastoplastic incremental analysis (EPIA) and elastic modulus adjustment procedure (EMAP). EPIA[2] simulates the strain hardening process of materials and the damage process of structures through the iterative analysis of the incremental load, but EPIA has such problems as great sensitivity of calculations to parameters when considering the strain hardening effect of materials. In order to overcome the shortcomings of EPIA[3], EMAP obtains the ultimate bearing capacity of the structure considering the strain hardening of materials by considering the strain hardening property of materials in the ultimate load multiplier[4]. According to the bound theorem of limit analysis, EMAP can be divided into lower bound method[5] and up bound method[6]. EMAP lower bound method is widely used[7]; EMAP higher bound method is suitable for plastic pressure machining and geotechnical engineering, and can be combined with the lower bound method to estimate the value of true ultimate bearing capacity. However, the strain hardening effect of materials is mainly used in the lower bound limit analysis of EMAP[8], and has not been introduced into the upper bound limit analysis of structures.

In view of this, in this paper, the study on elastic modulus reduction method[9] (EMRM) for upper bound limit analysis of pressure vessels considering strain hardening effect of materials was carried out with EMRM as a typical EMAP. The element bearing ratio (EBR) expressed as a generalized yield function was defined. The threshold of EMAP was determined. The ultimate load multiplier considering the strain hardening effect was determined according to the upper bound theorem of plastic limit analysis. This method relieves the limitation of the upper bound analysis method by the ideal elastoplastic constitutive model.

2. Equivalent yield strength

For the problem that the EMRM for the upper bound limit analysis of pressure vessels fails to consider the strain hardening effect of materials, the equivalent yield strength considering the strain hardening of materials was given on the basis of the principle of strain energy equivalence; and the equivalent ideal elastoplastic model for bilinear, power hardening and Ramberg-Osgood materials was constructed.

2.1. Equivalent ideal elastoplastic model based on principle of strain energy equivalence

In order to consider the strain hardening effect of materials upon the upper bound limit analysis for shell structures by EMRM, the simplified elastoplastic constitutive model reflecting the real stress-strain relationship should be converted into an equivalent ideal elastoplastic model. In Figure 1, the equivalent yield strength corresponding to the equivalent ideal elastoplastic model is obtained according to the principle of strain energy reciprocity (that is, areas A_1 and A_2 are equivalent) [8].

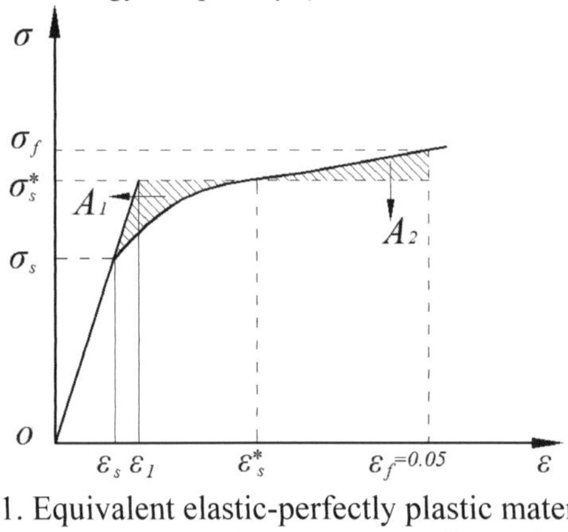

Figure 1. Equivalent elastic-perfectly plastic material model

The constitutive model considering the strain hardening effect of materials is characterized by Formula (1):

$$\varepsilon = g(\sigma_s, \sigma) \tag{1}$$

The equivalent yield strength under the equivalent ideal elastoplastic model σ_s^* is obtained by Formula (2):

$$\int_{\sigma_s}^{\sigma_s^*} \varepsilon \mathrm{d}\sigma - \frac{1}{2}\left(\sigma_s^* - \sigma_s\right)\left(\varepsilon_1 + \varepsilon_s\right) = \varepsilon_f\left(\sigma_f - \sigma_s^*\right) - \int_{\sigma_s^*}^{\sigma_f} \varepsilon \mathrm{d}\sigma$$
$$\varepsilon_s = \frac{\sigma_s}{E}, \ \varepsilon_1 = \frac{\sigma_s^*}{E} \tag{2}$$

2.2. Equivalent yield strength based on complete stress-strain data

In this paper, the equivalent yield strength corresponding to the common elastoplastic stress-strain model was solved on the basis of complete stress-strain data. The complete stress-strain data can be obtained by the static tensile test of materials. Different elastoplastic constitutive models can be fitted according to the stress-strain data. The common elastoplastic constitutive models are shown in Figure 2.

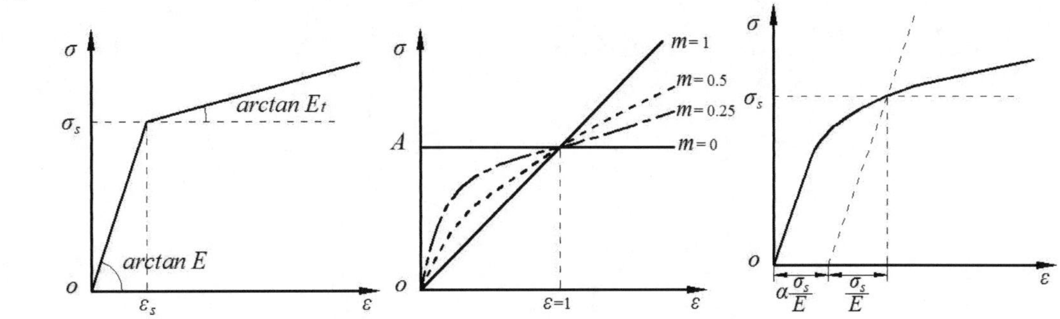

(a) bilinear material (b) power hardening material (c) Ramberg-Osgood material

Figure 2. Commonly used constitutive model

The corresponding equivalent yield strength σ_s^* can be obtained by substituting the basic parameters of the common elastoplastic constitutive models into Formula (2):

Bilinear:
$$\sigma_s^* = (1-\frac{E}{E_t})\sigma_s + \frac{E}{E_t}\sigma_f \pm (\sigma_f - \sigma_s)\sqrt{\frac{E}{E_t}(\frac{E}{E_t}-1)} \tag{3}$$

Power Hardening:
$$\sigma_s^* = E\left[\left(\frac{\sigma_f}{A}\right)^{\frac{1}{m}} \pm \sqrt{\left(\frac{\sigma_f}{A}\right)^{\frac{2}{m}} - \frac{2}{EA^{\frac{1}{m}}(1+\frac{1}{m})}(\sigma_s^{\frac{1}{m}+1} + \frac{1}{m}\sigma_f^{\frac{1}{m}+1}) + \frac{\sigma_s^2}{E^2}}\right] \tag{4}$$

Ramberg-Osgood:
$$\sigma_s^* = \left(\sigma_f + \frac{\alpha\sigma_f^n}{\sigma_s^{n-1}}\right) \pm \sqrt{\left(\sigma_f + \frac{\alpha\sigma_f^n}{\sigma_s^{n-1}}\right)^2 - \frac{2\alpha n\sigma_f^{n+1}}{(n+1)\sigma_s^{n-1}} - \sigma_f^2 - \frac{2\alpha\sigma_s^2}{n+1}} \tag{5}$$

Where, σ_f is the tensile strength of material. The simplified elastoplastic constitutive model is converted into the equivalent ideal elastoplastic constitutive model to obtain the equivalent yield strength considering the strain hardening effect. The upper bound load multiplier based on the equivalent yield strength can be obtained accordingly.

3. EMRM for upper bound limit analysis of pressure vessels considering strain hardening effect of materials

The EMRM for the upper bound limit analysis of shell structures considering the strain hardening effect of materials proposed herein takes EBR as a basic parameter of EMAP and the reference bearing ratio as the threshold of modulus adjustment, determines the ultimate load multiplier considering the strain hardening effect based on equivalent yield strength, and is combined with the EMAP strategy to form the maneuvering allowable strain field and displacement field gradually approaching the limit state by reducing the elastic modulus of the high bearing element, so as to obtain the ultimate bearing capacity of shell structures.

3.1. EBR

Element bearing ratio (EBR) is a control parameter that can simultaneously represent the internal force and resistance of the element in the EMRM modulus adjustment process. It is used to characterize the degree to which discrete elements approach plastic yield. For shell structures, the EBR r_k^e can be expressed as:

$$r_k^e = \sqrt{f} \tag{6}$$

where, the superscript e is the element number; the subscript k is an iterative step; f is a generalized yield function expressed as [10]:

$$\begin{cases} f = f_N + f_M + \dfrac{|f_{NM}|}{\sqrt{3}} \\ f_N = n_x^2 + n_y^2 - n_x n_y + 3n_{xy}^2 \\ f_M = m_x^2 + m_y^2 - m_x m_y + 3m_{xy}^2 \\ f_{NM} = n_x m_x - \dfrac{1}{2} n_x m_y - \dfrac{1}{2} n_y m_x + n_y m_y + 3n_{xy} m_{xy} \end{cases} \tag{7}$$

$$\begin{cases} n_x = \dfrac{N_x}{N_{px}};\ n_y = \dfrac{N_y}{N_{py}};\ n_{xy} = \dfrac{N_{xy}}{N_{pxy}} \\ m_x = \dfrac{M_x}{M_{px}};\ m_y = \dfrac{M_y}{M_{py}};\ m_{xy} = \dfrac{M_{xy}}{M_{pxy}} \end{cases} \tag{8}$$

where, N_x, N_y and N_{xy} are the internal forces of the film on the element section; M_x, M_y and M_{xy} are bending moments and torques on the element section; N_{px}, N_{py}, N_{pxy}, M_{px}, M_{py}, M_{pxy} are the section resistance under the action of each internal force alone. The diagram for internal forces of shell element with a thickness of T is shown in Figure 3.

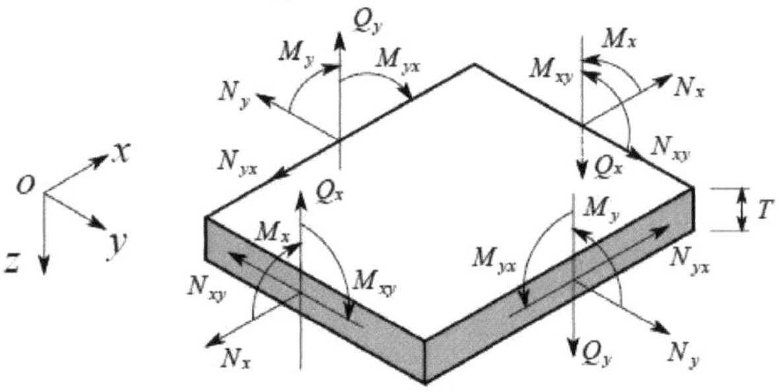

Figure 3. Diagram for internal force of plate and shell elements

3.2. EMAP strategy

The bearing ratio uniformity d_k can reflect the approaching degree of bearing capacity of all elements in the structure. d_k can be expressed as[7]:

$$d_k = \frac{\overline{r}_k + r_k^{\min}}{\overline{r}_k + r_k^{\max}},\ \overline{r}_k = \frac{1}{N}\sum_{e=1}^{N} r_k^e \tag{9}$$

where, $\overline{r}_k$, $r_k^{\max}$ and $r_k^{\min}$ are the mean, maximum and minimum of the EBR in the structure, respectively; N is the number of structural discrete elements.

As a reference value in the EMAP process, the reference bearing ratio r_k^0 can be expressed as [11]:

$$r_k^0 = r_k^{\max} - d_k \times \left(r_k^{\max} - r_k^{\min}\right) \tag{10}$$

According to Hooke's law, the EMRM strategy can be expressed as [7]:

$$E_{k+1}^e = \begin{cases} E_k^e \dfrac{2\left(r_k^0\right)^2}{\left(r_k^e\right)^2 + \left(r_k^0\right)^2}, & r_k^e > r_k^0 \\ E_k^e, & r_k^e \le r_k^0 \end{cases} \tag{11}$$

where, E_{k+1}^e and E_k^e are the elasticity modulus of element e at the k+1th and kth iteration, respectively; the elastic modulus of high bearing element with an EBR greater than r_k^0 will be reduced to simulate the damage evolution process of the structure.

3.3. Upper bound load multiplier considering strain hardening effect of materials

An optimization model for upper bound limit analysis can be constructed according to the condition that the work done by the external load is not less than the plastic dissipation work of the structure [10]:

$$P_{\mathrm{L}}^{u} = \min : \frac{\int_V \sigma_{\mathrm{s}} \varepsilon_{\mathrm{eq}}^{*} \mathrm{d}V}{\int_{S_\sigma} \eta_i u_i^{*} \mathrm{d}S}$$

$$\text{s.t.} \quad u_i^{*} = 0 \quad \text{on} \quad S_u \tag{12}$$

$$u_{i,i}^{*} = 0 \quad \text{in} \quad V$$

where, $P_{\mathrm{L}}^{\mathrm{u}}$ is the upper bound load multiplier; V is the volume of the structure; S_σ and S_u are the boundary surface of force and displacement, respectively; σ_{s} is the yield strength of the material; $\varepsilon_{\mathrm{eq}}^{*}$ is the equivalent strain field; u_i^{*} is the maneuvering allowable displacement field.

Combined with the EMRM strategy in Formula (11), this optimization model can simulate the process of structural plastic damage in the iterative analysis process to gradually obtain the failure mode approaching the plastic limit state of the structure, and then obtain the ultimate bearing capacity of the structure in combination of Formula (12) according to the principle of functional reciprocity:

$$P_{\mathrm{L}}^{\mathrm{u}} = \min\left(P_{\mathrm{L},1}^{\mathrm{u}}, P_{\mathrm{L},2}^{\mathrm{u}} \cdots P_{\mathrm{L},k}^{\mathrm{u}} \cdots P_{\mathrm{L},n}^{\mathrm{u}}\right)$$

$$P_{\mathrm{L},k}^{\mathrm{u}} = \frac{\int_V D_k^{*} \mathrm{d}V}{\int_V U_k^{*} \mathrm{d}V} = \frac{\int_V \sigma_{\mathrm{s}} \varepsilon_{\mathrm{eq},k} \mathrm{d}V}{\int_V \sigma_{\mathrm{eq},k} \varepsilon_{\mathrm{eq},k} \mathrm{d}V} \tag{13}$$

where, D_k^{*} is the plastic dissipation work of the structure at the kth iteration; U_k^{*} is the elastic strain energy at the kth iteration; $\sigma_{eq,k}$ and $\varepsilon_{\mathrm{eq},k}$ are the equivalent stress field and equivalent strain field at the kth iteration under the reference load.

The ultimate bearing capacity of the shell structure considering the strain hardening effect $P_{\mathrm{L},k}^{\mathrm{u}}$ can be obtained by considering structure discretization and substituting in the equivalent yield strength:

$$P_{\mathrm{L},k}^{\mathrm{u}} = \frac{\sum_{e=1}^{N} \sigma_{\mathrm{s}}^{*} \varepsilon_{\mathrm{eq},k}^{e} V^{e}}{\sum_{e=1}^{N} \sigma_{\mathrm{eq},k}^{e} \varepsilon_{\mathrm{eq},k}^{e} V^{e}}, \quad V = \sum_{e=1}^{N} V^{e} \tag{14}$$

where, V^e is the volume of element e. The mesh division of the shell structure is fine enough. Meanwhile, the thickness of the element is small. Therefore, the stress and strain at the centroid of the element can be approximated to represent the stress and strain of the whole element.

Formula (14) is solved iteratively until the ultimate bearing capacity of two adjacent iterative steps satisfies the convergence criterion:

$$\left|\left(P_{\mathrm{L},k}^{\mathrm{u}} - P_{\mathrm{L},k-1}^{\mathrm{u}}\right) / P_{\mathrm{L},k}^{\mathrm{u}}\right| \le \varepsilon \tag{15}$$

where, ε is the preset allowable error and takes 0.001 in the calculating example herein.

If the convergence of calculation results occurs after k iterations, the ultimate bearing capacity of the structure is:

$$P_{\mathrm{L}}^{\mathrm{u}} = P_{\mathrm{L},k}^{\mathrm{u}} \tag{16}$$

Accordingly, the equivalent yield strength is solved according to complete stress-strain data; the ultimate load multiplier is defined on the basis of equivalent yield strength; the ultimate load multiplier of pressure vessels considering the strain hardening of materials are solved in combination with the EMAP process.

4. Examples

4.1. Pressure vessel

4.1.1. Basic parameters

In order to verify the accuracy of the upper bound method of EMRM considering the strain hardening of materials proposed herein, the test data in the relevant study of Zhejiang University of Technology (ZJUT)[11] were selected for comparison. Figure 4 shows the test model structure of the pressure vessel, the material of which is Q235B. See Table 1 for steel parameters. See Table 2 for geometrical parameters of the vessel.

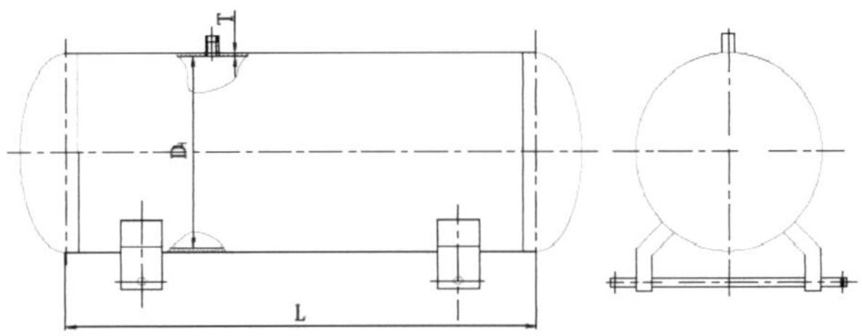

Figure 4. Schematic diagram of test model

Table 1. Steel parameters for the pressure vessel

Steel	Elastic modulus E / MPa	Yield strength σ_s / MPa	Tensile strength σ_f / MPa	Poisson's ration μ
Q235B	2.00×10^5	324	460	0.3

Table 2. Geometry parameters for the pressure vessel

Experiment object	D/ mm	L/ mm	T/ mm
Pressure vessel	160	380	1.5

4.1.2. Establishment of finite element model

The pressure vessel is under uniform internal pressure. The finite element analysis software ANSYS was needed for calculating the ultimate bearing capacity by the upper bound method of EMRM. The element SHELL181 was selected. See Figure 5 for the finite element mesh.

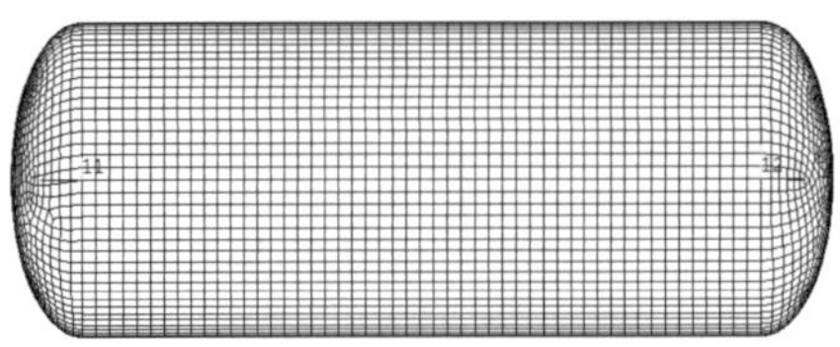

Figure 5. Finite element mesh for the pressure vessel

4.1.3. Calculation of equivalent yield strength
In order to compare the influence of different constitutive models on ultimate bearing capacity, three common elastoplastic constitutive models were selected. For the bilinear constitutive model, the tangent modulus is E_t=0.83×10^{5}MPa. For the power hardening constitutive model, the material parameter is A=545.93 and the power hardening coefficient m=0.096. For the Ramberg-Osgood constitutive model, the coefficient α=2.35; and n=10.08. See Table 3 for the equivalent yield strength corresponding to the common elastoplastic constitutive models.

Table 3. Equivalent yield strength / Mpa

Constitutive model	Bilinear	Power Hardening	Ramberg-Osgood
Equivalent yield strength based on complete stress-strain data	392.07	421.34	421.67

4.1.4. Influence of strain hardening effect of materials on ultimate bearing capacity
In this paper, the upper bound method of EMRM and the lower bound method of EMRM considering strain hardening effect were used to analyze the ultimate bearing capacity; and the results were compared with test results[11]. See Table 4 for calculations. The comparison of the ultimate bearing capacity calculated by the two numerical methods with the measured values obtained from the test is shown in Figure 6.

Table 4. Ultimate load bearing capacity of the pressure vessel calculated by lower bound EMRM and up bound EMRM

Constitutive model	Measured values / MPa	Lower bound EMRM			Up bound EMRM		
		Ultimate bearing capacity / MPa	Iterations / times	Error/ %	Ultimate bearing capacity / MPa	Iterations / times	Error/ %
Elastic-perfectly plastic	8.75	6.370	3	27.20	6.961	8	20.45
Bilinear	8.75	7.602	3	15.13	8.424	8	3.73
Power Hardening	8.75	8.160	3	7.23	9.053	8	3.46
Ramberg-Osgood	8.75	8.160	3	7.23	9.060	8	3.54

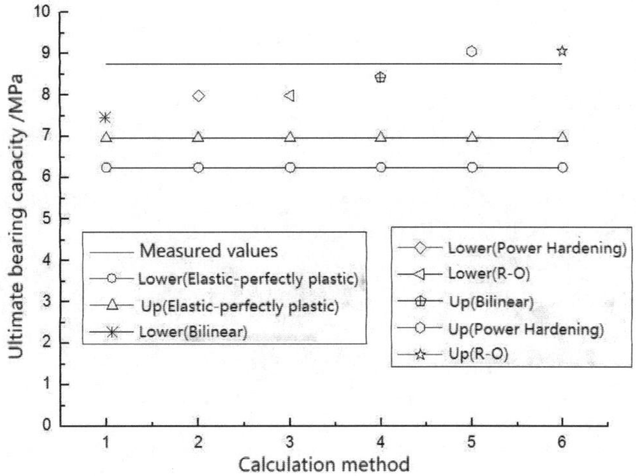

Figure 6. Comparison of ultimate load bearing capacity

According to Table 4, the error between the ultimate bearing capacity of the vessel calculated by the upper bound method of EMRM and the lower bound method of EMRM and with ideal elastoplastic constitutive models and the measured values obtained from the test [16] exceeded 20%. The reason is that there are some differences between the ideal elastoplastic constitutive models and the real stress-strain relationship of the pressure vessel. Compared with the lower bound method of EMRM, the calculation of the ultimate bearing capacity of the pressure vessel by the upper bound method of EMRM is more accurate with an error with the measured values within 4%. It can also be seen from Figure 6 that the calculation results by the upper bound method of EMRM considering the strain hardening effect of materials are closer to the measured values. Although constitutive models are different, the numerical values of the ultimate bearing capacity considering the strain hardening effect of materials are close. The reason is that these constitutive models are based on a real stress-strain relationship and can reflect the important properties of the material after reaching yield.

This pressure vessel calculating example verified the effectiveness of the calculation of the ultimate bearing capacity of the shell structure by the upper bound method of EMRM by comparing the calculation results by this method with the test results. In this paper, the EPIA with recognized effectiveness was used for the limit analysis of shell structures with different constitutive models and the comparison with the calculation results by the upper bound method of EMRM to further verify the effectiveness and applicability of the upper bound method of EMRM in shell structures.

5. Conclusion
In this paper, the equivalent yield strength was given on the basis of the principle of strain energy equivalence; the ultimate load multiplier based on equivalent yield strength was determined according to the upper bound theorem of plastic limit analysis; the EMRM for the upper bound limit analysis of pressure vessels considering the strain hardening effect of materials was established in combination with EMAP strategy. This method relieves the limitation of the upper bound analysis method by the ideal elastoplastic constitutive model and provides an efficient way to calculate the ultimate bearing capacity of pressure vessels. Based on the method proposed in this paper， A fast method for safety evaluation of pressure vessels can be presented.

Acknowledgments
This study was supported by the Science and Technology Project funded by the Department of Education of Jiangxi Province (GJJ180951) and the Guangxi Natural Science Foundation (2018GXNSFBA281186).

References
[1] Galambos, T.V. (1998) Guide to stability design criteria for metal structures. John Wiley and Sons, New York.
[2] Tong, R.C., Wang, X.C. (1997) Simplified method based on the deformation theory for structural limit analysis—I. Theory and formulation. International Journal of Pressure Vessels and Piping, 70: 43-49.
[3] Adibi-Asl, R., Seshadri, R. (2007) Local limit-load analysis using mβ method. Journal of Pressure Vessel Technology, 129: 296-305.
[4] Mahmood, S.L., Adibi-Asl R., Daley C.G. (2013) Plastic response estimation in repeated elastic analyses for strain hardening material model. Journal of Pressure Vessel Technology, 135:051201.
[5] Mahmood, S., Haddara, M., Seshadri, R. (2010) Lower bound limit loads of ship structure components using the mα-tangent method based on single linear elastic analysis[J]. Ocean Engineering, 37: 1139-1148.
[6] Wu, W.L., Yang, L.F., Zhang, W. (2012) Elastic modulus reduction method for the upper bound limit analysis of beam structures. Chinese Journal of Applied Mechanics, 29: 687-691.
[7] Yu B., Yang L.F. (2010) Elastic modulus reduction method for limit analysis considering initial

constant and proportional loadings. Finite Elements in Analysis and Design, 46: 1086-1092.
[8] Zhang W., Zhang Y., Ye Z.Y. (2017) Elastic modulus reduction method for limit analysis of structures consideration of material strain hardening effect. Journal of Basic Science and Engineering, 4: 109-118.
[9] Yu B., Yang L. (2010) Elastic modulus reduction method for limit analysis of thin plate and shell structures. Thin-Walled Structures, 48: 291-298.
[10] Yang L., Zhang W., Yu B. (2012) Safety evaluation of branch pipe in hydropower station using elastic modulus reduction method. Journal of Pressure Vessel Technology, 134: 041202.
[11] Jiang, Y.Z. (2015) Numerical Simulation and Experimental Study of Burst Pressure of Pressure Vessel. Zhejiang University of Technology, Hangzhou.

Experimental Study on Silt Liquefaction by Shaking Table Test

Qiang Luo[1*]

[1] Department of Civil and Architectural Engineering, Panzhihua University, Panzhihua, Sichuan, 617000, CHINA

[*]Corresponding author's e-mail: luoqiang@pzhu.edu.cn

Abstract. Based on the shaking table test, the liquefaction of silt under earthquake load and the earthquake damage after liquefaction are studied. The test results show that the lower silt is relatively dense, the pore pressure increases slowly, and the pore pressure ratio can not reach the initial liquefaction state. However, under the stronger earthquake action, the dense silt may also produce liquefaction, which indicates that the increase of pore water pressure has a great relationship with the compactness and earthquake intensity under the earthquake load.Because the permeability coefficient of silt is relatively low, the increase and dissipation of pore pressure are relatively slow. After the earthquake, he pore pressure of silt continues to rise, and the liquefaction will be further aggravated.

1.Introduction

Shaking table test is a large-scale indoor dynamic test initially used to study the liquefaction behavior of soil layer, which can simulate the stress conditions of natural soil layer with large samples. It can input a random wave or given harmonic vibration from the bottom to the ground, and can simulate the actual effect of seismic wave to the maximum extent. The large scale shaking table test can measure the law of pore water pressure growth and residual displacement during liquefaction, and can reproduce the macro process of soil liquefaction. The disadvantage of shaking table test is that undisturbed soil can not be used, and the consolidation state of remolded soil is difficult to be the same as that of the field. Therefore, the liquefaction test of seismic simulation shaking table can only qualitatively or semi quantitatively simulate the liquefaction performance of the undisturbed soil on site[1].

Early studies on liquefaction focused on the evaluation of resistance and residual strength of sand triggered by liquefaction, while few studies on dynamic characteristics of saturated silt under earthquake action[2]. A series of dynamic triaxial liquefaction tests and triaxial shear tests were carried out on saturated silt after pore water pressure dissipated in different degrees. The results show that the anti liquefaction strength of saturated silt decreases with the increase of applied dynamic stress amplitude, the larger the dissipation degree of pore pressure, the greater the undrained shear strength and the smaller the tangent modulus[3]. Through cyclic torsional shear test by GDS hollow cylinder apparatus, the effect of initial shear stress on liquefaction characteristics of saturated silt was studied[4]. The purpose of this shaking table test is to test the liquefaction resistance of silt in the site by inputting the seismic time history of the site seismic risk analysis and simulating the seismic vibration according to certain similarity criteria, so as to obtain the visual data of seismic liquefaction of silt.

2.Model design and sensor arrangement

The size of the shaking table is 5×5 meters, the maximum bearing capacity is 30 tons, three directions of vibration load can be input at the same time. The maximum horizontal displacement of the table is 8 cm, and the maximum frequency of simulation is 50 Hz. The size of the test model box is 3.8×2×1.6 meters, and the model box and the vibration table are fixed with 8 bolts.

The two ends of the model box are thickened with sponge cushion of about 10 cm and sealed with impermeable membrane. The soil sample in the model box is formed by water sinking method, and 10 cm high clean water is added into the box. The crushed silt is evenly scattered into the water, and the water surface in the box is kept higher than the soil surface, so that a more uniform saturated silt layer can be obtained. When the soil sample reaches 130 cm high and the water level is about 10 cm high, the loading of soil is stopped. Keep for 2-3 days,let the residual air bubbles out of the soil, make the soil sample fully saturated,and then add silt to the height of 150 cm. The soil is pre vibrated by inputting about 50 gal seismic acceleration time history to accelerate the consolidation speed of soil samples. The pre vibrated soil samples are placed for another 4 days, and then the formal test is started. Table 1 shows the saturated bulk density of soil samples in three states.

15 pore pressure sensors, 8 acceleration sensors, 4 earth pressure sensors and 2 displacement sensors are used in this test. The 15 pore pressure gauges are buried in different positions in the soil. The acceleration sensors J01, J02 and J03 are buried near the same height of the pore pressure gauges K01, K02 and k03. The acceleration sensors G01 and G02 are placed on the soil surface. The acceleration sensors D02 and D03 are placed on the vibration table and the bottom surface of the model box. The acceleration sensor D01 is placed on the top surface of the model box, The displacement sensors w01 and W02 are placed outside the vibration table and the bottom of the box.The earth pressure gauge shall be buried near the same height of K01, K02 and K03. Figure 1 shows the location of the sensor. In Figure 1, 1 is the pore pressure sensor, 2 is the acceleration sensor, 3 is the displacement sensor, 4 is the sea cushion, and 5 is the vibration table.

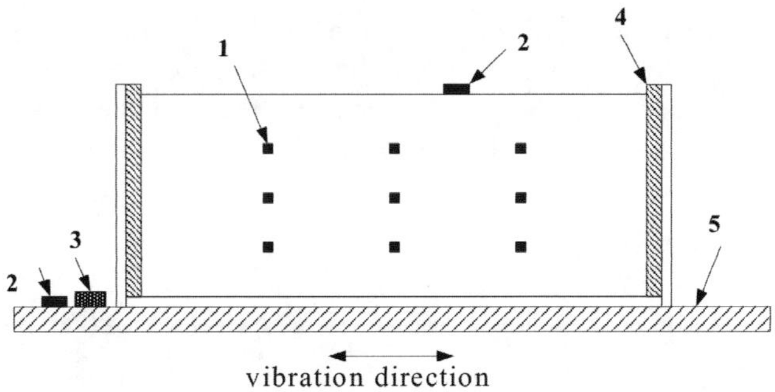

Figure 1. Schematic diagram of model box and some sensors

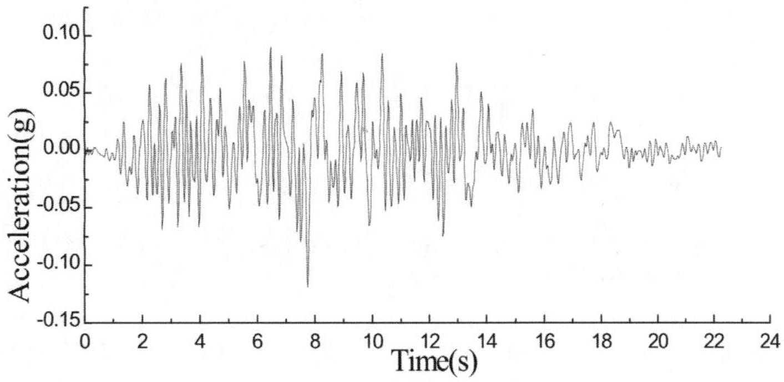

Figure 2. Original ground motion time history curve of test

3.Test ground motion input

In this experiment,the ground motion was only input in one horizontal direction, the maximum amplitude of the ground motion is 0.12g,and the predominant frequency is about 1Hz. Figure 2 shows the time history curve of the ground motion. According to the similarity law,the duration of the selected input seismic wave is compressed to 7.2 seconds, and the maximum amplitude of the input acceleration is adjusted to 0.254g, which is called XZ1.The input wave whose maximum acceleration amplitude is adjusted to 0.32g is called XZ2. Four vibration tests were completed. The first test is carried out one week after the soil sample is formed, and the time of the last three tests is respectively 30 minutes and 48 hours after the first vibration and 30 minutes after the third vibration.

4.Macroscopic phenomenon of the test

In the first test, water precipitated from the soil surface, the model subsided about 2 cm, and there was a clear gushing near the edge of the box.About 20 minutes after the test,about 0.5 cm high water precipitated from the soil surface, and more eruptive mounds similar to volcanic mounds appeared on the soil surface. These phenomena indicate that the soil in the model box has been liquefied in the process of vibration and a short time after vibration.

In the second test, a large area of gushing water appeared,the soil subsided about 2 cm, and the silt was completely liquefied. About 3 hours after vibration, water about 5 cm deep precipitated from the soil surface. After the water on the soil surface is drained away, it is found that the soil surface is covered with water spray holes similar to craters, which are very consistent with the liquefaction phenomenon at the earthquake site, indicating that the silt has undergone serious liquefaction.

After 48 hours, the last two tests were carried out, and the test phenomenon was basically the same as before, except that the water yield was reduced after vibration, and the spray hole was not as large as before.

5.Experimental analysis

Figure 3 shows that the acceleration of vibration table and box bottom is consistent. Figure 4 shows that the displacement records of vibration table and box bottom are consistent, and there is no relative motion. According to the location of the piezometer, the effective overburden pressure of each layer can be calculated, which is 11.05kpa for the first layer, 7.38kpa for the second layer and 2.78kpa for the third layer.

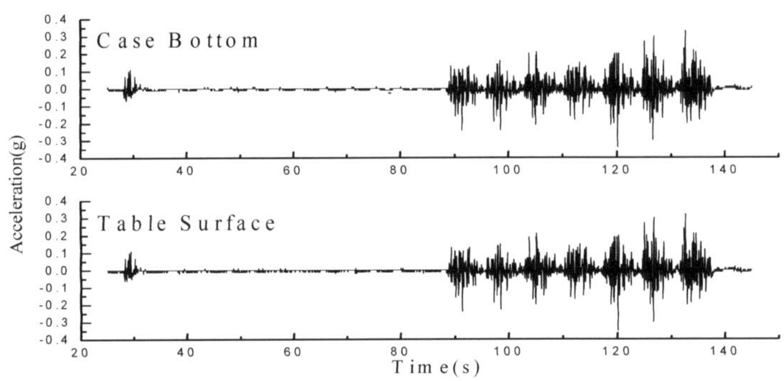

Figure3. Acceleration record of vibration table and box bottom

Figures 5 and 6 shows three pore water pressure recording curves on the central plane of the model box along the vibration direction. With the increase of silt burial depth, the pore water pressure gradually increases, and when the pore water pressure reaches a stable state, the pore water pressure of each layer is approximately equal. The average pore water pressure of the first layer is 10.21kpa when it is stable, the average pore water pressure of the second layer is 7.06kpa, and the average pore water pressure of the third layer is 3.13kpa. Therefore, it can be calculated that the pore pressure ratio of the first layer is 92.4%, that of the second layer is 95.6%, and that of the third layer is more than 1. Combined with the macroscopic phenomenon after vibration, it can be judged that the silt in the model box has been liquefied.

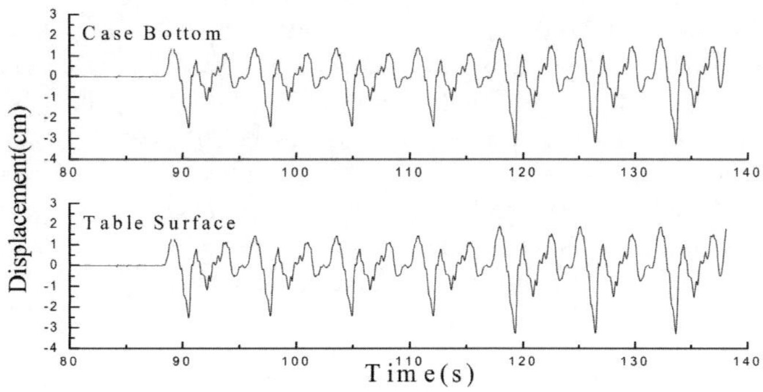

Figure 4. Displacement record of top and bottom of shaking table

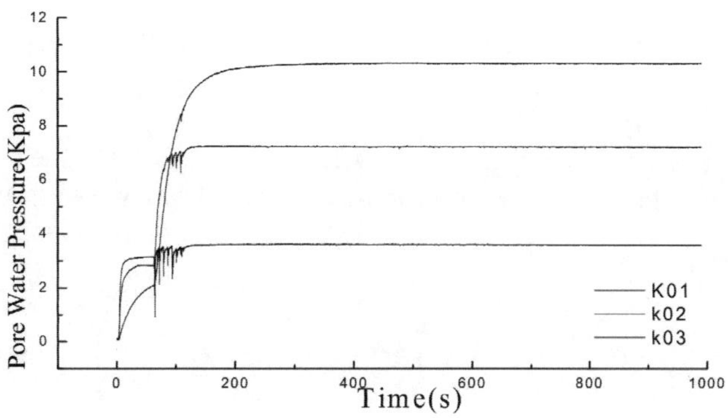

Figure 5. Time history curves of pore water pressure growth recorded by K01-K03

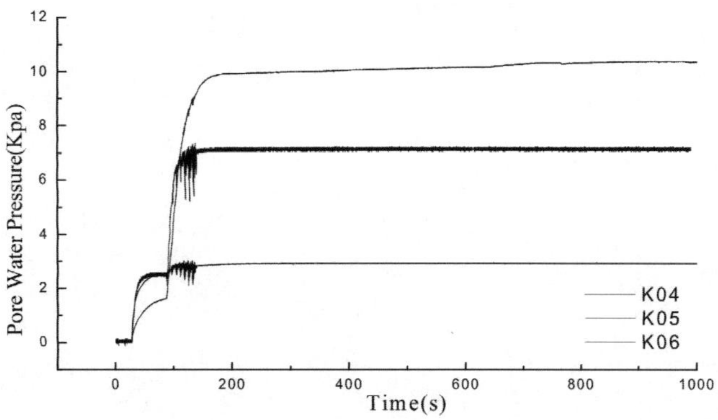

Figure 6. Time history curves of pore water pressure growth recorded by K04-K06

6.Conclusion

Through this test, we can get the intuitive experience about whether the silt liquefaction under the earthquake load and the earthquake damage after liquefaction. Through the analysis of the test records, the following conclusions are obtained:

(1) When the silt density is small, the whole saturated silt layer in the model box may liquefy. Under the same earthquake input, after pre vibration, the silt in the lower layer of the model box is relatively dense, the pore pressure increases slowly, and the pore pressure ratio can not reach the initial liquefaction state. However, under the stronger earthquake action, the dense silt layer may also produce liquefaction, which indicates that the increase of pore water pressure of silt layer under earthquake load has a great relationship with the compactness and earthquake intensity.

(2) It can be seen from the test results that liquefiable silt will not only liquefy during the earthquake, but also liquefy or aggravate the liquefaction due to the continuous increase of pore pressure after the earthquake. The liquefaction of silt will also produce serious water gushing phenomenon of sand blasting, and cause surface subsidence, resulting in foundation failure.

(3) The field conditions can not be fully simulated in the laboratory, so it is suggested to make a comprehensive analysis combined with the conclusions given by other methods.

References

[1]Kexu Zhang & Junfei Xie(1989), *Soil Dynamics*.In chinese: Beijing, ON: Seismological Press.

[2]Xiaofei Lv& Dong Li(2015), Experimental study on strength and deformation characteristics of silty soil after liquefaction in Hangzhou Bay. In Chinese: Journal of seismological Engineering, 37(43):857-861.

[3] Hui Dou&Yongxiang Yang etc(2019), Experimental Study on Liquefaction and Post-liquefaction Mechanical Properties of Saturated Silt Soil. In Chinese:Journal of Chongqing Architecture,2019(3):37-42.

[4] Zhenglong Zhou&Guoxing Chen etc(2016), Effect of initial static shear stress on liquefaction behavior of saturated silt. In Chinese: Journal of Geotechnical Engineering,2016(3):504-509.

Experimental research on dilatancy characteristics of coarse-grained soils by triaxial compression test

Jun Du*, Zhi-gang Qiu

College of Architecture and Civil Engineering, Kunming University, Kunming, Yunnan, 650214, China

*Corresponding author's e-mail: dujun0605@kmu.edu.cn

Abstract. Dilatancy is a basic behavior of soils which is closely related to strength and deformation behaviors of soils and makes soils significantly different from normal elastic materials. A large-scale triaxial compression testing machine was used to study the dilatancy characteristics of coarse-grained soils of earth rock dam. The results show that the confining pressure has a significant effect on the dilatancy of coarse-grained soils, the soil first shrinks and then expands under low confining pressure, and the soil mainly produces shear shrinkage deformation under high confining pressure. When the soil changes from shear shrinkage to dilatancy, the strain ratio of soil changes abruptly. Rowe dilatancy model can reflect the dilatancy deformation characteristics of coarse-grained soils. The parameter K_f of the model tends to be constant with the increase of axial strain, and the parameter K_f of coarse-grained soils is smaller when the confining pressure is larger.

1. Introduction

The volume change of soil under shear stress is called dilatancy, which is one of the basic characteristics of soil [1]. Different from the general elastic materials, the dilatancy of soil is very obvious, and the volume change caused by shear stress may be of the same order of magnitude as that caused by compressive stress, which cannot be ignored. The coarse-grained soils which is widely used in earth-rock dam filling is a kind of aggregate filled with soil particles of different sizes, shapes and properties. Due to the relatively large particle size of coarse-grained soils, the mechanism and characteristics of dilatancy and its influence on strength and deformation are different from those of sand and clay.

Several groups of triaxial compression tests on coarse-grained soils have shown that coarse-grained soils show an obvious dilatancy trend under low confining pressure. With the increase of confining pressure, the coarse-grained soils gradually transit from dilatancy to shrinkage until the dilatancy disappears [2-3]. The mutual occlusion between soil particles during shearing, as well as the overturning and rolling between particles, are the fundamental causes of soil dilatancy [4]. The dilatancy of coarse-grained soils is also affected by factors such as density [5], confining pressure [6], particle gradation [7], particle breakage [8], parent rock properties [9], and stress path [10]. Although the existing research results have pointed out the mechanism, law and influencing factors of dilatancy of coarse-grained soil, the dilatancy characteristics and quantitative description of soil under different stresses still need more in-depth research. Therefore, based on the conventional triaxial compression test, it is necessary to quantitatively analyze the dilatancy characteristics of coarse-grained soils from the strength and deformation characteristics of soil.

2. Triaxial compression test

2.1. Test instruments and test methods

The DJSZ-150 large-scale coarse-grained soils dynamic and static triaxial test machine was used in the test. The test machine included confining pressure servo system, dynamic and static loading system, pore pressure measurement system, volumetric strain measurement system and data acquisition system. The size of the sample was allowed to be Φ300 × 600 mm, the maximum axial load was 1500kN, and the maximum confining pressure could reach 3.0 MPa. According to the dry density, sample size and gradation of the test requirements, the soil required for the test is weighed, mixed evenly and filled in three layers. The test condition is consolidated drainage, shearing according to strain rate 1mm /min until the axial strain reaches 15 %.

2.2. Test material and test method

The test material is selected from the rock-fill of Daqiaopo reservoir in Lincang, Yunnan Province. Due to the limitation of the grain size and specimen size of the rock-fill, the indoor test required equal replacement of the original grade of the material for the scale reduction process, and the composition of the test material grade is shown in Table 1. The dry density of the sample is obtained from the relative density test, and the maximum dry density is determined by the shaking table test method, and then the minimum dry density is obtained by the method of the regulation. The dry density of the sample making is controlled according to the relative density Dr=0.9. In the triaxial compression test, the surrounding pressure is designed to be 200kPa, 400kPa, 600kPa, 800kPa and 1000kPa.

Table 1. Particle gradation composition of coarse-grained soils

grading type	Percentage content of granules /%						
	<2mm	2-5mm	5-10mm	10-20mm	20-40mm	40-60mm	>60mm
original	8.88	4.61	9.98	18.47	28.25	14.61	15.2
Shrinkage	8.88	4.61	12.12	22.41	34.27	17.71	-

3. Analysis of test results

3.1. Analysis of stress-strain characteristics

The curve of deviation stress-axial strain relationship of consolidation drainage test for coarse-grained soils is shown in figure 1. In figure1, the stress-strain curve is not smooth and fluctuates locally. When the axial strain is less than 5%, the soil presents linear elastic deformation, and the initial elastic modulus of the soil increases with the increase of confining pressure. Under low and medium confining pressure (200, 400, 600kpa), the stress-strain curve of soil has a peak value, and after the peak value, there is an obvious trend of stress softening. Under the high confining pressure (800, 1000kPa), the stress-strain curves of soil show weak hardening characteristics. Until the strain reaches 15% failure, the stress increases, but the strength increases slowly after the strain is more than 9%. The reason is that there is almost no relative movement between particles in the initial stage of soil shear, and the main deformation is elastic deformation. After that, due to the close contact between particles, some particles roll and friction, which need to overcome the larger bite force and show higher shear strength. In low or medium confining pressure, the soil structure becomes loose and the shear strength remains unchanged or decreases slightly. When the pressure is high, the contact point of particles breaks, and the fine particles fill into the original pores, which makes the structure more compact and can still bear large stress.

3.2. Analysis of deformation characteristics

Volume strain of soil consolidation drainage test εv and axial strain ε_a relationship curve is shown in figure 2, and the volume reduction is positive in the figure. In figure 2, at the initial stage of axial strain, the volume strain is positive and the soil is subject to shear shrinkage. With the increase of axial strain, the volume strain curve appears inflection point and develops to the direction of volume increase. When the confining pressure is 200 kPa and 400 kPa respectively, the volume strain of the soil becomes negative after the axial strain is 6% and 11%, and the dilatancy occurs. When the confining pressure is greater than 600kPa, the soil does not show dilatancy, but shows shear shrinkage. Therefore, under low confining pressure, the soil first shrinks and then expands, while under high confining pressure, the soil mainly shrinks. The reason is that there are still a certain number of pores in the soil after isotropic consolidation and drainage. In the initial stage of shear, the axial load increases, the soil is compacted, the pore water is discharged, and the volume of soil decreases, resulting in shear shrinkage. With the increase of axial load, the particles roll and bite gradually, and the volume of soil presents the tendency of swelling. When the confining pressure is low, the radial stress limit of the sample is small, and the stiffness of the soil particles is large, so the mechanical action between the particles is more significant, and the volume of the soil changes strongly until the shear failure, and the soil presents dilatancy. The radial stress of the sample is limited greatly when the confining pressure is high. With the increasing axial load, the particles will break gradually in the rolling, friction and bite action. The broken small particles will be further filled into the large particle pores until the shear failure, and the soil still presents shear shrinkage.

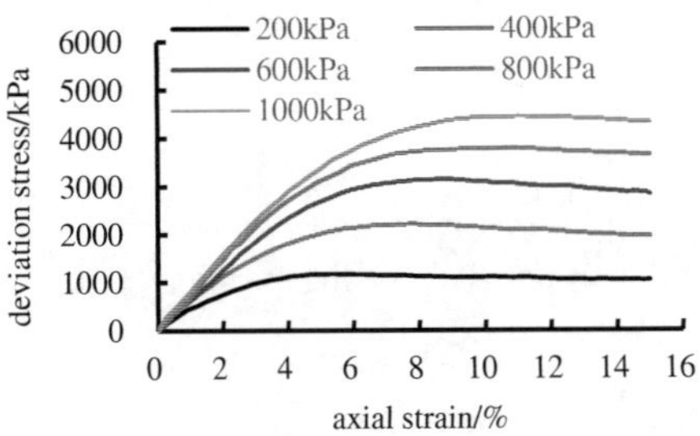

Figure 1. Deviation stress-axial strain curve

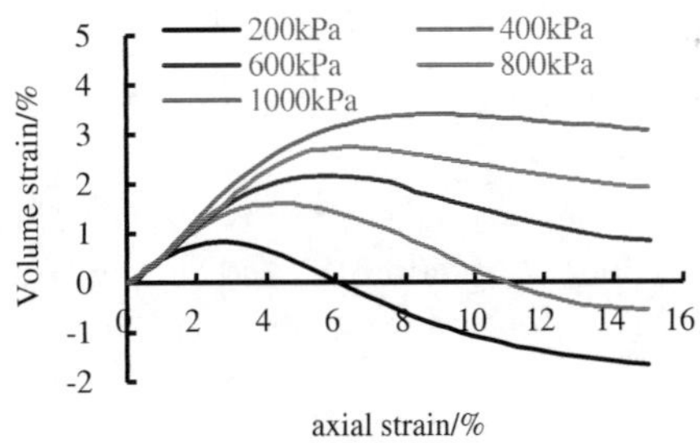

Figure 2. Volume strain curve

The ratio of axial strain to volume strain is defined as the strain ratio. The figure3 shows the variation of strain ratio under different confining pressures.

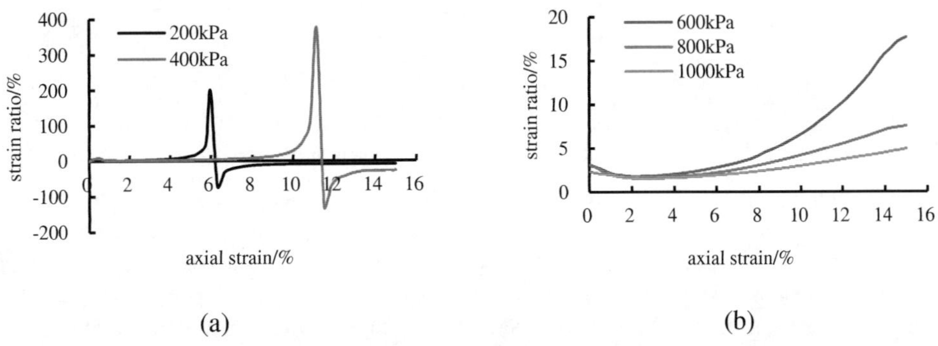

(a) (b)

Figure 3. Variation of strain ratio under different confining pressures. (a) low confining pressure; (b) medium and high confining pressure

In figure3(a), when the confining pressure is low (200, 400kPa), the strain ratio begins to be positive and increases slowly with the increase of shear strain, reaches an extreme point, then suddenly changes, decreases to a negative value, and finally slowly decreases, and gradually tends to a stable negative value. The location of the mutation point is consistent with the axial strain of the soil in figure2. Therefore, the mutation point can be defined as the critical point of the soil from shear shrinkage to dilatancy. The volume strain of soil is composed of axial strain and lateral strain. At the initial stage of the test, the soil sample mainly produces axial strain, and the lateral strain is small, so the strain ratio increases gradually, and the soil sample is mainly shear shrinkage. When the axial deformation develops to a certain extent, the lateral strain increases gradually, and the rolling between particles is more intense. The volumetric strain is mainly dilatancy. At this time, the strain ratio suddenly changes to a negative value, and the pores of soil particles show the phenomenon of inverted water absorption. Analysis of figure 3(b) shows that under high confining pressure (600, 800, 1000kPa), the strain ratio first decreases and then increases with the increase of shear strain, and the strain ratio is always positive, and the soil sample presents shear shrinkage state. The larger the confining pressure is, the more slowly the soil strain ratio increases.

3.3. Analysis of Rowe dilatancy model

The research on soil dilatancy is mostly based on Rowe dilatancy model, and mainly focuses on sand, while the research on coarse-grained soil is relatively less. Therefore, the applicability of Rowe dilatancy model for coarse-grained soil needs to be further analyzed. Under the condition of conventional triaxial compression test, the Rowe dilatancy equation can be expressed as follows:

$$\frac{\sigma_1}{\sigma_3} = 2K_f\left(-\frac{d\varepsilon_3}{d\varepsilon_1}\right) \tag{1}$$

Where σ_1 is large principal stress. σ_3 is small principal stress. K_f is Rowe dilatancy parameter. $d\varepsilon_1$ is axial strain increment. $d\varepsilon_3$ is lateral strain increment.

The expression of K_f can be obtained by formula (1), giving:

$$K_f = -\frac{\sigma_1 d\varepsilon_1}{2\sigma_3 d\varepsilon_3} \tag{2}$$

According to equation (2) and the test data, the relationship between the parameter K_f of Rowe dilatancy equation and the axial strain can be obtained, as shown in figure 4.

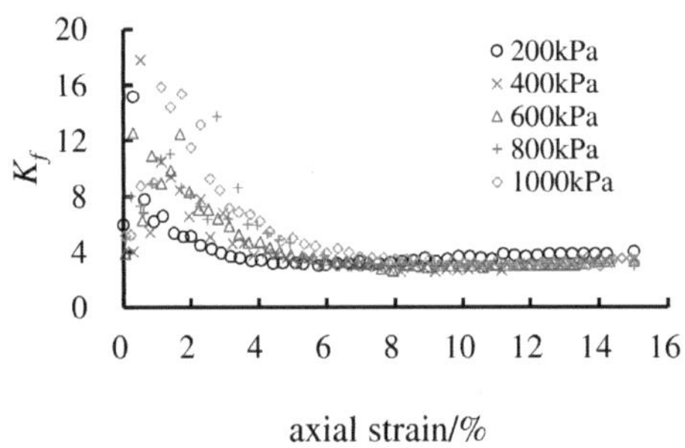

Figure 4. Relationship between dilatancy parameters K_f and axial strain

In figure 4, the parameter K_f of Rowe's dilatancy equation is discretized at the initial stage of shear (ε_1<3%). This is because Rowe's dilatancy model reflects the irreversible sliding deformation between particles, and the soil is similar to elastic deformation at the initial stage of loading deformation, and there is particle breakage phenomenon. With the increase of axial strain, the dispersion of parameter Kf decreases, and when the axial strain exceeds 5%, the parameter K_f shows significant normalization, which is characterized by the close filling between coarse-grained soils particles, and the deformation energy of coarse-grained and fine-grained soil gradually develops harmoniously. On the other hand, the Rowe dilatancy parameter K_f of coarse-grained soils is smaller when the confining pressure is larger. Therefore, Rowe equation which reflects the dilatancy of soil can be used to analyze the deformation characteristics of coarse-grained soils.

4. Conclusions

(1) Under the condition of low and medium confining pressure, the strength characteristics of coarse-grained soils tend to soften. Under high confining pressure, the stress-strain curve of coarse-grained soils is hardening.

(2) The coarse-grained soil presents a significant dilatancy trend in the shear process. Under low confining pressure, coarse-grained soil first shrinks and then expands. Under high confining pressure, coarse-grained soil mainly produces shear shrinkage deformation.

(3) Rowe dilatancy equation can describe the dilatancy of coarse-grained soils. When the shear strain of coarse-grained soils is over 5%, the Rowe dilatancy parameter K_f has the characteristic of normalization.

(4) Understanding the variation of dilatancy of coarse-grained soil and the relationship between dilatancy and stress-strain properties has important application value for further understanding the engineering characteristics of coarse-grained soil and providing reasonable basis for engineering analysis.

Acknowledgments
This research is supported by the Joint special fund project of basic research in local universities of Yunnan Province(202001BA070001-066).

References

[1] Simoni A., Houlsby G.T. (2006) The direct shear strength and dilatancy of sand-gravel mixtures. Geotechnical and Geological Engineering., 24(3): 523-549.
[2] Chu Fu-yong., Zhu Jun-gao., Yin Jian-hua. (2013) Study of dilatancy behaviors of coarse-grained soils in large-scale triaxial test. Rock and Soil Mechanics., 34(8) : 2249-2254.
[3] Been K., Jefferies M. (2004) Stress-dilatancy in very loose sand. Canadian Geotechnical Journal., 41(5): 972-989.
[4] Rowe P.W. (1962) The stress-dilatancy relation for static equilibrium of an assembly of panicles in contact. Proceedings of The Royal Society (Series A)., 269: 500-527.
[5] Jiang Jing-shan., Liu Han-long., Cheng Zhan-lin., et al. (2009) Influences of density and confining pressure on mechanical properties for coarse-grained soils. Journal of Yangtze River Scientific Research Institute., 26(8): 46-50.
[6] Wang Guang-jin., Yang Chun-he., Zhang Chao., et al. (2009) Experimental research on particle breakage and strength characteristics of rock and soil materials with different coarse-grain contents. Rock and Soil Mechanics., 30(12): 3649–3654.
[7] Konrad J.M., Lemieux N. (2005) Influence of fines on frost heave characteristics of a well-graded base-course material. Canadian Geotechnical Journal., 42(2): 515-527.
[8] Miura N., Ohara S. (1979) Particle Crushing of a decomposed Grained Soil Under Shear. Soils and Foundation., 19(3):61-76.
[9] Qin Hong-yu., Liu Han-long., Gao Yu-Feng., et, al. (2004) Research on strength and deformation behavior of coarse aggregates based on large-scale triaxial tests. Rock and Soil Mechanics., 25(10): 1575-1580.
[10] Chu Fu-yong., Zhu Jun-gao., Wang Ping., et, al. (2012) Study of deformation and strength characteristics of coarse grained soil under K0-consolidation condition. Rock and Soil Mechanics., 33(6):1625-1630.

Construction of $Fe(OH)_3$-ZnO/NF nanotube arrays with enhanced performance for water splitting

Yueyue Fu [1], Jiaqi Pan[1], Guangsheng Xiao [1], and Chaorong Li[1*]

[1]Department of Physics and Key Laboratory of ATMMT Ministry of Education Zhejiang, Sci-Tech University, Hangzhou 310018, China

[*]Corresponding author's e-mail: crli@zstu.edu.cn

Abstract: Well-aligned hierarchical hollow nanoarrays containing ZnO core and $Fe(OH)_3$ nanotube shell have been synthesized via simple one-step hydrothermal method and etching at room temperature. The obtained catalyst has a high specific surface area, a large number of catalytic active sites and good chemical stability. The resulting $Fe(OH)_3$-ZnO/NF nanotubes exhibits promising behavior in alkaline solution. When the current density is 10 mA cm^{-2}, the HER and OER overpotentials (η_{10}) of $Fe(OH)_3$-ZnO/NF are 190 and 388 mV, respectively.

1. Introduction

Energy and environment are two prominent issues that restrict the development of human society. In current, the traditional fossil fuels dominate the main energy use structure, despites it also brings abundant environmental pollution[1]. Numerous efforts are being strived for appropriate alternative power sources. As a renewable clean energy, hydrogen is considered to be the best alternative to traditional fossil energy[2]. Electrocatalytic water splitting reaction provides an uncomplicated and feasible method for preparing hydrogen, has attracted lots of attentions[3-4]. Due to the sluggish kinetics and slow rates of gas release at a low overpotential of both OER and HER, which are expected to be improved further. The electrocatalysts that can reduce overpotentials and accelerate the reaction rate at the same time are regarded as an ideal method[5-6]. Hence, it is extremely urgent to develop the efficiency, low-cost and sustainability of electrocatalysts.

Considerable research has been used to develop materials based on transition metal compounds, including phosphides[7], sulfides[8] and hydroxide[9]. As reported, the hydroxide of iron can promote the oxidation of water through the change of valence state of Fe^{3+}[10]. The hollow structure has a better specific surface area and also exposeing more active sites[11].

Herein, we have prepared the ZnO/NF hollow nanotubes via one-step hydrothermal method. Then, the $Fe(OH)_3$-ZnO/NF was obtained by etching ZnO through in-situ reaction ($ZnO+Mn^{+} + H_2O$ - Zn^{2+} $+M(OH)_N$, M=Fe or Ni)[10]. As expected, the prepared $Fe(OH)_3$-ZnO/NF with above features exhibit lower overpotential (190 mV for HER and 388 mV for OER at 10 mA cm^{-2}), excellent activity and durability for water splitting in alkaline electrolyte.

2. Experiment

2.1 Synthesis of ZnO/NF

At the beginning of material preparation, a piece of commercial NF (1×3 cm) was sonicated in HCl solution (1 M) and ethanol solution for 30 minutes to remove surface stains. The nickel foam was soaked in potassium permanganate solution for 20 minutes to activate the surface of the nickel foam. Next, disperse 20 mg $Zn(N0_3)_2{\cdot}6H_20$ in a mixture of 50 ul Nafion, 100 ul isopropanol and 850 uL distilled water. The nickel foam was placed in the above solution and soaked for 10 min, waiting for it to dry and then calcining at 400 °C for 2 hours in a nitrogen atmosphere. Finally, the nickel foam loaded with the ZnO seed layer was grown in an autoclave at 95°C in a mixed solution of ammonia water and $Zn(N0_3)_2{\cdot}6H_20$. After taking ZnO/NF out of the reaction kettle, rinse with ethanol and deionized water respectively, and put it in an oven to dry for using.

2.2 Synthesis of nanorod $Fe\ (OH)_3$-ZnO/NF

The ZnO/NF was soaked in a 0.005 mol/L $FeCl_3$ solution for 10 min, then taking the sample out, rinsing and dring it with distilled water for later using.

2.3 Characterzation

The phase composition of the prepared sample was tested by X-ray diffraction (XRD, Bruker D8 Discover). The morphology and microstructure of the sample were observed with a scanning electron microscope (FESEM Hitachi S-4800) and transmission electron microscope (TEM JEM-2100).

2.4 Electrochemical measurments

The complete hydrolysis performance of the material is tested using an electrochemical workstation (CHI660E), and all electrochemical tests are performed in 1 mol/L KOH aqueous solution. The three electrodes used in the test are: $Fe(OH)_3$-ZnO/NF working electrode with a size of 1 cm × 1 cm; platinum plate counter electrode; Ag/AgCl reference electrode. Use the following formula for potential calibration: $E_{RHE} = E_{Ag/AgCl} + 0.194 + 0.059 \times pH$.

3. Results and discussions

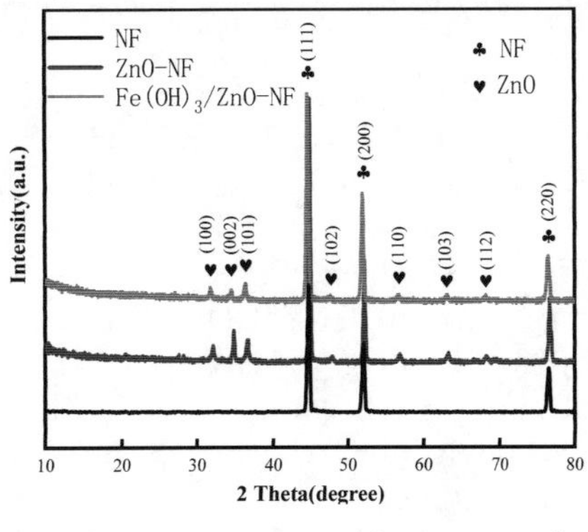

Fig.1. The XRD spectra of as-prepared samples

X-ray diffraction patterns of the $Fe(OH)_3$-ZnO/NF nanotube are shown in Fig. 1. The obvious diffraction peaks at 44.5°, 51.85° and 76.37° peak correspond to the (111), (200) and (220) crystal planes of Ni. What's more, the new peaks at 31.77 °, 34.42 °, 36.25 °, 47.54 °, 56.60 °, 62.86° and 67.96 ° can be observed in as-prepared ZnO-NF are ascribed to the (100), (002), (101), (102), (110), (103) and (112) planes of ZnO. In $Fe(OH)_3$-ZnO/NF, the characteristic peaks of ZnO can still be observed but there is no obvious peak of $Fe(OH)_3$, which can be ascribed to that the ZnO cannot be completely etched away and the crystallinity of $Fe(OH)_3$ is weak.

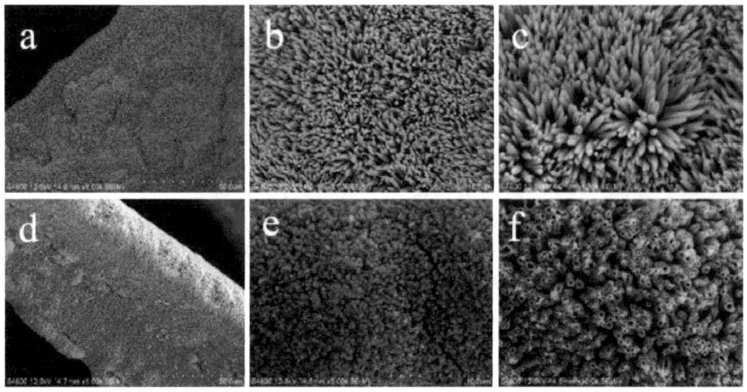

Fig.2. SEM of (a-c) ZnO/NF with magnification of 1K, 5K and 10K (d-f) $Fe(OH)_3$-ZnO/NF with magnification of 1K, 5K and 10K.

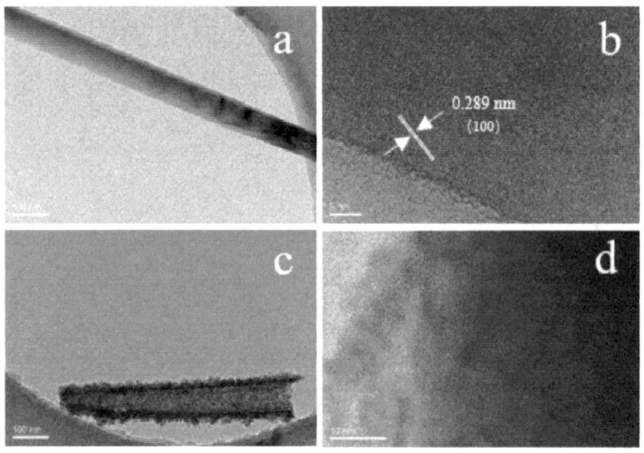

Fig.3. (a) TEM of ZnO nanorod (b) HR-TEM of ZnO nanorod(c) TEM of $Fe(OH)_3$ nanotube (d) HR-TEM of $Fe(OH)_3$ nanotube

The morphology of samples is showed in Fig.2 As revealed in Fig.2a, a large amount of ZnO nanorods are uniformly loaded on the foamed nickel. The diameter of the dense nanorods is about ~100 nm, and there is no adhesion between them. Fig.3.(b) shows that the interplanar spacing of the nanorods is 0.289 nm, which is consistent with the ZnO(100) plane. After immersing ZnO-NF in $FeCl_3$ aqueous solution for etching reaction, according to the reaction: $3ZnO + 2FeCl_3 + 3H_2O \rightarrow 2Fe(OH)_3 + 3ZnCl_2$, we can get the nanotube array as shown in Figure 2d. As shown in Fig.2f, it's clearly that the composite sample presents a hollow tubular structure. In addition, this tubular structure can provide more channels for electron migration during hydrogen evolution and oxygen evolution. The TEM image in Fig.3c shows that the nanotube wall is composed of many nanoparticles with a size of ~5 nm.

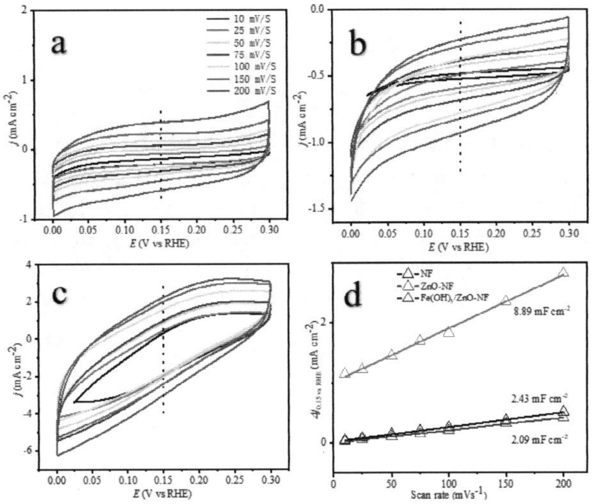

Fig.4. CV curves at different scan rates, (a) NF, (b) ZnO/NF, (c) $Fe(OH)_3$-ZnO/NF, (d) Fitting Cdl by current density change

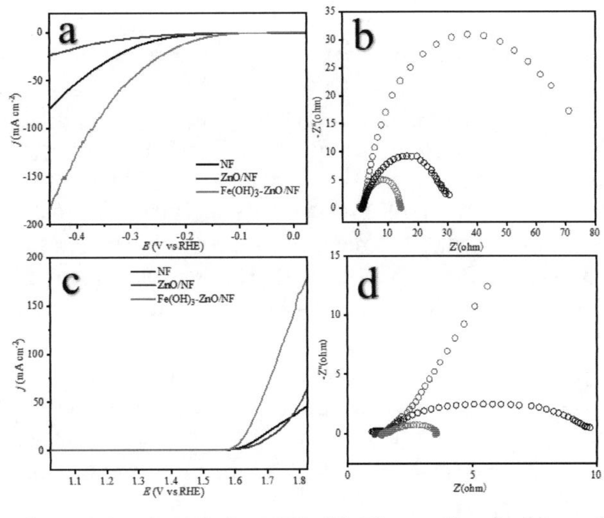

Fig.5. (a) Liner sweep coltammetry (LSV) for HER (b) Electrochemical impedance spectroscopy (EIS) for HER, (c) LSV for OER (d) EIS for OER

We further analyzed the active sites of composite materials and their contribution to electrocatalysis. Therefore, the electric double layer capacitance (Cdl), which is proportional to the electrochemical surface area (ECSA), is used for reasonable estimation and comparison. According to the cyclic voltammograms obtained at different scan rates (Figures 4a, b and c), the fitted Cdl of $Fe(OH)_3$-ZnO/NF is 8.89 mF cm^{-2} (Figure 4d), which is much higher than NF (2.43 mF cm^{-2}). The high C_{dl} value means that the catalyst obtains more active sites for complete hydrolysis.

The performance of the HER and OER was evaluated via the linear sweep coltammetry (LSV). In order to comparison, bare NF and ZnO/NF were also tested. Fig.5a and 5c display their polarization curves with iR correction. As shown in the Fig.5a, $Fe(OH)_3$-ZnO/NF has good hydrogen evolution performance, and as the overpotential increases, the prepared hollow nanotube catalyst performs better. When the current density reaches -10 mA cm^{-2}, the required overpotential is as low as 190 mV, which is better than NF (η_{10} =264 mV) and ZnO/NF (η_{10} = 350 mV). Compared with bare NF, the activity of ZnO/NF is slightly decreased, which indicates that the OER activity of ZnO is lower, which is consistent with the properties of ZnO semiconductor materials. The η_{10} of $Fe(OH)_3$-ZnO/NF is 388 mV, which is better than NF (η_{10} = 431 mV) and ZnO/NF (η_{10} = 450 mV)

4. Conclusion

In this work, the $Fe(OH)_3$-ZnO nanotube core-shell catalyst for water splitting was prepared by one-step hydrothermal and etching methods. Herein, the ZnO nanotubes directly grown on nickel foam can strengthen the contact between the active material and the conductive substrate and enhance the long-term stability of the electrocatalytic process. Subsequently, the $Fe(OH)_3$ nanoparticles accumulate on the surface of the nanorods. Due to the incomplete etching of ZnO, the catalyst can gradually form a core-shell nanotube structure. By evaluation, the composite catalyst exhibits a good performance in HER and OER.

Acknowledgments

This work was supported by the Natural Science Foundation of China (Nos. 51802282 and 51672249).

References

[1] Hadi, K., Nicholas, S.G., Wipf, D.O., Goodenough, J.B. (2019) Graphitic-Shell Encapsulation of Metal Electrocatalysts for Oxygen Evolution, Oxygen Reduction, and Hydrogen.AEM,1903215.

[2] Pan, J.Q., Chen, Z.F., Wang, P.H., Wang, P.P., Yu, Q., Zhao, W.J., Wang, J.J., Zhu, M., Zheng, Y.Y., Li, C.R. (2021) The overall water splitting of CdS/Ti3+-SrTiO3 core–shell heterojunction via OER enhancement of MnOx nanoparticles, Chem. Eng. J., 424: 130357.

[3] Yang, Y.Q., Zhang, K., Lin, H.L., Li, X., Chan, H.C., Yang, L.C. (2014) Hierarchical Nanowire Arrays Based on ZnO Core−Layered Double Hydroxide Shell for Largely Enhanced Photoelectrochemical Water Splitting, Advanced Functional Materials,24:580-586.

[4] Shao, M.F., Ning, F.Y., Wei, M., Evans, D.G., Duan, X., (2021) The photocatalytic overall water splitting hydrogen production of g-C3N4/CdS hollow core–shell heterojunction via the HER/OER matching of Pt/MnOx, Chem. Eng. J, 2405: 126622.

[5] Lu, F., Zhou, M., Zhou, Y.X., Zeng, X.H. (2017) First-Row Transition Metal Based Catalysts for the Oxygen Evolution Reaction under Alkaline Conditions: Basic Principles and Recent Advances, Small，1701931:1-18.

[6] Zeng, K., Zhang, D.k. (2010) Recent progress in alkaline water electrolysis for hydrogen production and applications, Progress in Energy and Combustion Science,36:307-326.

[7] Liu, H.T., Guan, J.Y., Yang, S.X., Yu, Y.H., Shao, R., Zhang, Z.P., Dou, M.L., Wang, F. (2020) Metal–Organic-Framework-Derived Co2P Nanoparticle/Multi-Doped Porous Carbon as a Trifunctional Electrocatalyst, AM,32:2003649.

[8] Yang, Y., Yao, H.Q., Yu, Z.H., Islam, S.M., He, H.Y., Yuan, M.W., Yue, Y.H., Xu, K., Hao, W.C., Sun, G.B., Li, H.F., Ma, S.L., Zapol, P., Kanatzidis, M.G. (2019) Hierarchical Nanoassembly of MoS2/Co9S8/Ni3S2/Ni as a Highly Efficient Electrocatalyst for Overall Water Splitting in a Wide PH Range, J. Am. Chem. Soc., 141:10417-10430.

[9] Yu, N., Cao, W., Huttula, M., Kayser, Y., Hoenicke, P., Beckhoff, B., Lai, F.Y., Dong, R.H., Sun, H.X., Geng, B.Y. (2020) Fabrication of FeNi hydroxides double-shell nanotube arrays with enhanced performance for oxygen evolution reaction, Applied Catalysis B: Environmental, 261:11893.

[10] Lei Dai, Zhe-Ning Chen, Liuxiao Li, Peiqun Yin, Zhengqing Liu, and Hua Zhang (2019) Ultrathin Ni(0)-Embedded Ni(OH)2 Heterostructured Nanosheets with Enhanced Electrochemical Overall Water Splitting, AM,1906915:1-7.

[11] Pan, J.Q., Liu, Y.Y., Li, H.L., Ou, W., Li, S., Wang, J.J., Zheng, Y.Y., Song, C.S., Y.Y., Li, C.R. (2020) The photocatalytic hydrogen evolution enhancement of the MoS_2 lamellas modified $g\text{-}C_3N_4/SrTiO_3$ core-shell heterojunction, Renew. Energy, 161: 340-349.

Changes in the Spectral Characteristics of Hydrophilic Fraction in Compost Organic Matter

Zhang Zilong[1], Xiao Yu[1,2*]

[1] College of environmental science and engineering, Guilin University of Technology, Guilin, Guangxi, China

[2] Guangdong University of Petrochemical Technology, Maoming, Guangdong, China

Fisrt author's email: 1584125859@qq.com

*Corresponding author's e-mail: xiaoyu@gdupt.edu.cn

Abstract. The hydrophilic fraction (Hyl) is one of the important components of humus, and its structure and composition are important factors for evaluating the stability and maturity of compost. This study used spectroscopy to analyze the composition and structure of Hyl. The results showed that with the passage of composting time, the UV–Vis curve had a red shift, indicating that the conjugated structure of the substance increased and the aromaticity is enhanced. The three-dimensional fluorescence excitation–emission spectrum (3D-EEM) showed that the fluorescence intensity of humus-like substances in HYI was significantly enhanced, and microbial metabolites such as protein-like substances, carbohydrates and other small molecules were continuously degraded, and the relative content decreased from 21% to 8%, the humus-like substances rose from 54% to 68%, which showed that the organic matter in the compost is continuously condensing and polymerizing, and the maturity and stability of the compost product are deepened.

1. Introduction

The sludge contains a lot of organic matter and nutrient elements such as nitrogen, phosphorus and potassium, Composting is one of the important ways to realize the resource utilization of sludge. Composting products can improve soil fertility and microbial community structure [1], and can be used for agricultural land, garden land and degraded soil remediation [2-3]. The material transformation of compost mainly occurs in the water phase. Dissolved organic matter (DOM) is the most active component of organic matter, which is regarded as the binding center of organic pollutants, heavy metals and other toxic substances [4], It is also an important interface of microbial metabolism and transformation, and its main components include humic acid, fulvic acid and Hyl. Hyl is a typical heterogeneous mixture, and its structure and composition are significantly different in each composting stage. At present, there are relatively few studies on the transformation of hydrophilic component in composting process. Based on this, this study separated and purified the hydrophilic components of the compost at different time periods, and used UV–Vis and 3D-EEM to explore the material structure of Hyl, reveal their structural transformation characteristics during the composting process. It provides a new strategy for the use of spectroscopy to evaluate compost maturity, and also provides a scientific basis for the subsequent compost products entering the soil.

2. Materials and methods

The compost material used 15kg municipal sludge and 5kg auxiliary material rice bran, mixed evenly and put it into a 50L fermentation tank, the particle size is about 10~20mm, and the initial moisture content is 62%. 150g samples of upper, middle and lower layers were collected at 0, 3, 6, 9, 14, 19, 24 and 29 days. After freeze-drying, grinding and sieving, the samples were stored in the refrigerator for testing. For the separation and purification of HyI, refer to the literature [5], using NaOH+ $Na_4P_2O_4$ mixed solution leaching, through alkaline extraction, acid precipitation, elution resin column and so on. The data are processed by Excel, the toolbox toolkit of Matlab software is used for parallel factor analysis, and the graph is drawn by Origin software.

3. Results & Discussion

3.1. Analysis of the UV–Vis spectra of HA

Diluted the DOC concentration of each hydrophilic component to 20 mg/L, eliminating the interference error of organic carbon, selected the data at the beginning and end of the compost, the high temperature period, and the maturity period to draw the graph. As shown in Figure 1, the UV–Vis curve has a red shift, and a shoulder peak appears at 270nm, which indicated that the humicity and aromaticity of the hydrophilic component are continuously increasing. Conjugated structures such as unsaturated aromatic substances absorb ultraviolet light [6], caused electrons to undergo a $\pi \rightarrow \pi^*$ transition, this also showed that the content of humic substances in HyI, is increased, and the structure is more complex.

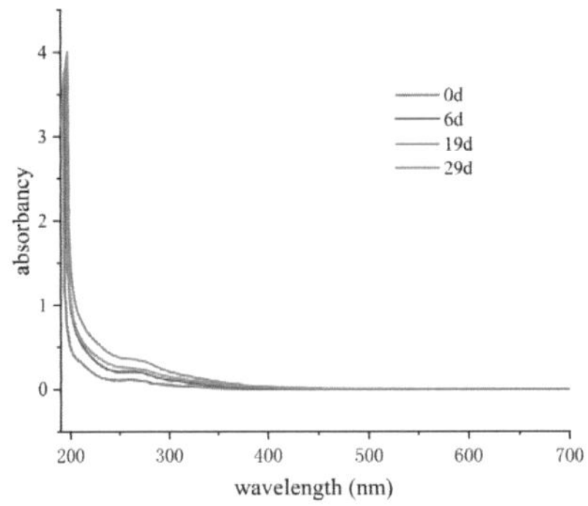

Figure 1 UV-Visible spectra of hydrophilic components.

3.2. Evolution of HyI components based on fluorescence spectroscopy

The organic components change drastically during the composting process. 3D-EEM is often used to study the structural characteristics of mixtures such as humus, which can effectively reflect the composting process and maturity. The DOC concentration of HyI samples in different time periods was uniformly diluted to 2mg/L, and then the fluorescence spectrum was scanned with the F98 fluorophotometer. As shown in Figure 2, SONG divided the fluorescence image into 5 regions [7]. The HyI mainly include by-products of microbial metabolism in region IV and humus-like substances in region V. In addition, the fluorescence peak shows a tendency to move to the V region, and the fluorescence intensity continues to increase, indicated that the concentration of humus-like substances has increased significantly, the degree of humification of hydrophilic components has increased, and the structure of the substance has become more stable. This is consistent with the results of UV-Vis spectra.

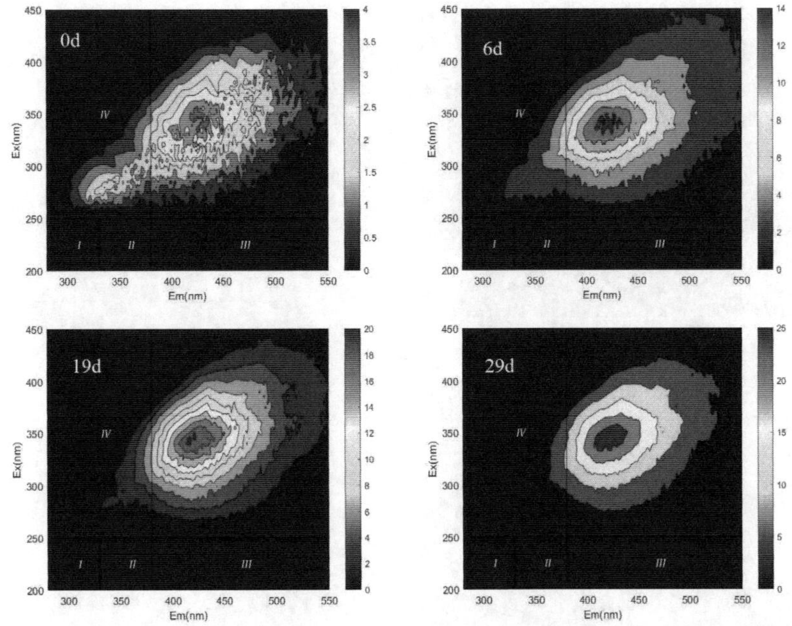

Figure 2 Three-dimensional fluorescence spectra of each stage of the hydrophilic component.

The parallel factor analysis (PARAFAC) method can perform qualitative and quantitative analysis of organic matter, it also can effectively avoid the overlap of fluorescence peaks and identify a single fluorescent component by three-dimensional fluorescence data analysis. Each fluorescent component represents a class of substances. Therefore, this method is more representative and scientific in quantitative characterization [8]. As shown in Figure 3, the PARAFAC divided the hydrophilic component into three components. Component 1 has a fluorescence peak at Ex=340nm, Em=415nm, which is closely related to the humus-like substances in the organic matter. The structure of humus-like substances is complex and difficult to decompose, which can prolong the residence time of organic matter in the soil and strengthen the soil Fertility. Component 2 has a strong and weak double peak at Ex=360nm, 395nm, Em=475nm, which is caused by the fluorescent group of the humic acid substance. The fluorescence peak of Component 3 appeared at Ex=310nm, Em=380nm, which is caused by protein-like substances such as microbial metabolites and their residues.

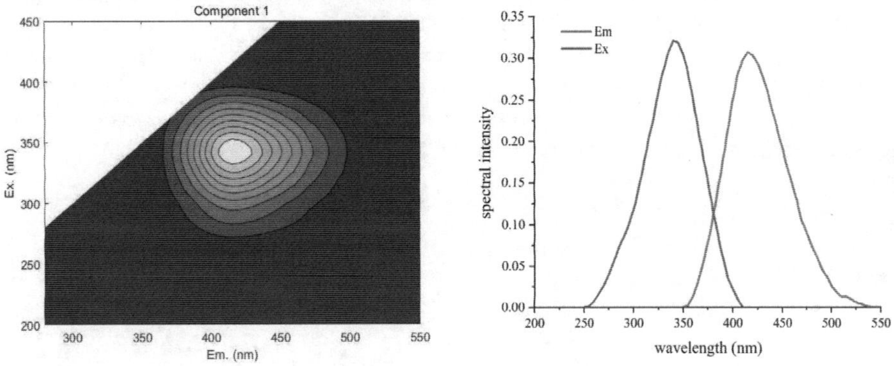

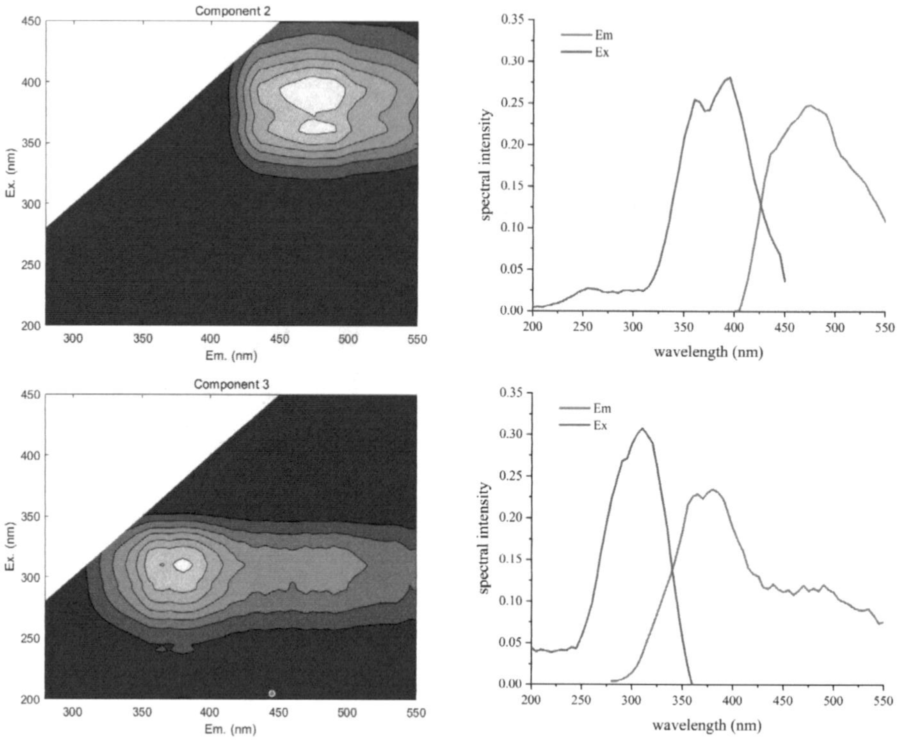

Figure 3 Diagram of HYI components analyzed by PARAFAC.

The content of the three components of HyI during the composting process is shown in Figure 4. The content of each component presents different trends. With the passage of composting time, the content of humus of Component 1 increased from 54% to 68%, which indicated that the hydrophilic components in the compost organic matter are gradually humified, and the carbon skeleton continues to condense to form humus-like substances with more complex components and larger molecular weights [9]. The content of Component 2 decreases first and then increases. This indicated that the microbes are active in the first and middle stages of composting, and the macromolecular humic acid will also be mineralized and decomposed. During this period, the rate of generation of humic acid is less than the degradation rate. After the compost temperature drops and enters the decomposing period, Microbial activity declines, small molecular substances gradually polymerize into humic acid, and the relative content of the substance increases. Studies have shown that protein substances are precursors of synthetic humus-like substances [10], and the initial content of Component 3 has been reduced from 21% to 8%, indicated that microbial metabolites and their residues are easily degradable protein-like substances (such as tyrosine, tryptophan, carbohydrates, etc.) are gradually degraded and converted into humus-like substances that are more stable and more complex in structure.

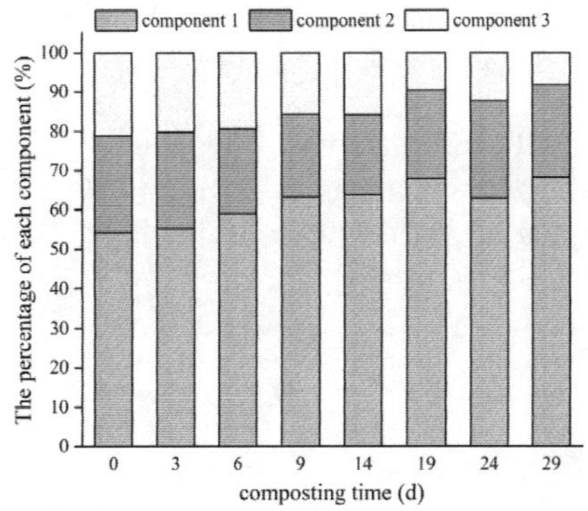

Figure 4 The relative content percentage diagram of each component.

4. Conclusion

The spectroscopy method can effectively characterize the change characteristics of organic matter during the composting process. With the progress of composting, the UV-Vis curve has a red shift, the conjugated structure of the substance increased, and the aromaticity enhanced. 3D-EEM combined with PARAFAC showed that HyI is continuously humified, and the protein-like substances in microbial residues and their metabolites are degraded in a large amount, and the relative content is reduced from 21% to 8%, while the high aromatization and high humus-like substances increased from 54% to 68% at the beginning, and the material structure became more stable and complicated, and the compost maturity was deepened.

Acknowledgments

Thanks for the scientific research funding of Guangxi Innovation Research Team Project (2018GXNSFGA281001) and Projects of Talents Recruitment of GDUPT 2020rc035.

References

[1] Awasthi S K, Sarsaiya S, Awasthi M K, et al. (2020) Changes in global trends in food waste composting: Research challenges and opportunities [J]. Bioresource Technology, 299: 122555.

[2] Titova J, Baltrenaite E. (2021) Physical and Chemical Properties of Biochar Produced from Sewage Sludge Compost and Plants Biomass, Fertilized with that Compost, Important for Soil Improvement [J]. Waste And Biomass Valorization, 12(7): 3781-3800.

[3] Pandey P K, Cao W, Biswas S, et al. (2016) A new closed loop heating system for composting of green and food wastes [J]. Journal of Cleaner Production, 133: 1252-1259.

[4] Zhao Y, Wei Y, Zhang Y, et al. (2017) Roles of composts in soil based on the assessment of humification degree of fulvic acids [J]. Ecological Indicators, 72: 473-480.

[5] Yu M, He X, Liu J, et al. (2018) Characterization of isolated fractions of dissolved organicmatter derived from municipal solid waste compost [J]. Science Of the Total Environment, 635: 275-283.

[6] He X S, Xi B D, Jiang Y H, et al. (2013) Structural transformation study of water-extractable organic matter during the industrial composting of cattle manure [J]. Microchemical Journal, 106: 160-166.

[7] Song C, Li M, Wei Z, et al. (2016) Effect of inoculation with multiple composite microorganisms on characteristics of humic fractions and bacterial community structure during biogas residue and livestock manure co-composting [J]. Journal Of Chemical Technology And Biotechnology, 91(1): 155-164.

[8] Song C, Li M, Xi B, et al. (2015) Characterisation of dissolved organic matter extracted from the bio-oxidative phase of co-composting of biogas residues and livestock manure using spectroscopic techniques [J]. International Biodeterioration & Biodegradation, 103: 38-50.

[9] Yu G H, He P J, Shao L M. (2010) Novel insights into sludge dewaterability by fluorescence excitation-emission matrix combined with parallel factor analysis [J]. Water Research 44(3): 797-806.
[10] Che J, Lin W, Ye J, et al. (2020) Insights into compositional changes of dissolved organic matter during a full-scale vermicomposting of cow dung by combined spectroscopic and electrochemical techniques [J]. Bioresource Technology, 301:122757.

Study on Permanent Thermal Insulation Formwork of Mass Concrete Prepared by Different Types of Nano-modified Lightweight Aggregate

Xiaojian Jiang[1], Sijia Zhang[2*], Zhifeng Gu[1], Licheng Wang[3], Yunfei Zhao[1]

[1]Sartohai Reservoir Management Office, Altay, Xinjiang, 836500, China

[2]China Institute of Water Resources and Hydropower Research, Beijing IWHR-KHL Co., Ltd., Beijing, 100038, China

[3]Bei Fang Investigation, Design & Research Co., Ltd, Tianjin, 300222, China

[*]Corresponding author's e-mail: zhangsj@iwhr.com

Abstract: Aiming at the problems of low strength, flammability, easy falling off, and aging of traditional organic thermal insulation materials, inorganic thermal insulation materials of different types of nano-modified lightweight aggregates were prepared in this paper. The feasibility of preparing permanent thermal insulation formwork of mass concrete with nano-modified lightweight aggregate was studied through experiments and calculation analysis of mechanical, durability, and thermal properties. The results show that the water absorption of shale ceramsite modified by nano-permeable hydrophobic agent is reduced to 1.21%, and the water absorption rate of inorganic precast hollow aggregate produced by nano-silica powder is lower, which is 0.79%. Inorganic precast hollow aggregate concrete, modified shale ceramsite concrete, and organic foamed aggregate concrete have the characteristics of high strength, fire resistance, aging resistance and low thermal conductivity. Inorganic precast hollow aggregate concrete and modified shale ceramsite concrete have excellent impermeability and frost resistance. Organic foamed aggregate concrete has the lowest thermal conductivity and good thermal insulation effect and its durability can be improved by surface protection technology. According to the principle of equivalent surface heat release coefficient, it is feasible to use lightweight aggregate precast formwork as permanent insulation formwork of mass concrete with thickness between 50 and 100 cm. The permanent thermal insulation formwork of nano-modified lightweight aggregate concrete can also be used as the formwork for concrete construction to realize the temperature control measures of tracking thermal insulation and slow cooling of mass concrete, reduce the risk of cracking, and improve the construction efficiency.

1. Introduction

In order to optimize the energy structure and rationally develop hydropower and water resources, China will plan and develop a number of water conservancy and hydropower projects in cold regions such as Xinjiang and Tibet. These areas are characterized by extreme temperature difference, large temperature difference between day and night, low annual average temperature, and dry and windy climate. Dam concrete is prone to produce harmful temperature cracks due to temperature drop shrinkage and dry shrinkage deformation, which seriously threatens the operation safety of the dam[1,2].

The technology of temperature control and crack prevention for mass concrete is still in the stage of exploration and research under severe climatic conditions such as high cold and large temperature difference [3]. Thermal insulation of dam surface during construction and operation is one of the main measures to reduce the temperature difference between inside and outside, reduce the temperature stress and prevent the concrete cracking of dam body [4,5]. At present, polyurethane, polystyrene board, and other organic thermal insulation materials are mainly used for the concrete dam surface thermal insulation [6-8]. From the perspective of application effect, these organic thermal insulation materials have many problems, such as flammability, low safety, short service life, poor ice pull-out resistance and water durability, easy to fall off and so on [9]. These problems have appeared in the application of concrete for dams in many cold areas in China, especially in the upstream water level change area of the dam body. As a result, it is difficult to achieve the original design of permanent insulation. Inorganic thermal insulation materials such as foamed concrete and lightweight aggregate concrete have the characteristics of light weight, heat insulation, and fire resistance. They have been widely used in construction. However, they have not been applied in the thermal insulation of concrete in dam body due to their low strength, unclosed voids, high water absorption, poor impermeability, and frost resistance [10-12].

The existing research data [13,14] indicate that the Baishan Hydropower Station in Jilin Province, which was built in the late 1980s in China, has a concrete gravity arch dam with a height of 149.5m and a non-dismantling wooden formwork for dam construction; the overflow weir of Qinglong Hydropower Station has adopted reinforced concrete precast non-dismantling formwork. The precast concrete non-demolition formwork technology has the characteristics of fast construction, safety, and beautiful appearance of the concrete and has significant advantages in concrete insulation, moisture retention, and crack resistance. The permanent thermal insulation formwork of mass concrete should have the characteristics of low thermal conductivity, high strength, excellent impermeability and frost resistance durability, and convenient construction.

In view of the shortcomings of traditional thermal insulation materials, the surface of ceramsite was modified by hydrophobic method in this paper. The mechanical properties, durability, and thermal properties of different types of lightweight aggregate concrete, such as modified shale ceramsite, inorganic precast hollow aggregate, and organic foamed aggregate, were studied. The feasibility of preparing permanent thermal insulation formwork of mass concrete with lightweight aggregate was analyzed. It is expected to solve the problems of cracking, aging and falling off of surface thermal insulation materials during the construction and operation of mass concrete in alpine region.

2. Experimental Section

2.1 Raw Materials

The raw materials used in the experiment are P·O42.5 ordinary Portland cement, class F grade II fly ash, polycarboxylic acid high-performance water reducing agent, and polyether sulfonate air entraining agent.

Modified Shale ceramsite: the shale ceramsite was immersed in permeable hydrophobic agent for 24 h and then taken out and dried to form a hydrophobic membrane on the surface of the modified shale ceramsite. The water absorption was reduced to 1.21%.

Inorganic precast hollow aggregate: the inorganic hollow aggregate is formed by calcining alumina, mullite, kaolin, and nano-silica powder at high temperature, with a water absorption rate of 0.79%, as shown in Figure 1.

Organic foamed aggregate: it is made of special organic materials and internal foaming process. The section structure of organic foamed aggregate is shown in Figure 2. The water absorption rate is 1.95%.

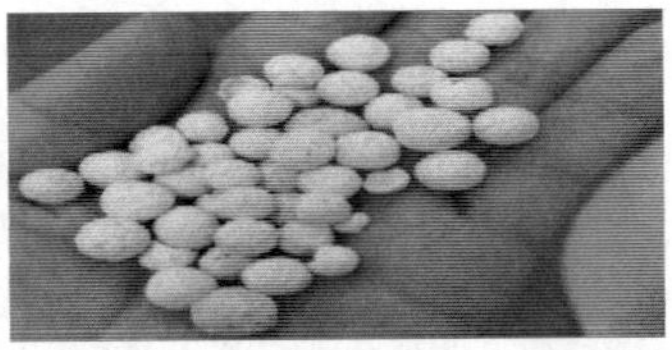

Figure 1 Inorganic precast aggregate

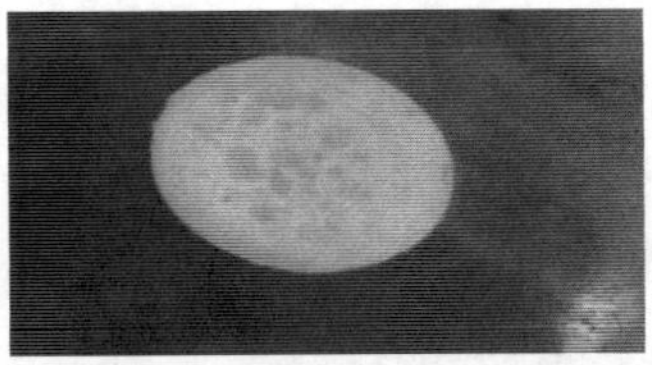

Figure 2 Cross-section of organic foamed aggregate

2.2 Methods

The compressive strength, tensile strength, impermeability and frost resistance tests of lightweight aggregate concrete were carried out according to SL 352-2006 "Test Code for Hydraulic Concrete". The size of the compressive strength and tensile strength specimens is 100mm × 100mm × 100mm.

The thermal conductivity test of lightweight aggregate concrete was conducted using the protective hot plate method in the steady-state method, which was in accordance with GB/T 10294-2008 "Thermal insulation–Determination of steady-state thermal resistance and related properties–Guarded hot plate apparatus". The test period was 28 days.

3. Results and Analysis

3.1 Mechanical properties

The compressive strength and tensile strength of different types of lightweight aggregate concrete are shown in Table 1. At the age of 28 days, the compressive strength of different types of lightweight aggregate concrete is between 10.6 and 21.9MPa, and the tensile strength is between 1.10 and 1.91MPa. The compressive strength and tensile strength of inorganic precast hollow aggregate concrete are the highest, followed by modified shale ceramsite concrete. Those of organic foamed aggregate concrete are the lowest. This may be due to the relatively smooth surface of the organic foamed aggregate and the low bonding strength of the interfacial transition zone between the aggregate and the cement stone, which leads to the decrease of the strength of the organic foamed aggregate concrete.

Table 1 Mechanical properties of different types of lightweight aggregate concrete

Aggregate type	Test Number	Water-Binder Ratio	Volume Ratio of Lightweight Aggregate（%）	Compressive Strength（MPa）		Tensile Strength（MPa）	
				7d	28d	7d	28d
Inorganic Precast Hollow Aggregate	WJYZ	0.40	70	10.7	21.9	1.32	1.91
Modified Shale Ceramsite	GXYY			11.4	17.9	0.97	1.75
Organic Foamed Aggregate	YJFP			7.7	10.6	0.58	1.10

3.2 Impermeability

Improving the impermeability of concrete is one of the important measures to improve the durability of concrete. The test results of impermeability of different types of lightweight aggregate concrete are shown in Table 2. During the test, the water pressure started from 0.1MPa and increased by 0.1MPa every 8h. The water seepage at the end face of the specimen was observed at any time. When the pressure reached 0.7MPa or there were more than two specimens with water seepage on the surface, the water pressure was recorded and the machine was shut down. The test results show that the impermeability grades of the inorganic precast hollow aggregate concrete and the modified shale ceramsite concrete are both≥W6, and the water seepage height of the modified ceramsite concrete is the lowest, which is 2.3cm. Therefore, the impermeability of concrete can be significantly improved by surface modification of porous shale aggregate. When the water pressure was increased to 0.6MPa, water seepage was found on the surface of the organic foamed aggregate concrete specimen, so the impermeability grade of the organic foamed aggregate concrete is ≥W4.

Table 2 Impermeability of different types of lightweight aggregate concrete

Aggregate type	Test Number	Test Period (d)	Maximum Water Pressure (MPa)	Seepage Height (cm)	Impermeability grade
Inorganic Precast Hollow Aggregate	WJYZ	28	0.7	4.0	≥W6
Modified Shale Ceramsite	GXYY		0.7	2.3	≥W6
Organic Foamed Aggregate	YJFP		0.5	/	≥W4

3.3 Frost resistance

The frost resistance of concrete is one of the important factors that affect the durability, especially in the cold area. Therefore, improving the frost resistance of concrete plays an important role in improving the structural durability. The frost resistance test results of different types of lightweight aggregate concrete are shown in Table 3, Figure 3, and Figure 4. The results show that the aggregate type has a significant effect on the frost resistance of concrete. The frost resistance of modified shale ceramsite concrete is excellent, reaching F300 frost resistance grade. The frost resistance grade of inorganic precast hollow aggregate concrete is F200. The relative dynamic elastic modulus of organic foamed aggregate concrete is less than 60% after 100 freeze-thaw cycles, which reaches the frost resistance grade of F100. The relative dynamic elastic modulus of concrete decreases with the increase of freeze-thaw cycles, but the quality variation of different types of aggregate concrete is different. The quality of inorganic precast hollow aggregate concrete decreases with the increase of freeze-thaw cycles because the surface mortar falls off during the freeze-thaw process. Since the modified shale ceramsite is a porous aggregate, it can absorb water during the freeze-thaw process. The quality of concrete increases with the increase of the number of freeze-thaw cycles. After reaching the saturation state, the surface quality of concrete is slightly lost because of freeze-thaw damage. Therefore, the quality of concrete decreases with the increase in the number of freeze-thaw cycles, but the total quality and appearance of the concrete does not change much. Because the organic foamed aggregate is also a porous material and has a certain water absorption effect, the quality of concrete first increases with the increase of freeze-thaw cycles. However, due to the low bond strength between the organic aggregate and the cement stone, the concrete is easy to be damaged by freezing. As shown in Figure 5, after 100 freeze-thaw cycles, concrete spalling occurs at the aggregate interface and the quality decreases significantly.

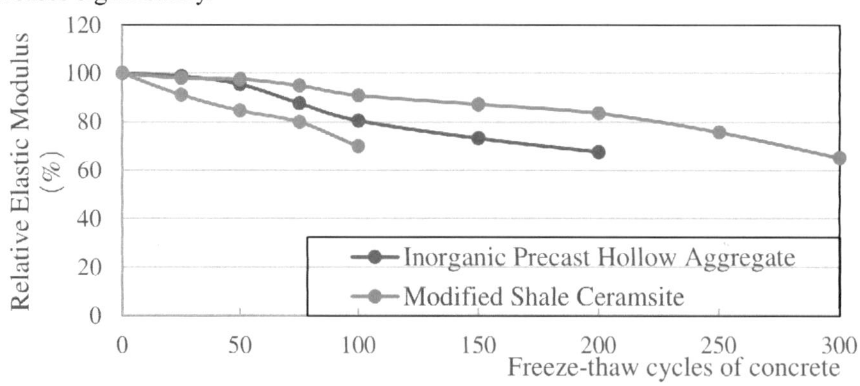

Figure 3 The relative dynamic elastic modulus of different types of lightweight aggregate concrete

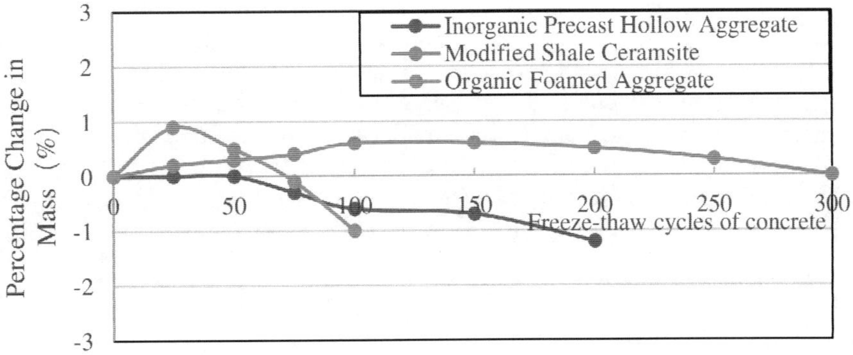

Figure 4 The percentage change in the mass of different types of lightweight aggregate concrete

Figure 5 Organic foamed aggregate concrete subjected to freezing

Table 3 Frost resistance of different types of lightweight aggregate concrete

Aggregate type	Test Number	Test Period (d)	Relative Elastic Modulus (%) Percentage Change in Mass (%)									Frost Resistance Grade
			0	25	50	75	100	150	200	250	300	
Inorganic Precast Hollow Aggregate	WJYZ	28	100 0.0	98.8 0.0	95.5 0.0	87.6 -0.3	80.5 -0.6	73.2 -0.7	67.6 -1.2	/	/	F200
Modified Shale Ceramsite	GXYY		100 0.0	98.0 0.2	97.6 0.3	94.9 0.4	90.9 0.6	87.2 0.6	83.5 0.5	75.7 0.3	65.2 0.0	F300
Organic Foamed Aggregate	YJFP		100 0.0	91.1 0.9	84.6 0.5	79.9 -0.1	69.9 -1.0	/	/	/	/	F100

3.4 Thermal Properties

The smaller the thermal conductivity of a material is, the better its thermal insulation performance is. The test results of thermal conductivity of different types of lightweight aggregate concrete are shown in Table 4. The test results show that the bulk density and thermal conductivity of the three types of lightweight aggregate concrete are within the normal range of thermal insulation materials. The type of aggregate is closely related to the thermal conductivity of concrete. The thermal conductivity of organic foamed aggregate concrete is the lowest, that of modified shale ceramsite concrete is slightly higher, and that of inorganic precast hollow aggregate concrete is the highest. This is mainly related to the pore structure of aggregate. Previous studies [12, 15, 16] have shown that when the porosity of the material is constant, the closer the pore structure is to the spherical shape, the smaller the equivalent radius is, and the lower the thermal conductivity of material is. As can be seen from Fig. 2, the inside of the organic foamed aggregate are all small closed circular air holes. The smaller the size of the air holes, the greater the resistance to the air convection amplitude. As a result, the reduces the efficiency of convection heat transfer and the thermal conductivity are reduced and the thermal insulation performance of concrete is improved. Shale ceramsite is made by firing clay minerals at high temperature. The pore size and connectivity degree of the aggregate are uncertain. Although the inorganic precast hollow aggregate has a closed circular pore size, the pore size is larger than that of organic foamed aggregate.

Table 4 Thermal conductivity of different types of lightweight aggregate concrete

Aggregate type	Test Number	Size (cm)	Bulk Density（kg/m³）	Thermal Conductivity W/(m·K)
Inorganic Precast Hollow Aggregate	WJYZ	30×30×3	1376	0.394
Modified Shale Ceramsite	GXYY		1108	0.307
Organic Foamed Aggregate	YJFP		752	0.269

3.5 Permanent thermal insulation formwork of lightweight aggregate concrete

As inorganic thermal insulation materials, the three types of lightweight aggregate concrete have the characteristics of high strength, aging resistance, and low thermal conductivity. Among them, the inorganic precast hollow aggregate concrete and the modified shale ceramsite concrete have excellent impermeability and frost resistance. For organic foamed aggregate concrete, the durability can be improved through surface protection.

According to the calculation method of the equivalent surface heat release coefficient[17] (formula (1)) , the equivalent heat release coefficient (β_s) of the concrete surface through the insulation layer to the surrounding medium is inversely proportional to the thickness of the insulation material and directly proportional to the thermal conductivity of the insulation material, that is, the thicker the insulation material, the lower the thermal conductivity, and the better the insulation effect. The available data show that the thermal conductivity of polyurethane material is about 0.03w/(m·K) and the thickness of thermal insulation layer of mass concrete is between 5 and 8 cm. According to formula (1), if lightweight aggregate concrete is used as permanent thermal insulation formwork, the thickness of formwork should be between 50 and 100 cm if it achieves the same thermal insulation effect as polyurethane. Moreover, the volume weight of lightweight aggregate concrete is only 1/3 to 1/2 of ordinary concrete. Since the prefabricated formwork is easy to be lifted and assembled, it is feasible for construction.

$$\beta s = \frac{1}{\frac{1}{\beta o}+\Sigma\frac{hi}{\lambda i}} \tag{1}$$

Where β_0 is the equivalent heat release coefficient of the uninsulated concrete surface, h_i is the thickness of the insulation material i, and λ_i is the thermal conductivity of the insulation material i.

Lightweight aggregate concrete precast formwork can be used not only as the formwork for mass concrete pouring construction, but also as the permanent insulation formwork of concrete. It can not only realize the permanent heat insulation of the concrete surface and improve the overall aesthetics, but also facilitate the implementation of the process control of "early thermal insulation, small temperature difference, and slow cooling" of mass concrete temperature control measures. As a result, the risk of concrete cracking can be reduced and the construction efficiency can be improved.

4. Conclusion

(1) The water absorption of shale ceramsite modified by nano-permeable hydrophobic agent is reduced to 1.21%. The water absorption rate of inorganic precast hollow aggregate produced by nano-silica powder is lower than modified shale ceramsite, which is 0.79%. The water absorption of porous lightweight aggregate can be effectively reduced by modifying or preparing aggregate with nano

materials.

(2) For the 28-day-old inorganic precast hollow aggregate concrete, the compressive strength and tensile strength of the are 21.9MPa and 1.91MPa respectively, the thermal conductivity is 0.394W/(m·K), the impermeability grade is ≥W6, and the frost resistance grade is F200. For the 28-day-old modified shale ceramsite concrete, the compressive strength and tensile strength are 17.9MPa and 1.75MPa respectively, the thermal conductivity is 0.307W/(m·K), the impermeability grade is ≥W6, and the frost resistance grade reaches F300. For the 28-day-old organic foamed aggregate concrete, the compressive strength and tensile strength are 10.6MPa and 1.10MPa respectively, the thermal conductivity is 0.269W/(m·K), the impermeability grade is ≥W4, and the frost resistance grade is F100.

(3) Compared with the traditional organic insulation materials, the inorganic precast hollow aggregate concrete, the modified shale ceramsite concrete and the organic foamed aggregate concrete have the characteristics of high strength, fire resistance, aging resistance, and low thermal conductivity. They can be made into prefabricated formwork through structural design and construction process optimization, taking into account the dual functions of permanent thermal insulation on the surface of mass concrete and the formwork for pouring construction. It is not only conducive to the implementation of tracking thermal insulation and slow cooling temperature control measures for mass concrete, solving the problems of cracking, aging, and falling off of surface thermal insulation materials of mass concrete in cold areas, but also improving the construction efficiency, which has good economic and social benefits.

(4) Inorganic precast hollow aggregate concrete and modified shale ceramsite concrete have excellent impermeability and frost resistance. Due to the low bonding strength between the organic foamed aggregate and the cement stone, the impermeability and frost resistance of concrete are limited. However, the organic foamed aggregate concrete has the lowest thermal conductivity and good thermal insulation effect. Its durability performance can be improved by surface protection.

(5) The use of lightweight aggregate precast formwork as permanent insulation and construction formwork of mass concrete can not only improve the construction quality and operation efficiency, but also innovate the dam construction technology in cold areas. The problems such as the design of light aggregate precast thermal insulation formwork, the coordination and safety of splicing and mass concrete construction need to be further examined and solved in practice.

Acknowledgments
The authors would like to acknowledge the financial support of Science and Technology Project of Xinjiang Water Conservancy (XSKJ-2021-10), IWHR Fund for Fundamental Science Research (SM0145B032021) and Open project of Guangxi Key Laboratory of Hydraulic Materials and Structures (GXHRI-WEMS-2020-02).

References
[1] Liu Y, Du L, Qian W, Ren B, Wang J, Wan L, Li S, Yan C. Study on key technical problems of high concrete dam construction in alpine region. Water Resour. Hydr. Eng. 2020; 51(03):45-52.
[2] Zhou Z, Liu C, Wang W. Durability problem of concrete in alpine region of Tibet and its countermeasures. Water Resour. Hydr. Eng. 2013; 44(03): 82-86.
[3] Zhang P. Dam construction technology of precast concrete without dismantling thermal insulation formwork in cold region. Water Resour. Hydr. Northeast Chin. 2020; (11): 14-16.
[4] Zhang Y. Research on temperature control and crack prevention measures of mass concrete in water conservancy project. Water Resources Planning and Design. 2015; (4): 60-62+110.
[5] Li S, Zhang G, Liu Y, Zhang Y. Mass concrete crack prevention intelligent monitoring technology and engineering application. Journal of China Institute of Water Resources and Hydropower Research. 2018; 16(1): 9-15.
[6] Qian Y, Zhao L. Application of polyurethane thermal insulation material in water conservancy project in severe cold area. Gansu Water Conservancy and Hydropower Technology. 2010;

46(7): 23-24.

[7] Liu Y, Huang G. Application of spraying polyurethane hard foam in concrete dam. 2005 RCC material and quality inspection Conference. Nanjing, 2005.

[8] Al-Homoud MS. Performance characteristics and practical applications of common building thermal insulation materials[J]. Building and Environment. 2005; (40): 353-366.

[9] Zhu B, Mai S. Permanent compound plate of thermal and water insulation for concrete dams. Water Resour. Hydr. Eng. 2006; 37(4): 13-18.

[10] Lei D, Guo L, Liu J, Miao C, Sun W. State of study and applicaiotn of foamed concrete. J Funct. Mater. 2017; 11(48): 37-53.

[11] Ramamurthy K, Nambiar EKK, Ranjani GIS. A classification of studies on properties of foam concrete. Cement and Concrete Composites. 2009; 32(31): 388-396.

[12] Sun Y. Development and applicaiotn of lightweight aggregate and high strength lightweight concrete. Nanjing: Dissertation of Southeast University. 2017.

[13] Du L, Xi Y. Key technologies and countermeasures for construction of super-high arch dam in severe cold and strong seismic region. Water Resour. Hydr. Eng. 2017; 48(10): 10-17.

[14] Liu Z. Application of the precast concrete form in the construction of overflow weirs in Qinglong hydropower station. Hydropower and New Engergy. 2015; 133(7): 40-50.

[15] Liu X, Zheng C, Huang C. Analysis of factors affecting thermal conductivity of porous materials. Low Temp. Archit. Technol. 2009; (9): 121-122.

[16] Tang W C, Lo Y, Nadeem A. Mechanical and drying shrinkage properties of structural-graded polystyrene aggregate concrete. Cem. Concr. Compos. 2008; 30(5): 403-409.

[17] Zhu B. Temperature stress and temperature control of mass concrete. 2nd Edition. Beijing: China Water&Power Press; 2012.

Component distribution of nano-carbon materials assisted by Time of Flight-Secondary Ion Mass Spectrometer

Ying Dong[1,2], Fan Xu[1], Yan Li[3], Tinglu Song[1*] and Guoqiang Tan[1,2*]

[1] Experimental Center of Advanced Materials School of Materials Science & Engineering, Beijing Institute of Technology, Beijing, 100081, China

[2] Beijing Key Laboratory of Environmental Science and Engineering, School of Materials Science & Engineering, Beijing Institute of Technology, Beijing, 100081, China

[3] Shenzhen Ori Technology Co., Ltd, Shenzhen, Guangdong, 518107, China

*Corresponding author's e-mail: song@bit.edu.cn; tan123@bit.edu.cn

Abstract. Currently, there exist various materials surface analysis and characterization techniques, but it is difficult to measure their component distribution, in particular from a three-dimensional prospect. In this paper, the component distribution of nano-carbon materials in 2D and 3D views were both characterized via Time of Flight-Secondary Ion Mass Spectrometer (TOF-SIMS). The results indicate that TOF-SIMS could provide comprehensive information about component distribution, which is expected to be developed as an advanced technology in material analysis and characterization to the research community.

1. Introduction

The completion of materials analysis and characterization depends on advanced technologies[1]. In recent years, more advanced demands have been required for materials analysis and characterization, such as the detection limit, the spatial resolution and sensitivity, etc., which promotes the development of materials analysis and characterization technology. Among them, TOF-SIMS technology has attracted much attention and has been applied into various fields.

TOF-SIMS is a technology that combines time of flight (TOF) and secondary ion mass spectrometry (SIMS). The basic principle is to employ high energy primary ion sources (e.g., Ga^+, Bi^{3+}, etc.) to shine on the sample surface and excite secondary ions with the same initial kinetic energy. Ions that generated at the surface with different m/z values will then reach the detector at different times. The information about elements, isotopes, molecular structure and chemical bonding could be obtained. In addition, by using TOF-SIMS technology, not only the surface component information can be acquired, but also in the depth direction through continuous sputtering, which is illustrated in depth profiles, 2D and 3D images[2-4].

Nowadays, TOF-SIMS has been widely employed in various fields, such as semiconductor devices, nano devices, biomedicine, quantum structure, high-energy secondary batteries, high-tech materials including polymer materials, metals, semiconductors, glass ceramics, nano coating, paper, film, fiber, etc[5-7]. Compared with other surface analysis and characterization techniques, TOF-SIMS exhibits several advantages, such as a high spatial resolution of 70 nm, a ppb level detection limit, and excellent sensitivity. The main application capabilities of some popular surface analysis and characterization techniques were summarized in Table 1.

Table 1. Comparison of some popular surface analysis and characterization techniques in main application capabilities[8].

	XPS	AES	TOF-SIMS	SEM-EDS	XRF
Signal type	electron	electron	secondary ion	X ray	X ray fluorescence
Detection depth	~ 10 nm	~ 5 nm	1 ~ 2 nm	~ 1 μm	~ 2 μm
Detectable element	Li ~ U	Li ~ U	H ~ U	B ~ U	F ~ U
Detection limit	0.01%	0.1%	ppb	0.1%	ppm
Chemical analysis	excellent	general	excellent	-	-
Depth profiling	excellent	excellent	excellent	-	-
Insulation materials	excellent	general	excellent	need to spray gold	excellent
Spatial resolution	10 μm	8 nm	70 nm	1 μm	1 mm

With the rapid development of nanomaterials, its surface analysis and characterization are particularly important. This paper aims to analyze and characterize nanomaterials assisted by TOF-SIMS, and obtain rich and comprehensive information about the component distribution, so as to show that TOF-SIMS could be considered as a simple, fast and efficient method for nanomaterials analysis and characterization. It's necessary to introduce this technology because it could provide component distribution of nanomaterials from a 3D prospect, which could not be achieved via other technologies.

2. Materials and methods

Samples: Three samples were involved in this paper. The sample 1# was carbon paper produced by Toray Company of Japan. The sample 2# was carbon paper after cycling in a half cell with lithium metal. The 3# sample was obtained when the 2# sample used as the cathode material of the lithium-oxygen battery after discharging.

Equipment: ULVAC-PHI, Inc, Equipment type: PHI nano TOF II.

Each image had 512×512 pixels. TOF-SIMS depth profiling used 3 keV Ar^+ ion sputtering with an area of 400um×400um. The beam current was approximately 100 nA.

3. Results and discussion

3.1. Images of surface component distribution

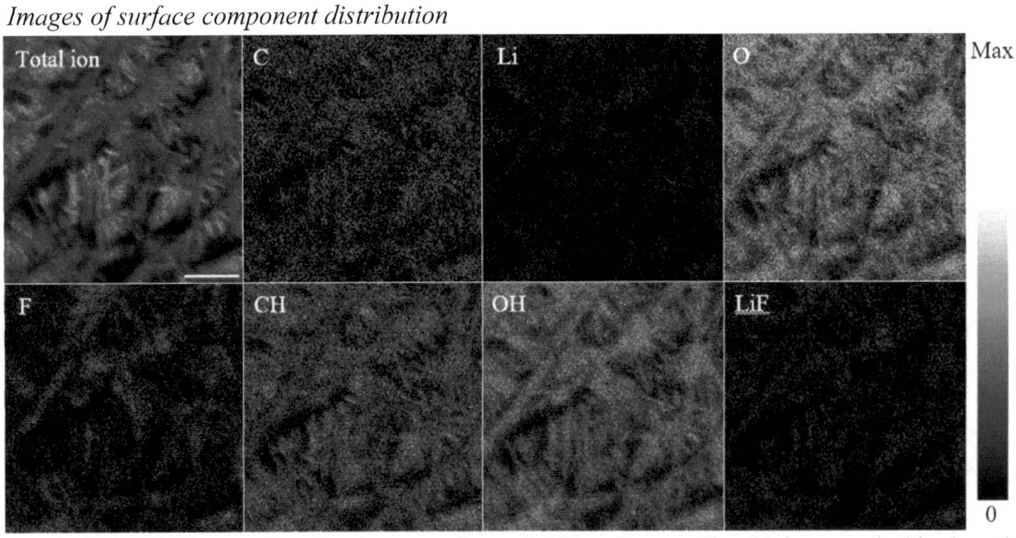

Figure 1. TOF-SIMS secondary ion images of sample 2# under negative (-) ions mode (the length of scale bar is 100 μm).

Journal of Physics: Conference Series **2011** (2021) 012071 doi:10.1088/1742-6596/2011/1/012071

TOF-SIMS secondary ion images of sample 2# are shown in figure 1, which directly reflects the distribution of chemical species on the surface of the sample. The essential form of carbon fibers with sub-nano scales could be observed in these images. In addition, the chemical species on the surface of the sample grow uniformly attached to the carbon fibers. Among all, the intensity of O element is the highest[9].

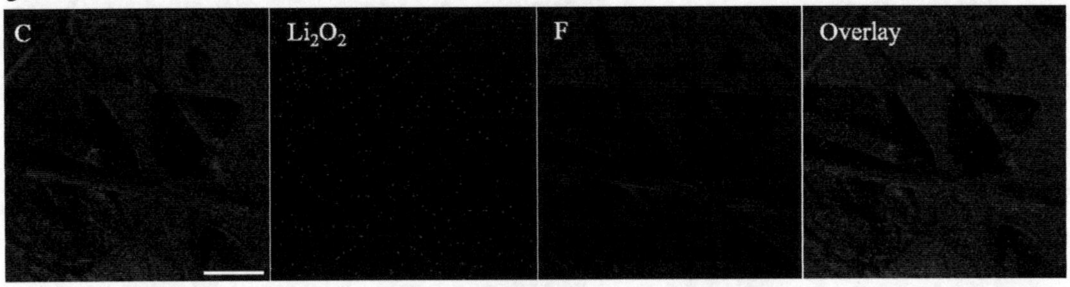

Figure 2. TOF-SIMS secondary ion images of sample 3# under negative (-) ions of UB mode (the length of scale bar is 50 μm).

Figure 2 shows the distribution of several major chemical species (C, F, Li_2O_2) on the surface of sample 3#. The C and F elements grow along the carbon fibers, which exhibit a relatively high intensity. On the contrary, the discharging product of lithium-oxygen batteries, Li_2O_2, is usually hard to be probed via other surface analysis and characterization techniques. However, the existence of Li_2O_2 could be detected by conducting TOF-SIMS measurement, which further indicates its high sensitivity and low detection limit[10].

3.2. Mass spectrum

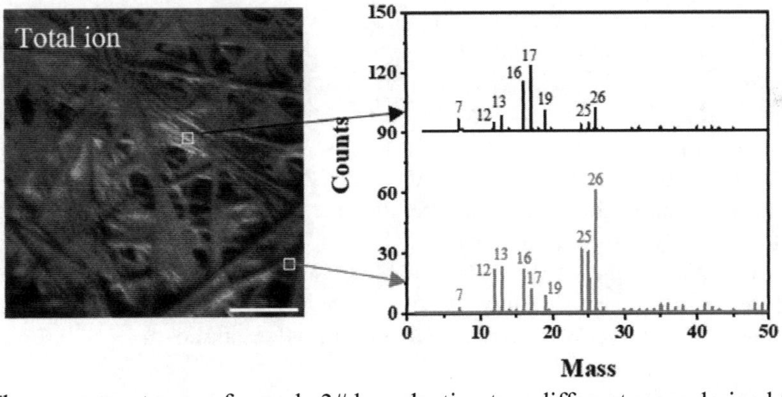

Figure 3. The mass spectrums of sample 3# by selecting two different areas derived from TOF-SIMS under negative (-) ions mode (the length of scale bar is 50 μm).

As can be observed from the left part of figure 3, the sample with different areas exhibit different light and dark degrees, which may be caused by disparate component intensity. To explore the reason, the mass spectrum of the region can be measured via TOF-SIMS to analyze the distribution and intensity of its components. The results indicate that the distribution of components in different regions are different. To be more specific, the brighter region contains more O and F elements, while more C element, C-H chemical bonding, C_2H and LiF components exist in the darker region[11].

3.3. Depth profiling

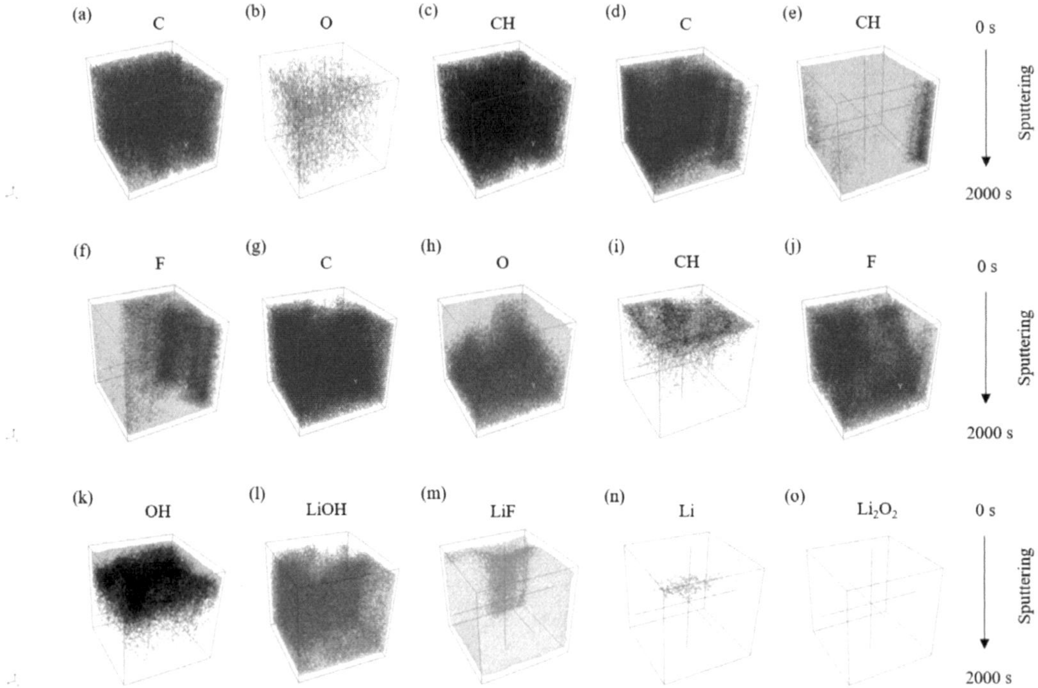

Figure 4. Three-dimensional view of component distribution for three samples in the TOF-SIMS sputtered volumes, 1# (Fig 4a-4c), 2# (Fig 4d-4f), 3# (Fig 4g-4o).

As is shown in figure 4, the 3D images obtained by TOF-SIMS show a relatively uniform distribution of C element within the material in all three samples, while the opposite trend is observed for most of other species. In sample 1#, the O element and C-H chemical bonding are evenly distributed. As for sample 2#, after cycling in the lithium-ion battery, it is found that the distribution of C-H chemical bonding and F element in the material are different from each other. The intensity of the former is high at both sides of the material, while the latter is high only at one side. The intensity of O element in sample 3# is higher than sample 2#, which may be attributed to the generation of oxygen-containing compounds LiOH and Li_2O_2 during the discharging behavior of lithium-oxygen battery. In addition, the spatial distribution of C-H chemical bonding also changes, which exhibits a reduced trend from surface to the bulk in terms of intensity. However, LiF is mainly distributed at one side of the sample, while Li and Li_2O_2 exhibit a relatively even distribution within the sample[12].

4. Conclusions

In summary, we have characterized the component distribution of three nanoscale samples assisted by TOF-SIMS and discussed about it from three aspects of 2D/3D images and mass spectrums. As an advanced technique, TOF-SIMS exhibits various merits including high detection limit, high spatial resolution, good sensitivity, etc. Moreover, TOF-SIMS has an incomparable advantage over other technologies, that is providing 3D images of material component distribution, which confirms the great necessity of this paper. In future studies, we hope to gain more information about the material at smaller nanoscale. A broader application of TOF-SIMS technique in nanomaterials is expected in the near future.

Acknowledgments

The authors gratefully acknowledge financial support from the National Natural Science Foundation of China (22075026). G.T. acknowledges the support from Beijing Institute of Technology Teli Young Fellow Program.

Reference

[1] Tan, G. Q., Chong, L., Zhan, C., et al. (2019). Insights into structural evolution of lithium peroxides with reduced charge overpotential in Li_2O_2 system[J]. Advanced Energy Materials, 9: 1900662.1-1900662.12

[2] Mamyrin, B. A., Karataev, V. I., Shmikk, D. V., et al. (1973). The mass-reflectron, a new nonmagnetic time-of-flight mass spectrometer with high resolution[J]. Journal of Experimental & Theoretical Physics, 37: 45-48.

[3] Stephan, T. (2001). TOF-SIMS in cosmochemistry. Planetary and Space Science, 49: 859-906.

[4] Fearn, S. (2015). An introduction to time-of-flight secondary ion mass spectrometry (TOF-SIMS) and its application to materials science. Morgan & Claypool Publishers, San Rafael, CA, USA.

[5] Fisher, G. L., Belu, A. M., Mahoney, C. M., et al. (2009). Three-dimensional time-of-flight secondary ion mass spectrometry imaging of a pharmaceutical in a coronary stent coating as a function of elution time. Analytical Chemistry, 81: 9930-9940.

[6] Liu, Y., Green, T. J., Kitts, D. D. (2015). Stability of microencapsulated L-5-methyltetrahydrofolate in fortified noodles. Food Chemistry, 171: 206-211.

[7] Chaudhari, P., Singh, A., Topkar, A., et al. (2015). Fabrication and characterization of silicon based thermal neutron detector with hot wire chemical vapor deposited boron carbide converter. Nuclear Instruments and Methods in Physics Research Section A: Accelerators, Spectrometers, Detectors and Associated Equipment, 779: 33-38.

[8] Song, T. L., Xu, F., Chen, H. Y., et al. (2020). Application of X-ray photoelectron spectroscopy in the study of flame retardant materials. In China Flame-retardant Academic Conference. Zhu Hai. pp. 226-233.

[9] Walther, F., Koerver, R., Fuchs, T., et al. (2019). Visualization of the interfacial decomposition of composite cathodes in argyrodite-based all-solid-state batteries using time-of-flight secondary-ion mass spectrometry. Chemistry of Materials, 31: 3745-3755.

[10] Wang, S., Xu, H., Li, W., et al. (2018). Interfacial chemistry in solid-state batteries: formation of interphase and its consequences. Journal of the American Chemical Society, 140: 250-257.

[11] Zhang, X., Wang, S., Xue, C., et al. (2019). Self-suppression of lithium dendrite in all-solid-state lithium metal batteries with poly (vinylidene difluoride)-based solid electrolytes. Advanced Materials, 31: 1806082.

[12] Zhao, F., Sun, Q., Yu, C., et al. (2020). Ultrastable anode interface achieved by fluorinating electrolytes for all-solid-state Li metal batteries. ACS Energy Letters, 5: 1035-1043.

Template Preparation and Experimental Study on Superhydrophilic-Superoleophobic Nano-structures Surfaces Based on Titanium Mesh Membrane

Wang Jian[1*,2], Zhao Xuezeng[2], Chen lei[1] and Chen Mingying[1]

[1]Department of Mechatronics Engineering, East University of Heilongjiang, Harbin, Heilongjiang, China

[2]School of Mechatronics Engineering, Harbin Institute of Technology, Harbin, Heilongjiang, China

Corresponding author's e-mail: 19bg08153@stu.hit.edu.cn

Abstract. With the development of industrial process, the crude oil leakage of industrial oil wastewater is increasingly serious. The development of efficient and environmentally friendly separation mesh membrane technology is the best way to solve the water pollution problem. With the rapid development of micro-nanotechnology, the wetting change caused by surface structure (micro-nano scale) that can solve the problem of fabrication for oil-water separation mesh. If the liquid contact angle at the surface is less than 10°, the surfaces are called superhydrophilic surfaces. Meanwhile, the liquid contact angle is greater than 150°, the surfaces are called superhydrophobic surface. Surface with superhydrophilic/superhydrophobic is also known as special wetting surfaces. The special wetting surface is fabricated by coating / spraying. The process is simple and easy to operate. By constructing the multi-stage surface roughness and reducing the surface energy, the titanium mesh with different orders are used as the base. the mesh membrane with special superhydrophilic-superoleophobic properties is constructed. Comparing the influence of titanium mesh with different molybdenum numbers (mesh size) on hydrophilic/ hydrophobicity, the mechanism of the macroscopic size of the pore spacing substrate on the special surface wetting is analyzed. The preparation of ultra-hydrophobic multi-stage surface roughness titanium mesh is realized. The experimental analysis of oil and water separation effect is carried out. For the preparation of high-efficiency oil-water separation mesh film that can be used in industrial production.The oil and water separation mesh membrane was prepared by the titanium mesh substrate to achieve effective oil and water separation.

1. Introduction

With the development of economy, it is increasingly prominent that the environmental pollution problems of oil wastewater and crude oil exploitation leakage in industrial production. The separation of oil wastewater mainly includes gravity separation, centrifugal separation, ultrasonic separation, gas float method, electric field method, condensation method, adsorption method, biological method and membrane separation method [1,2]. By comparing the above methods, the most popular method is the membrane separation method. The mesh membranes commonly used for oil-water separation include: superhydrophobic/super-oleophilicity type and superhydrophilic/superoleophobic type.The preparation process of superhydrophobic/super-oleophilicity is relatively simple and the preparation materials are easy to obtain. But there are many shortcomings in the practical engineering application process[3,4]. For the oil-water separation by filtration method, the oil droplets are easy to attach or blocked in the filter mesh hole. The oil-water separation mesh film is easy to suffer pollution and cause failure. And it needs high requirements for the viscosity of oil in the oil-water mixture[5]. Due to the different proportion of water and oil, the small proportion of oil is layered above the mixture, resulting in the inability to use gravity driving method for separation of oil and water. Requiring special processes for auxiliary driving and increasing the cost of separation. Relatively by contrast, the superhydrophilic/superoleophobic type can realize the adsorption and filtration of water. The mesh material is not easy to pollution.The use and circulation performance is good. The service life is greatly extended. Since the density of water is generally larger than oil, gravity driving can be used as the power of oil and water separation, energy saving, consumption reduction and simple operation. Therefore, superhydrophilic/superhydrophobic material mesh membrane is one of the directions of research. However, the surface energy of water can be larger than that of oil. The surface of hydrobic material is also hydrophobic. So the preparation of hydrobic material is difficult[6].

Current reliable methods to constructing such mesh membranes include two methods:One is to modify the surface properties, reducing the surface energy to simultaneously construct the hydrophobic surface simultaneously; The second is to establish graded micro-nanostructures on the surface to increase the surface roughness. The study of membrane surface wetting begins with bionics, self-cleaning of the surface of lotus leaves in nature. Fish remain clean and unpolluted in oil-polluted rivers, and the special phenomenon that water striders can walk freely on the water surface. The above are superhydrophobic to water. it provides design ideas for special wetting surface[7]. Scholars from various countries have conducted extensive theoretical and experimental research on superhydrophobic/superhydrophilic oil separation mesh membrane. However, the preparation of superhydrophilic/superhydrophobic mesh membrane is still being explored. A variety of superhydrophilic/hyperhydrophobic mesh separation media with different preparation methods, different substrates and different properties are explored for the research[8].

In this paper, a metallic titanium mesh is used as a substrate, by using titanium mesh number with 40/100/200 items as mesh infrastructure. The surface of the titanium mesh is modified by constructing a hydrophilic group of multi-stage nanostructures by using spraying/immersion. Three different numbers of titanium mesh are obtained. Comparative analysis of oil and water contact angle of metal titanium mesh with different numbers. Making a simple device for oil and water separation for oil and water separation test that provide technical feasibility for the next engineering application.

2. Fabrication of microstructures on titanium mesh

2.1. Materials

Titanium with mesh number40/100/200 (purchased from Rui Metal Store), TiO_2 Nanoparticles (P5~10)(Aladdin,99.8%), Hexadecane(Sigma Aldrich, 99%), Ethanol absolute(Aladdin,99.5%), FS-50(Capstone), Ultra-pure water was used in all experiments.

2.2. Laboratory apparatus

Table 1. Laboratory apparatus.

	Model / Features	Manufacturer
Spray gun	116BK	Hangzhou Shenchao Co., Ltd.
Magnetic stirrer	85-2	Droide Co., Ltd.
Ultrasonic cleaner	PS-20	Whale Control instruments Co., Ltd
Contact angle measuring meter	DropMeter A-60	Haishu Mai Testing Co., Ltd
Vacuum drying oven	DZF	Shanghai Yiheng Co., Ltd.

2.3. Methods
Cut and sample the titanium mesh with different numbers of 40/100/200 respectively, then polished with sandpaper to remove impurities such as oxides on the titanium mesh surface. The polished titanium mesh was cleaned with acetone, anhydrous ethanol and deionized water ultrasound for 20 minutes. The cleaned titanium mesh passes 60 °C for constant temperature drying for 1 hour. Next, a hydrophilic TiO_2 nano-solution is prepared. 50ml anhydrous ethanol is taken aside. 2g FS-50 added to ethanol solution, mixing for 2 hours with a magnetic mixer. then add 1g TiO_2 nanoparticles to the stirred solution. Continue stirring fully for 2 hours. Finally, the prepared solution is sprayed to the titanium mesh surface, keeping spray nozzle 15cm from titanium mesh substrate. The spraying uniform spraying time is not limited in the spraying process. Or that an immersion method may be used, diaching the titanium mesh substrate in the prepared solution. Metal mesh is need to soak in solution by 60s-120s. Finally, sprayed / dipped titanium mesh is placed in vacuum drying oven for 4 hours at temperature:60°C. A titanium mesh with different mesh numbers for 40/100/200 with superhydrophilic/superhydrophobic characteristics is obtained after cooling.

2.4. Characterization
The TiO_2 nanoparticles selected in the experiment were 50nm diameter particles. It can increase the surface roughness by preparing a nanoparticle suspension solution for surface modification with spraying/infiltration to make it attach to the metal titanium mesh substrate. The construction of surface multi-scale micronano roughness is realized by increasing the surface roughness structure of the metal mesh. According to the influence of infiltration characteristics of contact surface, the constructed multi-stage roughness structure can realize the liquid drainage characteristics of the surface interface. Also its surface energy is reduced by FS-50 fluoride. Both ways have achieved improving the characteristics of network drainage. Surfaces with hydrophobic properties can be made hydrophilic by grafting of hydrophilic substances. Ethanol solution containing FS-50/TiO_2 include fluoride reduced surface energy and mass material with hydrophilic characteristics. By attaching it to the titanium substrate network with a different number of entries later. A titanium base filter mesh membrane with hydrophilic/hydrophobic can be constructed. The comparison of different number of titanium mesh substrate is made. Its hydrophobic characteristics of titanium mesh base were obtained by experimental analysis.

3. Hydrophilic-hyperhydrophobic of titanium mesh substrate

3.1. Contact angle measurement
According to the special wetting interface theory, when the water/oil contact angle at the contact surface is less than 10°, such surfaces are called superhydrophilic / liquid surfaces; when the contact angle of the water/oil on the contact surface is greater than 150, such surfaces become hyper-hydrophobic/oil surfaces. For a titanium mesh of three orders prepared in this paper, the oil-water contact angle was measured on the three titanium mesh 40 / 100 / 200 respectively by using contact angle meters. Using pure water and positive-hexane on the base of three orders to get contact angle measurement using a contact angle tester. According to the results, three kinds of titanium mesh substrate mesh membrane have a certain oil thinning. Drop orthohexane on a titanium net of 40/100/ 200, oil droplets gather into spherical. As shown in Figure 1. oil and water separation mesh membrane with 200 mesh number has the hydrophilic and oleophobic properties:contact angle of the oil and titanium mesh is 152.4°, contact angle of water and titanium mesh is 0°. As shown in Figure 2.oil and water separation mesh membrane with 100 mesh number has the hydrophilic and oleophobic properties:contact angle of the oil and titanium mesh is 110.1°, contact angle of water and titanium mesh is 0°. As shown in Figure 3. oil and water separation mesh membrane with 200 mesh number has the hydrophilic and oleophobic properties:contact angle of the oil and titanium mesh is 152°, contact angle of water and titanium mesh is 0°.

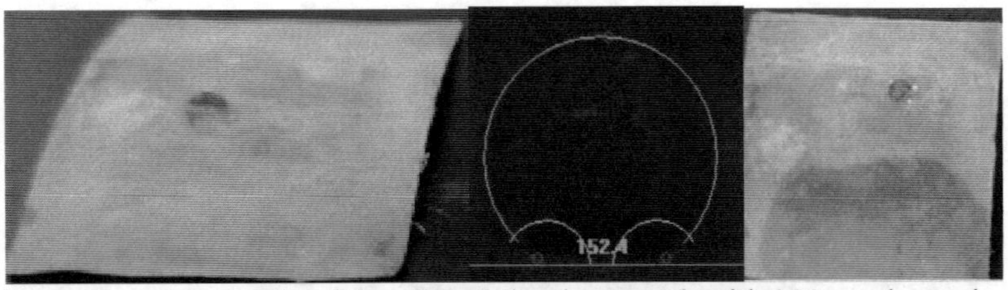

Figure 1. contact angle of the oil and titanium mesh with 200 mesh number.

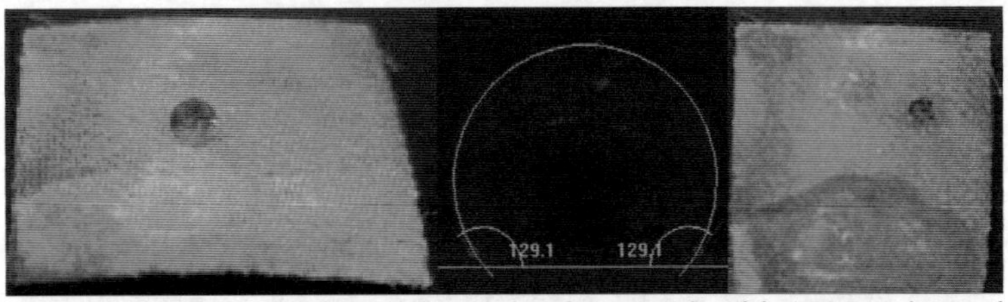

Figure 2. contact angle of the oil and titanium mesh with 100 mesh number.

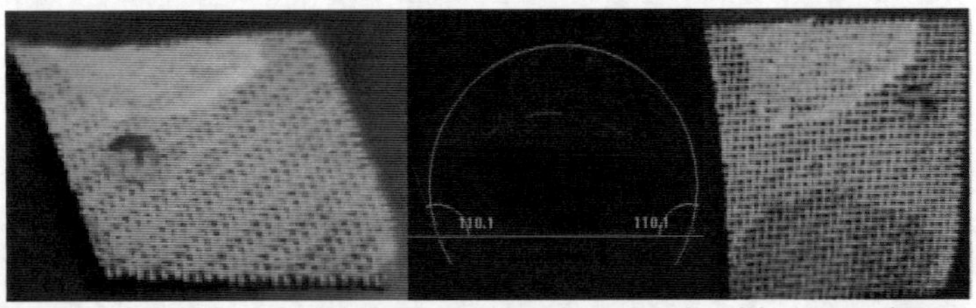

Figure 3. contact angle of the oil and titanium mesh with 40 mesh number.

According to hydrophilic test, water droplets are observed to fully infiltrate the titanium mesh after placing water droplets on the net base of three counts. Comparison of titanium mesh substrate with hydrophilic/oil thinning effect with three mesh numbers.Titanium mesh substrate is better hydrophilic to water with fully spread. Oil-water separation mesh film prepared by titanium mesh base of 200 objective has super oil thinning characteristics.

3.2. Oil and water separation experiment

By measuring the contact angle of three order titanium mesh, the 200 mesh film is super hydrophilic/ superhydrophobic in three orders. Therefore, 200 mesh number is selected as the mesh film of the oil-water separation experiment. First, the oil-water mixture is prepared by mixing 30ml orthohexane with 30ml purified water stained with methyl blue. The oil-water mixture is placed in the beaker for reserve. Oil and water separation device using double pass barrel/test bench/200 titanium mesh film. Process steps are: Cutting the 200-eye titanium mesh into a round with same-section diameter as the cylindrical section. Attaching it to the double cylinder side, securing the cylinder to the metal stand and keep the open portion upward. It is placed that a 100ml empty beaker directly below the titanium mesh. Then the oil-water mixture is slowly poured from the opening into the cylinder. The process of separation of oil and water separation device is shown in Figure 4. After the oil-water mixture is completely poured into the cylinder. The stained blue water flows through the titanium mesh into the lower beaker. Orthohexane is blocked by the titanium mesh in the upper cylinder. Reliable oil and water separation is achieved.

Figure 4. The separation process of oil and water .

4. Conclusion

This research object is titanium mesh metal as the base. A layer of multistage roughness structure with nanostructure with three different orders. The surface energy of structure is reduced by the surface modified by fluoride. Multi-stage nanostructure also improves the surface roughness.Thus hydrohydroproperties is produced. Since the presence of a hydrophilic group material is achieved as hydrophilic by constructing the titanium mesh surface modified by fluoride and titanium dioxide. The hydrophilic and oily mesh membrane is realized. According to the experimental results: titanium mesh with 200 mesh number can realize the functional characteristics of Superhydrophilic/Superoleophobic nanostructures surfaces. The titanium mesh in the remaining two orders also exhibit superhydrophilic / hydrophobic properties. It shows that the base order number of metallic titanium mesh has a great influence on superhydrophilicity. It is also proved that the superhydrophilic/superhydrophobic mesh membrane is achieved by the oil-water separation test. Superhydrophilic/Superoleophobic nanostructures surfaces play an important role in pollution control of industrial crude oil wastewater. Further study is necessary on the mesh membrane service life and separation efficiency of oil and water.

Acknowledgments

This project is supported by School level scientific research project of East University of Heilongjiang (Grant No. HDFKY200130).

References

[1] Song S K,Shon Z H,Kim Y K,et al. (2011)An oil spill accident and its impact on ozone levels in the surrounding coastal regions. Atmospheric Environment,45: 1312-1322.
[2] Ma Q, Cheng H, Fane A G, et al.(2016)Recent development of advanced materials with special wettability for selective oil / water separation,12:2186-2202.
[3] Arabtan T,Fang X,Shen W.(2011)Superhydrophobic ando- leophilic calcium carbonate powder as a selective oil sorb-ent with potential use in oil spill clean-ups. Chemical Engineering Journal,166:787-791.
[4] Xue Z, Cao Y, Liu N, et al. (2014)Special wettable materials for oil /water separation. Journal of Materials Chemistry A,2:2445-2460.
[5] Yang H C, Xie Y S, Chan H, et al.(2018)Crude-oil-repellent membranes by atomic layer deposition oxide interface engineering.ACS Nano, 12:8678-8685.
[6] Hyeok, Zhang K, et al.(2015)Effect of permeate flux and tangential flow on membrane fouling for wastewater treatment.Separation& Purification Technology,45:68-78.
[7] Ni W, Miao X, Yang X, et al.(2018)An alternative fabrication of under oil superhydrophobic or underwater superoleophobic stainless steel meshes for oil-water separation: originating from one-step vapor deposition of polydimethylsiloxane.Separation&Purification Technology,204:116-126.
[8] Zheng J,Zhang H,Zhao Z,et al.(2012) Construction of hierarchical structures by electro spinning or electro spraying.Polymer,53:546-55.

Experimental study on the evolution law of residual stress in the fatigue process of inductively quenched 42CrMo steel

Yongfang Zhang[1], **Sifan Kang**[1], **Lihong Dong**[2], **Haidou Wang**[2], **Yuelan Di**[2], **Shaofan Li**[3]

[1] Xi'an University of Technology, Xi'an 710048

[2] Academy of Armored Forces Engineering, Beijing 100072

[3] Hebei University of Technology, Tianjin 300401

zhangyf@xaut.edu.cn

Abstract. In the process of remanufacturing a crankshaft with a history of service, it is found that the residual stress of the crankshaft will relax during operation. Therefore, the relaxation of residual stress must be considered in the study of crankshaft remanufacturing. X-ray diffraction is used to measure the initial and final residual stresses after each loading cycle, and the changes in the microstructure of the material after induction quenching were also measured by scanning electron microscopy. The results show that the maximum residual compressive stress appears inside the sample, and its size decreases with increasing depth. The relaxation of residual stress on the surface of the sample with hardened layer can be roughly divided into three stages: rapid relaxation, stable relaxation, and instantaneous release of the final fracture residual stress. An empirical model is proposed to estimate the residual stress relaxation, considering the depth of the hardened layer and the external load during the fatigue process to predict the relaxation law of the residual stress.

1. Introduction

Crankshaft is one of the most typical and important core components of heavy-duty engines. In the service process, fillet bending fatigue fracture is the most common cause of crankshaft failure, accounting for about 80% of the total failure[1-9]. In order to improve the surface performance of the crankshaft and increase the fatigue strength, carburizing[10], nitriding[11], induction hardening[12], shot peening[13] and other surface strengthening technologies are usually used at home and abroad. At present, induction hardening has become one of the most widely used surface treatment technologies due to its advantages such as easier control of the hardened layer, short heating time, and small deformation[14-16].

Heavy-duty crankshaft has become one of the main research objects of remanufacturing technology due to its complex structure, difficult processing technology and high processing cost[7]. Compared with new products, the stress state of induction hardened crankshafts with service history will change, with both stress relaxation[17] and changes in stress concentration positions. Therefore, precise control of the residual stress state during service will become a powerful challenge in the crankshaft remanufacturing process.

It is the core of the research work to reveal the nature and size of the residual stress and the law of change along the surface depth. Some documents show that surface strengthening technology has a greater impact on residual stress. Yan[18] studied the effect of shot peening residual stress on the fatigue life of TC11 titanium alloy. The results showed that a significant residual compressive stress field would be formed on the surface of the material after shot peening, and the fatigue strength of the material increased by 37.6% after shot peening. Yin[19] studied the influence of induction heating and quenching process on the residual stress distribution and fatigue properties of 45 steel pin shaft. The results show that the residual stress distribution of hardened layer is mainly compressive stress, and its value decreases first and then increases with the increase of depth. Using laser shot peening to introduce gradient residual stress on the surface of titanium alloy, Xian[20] found that the residual stress relaxation rate of high temperature shot peening (WLP) was about 66% after 10 fatigue cycles, while the relaxation rate of room temperature shot peening (RT-LP) was as high as 91%.Lu[21] established a residual stress relaxation model. The results show that the residual stress relaxation of the cyclic softening material under fatigue load increases with the increase of the cyclic stress amplitude and the number of cycles, while the residual stress relaxation of the cyclic hardening material mainly occurs in the first few fatigue cycles. The purpose of this research is to explore the change law of residual stress of specimens in different cycle stages through three-point bending fatigue experiment, and establish a residual stress relaxation model to explore the main reasons that affect the change of residual stress relaxation.

2. Experiments

2.1. Substrate materials and specimens

The experimental research object is the 6DL induction hardened crankshaft manufactured by FAW. The material is 42CrMo steel. This material has good hardenability, no obvious temper brittleness, and has high strength and toughness. The matrix material composition and mechanical properties are shown in Table 1 and Table 2. A sample of 200mm×30 mm×10 mm was cut with a wire cutter, as shown in Fig. 1

Table 1 Chemical composition of the 42CrMo (wt%)

element	C	Si	Mn	P	S	Cr	Mo
percentage	0.38~0.45	0.17~0.37	0.50~0.80	≤0.035	≤0.035	0.90~1.20	0.15~0.25

Table 2 Mechanical properties of 42CrMo substrate (wt%)

Strength of extension/Rm	yield strength /$Rp_{0.2}$	elasticity modulus /E	Density /ρ	hardness
830MPa	510MPa	207.7GPa	7.85g/cm^3	305.6$HV_{0.2}$

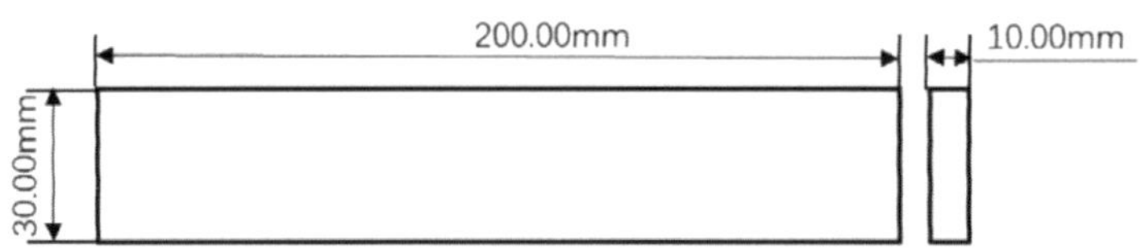

Figure 1 Size of matrix sample

2.2. Induction hardening processing

The induction heating of parts mainly relies on the phenomenon of electromagnetic induction, which is essentially an alternating magnetic field that can cause an alternating electric field[22]. The research object of this experiment is 6DL induction hardened crankshaft, the depth of the fillet hardened layer is 3.6mm, the shaft diameter is 76mm, and the hardening ratio is about 5%. The thickness of the sample used in this experiment is 10 mm, so there are two main considerations. The depth of the hardened layer is 0.5 mm, 1.6 mm.

The induction hardening equipment uses a super audio GCT-1200 vertical induction hardening machine tool, as shown in Figure 2. This equipment mainly completes the preparation of samples with different hardened layer thicknesses by controlling the power output and operating speed. The sample is single-sided induction quenching, and the quenching area is 100 mm in the center of the sample. The quenching process parameters are shown in Table 3. The quenching coolant is water. After quenching, it is tempered at 180°C for 1 hour and cooled at room temperature.

Table 3 Induction quenching process parameters

technological process	Voltage /V	Anode flow /A	Grid current /A	Running speed mm/min	Depth of hardening zone/mm
1	10000	5	1	1200	0.51
2	8000	4	0.8	300	1.6

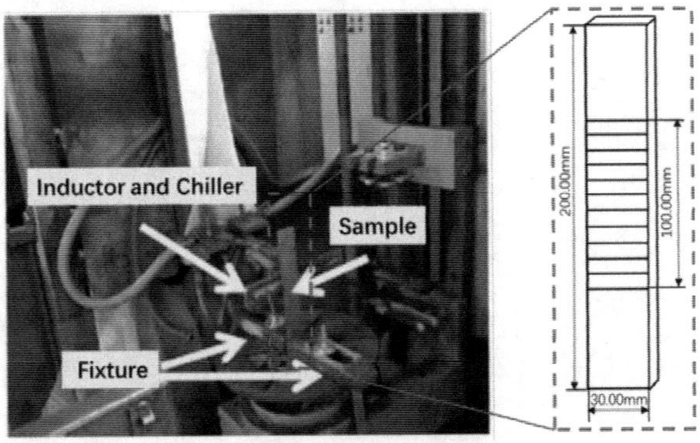

Figure 2 GCT-1200 vertical induction quenching machine tool and schematic diagram of sample quenching area

2.3. Three-point bending fatigue test

The equipment used in the experiment is a GPS-20 high-frequency fatigue testing machine, as shown in Figure 3. The maximum static load of the equipment is 20kN, and the maximum alternating load is 10kN.The side of the quenched surface is the tensile side, and the stress ratio is R=0.1.

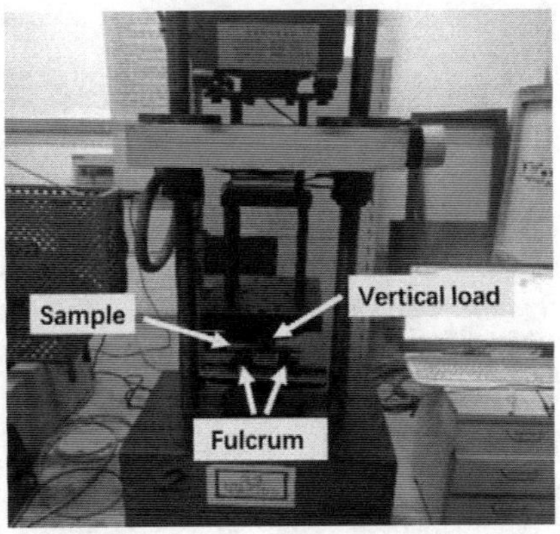

Figure 3 GPS-20 high frequency fatigue testing machine

The final test loads are determined to be 540 MPa and 450 MPa respectively. Two kinds of loads are applied to the samples of the two induction hardening processes respectively until the samples are broken, and the samples under different fatigue cycles are obtained. For the matrix samples, the first 100×10^4 weeks are mainly studied.

3. Establishment of stress relaxation model and analysis of influencing factors

3.1. Residual stress relaxation

Figure 4 shows the surface residual stress relaxation curves of the two kinds of induction hardening process specimens in the fatigue process under two kinds of loads.

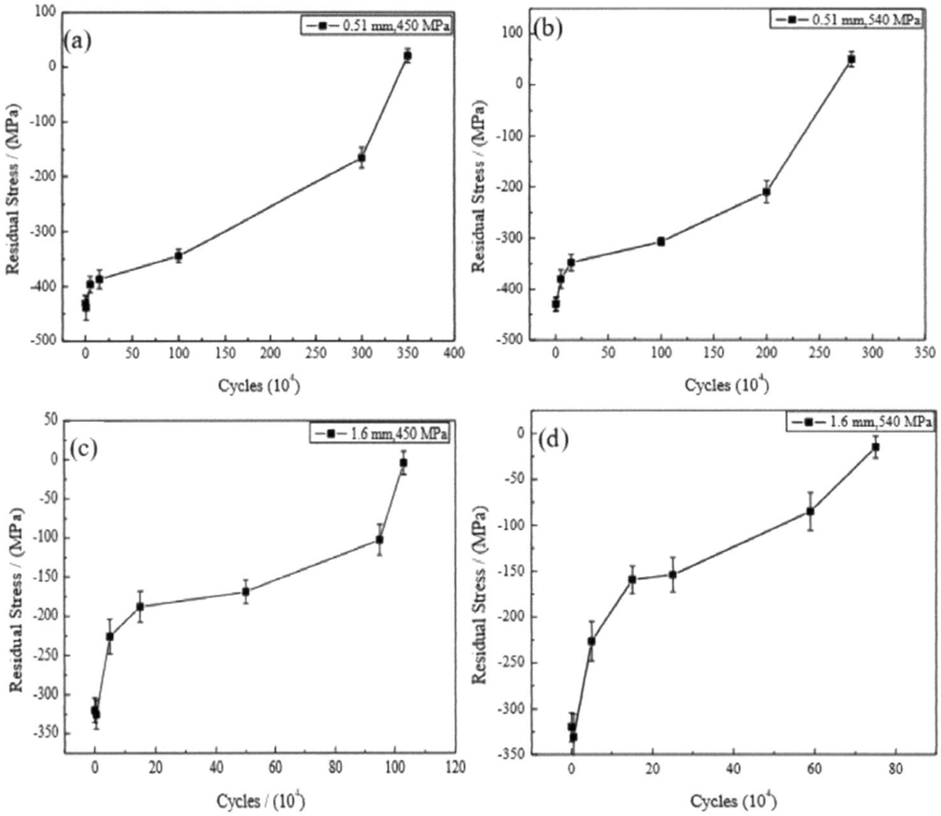

Figure 4 Relaxation curve of residual stress of induction hardening sample. (a) 0.51mm, 450Mpa; (b) 0.51mm, 540Mpa; (c) 1.6mm, 450Mpa; (d) 1.6mm, 540Mpa

According to Figure 4, the residual stress relaxation of the sample containing the hardened layer can be roughly divided into three stages: the rapid relaxation of the initial residual stress, the stable relaxation of the mid-term residual stress, and the instantaneous residual stress around the crack after the final sample fractures. freed. The initial rapid relaxation belongs to static relaxation, while the mid-term residual stress relaxation belongs to stable relaxation. Afterwards, in order to obtain the relaxation model, the residual stress value when the cycle cycle is 0 is removed.

After taking the logarithm of the abscissa cycle cycles, the results are shown in Figure 5. It can be clearly seen that in the early stage of fatigue, the residual stress value after relaxation of the sample containing the hardened layer exhibits an obvious power function relationship with lgN. Power function fitting is performed on the data, followed by data regression processing, the fitting results are shown in Figure 6.

 2011 (2021) 012073 doi:10.1088/1742-6596/2011/1/012073

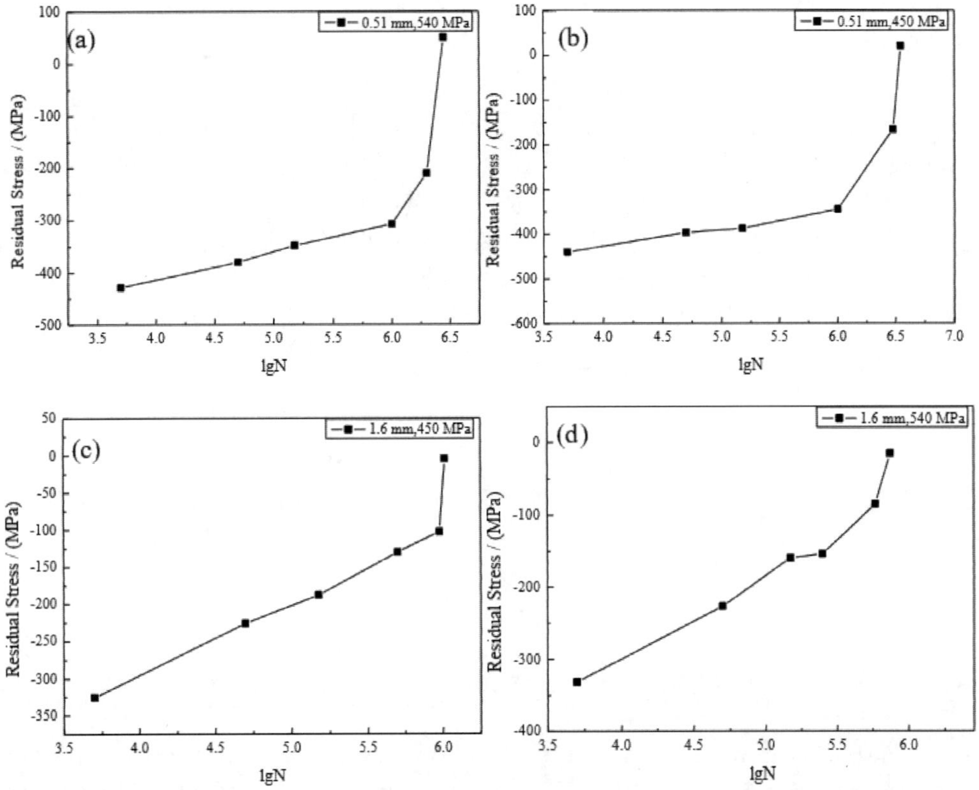

Figure 5 Residual stress relaxation curve after logarithm of abscissa. (a) Hardened layer 0.51 mm, Maximum stress 450 MPa;(b)Hardened layer 0.51 mm, Maximum stress 540 MPa; (c)Hardened layer 1.6 mm, Maximum stress 450 MPa;(d)Hardened layer 1.6 mm, Maximum stress 540 Mpa

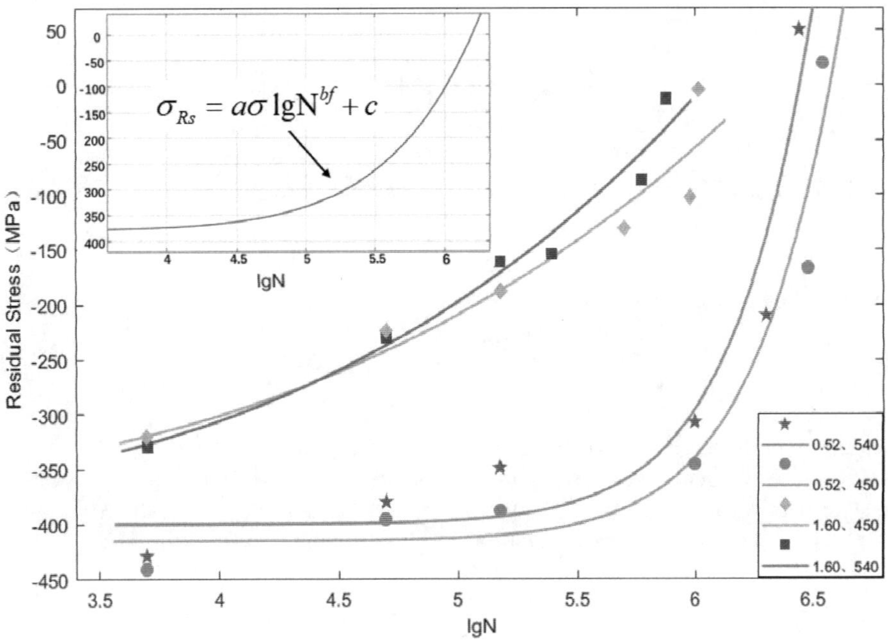

Figure 6 Residual stress and lgN fitting curve

The main variables of the experiment are the depth b of the hardened layer and the applied load σ during the fatigue process, so there are

$$\sigma_{Rs} = a\sigma \lg \mathrm{N}^{bf} + c \tag{1}$$

Make $\sigma_{Rs} = f(A, B, C, x)$、 $A = a\sigma$、 $B = bf$ 、 $x = \lg \mathrm{N}$、 $C = c$

Power function

$$\sigma_{Rs} = f(A, B, C, x) = Ax^B + C \tag{2}$$

Use this power function as a fitting function to fit the data. In the formula, A, B, and C are the fitting coefficients, and the method of solving the coefficients is given below.

Find the first derivative on both sides of equation (2) to get

$$x\frac{d(A、B、Cx)}{dx} = ABx^B \tag{3}$$

which is

$$\frac{x}{B}\frac{d(A、B、Cx)}{dx} = Ax^B \tag{4}$$

Bring formula (3) into formula (4) and organize it to get

$$x\frac{df(A, B, C, x)}{dx} = -BC + Bf(A, B, C, x) \tag{5}$$

Make $D = -BC$, then there are

$$x\frac{df(A, B, C, x)}{dx} = D + Bf(A, B, C, x) \tag{6}$$

In this way, the nonlinear fitting problem in equation (5) is transformed into a linear fitting problem. According to the measured data (x_i, y_i) (i=1,2, …, m), The numerical differential formula is used to find the corresponding $(\frac{dy}{dx})_i$ value of data (x_i, y_i), and then according to the least squares method, the equations for determining the coefficients D and B in the formula (6) are:

$$\begin{vmatrix} \sum 1 & \sum y_i \\ \sum y_i & \sum y_i^2 \end{vmatrix} \bullet \begin{vmatrix} D \\ B \end{vmatrix} = \begin{vmatrix} \sum x_i (\frac{dy}{dx})_i \\ \sum x_i (\frac{dy}{dx})_i y_i \end{vmatrix} \tag{7}$$

Solve the equations (7) to obtain B=0.281, C=-5240.818, and then use the least square method to determine A=3166.207, the fitting function is

$$y = 3166.207x^{0.281} - 5240.818 \tag{8}$$

Then the final data model can be expressed as:

$$y = \frac{3166.207}{\sigma} x^{\frac{0.281}{f}} - 5240.818 \tag{9}$$

3.2. *Study on the influence factors of residual stress relaxation*

Table 4 shows the residual stress relaxation of the two induction hardening process samples (the hardened layer depth is 1.6 mm and 0.51 mm respectively) under two loads (450 MPa, 540 MPa) in the three-point bending fatigue test. The calculation formula is shown in formula (10):

$$m = \frac{\sigma_{N1} - \sigma_{N2}}{\sigma_0} \tag{10}$$

Among them, σ_{N1} and σ_{N2} are the residual stress values on the surface of the sample after fatigue cycles of N_1 and N_2, where N_1 is greater than N_2; σ_0 is the residual stress value when the fatigue cycle of the sample is 0.

Table 4 Three-point bending fatigue residual stress relaxation of specimen

Serial number	Hardened layer depth/mm	Maximum stress/MPa	Residual stress relaxation in the first stage	Residual stress relaxation in the second stage	Total slack	fatigue life ($\times10^4$)
1	1.6	450	40%	26.7%	66.7%	103
2	1.6	540	50.1%	23.3%	73.4%	75
3	0.51	450	10%	51.39%	61.39%	350
4	0.51	540	19%	39.5%	58.5%	280

According to the data in Table 4, comparing 1 group of samples and 2 groups of samples and 3 groups of samples and 4 groups of samples, it can be seen that as the fatigue load increases, the residual stress relaxation in the first stage of the sample surface is Increasing trend, while for the second stage, the amount of relaxation has a decreasing trend. For the total slack, there is a big difference between the first group of samples and the second group of samples. As the load increases, the amount of slack also increases; while for the third group of samples and the fourth group of samples, the difference is not obvious. It can be seen that the effect of the change of fatigue load on the relaxation of residual stress is mainly reflected in the first stage. The greater the load, the greater the amount of relaxation of residual stress.

4. Conclusions

The following conclusions can be drawn from the experimental results:

(1) After surface quenching, the maximum residual compressive stress of 42CrMo material samples appears in the inside of the sample. The deeper the hardened layer is the closer the maximum residual stress point is to the surface. After the maximum residual stress point, the magnitude of residual stress decreases with the increase of depth.

(2) The residual stress relaxation model is proposed, which is a power exponential relationship. The depth of hardened layer and the applied load in the fatigue process are taken into account to predict the relaxation law of residual stress. It is found that the estimation is in good agreement with the experimental data.

(3) The residual stress relaxation on the surface of the sample containing the hardened layer can be roughly divided into three stages: the rapid relaxation of the initial residual stress, the stable relaxation of the intermediate residual stress, and the instantaneous release of the residual stress around the crack after the final specimen fracture.

Author

Biography: ZHANG Yong-fang (1975-), Female, Ph.D. Professor. Main research interests are intelligent monitoring and remanunfacturing life assessment.

Corresponding author: DONG Li-hong, Female, Ph.D. Associate researcher, Research focus: non-destructive testing and life assessment.

Acknowledgments

This work is supported by Key Project of the National Natural Science Foundation of China (Grant No. 51535011); General Program of the National Natural Science Foundation of China (Grant No.

51675532): Key Research and Development Program of Shaanxi Province of China (Grant No. 2020GY-106).

References

[1] S.K. Bhaumik, R. Rangaragu, M.A. Venkataswamy, T.A. Bhaskaran, M.A. Parameswara. Fatigue fracture of crankshaft of an aircraft engine[J]. Eng Fail Anal, 9 (2001), pp. 255-263.
[2] M. Fonte, M. de Freitas. Marine main engine crankshaft failure analysis: a case study[J]. Eng Fail Anal, 16 (2009), pp. 1940-1947.
[3] F.J. Espadafor, J.B. Villanueva, M.T. Garcia. Analysis of a diesel generator crankshaft failure[J]. Eng Fail Anal, 16 (2009), pp. 2333-2341.
[4] F.S. Silva. Analysis of vehicle crankshaft failure[J]. Eng Fail Anal, 10 (2003), pp. 605-616.
[5] D. Taylor, W. Zhou, A.J. Ciepalowicz, J. Devlukia. Mixed-mode fatigue from stress concentrations: an approach based on equivalent stress intensity[J]. Int J Fatigue, 21 (1999), pp. 173-178.
[6] M.M.I. Hammouda, H.E.M. Sallam, H.G. Osman. Significance of crack tip plasticity to early notch fatigue crack growth[J]. Int. J. Fatigue, 26 (2004), pp. 173-182.
[7] M.M.I. Hammouda, H.G. Osman, H.E.M. Sallam. Mode I notch fatigue crack growth behaviour under constant amplitude loading and due to the application of a single tensile overload[J]. Int. J. Fatigue, 26 (2004), pp. 183-192.
[8] Hammouda, Sherbini, Sallam. Deformation behavior at the tip of a physically short fatigue crack due to a single overload[J]. Fatigue Fract. Eng. Mater. Struct., 22 (1999), pp. 145-151.
[9] W. Chien, J. Pan, D. Close, S. Ho. Fatigue analysis of crankshaft sections under bending with consideration of residual stresses[J]. Int. J. Fatigue, 27 (2005), pp. 1-19.
[10] K. Farrell, E.D. Specht, J. Pang, L.R. Walker, A. Rar, J.R. Mayotte. Characterization of a carburized surface layer on an austenitic stainless steel[J]. J. Nucl. Mater., 343 (2005), pp. 123-133.
[11] E. Menthe, K.T. Rie. Further investigation of the structure and properties of austenitic stainless steel after plasma nitriding[J]. Surf. Coat. Technol., 116–119 (1999), pp. 199-204.
[12] Cryderman R , Shamsaei N , Fatemi A . Effects of continuous cast section size on torsion deformation and fatigue of induction hardened 1050 steel shafts[J]. Journal of Materials Processing Tech, 2011, 211(1):66-77.
[13] V. Schulze, F. Bleicher, P. Groche, Y. Guo, Y. Pyun. Surface modification by machine hammer peening and burnishing[J]. CIRP Annals, 65 (2016), pp. 809-832.
[14] H. Kristoffersen, P. Vomacka. Influence of process parameters for induction hardening on residual stresses[J]. Mater Des, 22 (2001), pp. 637-644.
[15] J. Grum. Measuring and analysis of residual stresses after induction hardening and grinding[J]. Mater Sci Forum, 347–349 (2000), pp. 453-458.
[16] L. Bertini, V. Fontanari. Fatigue behaviour of induction hardened notched components[J]. Int J Fatigue, 21 (1999), pp. 611-617.
[17] Hanagarth H, Vohringer O, Macherauch E, et al. Relaxation of shot peening residual stresses of the steel 42CrMo 4 by tensile or compressive deformation[J]. The japan society of precision Engineering, 1990, 4:337-345.
[18] YAN Zhen, LIANG Yilong, ZHANG Zejun, et al. Effect of Shot Peening on High Cycle Fatigue Properties of TC11 Titanium Alloy [J]. Rare Metals, 2014, 38(004):554-560.
[19] YIN Yingyue, ZHANG Jianjie, SUN Wenlei, et al. Effect of Induction Heating Quenching on Residual Stress Distribution and Fatigue Performance of 45 Steel Pin Shaft [J]. Heat Treatment of Metals, 2018, 43(3):230-232.
[20] Meng X K, Zhou J Z, Su C, et al. Residual stress relaxation and its effects on the fatigue properties of Ti6Al4V alloy strengthened by warm laser peening[J]. Materials science and engineering: A, 2017, 680:297-304.

[21] Lu J, Flavenot J F, Turbat A. Prediction of residual stress relaxation during fatigue loading and taking residual stresses in a multiaxial fatigue criteria into account[M]. International conference on residual stresses. 1989, 784-790.

[22] Komotori J, Shimizu M, Misaka Y, et al. Fatigue strength and fracture mechanism of steel modified by super-rapid in duction heating and quenching[J]. International journalof fatigue, 2001, 23(1):225-230.

Parameters Extraction Methods of Compound Semiconductor Photovoltaic Modules

Chuxiang Zhou[a], Yu Wang[b], Jie Xu[c*]

College of Electronic and Optical Engineering, College of Microelectronics, Nanjing University of Posts and Telecommunications, Nanjing

[a]email: 1220024821@njupt.edu.cn, [b]email: 1220024536@njupt.edu.cn, [c*]email: jiexu@njupt.edu.cn

Abstract. The parameters extraction method of the single diode model is usually used to analyse silicon-based solar cells to predict their behaviour from IV curves. However, the film solar cells based on compound semiconductor, for example the CdTe and CIGS, which have been widely used in the market, have few relevant studies still now. This paper presents two parameters extraction methods and applies to two photovoltaic modules, then we use a function in MATLAB to refine the results. Finally, the error of the two methods has been compared to find the best fitting result. Final fitting results show that these methods have their own advantages and disadvantages and both have great fitting accuracy on the two modules, and they can be used in various irradiance conditions.

1. Introduction

The IV characteristics of photovoltaic (PV) modules are an important factor during the process of production and use. In recent years, many related fitting studies have been done [1,2,3]. It is a typical regression analysis problem and the main research method is curve fitting. Research objects are usually the silicon-based solar cells, and the average errors are usually between 10^{-3} and 10^{-4}. But it is seldom used in CdTe and CIGS solar cells.

In this paper, the single diode model (SDM) is applied to the CdTe and CIGS PV modules. As shown in figure 1. It contains five parameters [4,5], photovoltaic current (I_{ph}), diode reverse saturation current (I_0), diode ideal factor (n), series resistor (R_s) and parallel conductance (G_{sh}). We reduce the number of parameters from 5 to 2 (n and R_s) with the conditions of open circuit, short circuit and the maximum power, in order to simplify calculation [6]. Then, Newton-Raphson method (NR) and particle swarm optimization (PSO) are used to find out the optimal solution of n and Rs under the condition of the least squares fitting. Finally, five parameters of SDM can be extracted out by back substitution.

Tests on PV module of CdTe and CIGS had been done under various environmental conditions. The fitting results show that NR has faster calculation speed, the PSO has smaller error. They both show better fitting results after refinement, and finally results are not inferior to other methods It indicates they are both appropriate parameters extraction methods for compound semiconductor modules.

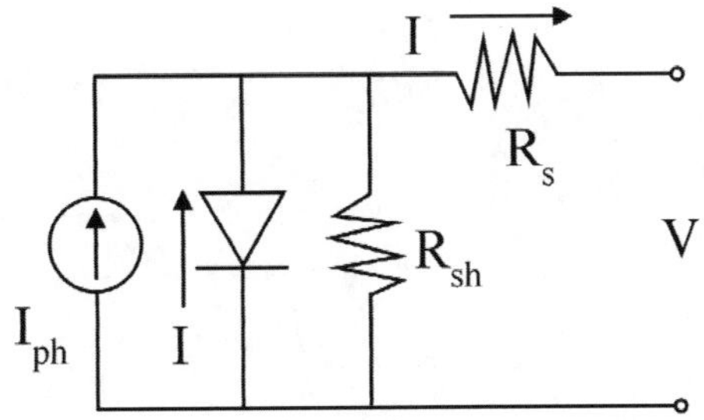

Figure 1. The SDM and five parameters of PV module.

2. Parameter extraction methods

2.1. Parameters reduction

The mathematical expression of PV module can be gotten by doing circuit analysis in figure 1 with Kirchhoff's current law. As shown in equation (1), and thermal voltage $V_t = kT/q$.

$$I = I_{ph} - I_o \left[exp\left(\frac{V + IR_s}{nV_t}\right) - 1\right] - G_{sh}(V + I \cdot R_s) \tag{1}$$

It is too complicated to deal with the five parameters directly. We find a method to reduce parameters [7]. According to the open circuit ($I=0$, $V=V_{oc}$), short circuit ($I=I_{sc}$, $V=0$) and maximum power ($I=I_{mp}$, $V=V_{mp}$) conditions, I_{ph}, I_0 and R_{sh} can be expressed with n and Rs. As shown in equations (2)-(4).

$$I_0 = \frac{V_{oc}(I_{sc} - I_{mp}) - V_{mp}I_{sc}}{A_1EXP_{sc} + A_2EXP_{mp} + A_3EXP_{oc}} \tag{2}$$

$$G_{sh} = \frac{EXP_{oc}(I_{mp} - I_{sc}) + EXP_{mp}I_{sc} - EXP_{sc}I_{mp}}{A_1EXP_{sc} + A_2EXP_{mp} + A_3EXP_{oc}} \tag{3}$$

$$I_{ph} = \frac{I_{sc}V_{oc}(EXP_{mp} - 1) + I_{sc}V_{mp}(1 - EXP_{oc}) + I_{mp}V_{oc}(1 - EXP_{sc})}{A_1EXP_{sc} + A_2EXP_{mp} + A_3EXP_{oc}} \tag{4}$$

Among them, abbreviations are as following equations (5)-(10).

$$EXP_{oc} = exp(V_{oc}/nV_t) \tag{5}$$

$$EXP_{mp} = exp[(V_{mp} - R_sI_{mp})/nV_t] \tag{6}$$

$$EXP_{sc} = exp(R_sI_{sc}/nV_t) \tag{7}$$

$$A_1 = V_{mp} + R_sI_{mp} - V_{oc} \tag{8}$$

$$A_2 = V_{oc} + R_sI_{sc} \tag{9}$$

$$A_3 = R_sI_{sc} - R_sI_{mp} - V_{mp} \tag{10}$$

2.2. NR method

Firstly, a function SE has been constructed to evaluate the effect of curve fitting. As shown in equation (11).

$$SE = \sum_{i=1}^{N}\left[I_i - I_{ph} + I_o exp\left(\frac{V_i + I_iR_s}{nV_t} - 1\right) + G_{sh}(V_i + I_iR_s)\right]^2 \tag{11}$$

The best fitting means that SE needs to be the minimum. For a binary function, studying two points whose partial derivatives are zero can find the minimum. Thus, I choose the NR method to solve this problem [8]. The related expressions and iterative equations are shown in equations (12)-(14).

$$f_{R_s} = \frac{\partial SE}{\partial R_s} = -2\sum_{i=1}^{N}\left[\frac{I_0 I_i}{nV_t}exp\left(\frac{V_i + I_i R_s}{nV_t}\right) + G_{sh}I_i\right]\left[I_i - I_{ph} + I_o exp\left(\frac{V_i + I_i R_s}{nV_t} - 1\right) + G_{sh}(V_i + I_i R_s)\right] \tag{12}$$

$$f_n = \frac{\partial SE}{\partial R_s} = 2\sum_{i=1}^{N}\left[\frac{I_0(V_i + I_i R_s)}{n^2 V_t}\right]\left[I_i - I_{ph} + I_o exp\left(\frac{V_i + I_i R_s}{nV_t} - 1\right) + G_{sh}(V_i + I_i R_s)\right] \tag{13}$$

$$\begin{bmatrix} R_s^{t+1} \\ n^{t+1} \end{bmatrix} = \begin{bmatrix} R_s^{t} \\ n^{t} \end{bmatrix} - \begin{bmatrix} \frac{\partial f_{R_s}}{\partial R_s} & \frac{\partial f_{R_s}}{\partial n} \\ \frac{\partial f_n}{\partial R_s} & \frac{\partial f_n}{\partial n} \end{bmatrix}^{-1} \begin{bmatrix} f_{R_s}(R_s^t, n^t) \\ f_n(R_s^t, n^t) \end{bmatrix} \tag{14}$$

The experiment proves that SE decreases rapidly with the number of iterations, and iteration will be ceased when SE changes very little. However, this method needs accurate initial values. Otherwise, it will fall into local optimization and get a bad fitting.

2.3. PSO method

Considering the problem of local optimization, as a common global optimization algorithm, PSO can be used to solve it [9]. We can assume the problem as a point with coordinates n and R_s, looking for the deepest pit on the plane with n and R_s, in which the deepest pit means the smallest SE.

This method can let different number of points search the deep pit along with different directions, in different velocities. Recording the depth every time to get the deepest pit which means the global optimal solution with compare. The logic chart is shown in figure 2.

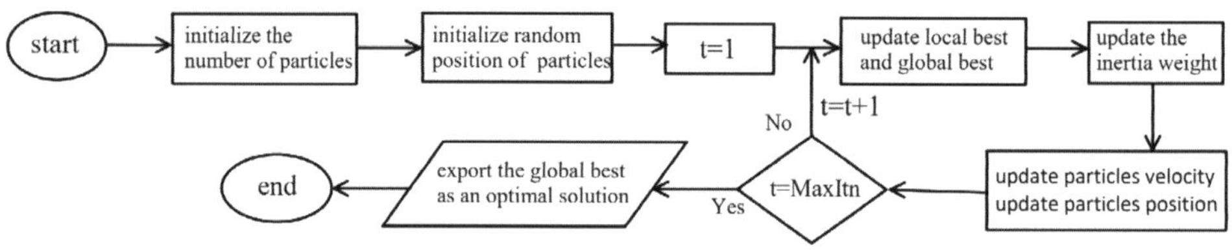

Figure 2. The logic chart of PSO method.

This method can choose the initial values in a certain range. It depends on the number of cells in series, the V_{oc} and the I_{sc} of the module. Usually, the inertia weight in the iterative equation and the number of particles have an influence on solution speed and fitting results.

3. Parameters extraction and fitting results

We choose two kinds of PV modules' data, which is from National Renewable Energy Laboratory, Golden, CO, USA [10], to test the two methods. They are two practical film solar cells, and one is based on CdTe, the other one is based on CIGS. In order to get the more accurate results, a refinement step has been added after that. The optimization function "lsqnonlin" in MATLAB is good for the least square problem, and the previous results can be substituted in it as initial values to get refined results.

Finally, the analyses of the results have been done, including the root mean squared error (RMSE), the mean absolute error (MAE) and the calculation steps in different irradiance conditions. Due to the large amount of data, we list the five parameters extracted under $1000 W/m^2$ and average analysis results to form a table, which is shown in table 1.

Table 1. Five parameters under 1000W/m^2, average error analysis and iteration steps of all methods.

module	CdTe				CIGS			
method	NR	PSO	NR_rf	PSO_rf	NR	PSO	NR_rf	PSO_rf
parameters	1000w/m^2							
n	239.305	239.794	239.788	239.794	105.853	106.301	106.301	106.301
R_s	7.199	7.173	7.174	7.173	0.975	0.968	0.968	0.968
I_{ph}	1.181	1.181	1.181	1.181	2.491	2.491	2.491	2.491
I_0	4.035E-06	4.140E-06	4.139E-06	4.140E-06	1.004E-06	1.068E-06	1.068E-06	1.068E-06
G_{sh}	5.294E-04	5.279E-04	5.280E-04	5.279E-04	2.413E-03	2.379E-03	2.379E-03	2.379E-03
error	average							
RMSE	1.254E-03	1.243E-03	1.243E-03	1.243E-03	5.470E-03	5.398E-03	5.398E-03	5.398E-03
MAE	1.050E-03	1.043E-03	1.043E-03	1.043E-03	4.815E-03	4.750E-03	4.750E-03	4.750E-03
steps	8	78	1	0	14	84	2	0

Under different irradiance conditions (400, 600, 800,1000 and 1200 W/m^2), the best fitting results are shown in figure 3.

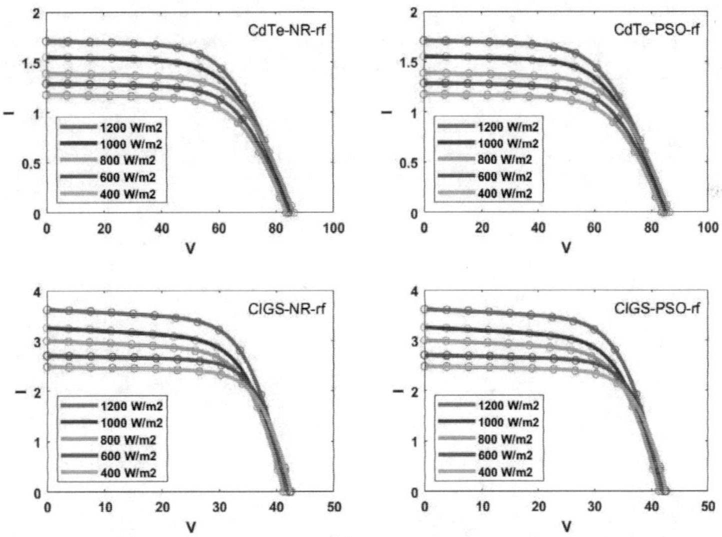

Figure 3. Fitting results under various irradiance conditions after refinement.

4. Conclusion

The fitting results and error analysis show that: (1) Five parameters of compound semiconductor PV module can be extracted out accurately by the three methods; (2) No negative values appear, which means that results are in accords with their physical meanings, SDM is also suitable for CdTe and CIGS solar cells; (3) NR has faster solution, but the accuracy of PSO is better than NR; (4) CIGS module has worse fitting result than CdTe module, maybe other algorithm can solve this problem.

Acknowledgment
This work was supported by NSFC (No. 61804080).

References

[1] Zhang C.F., Zhang J.C., Hao Y., et al. (2011) A simple and efficient solar cell parameter extraction method from a single current-voltage curve. Journal of Applied Physics, 110: 064504.
[2] Cardenas A.A, Carrasco M., David F.M., et al. (2017) Experimental parameter extraction in the single-diode photovoltaic model via a reduced-space search. IEEE Transactions on Industrial Electronics, 64: 1468-1476.
[3] Panchal A. K. (2020) A per-unit-single-diode-model parameter extraction algorithm: A high-quality solution without reduced-dimensions sea. Solar Energy, 207: 1070-1077.
[4] Toledo F.J., Blanes J.M. (2016) Analytical and quasi-explicit four arbitrary point method for extraction of solar cell single-diode model parameters. Renewable Energy, 2: 346-356.
[5] Ayodele T.R., Ogunjuyigbe A.S.O., Ekoh E.E. (2016) Evaluation of numerical algorithms used in extracting the parameters of a single-diode photovoltaic model. Sustainable Energy Technologies and Assessments, 13: 51-59.
[6] Laudani A, David F.M., Fulginei F.R., et al. (2013) Reduced-form of the photovoltaic five-parameter model for efficient computation of parameters. Solar Energy, 97: 122-127.
[7] Laudani A., Fulginei F.R., Salvini A. (2014) High performing extraction procedure for the one-diode model of a photovoltaic panel from experimental I-V curves by using reduced forms. Solar Energy, 103: 316-326.
[8] Abdulrazzaqa A.K., Bognára G., Plesz B. (2020) Evaluation of different methods for solar cells/modules parameters extraction. Solar Energy, 196: 183-195.
[9] Jarboui B., Damak N., Siarry P., et al. (2008) A combinatorial particle swarm optimization for solving multi-mode resource-constrained project scheduling problems. Applied Mathematics and Computation, 195: 299–308.
[10] Marion B., Anderberg A., Deline C., et al. (2014) New data set for validating PV module performance models, IEEE 40th Photovoltaic Specialist Conference (PVSC), 1362-1366.

Study on the characteristics of V_2O_5-WO_3/TiO_2 denitration catalyzer synergistic mercury removal

LU Zhiyong[1], QU Litao[2]

[1] Tianjin Huadian Nanjiang Thermal Power Co., Ltd., Tianjin, 300000, China,

[2] Huadian Power Science Research Institute Co., Ltd., Shenyang, 110000, China;

[*]Corresponding author's e-mail: litao-qu@chder.com

Abstract. Ti-V_1W_9 catalyzer obtained the best 68.18 % elemental mercury (Hg^0) oxidation efficiency at the temperature of 250 °C,The elemental mercury (Hg^0) oxidation efficiency can be maintained above 60 % in the reaction temperature range of 250-350 °C, indicating that V_2O_5-WO_3/TiO_2catalyzer has certain oxidizing ability to elemental mercury (Hg0)； SO_2 inhibits Ti-V_1W_9catalyzer's ability to oxidize elemental mercury (Hg^0), and the toxicity of Ti-V_1W_9catalyzer cannot be reversed to some extent； O_2 has a very obvious promoting effect on the oxidation performance of Ti-V_1W_9catalyzerelemental mercury(Hg^0). Increasing the concentration of O_2 can significantly improve the elemental mercury(Hg^0) oxidation efficiency； Low concentration of NO contributes to the adsorption oxidation of elemental mercury (Hg^0) by Ti-V_1W_9 catalyzer, while high concentration of NO significantly inhibits oxidation of elemental mercury (Hg^0)； NH_3 inhibits the adsorption and oxidation of elemental mercury (Hg^0) on the surface of catalyst； H_2O(g) has very obvious inhibitory effect on elemental mercury (Hg^0) adsorption oxidation.

1. Introduction

In order to achieve sustainable development of the environment, the former Ministry of Environmental Protection promulgated the "Emission Standards of Air Pollutants for Thermal Power Plants" (GB13223-2011) for the first time clarified the emission limit of 0.03 mg/m^3 of mercury and its compounds from coal-fired boilers in thermal power plants. Implemented on January 1, 2015 [1-2].

Mercury and its compounds in flue gas are transformed from mercury in coal during combustion, and exist in three forms of Hg^0, Hg^{2+} and Hg^P in flue gas. Hg^0 is highly volatile and insoluble in water, so it cannot be removed in subsequent flue gas treatment facilities; Hg^{2+} has strong water solubility and can be removed in the subsequent wet flue gas treatment facilities; Hg^P can be removed together with the smoke and dust in the dust collector system[3-4]. Therefore, mastering the transformation law of Mercury forms in flue gas becomes the key factor and effective control method to control mercury emission concentration. Studies have proved[5-8], that the SCR denitration catalyst can convert Hg^0 to Hg^{2+}, so that it can be removed synergistically in subsequent wet desulfurization facilities. Therefore, researching and improving the Hg^0 oxidation performance of SCR denitration catalysts can efficiently remove mercury and its compounds in flue gas. This experiment is based on the research of the removal characteristics of Hg^0 by V_2O_5-WO_3/TiO_2 catalyst.

2. Experimental method

2.1. Preparation of catalyst sample

The catalyst prepared in this experiment is supported by TiO_2 and supported by V_2O_5 and WO_3 as active components. In the laboratory, V_2O_5-WO_3/TiO_2 catalyst is prepared by ultrasonic assisted equal amount impregnation method. The preparation of mixed samples of V_2O_5 and 9% WO_3 is marked Ti-V_1W_9; The preparation load content of V_2O_5 sample was marked Ti-V_1; The sample with 9% WO_3 load was labeled Ti-W_9; The pure TiO_2 was prepared as a sample labeled TiO_2.

2.2. Oxidation performance test of Hg^0

The oxidation performance test of Hg^0 catalyst is completed in the gas-solid micro reactor. The catalyst test system is shown in Figure 1. The simulated flue gas components are 400 ppm-1200 ppm SO_2, 0-300 ppm NO, 5% O_2 (volume fraction, the same below), about 8% H_2O (g) and 0-530 ppm NH_3 (when required), N_2 (purity $\geq$ 99.99%), Hg^0 is provided by mercury vapor generating device.

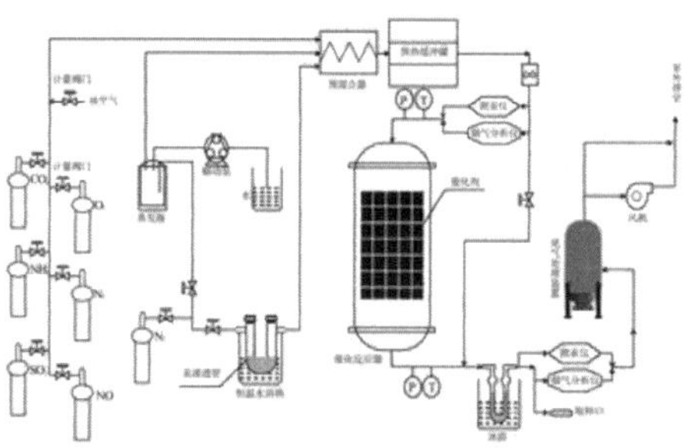

Fig 1. Catalyst test system diagram.

Each component of the simulated flue gas is controlled by a mass flow meter. During the experiment, first turn on N_2, and control the Hg^0 content in the system by adjusting the carrier gas flow of the Hg^0 vapor generator. After the Hg^0 content in the inlet and outlet of the reaction device is stable, then according to the operating conditions required by the experiment, adjust the N_2, SO_2, NO, O_2 gas content in the simulated flue gas, and control the total flow of the simulated flue gas at 1.0 L/min; adjust the temperature according to the experimental requirements, and start testing the Hg^0 inlet and outlet concentration after the program is heated to the set temperature .

3. Experimental results and discussion

3.1. Hg^0 efficiency test of series catalyst samples

At 250 °C, 5% O_2, 400 ppm SO_2, 300 ppm NO, mercury vapor (100 ml/min high purity N_2 as carrier gas) and equilibrium gas N_2 were introduced into the system. The oxidation performance of V_2O_5-WO_3/TiO_2 catalyst series samples prepared in the laboratory for Hg^0 was tested under simulated flue gas experimental conditions. The comparison results are shown in Figure 2.

The experimental results show that the oxidation efficiency of Hg^0 is less than 20% and stable when TiO_2 is used as catalyst alone. The oxidation efficiency of Ti-V_1 and Ti-W_9 increased to a certain extent when V_2O_5 and WO_3 were loaded on TiO_2, but the oxidation efficiency of both Ti-V_1 and Ti-W_9 was still lower than 50%. Ti-V_1W_9 loaded with V_2O_5 and WO_3 at the same time showed good Hg^0 oxidation efficiency, and more than 60% of Hg^0 oxidation efficiency lasted for 3h. This may be due to the synergistic effect of V_2O_5 and WO_3 on Hg^0 adsorption and oxidation.

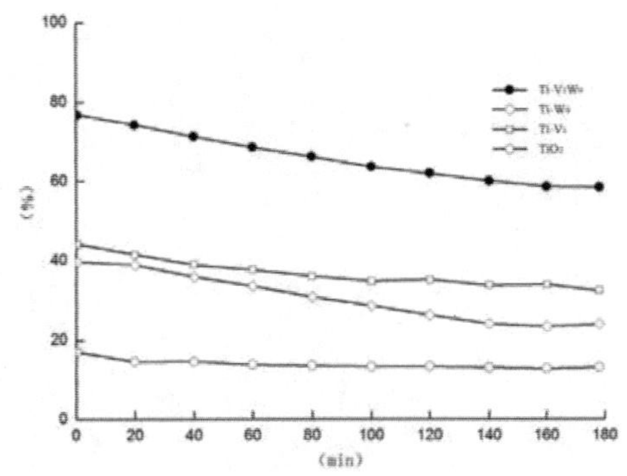

Fig 2. Different catalyst samples for Hg^0 oxidation efficiency.

3.2. Effect of reaction temperature on Hg0 oxidation efficiency of catalyst

The oxidation properties of V_2O_5-WO_3/TiO_2 series samples for Hg^0 were tested in the temperature range of 150-400 °C. According to the experimental results in Figure 3, the oxidation performance of TiO_2 to Hg^0 is poor, and the oxidation efficiency of Hg^0 is maintained at 10%. The oxidation efficiency of Ti-V_1W_9, Ti-V_1 and Ti-W_9 catalysts increased with the increase of temperature from 150 °C to 250 °C, and the best performance of Hg^0 oxidation was achieved at 250 °C. The results show that the oxidation efficiency of Ti-V_1W_9 reaches 68.18% at 250 °C, which is the highest oxidation efficiency of Hg^0 in this series of samples. After that, the oxidation efficiency of Ti-V_1W_9 decreases. However, Ti-V_1W_9 has good activity in the reaction temperature range of 250 ~ 350 °C, and the oxidation efficiency of Hg^0 can still maintain above 60%.

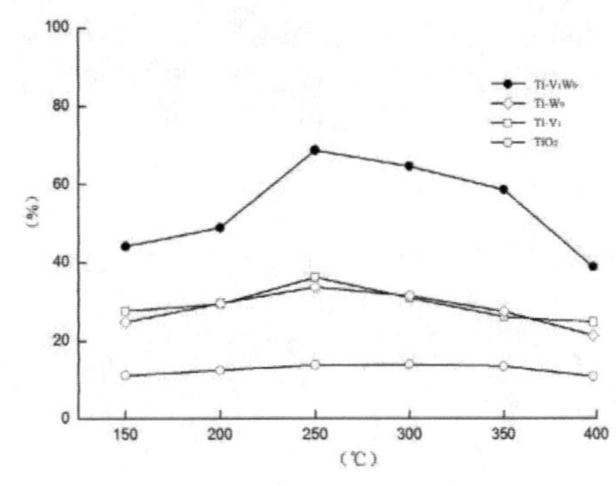

Fig 3. Effect of Different Reaction Temperatures on Oxidation Efficiency of Catalyst Hg^0.

3.3. Effect of flue gas composition on Hg^0 oxidation efficiency of catalyst

Effect of No: the effect of no on the Hg^0 oxidation removal efficiency of the catalyst is shown in Figure 4. It can be seen that when 100 ppm NO is injected into the simulated flue gas, the oxidation efficiency of Ti-V_1W_9 catalyst for Hg^0 is significantly higher than that of pure N_2. However, when the concentration of no is continuously increased from 100 ppm to 300 ppm, the oxidation efficiency of Ti-V_1W_9 catalyst for Hg^0 is significantly decreased, even lower than that of pure N_2, indicating that low concentration of no is conducive to the adsorption and oxidation of Hg^0, The oxidation of Hg^0 was inhibited by high concentration of No; When 5% O_2 was added into 300 ppm NO flue gas, it was observed that the oxidation efficiency of Ti-V_1W_9 catalyst for Hg^0 increased. This was mainly due to the existence of O_2. NO was adsorbed on the surface of Ti-V_1W_9 catalyst in molecular form and reacted with O_2 to form intermediate active substances such as NO_2, NO^+, NO^{2+}, etc. these active substances were more likely to react with Hg^0, thus increasing the oxidation efficiency of Hg^0.

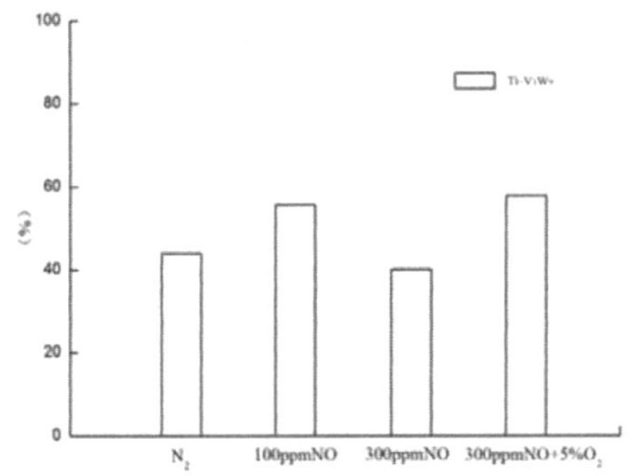

Fig 4. Effect of NO on Oxidation Efficiency of Catalyst Hg^0.

Effect of SO_2: the oxidation performance of Ti-V_1W_9 catalyst for Hg^0 was investigated under the conditions of pure N_2 and 400-1200 ppm SO_2 at 250 °C, and the results are shown in Figure 5. Figure 5 (a) shows that with the increase of SO_2 concentration, the oxidation efficiency of Hg^0 gradually decreases, indicating that the catalytic oxidation activity of Ti-V_1W_9 catalyst is obviously inhibited. When 5% O_2 was injected at 1200 ppm SO_2 concentration, it was observed that the oxidation efficiency of Hg^0 increased significantly, indicating that O_2 had a positive role in promoting the adsorption and oxidation of Hg^0; The sulfur resistance of Ti-V_1W_9 catalyst was tested in Figure 5 (b). After adding 400 ppm SO_2, the oxidation efficiency of Hg^0 decreased rapidly. When the oxidation efficiency of Hg^0 decreased to 40% and was basically stable, SO_2 was shut down. It can be seen that the oxidation efficiency of Hg^0 increased to a certain extent, but could not return to the initial level in a short time, The results showed that the toxicity of SO_2 to Ti-V_1W_9 catalyst could not be reversed to some extent. The results show that [14], there is a competitive relationship between SO_2 and SO_3, SO_3 can promote the oxidation of Hg^0 to a certain extent, and high concentration of SO_2 can significantly inhibit the oxidation of Hg^0, reducing the oxidation efficiency of Hg^0.

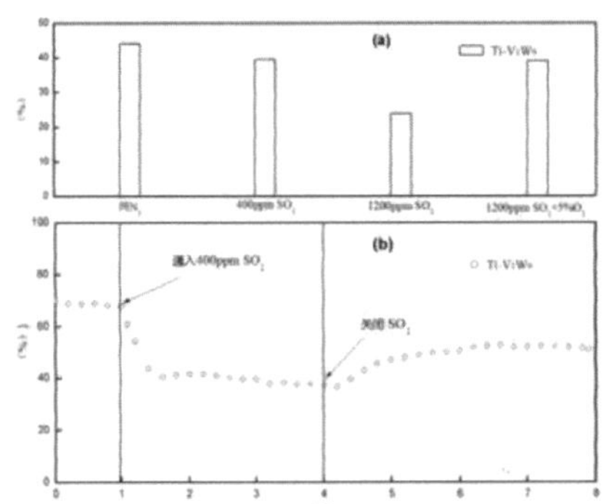

Fig 5. Effect of SO_2 on Hg^0 oxidation efficiency .

Influence of H_2O: water vapor is introduced into dry flue gas to form wet flue gas with humidity of 8%. The Hg^0 oxidation efficiency of Ti-V_1W_9 catalyst is tested, and the results are shown in Figure 6. Comparing and analyzing the Hg^0 oxidation efficiency of dry and wet flue gas, it can be seen that the Hg^0 oxidation efficiency of Ti-V_1W_9 catalyst in wet flue gas is significantly reduced, indicating that H_2O (g) in flue gas has a very obvious inhibition effect on Hg^0 adsorption and oxidation, because H_2O

(g) on the catalyst surface affects the physical and chemical adsorption process of Hg^0, hinders the reaction process of Hg^0 on the catalyst surface, resulting in catalyst deactivation [7].

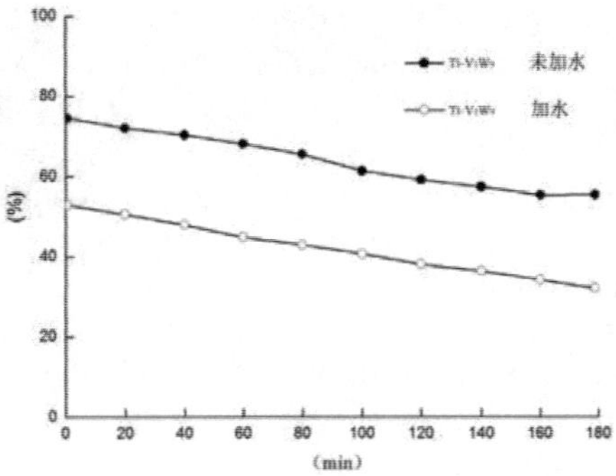

Fig 6. Effect of H_2O on Oxidation Efficiency of Hg^0 Oxidation Efficiency.

Influence of NH_3: NH_3 is the most commonly used reducing agent in SCR Flue Gas Denitration Process. In this experiment, the activity of catalyst in the presence of NH_3, NH_3、NO and O_2 was investigated, and the results are shown in Figure 7. In the absence of O_2 and NO, the Hg0 oxidation efficiency of Ti-V_1W_9 catalyst decreased gradually with the increase of NH_3 concentration. When NH_3 concentration reached 300 ppm, the Hg^0 oxidation ability of Ti-V_1W_9 catalyst was obviously inhibited, and then 5% O_2 was added, the Hg^0 oxidation efficiency increased obviously; When 300 ppm NO is added, the activity of Ti-V_1W_9 decreases slightly with the increase of NH_3/NO ratio from 1:3 to 1:1, which is mainly caused by the inhibition of Hg^0 adsorption and oxidation by NH_3 on the catalyst surface. The oxidation efficiency of Hg^0 increases significantly after 5% O_2 is added.

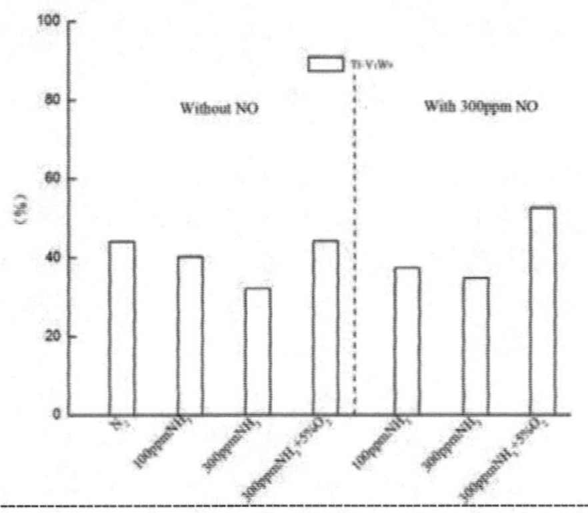

Fig 7. Effect of NH_3 on Hg^0 oxidation efficiency.

Through the above conditions, it can be found that the addition of O_2 has a positive effect on the Hg^0 adsorption and oxidation efficiency. After the addition of 5% O_2, the Hg^0 oxidation efficiency of the samples under the above conditions is slightly improved. The reason may be that O_2 provides oxygen atoms for the oxidation reaction, or supplements the oxygen vacancies formed after the oxidation of Hg^0. Hg^0 can be oxidized by O_2 or lattice oxygen, but the reaction energy barrier is very high, and the reaction is difficult to occur [11], However, O_2 still has a positive effect on the catalytic oxidation of Hg^0.

3.4. Analysis of Hg^0 oxidation mechanism of catalyst

According to the above test results of Hg0 oxidation efficiency of the catalyst, based on the Mars Maessen mechanism, the following mechanism analysis of Hg0 adsorption and catalytic oxidation reaction process under the action of Ti-V1W9 sample in this experimental study was made: firstly, after Hg0 in the flue gas was adsorbed on the surface of the catalyst, Hg0 in the gas phase was adsorbed on the surface of the catalyst, The active component VOx in the catalyst can provide lattice oxygen or oxygen atoms in the valence reduction of V5+ and V4+, and the latter can react with Hg0 adsorbed to produce stable oxidized Hg0. The supply of oxidant components such as O atoms in the catalytic oxidation process is the focus of the catalytic oxidation process of elemental mercury Hg0.

In the above experiments, the oxidation ability of Ti-V_1W_9 was obviously inhibited after adding certain concentration of SO_2, NH_3 and other gases into the system. After adding 5% O_2, it was observed that the O_2 in the simulated gas filled the o consumed in the oxidation reaction of Hg^0, which increased the catalytic oxidation ability of Ti-V_1W_9 to a certain extent and improved the oxidation efficiency of Hg^0, which confirmed the above mechanism.

4. Conclusion

(1) Under the experimental conditions, the Ti-V_1W_9 catalyst prepared in the laboratory has the best Hg^0 oxidation efficiency of 68.18% in the series of samples at 250 °C. It has good activity in the reaction temperature range of 250 ~ 350 °C, and the Hg^0 oxidation efficiency can maintain more than 60%. It shows that V_2O_5-WO_3/TiO_2 catalyst has a certain oxidation ability for Hg^0.

(2) The results show that the influence of flue gas components on the Hg^0 oxidation ability of Ti-V_1W_9 catalyst is quite different, and SO_2 can obviously inhibit the Hg^0 oxidation of Ti-V_1W_9 catalyst; Low concentration of no is helpful to the adsorption and oxidation of Hg^0 on Ti-V_1W_9 catalyst, while high concentration of no obviously inhibits the oxidation of Hg^0; The oxidation efficiency of Ti-V_1W_9 catalyst can be improved by increasing the concentration of O_2.

(3) $H_2O(g)$ on the catalyst surface affects the physical and chemical adsorption process of Hg^0, hinders the reaction process of Hg^0 on the catalyst surface, resulting in the deactivation of the catalyst.

References

[1] Wang Lei, Wang Lijuan. Status quo and suggestions of mercury pollution control in coal-fired power plants in China [J]. Environmental science and technology, 2014 (S2): 285-289.
[2] Chinese Academy of Environmental Sciences. Emission standard of air pollutants for thermal power plants: GB 13223-2011 [M]. China Environmental Science Press, 2012 .
[3] LIU Fasheng, XIA Yongjun, XU Rui,*et al.* Experimental Study on Mercury Removal Effect and Mercury Emission Characteristics of Pollution Control Equipment in Coal-Fired Power Plants[J]. Electric Power, 2017, 50(4): 162-166.
[4] YANG Liguo, DUAN Yufeng, YANG Xianghua, *et al.* Mercury emission characteristics from coal-fired power plants [J]. Journal of Southeast University(Natural Science Edition), 2007, 37(5):817-821.
[5] CAI Xingfei, NIU Tianwen, ZHANG Yubin, *et al.* Progress in the Field of Low-temperature SCR of Flue Gas Denitr Ification Technology [J]. Boiler Technology, 2014, 45(2):77-80.
[6] XU Yueyang, XUE Jianming, WANG Hongliang, *et al.* Research on Mercury Collaborative Control by Conventional Pollutants Purification Facilities of Coal-fired Power Plants [J]. Proceedings of the CSEE, 2014, 34(23):3924-3931.
[7] TAO Li, ZHANG Xunan, LI Caiting, *et al.* Oxidation and capture of elemental mercury over SCR catalyst in simulated flue gas [J]. Chinese Journal of Environmental Engineering, 2015, 9(6):2925-2932.
[8] ZHENG Yiwu, DUAN Yufeng, TANG Hongjian, *et al.* Characteristics of the existing air pollutant control devices on Hg synergistic removal in a coal-fired power plant [J]. China Environmental Science, 2018, 38(3): 862-870.

[9] ZHANG Xunan. Experimental Study on Simultaneous Removal of HgO and NO in Flue Gas with CeO2 Modified SCR Catalyst[D]. Hunan University, 2015.

[10] YANG Jian, YANG iang, SUN Jian, et al. Effects of mercury oxidation on V_2O_5-WO_3/TiO_2 catalyst properties in NH_3-SCR process[J]. Catalysis Communications, 2015, 59: 78–82.

[11] Gao W, Liu Q, Wu C Y, et al. Kinetics of mercury oxidation in the presence of hydrochloric acid and oxygen over a commercial SCR catalyst[J]. Chemical Engineering Journal, 2013, 220(11):53-60.

[12] SUAREZ NEGREIRA A, WILCOX J. DFT study of Hg oxidation across vanadia-titania SCR catalyst under flue gas conditions[J]. The Journal of Physical Chemistry C, 2013, 117(4): 1761–1772.

[13] ZHAO Li, LIU Yu, WU Yangwen, *et al.* Review on Theoretical Research of the Catalytic Oxidation of Hg0 in Coal-Fired Flue Gas [J]. Electric Power, 2018, 51(3): 170-176.

[14] Ye Z, Laumb J, Liggett R, et al. Impacts of acid gases on mercury oxidation across SCR catalyst[J]. Fuel Processing Technology, 2007, 88(10):929-934.

[15] Pudasainee D, Kim J H, Yoon Y S, *et al.* Oxidation, reemission and mass distribution of mercury in bituminous coal-fired power plants with SCR, CS-ESP and wet FGD[J]. Fuel, 2012, 93(1):312-318.

[16] WANG Zhen, LIU Jing, ZHANG Bingkai, *et al.* Mechanism of heterogeneous mercury oxidation by HBr over V_2O_5/TiO_2 catalyst[J]. Environmental Science & Technology, 2016, 50(10): 5398–5404.

[17] JIANG Ye, GAO Xiang, WU Weihong, *et al.* Review of the deactivation of selective catalytic reduction De NOx catalysts[J]. Proceedings of the CSEE, 2013, 33(14):18–31.

[18] QU Litao, LI Chao, YU Honghai, *et al.* I Impact Analysis and Countermeasures on Coal Fired Units In-depth Peak Regulation of Environmental Protection Facilities [J]. Northeast Electric Power Technology, 2016, 37(10):38-40.

[19] LI H, YING L, WU C Y, *et al.* Oxidation and capture of elemental mercury over SiO_2 –TiO_2 –V_2O_5, catalysts in simulated low-rank coal combustion flue gas[J]. Chemical Engineering Journal, 2011, 169(1):186-193.

[20] LI H, WU C Y, LI Y, *et al.* Superior activity of MnO_X -CeO_2 /TiO_2, catalyst for catalytic oxidation of elemental mercury at low flue gas temperatures[J]. Applied Catalysis B Environmental, 2012, 111(3):381-388.

[21] YAN N, CHEN W, CHEN J, *et al.* Significance of RuO_2 Modified SCR Catalyst for Elemental Mercury Oxidation in Coal-fired Flue Gas[J]. Environmental Science & Technology, 2011, 45(13):5725-30.

[22] Zhao Lingkui. Experimental and theoretical study on simultaneous denitrification and mercury removal by selective catalytic reduction catalyst [D]. Hunan University, 2017.

[23] GUO Pan, GUO Xin, ZHENG Chuguan. Computational insights into interactions between Hg species and α-Fe_2O_3 (001)[J]. Fuel, 2011, 90(5): 1840–1846.

Three-layer structure hot bending/diffusion bonding process of reactive sintering Ti-22Al-25Nb alloy and its load response at high temperature

YU Long-hai[1*], Wang Yuan-xin[2]

[1] Engineering Training Center, Xinyu University, Xinyu,338004, China.

[2] ZheJiang HengJi YongXin new materials Co.,Ltd, Ningbo,315500, China

[1*] yulonghai@xyc.edu.cn

Abstract Three-layer structure part as a ideal structure part, that had light weight and enough strength to be widely used, was prepared combining hot bending and diffusion bonding processes in this paper. The Ti-22Al-25Nb sintered alloy that owned better using temperature was applied to manufacture three-layer structure part. The forming processes of manufacturing three-layer structure was divided into hot bending forming of the core sheet and diffusion bonding of three layers sheets. Firstly, the hot bending of core sheet was performed at 1423K for 10MPa in a vacuum furnace. Secondly, the diffusion bonding of three layers sheets was carried out at 1473K for 120min and 10MPa. Furthermore, in order to evaluate the quality and performances of three-layer structure, the microstructure of bonding joint of three layers sheets was analyzed by SEM, and the properties of three-layer structure part were tested at 1073K by compressing test and three-point bending test. Compressing load and compressing strength were 9014N and 10MPa, respectively. Bending strength was up to 54MPa. In addition, the experimental results also indicated that three-layer structure that prepared using the Ti-22Al-25Nb sintered alloy owned good properties.

1. Introduction

With the development of aeronautic and astronautic engineering, lower weight and higher efficiency had been already considered as a standard of structural components. Thus, light alloys, such as Ti, TiAl and Ti_2AlNb alloys, had been widely developed and applied depending on its outstanding properties at elevated temperature[1-3]. Especially, Ti_2AlNb alloy, as a promising structural material at elevated temperature, had been widely studied since it was found by Banerjee D[4]. It had been regarded as a promising candidate of a structural material substituting for Ni-based alloy, which was expected to apply for higher temperature than that of Ti-based alloy. Moreover, as an excellent structural part, multi-layer structure design had the advantages of lower weight, higher specific strength and good integrity[5,6]. At present, however, the studies on Ti_2AlNb-based alloys mainly focused on the microstructures and basic mechanical properties[7-11]. As to the applications of Ti_2AlNb alloy serving as structural parts, it rarely reported. So, in this paper, the Ti-22Al-25Nb alloy was prepared by reactive sintering with element powders, and then three-layer structure had been fabricated with the Ti-22Al-25Nb sintered alloy. Combining with their respective advantages, it expected to obtain the structural component in order to meet better requirements.

2. Material and experimental procedure

The as-received material was prepared by step reactive sintering with element powders. And then sheets to be forming were grinded and polished prior to forming. The forming process of three-layer structure was divided into two stages: bending forming of the core sheet and diffusion bonding of three layers sheets. Firstly, bending forming of the core sheet was carried out at 1423K in vacuum furnace. Secondly, diffusion bonding of three layers sheets was implemented at 1473K, holding 10MPa for 120min in order to ensure to obtain good diffusion joint. The forming molds were manufactured with graphite. The microstructure of diffusion bonding joint of as-received three-layer structure were observed by SEM(with BSE). In addition, the microstructures of the as-received material in pre-forming and post-forming were analyzed. In order to estimate the basic properties of three-layer structure during using temperature, compressing test and three-point bending test were conducted at 1073K on Instron 5500R with a high temperature furnace, the rate was used at 1mm/min.

3. Results and discussion

3.1. The as-sintered alloy and its microstructure analysis

Fig.1 showed the sintered alloy and its microstructure at sintered state. It was observed in Fig.1 that the as-sintered alloy owned good quality(the final density was 5.218g.cm^{-3}, closing to the density of 5.3g.cm^{-3} in as-casted alloy)[12]. The microstructure of the sintered alloy indicated that no obvious holes were observed. The laths of O phases(the grey represented O phase) uniform dispersed in B2(the white indicated B2 phase) matrix (phases identification by XRD in previous experiments by authors). Despite of ideal phases of O and B2 phases in the as-sintered alloy, it was found that the microstructure of the sintered alloy seem to be different from the literatures reported which mentioned and described the acicular O phases in B2 substrates. The reason that led to the coarse laths of O phases could be cooling condition after sintering. In addition, the reactive mechanism and diffusion condition during reactive sintering course were also different from that of as-casted alloy. On the other hand, O phase owning better strength(as orthorhombic structure) and B2 phase having better ductility(that owned b.c.c structure) had been studied by many researchers[13]. Therefore, lots of O phases crisscross distributing in B2 matrix would benefit the forming property and further contribute to the followed fabrication of three-layer structure.

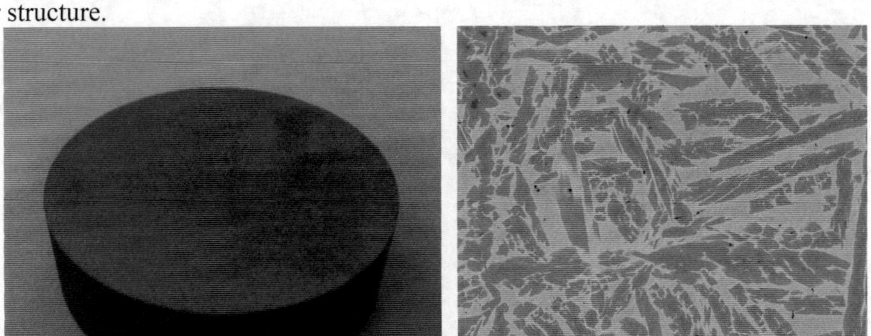

Figure 1 The sintered alloy and its microstructure at sintered state

3.2. The forming process of three-layer structure

Fig.2 showed bending mold and bonding schematics. As can be seen from Fig.2, the forming process of three-layer structure was divided into two stages. Firstly, the bending formation of the core sheet was performed at 1423K in a vacuum furnace, holding 10min and 10MPa. Secondly, the forming process of three-layer structure, that is, diffusion bonding was carried out at 1473K between top-face sheet and core sheet and bottom-face sheet, holding 120min and 10MPa. Graphite foils were selected as separation between sheets and cushion blocks in order that it was easy to demould after post-forming.

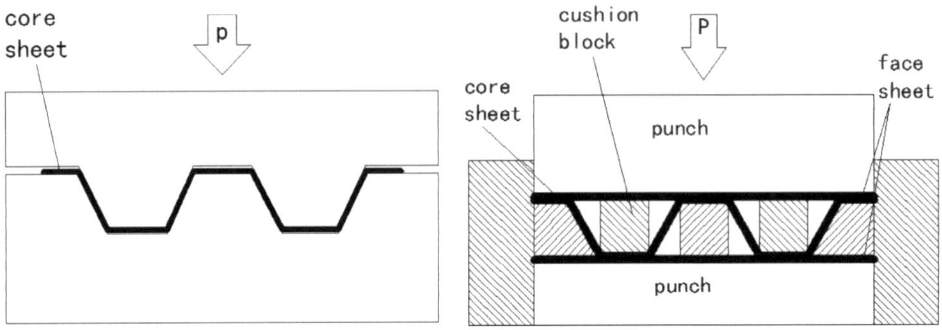

Figure 2 Bending forming and diffusion bonding schematics

Fig.3 showed the products pictures of the core sheet after bending forming and as-fabricated three-layer structure part. The bending forming object of the core sheet was presented in Fig.3(a). Good quality surface and precise shape were obtained in the bending forming. The bending forming of core sheet was performed at 1423K in a vacuum furnace. After bending forming, the core sheet was ground and polished for followed diffusion bonding meeting good effect. Diffusion bonding of three-layer structure was a crucial process, so, the temperature of diffusion bonding was set at 1473K in order to have enough diffusing time, higher than that of bending forming of the core sheet. The more appropriate process parameters were selected by means of previous optimization experiments of diffusion bonding at different conditions by authors. Therefore, diffusion bonding between top-face sheet and core sheet and bottom-face sheet was conducted at 1473K in a vacuum furnace, the bonding positions were holden pressure of 8MPa for 2h. As can be seen in Fig.3(b), three-layer structure part was successfully fabricated. Its shape and size met design requirements. Next, as to its bonding qualities and mechanical properties during the range of using temperature, the microstructure of bonding joint was analyzed, moreover, the compressive strength and flexural strength were tested.

(a) (b)

Figure 3 The products pictures: (a) the core sheet after bending part ;(b) as-fabricated three-layer structure part

Fig.4 showed the cross-sectional microstructures of diffusion bonding joint and non-bonding position. Diffusion bonding had a key effect on three-layer structure. In addition, quality of bonding joint would make sure if three-layer structure meeting good standard and owning quality qualified was fabricated. That is, quality of diffusion bonding is a key factor on evaluating properties of three-layer structure. In Fig.4(a), the microstructure of diffusion bonding joint was observed. It was found that bonding interface of three-layer structure presented good bonding quality and homogeneity, the phases in two different sheets seemed to be full contact and growth until full fusion. The flaw between two layers sheets was hardly observed. But, the presence of bonding interface seemed to be observed by means of the directions of O phases in two different sheets. From the morphology of the bonding joint, it indicated that the Ti-22Al-25Nb sintered alloy had good diffusion bonding ability, even if there no filler metal in the boundary of bonding joint. Thus, no differences between the microstructures of bonding joint and substrates would contribute to homogeneity and compatibility in the whole part.

In addition, it can be seen in Fig.4(b) that the microstructure of O phases of non-bonding positions became coarser than that of bonding joint in three-layer structure. The positions of bonding joint under pressure bore biaxial compressive stress could result in restricting precipitation and growth of O phases. However, non-bonding positions was at free condition, and the final forming temperature was up to 1473K, moreover, a slower cooling after forming also provided more time for the growth of O phases. All these during forming course would lead to precipitation and growth of O phases having conditions. Meanwhile, compared to the microstructure of the as-sintered alloy, it seemed that there no obvious difference between the two. The reason that could undergo the almost same environmental conditions gave rise to no obvious difference in the final morphologies.

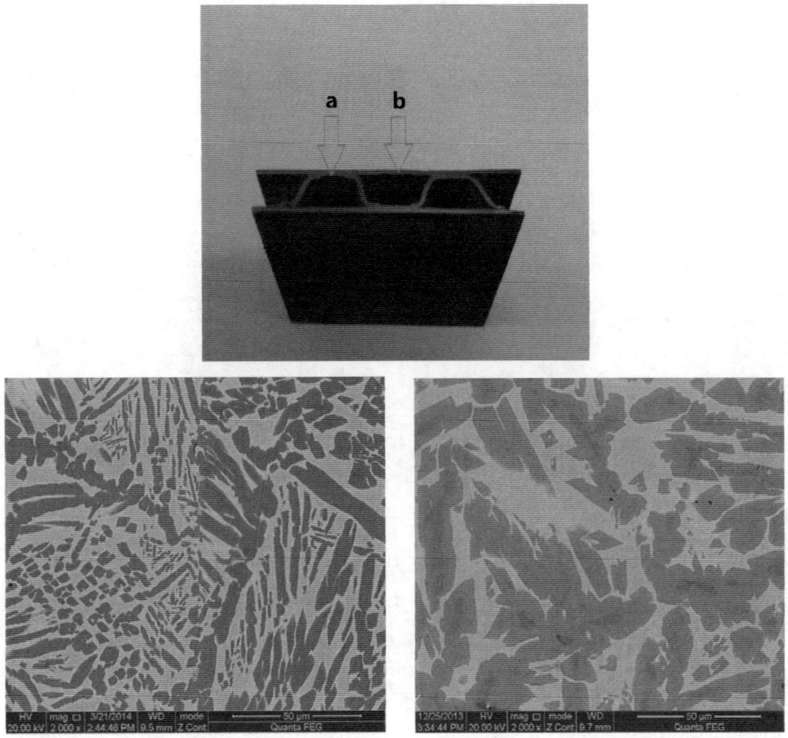

(a) (b)

Figure 4 Cross-sectional microstructures of diffusion bonding joint(a) and non-bonding position(b)

3.3. Compression test and three-point bending test

As a structural part, the basic mechanical performances of three-layer structure during using temperature range would verify its applicability. So, compressing test and three-point bending test were conducted in order to evaluate the basic properties of three-layer structure part. The three-layer structure parts were machined with a length of 35mm and a width of 35mm and a height of 6.6mm.

Fig.5 showed schematic of compression test(a) and compression curve of three-layer structure at 1073K(b).

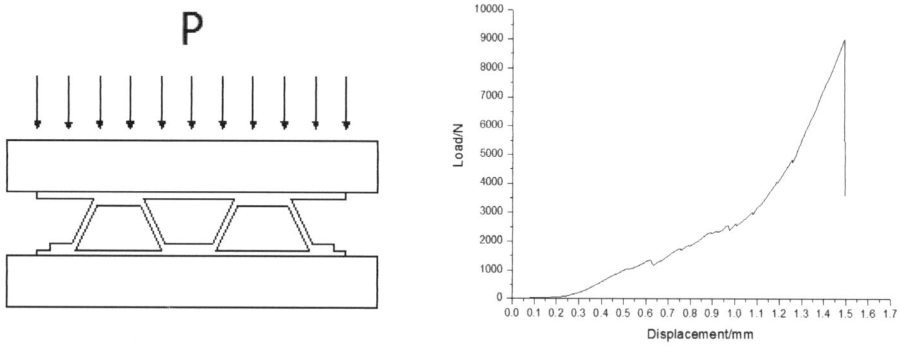

(a) (b)

Figure 5 schematic of compression test(a) and compression curve of three-layer structure at 1073K(b): the load VS. displacement

Compression test schematic was given in Fig.5(a). Compressing test was performed at 1073K with the rate of 1mm/min. The load increased with displacement increasing. The maximum load was 9014N while failure behavior of three-layer structure part occurred. From compression curve of three-layer structure at 1073K in Fig.5(b), no obvious plastic deformation behavior occurred. The compressing strength was up to 10MPa(the effective area is projected area of the core sheet). The compressing strain was 22.63% until three-layer structure broken. The three-layer structure part had no obvious deformation due to the sintered alloy owning poor plasticity, even if the temperature was kept 1073K. Moreover, three-point bending test was conducted at the same conditions. The maximum load was up to 1800N. The maximum bending strength was up to 54.8MPa before brittle failure.

4. Conclusions

In this paper, three-layer structure was successful fabricated by the processes of bending forming and diffusion bonding, using the Ti-22Al-25Nb sintered alloy which was prepared by reactive sintering with element powders. The qualities of bending and diffusion bonding were examined, and then the microstructures of the three-layer structure were observed and analyzed, compared to the microstructure of the as-sintered alloy. Furthermore, in order to evaluate the property and homogeneity for applications, the compressing testing and three points bending testing were performed at 1073K. The maximum compressive load and compressive strength of 10MPa were obtained. And the bending strength was up to 54MPa. From the experimental results, it was found that the sintered Ti22Al25Nb alloy owned good shaping ability and bonding ability for its applications at elevated temperature. Especially, three-layer structure that was fabricated with the Ti-22Al-25Nb sintered alloy, still owned good resistances of compression and bending by the tests of compression and three-point bending at high temperature.

References

[1] C.Leyens , M.Peters. Titanium and Titanium alloys, Wiley, Koln, 2003.
[2] He P, Feng J.C, Zhang B.G. Material Characterization, 48(2002):401-406.
[3] O.A Kaibyshev. Materials Science Technology, 22(2006):343-348.
[4] D. Banerjee, A.K. Gogia, T.K. Nandi, Acta Metallurgica, 36(1988):135-158.
[5] Xun Y.W, Tan M.J. Journal Processing Technology, 99(2000):265-268.
[6] Z.Q. Li, H.P. Guo. Aeronautical Manufacturing Technology, 8(2010):32-35.
[7] Seung Jin Yang, Soo Woo Nam, Masuo Hagiwara. Journal of Alloys and Compounds, 368(2004):197-210.
[8] Peng Lin, Zhubin He, Shijian Yuan, Jun Shen, Yongjiang Huang, Xiaobo Liang. Journal of Alloys and Compounds, 578(2013):96-102.
[9] Jorg Kumpfert, Advanced Engineering Materials, 11(2001):851-863.
[10] C.J. Boehlert, Materials Science and Engineering A, 279(2000):118-129.

[11] Satoshi Emura, Kaneaki Tsuzaki, Koichi Tsuchiya, Materials Science and Engineering A, 528(2010):355-362.
[12] Zhang Jianwei, Li Shiqiong, Liang Xiaobo, Cheng Yunjun, The Chinese Journal of Nonferrous Metals, 20(2010)S1:336-422.
[13] Chen Xue, Weidong Zeng, Bin Xu, Xiaobo Liang, Jianwei Zhang, Shiqiong Li, Intermetallics, 29(2012):41-47.

Butane Tetracarboxylic Acid as Green Modifier for Tuned Hydrophilicity of Surface Modified Cellulose

Shuxian Yu[1,2], Jie Wei[1], Jia Shuai[1], Shaoyi Lyu[2*], Chen Pan[1], Ziqiang Shao[1*]

[1]Beijing Engineering Research Center of Cellulose and Its Derivatives, School of Materials Science and Engineering, Beijing Institute of Technology, Beijing, China

[2] Research Institute of Wood Industry, Chinese Academy of Forestry, Beijing, 100091, China

*Corresponding author E-mail: lvsy@caf.ac.cn, shaoziqiang@263.net

Abstract. This study proposes one straightforward way for butane tetracarboxylic acid modification of cellulose nanofibrils (CNF). Esterified cellulose nanofibrils (ECNF) were characterized using Fourier Transform Infrared Spectroscopy (FT-IR), scanning electron microscopy (SEM), transmission electron microscopy (TEM). The reaction mechanism between tetracarboxylic acid modification and CNF was discussed. The resultant ECNF exhibited improved hydrophobicity in water compared with pure CNF. These results confirmed how esterified (ECNF) biobased nanoparticles with tuned hydrophilicity, obtained by treatment with a low cost, sustainable and easily soluble cross-linker, have the potential for widespread applicability in the field polymeric based nanocomposites having different polarity.

1. Introduction

Nanocellulose is a general term for cellulose with one or more dimensions on the nanometer scale. Not only nanocellulose retains regenerative and biodegradable characteristics of natural cellulose, but it also has the advantages of large specific surface area, high strength (110-220 GPa), low density[1] (1.6 g/cm ^{3}) high aspect ratio, extensive chemical modification, and is excellent candidate materials as reinforcing fillers. The high aspect ratio enables excellent stress transfer between the filler and the matrix. However, nanocellulose is rich in abundant hydroxyl, resulting in incompatible hydrophobic thermoplastics and poor interfacial adhesion between the fillers and polymer matrix. This dramatically limits the utilization of nanocellulose fillers for reinforcement.

To enhance the interactions between the cellulose surface and polymers, thereby improving its dispersibility among the composite properties, chemical modifications such as acidification[2], etherification[3], urbanization, silane coupling,[4] and polymer grafting through surface-initiated radical polymerization and ring-opening polymerization, have been performed. Meanwhile, hydrogen bonding, electrostatic interactions, and physisorption have also been used. However, the chemical modification of nanocellulose is mainly carried out in solvent system[5], which requires multiple solvent replacement, resulting in a complex modification process, longer treatment cycle, more waste liquid, and severe environmental pollution, which is not suitable for the large-scale and industrial preparation of nanocellulose reinforced polymer[6].

In this study, a combination of freeze-drying and chemical modification can be used to prepare large quantities of modified nano-cellulose powder. Specifically, the chemical modifier BTCA and nanocellulose were mixed uniformly in a water/tert-butanol system and then freeze-dried to obtain nanocellulose aerogel. Then, the aerogel was placed at 170°C for 3min to bring the surface esterification modified nanocellulose aerogel powder. BTCA is a polycarboxylic acid compound, which can be easily esterified with the hydroxyl group of nanocellulose. By controlling the addition amount of BTCA[7], the degree of modification can be well controlled. In addition, the modification method is carried out in the dry state, which can reduce the use of a large number of organic solvents for the modification of conventional nanocellulose. The addition of tert-butyl alcohol can affect the growth of ice crystals and the aggregation of nanocellulose, which ensures that the nanocellulose aerogel powder has a suitable microstructure.

2. Experiment

2.1. Experimental material and setups

The nanocellulose was pretreated with inorganic acid and then homogenized under high pressure. 1,2,3,4-butane tetracarboxylic acid (BTCA, 99%) was purchased from Sigma-Aldrich Co. Sodium hypophosphite monohydrate (SHP, 98%) as a catalyst for the reaction of BTCA with CNF. All the reagents used are analytical grade and without further purification.

2.2. Preparation of CNF aerogels

The CNF aqueous suspensions (0.2 wt%) was mixed with t-BuOH and water at different mass ratios (5/5), as shown in Table 1. Then they were poured into a high-speed disperser (Ultra-Turrax, IKA) by vigorous stirring for 12 h to get the homogenous suspensions. And the amount of SHP was always the same as the BTCA. The mixed suspensions were poured into cylindrical or rectangular moulds and freeze-dried in a lyophilizer (TELSTAR LYOBETA MINI, Spain) at a condenser temperature of −85 °C for three days. The crosslinking of aerogels were achieved by curing the freeze-dried samples containing the cross-linker in a vacuum oven at 170 °C for 3 min to perform sufficient covalent crosslinking. For comparison, neat CNF aerogels and BTCA crosslinked CNF aerogels were also fabricated according to similar procedures as mentioned above. The macroscopic morphologies of all the prepared aerogels were white with well-defined shapes, as shown in Figure 1.

Table 1.Formula of CNF aerogel

Sample	t-BuOH(ml)	Water(ml)	BTCA(g)	SHP(g)	CNF(g)
1	25	25	0.003	0.0015	1.5

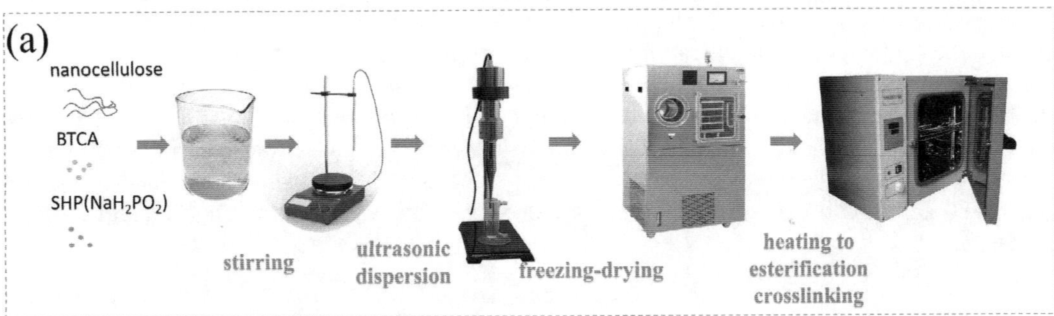

Figure 1. Schematic illustration for the experimental process.

2.3. Characterizations

The nanocellulose suspension sample was diluted with deionized water to a mass fraction of 0.1% and ultrasonic for two minutes with a Sonics(VCX 1500HV) ultrasonic dispersion instrument. The sample was dropped onto a TEM network and stained with uranium dioxide acetate. Shot by Tecnai G2 F20 S-TWIN (200KV).

Morphology of freeze-drying CNF aerogel was conducted under 2500 magnification using an SEM system (S-4800, Hitachi, Japan). All samples were sputter-coated with gold before testing.

The chemical compositions of the neat CNF aerogel and esterified crosslinking CNF aerogel were analyzed via Fourier transform infrared (FTIR) spectroscopy (Nicolet IS10, Madison, WI) in the range of 4000–500 cm^{-1} with a resolution of 4 cm^{-1}. All tests were repeated three times.

3. Results and Discussions

3.1. cellulose nanofibrils structure analysis

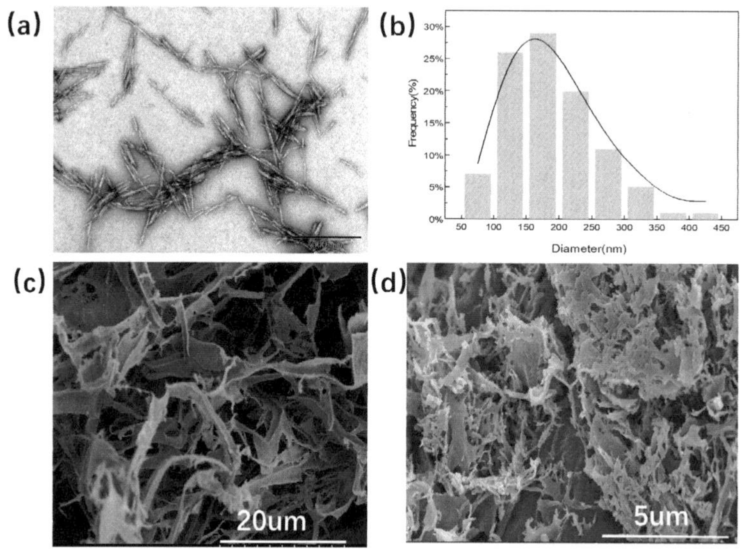

Figure 2. (a)TEM figure of cellulose nanofibrils suspension (b)Diameter distribution of cellulose nanofibrils(c)(d)SEM figure of freeze-dried cellulose nanofibril aerogel.

Fig2(a) shows the morphology of the prepared cellulose nanofibrils, and the average length of cellulose nanofibrils is between 100nm and 250nm. The addition of tert-butyl alcohol effectively prevented the agglomeration and keratinization of nanocellulose, as shown in **Fig(c-d)**.

Similar to previous reports in the literature, the possible reasons for this phenomenon are tert-butyl alcohol can accelerate this drying rate due to its higher vapour pressure (4.1 KPa at 20 °C) and lower boiling point (82.2 °C) firstly, and reduce the shrinkage degree of nanocellulose during drying process because of its low surface tension. More critical,water/t-BuOH[8] mixture can reduce interfibrillar contacts and adhesion by giving more interfibrillar connections and bonding due to larger molecular size than water.

3.2. cellulose nanofibrils structure analysis

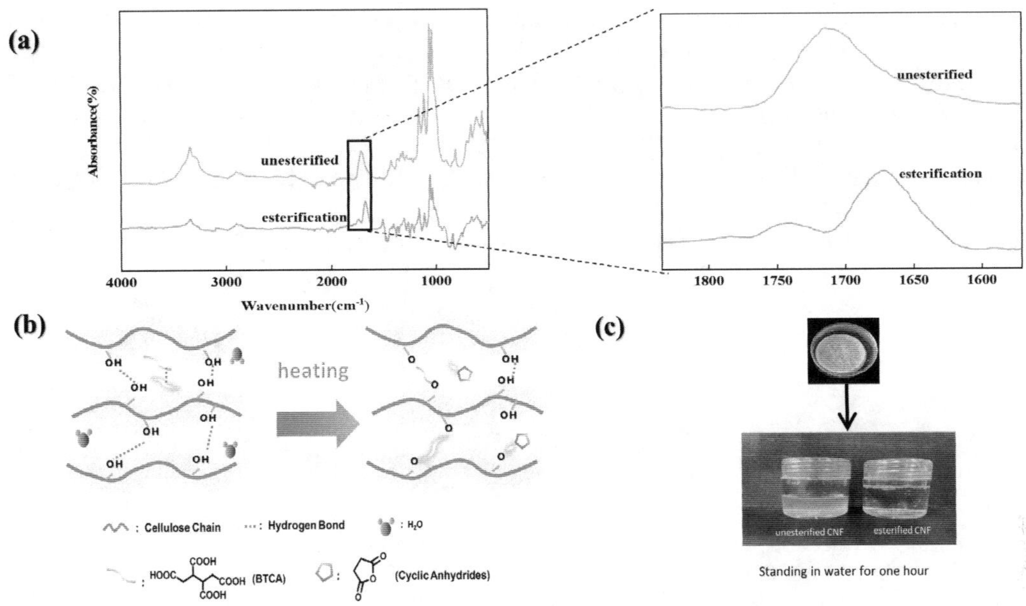

Figure 3. (a)FTIR spectra of pure and modified cellulose nanofibrils (b)Schematic diagram of esterification mechanism(c)Aerogel holds its shape in the water.

Fig.3(a) shows the FTIR spectra of the neat CNF aerogel and CNF/BTCA/SHP aerogel. The FTIR spectra of all of the samples exhibited broadband from 3100 to 3400 cm^{-1} and a strong peak at 2910 cm^{-1}, which were assigned to the O-H and C-H stretching vibrations, respectively. The crosslinking reaction between the cellulose nanofibrils and BTCA occurs through transesterification reactions between the hydroxyl groups of cellulose and carboxylic acid groups of BTCA.

And the specific process, as shown in **Fig.3(b)**, is that Na^+ react with carboxylic acid groups and transform them into corresponding carboxylates with the catalyst SHP. The strong hydrogen bonding is weakened because of the polar groups such as C=O and O-H[9]. With the temperature increasing, cyclic anhydrides gradually form from free carboxylate and carboxylic acid groups. They immediately undergo esterification with O-H groups in cellulose and achieve crosslinking.

The aerogel prepared by the modified crosslinking method has certain hydrophobicity and can maintain its shape in the water.

4. Conclusions

In summary, we successfully prepared a nanocellulose and changed its hydrophilicity by BTCA esterification and crosslinking. FTIR shows that we have successfully modified the aerogel and retains its shape in water rather than dispersed out. Compared with other modification methods, our one-step modification method does not use organic solvents, and the process is simple, environmentally friendly and suitable for industrial production.

References

[1] Wang, J.; Liu, X.; Jin, T.; He, H.; Liu, L., Journal of Biomaterials Science, Polymer Edition 30, 919 2019.
[2] Cui, X.; Honda, T.; Asoh, T.-A.; Uyama, H., Carbohydrate Polymers 230, 115662 2020.
[3] Mo, Y.; Yang, L.; Hou, W.; Zou, T.; Huang, Y.; Liao, R., Macromolecular Materials and Engineering 305, 2000063 2020.

[4] Nagamine, S.; Mizuno, Y.; Hikima, Y.; Okada, K.; Wang, L.; Ohshima, M., Journal of Applied Polymer Science 138, 49851 2021.
[5] Mohrmann, S.; Linganiso, L.; Song, S. P.; Motloung, S. V.; Koao, L.; Motaung, T. E., Plastics, Rubber and Composites 48, 401 2019.
[6] He, X.; Luzi, F.; Yang, W.; Xiao, Z.; Torre, L.; Xie, Y.; Puglia, D., ACS Sustainable Chemistry & Engineering 6, 9966 2018.
[7] Hou, L.; Wu, P., Cellulose 26, 2759 2019.
[8] Noguchi, T.; Niihara, K.-i.; Iwamoto, R.; Matsuda, G.-i.; Endo, M.; Isogai, A., Composites Science and Technology 207, 108734 2021.
[9] Guo, L.; Chen, Z.; Liu, S.; Fu, F.; Wang, S., Carbohydrate Polymers 179, 333 2018.

Effect of particle size on electrochemical performance of lithium iron phosphate materials

Rui He[1, 2], Aijia Wei[1, 2], Xue Bai[1, 2], Lihui Zhang[1, 2], Xiaohui Li[1, 2], Zhenfa Liu[1, 2]*

[1]Institute of Energy Resources, Hebei Academy of Science, Shijiazhuang, Hebei Province, P.R. China, 050081

[2]Hebei Engineer Research Center for Water Saving in Industry, Shijiazhuang, Hebei Province, P.R. China, 050081

*Corresponding author's e-mail: lzf63@heb-as.com

Abstract: $LiFePO_4$ material has many virtues, such as rich raw materials, low price, excellent high temperature cycling performance and safety property. Lithium ion battery with $LiFePO_4$ as positive electrode active material is the most promising. However, compared with other positive electrode active materials, $LiFePO_4$ material has the inherent disadvantage of poor electrical conductivity, it greatly limits its dynamic characteristics at low temperature. In this paper, the effect of particle size of $LiFePO_4$/C material on the low temperature performance prepared via solid-state method is systematically investigated. The structure and morphology of the samples were characterized by X-ray diffraction (XRD) and scanning electron microscope (SEM). The electrochemical properties of cathode materials were studied by Land2001.

1. Introduction

With the increasing prominence of energy and environmental problems, the new generation of safe and green lithium ion batteries are widely used [1-4]. Relative to the traditional lithium ion battery materials, Lithium iron phosphate (LFP) has the advantages of low price, environment friendly, good thermal stability, abundant resources, good safety and so on [5,6,7]. At present, the synthesis methods of lithium iron phosphate have solid state method with high temperature calcine, carbothermal way, microwave synthesis, sol-gel, hydrothermal method and so on. The application of LFP materials is restricted by its low electron conductivity and low lithium ion diffusion coefficient. In addition, the low temperature performance can not meet the current requirements.

Based on the above problems, coating and doping methods are generally used for modification $LiFePO_4$. In this paper, nanocrystallization is used to enhance the properties of LFP. Using LiH_2PO_4 and Fe_2O_3 as raw materials, the preparation process of low cost, high efficiency and clean $LiFePO_4$ for industrial application was studied. And citric acid is added as carbon source. The particle size of the material is controlled by ball milling time and milling process.

2. Experimental

2.1 Apparatus and Reductant

Lithium dihydrogen phosphate was manufacted by Tangshan aboundant south area into feng technology co., LTD. Iron oxide is produced by Tianjin zhengyuan haoye. Citric acid is produced by Tuosheng Chemical Co., Ltd.

Scanning electron microscopy (SU8020, Hitachi), X-ray diffraction (BRUCKER D8 ADVANCE), Electrothermal blowing dry box (150-1B, Shanghai), Electronic balance (BSA423S-CW, Sartorius), Glovebox (FG1200/750 Vigor (CHINA) CO., LTD), MICROTRAC HRA (Topsizer plus, OMCC)

2.2 Synthesis and characterization

Lithium iron phosphate was prepared by carbothermal reduction using lithium dihydrogen phosphate and iron oxide as raw materials. Add citric acid as carbon source. According to the following synthesis steps to prepare. The molar ratio of lithium dihydrogen phosphate (AR), Fe_2O_3(AR), citric acid (AR) were blended in acetone and ball milled under the speed of nearly 400r/minute for 4h, 6h, 8h and 10h, resbectivily. Then, the mixture was dried at 105°C in the oven for 5h. Then the drying materials were put in an atmosphere furnace and was calcined in Ar atmosphere at 300°C for 5h, then 700°C for 10h, the rate of elevated temperature was 3°C/min. The composite electrode material was prepared by atmosphere tube furnace (GELISI Luoyang). Then the particle size of materials was measured by laser particle size analyzer.

The pureness and construction of the samples were tested by X-ray diffraction (XRD) (BRUCKER D8 ADVANCE) using Cu Kα radiation. Scanning electron microscopy (SU8020, Hitachi) was used to observe the particle morphology.

3. Results

3.1 Particle size analysis

Table.1 illustrates the particle size of the $LiFePO_4$/C samples at different balling mix time. It can be seen from Table 1 that the particle size reaches the minimum when the milling time is 4h. And the D50 is 1.50μm at 4h. The reduction of particle size is beneficial to shorten the transport path of lithium ions and allow more facile Li+ ions intercalation/deintercalation process.

Table 1. Particle size of $LiFePO_4$/C samples

Samples	Particle size (μm)		
	D10	D50	D90
LFP-4h	0.78	1.50	1.94
LFP-6h	1.10	2.04	2.12
LFP-8h	1.05	1.90	2.01
LFP-10h	1.24	2.31	2.54

3.2 X-ray diffraction Discussion

Fig. 1 demonstrates the XRD spectra of $LiFePO_4$/C samples prepared at four different ball milling times. Each diffraction peak is consistent with the position that corresponding to the diffraction angle of standard card JCPDF (40-1499). There is no impurity peak in XRD pattern, indicating that the composite electrode materials are pure. In addition, there is no diffraction peak of carbon in the figure, indicating that the amorphous carbon formed after high temperature pyrolysis of citric acid, and the crystal structure of $LiFePO_4$ olivine was not affected.

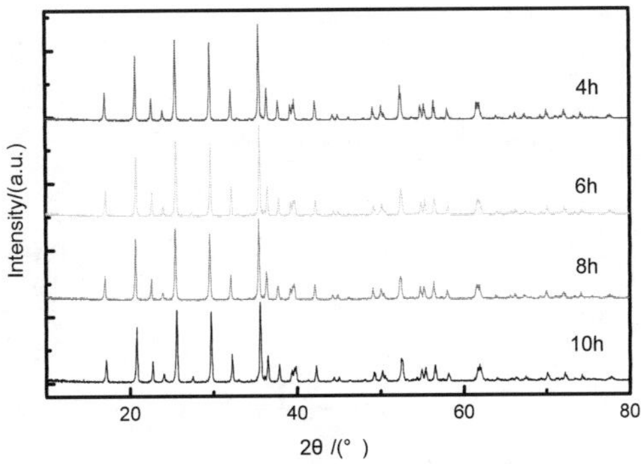

Fig.1 XRD patterns of $LiFePO_4$ at various balling time

3.3 SEM analysis of the material

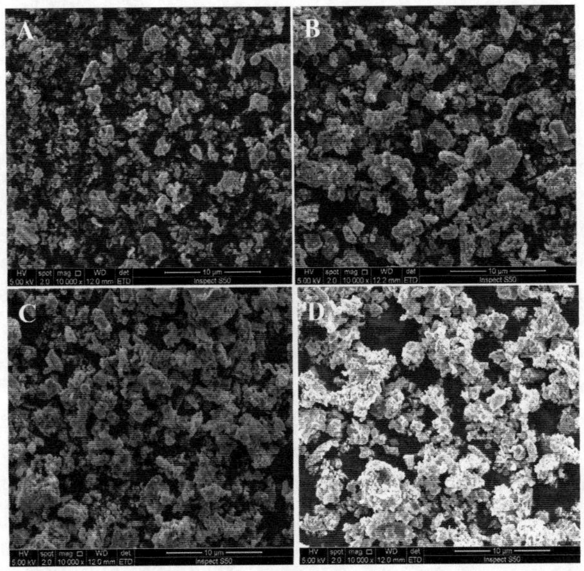

Fig. 2 SEM images of the material (A-4h, B-6h, C-8h, D-10h)

Fig. 2 shows the SEM images of materials obtained at various milling time. The SEM images of $LiFePO_4$ obtained by various ball milling times are composed of irregular particles. From Fig. 2, we can see that the particles in Figure 2A are smaller than those in Fig. 2B 2C and 2D. And the particles in Figure 2A are relatively uniform than other samples.

3.4 Electrochemical performance of the material

Figure. 3 illustrates the initial charge and discharge curves of $LiFePO_4$/C material at different milling time. From the Fig. 3 it can be seen that all samples at 0.2C have a long flat discharge plateau at around 3.4V, which indicates that the two-phase redox reaction proceeds via a first-order transition between $LiFePO_4$ and $FePO_4$[8,9]. The specific discharge capacities of the samples obtained at different milling times are 158.6mAh/g, 157.3mAh/g, 155.6mAh/ g and 152.4mAh/g, respectively. The best milling time is 4h. These results indicate that the ball milling time has a certain effect on the granular size of the sample, thus affecting the electrochemical properties of the sample.

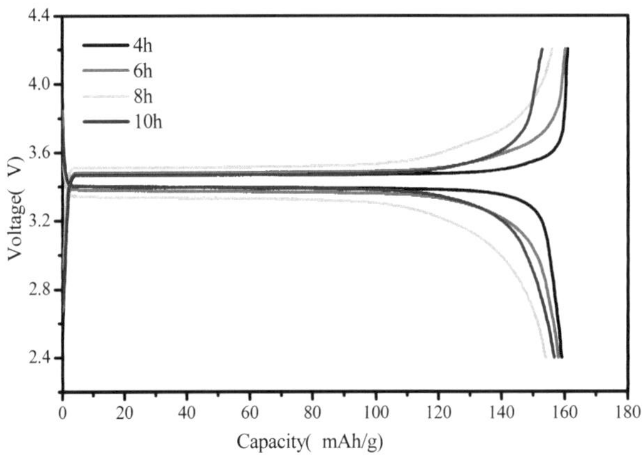

Fig. 3 Fist charge-discharge capacity of the material

4. Conclusion

In this paper, the influence of different ball milling time on the particle size of the material was researched by high temperature solid phase method, using lithium dihydrogen phosphate and iron oxide as raw materials and citric acid as carbon source. The results show that the microstructure and size of the particles are affected by different milling time. And the smaller of the particle size, the more conducive to the deintercalation of the lithium ions. Thus the material with small particle size has better electrochemical properties.

Acknowledgements

This work was supported by the Hebei Province Major Scientific and Technological Achievements Transformation Project (20284401Z) and the Project of Hebei Academy of Science (21708 and 21709).

References

[1] Ciurduc, D. E., Boaretto, N., Fernández-Blázquez, J. P., & Marcilla, R. (2021). Development of high performing polymer electrolytes based on superconcentrated solutions. Journal of Power Sources, 506, 230220.

[2] Lindén, H., Tetzlaff, T., Potjans, T. C., Pettersen, K. H., Grün, S., Diesmann, M., & Einevoll, G. T. (2011). Modeling the spatial reach of the LFP. Neuron, 72(5), 859-872.

[3] Zhao, R., & Zhu, Q. (2021). (Local) Non-Asymptotic Analysis of Logistic Fictitious Play for Two-Player Zero-Sum Games and Its Deterministic Variant. arXiv preprint arXiv:2106.15786.

[4] Choi, N. S., Chen, Z., Freunberger, S. A., et al. (2012) Challenges Facing Lithium Batteries and Electrical Double-Layer Capacitors[J]. Angewandte Chemie International Edition., 51(40): 9994-10024.

[5] Barban, F., Averna, A., Murphy, M. D., Carè, M., Nudo, R. J., Guggenmos, D. J., & Chiappalone, M. (2021, May). LFP based analysis of brain injured anesthetized animals undergoing closed-loop intracortical stimulation. In 2021 10th International IEEE/EMBS Conference on Neural Engineering (NER) (pp. 293-296). IEEE.

[6] Yuan, L. X., Wang, Z.H., Zhang, W. X., et al. (2011) Development and challenges of $LiFePO_4$ cathode material for lithium-ion batteries. Energy Environ. Sci., 4: 269.

[7] Gu, Y., Zhang, X. J., Lu, S. G. (2015) High rate performance of LiF modified $LiFePO_4/C$ cathode material. J. Solid State lonics., 269: 30-36.

[8] Henrie, J. A., & Shapley, R. (2005). LFP power spectra in V1 cortex: the graded effect of stimulus contrast. Journal of neurophysiology, 94(1), 479-490.

[9] Pan, W. J., Thompson, G. J., Magnuson, M. E., Jaeger, D., & Keilholz, S. (2013). Infraslow LFP correlates to resting-state fMRI BOLD signals. Neuroimage, 74, 288-297.

Study on ionic liquid treatment and enzymatic hydrolysis of corn stover after activation of zinc chloride solution

Guoming Zeng[1,2,3], Xin Huang[1], Yan Li[4,a], Da Sun[5*] and Bin Zhou[6,b*]

[1]Chongqing University of Science and Technology, Chongqing 401331, China

[2]China Metallurgical Construction Engineering Group Co., Ltd., Chongqing 643000, China

[3]Chongqing University， Chongqing 40044, China

[4]School of Pharmacy, Taizhou polytechnic college, Taizhou 225300, China

[5]Wenzhou University ,Wenzhou 325035, China

[6] GENERAL RESEARCH INSTITUTE OF ARCHITECTURE & PLANNING DESIGN CO.,LTD,CHONGQING UNIVERSITY, Chongqing 400045, China

*Corresponding author: 2017015@cqust.edu.cn

Abstract. With the intensification of the energy crisis, this thesis uses corn stalk cellulose as the raw material to carry out research on biomass energy. By designing the response surface experiment, a multiple quadratic regression equation prediction model of the response surface factors and the yield of enzymatic reducing sugars after the activation of the zinc chloride solution treated by the ionic liquid was established, which provided the technology for the ionic liquid treatment of the metal ion solution to activate the corn stalk raw materia reference. The experimental results show that the yield of enzymatically hydrolyzed reducing sugars of corn stalks after being activated by metal ions and then treated with ionic liquids is greatly improved compared to the raw materials.

1. Introduction

With the increasingly tight global energy supply, lignocellulosic resources are considered to be an ideal third-generation energy source that can replace traditional fossil energy[1]. Approximately 640 million tons of crop stalks are left on arable land in China each year. Therefore, the use of high-lignocellulose biomass to develop biomass energy conversion technology has great potential. The special structure of lignocellulosic biomass and the highly ordered dense crystalline structure seriously restrict its enzymatic saccharification rate[2]. As a new type of environmentally friendly organic solvent, ionic liquid has special physical and chemical properties, which can destroy the inherent compact structure of lignocellulosic biomass and improve its biodegradability[3]. It has broad development prospects in the pretreatment of lignocellulosic biomass.

In order to improve the effect of the ionic liquid in treating cellulose, the cellulose raw material must undergo activation treatment. Metal ion solution activation pretreatment technology is a method that has been studied more in recent years[4]. This method has high efficiency and low energy, and can effectively separate the components in the pretreatment process. A single pretreatment method has limited damage to the corn stalk structure and cellulose crystallinity. Therefore, the response surface

Journal of Physics: Conference Series **2011** (2021) 012079 doi:10.1088/1742-6596/2011/1/012079

method was used to optimize the ionic liquid treatment of corn stover cellulose after the activation of the metal ion solution. Through statistical optimization experiments on the three factors, a better pretreatment effect was obtained and basic parameter guidance was provided for the amplification of related processes[5].

2. Materials and methods

2.1. Ionic liquid treatment experiment

The corn stalk raw materials are air-dried naturally and set aside. The corn stover raw materials with a particle size of 60 meshes are processed according to the optimal conditions of the metal ion solution pretreatment process. After filtration, the materials treated by the metal ion solution are collected, rinsed with deionized water five times, and then dried at 105°C for use. Finally, the dried material is processed by the newly developed ionic liquid method.

2.2. Optimization of enzymatic hydrolysis experiments by response surface method

In the reaction process of ionic liquid treating metal ion solution to activate corn stalk raw materials, since the yield of reducing sugar after enzymatic hydrolysis is related to the mass fraction of substrate, reaction temperature, and reaction time, the three of the three are selected as the optimization experiment of enzymatic response surface three factors. First of all, single-factor experiments are carried out on these three factors to determine their appropriate scope of action. According to the results of the single factor experiment, the enzymatic hydrolysis experiment was carried out on the corn stalk raw material after the activation of the metal ion solution by the ionic liquid according to the response surface experiment design of 3 factors and 3 levels.

3. Results and discussion

3.1. Response surface method to optimize experimental results

Taking reducing sugar yield as the response value, according to the response surface design experiment scheme, the ionic liquid treatment of the activated metal ion solution was carried out for enzymatic hydrolysis experiments. The experimental results are shown in Tables 3.1, 3.2, and 3.3.

Table 1. Experimental design and response of RSM

Test Numbering	Factor			Reducing sugar yield (mg·g-1)
	X1	X2	X3	
1	-1	1	0	614.67
2	0	0	0	689.74
3	-1	-1	0	570.12
4	0	1	1	679.52
5	1	-1	0	660.57
6	1	1	0	669.73
7	-1	0	-1	576.92
8	0	-1	1	648.56
9	0	0	0	690.12
10	-1	0	1	616.47
11	0	0	0	691.23

12	1	0	1	668.93
13	0	1	-1	657.89
14	0	-1	-1	579.64
15	1	0	-1	667.82

Table 2. Significance test of each coefficient of developed quadratic regression model

Item	Coefficient	Standard error	t	p
Constant	690.363	6.821	101.208	0.0001
X1	36.109	4.177	8.644	0.0001
X2	20.365	4.177	4.875	0.0005
X3	16.401	4.177	3.926	0.0112
X1X1	-35.229	6.149	-5.730	0.0021
X2X2	-26.362	6.149	-4.287	0.0082
X3X3	-22.599	6.149	-3.676	0.0141
X1X2	-8.847	5.907	-1.498	0.0193
X1X3	-9.610	5.907	-1.627	0.0454
X2X3	-11.822	5.907	-2.001	0.0323

Table 3. ANOVA analysis for regression equation

Source of Variance	Degree of freedom	sum of squares	Corrected sum of squares	Modified variance	F	p
return	9	25039.1	25039.1	2782.12	19.93	0.002
Linear	3	15900.6	15900.6	5300.20	37.97	0.001
square	3	7896.9	7896.9	2632.30	18.86	0.004
Interaction	3	1241.6	1241.6	413.87	2.96	0.136
Residual error	5	697.9	697.9	139.59		
Lack of fit	3	696.7	696.7	232.25	387.44	0.003
Pure error	2	1.2	1.2	0.60		
total	14	25737.0				

Take the enzymatic hydrolysis reducing sugar yield Y as the response value, X1 (substrate mass fraction), X2 (reaction temperature), and X3 (reaction time) as independent variables. After processing by Design-Expert.8.0 software, ionic liquid treated chlorine is obtained. The multiple quadratic regression equation of the response surface factors and the yield of enzymatically hydrolyzed reducing sugar after activation of the zinc chloride solution is:

$$Y=690.36+36.11X1+20.37X2+16.40X3-35.23X1X1-26.36X2X2-22.60X3X3-8.85X1X2-9.61X1X3-11.82X2X3$$

4. Conclusion

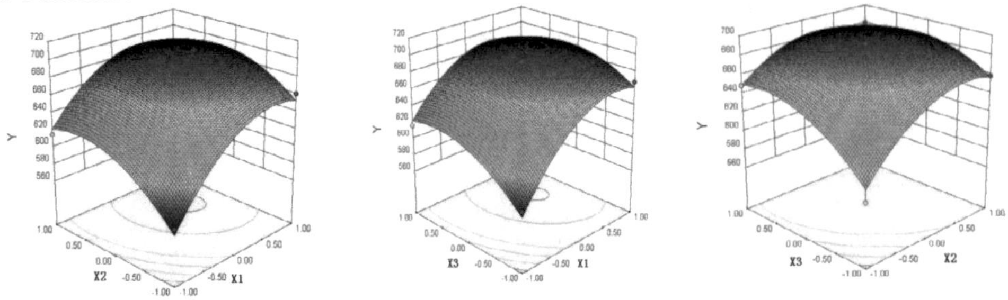

Figure 1. Response surface plot on the cellulose solution of pretreated corn stover
(a)Y and X2，X1 surface map (b)Y and X1，X3 surface map (c)Y and X2，X3 surface map

In summary, low-level or high-level influencing factors are not conducive to the improvement of enzymatic hydrolysis yield. When the substrate mass fraction is 2%-4%, the reaction temperature is 80-100°C, and the reaction time is 4-6h. From time to time, the enzymatic hydrolysis yield is the highest. The optimal conditions are: when the substrate mass fraction is 3.8%, the reaction temperature is 103°C, and the reaction time is 357min, the maximum theoretical value (Y) of reducing sugar yield is 702.85mg/g. In order to verify the accuracy of the model, a verification test was carried out under these conditions, and the yield of reducing sugar after enzymatic hydrolysis was 696.34mg/g, which was close to the predicted result of the model, indicating that the model was effective. At the same time, compared with the untreated raw material, the enzymatic hydrolysis reducing sugar yield increased by 154.14%.

Acknowledgments
This work was supported by the National Natural Science Foundation of China (Grant No 51808086),China Postdoctoral Science Foundation(Grant No 2020M673112), Taizhou Science and Technology Support Program (Agriculture) Project (TN202012), Wenzhou Science and Technology Bureau Project(Grant No S2020005) and Provincial and Ministerial Co-constructive of Collaborative Innovation Center for MSW Comprehensive Utilization.

References
[1] Prasad, A., Sotenko, M., Blenkinsopp, T and Coles, S.R. (2016) Life cycle assessment of lignocellulosic biomass pretreatment methods in biofuel production. The International Journal of Life Cycle Assessment, 21(1):44-50.

[2] Satari, B., Karimi, K. and Zamani, A. (2016) Oil, chitosan, and ethanol production by dimorphic fungus Mucor indicus from different lignocelluloses[J]. Journal of Chemical Technology & Biotechnology, 91(6):1835-1843.

[3] Zhang H,Wu J,Zhang J,He J, et al.1-allyl-3-methylimidazolium chloride room temperature ionic liquid:a new and powerful nonderivatizing solvent for cellulose[J], Macromolecules, 2005, 38,8272-8277.

[4] Kubisa P.Application of ionic liquids as solvents for polymerization processes[J]，Progress in Polymer Science,2004,29:3-12.

[5] Yukinobu F，Kensaku H，Masahisa W，Hiroyuki O, et al.Cellulose dissolution with polar ionic liquids under mild conditions:required factors for anions[J],Greenchem,2008,10:44-46.

Study on Corrosion Characteristics of Grounding Materials in High Corrosion Area

Mingxiang Huang[1*] Jianxun Zhang[1]

[1] State Grid Fujian Electric Power Co., Ltd. Construction Branch, Fuzhou, Fujian Province, 350012, China

[*]Corresponding author's e-mail: mingxianghuang@fj.sgcc.com.cn

Abstract. Due to the differences in the causes of corrosion, the soil corrosion characteristics in coastal mountains and coastal areas may be quite different. As a result, there are differences in the corrosion causes, corrosion status and service life of the grounding grids of the transmission line towers built in coastal areas. Based on the difference in apparent resistivity of the soil under the transmission line towers in a coastal area with high corrosion, this paper selects 12 soil samples for physical and chemical analysis. The analysis results show that there are two typical soils of acid and weak alkaline in the transmission line erection area, which are highly corrosive but with different corrosion mechanisms. Then the soil liquids of two typical soils were extracted, and the polarization curve test of carbon steel and galvanized steel, which are commonly used in the grounding grid of poles and towers, was carried out for 7 consecutive days, and the corrosion laws of carbon steel and galvanized steel were obtained. The results show that the corrosiveness of acidic soil is much greater than that of weakly alkaline soil. The corrosion rates of carbon steel and galvanized steel in acidic soil are 3.7 times and 5.2 times that of weakly alkaline soil, respectively. In addition, using the extracted soil liquid, a 30-day accelerated corrosion experiment was performed on carbon steel and galvanized steel, and an empirical formula for the change of corrosion weight with time was obtained. Combined with the surface morphology characteristics and corrosion rate after corrosion, this paper analyzed corrosion mechanism of two grounding materials. The research results of this paper provide a reference for the selection and application of grounding materials.

1. Introduction

The grounding grid is an important part of lightning protection. The grounding grids of substations and transmission line towers are responsible for the lightning protection of the equipment and lines in the station, and are valued by the design and production and operation departments [1]. The grounding grid buried in the soil is affected by different types of soil corrosion, including chemical corrosion, electrochemical corrosion, microbial corrosion, and stray current corrosion caused by power equipment leakage [2,3]. Corrosion products often degrade the flow performance of the grounding grid, and even cause the grounding grid to break [4-6]. The survey results show that the annual corrosion rate of the grounding grid metal can reach 1.0～2.0mm/a in the corrosive soil, and the highest is 3.0～4.0mm/a in the corrosive soil [7,8]. The corrosive soil medium can be as high as 8.0mm/a. In these areas, the corrosion of the grounding grid is an important factor affecting the safe operation of the power system [9.10].

At present, there are a large number of researches on the corrosion performance of grounding materials, and the research object is mainly carbon steel. There is not much research on the corrosion mechanism and behavior of galvanized steel. Literature [11-13] studied the effects of temperature and humidity on the self-corrosion potential, corrosion products and charge transfer resistance of carbon steel, and established the law of the influence of various physical and chemical properties of soil on the corrosion of carbon steel. Literature [14-17] compared the corrosion rate of galvanized steel with other materials, and studied the electrochemical corrosion process of galvanized steel, but the corrosion mechanism of galvanized steel in the soil environment has not been fully studied. Due to the large differences in soil composition in different regions, the differences in corrosion performance are also obvious.

In most coastal areas, thunderstorms are frequent, and the grounding grid is more important. However, for areas with more coastal and mountainous areas, the same transmission line tower will pass through beaches and mountains. Due to different soil causes, the corrosion mechanism is different, and the corrosion rate and service life of the grounding grid of the tower in different areas are also different.

According to the distribution of apparent resistivity of the soil under each base pole and tower, this paper selects 12 samples for soil sampling and analyzes the physical and chemical properties of the soil samples in the area passing through a transmission line along the coast; Two typical soils were selected based on the analysis results of the physical and chemical properties of 12 samples. The soil corrosion liquid was extracted from the soil, and the polarization curve experiment was carried out on carbon steel and galvanized steel to study the corrosion characteristics of the two grounding materials in the soil liquid; A 30-day indoor accelerated corrosion experiment was carried out on carbon steel and galvanized steel in two soils, and the weight loss analysis was carried out to obtain the corrosion rate, and the surface morphology of the corroded sample was analyzed. Finally, the corrosive behavior of two soils on grounding materials is summarized, which provides data support for the design and protection of grounding grids in coastal areas with high corrosion.

2. Soil sample analysis and selection

According to the actual measurement results of soil resistivity in a coastal area, the transmission line passes through areas such as coastal beaches and coastal mountains, and the soil resistivity is between 4 Ω·m and 4345 Ω·m. According to the distribution of soil apparent resistivity and topographical characteristics, this paper selected 12 sampling points for soil sampling. When sampling the soil, the ring knife method [17] was used to sample the soil 80 cm away from the surface of the sampling point. Analyze the physical and chemical characteristics of the collected 12 groups of soils in accordance with the standards of soil testing [18], and analyze the content of nitrate nitrogen, carbonate, bicarbonate, chloride, sulfate, water content and pH as shown in Table 1.

Table 1. Analysis results of physical and chemical properties of soil samples

Sample number	Soil resistivity Ω·m	Nitrate mg/kg	Carbonate mg/kg	Bicarbonate mg/kg	Chloride mg/kg	Sulfate mg/kg	Water content %	pH
1	1181	4.38	<3	68.8	0.02	344.4	21.5	4.5
2	1634	2.73	<3	62.5	0.02	35.6	23.7	4.8
3	450	6.87	<3	48.8	0.03	51.5	20.7	4.7
4	751	0.58	<3	46.3	0.02	31.7	21.4	4.9
5	452	<0.25	<3	60	0.03	91	19.9	4.8
6	452	1.47	<3	51.3	0.03	112.9	19.5	4.8
7	314	7.58	<3	61.3	0.03	126.6	19.2	5.1
8	7	<0.25	<3	169	4.09	722.5	29.4	8.5

9	2393	<0.25	<3	83.7	0.03	63.3	7.2	7.2
10	4	0.37	<3	158	0.48	245.6	19.1	7.8
11	4345	0.33	<3	157	0.48	146.5	23.9	7.6
12	4345	4.63	<3	158	0.11	91.1	27.8	7.8

According to the analysis results of soil physical and chemical properties, the 12 samples can be divided into two types: acidic soil (No. 1-7) and weakly alkaline soil (No. 8-12) according to the pH value. Acidic soils are mainly distributed in mountainous areas, hills and areas far away from the sea, and weakly alkaline soils are mainly distributed in offshore areas. Except for No. 5 soil and No. 12 soil, the nitrate nitrogen content of acid soil is basically higher than that of weak alkaline soil. The content of carbonate ions in the 12 soil samples was roughly the same. The bicarbonate ion content of acidic soil is significantly lower than that of weakly alkaline soil. The chloride ion content of acidic soil is also significantly lower than that of weakly alkaline soil. Soil No. 8 is closest to the sea. The seawater can inundate this area at high tide, so its chloride ion content is much higher than other soils. Except for the No. 9 soil sample, the soil moisture content is about 20%.

Among the 12 soil samples, No. 8 soil has the highest pH value, the strongest alkalinity, and contains the most carbonate ions, chloride ions, and sulfate ions, while No. 1 soil has the lowest pH value. This paper selects No. 1 soil and No. 8 soil as the test soil for grounding materials.

3. Analysis of polarization characteristics of grounding materials

The polarization characteristics of metal materials can effectively reflect the electrochemical reaction rate of metals in the electrolytic cell. This paper extracts the soil solution and uses an electrochemical workstation to test the polarization characteristics of the materials.

3.1. Principle of the experiment

For electrochemical corrosion, according to the principle of corrosion potential, when the potentials of the two local reactions are balanced, the relationship between the polarization current I and the electrode potential E is shown in equation (1) [7].

$$I = I_k \{\exp[\frac{2.3(E - E_k)}{b_a}] - \exp[\frac{2.3(E_k - E)}{b_c}]\} \tag{1}$$

Among them, I is the measured polarization current; E is the polarization potential; I_k is the natural corrosion current; E_k is the natural corrosion potential; b_a is the anode Tafel constant; b_c is the cathode Tafel constant. Taking the logarithm of I, in the linear section of the polarization curve, the polarization potential has a linear relationship with the applied current, and two straight lines, namely the Tafel straight line, can be obtained. The self-corrosion potential and self-corrosion current at the intersection point are obtained by extrapolating the intersection of Tafel straight lines in the strong polarization section of the polarization curve [19].

3.2. Polarization curve experiment platform

Most of the metal corrosion process is electrochemical corrosion. The electrochemical test of metal corrosion is fast and sensitive, can reflect the instantaneous information of metal corrosion, and can also detect the surface condition of metal electrodes for a long time. Before the experiment, we wipe the corrosion test piece with absolute alcohol, dry it, weigh it, and keep it sealed. The size of the corrosion piece is 50×25×2mm, and the type is Q235 carbon steel and galvanized steel coupons (the thickness of the galvanized layer is 8~10μm). The acid soil and the weakly alkaline soil are crushed, sieved, dried in a constant temperature drying oven at 70°C for 24 hours, and sealed for storage. According to the standard, we take the dried soil, add deionized water according to the ratio of soil and deionized water 1:5, stir evenly, take the supernatant after standing, and put them in two containers A1 and A2 respectively for experiment. The carbon steel sheet and the galvanized steel sheet are placed in a sealed container, and the sealed container is placed in a constant temperature and humidity box with a temperature of 70°C and a relative humidity of 20%, and the corrosion test is carried out for 7 consecutive days. The corroded sheet was taken out at a fixed time every day, and the polarization curve experiment was performed on the test bench shown in Figure 1.

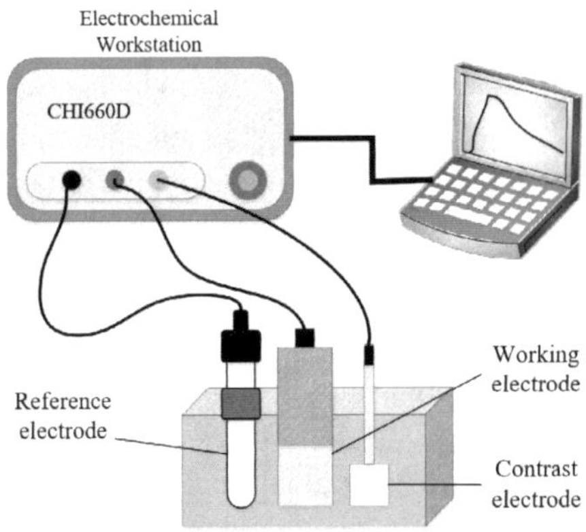

Figure 1. The schematic figure of experiment platform for polarization curve

The working electrode is the piece to be tested. The diameter of the wire is 2.5mm, and the copper wire is wound around the missing hole of the test piece. The wire and the entire test piece are wrapped by the insulating tape, and cuts the electrode working surface of 10mm×10mm. The auxiliary electrode is a platinum electrode with a size of 20mm×20mm×0.1mm, which provides a current loop for the external excitation signal. The reference electrode is a saturated calomel electrode [20] (the salt bridge solution is saturated potassium chloride), which provides a stable potential reference point for the three-electrode system. The working surface of the working electrode is horizontally opposite to the platinum electrode at a distance of 5mm. The saturated calomel electrode of the reference electrode is placed in the third position at a distance of 5mm in parallel. Put the working electrode, platinum chip electrode and saturated calomel electrode with a working area of 1cm2 into a customized 250ml electrolytic cell with an electrode slot reserved on the electrolytic cell.

3.3. Experimental results and analysis

A 7-day polarization curve experiment was performed on carbon steel sheets and galvanized steel sheets of the same size in acidic soil liquid and weakly alkaline soil liquid, respectively. Figure 2 shows the polarization curves of two grounding materials on the first and sixth days of corrosion in acidic soil liquid. Figure 3 shows the polarization curves of two grounding materials on the first and sixth days of corrosion in a weakly alkaline soil solution. The solid line is carbon steel and the dotted line is galvanized steel.

In the acidic solution, in terms of self-corrosion potential, on the first day of corrosion, carbon steel is the highest and galvanized steel is the lowest. On the 6th day of corrosion, galvanized steel was the highest and carbon steel was the lowest. As the corrosion progresses, the self-corrosion potential of carbon steel gradually decreases, and the self-corrosion potential of galvanized steel is basically similar. The lower the self-corrosion potential, the easier the corrosion is. In the self-dissolution process of galvanized steel, the standard electrode potential of the oxidation-reduction reaction of the oxidant Zn is lower than that of Fe. Therefore, the metal oxidation reaction on Zn is dominant during the corrosion process and becomes the anode zone. The dissolution reaction of zinc occurs first. Therefore, the self-corrosion potential of galvanized steel is lower than that of carbon steel on the first day. At this time, the corrosion reaction of galvanized steel is the galvanized layer. As the corrosion progresses, the galvanized layer of the galvanized steel is gradually corroded, and the exposed carbon steel in the galvanized steel begins to corrode in the acid soil solution. Since the self-corrosion potential of carbon steel is higher than that of galvanized steel, this increases the self-corrosion potential of galvanized steel to some extent, and the self-corrosion potential of galvanized steel gradually approaches carbon steel. When the galvanized layer is completely dissolved, only carbon steel is left in the galvanized steel that participates in the corrosion reaction. And at this time, the iron in the galvanized steel participates in corrosion for a shorter time than that in the carbon steel. Therefore, on the sixth day, the self-corrosion potential of the galvanized steel is already higher than that of the carbon steel.

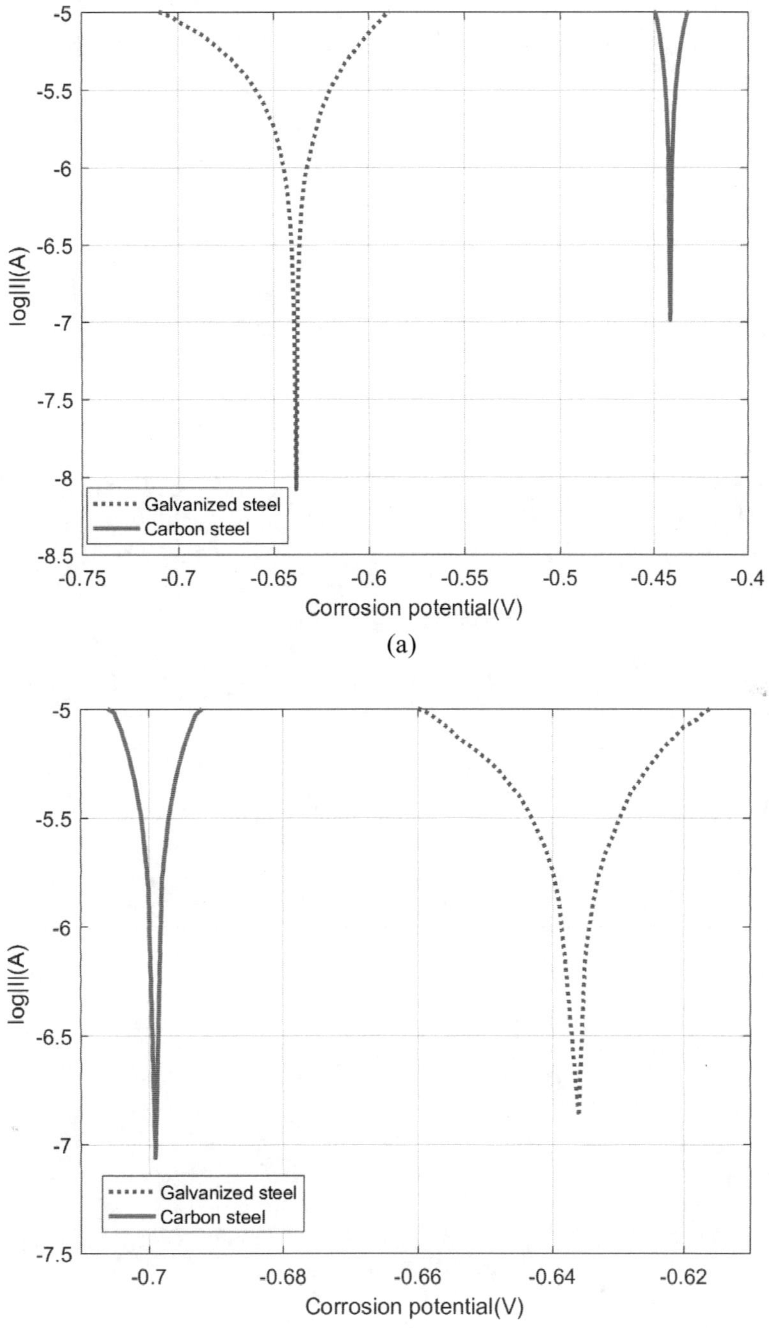

Figure 2. Polarization curves of two types of grounding materials in acidic soil fluids, (a) the first day, (b) the sixth day

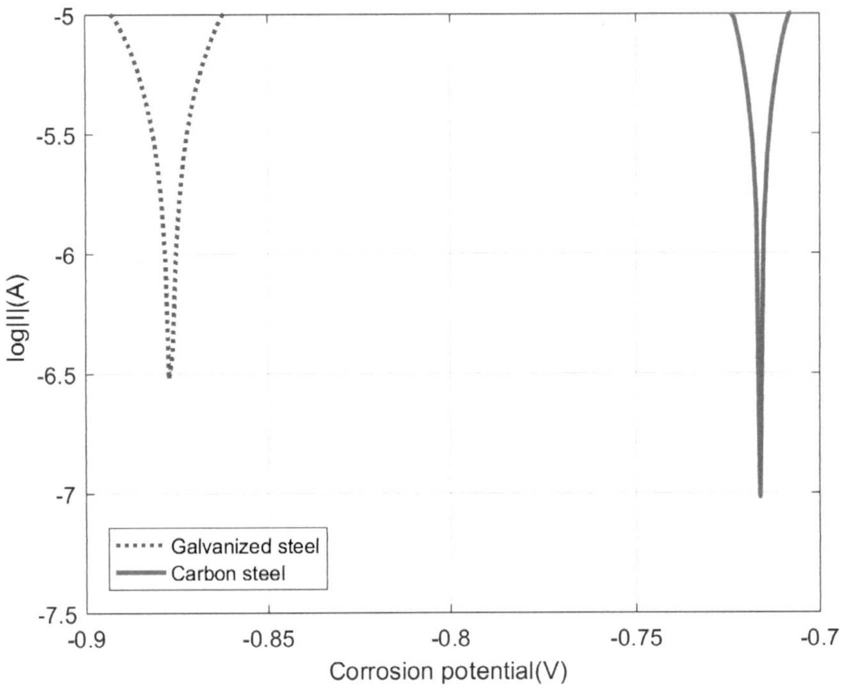

(a)

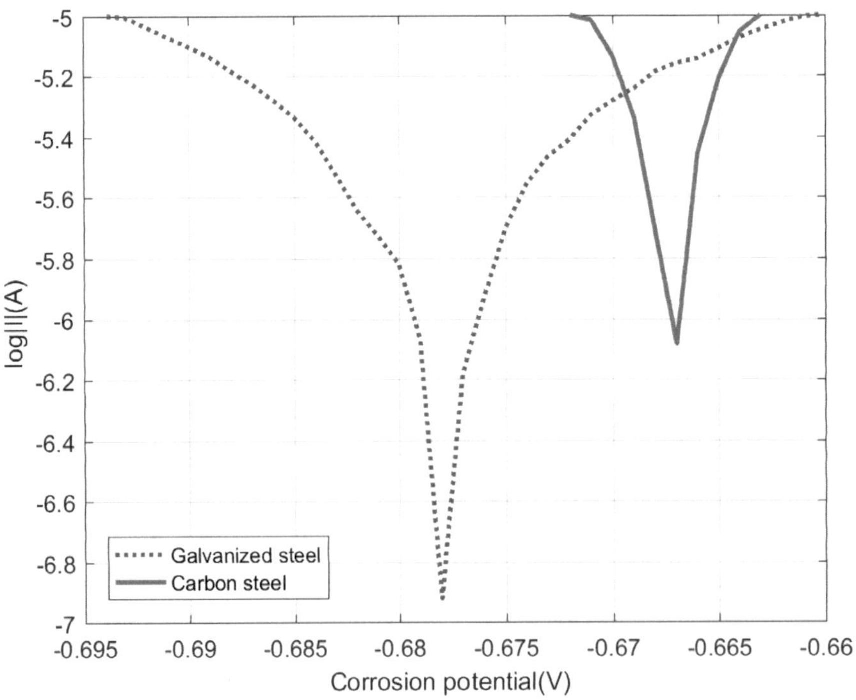

(b)

Figure 3. Polarization curves of two types of grounding materials in weak alkaline soil fluids, (a) the first day, (b) the sixth day

In the weak alkaline solution, the self-corrosion potential of carbon steel is higher than that of galvanized steel from day 1 to day 6 of corrosion. As the corrosion progresses, the self-corrosion potential of both carbon steel and galvanized steel increases. This is because the galvanized layer is not completely corroded during the corrosion process, that is, the corrosion rate in weakly alkaline soil is lower than that in acid soil. The increased self-corrosion potential of carbon steel may be due to the corrosion products of carbon steel in the weakly alkaline soil solution inhibiting corrosion.

According to the self-corrosion potential, whether in acidic soil liquid or alkaline soil liquid, the order of corrosion is zinc before iron. The anti-corrosion principle of galvanized steel is to sacrifice the anode to protect the cathode, that is, to let the galvanized layer undergo corrosion reaction first to protect the inner steel.

4. Weightlessness analysis of accelerated corrosion of grounding materials

4.1. Principles of Weightlessness Analysis

The weight loss method calculates the metal corrosion rate according to the change in the mass of the metal sample measured before and after corrosion, and the average corrosion rate can be obtained by the weight loss method [21]:

$$v = \frac{m_0 - m_1}{S \cdot t} \tag{2}$$

In the formula, v represents the corrosion rate of the metal, g/(m2•h); m_0 is the mass of the specimen before corrosion, g; m_1 is the mass of the specimen to remove the corrosion products after corrosion, g; S is the surface area of the test piece in a corrosive environment, m^2; t is the corrosion time of the specimen, h.

4.2. Indoor accelerated corrosion experiment design

The weightlessness method is a classic method suitable for laboratory and field corrosion tests. It is the most reliable and closest method to the actual conditions for determining the corrosion rate of metals. The weightlessness method calculates the average corrosion rate of metals based on the changes in the mass of the sample before and after corrosion. Before the experiment, we wipe the corrosion test piece with absolute alcohol, dry it, weigh it, and keep it sealed. The size of the corrosion piece is 50×25×2mm, and the type is Q235 carbon steel and galvanized steel coupons (the thickness of the galvanized layer is 8~10μm). The acid soil and the weakly alkaline soil are crushed, sieved, dried in a constant temperature drying oven at 70°C for 24 hours, and sealed for storage. According to the standard, we take the dried soil, add deionized water according to the ratio of soil and deionized water 1:5, stir evenly, take the supernatant after standing, and put them in two containers A1 and A2 respectively for experiment. The carbon steel sheet and the galvanized steel sheet are placed in a sealed container, and the sealed container is placed in a constant temperature and humidity box with a temperature of 70°C and a relative humidity of 20%. The accelerated corrosion test is carried out, and the test period is 30 days. Add water to the soil at a fixed time every day to ensure that the soil moisture content is 20% every day. The corroded pieces are taken out and photographed and kept on the 1st, 2nd, 3rd, 4th, 5th, 6th, 7th, 10th, 15th, 20th, 25th, and 30th days. All corroded pieces are derusted in a mixture of urotropine and 50% hydrochloric acid [5]. The corroded pieces after rust removal are washed with deionized water, dehydrated in absolute ethanol, and then dried in a constant temperature oven at 70°C and weighed. According to the weight before and after corrosion, the corrosion rate of each corroded piece is calculated. The schematic diagram of the experimental platform of the indoor accelerated corrosion experiment is shown in Figure 4.

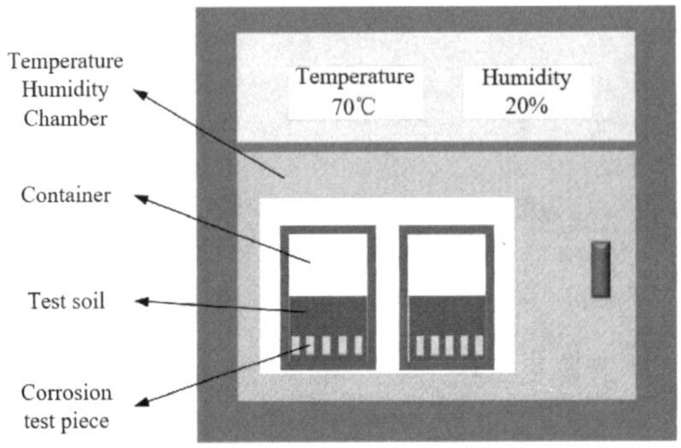

Figure 4. Schematic diagram of indoor accelerated corrosion test platform

4.3. Experimental results and analysis

Keeping the temperature in the constant temperature and humidity box at 70°C and the soil moisture content of 20%, after 30 days of accelerated corrosion experiments in the room, the carbon steel sheets and galvanized steel sheets were subjected to rust removal and weight loss analysis. The results are shown in Figure 5. Among them, the red is carbon steel; the green is galvanized steel; the scattered points are the actual measurement results of weight loss; the solid line is the curve fitting according to the power exponential function. The figure (a) is the result in acidic soil, and the figure (b) is the results in alkaline soil. The corrosion weight of metals in the soil can be fitted with a power exponential function [22], and the fitting formula is:

$$M = Dt^n \quad (3)$$

Among them, M is the reduced mass, the unit is g; t is the corrosion time, the unit is d; D and n are both fitting constants, and the fitting parameters are shown in Table 2.

Table 2. Fitting parameters of grounding materials weightlessness after accelerated corrosion experiment

Soil	Material	D	n	Goodness of fit R
Acid soil	Carbon steel	0.05379	1.009	0.9875
	Galvanized steel	0.015254	1.473	0.9981
Weakly alkaline soil	Carbon steel	0.04788	0.855	0.9975
	Galvanized steel	0.03424	0.7587	0.9939

According to the fitting results, the weight loss of carbon steel and galvanized steel after accelerated corrosion in acid soil and alkaline soil satisfies the power exponential function, and the goodness of fit is good. For carbon steel, the coefficient D and power exponent n of the fitting function in acid soil are both larger than those in weak alkaline soil, which indicates that the corrosion rate is faster in acid soil; similar results are also obtained for galvanized steel. The weight loss of carbon steel is greater than that of galvanized steel from the beginning, whether it is in acid soil or weakly alkaline soil. That is to say, carbon steel is more prone to corrosion than galvanized steel in acid soil or weak alkaline soil.

After 30 days of indoor accelerated corrosion experiment, total weight loss (g), average corrosion rate (g/m2·h), and annual corrosion depth (mm/a) of the two grounding materials of carbon steel and galvanized steel are shown in Table 3 respectively.

Table 3. Comparison of grounding materials corrosion rate after accelerated corrosion experiment

Soil	Material	Total weightlessness (g)	Average corrosion rate ($g/m^2 \cdot h$)	Annual average corrosion depth (mm/a)
Acid soil	Carbon steel	3.1576	1.5663	1.7933
	Galvanized steel	2.2590	1.1205	1.2219
Weakly alkaline soil	Carbon steel	0.8623	0.4277	0.4988
	Galvanized steel	0.4326	0.2146	0.2333

No matter in acid soil or weak alkaline soil, the average annual corrosion depth and average corrosion rate of carbon steel are higher than the corresponding values of galvanized steel. The corrosion rate of carbon steel in acid soil is about 1.4 times that of galvanized steel. The corrosion rate of carbon steel in weak alkaline soil is about twice that of galvanized steel. The corrosion rate of carbon steel in acid soil is about 3.7 times that of alkaline soil, and the corrosion rate of galvanized steel in acid soil is about 5.2 times that of alkaline soil. Therefore, the corrosion resistance of galvanized steel is better than that of carbon steel no matter in acid soil or weak alkaline soil, and whether it is carbon steel or galvanized steel, the corrosion resistance of acid soil is much stronger than that of weak alkaline soil.

5. Analysis of products after accelerated corrosion of grounding materials

After 30 days of accelerated corrosion tests on carbon steel and galvanized steel in acidic and weakly alkaline soils with a temperature of 70°C and a soil humidity of 20%, the types and contents of the corrosion products on the surface of the corroded sheet are analyzed respectively by the 550i X-rays energy spectrum analyzer (EDS), and the corrosion product composition of the corroded piece is analyzed by D8 Advance X-ray diffractometer (XRD). The element content of surface corrosion products is shown in Table 4.

Table 4. Element content (%) of corrosion products on the surface of grounded materials after indoor accelerated corrosion test

Soil	Material	C	O	Al	Si	S	Cl	K	Ca	Mn	Fe	Zn
Acid soil	Carbon steel	2.272	20.206	3.453	7.980	0	0	0.492	0.383	4.384	60.829	0
	Galvanized steel	14.761	45.335	4.187	7.878	0.222	0.345	0.476	0.263	0.739	19.125	7.560
Weakly alkaline soil	Carbon steel	3.549	26.058	2.744	5.687	0	0	0.569	14.091	1.052	54.106	0
	Galvanized steel	2.024	17.190	1.171	2.009	0	0.274	0	0	0.788	56.451	20.232

Among the corrosion products of carbon steel, the most abundant elements are Fe and O, and a small amount of C, Al, Si, K, Ca, Mn and other elements from the soil or matrix. No matter in acid soil or weak alkaline soil, the content of Fe and O together exceeds 80%, so the corrosion products of carbon steel are mainly Fe oxides. Among the corrosion products of galvanized steel, the most content elements are Fe, Zn, O, and a small amount of C, Al, Si, K, Ca, Mn, Cl, S and other elements from the soil or matrix. In acid soils, the content of O is the highest, while the content of Fe in weakly alkaline soils is the highest. This indicates that the corrosion products in acid soils are mainly Fe oxides with higher oxygen content, while Fe oxides and Zn oxides dominate in weakly alkaline soils.

6. Conclusion

In this paper, two types of highly corrosive typical soils in a coastal area are tested on the corrosion characteristics of two grounding materials. The electrochemical characteristics of the soil extract are measured, and the accelerated corrosion experiment of grounding materials is carried out. The weightlessness analysis, surface morphology analysis and corrosion product analysis of the corroded grounding materials are carried out. The main conclusions of this paper are as follows:

(1) These two typical highly corrosive soils have obvious differences in physical and chemical properties, so the corrosiveness of grounding materials is obviously different. The corrosiveness of

acidic soil is much greater than that of weakly alkaline soil. The corrosion rates of carbon steel and galvanized steel in acid soil are 3.7 times and 5.2 times that in weak alkaline soil, respectively.

(2) The results of accelerated corrosion experiments on carbon steel and galvanized steel show that in acid soil, the average corrosion rate of carbon steel is about 1.7 times that of galvanized steel, and in weak alkaline soil, the average corrosion rate of carbon steel is about twice that of galvanized steel. On this basis, the fitting curve of corrosion weight loss of galvanized steel and carbon steel in typical soil is given.

(3) After being corroded by acid soil, the corrosion products of carbon steel appear as large lumps and layers, which are uniform corrosion, while the corrosion products of galvanized steel show larger and deeper pores locally, and the type of corrosion is between uniform corrosion and pitting corrosion. After being corroded in weakly alkaline soil, the corrosion products of carbon steel show partial stratification, which belongs to incomplete uniform corrosion, while the corrosion products of galvanized steel show dense small cavities, which belong to pitting corrosion. The corrosion products of carbon steel are mainly Fe_2O_3 and its water compound FeOOH. In addition to Fe_2O_3 and its water compound FeOOH, the corrosion products of galvanized steel also include ZnO and its hydrate $Zn(OH)_2$.

Acknowledgement
This work was supported by Science and technology project of State Grid Fujian Electric Power Co. "Application of new grounding materials in high corrosive soil environment of Fujian coastal area".

Reference
[1] Zhao Z. (2006) Analysis of grounding grids buried in complex soil. North China Electric Power University. Beijing.
[2] YANG T. (2011) The Research on Corrosion Detection and Life Prediction for Grounding Grids. Hunan University. Changsha.
[3] Xiao W., Hu Y., Ruan J., Zhan Q., Huang D. Flexible Graphite Composite Electrical Grounding Material and Its Grounding Application Features. (2017) Transactions of China Electrotechnical Society, 32: 85-94.
[4] Xu L., Li L. Fault diagnosis for grounding grids based on electric network theory . (2012)Transactions of China Electrotechnical Society, 27: 270-276.
[5] Chen Y. (2009) Soil Accelerated Corrosion And Protection Of Grounding Net. Xi'an University of Technology, Xi'an.
[6] Ji Y. (2013) Research on the Corrosion of Substation Grounding Grids by Electrochemical Measurement Methods. North China Electric Power University, Beijing.
[7] Lu P. J., Huang S.B., Dou P. Corrosionstatus of substation grounding grid in Foshan area. (2008) High Voltage Engineering, 34: 1996-1999.
[8] Wen X. S., Song Z., Tan B. Electrical properties of copper-clad steel grounding conductors. (2015) High Voltage Engineering, 41: 608-614.
[9] Yuan T, Lei C. P., Sima W.X. Analysis of grounding resistance reduction effect based on enhancing impulse current leakage efficiency. (2012) Transactions of China Electrotechnical Society, 27: 278-284.
[10] Liu Y., Cui X., Zhao Z. Design and application of testing magnetic field system for corrosion diagnosis of grounding grids in substation. (2009) Transactions of China Electrotechnical Society, 24: 176-182.
[11] Cao J. F. (2007) Effects of temperature and water content on soil corrosion behavior of carbon steel at different exposure time. Huazhong University of Science and Technology, Wuhan.
[12] Li J. (2015) Study on low alloy steel designed for grounding grids and its corrosion behavior in acidic soil. University of Science & Technology Beijing, Beijing.
[13] Huang T. (2014) Study on corrosion behavior of ground connection network materials in simulated soil of substation. Kunming university of science and technology, Kunming.

[14] Densioni A., Romanoff M. Corrosion of galvanized steel in soils. (1952) Journal of Research of the National Bureau of Standards, 49: 299-316.
[15] Aijun Y. , Yi C. , Lajun F. Soil Corrosion Performance of Several Grounding Net Materials. (2010) Corrosion Science & Protection Technology, 22:197-199.
[16] Liu S., Sun H.Y., Fan H.J. Research progress on corrosion behavior of galvanized steel. (2012) Material protection, 45: 42-5+1.
[17] Zhan M., Li J., Nie X. Corrosion Behavior of Galvanized Q235 Steel for Grounding in Acid Soils. (2015) Journal of Chinese Society for Corrosion & Protection, 35:27-32.
[18] Nie X, Li X, Li Y. Study on the correlation between acceleration ratio and kinetics of soil corrosion acceleration test. (2011) Journal of Chinese Society for Corrosion and Protection, 31: 208-13.
[19] Shao Y., Yan A., Li B. Corrosion Behavior of Q235 Carbon Steel for Grounding Grids in Soil at Shanghai District. (2015) Corrosion Science And Protection Technology, 4: 333-8.
[20] Su Y.G., Fang B.L., Lv P.H. Initial Corrosion Behavior of Grounding Materials in Three Typical Soils in Shaanxi Province. (2018) Corrosion Science And Protection Technology, 30: 279-84.
[21] Cai Z. Z., Cai Y. X., Tao S. Y. Corrosion Performance on Several Grounding Net Materials in Soils of Qinghai-Tibet Plateau. (2017) Corrosion Science And Protection Technology, 29: 183-187.
[22] Fan X., Wang J. G., Zhou M. Evaluation method of weak polarization curve of corrosion velocity of grounding materials. (2012) Proceedings of the CSEE, 28: 192-198,202.

Study of structure and electrochemical properties of Fe-doped carbon aerogels

Shasha Wang[1], Bin Ren[1], Lihui Zhang[1], and Zhenfa Liu[1*]

[1] Institute of Energy Resources, Hebei Academy of Sciences, Shijiazhuang, 050081, China

[*]Corresponding author's e-mail: lzf63@heb-as.com

Abstract. A nanocomposite material CAFe was obtained by assembling of Fe and carbon aerogels. The physical characterizations of CAFe were compared using XRD, Raman, XPS, and SEM. The results showed that Fe was successfully loaded onto carbon aerogels. The results of the electrochemical performance revealed CAFe exhibited high specific capacitance (78.7 F g^{-1}) and excellent stability. Therefore, Fe-doped carbon aerogel is a good electrode material.

1. Introduction

Electrical double-layer capacitance (EDLC) supercapacitor, has received increasing attention, due to its greater power density, longer cycle-life, and lower cost compared to other conventional capacitors [1]. EDLC focused on the ion deposition at electrode-electrolyte interface, which relies upon electrode porosity and surface area. Carbon materials, such as carbon aerogels, nanotubes and activated carbons have been widely studied as electrode materials for EDLC supercapacitor, due to their controllable pore structure, great surface area, high chemical stability and cost-effectiveness [2-3].

In the present study, nanocomposite material CAFe and carbon aerogel (CA) were obtained. Their differences are compared by X-ray diffraction (XRD), Raman spectroscopy (Raman), X-ray photoelectron spectroscopy (XPS) techniques. Electrochemical performance of the composite electrode was assessed using cyclic voltammetry (CV), galvanostatic charge-discharge test, and electrochemical impedance spectroscopy (EIS).

2. Methods

2.1. Synthesis of carbon materials

CAFe was prepared by a sol-gel approach. Resorcinol (R), phloroglucinol (P), formaldehyde (F, 37% aqueous solution) and $Fe(NO_3)_3 \cdot 9H_2O$ (Fe) were all purchased from Aladin, and then dissolved in distilled water at a R/F/P/Fe molar ratio of 7:16:1:1 and initial concentration of 45% (w/w). All reagents were mixed through 30-min ultrasonication, and then heated at 80 °C in a water bath for 72 h in order to attain wet gels. After that, the obtained wet gels were dried at vacuum condition, followed by pyrolysis at 900 °C in a N_2 atmosphere for 3 h in order to synthesize CAFe.

2.2. Characterization

The surface composition and structure of CAFe and CA was evaluated by XRD (Ultima IV XRD system, Rigaku, Japan) with CuKa radiation (λ=1.54 Å), Raman (DXR Raman microscope, Thermo Fisher, USA) and XPS (PHI5600, Physical ElectronicsInc, USA). Scanning electron microscopy (SEM; Tescan Maia3 MHL) were applied to characterize the morphology of as-prepared samples.

3. Results and discussion

The structures of CA and CAFe were assessed by XRD and Raman. The XRD patterns are summarized in Fig. 1(a). Notably, the XRD patterns indicated 2 peaks at approximately 24° and 44°, accounting for the (002) and (101) diffraction peaks of graphitic carbon, respectively [4]. In addition, the existence of Fe_3O_4 was confirmed.

The Raman spectrum (Fig. 1b) obviously displayed D-band and G-band peaks centered at 1350 cm^{-1} and 1590 cm^{-1}, respectively. The D band is assigned to edge defects and highly disordered carbon, whereas the G band is attributed to C=C backbone stretching and highly ordered carbon [5]. The relative intensity ratios (R) of D band to G band ($R=I_D/I_G$) are listed in Fig 1(b). It further noticed that, the R value of CAFe (R=0.90) is lower than that of CA (R=0.98). Such low ratio reflects a high degree of graphitization of CAFe [6], which can markedly increase its electrical conductivity. These findings were consistent with the impedance behavior data (Fig. 4).

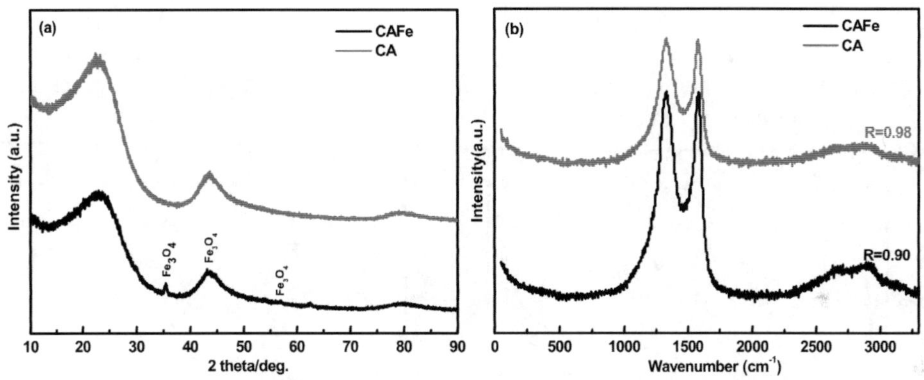

Fig. 1 (a) X-ray diffraction patterns, (b) Raman spectra of all samples.

To attain more information regarding the elements of CAFe, XPS analysis was carried out and the obtained data are demonstrated in Fig. 2. As presented in Fig. 2, the full XPS spectrum of CAFe confirmed the presence of heteroatoms (Fe, O).

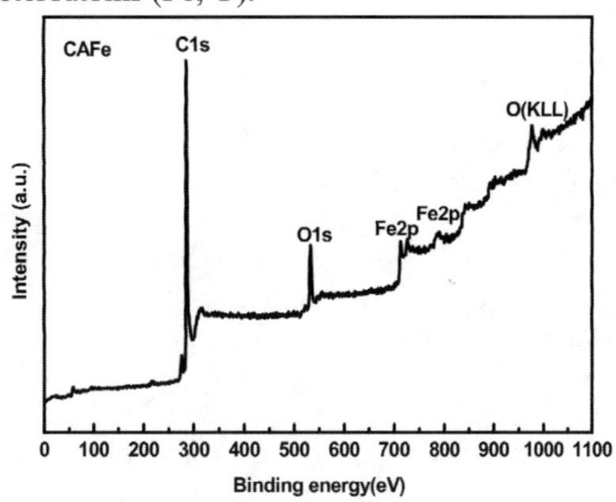

Fig. 2. Full XPS spectrum of CAFe.

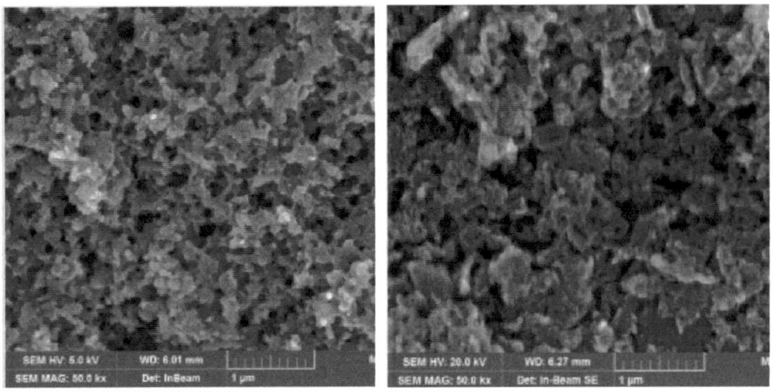

(a) (b)

Fig. 3 SEM images of CA (a) and CAFe (b).

To further assess the electrochemical performances, CV test and galvanostatic charge–discharge tests were performed in 6 M KOH under a three-electrode configuration, and the results are shown in Fig. 4.

Fig. 4 (a) demonstrates the CV curves of CAFe and CA at 1mV s^{-1}. The CV curves revealed apparently well-defined redox peaks at approximately -0.004 A and 0.004 A for CAFe electrodes, indicating a major capacitive role of high pseudocapacity. It is noted that the CV curve shape of CA is comparatively different from that of CAFe, which is reasonably symmetrical nature either anodic or cathodic direction. These findings suggest the dominant EDLC behavior [7] of CA, and almost no faradaic contribution from the foreign dopant.

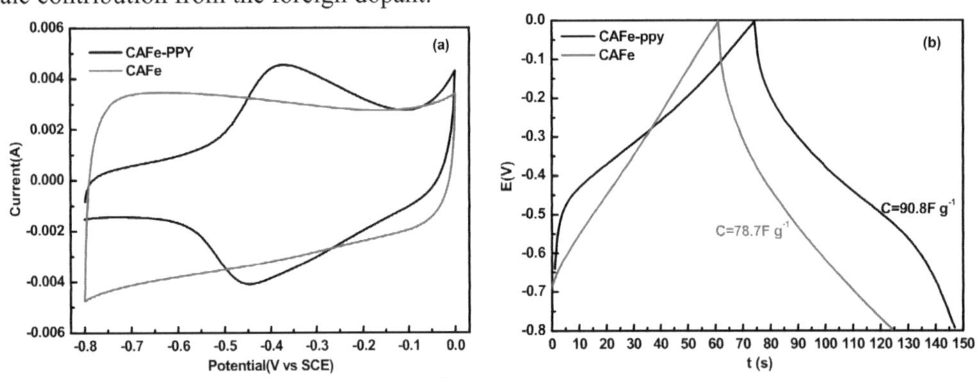

Fig. 4 (a) Charge/discharge curves at 1 A g^{-1}, (b) Cyclic voltammograms curves (scan rate =1 mV s^{-1}).

Fig. 4(b) displays the specific capacitance (C) and galvanostatic charge–discharge curves of CAFe and CA at 1 Ag^{-1}. The specific capacitances of 90.8 F g^{-1} and 78.7 F g^{-1} were obtained at 1 A g^{-1} current density for CAFe and CA electrodes, respectively. The galvanostatic charge–discharge curve of CA was almost linear with asymmetrical triangular shape, indicating that this sample displays good electrochemical reversibility and high coulombic efficiency. Considering the pseudocapacitive behavior of CAFe's functional groups, its galvanostatic charge-discharge curve is relatively unsymmetrical but slightly distorted, as consistent with its CV curve.

4. Conclusion

CA and CAFe were prepared. XRD, Raman, XPS, and N_2 adsorption-desorption were adopted to characterize their physical structure. The results showed that Fe, N and O were successfully loaded onto CAFe. Electrochemical performance tests revealed that CAFe exhibited obvious EDLC behavior and faradaic contribution.

Acknowledgments

This work was financially supported by the Foundation of Key R&D Program of Hebei Province (18393616D) and Hebei Province Applied Basic Research Program-Key Basic Research Project (18964005D).

References

[1] Guo, S. Lu, G. Qiu, S. Liu, J. Wang, X. He, C. Wei, H. Yan, X. Guo, Z. (2014) Carbon-coated MnO microparticulate porous nanocomposites serving as anode materials with enhanced electrochemical performances. Nano Energy, 9: 41-49.

[2] Liu, R. Wan, L. Liu, S. Pan, L. Wu, D. Zhao, D. (2015) Supercapacitors: An Interface Induced Co-Assembly Approach Towards Ordered Mesoporous Carbon/ Graphene Aerogel for High-Performance Supercapacitors. Adv. Funct. Mater, 25: 615-651.

[3] Wang, J. Chen, M.M. Wang, C.Y. Wang, J.Z. Zheng, J.M. (2011) Preparation of mesoporous carbons from amphiphilic carbonaceous material for high-performance electric double-layer capacitors. Journal of Power Sources, 196: 550-558.

[4] Lang, J.W. Bi, X. Yan, X.Y. Yang, J. Xue, Q.J. (2011) Study on the electrochemical properties of cubic ordered mesoporous carbon for supercapacitors. Journal of Power Sources, 196:10472-10478.

[5] Wu, J. Zhang, D. Wang, Y. Hou, B. (2013) Electrocatalytic activity of nitrogen-doped graphene synthesized via a one-pot hydrothermal process towards oxygen reduction reaction. J. Power Sources, 227: 185-189.

[6] Gao, W. Wan, Y. Dou, Zhao,Y.D. (2011) Synthesis of partially graphitic ordered mesoporous carbons with high surface areas. Adv. Energy Mater, 1: 115-123.

[7] Ma, X. Liu, M. Gan, L. Zhao, Y. Chen, L. (2013) Synthesis of micro-and mesoporous carbon spheres for supercapacitor electrode. J. Solid State Electr., 17: 2293-2301.

Research on capacity degradation and aging state estimation of lithium-ion battery

Xu Li, Qi Zhang, Bowen Yang, Dafang Wang*, Shengmin Cui

School of Automotive Engineering, Harbin Institute of Technology, Weihai, Shandong, 264209, China

*Corresponding author's e-mail: wangdf@hit.edu.cn

Abstract. The aging state estimation of the power battery is very important for the battery management system (BMS). Regular estimation of the battery' state of health (SOH) can improve the service life of electric vehicles. The reasons for the aging of lithium-ion battery are briefly introduced in this paper. What's more, the estimation methods of battery aging state are summarized and described from both the model-based and data-driven aspects.

1. Introduction

Recently, the share of electric vehicles in the automotive market has seen explosive growth. The increase in the number of electric vehicles has put forward higher requirements for the cycle life and reliability of power batteries. However, capacity degradation is inevitable. Some external factors such as extreme temperature and abnormal charging and discharging conditions can accelerate battery aging and even cause battery failure. Capacity degradation of the battery will not only reduce the car's cruising range, but also affect the accuracy of state of charge (SOC) estimated by BMS. The battery may be in a state of overcharging and overdischarging, causing more serious failures. Therefore, it is of great significance to estimate SOH of the battery regularly. It can not only improve battery performance and battery life, but also effectively avoid safety problems caused by battery aging failure.

2. Reasons for lithium-ion battery aging

2.1 Internal reasons

Internal reasons mainly include loss of lithium inventory (LLI), conductivity loss (CL) and loss of active material (LAM). Available lithium ions mainly come from the cathode material. When the battery is working, lithium ions are intercalated back and forth between the positive and negative electrodes, and the stability of the positive and negative electrode structures decreases or even collapses. As a result, not all lithium ions can be inserted back into the structures, resulting in the loss of lithium ions. The formation, destruction and regeneration of solid electrolyte interphase (SEI) can also lead to the loss of lithium ions. CL is mainly due to the aging of electronic components, such as the corrosion of the collector. LAM usually related to the dissolution and consumption of the positive and negative active materials of the battery and the loss of electrolyte[1-2].

2.2 *External reasons*

Different external factors will affect the internal reaction process of the battery, such as ambient temperature, charge and discharge rate and charge method (overcharge and overdischarge), etc.

Charging and discharging at too low a temperature will cause lithium deposition reaction in the battery, which will seriously damage the health of the battery. When charging and discharging at a high rate for a long time, the interface polarization of the electrode and electrolyte and the reaction polarization of the active material will increase significantly, making the battery impedance also increases. The charge-discharge cut-off voltage will also cause irreversible attenuation of the battery capacity. Too high or too low cut-off voltage will damage electrode structures of the battery, making it impossible to insert lithium ions back into the electrode fully, causing lithium deposition, and even cause safety hazards such as short circuits.

3. Estimation of battery aging state

3.1 Model-based estimation method

There are currently different classification standards for model-based battery aging state estimation research, such as equivalent circuit model (ECM), electrochemical mechanism-based model (EMM), and hybrid model.

3.1.1 Equivalent circuit model

The ECM does not consider the internal chemical reaction process and mechanism of the battery, only uses electronic components such as resistors and capacitors to build the model based on the electrical characteristics. The common ones are Rint model, Thevenin model, PNGV model and more complex models improved based on the above models[3], as shown in Fig.1. Qian K et al. used a second-order RC equivalent circuit to describe the voltage relaxation, and obtained characteristic parameters that can distinguish ohmic polarization, reaction polarization, and concentration polarization by simulating the voltage relaxation curve, and used parameter sensitive changes to evaluate battery aging state[4]. Fang L L et al. used the forgetting factor recursive least squares method to identify model parameters on the basis of the ECM, and proposed a joint estimation method. Double extended Kalman filter algorithm is used to analyse the battery SOC-OCV curve, battery capacity and parameters by building an error estimation model, through which SOC and SOH are jointly estimated[5].

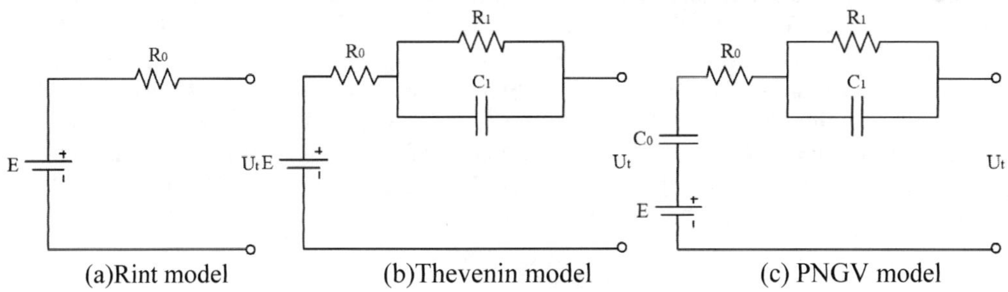

(a)Rint model (b)Thevenin model (c) PNGV model

Fig.1. The commonly used ECM

3.1.2 Electrochemical mechanism-based model

The EMM describes the microscopic physical and chemical reaction process based on the complex electrochemical phenomenon and aging mechanism inside the battery, and can characterize the essence of the battery, but the modeling process is quite complicated. A. Lamorgese et al. simulated a simple lithium-ion battery aging model based on electrochemical thermally coupled pseudo two-dimensional (P2D) theory, and estimated capacity attenuation by evaluating the loss of recyclable lithium ions at each time point[6]. Yang S C et al. coupled the EMM with the thermodynamic model to study the battery degradation and aging. The model describes the anode side reaction and the loss of cathode active materials, and studied the battery aging behavior under different discharge rates and ambient temperatures[7] .

3.1.3 Hybrid model

Hybrid model generally refers to the model based on the ECM combined with the battery chemistry mechanism to improve the accuracy of the model. An electrochemical-equivalent circuit unified parameter model based on the electrochemical mechanism and ECM is established to study the side reactions of the battery negative electrode, and established the battery aging model through the side reaction control equation to study the aging of the battery. Typical aging values such as the internal resistance, growth of the deposition layer and SEI film and are used to characterize the aging state[8].

3.2 Data-driven estimation method

This method mainly focus on the estimation algorithm or parameters which can characterize battery aging. A large amount of data are used for training to make the model result consistent with the measured data, instead of studying the battery itself. Current, voltage, capacity, impedance, and temperature, etc are collected to estimate SOH with the help of algorithms. There is no need considering the internal electrochemical reaction and aging mechanism of the battery, let alone the establishment of a complex electrochemical system model, so it has gradually become a research hotspot in recent years.

3.2.1 SOH estimation based on machine learning algorithm

Common algorithms include artificial neural network (ANN), support vector machines (SVM), relevance vector machine (RVM), fuzzy algorithm (FA), gaussian process regression (PGR), etc. Shehzar Shahzad Sheikh et al. proposed a method to study battery health and state of charge, using machine learning algorithms to extract important features from the discharge curve to estimate SOH and SOC, mainly relying on powerful learning algorithm to achieve state estimation[9]. T-S fuzzy control is used to establish a dynamic prediction model. With the number of cycles, voltage drop and internal resistance change as the input, and the corresponding SOH weight as the output, the SOH is calculated according to the SOH weight and boundary conditions[10]. The wavelet denoising (WD) method is also combined with the PGR to predict the remaining battery life[11].

3.2.2 SOH estimation based on filtering algorithm

Filter-based estimation algorithms usually include two steps: prediction and correction. A state equation is supposed to be constructed, which takes the parameters (internal resistance, capacity, etc.) representing SOH as state variables to establish a state space model, to realize dynamic tracking and prediction of running conditions. Then the filtering algorithm is used to solve SOH iteratively[12]. Commonly used filtering algorithms include particle filter (PF), Kalman Filter (KF), extended Kalman Filter (EKF), unscented Kalman Filter (UKF) and so on. Chen Z P established an 18650 battery state of health estimation model, proposing an improved battery SOH estimation method based on the UKF algorithm[13].

3.2.3 SOH estimation based on time series algorithm

This method is based on past trends to predict future development. Simple sequence average method, weighted sequence average method, moving average method, and weighted moving average method are commonly used. These methods assign different weights according to the degree to which the data affects the predicted value at different times in the moving segment, and then use the average value to predict the future value. Auto-regressive moving average model (ARMA), long short-term memory (LSTM) and recurrent neural network (RNN) are also important methods for time series forecasting[12]. Empirical mode decomposition (EMD) was used to separate the global deterioration trend and capacity regeneration from the health state time series, which were used in the ARIMA model to predict the global deterioration trend and capacity regeneration respectively[14]. Qu J T et al. proposed to combine LSTM networks with particle swarm optimization and attention mechanisms to estimate the RUL and SOH of lithium-ion batteries[15].

3.2.4 SOH estimation based on capacity increment curve

The capacity increment (IC) curve is derived from the battery voltage curve and represents the capacity change corresponding to the change in unit voltage. The voltage curve changes relatively smoothly, make it difficult to detect weak characteristics changes. The IC curve has multiple peaks and valleys, which can characterize the phase change reaction of the battery, and the peaks of the waves will move in height and position as the battery ages, as shown in Fig.2. The battery SOH can be estimated quantitatively combined with the regression algorithm. Weng C H et al. used IC analysis to identify reliable features related to battery aging to predict SOH[16]. Li X Y et al. used gray relational analysis (GRA) with entropy weight method to extract health indicators based on IC curve. The entropy weight method is used to evaluate the significance of each health index and SOH is evaluated by calculating the gray correlation degree between the reference sequence and the comparison sequence[17]. IC curve can also be combined with PGR to predict the health of the battery[18].

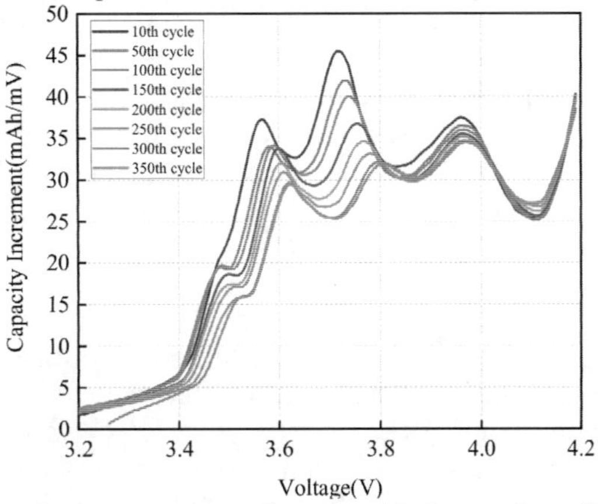

Fig.2. The capacity increase curve changes with the number of battery cycles

3.2.5 SOH estimation based on linear relationship between parameters and capacity decay

Some scholars predict aging state by looking for the linear relationship between certain variables and the battery SOH. There is no need to build a complex model and no need for powerful algorithms. For example, Tang X P et al. introduced the concepts of regional capacity and regional voltage to predict the SOH of four batteries based on the IC curve, finding that SOH and regional capacity present a simple linear relationship, which can directly predict SOH. This method is efficient and novel, insensitive to noise and filtering algorithms[19]. Through the analysis of battery cycle life test data, the instantaneous discharge voltage V and its unit time voltage drop V' can be used as model parameters to estimate SOH, which is linearly related to 1/V' multiplied by the correction factor[20].

4. Conclusion

The reasons for the battery aging and the estimation methods of SOH are discussed. In terms of battery SOH estimation, the model-based SOH estimation method has high accuracy and can be effectively combined with the chemical mechanism of the battery to explain the aging mechanism of the battery, but the modeling and parameter identification process is more complicated. Data-driven SOH estimation method can also achieve very good estimation results with the improvement of estimation algorithms and it has attracted widespread attention for it's efficient and convenient application in vehicles BMS.

Acknowledgments

The author is very grateful to Professor Dafang Wang for his guidance and the support of the doctors in the laboratory!

References

[1] Dubarry M, Truchot C, et al. Synthesize battery degradation modes via a diagnostic and prognostic model[J]. Journal of Power Sources, 2012,219(DEC.1): 204-216.
[2] Vetter J, P Novák, Wagner M R, et al. Ageing mechanisms in lithium-ion batteries[J]. Journal of Power Sources, 2005,147(1/2):269-281.
[3] Jiang L, Li Q, Chen WR, et al. Study on parameter identification of third-order RQ equivalent circuit of PEMFC based on Nelder-Mead optimization[J]. Power Supply, 2019,17(2):12-18, 25.
[4] Qian K, Huang B H, Ran A H, et al. State-of-health (SOH) evaluation on lithium-ion battery by simulating the voltage relaxation curves[J]. Electrochimica Acta, 2019(303):183-191.
[5] Fang L L, Li J Q, Peng B. Online Estimation and Error Analysis of both SOC and SOH of Lithium-ion Battery based on DEKF Method[J]. Energy Procedia, 2019(158):3008-3013.
[6] Lamorgese A, Mauri R, Tellini B. Electrochemical-thermal P2D aging model of a LiCoO 2 /graphite cell: Capacity fade simulations, 2018(20):289-297.
[7] Yang S C, Hua Y, Qiao D, et al. A coupled electrochemical-thermal-mechanical degradation modelling approach for lifetime assessment of lithium-ion batteries, 2019(326): 134928.
[8] Zhang X, Lu J L, Yuan S F, Yang J, Zhou, X. A novel method for identification of lithium-ion battery equivalent circuit model paramameters considering electrochemical properties[J]. Journal of Power Sources, 2017,345:21-29.
[9] Sheikh S S, Anjum M, Khan M A, et al. A Battery Health Monitoring Method Using Machine Learning: A Data-Driven Approach[J]. Energies, 2020(13):3658.
[10] Xia Z, Qahouq J. Adaptive and fast state of health estimation method or lithium-ion batteries using online complex impedance and artificial neural network[J]. Applied Power Electronics Conference and Exposition, 2019:17-21.
[11] Peng Y, Hou Y D, Song Y C, et al. Lithium-Ion Battery Prognostics with Hybrid Gaussian Process Function Regression[J]. Energies, 2018, 11(6):1420.
[12] Tian H, Qin P, Li K, et al. A review of the state of health for lithium-ion batteries: Research status and suggestions[J]. Journal of Cleaner Production, 2020, 261:120813.
[13] Chen Z P, Wang Q T. The Application of UKF Algorithm for 18650-type Lithium Battery SOH Estimation[J]. Applied Mechanics & Materials, 2014, 519-520:1079-1084.
[14] Zhou Y P, Miao H, et al. Lithium-ion batteries remaining useful life prediction based on a mixture of empirical mode decomposition and ARIMA model[J]. Microelectronics Reliability, 2016(65):265-273.
[15] Qu J T, Liu F, Ma Y, et al. A Neural-Network-Based Method for RUL Prediction and SOH Monitoring of Lithium-Ion Battery[J]. IEEE Access, 2019, 7(99):87178-87191.
[16] Weng C H, Cui Y J, Sun J, et al. On-board state of health monitoring of lithium-ion batteries using incremental capacity analysis with support vector regression[J]. Journal of Power Sources, 2013(235):36-44.
[17] Li X Y, Wang Z P, Zhang L, et al. State-of-health estimation for Li-ion batteries by combing the incremental capacity analysis method with grey relational analysis[J]. Journal of Power Sources,2019(410-411):106-114
[18] Li X Y, Wang Z P, Yan J Y. Prognostic health condition for lithium battery using the partial incremental capacity and Gaussian process regression[J]. Journal of power sources, 2019, 421(MAY 1):56-67.
[19] Tang X P, Zou C F, Yao K, et al. A fast estimation algorithm for lithium-ion battery state of health [J]. Journal of Power Sources, 2018(396):453-458.
[20] Huang S C, Tseng K H, Liang J W, et al. An Online SOC and SOH Estimation Model for Lithium-Ion Batteries[J]. Energies, 2017(10):512.

Properties of carbon-doped GaN using isobutane as a dopant

Xun Li[1,a*], Songran Zhu[1,b]

[1]School of Electronics and Information Engineering, Hebei University of Technology, 300401 Tianjin, China

*Corresponding author: [a]email: lixun@hebut.edu.cn

[b]email: 201921901026@stu.hebut.edu.cn

Abstract. Carbon doping is an effective method to obtain semi-insulating GaN buffer, which is a necessity to prevent current leakage, in the high-electron-mobility-transistor device structure. The properties of intentionally carbon-doped GaN using isobutane gas as a dopant has been studied in detail. The carbon incorporation efficiency has been measured by secondary ion mass spectrometry. It is found that the carbon concentration could be directly controlled by the flow rate of isobutane precursor. The surface morphology of carbon-doped gallium nitride epitaxial layers has been investigated by optical microscopy and atomic force microscopy. The growth mode of GaN layers changes from step-flow to island growth, when the incorporated carbon concentration is higher than 1×10^{19} cm^{-3}. In order to evaluate the structural quality of intentionally carbon-doped GaN, the full-width-at-half-maximum values are extracted from the rocking curves in six different reflections measured by high resolution X-ray diffraction. Raman spectroscopy is utilized to evaluate the physical properties of the carbon-doped GaN epitaxial layer.

1. Introduction

Wide band gap semiconductor gallium nitride (GaN) is considered to be a very promising material in the applications of optoelectronics [1] and high-frequency and high-power devices [2] due to its remarkable properties.

In the GaN-based high-electron-mobility-transistors (HEMT) structure, a semi-insulating buffer is necessary in order to decrease leakage current [3]. Nowadays, there are two common ways to obtain the semi-insulating GaN buffer layers, i.e., iron doping and carbon doping [4-7]. The drawback of iron doping is the memory effect during the growth. Iron will exist on the surface of GaN epitaxial layer for a while, even if the iron precursor is shut down [8]. In this point of view, the growth of semi-insulating GaN by carbon doping is preferred.

For the chemical vapour deposition (CVD) technique, the conventional way of incorporating carbon into GaN film is using the carbon atoms from the gallium precursor [9]. By varying the growth condition, the carbon doping rate could be controlled. However, this kind of method has difficulties in maintaining growth uniformity and reproducibility when process temperature and pressure are changed. Therefore, a new way of carbon doping is emerged and studied. In this new method, an optimal unintentionally doped GaN growth condition is selected first. At this condition, the concentration of carbon source from the gallium precursor is suppressed under the detection limit of SIMS. Then a carbon gas precursor is put into the reactor to dope GaN. The carbon incorporation efficiency could be easily managed by tuning the flow of the carbon precursor. Few studies have been made in this field. Sweden researchers have discussed the device performance of a HEMT structure, which buffer layer was doped by the propane gas [10].

In this paper, a detailed study of intentionally carbon-doped GaN by isobutane (i-C_4H_{10}) precursor is reported. Morphological, structural and physical properties of the carbon-doped GaN epitaxial layers are investigated.

2. Experimental details

In this section, detailed information of sample preparation and characterization techniques is described.

2.1. Sample preparation

All the experiments were carried out in a metal-organic CVD (MOCVD) reactor with substrate rotation. The GaN epitaxial layer was grown on the 4H-SiC on-axis substrates with a 100 nm thick aluminum nitride (AlN) interlayer. The process pressure used in the present study was kept at 5000 Pa. Trimethylgallium (TMGa), trimethlylaluminum (TMAl), and ammonia (NH_3) were used as precursors for Ga, Al, and N, respectively. A mixture of hydrogen (H_2) and nitrogen (N_2) was used as carrier gas. The growth temperature of GaN layers was fixed at 1060 °C. The growth rate of all the GaN samples is around 0.8 m/h at this temperature. Isobutane gas is a gaseous precursor, which was chosen as a carbon dopant during the growth of GaN layers. The V/III ratio used in this study was kept at 1200.

2.2. Characterization technique

Secondary ion mass spectrometry (SIMS) measurement was performed in order to detect the impurity concentrations. The detect limit for carbon (C) impurity is around 1×10^{16} cm^{-3}. Optical microscopy and atomic force microscopy (AFM) were utilized to examine the morphology and surface roughness of the grown epitaxial layers. Root-mean-square (RMS) value is used in this study to quantify the surface roughness in AFM measurements. High resolution X-ray diffraction (HRXRD) equipped with a triple-axis configuration was used to measure the rocking curves. Raman spectroscopy was done at room temperature with excitation laser wavelength of 532 nm.

3. Results and discussion

In this section, the experimental results and discussion regarding carbon incorporation efficiency, surface morphology, crystalline quality as well as Raman spectroscopy are presented.

3.1. Carbon incorporation efficiency

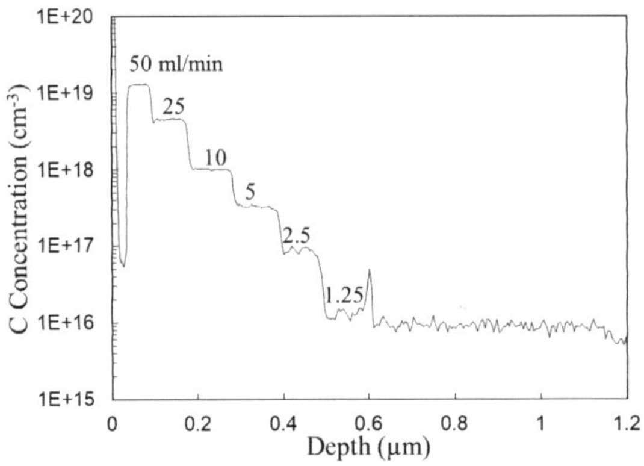

Figure 1 Carbon concentration versus depth of an isobutane-doped GaN sample measured by SIMS.

The optimal growth condition to obtain a high quality unintentionally doped GaN epitaxial layer in our reactor is at a temperature of 1060 °C with a reduced pressure of 5000 Pa. Carbon concentration of undoped GaN layers is under the SIMS detection limit at this temperature. Therefore, i-C_4H_{10} gas precursor is added at the abovementioned process condition when growing carbon-doped GaN layers. In order to study the relation between carbon incorporation efficiency and flow rate, a sample with several GaN layers were grown. Firstly, a 0.6 μm undoped GaN epitaxial layer was grown, followed by six carbon-doped layers with different flow rates of i-C_4H_{10} precursor. Finally, the carbon gas precursor was shut down and a thin undoped GaN was grown. Figure 1 shows the SIMS measurement done in the center of this sample. There's a variation at the depth of 0.6 μm in the figure due to the carbon precursor was added into the reactor at this point. The used flow rate of i-C_4H_{10} gas is given in the figure. When the flow rate of i-C_4H_{10} precursor is higher than 5 ml/min, the carbon concentration profile is very stable.

3.2. Surface morphology

In order to study the surface morphology, a few 1 μm thick GaN samples intentionally doped by i-C_4H_{10} precursor with only one concentration were deposited at 1060 °C. Figure 2 is the AFM images performed on the surface of these samples. An unintentionally doped GaN sample is shown in figure 2(a) for comparison. Its surface roughness is 0.178 nm over a 3×3 μm^2 area. The surface roughness of GaN layer increases to 0.212 nm, when increasing the flow rate of i-C_4H_{10} precursor to 10 ml/min, shown in figure 2(b). Figure 2(c) is the AFM image of GaN epitaxial layer doped by 25 ml/min i-C_4H_{10} gas. Its surface becomes rougher but still remains to step-flow growth mode. Its surface roughness is 0.272 nm over a 3×3 μm^2 area and 0.765 nm over a 10×10 μm^2. However, when further increase the flow rate of carbon precursor to 50 ml/min, the GaN epitaxial layer change to island growth, as shown in figure 2(d).

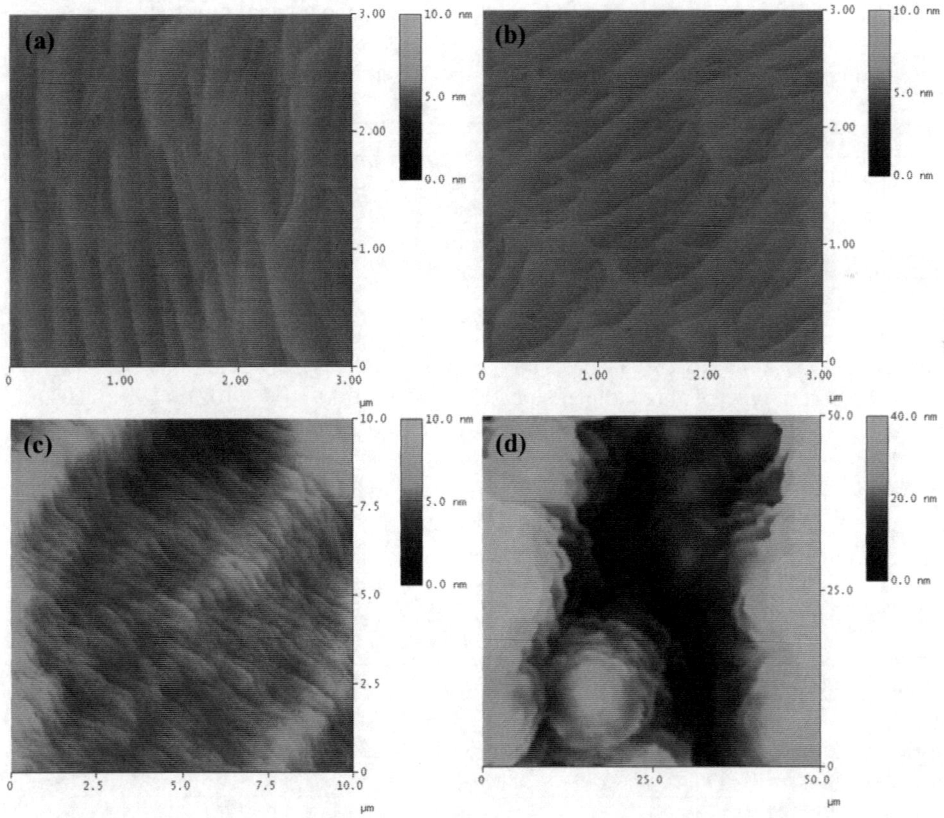

Figure 2 AFM images of carbon-doped GaN epitaxial layers with i-C_4H_{10} precursor flow rate of (a) 0 ml/min, (b) 10 ml/min, (c) 25 ml/min, and (d) 50 ml/min, respectively.

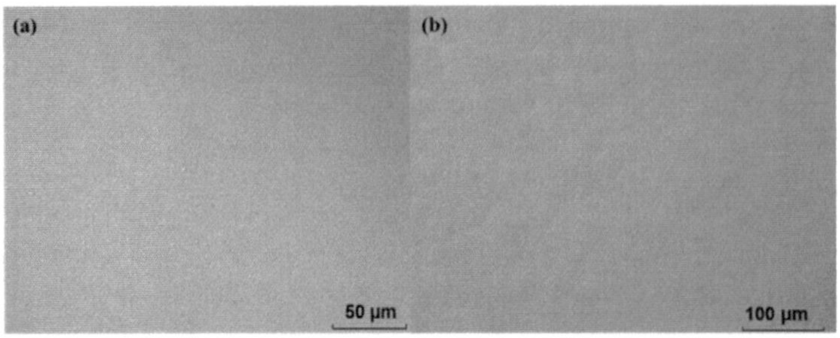

Figure 3 Optical microscope images of (a) an unintentionally doped GaN epitaxial layer, and (b) a carbon-doped GaN layer using 50 ml/min i-C_4H_{10} gas.

Figure 3(a) is the typical optical microscope image for unintentionally doped samples. Its surface is very smooth, which could hardly see any feature. Figure 3(b) is the optical image of the same sample presented in figure 2(d). The surface of this sample becomes rougher and has decorative pattern. The information of i-C_4H_{10} flow rate, C concentration and related surface roughness measured in the area of 3×3 µm^2 are summarized in table 1. When the carbon precursor are 5 ml/min, 10 ml/min, 25 ml/min and 50 ml/min, the incorporated carbon concentration could reach to the lever of 3.5×10^{17} cm^{-3}, 1×10^{18} cm^{-3}, 4.5×10^{18} cm^{-3} and 1.3×10^{19} cm^{-3}, respectively. Meanwhile, the surface of the GaN epitaxial layers doped by i-C_4H_{10} gas becomes rougher as the incorporated carbon concentration increases.

Table 1 Carbon precursor flow rate, carbon concentration and corresponding surface roughness in an area of 3×3 µm^2.

i-C_4H_{10} flow rate (ml/min)	0	5	10	25	50
C concentration (cm^{-3})	undoped	3.5E17	1E18	4.5E18	1.3E19
RMS (nm)	0.178	0.2	0.212	0.272	0.388

3.3. Structural quality

HRXRD was measured on the centre part of the intentionally carbon-doped GaN samples with different i-C_4H_{10} gas flow. In order to evaluate the structural quality, the full-width-at-half-maximum (FWHM) values were extracted from the rocking curves and listed in table 2. The (002), (004) and (006) reflections could reveal the screw type of threading dislocations, while the (101), (102), (103), (104) and (302) reflections are related to the edge type of threading dislocations. Generally, the carbon doping will deteriorate the structural quality of GaN epitaxial layers. The FWHM values increase as the flow rate of i-C_4H_{10} gas goes up from 0 ml/min to 25 ml/min. However, when further increase the carbon precursor flow rate, the FWHM values decrease for all the reflections. This may because the growth mode has been changed from step-flow growth to island growth as observed in the AFM pictures.

Table 2 FWHM values extracted from XRD rocking curves with different reflections for i-C_4H_{10} doped GaN.

reflection	(002)	(004)	(006)	(101)	(102)	(103)	(104)	(302)
undoped	223	211	210	297	69	237	222	298
5 ml/min	259	235	234	362	313	288	273	362
25 ml/min	300	277	270	393	349	325	302	385
50 ml/min	171	161	155	380	313	270	237	373

3.4. Raman spectroscopy

Raman measurement has been carried out at room temperature at the centre part of an i-C_4H_{10} doped GaN sample with a carbon concentration of 1×10^{18} cm^{-3}, shown in the figure 4. The GaN E2 (high) peak positioned at 567.5 cm^{-1} is quite strong. The relatively weak GaN E2 (low) and A1 (LO) phonon modes are at 143.5 cm^{-1} and 733.6 cm^{-1}, respectively. The weak peaks at 266 cm^{-1}, 610 cm^{-1}, 799 cm^{-1} and strong peaks at 776 cm^{-1} are all related to the SiC substrate. There is also a very weak E2 (high) peak located at 657 cm^{-1}, which is contributed from the 100 nm thick AlN interlayer.

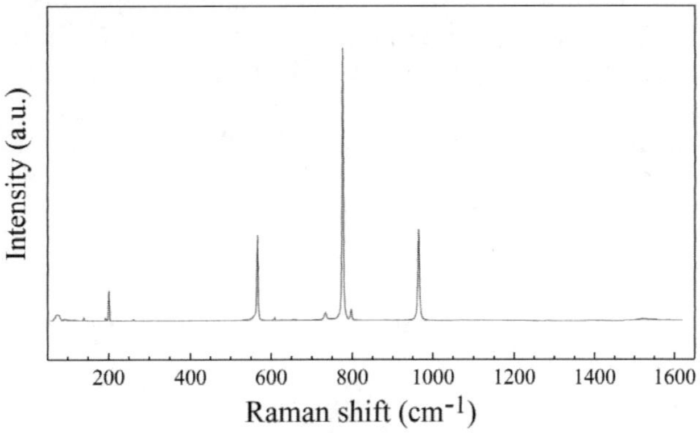

Figure 4 Raman spectrum of the i-C_4H_{10} doped GaN sample with a C concentration of 1×10^{18} cm^{-3}.

4. Conclusions

In this paper, the morphological, structural and physical properties of intentionally carbon-doped GaN epitaxial layers by i-C_4H_{10} precursor were studied. The carbon precursor was added into the reactor at the optimized growth condition for undoped GaN. The incorporated carbon concentration could be directly controlled by the flow rate of i-C_4H_{10} gas. The carbon incorporation efficiency was studied by SIMS measurements. The growth mode of i-C_4H_{10} doped GaN layers, observed from AFM images, changed from step-flow growth to island growth when the doped carbon concentration was higher than 1×10^{19} cm^{-3}. The structure quality of GaN layers with different carbon doping levels was studied by HRXRD. In general, the structure quality of the intentionally carbon-doped GaN film would decline with the increased flow rate of i-C_4H_{10} precursor. Nevertheless, the crystal quality got better at the carbon concentration of 1×10^{19} cm^{-3}, which could be explained by the alternation of growth mode. Raman measurement was performed on a GaN sample with a carbon concentration of 1×10^{18} cm^{-3} in order to study its physical property.

Acknowledgments

This work was financially supported by National Natural Science Foundation of China (Grant No. 61804044) and Natural Science Foundation of Hebei Province (Grant No. F2018202234).

References

[1] Wang, K., Chen, P., Chen, J., Liu, Y., Wu, C., Sun, J., Zhou, X., Zhang, Y., Guo, T. (2021) Alternating current electroluminescence from GaN-based nanorod light-emitting diodes. Opt. Laser Technol., 140: 107044-1–107044-7.

[2] Wang, Y., Chen, T., Hua, M., Wei, J., Zheng, Z., Song, W., Yang, S., Zhong, K., Chen, K.J. (2021) A physics-based empirical model of dynamic I_{OFF} under switching operation in p-GaN gate power HEMTs. IEEE Trans. Power Electron., 36: 9796–9804.

[3] Huang, C., Liu, C., Kao, H., Chiu, H., Chen, C., Chang, K. (2021) The characteristics of 6-inch GaN on Si RF HEMT with high isolation composited buffer layer design. Electronics, 10: 46.

[4] Uren, M.J., Kuball, M. (2021) Impact of carbon in the buffer on power switching GaN-on-Si and RF GaN-on-SiC HEMTs. Jpn. J. Appl. Phys., 60: SB0802-1–SB0802-14.

[5] Noh, Y., Lee, S., Kim, M., Oh, J. (2019) High electron mobility transistors with Fe-doped semi-insulating GaN buffers on (1 1 0) Si substrates grown by ammonia molecular beam epitaxy. J. Cryst. Growth, 509: 141–145.

[6] Hao, M., Wang, Q., Jiang, L., Feng, C., Chen, C., Wang, C., Xiao, H., Liu, F., Xu, X., Wang, X., Wang, Z. (2018) Trapping effects induced by gate OFF-State stress in AlGaN/GaN high-electron-mobility transistors with Fe-doped buffer. J. Nanosci. Nanotechno., 18: 7479–7483.

[7] Challa, S.R., Vega, N.A., Mueller, N.A., Kristukat, C., Debray, M.E., Witte, H., Dadgar, A., Strittmatter, A. (2021) Understanding high-energy 75-MeV sulfur-ion irradiation-induced degradation in GaN-Based heterostructures: The role of the GaN channel layer. IEEE Trans. Electron Devices, 68: 24–28.

[8] Lee, W., Ryou, J., Yoo, D., Limb, J., Dupuis, R.D., Hanser, D., Preble, E., Williams, N.M., Evans, K. (2007) Optimization of Fe doping at the regrowth interface of GaN for applications to III-nitride based heterostructure field-effect transistors. Appl. Phys. Lett., 90: 093509-1–093509-3.

[9] Kangawa, Y., Kusaba, A., Kempisty, P., Shiraishi, K., Nitta, S., Amano, H. (2021) Progress in modeling compound semiconductor epitaxy: Unintentional doping in GaN MOVPE. Cryst. Growth Des., 21: 1878–1890.

[10] Li, X., Bergsten, J., Nilsson, D., Danielsson, O., Pedersen, H., Rorsman, N., Janzen, E., Forsberg, U. (2015) Carbon doped GaN buffer layer using propane for high electron mobility transistor applications: Growth and device results. Appl. Phys. Lett., 107: 262105-1–262105-5.

Research on the Anode Protection of Lithium

Chen Tao [1, a *], Li Jingze [2, b], Liu Hongli [1, c], Shi Jianjian [1, d]

[1] Chengdu Institute of technology, Pidu District, Chengdu, China

[2] School of materials and energy, University of Electronic Science and technology, Pidu District, Chengdu, China

[a] email: chentao@cdtu.edu.cn, [b] email: lijingze@uestc.edu.cn, [c] 945634836@163.com, [d] 1776573169@163.com

*Corresponding Author: [a] email: chentao@cdtu.edu.cn

Abstract: The anode of Lithium serves as the most ideal substitute for graphite anode. However, uncontrolled growth of lithium dendrite brings about service life and safety problems, which limit its commercial application. According to the thermodynamic and kinetic factors of lithium dendrite growth, many strategies have been proposed to inhibit the growth of lithium dendrite, such as structural composite lithium metal anode, optimizing SEI film, using solid electrolyte, etc. This paper reviewed the development process of different strategies, and analyzed their advantages and limitations, looking forward to the possible direction of inhibiting the growth of lithium dendrite.

1. INTRODUCTION

The energy density of lithium-ion battery with graphite as negative electrode has reached the limit. Lithium metal has ultra-high theoretical specific capacity (3860 $mAh \cdot g^{-1}$) and the lowest oxidation-reduction potential (-3.040 V *Vs.* standard hydrogen electrode), becoming the most promising alternative [1, 2]. However, the electrochemical behavior of lithium ion in the lithium metal electrode is uncontrollable, which leads to dendrite growth in the lithium metal anode during charging and discharging, resulting in low coulomb efficiency, short cycle life and even safety problems [2]. Therefore, how to inhibit the growth of lithium dendrite has become a hot topic in the academic and industrial circles. In this paper, different strategies to inhibit the growth of lithium dendrite are reviewed, including the modification of solid electrolyte interface (SEI) membrane, the use of solid electrolyte instead of liquid electrolyte and structured composite lithium metal negative electrode. The advantages and limitations of different strategies to inhibit the growth of lithium dendrite are discussed. On this basis, the possible effective direction to inhibit the growth of lithium dendrite is discussed.

2. Structured Composite Lithium Metal Anode

Cui Yi's team [3] found that the growth of lithium dendrite is closely related to the initial nucleation of lithium ion. When lithium ion has no heterogeneous nucleation barrier on the substrate, this material can induce more uniform nucleation of lithium ion. They found that when Li nucleates on Ag, Au, Mg, Zn and other materials, there is no barrier (Figure1), which is called lithium affinity, Therefore, the initial homogeneous nucleation of lithium can be induced by adding lithium hydrophilic material into the electrode. According to the sand time, the growth rate of lithium dendrite is related to the effective current density. Therefore, increasing the contact area between the electrode and electrolyte can reduce the effective current density on the electrode surface and slow down the growth of lithium dendrite. Based on the above understanding, it is an important strategy to construct electron conduction network and to contain lithium layer or lithium point in the electrode to inhibit the growth of lithium dendrite.

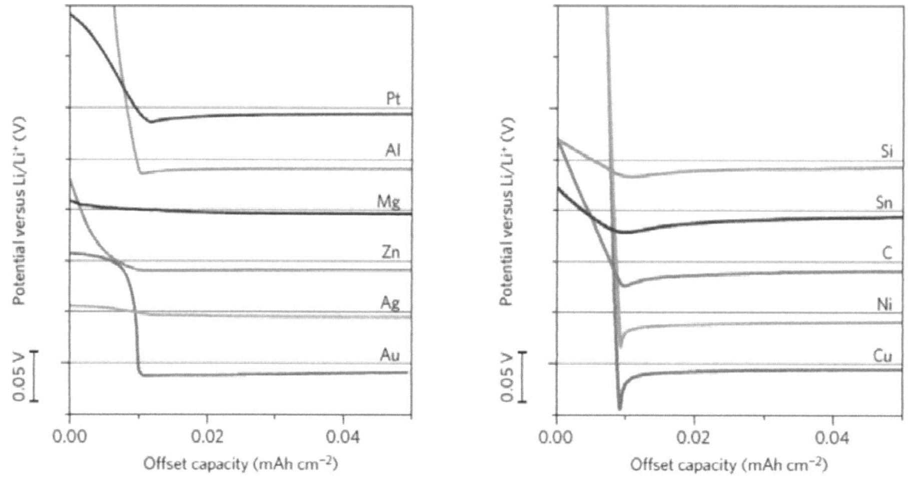

Figure 1 10 μA·cm^{-2} voltage curves of electrochemical precipitation of lithium on different materials at current density [3]

Liang Jiajie's research group [4] prepared three-dimensional graphene framework containing Ag nanowires by freeze-drying method. The three-dimensional graphene framework in the electrode can increase the electrochemical reaction area and reduce the effective current density. Ag has lithium affinity, which can induce uniform precipitation of lithium ions and effectively inhibit the growth of lithium dendrites (Figure2).

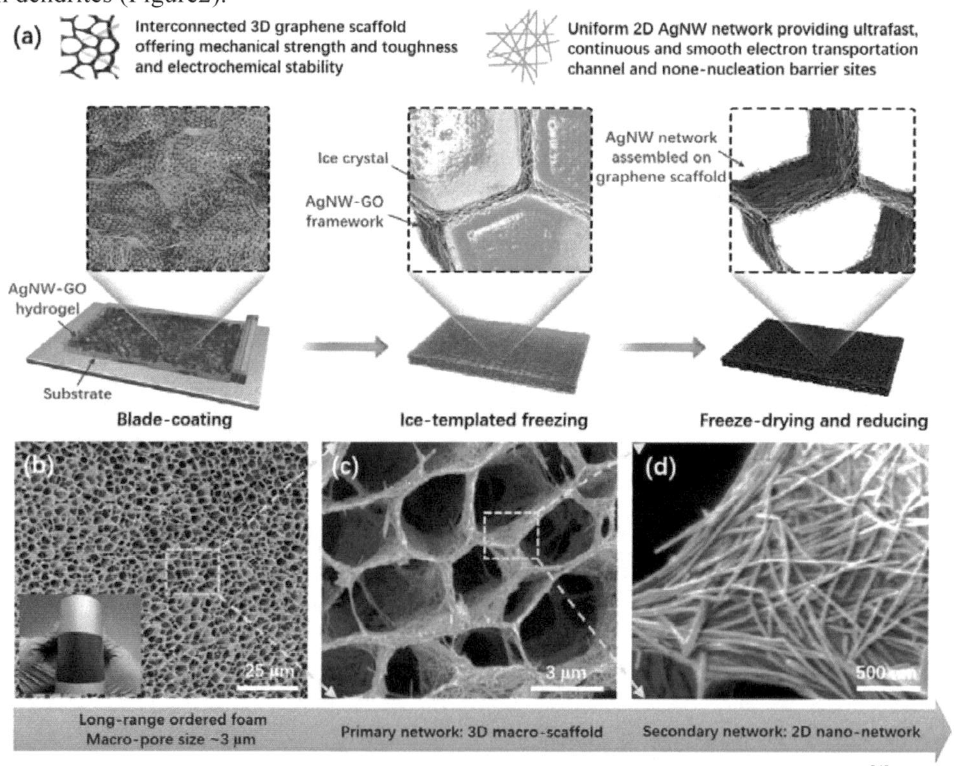

Figure 2 preparation of 3D graphene framework containing Ag nanowires [4]

Although the 3D electron conduction network and lithium sites increase the effective active area and reduce the current density, the increase of the electrochemical active area leads to more irreversible side reactions [5]. Moreover, because the electrode/ electrolyte interface is inside the electrode, this strategy still cannot solve the problem of volume change during the charging and discharging process, as well as the problems of repeated rupture of SEI film and repeated consumption of lithium and electrolyte. The above problems can be solved by adding 3D solid-state lithium-ion conduction network. Wang Jiulin team of Shanghai Jiao Tong University [6] prepared a composite lithium electrode by mixing the micron particles of Li, carbon nanotubes, sulfurized polyacrylonitrile and MgO particles. In this electrode, carbon nanotubes were used as conductive framework and MgO particles were used as lithium affinity points. Lithium micron particles spontaneously reacted with sulfurized polyacrylonitrile to form Li_2S and Li_3N as solid Li^+ conductive framework. It shows stable electrochemical performance in Li-S battery test. Figure 3 is a schematic diagram of the working mechanism and structural composition of the electrode.

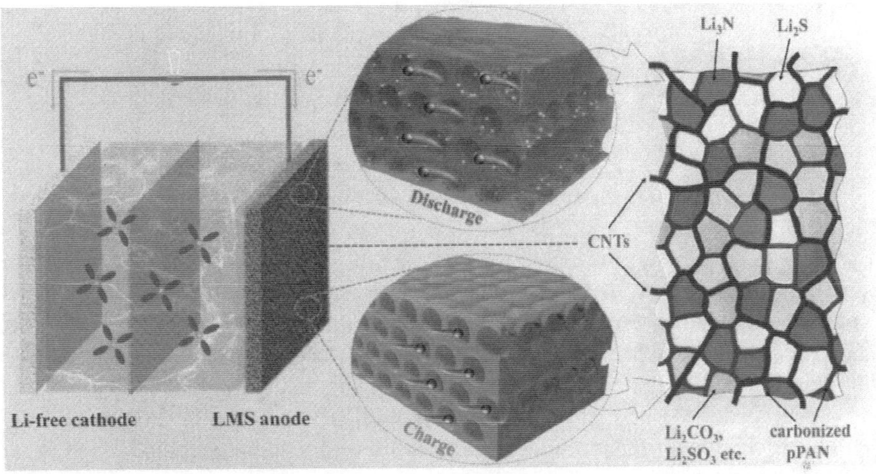

Figure 3 schematic diagram of LMS anode electrochemical lithium plating/lithium removal process and composition of 3D ion/electron hybrid framework [6]

3. Artificial SEI Membrane

In case lithium metal anode contacts with electrolyte, irreversible reaction occurs between them. Solid electrolyte interface (SEI) film with solid electrolyte properties is formed on the surface of lithium metal. The film is an ion conductor and an electronic insulator. It has a double-layer structure. Near the surface of lithium metal is an inorganic layer composed of Li_2O, LiF, LiOH and Li_2CO_3, Near the electrolyte is an organic layer composed of $ROCO_2Li$, ROLi and $RCOO_2Li$ (R is an organic group related to solvent). The organic and inorganic components are competitively balanced and coexisting, with mosaic morphology [7], as shown in Figure 4.

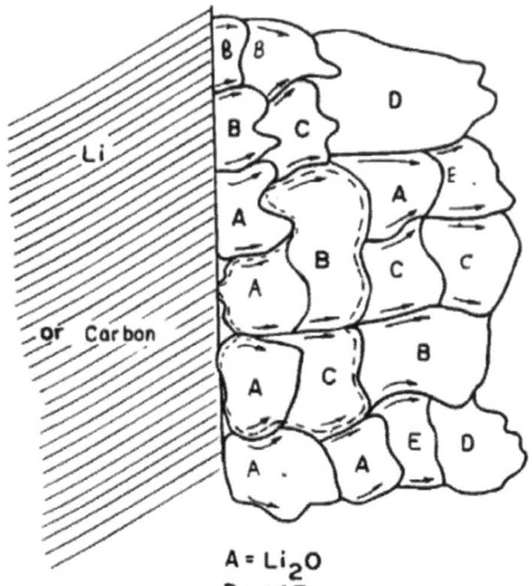

Figure 4 schematic diagram of SEI mosaic structure [7]

Li^+ exists in the form of solvation in the electrolyte. When Li^+ migrates in the SEI film, the desolvation process occurs first, and then Li + migrates in the SEI film [8]. The SEI film is the last medium for Li^+ to reach the electrode surface. Therefore, the morphology of the SEI film directly affects the morphology of the electrochemical precipitation of Li on the electrode surface. According to the structure of SEI described in reference [7]. The distribution of chemical components and the thickness of SEI film are not uniform, and there is obvious grain boundary structure. Single et al. [9] found that Li^+ preferentially passes through the SEI film at the grain boundary between the components, which leads to the uneven electrochemical precipitation of lithium ion on the electrode surface. In order to further understand the influence of each inorganic component on the migration of Li^+, Y. C. Chen et al. [10] studied the migration barrier of Li^+ in the main inorganic components. It was found that the migration barrier of Li^+ was the smallest in Li_2O (0.152 eV), the lowest in Li_2CO_3 (0.227-0.491 eV), and the highest in LiF (0.729 eV). The mobility of Li^+ in Li2O was the highest. Jie Xiao [11] pointed out that the concentration gradient of Li^+ is an important factor affecting the morphology of lithium-ion electrochemical precipitation. Suppose the conduction velocity of Li^+ is slow, a large concentration gradient will be formed on the electrode surface to promote the growth of lithium dendrite. Therefore, the different conductivity of Li^+ in each component is also an important factor leading to the growth of lithium dendrite. Therefore, the construction of SEI film with uniform chemical composition distribution, no grain boundary and high Li^+ conductivity becomes a main direction to inhibit the growth of lithium dendrite. According to the above analysis results, Cui Yi's research team [12] heated the sulfur powder to 240 °C at the bottom of the stainless-steel container to generate sulfur vapor, and placed metal lithium on the top to make lithium at 170 °C. The Li + conductivity ($\approx 10^{-5}$ $S \cdot cm^{-1}$) of this layer is much higher than that of Li_2CO_3, RCO_2Li ($\approx 10^{-8}$ $S \cdot cm^{-1}$) and LiF ($\approx 10^{-9}$ $S \cdot cm^{-1}$), which can effectively inhibit the growth of lithium dendrite caused by the Li^+ concentration gradient on the electrode surface. The single chemical composition makes the layer without grain boundary, which can effectively inhibit the growth of lithium dendrite at the grain boundary. Under the condition of current density of 2 $mA \cdot cm^{-2}$ and specific capacity of 5 $mAh \cdot cm^{-2}$, the symmetrical battery shows stable electrochemical performance in constant current test. Although Li_2S has high Li^+ conductivity, it has poor flexibility and poor contact effect with the electrode surface. Ding Yi's team [13] formed a Hg-Li alloy layer on the surface of lithium metal in situ through the direct reaction of Hg and Li. This kind of alloy layer has good Li^+ conductivity and flexibility. It can provide uniform Li^+ flux and keep good contact with the surface of lithium metal. It has good capacity retention in the whole battery test composed of $LiFePO^4$ (LFP) and

$LiNi_{0.6}Co_{0.2}Mn_{0.2}O_2$ (NCM622). This method provides a new idea for the construction of artificial SEI membrane. Xiong Jie et al. [14] of the University of Electronic Science and technology of China revealed the diffusion mechanism of Li^+ in the mixed SEI by constructing an artificial mixed SEI layer composed of lithium antimony alloy (Li_3S_b) and lithium fluoride (LiF) on the Li metal anode, which provided an idea for stabilizing the Li metal anode under actual high-rate conditions (Figure5).

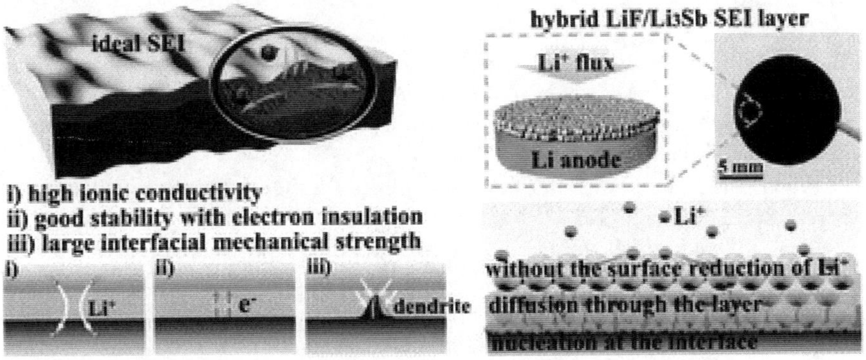

Figure 5 Li^+ diffusion in Li_3S_b/LiF composite SEI [14]

4. Solid Electrolyte

Solid electrolyte can not only solve the problems of leakage, poor chemical stability and explosion of liquid electrolyte, but also effectively inhibit the growth of lithium dendrite. At present, sulfides, oxides and polymers are mainly used as solid electrolytes. Kamaya, n et al. [15] prepared a new sulfur-based solid electrolyte $Li_{10}GeP_2S_{12}$ (LGPS) by mixing Li_2S, GeS_2 and P_2S_5 in a vacuum quartz tube at 550 °C. The lithium-ion conductivity of the material is 12 $mS\cdot cm^{-1}$ at room temperature, which is equivalent to that of liquid electrolyte. However, the material is unstable to lithium, which limits its use. Murugan, R. et al. [16] prepared a new garnet type solid electrolyte $Li_7La_3Zr_2O_{12}$ (LLZO), which is more stable than the former, but the lithium-ion conductivity is significantly lower than the former (3×10^{-4} $S\cdot cm^{-1}$). In addition, it cannot adapt to the volume change of electrode materials. In view of the instability of sulfide solid electrolyte, Lee, Y. g. et al. [17] of SamSung company prepared lithium free anode containing Ag-C composite layer. Through Ag alloying and dealloying and good mobility of Li^+ in nano carbon, lithium metal anode without dendrite morphology was realized (Figure 6). Professor Li Xin's team of Harvard University [18] creatively proposed a sandwich structure solid-state electrolyte design scheme, which sandwiched the less stable $Li_{5.5}PS_{4.5}CL_{1.5}$ (LPSCl) electrolyte between the more stable graphite, and prevented the growth of lithium dendrite through good local decomposition in the less stable electrolyte layer. In the whole cell test with $LiNi_{0.8}Mn_{0.1}Co_{0.1}O_2$（NMC811）, the cycle performance is very stable, and the capacity retention rate is 82% after 10,000 cycles at 20 C.

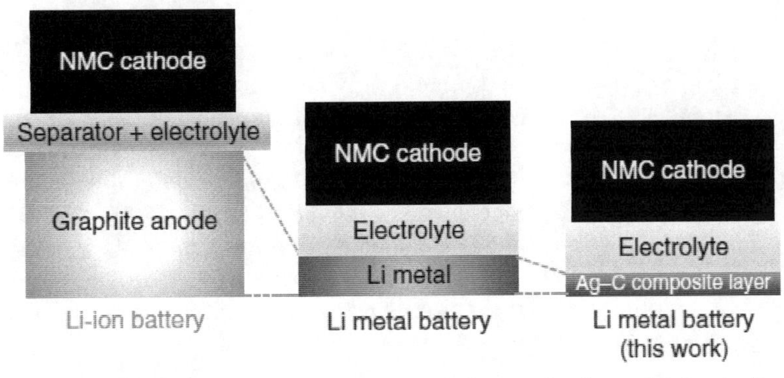

Figure 6 Lithium metal battery with Ag-C nanocomposite anode shows higher bulk energy density than traditional lithium-ion battery and lithium metal battery [18].

5. Conclusion

This paper introduces the research progress of lithium metal anode protection, mainly introduces the strategies of structured lithium metal anode, artificial SEI film and solid electrolyte, and analyzes the different characteristics of these strategies. Through the analysis in this paper, the combination of composite lithium metal anode and solid electrolyte can be an effective strategy for lithium metal protection in the future.

References

[1] Tarascon, J.M., Armand. M. *Issues and Challenges Facing Rechargeable Lithium Batteries* [J]. Materials for sustainable energy: a collection of peer-reviewed research and review articles from Nature Publishing Group, 2011: 171-179.

[2] Xu, W., Wang, J.L., Ding. F., et al. *Lithium Metal Anodes for Rechargeable Batteries* [J]. Energy Environ. Sci, 2014, 7: 513–537.

[3] Yan, K., Lu, Z., Lee, H.W., et al. *Selective Deposition and Stable Encapsulation of Lithium through Heterogeneous Seeded Growth*[J]. Nature Energy, 2016, 1(3): 16010.

[4] Xue, P., Liu, S., Shi, X., et al. *A Hierarchical Silver-Nanowire–Graphene Host Enabling Ultrahigh Rates and Superior Long-Term Cycling of Lithium-Metal Composite Anodes* [J]. Adv. Mater. 2018, 1804165.

[5] AI Xinping. *Where is the Way to Rechargeable Lithium Anode* [J]. energy storage science and technology, 2018, 7 (1): 37

[6] Lu, H., Zhang, T., Kuai, Y., et al. *A Superb 3D Composite Lithium Metal Anode Prepared by In-Situ Lithiation of Sulfurized Polyacrylonitrile* [J]. Energy Storage Materials, 2020, 33: 452-459.

[7] Peled, E. *Advanced Model for Solid Electrolyte Interphase [SEI] Electrodes in Liquid and Polymer Electrolytes*[J]. Journal of the Electrochemical Society, 1997, 144(8): L208-L210.

[8] Xu, K., Cresce, V. A., Lee, U. *Differentiating Contributions to "Ion Transfer" Barrier from Interphasial Resistance and Li^+ Desolvation at Electrolyte/Graphite Interface*[J]. Langmuir,2010, 26, 11538-11543.

[9] Single, F., Horstmann, B., Latz, A. *Revealing SEI Morphology: In-Depth Analysis of a Modeling Approach* [J]. Journal of the Electrochemical Society, 2017, 164(11): E3132-E3145.

[10] CHEN, Y.C., OUYANG, C.Y., SONG, L. J., et al. *Electrical and Lithiumion Dynamics in Three Main Components of Solid Electrolyte Interphase from Density Functional Theory Study* [J]. The Journal of Physical Chemistry C, 2011, 115(14): 7044-7049.

[11] Xiao, J. *How Lithium Dendrites Form in Liquid Batteries* [J]. Science 366 (6464), 426-427.

[12] Chen. H, Pei, A, Lin, D., et al. *Uniform High Ionic Conducting Lithium Sulfide Protection Layer for Stable Lithium Metal Anode*[J]. Advanced Energy Materials, 2019, 9(22): 1900858.

[13] He, G., Li, Q.W., Shen, Y. L., et al. *Flexible Amalgam Film Enables Stable Lithium Metal Anodes with High Capacities*[J]. Angewandte Chemie International Edition, 2019, 58(51), 18637-18641.

[14] Hu, A., Chen, W., Du, X., et al. *An Artificial Hybrid Interphase Endowing Ultrahigh-Rate and Practical Lithium Metal Anode*[J]. Energy Environ. Sci., 2021, DOI: 10.1039/d1ee00508a.

[15] Kamaya, N., Homma, K., Yamakawa, Y., et al. *A Lithium Superionic Conductor* [J]. Nature Materials, 2011, 10(9), 682−686.

[16] Murugan, R., Thangadurai, V., Weppner, W. *Fast Lithium-Ion Conduction in Garnet-Type Li7La3Zr2O12* [J]. Angewandte Chemie International Edition, 2007, 46, 7778−7781.

[17] Lee, Y. G., et al. *High-Energy Long-Cycling All-Solid-State Lithium Metal Batteries Enabled by Silver–Carbon Composite Anodes*. Nat Energy (2020). DOI: 10.1038/s41560-020-0575-z.

[18] Ye, L. H., Li, X. *A Dynamic Stability Design Strategy for Lithium Metal Solid State Batteries* [J]. Nature, 2021, 593: 218–222.

Low frequency electromagnetic casting of large size ZA21 magnesium alloy slab ingot

Yanxia Niu[1,2,a], Yan Shi[1,b*], Qichi Le[3,c*], Yixuan Wang[1], Shuai Ren[1], Na Wang[1]

[1] School of Automation Science and Electrical, Beihang University, Beijing 100083, China

[2]Jingdezhen Research Institute of Beihang University, Jingdezhen 333000, China

[3] Key Laboratory of Electromagnetic Processing of Materials, Ministry of Education, Northeastern University, Shenyang 110819, China

[b]*Corresponding author. Prof., Ph.D.; Tel.: +86-1082315821
E-mail address:shiyan@buaa.edu.cn (Yan Shi).

[c]*Corresponding author. Prof., Ph.D.; Tel.: +86-2483683312
E-mail address: qichil@mail.neu.edu.cn (Qichi Le).

Abstract. Surface quality and microstructure of ZA21 magnesium slab ingots by the direct-chill (DC) casting and low frequency electromagnetic direct chill casting (LFEC)were Deeply and systematically investigated. The surface quality of alloy ingot by LFEC casting is better than that without electromagnetic casting. The crack of ingot was avoided by applying electromagnetic field. The microstructure of ingot without electromagnetic field is coarse columnar grains, on the contrary, the microstructure of ingot by applying electromagnetic field is uniform equiaxed grains. The electromagnetic field can accelerate the flow of the melt and make the dendrite arm break to form equiaxed grains. In addition, the flow of the melt reduces the stress concentration in the solidification process, so that the crack can be avoided.

1. Introduction

The density of Mg alloys is about 1.74 g/cm^3[1]. It is obviously lower than that of traditional alloys, such as Al alloys (~2.70 g/cm^3) and ferrous materials (~7.87 g/cm^3) [2]. Therefore, Mg and its alloys are recognized as the next generation of light-weight metallic materials[3-4]. In addition, Mg alloys has the advantages of good castability and high specific strength (strength-to-weight ratio). Nevertheless, Mg and its alloys are not widely used as structural materials. This is due to the close packed hexagonal (HCP) structure of magnesium alloys[5]. The HCP structure leads to its difficult forming property, low strength and plasticity[6, 7]. For magnesium alloys, grain refinement is an effective method to improve its formability and mechanical properties[8].

It is an effective way to refine grains by applying external physical fields such as magnetic and ultrasonic during the casting process[9, 10]. But the ultrasonic field can make the ultrasonic rod extend into the melt and lead to the melt pollution[11]. In contrast, the application of electromagnetic field has the advantages of no pollution and easy control of intensity[12]. Direct chill (DC) casting technology can produce large magnesium alloy billets. However, the magnesium alloy ingots produced by traditional DC casting often have the disadvantages of coarse microstructure, serious chemical segregation, hot tearing porosity and so on[13]. These characteristics not only impose a burden on the next hot forming process, but also it is not conducive to improve the mechanical properties of the alloy. Low frequency electromagnetic direct chill casting (LFEC) can well refine the grains of as-cast magnesium alloys, reduce elemental segregation and improve the quality of casting products[14]. LFEC is mostly used in the production of round ingots but seldom used in the production of magnesium alloy flat ingots especially large-scale flat ingots.

In the present work, LFEC is applied to the production of large scale ZA21 magnesium alloy flat ingots. And the metallurgical quality of magnesium alloy ingots obtained by applied and unapplied electromagnetic fields is compared to explore the effect of the application of electromagnetic field on grain refinement.

2. Experimental procedures

ZA21 magnesium alloy is used in this study. The nominal composition was 2.0% Zn, 1.0% Al with balance Mg (in wt.%). The ZA21 magnesium alloy was synthesized with pure Mg ingot, pure Al ingot, and pure Zn ingot. The ZA21 magnesium alloy ingots can be obtained by traditional DC and LFEC methods. The first step, to heat pure Al, Zn and Mg ingots to melt by the resistance melting furnace. Second, to cool it to 700~710°C. Finally, to clear Fe by $MnCl_2$. When the temperature of the melt dropped to 680~700°C, to do DC and LFEC treatments respectively under gas protection. The shielding gas is a mixture of CO_2 and SF_6, and the volume ratio is 99:1. The DC method was to keep the water flow at 200L/min and the casting speed at 60mm/min. The LFEC conditions was to set ampere turns of excitation coil 8000AN and electromagnetic frequency 20Hz, the water flow rate was 200L/min, the casting speed is 60mm/min. The obtained ingot section size is 300mm×200mm.

Fig. 1 shows the casting process by the DC and LFEC treatments. In order to study the effect of electromagnetic field on the microstructure of alloy ingot, the microstructure of ingots was investigated, and Fig. 2 shows the specific sampling method.

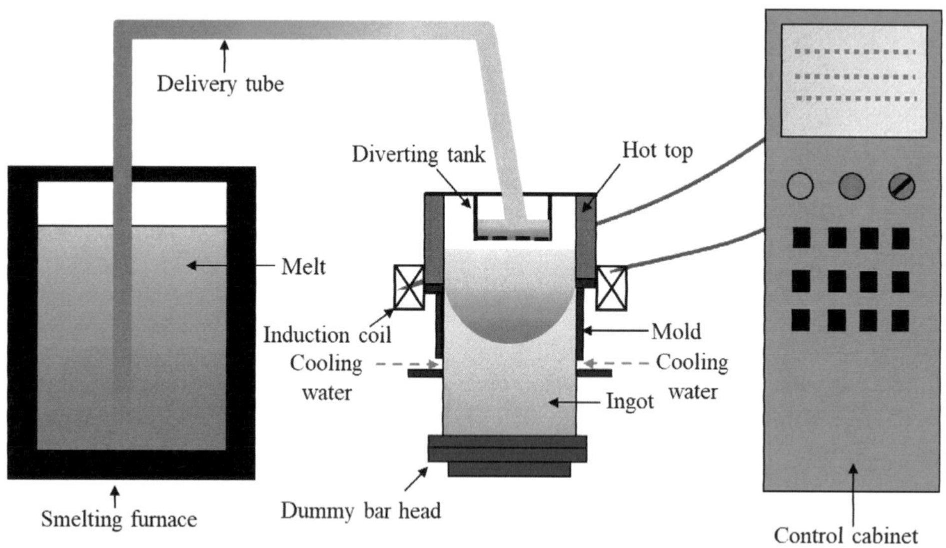

Fig.1 Schematic diagram of DC and LFEC casting

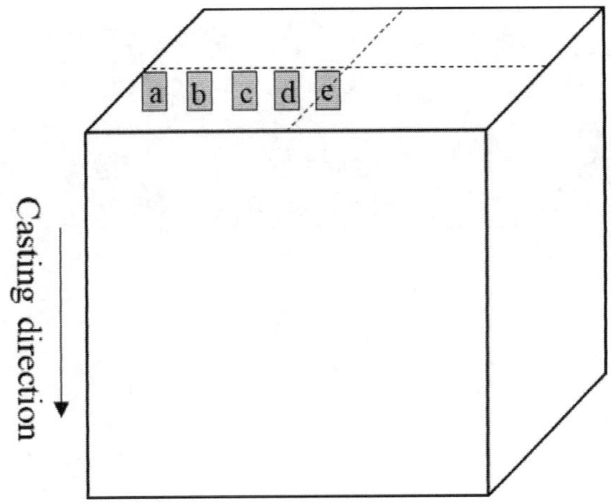

Fig.2 Schematic diagram of microstructure observation position

3. Results and discussions

The surface photos of the ZA21 alloy ingot obtained by DC and LFEC in the experiment are shown in Fig.3. It can be seen that the surface quality of the two ingots is quite different. The surface of DC ingot is very rough and obvious crack can be observed. In the contrast, the surface of LFEC ingot is relatively flat and free of cracks. This is because the Lorentz force generated by the electromagnetic field makes the melt to flow fully so that the melt can be better filled in the inner wall of the mold. Therefore, the surface of the ingot by LFEC is smoother than the ingot by DC. On the other hand, the full flow of the melt can release the internal stress of the melt and reduce the stress concentration during the solidification process, thus cracks can be avoided by LFEC. The result shows that the surface quality of ingot can be improved by applying electromagnetic field during DC casting.

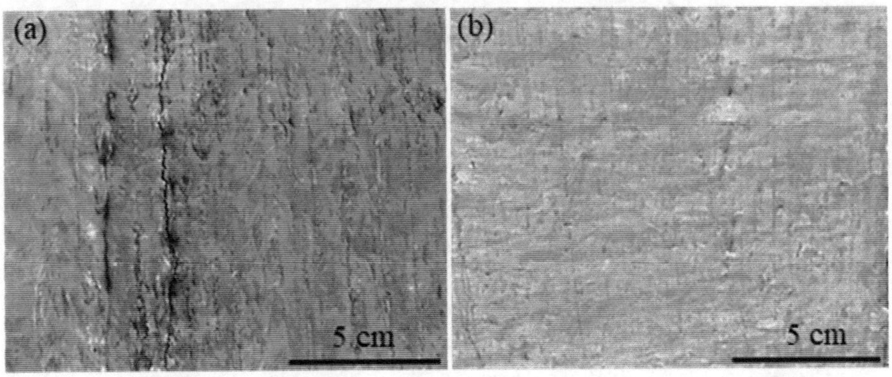

Fig.3 Surface of ingots by different casting method (a) DC; (b)LFEC

The microstructures of the ingot by DC can be seen in Fig.3. Fig. 2 shows the location of microstructure observation. It can be seen that from Fig.4(a) there is a thin layer of chill grain zone on the edge. Columnar grain zone is next to the chill grain zone, and the long axis direction of the columnar grains is about 45° with the casting direction, which can be seen in Fig.4(a). There is a large columnar grain zone, which can be seen in Fig.4(b) (c) (d), and the long axis direction of these columnar grains is parallel to the casting direction. The central position of ingot is coarse equiaxed grains which is shown in fig.4(e). The formation of chill grain zone is due to the rapid cooling effect of cooling water, and the edge region quickly crystallizes and solidifies to form fine grains. High temperature gradient and the specific shape of solid-liquid boundary leads to the formation of columnar crystals. The coarse grains are formed because of the slow heat dissipation at the center of the ingot.

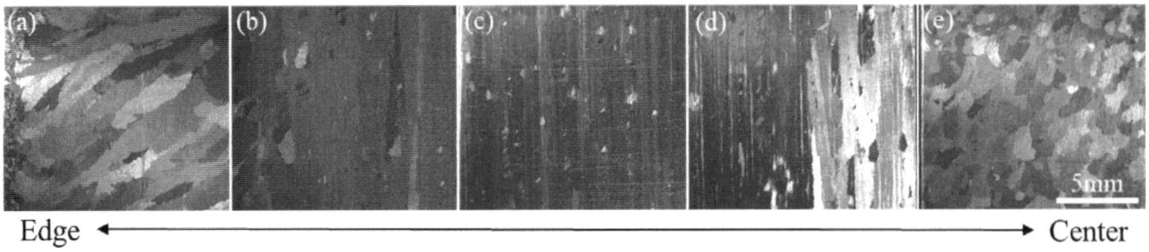

Fig.4 Microstructures of ingot by DC

Fig. 5 is the microstructures of ingot by LFEC. The central part of the ingot is the mixed grain state of coarse grains and columnar grains(see Fig.5e). The other part of the ingot is relatively uniform equiaxed grains (see Fig.5a, b, c and d). In addition to the uniformity of grains, the size of grains is smaller than that of DC ingot. The Lorentz force generated by the electromagnetic field makes the melt to flow and a large number of dendrite arms were broken to form equiaxed fine grains. From another perspective, the flow of the melt makes the temperature gradient in the solidification process more uniform, so the grains formed are also more uniform.

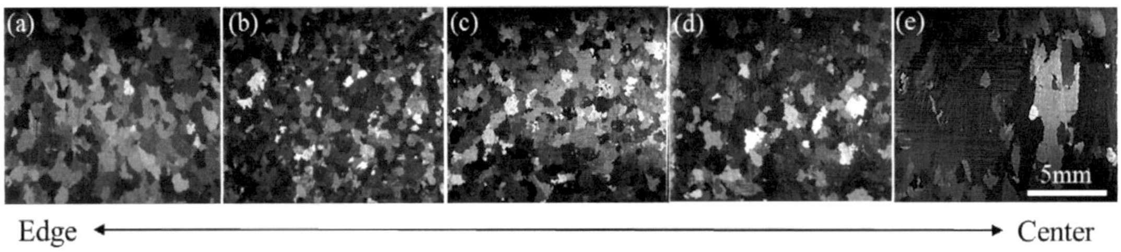

Fig.5 Microstructures of ingot by LFEC

4. Conclusions

In this work, the DC and LFEC were carried out to study the influence of electromagnetic field on surface quality and microstructures of large size slab ingot. The main conclusions are as follows.

(1) ZA21 magnesium alloy slab ingot with good surface quality was obtained by LFEC casting.
(2) The Loren magnetic force produced by the electromagnetic field can promote the melt flow and make the melt fill the inner wall of the mold better and the surface of the ingot smoother.
(3) The sufficient liquidity of the melt brought by Lorentz force can reduce the stress concentration and avoid the crack.
(4) Equiaxed fine grains can be obtained by LFEC, the reason is that Lorentz force makes the melt to flow and a large number of dendrite arms are broken to form equiaxed fine grains.

References

[1] X.S. Huang, Y. Chino, M. Yuasa, etal. Microstructure and mechanical properties of AZX912 magnesium alloy extruded at different temperatures, Mat Sci Eng a-Struct 679 (2017) 162-171.
[2] W.T. Jia, F.K. Ning, Y.P. Ding, etal. Cui, Role of pre-width reduction in deformation behavior of AZ31B alloy during break-down rolling and finish rolling, Mat Sci Eng a-Struct 720 (2018) 11-23.
[3] C. Xu, M. Zheng, S. Xu,etal. Improving strength and ductility of Mg–Gd–Y–Zn–Zr alloy simultaneously via extrusion, hot rolling and ageing, Materials Science and Engineering: A 643 (2015) 137-141.
[4] S.H. You, Y.D. Huang, K.U. Kainer, etal. Recent research and developments on wrought magnesium alloys, J Magnes Alloy 5(3) (2017) 239-253.

[5] S.Q. Zhu, H.G. Yan, J.H. Chen. Effect of twinning and dynamic recrystallization on the high strain rate rolling process, Scripta Mater 63(10) (2010) 985-988.
[6] S.H. Lu, D. Wu, R.S. Chen. The effect of twinning on dynamic recrystallization behavior of Mg-Gd-Y alloy during hot compression, Journal of Alloys and Compounds 803 (2019) 277-290.
[7] S.L. Yu, C.M. Liu, Y.H. Gao, etal.Dynamic recrystallization mechanism of Mg-8.5Gd-2.5Y-0.4Zr alloy during hot ring rolling, Mater Charact 131 (2017) 135-139.
[8] U.M. Chaudry, T.H. Kim, Y.S. Kim, K. Hamad, Y.G. Ko, J.G. Kim, Dynamic recrystallization behavior of AZ31-0.5Ca magnesium alloy during warm rolling, Mat Sci Eng a-Struct 762 (2019).
[9] X.R. Chen, Q.Y. Liao, Y.X. Niu, , etal. Comparison study of hot deformation behavior and processing map of AZ80 magnesium alloy casted with and without ultrasonic vibration, Journal of Alloys and Compounds 803 (2019) 585-596.
[10] J. Dong, J.Z. Cui, F.X. Yu, etal. Effect of low-frequency electromagnetic casting on the castability, microstructure, and tensile properties of direct-chill cast Al-Zn-Mg-Cu alloy, Metall Mater Trans A 35a(8) (2004) 2487-2494.
[11] X.R. Chen, F.K. Ning, J. Hou, etal. Tang, Dual-frequency ultrasonic treatment on microstructure and mechanical properties of ZK60 magnesium alloy, Ultrason Sonochem 40 (2018) 433-441.
[12] J.Z. Cui, Z.Q. Zhang, Q.C. Le. Direct-chilling casting of Mg alloy under electromagnetic and ultrasonic combined field, T Nonferr Metal Soc 20 (2010) S297-S305.
[13] Y.B. Zuo, B. Jiang, Z. Fan. Microstructures of DC cast light alloys under the influence of intensive melt shearing, Light Metals Technology V 690 (2011) 137-140.
[14] S.J. Guo, Q.C. Le, Z.H. Zhao, etal. Microstructural refinement of DC cast AZ80 Mg billets by low frequency electromagnetic vibration, Mat Sci Eng a-Struct 404(1-2) (2005) 323-329.

Study on Micro-arc Oxidation of Mg-Zn-Y-Nd-Ca Magnesium Alloys

Yuelai Dai*, Yulei Li, Juan Lv, Jingwei Li

Inner Mongolia Metallic Materials Research Institute, Ningbo, 315103, China

*Corresponding author's e-mail: dyl@norincogroup.com.cn

daiyuelai_1988@126.com

Abstract. To solve the problem of rapid degradation of magnesium alloy implanted in human body, the surface of magnesium alloy was modified by micro-arc oxidation treatment. In order to improve the comprehensive mechanical properties and corrosion resistance of magnesium alloy materials, a new type of Mg-Zn-Y-Nd-Ca degradable magnesium alloy materials was formed by adding a certain proportion of rare earth elements Y and Nd, as well as Zn and Ca. A micro - arc oxidation layer containing Ca, P and Si was formed on the surface of magnesium alloy by secondary micro - arc oxidation. The surface of the micro-arc oxidation film was covered by sol-gel technology to reduce the micro-crack and micro-pore structure of the film and improve the corrosion resistance of the film in the human body simulation liquid.

1. Introduction

Magnesium alloys have attracted wide attention as medical degradable implant materials due to their biodegradability and good mechanical properties[1]. Traditional biometallic materials such as 316L stainless steel, cobalt-chromium alloy and titanium alloy cannot decompose by themselves after being implanted into human body, resulting in many adverse reactions and side effects[2]. However, the rapid degradation rate of magnesium alloy limits its clinical application. The corrosion resistance was improved by alloying and micro-arc oxidation treatment on the alloy surface.

The properties of micro-arc oxidation coatings are affected by many factors, including the composition of magnesium alloy, electrolyte composition and electrical parameters.Liang et al. compared the effects of different electrolyte systems on the properties of micro-arc oxidation coatings on the surface of AM50 magnesium alloy[3]. Cai et al. prepared phosphorus film and silicon film on the surface of AZ91D magnesium alloy, and the experiment found that the silicon film had higher corrosion resistance than the phosphorus film[4].

2. Composition design of magnesium alloy

In this paper, Mg-Zn-Y-Nd-Ca magnesium alloy is taken as the research object, and the flux covering method is adopted to melt the magnesium alloy material. The addition of rare earth elements Y and Nd has a great effect on the micro-structure of magnesium alloys. Firstly, both elements can combine with magnesium to form binary phase, which becomes the core of heterogeneous nucleation of magnesium matrix.

Secondly, during the solidification process, Y and Nd gather in the front edge of the alloy to form a component of super cooling, which impedes the growth of the magnesium matrix, refine the as-cast micro-structure of the magnesium alloy, and improve the comprehensive mechanical properties of the alloy. The addition of Zn element can enhance the mechanical properties of magnesium alloy by solution strengthening[5]. The addition of Ca element can refine the grain[6]. The element addition is mainly in the form of magnesium - master alloy in the melting process of magnesium alloy. Mg master alloys are mainly Mg-20%Y, Mg-20%N, Mg-12%Zn and Mg-30%Ca, and the composition of the final Mg-Zn-Y-Nd-Ca magnesium alloy is shown in Table 1.

Table 1 Composition of Mg-Zn-Y-Nd-Ca magnesium alloy (wt%)

Y	Nd	Ca	Zn	Mg
3.5~4.2	2.4~2.7	0~0.6	0.5	Balance

3. Preparation and technological parameters of micro-arc oxidation electrolyte

The basic components of micro-arc oxidation electrolyte are mainly potassium hydroxide, ammonium hydrogen fluoride, glycerol, hydrogen peroxide and so on. Formulations for electrolytes containing calcium and phosphorus and electrolytes containing silicate are shown in Table 2 and Table 3, respectively.

Table 2 Chemical composition of calcium-phosphate electrolyte

Constitute	$C_3H_7CaO_6P{\cdot}H_2O$	$(NaPO_3)_6$	KOH	NH4HF2	$C_3H_8O_2$	H_2O_2	$N(CH_2CH_2OH)_3$
Concentration	7.00g/L	7.50g/L	4.5g/L	7.0g/L	5.0mL/L	7.5mL/L	5.0mL/L

Table 3 Chemical composition of silicate electrolyte

Constitute	$NaSiO_3{\cdot}9H_2O$	KOH	NH_4HF_2	$C_3H_8O_2$	H_2O_2	$N(CH_2CH_2OH)_3$
Concentration	15g/L	5.0g/L	7.0g/L	5.0mL/L	7.5mL/L	5.0mL/L

The process parameters used in the process of micro-arc oxidation, such as positive and negative voltage, oxidation time and duty cycle, are closely related to the thickness of the oxidized ceramic layer, the size and number of micro-pores, as well as the composition and phase composition of the film. Therefore, it is necessary to select reasonable process parameters. Constant pressure mode is used in the test, and the specific parameters are shown in Table 4.

Table 4 Electrical parameters setting of the prepared micro-arc oxidation film

Forward voltage (V)	Negative voltage (V)	Oxidation time (min)	Duty ratio (%)	Pulse frequency (Hz)
400	50	10	40	1000

4. Preparation of micro-arc oxidation layer

Electrolytic liquid systems commonly used in micro-arc oxidation of magnesium alloys include single electrolytic liquid system and composite electrolytic liquid system. Single electrolytic liquid system mainly includes silicate, phosphate and silicate composition[7].

Preparation of micro-arc oxidation coatings containing calcium and phosphorus. The pretreated magnesium alloy is connected to the positive electrode of the power supply with an aluminum wire, and the stainless steel container containing calcium and phosphorus electrolyte is poured into the container to be connected to the negative electrode of the power supply. The magnesium alloy sample is placed in the center of the stainless steel container to open the circulating cooling system. Turn on the micro - arc oxidation power supply and carry out the micro - arc oxidation experiment according to the set electrical parameters. At the end of the experiment, the samples were taken out, washed several times in deionized water and anhydrous ethanol, and dried to obtain the samples with micro-arc oxidation layer.

Preparation of micro - arc oxidation films containing calcium, phosphorus and silicon. The dried micro-arc oxidation film samples containing calcium and phosphorus are connected to the positive pole of the power supply, and the stainless steel container containing silicate electrolyte is poured into the container to connect to the negative pole of the power supply, and the magnesium alloy samples are placed in the center of the stainless steel container to open the circulating cooling system. Turn on the micro - arc oxidation power supply and carry out the micro - arc oxidation experiment according to the set parameters. After the end of the experiment, the samples were taken out, washed several times in deionized water and anhydrous ethanol, and then dried in a blast oven at 35°C for 3h to obtain the samples with calcium, phosphorus and silicon as micro-arc oxidation film.

The thickness of the films containing Ca and P is between 20 and 28μm, while the films containing Ca and P and Si are between 30 and 41μm. Figure 1 shows the microscopic images of the micro-arc oxidation films containing calcium phosphorus and calcium phosphorus silicon. There are micro-pores of different sizes in the two kinds of films. The micro-pores of the films containing Ca, P and Si are 2-5μm in diameter, while those containing Ca, P and Si are relatively small, about 1-2μm in diameter.

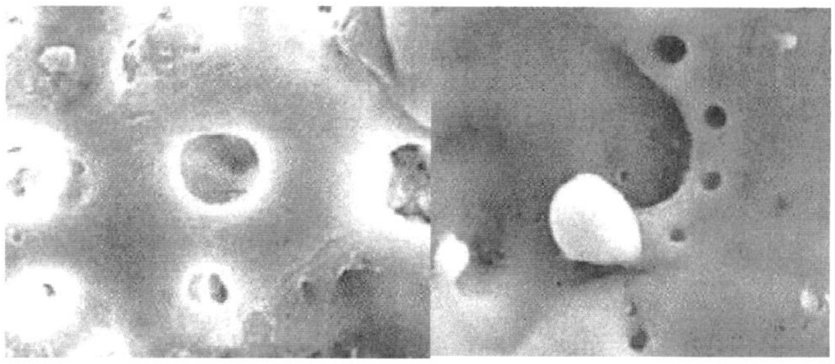

Ca-P Ca-P-Si

Figure 1 Microscopic image of micro-arc oxidation film

There are no obvious micro-cracks on the surface of the micro-arc oxidation film, and the micro-cracks mainly exist in the interior of the film. The formation process of the film layer is through local inward growth rather than uniform growth on the surface [8]. The presence of micro-pores and micro-cracks has a certain effect on the corrosion resistance of magnesium alloys.

5. Sol-gel technology

Micro-arc oxidation surface treatment can greatly improve the surface characteristics of the alloy, such as hardness, corrosion resistance, wear resistance and adhesion to the matrix. However, at the same time, the existence of micro-pores and micro-cracks in the micro-arc oxidation ceramic film reduces the protective power of the micro-arc oxidation film to the body. In order to improve the protective characteristics of the micro-arc oxidation film on the surface of magnesium alloy, it is necessary to seal the pores of the film. After immersion in the sol, a sol gel film appears on the surface of the micro-arc oxidation film. However, the sol-gel film did not completely cover the micro-pores of the micro-arc oxidation film after one and two times of dipping and pulling. With the increase of the number of dipping and pulling, the micro-pores of micro-arc oxidation are basically sealed. The magnesium alloy without micro-arc oxidation treatment, the magnesium alloy with Ca-P-Si film, and the sealed magnesium alloy were immersed into human body fluid for 3d,6d,12d, and 18d respectively. The weight loss rates were shown in Table 5, Table 6, and Table 7 respectively. Micro-arc oxidation film can effectively reduce the corrosion rate of magnesium alloy in human body fluid simulation, and sol-gel technology can make the micro-arc oxidation film have a lower corrosion rate.

Table 5 Weight loss rate of Mg-Zn-Y-Nd-Ca magnesium alloy immersed in simulated body fluids

Time	3d	6d	12d	18d
Weight loss(%)	1.06	2.03	5.27	8.43

Table 6 Weight loss rate of composite film containing Ca-P-Si in simulated body fluid immersion

Time	3d	6d	12d	18d
Weight loss(%)	0.64	1.22	2.35	3.27

Table 7 Weight loss rate of sol-gel treated composite film in simulated body fluid immersion

Time	3d	6d	12d	18d
Weight loss(%)	0.22	0.51	1.25	2.14

6. Conclusion

The Mg-Zn-Y-Nd-Ca magnesium alloy was used as the matrix, and the micro-arc oxidation film containing Ca, phosphorus and silicon was obtained by secondary micro-arc oxidation on the basis of micro-arc oxidation film containing Ca, phosphorus and silicon. On the basis of the micro-arc oxidation film containing Ca-P-Si, a composite film was prepared on the surface of the film by sol-gel technology to seal the micro-cracks and micro-pores in the film, and effectively reduce the degradation rate at the early stage of soaking.

Acknowledgments

This study was supported by Zhejiang Basic Public Welfare Research Project of China under Grant No.LGF19E010001; Natural Science Foundation of Ningbo, China(No.2018A610059).

References

[1] Li X, Liu X, Wu S, Yeung K W K, Zheng Y, Chu P K. Design of magnesium alloys with controllable degradation for biomedical implants: From bulk to surface[J]. Acta Biomaterialia, 2016, 45: 2-30.

[2] Gordon J. Mcdougall, Partricia Dobson, Pauline Smith, Alison Blake, Derke Stewart. Assessing potential bioavailability of raspberry anthocyanins using an in vitro digestion system[J]. Agric. Food Chem. 2005,53: 5896-5904.

[3] Liang J, Srinivasan P B, Blawert C, Stormer M, Dietzel W, Electrochemical corrosion behavior of plasma electrolytic oxidation coatings on AM50 magnesium alloy formed in silicate and phosphate based electrolytes[J]. Electrochimica Acta, 2009,54(14):3842-3850.

[4] Cai Q, Wang L, Wei Bo, Liu Q. Electrochemical performance of microarc oxidation films formed on AZ91D magnesium alloy in silicate and phosphate electrolytes[J]. Surface and Coatings Technology, 2006,200(12-13):3727-3733.

[5] Yamasaki Y, Yoshida Y, Okazaki M, et al. Action of FGMgCO 3 Ap-collagen composite in promoting bone formation[j]. Biomaterial, 2003, 24(27): 4913-4920.

[6] Ilich J Z, Kerstetter J E. Nutrition in bone health revisited: a story beyond calcium[J]. Journal of the American College of Nutrition, 2000, 19(6): 715-737.

[7] Sankara Narayanan T S N, Park I S, Lee M H. Strategies to improve the corrosion resistance of microarc oxidation (MAO) coated magnesium alloys for degradable implants: Prospects and challenges[J]. Progress in Materials Science, 2014,60: 1-71.

[8] Gu Y, Bandopadhyay S, Chen C, Ning C, Guo Y. Long-term corrosion inhibition mechanism of microarc oxidation coated AZ31 Mg alloys for biomedical applications[J]. Material & Design, 2013,46: 66-75.

Preparation of a Novel UHMWPE Lithium Battery Separator by Electrospraying Method

Taiqi Liu*, Yunteng Zhao, Ziye Han, Jincheng Wu, Rong Zhao, Xue An, Xiaomin Zhao, Yan Li, Lei Zhang

([1]Research Center of Ecomaterial , Beijing Institute of Petrochemical Technology, Beijing 102617, China)

*Corresponding author: TAIQI LIU， Dr., Prof.; Tel./Fax:86-010-81292073; E-mail:liutaiqi@bipt.edu.cn

Abstract: In this paper,an ultra high molecular weight polyethylene (UHMWPE) lithium ion battery composite separator was successfully prepared via the electrospraying technique. First, the optimal electrospraying parameters of Polyvinylidene fluoride (PVDF), mass fraction of PVDF 3wt% and electrospraying voltage 21kV, were successfully determined by adjusting the effect of voltage and concentration of PVDF solution. Then the composite separator was prepared by electrospraying PVDF particles on the UHMWPE film. Finally the porosity rate, thermal stability and charge and discharge properties of the separator were analyzed.The analytical results show that the porosity rate of the composite separator is increased from 46.5% to 73.1%, and the heat shrinkage of the separator in the longitudinal is reduced from 2.6% to 1.3%. The first specific discharge capacity of the composite separator increases by 5.8%, and the separator has a good cycle stability after 50 cycles.

1. Introduction

Because the lithium battery has excellent performance, such as high energy density, high operating voltage, long cycle life, low self-discharge rate [1], the battery is widely used in electronic products including phone, laptop and so on.

As an important component of the lithium battery, the separator which is between the positive and the negative has the effect to provide a microporous channels for the ion mobility [2]. The capacity, cycle stability and safety of the battery are directly affected by the quality of the separator, that is to say, if there is no high-quality separator, the excellent battery will not exist [3]. Electrospraying is a kind of processing method to prepare polymeric particle *via* a high-voltage electrostatic field. During the electrospraying process, polymeric parameter, solution parameter, voltage and receiving distance have influence on the size of the particle and morphology [4-6]. Hu prepared the biodegradable polymeric microspheres by the electrospraying which was used as the drug carrier [7]. Cao deposited the nano titanium dioxide on the surface of the purification materials to prepare the sandwich materials which possess a good antibacterial properties [8]. Yin et al applied the electrospraying technique to prepare the Si/C nanomicrospheres which can be used as the positive material of the lithium battery, his research results show that the nanomicrospheres have the high reversible specific capacity and excellent cycle property [9]. Volvo prepared a metal oxide/PVDF nanocomposition to modify the negative material of the lithium battery [10]. Nanoparticle-on-nanofiber hybrid membranes were prepared by electrospraying of SiO2 dispersions and electrospinning of polyvinylidene fluoride (PVDF) on PP separator simultaneously, as the separator of the lithium battery, the pore rate and thermal stability of this separator has got significant improvement [11].

The research by electrospraying of organics to modify the UHMWPE separator has not been reported. In this article, PVDF particles was electrosprayed on the surface of the commercial UHMWPE separator to prepare a novel PVDF/UHMWPE composite separator. The analytical results show that the porosity rate of the composite separator is increased from 46.5% to 73.1%, the heat shrinkage of the separator in the longitudinal is reduced from 2.6% to 1.3%, the first specific discharge capacity of the composite separator increases by 5.8%, and the separator has a good cycle stability after 50 cycles.

2. Experimental Section

2.1 Chemical

Polyvinylidene fluoride: product 44080, molecular 534000, Alfa Aesar; Acetone: AR, from Beijing Chemical Works, China; N, N- dimethylformamide: AR, from Beijing Chemical Works, China; UHMWPE separator: SG208, molecular weight 1800000, from HeBei Gellec New Energy Material Science&Technoloy Co., Ltd., China

2.2 Preparation of PVDF particles

The PVDF solution was prepared by dissolving 3wt% PVDF in DMF/acetone (2:3 by weight),which was stirred at 60°C for 8h. The PVDF mass fraction of the solution were 1 wt%, 3 wt%, 5 wt%, 7 wt%, 9 wt%, respectively. After ultrasound (power 150W) treatment for 30 min.s, PVDF particle was electrosprayed on the surface of UHMWPE film. The average diameter of particle was analyzed by the RISE-2008 laser particle size analyzer [12].

2.3 Preparation of composite separator

The UHMWPE composite separator was prepared via electrospraying the different contents of PVDF particle on commercial UHMWPE separator under the optimum condition.

2.4 Characterization and performance test of UHMWPE composite separator

The morphology of UHMWPE composite separator was determined by a S-4800 field emission scanning electron microscope (SEM, S-4800, Hitachi). The porosities of the separators were analysized by using n-butanol uptake tests. The porosity was calculated using the following equation:

$$\text{Porosity}(\%) = \frac{w_2 - w_1}{\rho V} \qquad (1)$$

Where w_2 and w_1 are the weights of wet and dry separators, respectively, ρ is the density of n-butanol, and V is the geometric volume of the separator.

The dimensional stability of the separators were determined by thermal shrinkage test at 120 °C for 1h. The shrinkage of separators was calculated according to the following equation:

$$\text{Shrinkage}(\%) = \frac{D_0 - D}{D_0} \times 100 \qquad (2)$$

Where D_0 and D are the diameters of separators before and after heat treatment, respectively.

The coin-type cells were assembled by sandwiching the composite separators between $LiFePO_4$ and Li metal electrodes in a glove box filled with dry argon. The cycle performance of cells fabricated with different separators were examined under the constant current mode at 0.2C with the operating voltage ranging from 2.5V to 4.2V using a cycle test instrument (CT2001A, Wuhan Jinnuo Electronics, China).

3. Results and discussion

3.1 Effect of concentration on the PVDF solution electrospraying

Concentration of a solution is an important parameter of electrospraying. The surface tension and viscosity can be adjusted by changing the concentration of PVDF , which will affect the size of the particles [7]. Due to the low entanglement of molecular chains in the 1 wt% PVDF solution, the jet was unstable under electrostatic force and surface tension, and the droplets occurred more frequently[8]. From the Figure.1, we can see that there were smaller and uniform particles in the concentration of 3 wt% PVDF, the particle size became larger in the concentration of 5 wt% PVDF, however, the concentration increased to 7-9 wt%, the electrostatic spraying process was difficult due to a high viscosity of the solution.

3.2 The influence of voltage on the electrospraying

According to the Scaling law of Hartman [13], the diameter of electrospraying particles is inversely proportional to the current, the greater the current, the smaller the diameter. And current is directly proportional to the voltage between the syringe and receiving device, the greater the voltage, the smaller the spray particle diameter. Table. 1 shows that when voltage increases from 13 kV to 21 kV, its average particle diameter reduce from 3.83μm to 1.70μm. We conclude from Figure.2 that there is reunion content at 13kV, and no reunion at 21kV. So we choose 21 kV as spraying voltage.

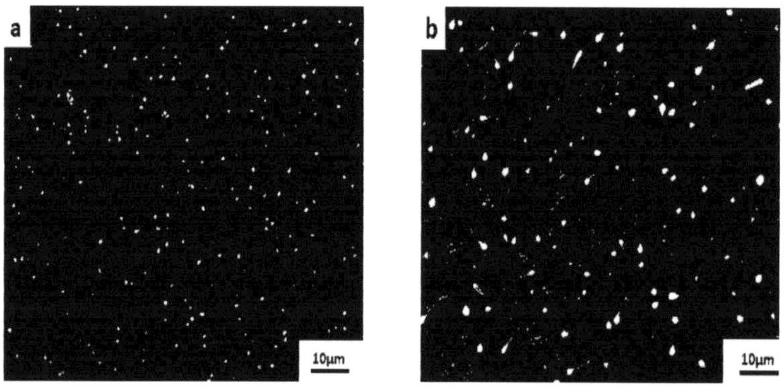

(a) 3 wt % (b) 5 wt %

Figure.1 SEM images of PVDF particles with different solution concentration

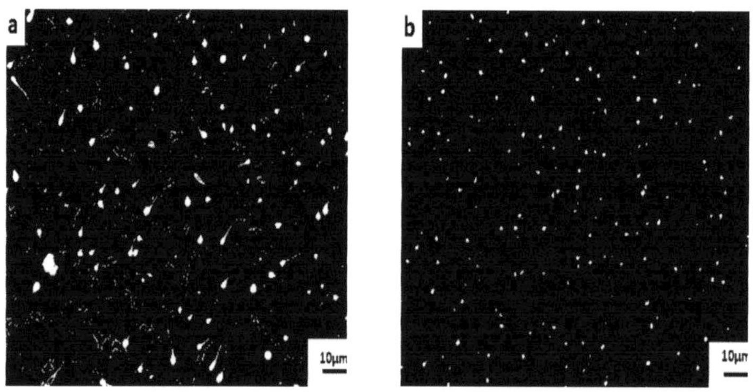

(a) 13kV (b) 21kV

Figure .2 SEM images of PVDF particles with different electric voltage

Table.1 Diameters of the electrospraying PVDF particles

Voltage (kV)	concentration (wt%)	Spraying distance(cm)	Average particle diameter(μm)
13	3	15	3.83
15	3	15	3.56
17	3	15	2.94
19	3	15	2.12
21	3	15	1.70

3.3 Porosity of composite separator

The test of composite separator porosity results with different quantities of PVDF particles spraying on UHMWPE separator surface are shown in Table.2. From Table.2, We can conclude that the porosity of composite separator is between 46% ~ 74%. When spraying time is 2 min, the maximum porosity is 73.1%, increased by 57.2 % compared with the UHMWPE separator, there is a maximum porosity because of pore jams formed with the increase of particles [11].

Table.2 The porosity of Composite Separator

Spraying time /min	The mass of dry separator M_1/g	The mass of wet separator M_2/g	The volume of dry separator V/cm^3	Porosity /%
PVDF 0min	0.0022	0.0036	0.0036	46.5
PVDF 2min	0.0023	0.0045	0.0036	73.1
PVDF 4min	0.0025	0.0043	0.0036	59.8
PVDF 6min	0.0028	0.0042	0.0036	46.5

3.4 Dimensional stability

We also studied the thermal stability of the PVDF/UHMWPE composite separator compared with commercial UHMWPE separator by calculating the dimensional change after heat treatment. Figure.3 shows that the UHMWPE separator exhibits relatively poor thermal stability and the longitudinally thermal shrinkage is about 2.6% at 120°C after 1h treatment. In contrast, the composite separator after spraying 6 min shows improved thermal stability and displays only 1.3% shrinkage at the same condition. The transverse thermal shrinkage of the commercial UHMWPE separator and the composite separator have no significant change.

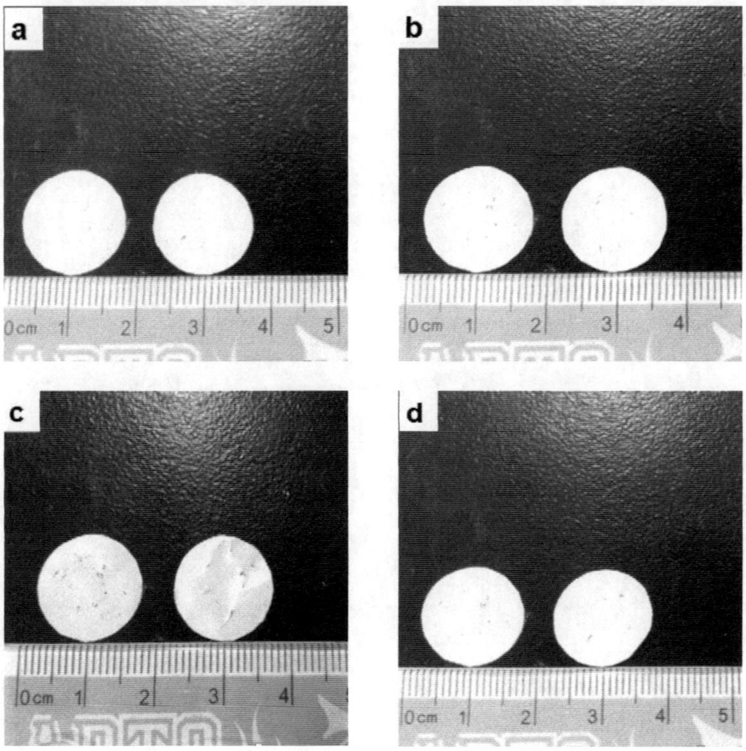

Figure .3 the Dimensional Stability of Composite Separator

Table.3 the Heat Shrinkage of Composite Separator

Spraying time /min	Transverse shrinkage /%	Longitudinally shrinkage /%
PVDF 0min	0.6	2.6
PVDF 2min	0.6	1.9
PVDF 4min	0.6	1.6
PVDF 6min	0.6	1.3

3.5 Charge and discharge performance of composite separator

The first discharge capacity curve of composite separator was shown in Figure.4. It shows that the first discharge capacity of UHMWPE separator is 116.67mAh.g^{-1}. When the spray time of PVDF particles is for 2 min, 4 min, 6 min, respectively, the first discharge capacity is 122.47 mAh.g^{-1}, 120.26 mAh.g^{-1}, 116.68 mAh.g^{-1}, respectively, we can see that the largest first discharge capacity appears in the spray time for 2 min, which increase by 5.8% than that of common UHMWPE separator. However, the first discharge capacity decreases with the increasing of spraying time. This is plausibly because that small amount of PVDF particles on the surface of UHMWPE can improve the porosity of the separator and let more lithium ions go across, more quantity of PVDF particles on UHMWPE surface not only will decrease the porosity of the separator and may clog pores on the surface of the separator, which results in the decrease of porosity and a reduction of discharge capacity, therefore, the best spray time is for 2 min [11].

Figure.5 is the cyclic stability of composite separator by electrostatic spraying PVDF particles on UHMWPE separator in different time. From which we see a good cyclic stability exists after 50 times charging and discharging test.

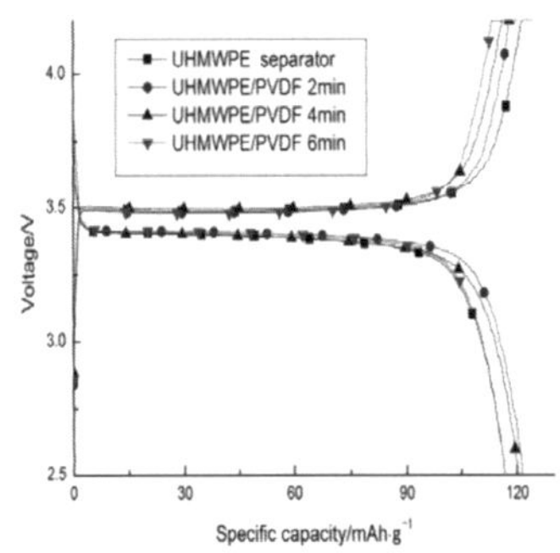

Figure.4 the First Discharge Capacity Curve of UHMWPE/PVDF Separator

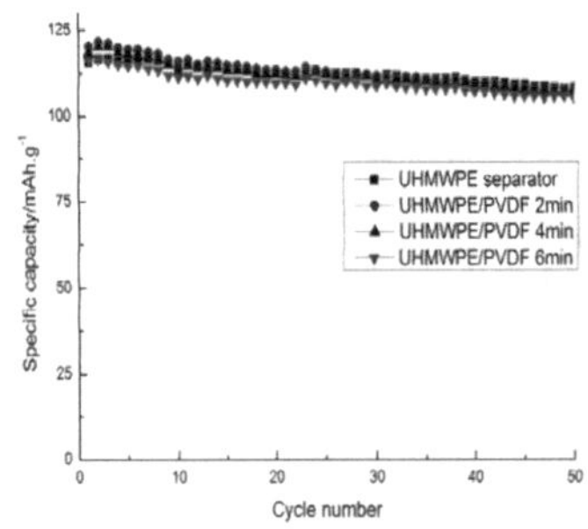

Figure.5 the Cyclic Performance Curve of UHMWPE/PVDF Separator

4. Conclusion

Ultra-high molecular weight polyethylene composite separator was made through electrospraying technology by loading PVDF particles on the surface of ultra-high molecular weight polyethylene separator. The separator has higher porosity and better dimensional stability than that of a commercial UHMWPE separator, The best parameters of spraying conditions are, a solution concentration of 3 wt %, a spraying voltage of 21 kV, a receiving distance of 15 cm. Tests of lithium ion battery bearing the composite separator show that various properties of composite separator is excellent, and the battery discharge capacity increased by 4.2% for the first time. After 50 times charging and discharging test, the battery exists a good cycle stability.

Acknowledgment:
The authors gratefully acknowledge the financial help of the project, Platform Construction of Sci-tech Innovation of Beijing Municipal Education Commission (PXM2016-014222-0000
15) for support of this work.

References

[1] Yanilmaz M, Lu Y, Li Y, Zhang X. J. Power Sources. 2015; 273:1114-1119.
[2] Cao J, Yu X, Tang D, Wu M, Wu D. Scientia Sinica (Chimica) .2014; 44 (7) :1125-1149.
[3] Huang YQ, Guan DA. Marine Electric & Electronic Engineering. 2011; 31 (1) :26-29.
[4] Yeo LY, Gagnon Z, Chang H. Biomaterials. 2005; 26 (31) :6122-6128.
[5] Jaworek A, Krupa A, Lackowski M, Sobczyk AT, Czech T, RamakrishnaS, SundarrajanS, PliszkaD. J. Electrostatics. 2009; 67 (2-3) :435–438.
[6] Jaworek A, Sobczyk AT. J. Electrostatics. 2008; 66 (3-4) :197-219.
[7] Hu CM. Fabrication of composite microspheres and fibers via electrospraying method for drug carriers. Dissertation of Shanghai Jiao Tong University, China, 2013.
[8] Cao BB, Liu TQ, Liu RX. Polymer Materials Science & Engineering. 2012; 28 (4) :117-120.
[9] Yin YX, Xin S, Wan LJ, Li CJ, Guo YG. J. Physical Chemistry C. 2011; 115 (29) :14148-14154.
[10] Valvo M, Garcia-tamayo E, Lafont U, Kelder EM. J. Power Sources. 2011; 196 (23) :10191-200.
[11] Yanilmaz M, Lu Y, Dirican M, Fu K, Zhang XW. J. Membrane Science. 2014; 456: 57-65.
[12] Reng F, Liu TQ, Filtration Property of Sandwich Structure Nanofiber Super-cleaning Material used to Remove Microzyme, Polymer Materials Science & Engineering. 2008; 24 (5) :135-138.
[13] Hartman RPA, Brunner DJ, Camelot DMA, Marijnissen JCM, Scarlett B. J. Aerosol Science. 2000; 31 (1) :65-95.

The photoelectric conversion enhancement of p-Cu_2O/n-ZnO heterojunction by deposition temperature

Jingjing Niu[1], Jiaqi Pan[1], Wendi Fu[1], and Chaorong Li[1*]

[1]Department of Physics and Key Laboratory of ATMMT Ministry of Education

Zhejiang, Sci-Tech University, Hangzhou 310018, China

[*]Corresponding author's e-mail: crli@zstu.edu.cn

Abstract: Dense ZnO films were prepared by RF magnetron sputtering, and PN junctions were formed with Cu_2O films prepared at different deposition temperatures. Different characterization methods show that the PN junction has better transmittance. Compared with the unheated Cu_2O, the Cu_2O / ZnO PN junction formed by the heated Cu_2O film shows obvious enhancement of photoelectric properties, which can be attributed to that the improved film crystallinity with the increased sputtering substrate temperature, can reduce the corresponding defects in the film.

1. Introduction

ZnO is regarded as one of the most promising optoelectronic materials, with high optical transmittance, piezoelectric and photoelectric properties.[1] It can be used to prepare zero-dimensional,one-dimensional and two-dimensional nanostructures according to requirements.[2-4]. Also, the preparation process is mature, and is expected being used as an ideal window material.[5] Up to now, a lot of research work on ZnO has been reported. Because of its excellent photoelectric performance, ZnO has been used in photoelectric detectors, light-emitting diodes, solar cells and other fields.[6,7] However, due to the wide band gap and intrinsic defects of ZnO, it would almost not respond in visible light and recombine rapidly. Obviously, the low solar energy utilization and high recombination rate of photo-generated electron hole pair would be the most important issue. Therefore, the key to preparation of transparent ZnO based PN junction is to select p-type materials with matching energy band.[8,9]

As known, the narrow-band gap p-type semiconductor, Cu_2O (2.2eV) is regarded as a potential light-absorbing layer material, because of its high absorption coefficient in the visible region and low preparation cost.[10,11] Especially combination of Cu_2O and ZnO to prepare PN junction can broaden the utilization rate of light by multilayered structure, enhance the absorption of light by Cu_2O based materials, inhibit the recombination of photogenerated carriers and enhance the performance of PEC via the Cu_2O/ZnO pn junction.[12,13]

In this work, the Cu_2O thin films were prepared by RF magnetron sputtering at room temperature and 275 °C to form PN junction with ZnO thin films. The results show that the p-Cu2O/n-ZnO film device obtains a good transmittance and a high photocurrent at 275 °C.

2.Experiment

2.1 Preparation

Preparation of the ZnO films: ZnO film was deposited on the glass via RF magnetron sputtering. The cavity background vacuum was 6×10^{-4} Pa, sputtering power was 100W, sputtering temperature was 350

°C, sputtering time was 30 min.

Preparation of the Cu_2O films: Cu_2O films was deposited on the glass via RF magnetron sputtering. The cavity background vacuum was 6×10^{-4} Pa, sputtering power was 90W, sputtering temperature was room temperature and 275 °C, sputtering time was 30 min.

2.2 Characterization

Micromorphology and microstructure of the samples were investigated by field emission scanning electron microscopy (FESEM, Hitachi S-4800, Japan) and transmission electron microscopy (TEM, JEM-2100, Japan). The crystallographic information and phase were characterized by X-ray diffraction (XRD, Bruker AXS D8-discover, Germany). Optical properties were determined from the ultraviolet-visible spectroscopy absorption (U-3900 Hitachi). The photoresponse was evaluated by electrochemical workstation (ZAHNER IM6, Germany). The illumination was obtained by a xenon lamp at 100mW/cm2. The photocurrent-voltage properties were performed by Keithley 4200.

3.Results and Discussion

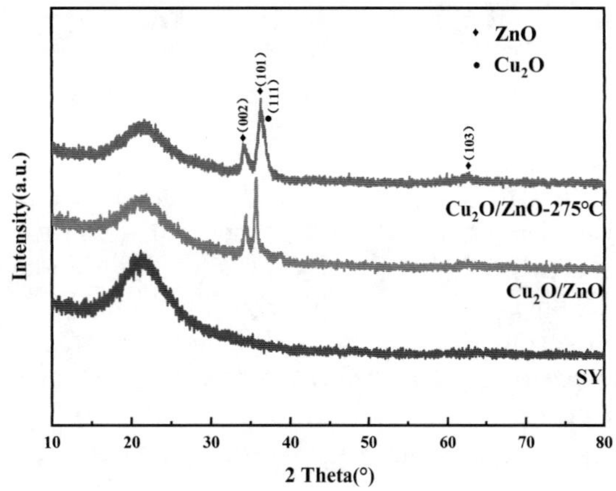

Fig.1. The XRD spectra of as-prepared samples

Fig. 1 illustrates the XRD of p-Cu_2O/n-ZnO device prepared at room temperature and 275 °C on the substrate. There, the diffraction peaks at 34.3°, 36.0°, 62.6°correspond to the(002), (101) and (103) crystalline planes of the ZnO(PDF#79-0208). The diffraction peaks at 36.5 ° correspond to the (111) plane of the Cu_2O(PDF#74-1230). However, the diffraction peak of Cu_2O with weak crystallinity could not be observed in p-Cu_2O/n-ZnO prepared at room temperature, but at 275 °C. The results show that 275 °C deposition temperature can improve the crystallinity of Cu_2O effectively.

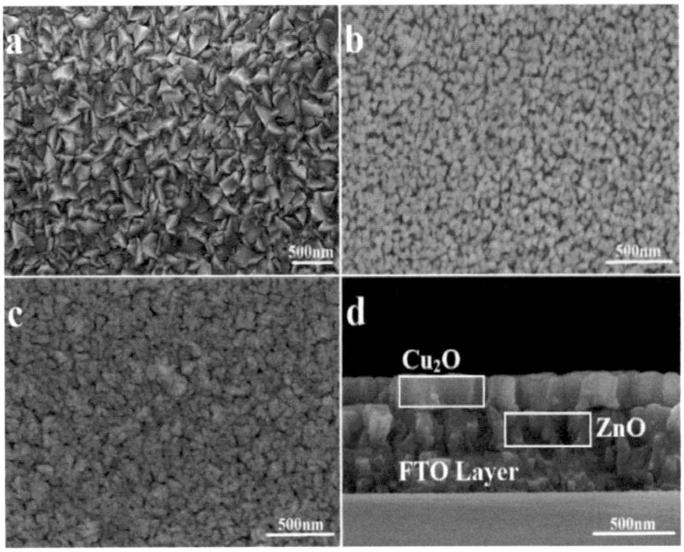

Fig.2. SEM and TEM of samples, (a) ZnO film, (b) Cu_2O film deposited at room temperature, (d) Cu_2O film deposited at 275 °C, (d) the cross section of the p-Cu_2O/n-ZnO device

Fig. 2 (a) shows the SEM of ZnO films. As revealed, the surface grain size of ZnO film is large, uniform, smooth, and relatively dense. Fig. 2 (b) and (c) shows SEM of Cu_2O films at room temperature and 275 °C, respectively. As shown in Fig. (b), the Cu_2O prepared at room temperature has a smaller grain size, and the surface is rough and porous. However, the grain size of Cu_2O prepared at 275 °C becomes larger and the film becomes denser. Fig. 2 (d) shows the cross section of as-prepared PN junction. As shown, the interface of Cu_2O and ZnO can be clearly distinguished, and both thicknesses are about ~200 nm, respectively.

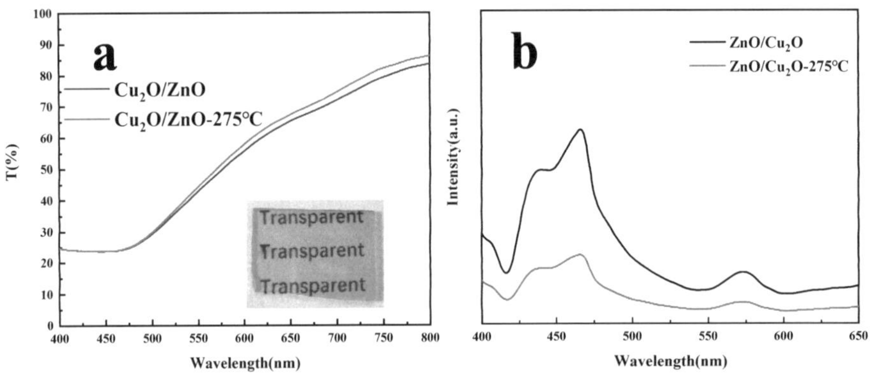

Fig.3. (a) The optical transmittance of different samples, (b) The PL of different samples

Fig. 3(a) presents the optical transmittance of as-prepared p-Cu_2O/n-ZnO device. It is obvious that the PN junction deposited at 275 °C exhibits a high transmittance of ~ 85%. It should be pointed out that the optical transmittance of p-Cu_2O/n-ZnO junction devices mainly depends on Cu_2O, because the band gap of ZnO is wider than that of Cu_2O. The inset is a physical photo of the p-Cu_2O/n-ZnO device. The transparent and distortion-free image shows that the P-N device obtains excellent transparency. Fig. 3(b) demonstrates PL of as-prepared p-Cu_2O/n-ZnO device. It illustrates that the PN junction device prepared at 275 °C exhibits lower PL. Above results can be attributed to that the reduced photo-generated carrier recombination and decreased defects in Cu_2O induced by the increased deposition temperature, which is conducive to improve the efficiency of photo-generated carriers.

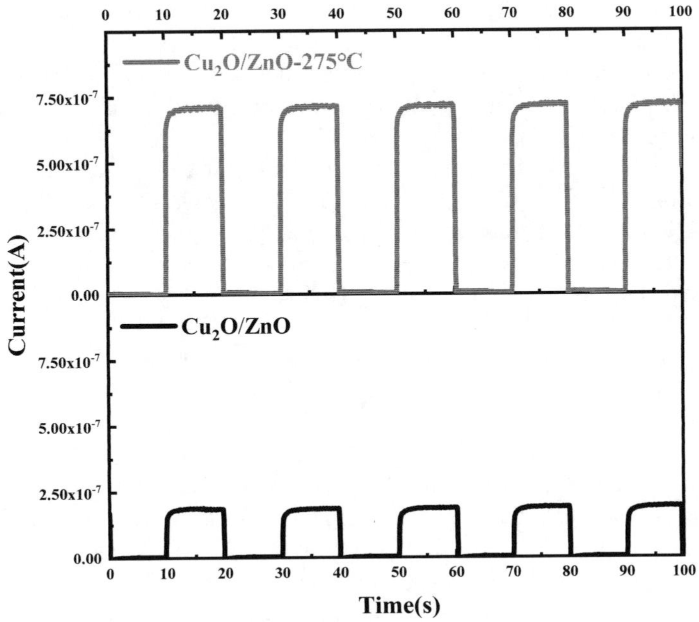

Fig.4. The photovoltaic conversion properties of the p-Cu_2O/n-ZnO device.

Fig. 4 is the photovoltaic conversion properties of the as-prepared p-Cu2O/n-ZnO device under 0 V bias. As displayed, the photocurrents of Cu_2O/ZnO and Cu_2O/ZnO-275 °C are about ~1.88×10^{-7} and~ 7.09×10^{-7} A, respectively. It's obvious that, the photovoltaic response of the p-Cu_2O/n-ZnO exhibits an obvious improvement after 275 °C high-temperature processing.

4. Conclusions

In this work, we have prepared the transparent p-Cu_2O/n-ZnO pn junction device by magnetron sputtering successfully. Herein, the PN junction device formed by Cu_2O thin film and ZnO thin film deposited at 275 °C shows a high transparency of about ~85%. Compared with the PN junction device prepared at room temperature, the photoelectric properties of Cu_2O/ZnO-275 °C exhibits an obvious enhancement. Which can be mainly attributed to that, with the increase of sputtering substrate temperature, the crystallinity of Cu_2O thin films obtain improvement, which makes the defects in Cu_2O thin films decreasing accordingly, and is beneficial to improve the efficiency of photo-generated carriers.

Acknowledgments

This work was supported by the Natural Science Foundation of China (Nos. 51802282 and 51672249).

References

[1] Pradel, KC.,Wu, W. Z.,Zhou, Y. S.,Wen, X. N.,Ding, Y.,Wang, Z. L. (2013) Piezotronic effect in solution-grown p-type ZnO nanowires and films. Nano Lett 13:2647e53.

[2] Pan, J. Q.,Chen Z. F.,Wang P. H.,Wang P. P.,Yu, Q.,Zhao W. J.,Wang J. J.,Zhu M.,Zheng Y. Y.,Li C. R.(2021) The overall water splitting of CdS/Ti^{3+}-$SrTiO_3$ core–shell heterojunction via OER enhancement of MnOx nanoparticles, Chem. Eng. J 424: 130357.

[3] Pan, J. Q.,Liu, Y. Y.,Ou, W.,Li, S.,Li,H. L.,Wang, J. J.,Song, C. S.,Zheng, Y. Y.,Li C. R. (2020) The photocatalytic hydrogen evolution enhancement of the MoS_2 lamellas modified g-C_3N_4/$SrTiO_3$ core-shell heterojunction, Renew. Energy 161: 340-349.

[4] Pan, J. Q.,Li, H. L.,Li, S.,Ou, W.,Liu, Y. Y.,Wang, J . J.,Song, C. S.,Zheng, Y. Y.,Li C. R. (2020) The enhanced photocatalytic hydrogen production of nickel-cobaltbimetals sulfide synergistic modified CdS nanorods with active facets, Renew. Energy 156: 469-477.

[5] Pan, J. Q.,Li, S.,Ou, W.,Liu, Y. Y.,Li, H. L.,Wang, J . J.,Song, C. S.,Zheng, Y. Y.,Li C. R. (2020) The photovoltaic conversion enhancement of NiO/Tm:CeO_2/SnO_2 transparent p-n junction device with dual-functional Tm:CeO_2 quantum dots, Chem. Eng. J 393: 124802.

[6] Cui, J.,Gibson, UJ. (2010) A simple two-step electrodeposition of Cu_2O/ZnO nanopillar solar cells. J Phys Chem C 114:6408e12.

[7] Yu, L.,Xiong L. B.,Yu, Y. (2015) Cu_2O homojunction solar cells: F-doped N-type thin film and highly improved efficiency. J Phys Chem C 119:22803e11.

[8] Kandjani, A. E.,Sabri, Y. M.,Periasamy, S. R.,Zohora, N.,Amin, M. H.,Nafady, A.,Bhargava, S. K. (2015) Controlling core/shell formation of nanocubic p-Cu_2O/n-ZnO toward enhanced photocatalytic performance. Langmuir 31:10922e30.

[9] Liu, Y.,Dong, R.,M, YJ.,Liu, W. W.,Zhu, A. Q.,Tan, P. F., Bian, Y.,Xiong, X.,Li, E.,Pan, J. (2019) Improved photocatalytic hydrogen evolution by facet engineering of core-shell structural CdS@ZnO. Int J Hydrogen Energy 44:25599e606.

[10] Pan, J. Q.,Wang P. H.,Wang P. P.,Yu, Q.,Wang, J . J.,Song, C. S.,Zheng, Y. Y.,Li C. R. (2021) The photocatalytic overall water splitting hydrogen production of g-C_3N_4/CdS hollow core–shell heterojunction via the HER/OER matching of Pt/MnO_x, Chem. Eng. J 405: 126622.

[11] Wang, Y. F.,Gao, J.,Wang, X. Z.,Jin, L. P.,Fang, L. L.,Zhang, M.,He, G.,Sun, Z. Q. (2018) Facile synthesis of core-shell ZnO/Cu_2O heterojunction with enhanced visible light-driven photocatalytic performance. J Sol-Gel Sci Technol 88:172e80.

[12] Kuang, P. Y.,Zheng, X. J.,Lin, J.,Huang, X. B.,Li, N.,Li, X.,Liu, Z. Q. (2017) Facile construction of dual p-n junctions in CdS/Cu_2O/ZnO photoanode with enhanced charge carrier separation and transfer ability. ACS Omega 2:852e63.

[13] Ma, C. H.,Liu, Z. F.,Cai, Q. J.,Han, C. C. ,Tong, Z. F. (2018) ZnO photoelectrode simultaneously modified with Cu_2O and Co-Pi based on broader light absorption and efficiently photogenerated carrier separation. Inorg Chem Front 5:2571e8.

Journal of Physics: Conference Series **2011** (2021) 012089 doi:10.1088/1742-6596/2011/1/012089

Metal-Organic-Frameworks-driven hollow core-shell Co_3O_4/TiO_2 Heterojunction for Enhancement of Photocatalytic Hydrogen generation

Guangsheng Xiao[1], Jiaqi Pan[2], Yueyue Fu[3], and Chaorong Li[1*]

Department of Physics and Key Laboratory of ATMMT Ministry of Education

Zhejiang, Sci-Tech University, Hangzhou 310018, China

*Corresponding author's e-mail: crli@zstu.edu.cn

Abstract: In recent years, how to further enhance the efficiency of photocatalytic hydrogen generation has become a research hotspot. At present, the metal-organic-frameworks (MOF) has attracted much attention in the field of catalysis due to its special physical and chemical properties. Herein, the hollow core-shell Co_3O_4/TiO_2 heterojunction is prepared by deposition and annealing method with ZIF-67 as self-sacrificial templates and metal precurous. The characterzation of XRD, SEM, and TEM indicate that the hollow core-shell Co_3O_4/TiO_2 heterojunction is synthesized successfully. The characterzation of UV, PL indicate that the Co_3O_4/TiO_2 show superior photoelectric performance. By evaluation, the Co_3O_4/TiO_2 heterogeneous catalyst exhibits a remarkable photocatalytic hydrogen production performance, which is ~11 and ~14 times higher than that of pure Co_3O_4 and TiO_2. Studies prove that the formation of heterojunction can promote charge carriers separation, the hollow structure can raise the solar efficiency, offer more surface areas and active sites, all these are beneficial for photocatalytic performance.

1. Introduction

Rencently, with the energy consuming crisis and environmental pollution issues, the development of renewable and pollution-free energy has attracted researchers' attentions increasingly such as hydrogen fule and solar energy et al.[1] Especially, the photocatalytic hydrogen generation(HER) is regarded as the most promising technology to convert solar fule into chemicial energy.[2] In past few decades, the TiO_2-based photocatalytic has been widely explored due to its suitable energy band structure, corrosion resistance and low cost.[3] However, the high recombination rate of photon-generated carriers and low active site area limit the application in potocatalytic water splitting.[4] Herein, the typical p-type semiconductor Co_3O_4, due to narrow band gap and abundant active sites, has been regarded as an ideal candidate material for construction of p-n heterostructures to improve the photocatalytic performance.[5]

It is worth mentioning that designing advanced structure is an efficient strategy to improve the photocatalytic activities.[6] In particular, the hollow structure has attracted great attention, because it can enhance solar light utilization, offer unique inner-external surface, and expose abundant active sites.[7] As reported, as one of the typical MOF, the zeolitic imidazolate framework(ZIF) exhibites novel properties, such diversity in structure, morphology and high specific surface area and themal stability.[8] The above physical and chemical properties can be used as metal precurous and self-sacrificial templates.[9] For the Photocatalysis, the ZIF is a effective route to fabricate the hollow structure, avoiding the complicated process of removing templates.[10]

Herein, we have prepared hollow core-shell Co_3O_4@TiO_2 heterostructures nanocompositics via using ZIF-67 as self-sacrificial templates and metal precurous. Compared with pure TiO_2, the Co_3O_4@TiO_2 nanocomposities exhibits obvious enhancement in phoyocatalytic HER. In additional, machanism of the heterojunction was also studied.

2. Experiment

2.1. Synthesis of ZIF-67

Firstly, the self-template and metal percusor of ZIF-67 was synthesised with some modification. Typically 0.6 g 2-methylimidazole was dissolved in 20 ml methanol solution. Subsequently, the 2-methylimidazole solution was droped into 15 ml solution combined with 0.3 g $Co(NO)_3 \cdot 6H_2O$ and 20 ml methanol with 20 min stiring. Then the mixed solution was incubated at 25 ° C for 3 hours. Finally, the prepared precipitate was washed by ethanol four times, dried in a hot vacuum drying oven at 70 °C overnight.

2.2 Synthesis of hollow core-shell Co_3O_4/TiO_2

Firstly, 0.5 ml titanium (IV) isopropoxide was dropped and mixed with 20 ml ethanol. 0.1 g ZIF-67 was put into 25 ml ethanol with sonicating for 5 min. Then, the above solutions were mixed with stiring for 6 h after droping 0.5 ml distilled water. Subsequently, the obtained precipitate was centrifuged with ethanol, dried in a hot vacuum oven at 60 °C for overnight. Finally, the hollow core-shell Co_3O_4/TiO_2 nanocomposities were prepared by calcination at 450 ° C (2°C /min) for 1 h in air.

2.3 Characterzation

The phase and crystal structure of the samples were tested by X-ray power diffraction. The morphology of the samples was characterized by FESEM (Hitachi S-4800) and TEM (JEM-2100). The UV and PL curves were characterized by spectrophotometer (Hitachi-U3900) and spectrophotomater (Hitachi F-7000). The HER characterization was obtained by Labsolar 6A photocatalytic system (Perfect Light Technology).

2.4 Photocatalytic activity

Photocatalytic HER perfrmances were measured in a gas-sealed glass reactor and evacuation system. Typicaly, 25 mg catalyst was mixed with 80 ml Na_2S (0.25 M) and Na_2SO_3 (0.15 M) solution (sacrificial agent). A 300W Xe arc lamp was used as light source. Firstly, the air in reactor was remove by the high purity argon gas for 40 min. Then the solution was ultrasound for 15 min to obtain a unifoem suspension. Finally, the suspension was put into the reactor for experiment. The actual production of H_2 was tested by Labsolar 6A photocatalytic system and gas chromatograph (Techcomp GC-7900).

3. Results and discussions

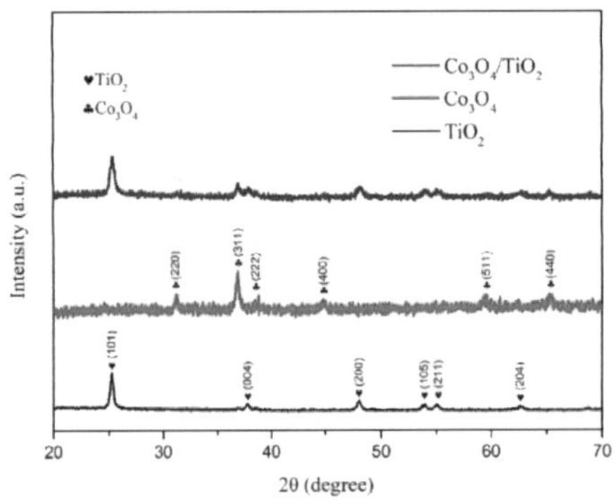

Fig.1. The XRD spectra of the samples

Journal of Physics: Conference Series **2011** (2021) 012089 doi:10.1088/1742-6596/2011/1/012089

As shown in Fig. 1, the XRD results of TiO2, Co3O4, and Co3O4/TiO2. For the TiO2, diffraction peaks located at 25.28°, 38.22°, 48.23°, 54.18, 55.28° and 63.01° are belong to the anatase phase TiO_2(PDF#84-1285). What's more, the new peaks at 31.30°, 36.91°, 38.58 °, 44.81°, 59.45 ° and 65.33 ° can be ascribedof the Co_3O_4/TiO_2 hollow core-shell heterojunction, which can be ascribed to the (202), (311), (222), (400), (511) and (440) planes of Co_3O_4 (PDF#74-2120). There is no extra peaks in all curves, that indicates the sample is composed of Co_3O_4 and TiO_2.

Fig.2. SEM and TEM of samples, (a) ZIF-67, (b,d) ZIF-67/Ti(OH)$_4$, (c,e) Co_3O_4/TiO_2; HRTEM of Co_3O_4/TiO_2 (f).

The morphology of samples is exhibited in Fig.2. As revealed in Fig. 2a, the as-prepared of ZIF-67 is a classical dodecahedral structure with a diameter of ~300 nm. Further, with the formation of ZIF-67/Ti(OH)$_4$ sold-polyhedron (Fig.2b, d), the surface turns to be more coarse, and the diameter increases to about ~380 nm. Subsequently, after calcination, the sold-polyhedron is transformed into a complete hollow structure and Co_3O_4 nanoparticles are distributed on the inner surface of hollow shells of TiO_2 (Fig.2c, e), which can provide efficient specific surface area and more active sites for HER. Finally, the high resolution TEM is shawn in images (Fig.2.f), the measured lattice length of 0.352 nm is assigned to the (101) plane of anatase TiO_2 and the 0.244 nm is belonged to the (311) plane of Co_3O_4. These above results indicate that the hollow Co_3O_4/TiO_2 heterojunction has been fabcricted successfully.

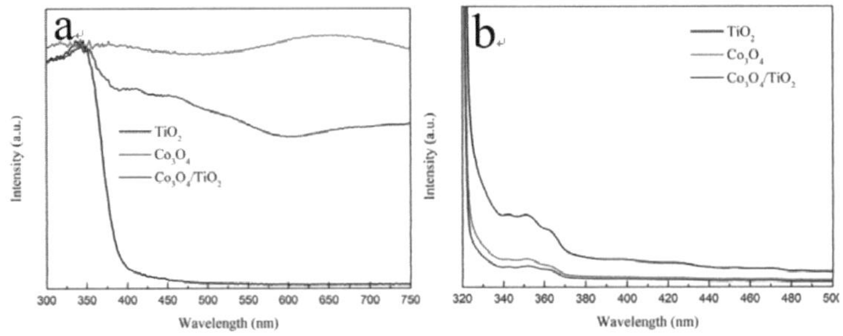

Fig.3. The UV-vis absorption spectra (a) and PL spectra (b) of as-prepared samples

As revealed in Fig. 3a, the pure TiO_2 exhibits the weakest light absorption and the absorption edge is located at ~380 nm. What's more, the sample of Co_3O_4/TiO_2 exhibits significant absorption of visible light, because the band gap of Co_3O_4 is narrower than pure TiO_2. As revealed in Fig. 3b, the PL of the as-prepared Co_3O_4/TiO_2 is lower than both pure TiO_2 and Co_3O_4, which indicates that, due to the formation of pn junction, the separation and imigragration of the photogenerated electron-hole pair become more effective. All above properties promote the performance of photocatalytic HER.

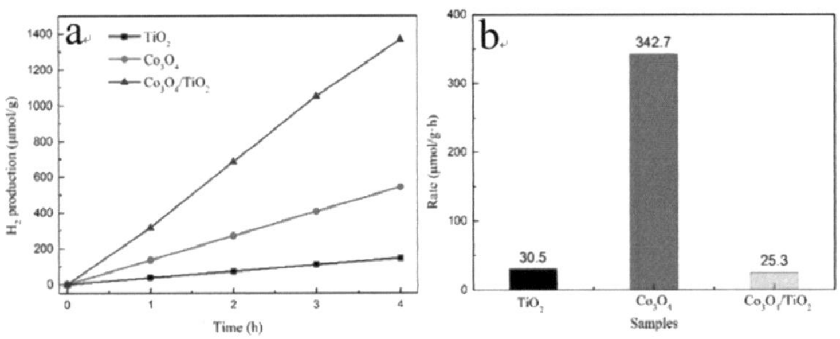

Fig.4. The photocatalytic hydrogen production activity of as-prepared samples

The photocatalytic hydrogen generation capacity of the samples are shown in image 4. As revealed in Fig.4a, the Co_3O_4/TiO_2 heterostructure exhibits a remarkably superior HRE performance than both pure TiO_2 and Co_3O_4 during 4 hours reaction. By calculation, the rate of HER is revealed in Fig. 4b, herein, the hydrogen production rate of composited sample is 342.7 μmol $(g{\cdot}h)^{-1}$, which is ~11 and ~14 times higher than that pure TiO_2 (30.5μmol $(g{\cdot}h)^{-1}$) and Co_3O_4 (25.3 μmol $(g{\cdot}h)^{-1}$), respectively. These results indicate that the hollow core-shell Co_3O_4/TiO_2 heterostructures can increase the hydrogen generation performance efficiently.

4. Conclusion

In this work, the hollow core-shell Co_3O_4/TiO_2 heterogeneous been prepared by deposition-annealing method with ZIF-67 as self-sacrificial templates and metal precurous. By evaluation, the photocatalytic hydrogen production of Co_3O_4/TiO_2 exhibits obvious enhancing, that is ascribed to the formation of heterojuncture can improve the efficiency of photogenerated electron-hole pairs separation, the hollow structure can raise the solar efficiency, offer more surface areas and active sites, all these are beneficial for photocatalytic performance. Consequently, this study provides an important basis to improve the photocatalytic performance by constructing heterojunction and regulating the morphology of the catalyst. Our research also extend the application of MOF in the field of photocatalysis.

Acknowledgments

This work was supported by the Natural Science Foundation of China (Nos. 51802282, 51672249).

References

[1] Kudo, A.,Miseki, Y. (2009) Heterogeneous photocatalyst materials for water splitting. *Chem Soc Rev 38* (1): 253-78.

[2] Low, J. X.,Yu, J. G.,Jaroniec, M.,Wageh, S.,Al-Ghamdi, A. A. (2017) Heterojunction Photocatalysts. *Adv. Mater. 29* (20): 20.

[3] Huang, M.,Chu, Y. T.,Xi, B. J.,Shi, N. X.,Duan, B.,Zhang, C. H.,Chen, W. H.,Feng, J. K.,Xiong, S. L. (2020) TiO_2-Based Heterostructures with Different Mechanism: A General Synergistic Effect toward High-Performance Sodium Storage. *Small 16* (42): 12.

[4] Guo, L. M.,Zhong, C. F.,Shi, L.,Ju, L. C.,Wang, X. H.,Yang, D. Q.,Bi, K.,Hao, Y. N.,Yang, Y. (2019) Phase and Defect Engineering of MoS_2 Stabilized in Periodic TiO_2 Nanoporous Film for Enhanced Solar Water Splitting. *Adv. Opt. Mater. 7* (5): 8.

[5] Chen, Z. F.,Pan, J. Q.,Mei, J.,Yu, Q.,Wang, P. H.,Wang, P. P.,Wang, J. J.,Song, C. S.,Zheng, Y. Y.,Li, C. R. (2021) Ternary $Co_3O_4/CdS/SrTiO_3$ core-shell pn junctions toward enhanced photocatalytic hydrogen production activity. *J. Environ. Chem. Eng. 9* (1): 10.

[6] Zhang, P.,Lou, X. W. (2019) Design of Heterostructured Hollow Photocatalysts for Solar-to-Chemical Energy Conversion. *Adv. Mater. 31* (29): 18.

[7] Lai, J.,Qin, Y.,Yu, L.,Zhang, C. (2016) GSH-assisted hydrothermal synthesis of $Mn_xCd_{1-x}S$ solid solution hollow spheres and their application in photocatalytic degradation. *Materials Science in Semiconductor Processing 52*: 82-90.

[8] Wang, J. Y.,Cui, Y.,Wang, D. (2019) Design of Hollow Nanostructures for Energy Storage, Conversion and Production. *Adv. Mater. 31* (38): 24.

[9] Chen, W. Y.,Han, B.,Tian, C.,Liu, X. M.,Liang, S. J.,Deng, H.,Lin, Z. (2019) MOFs-derived ultrathin holey Co_3O_4 nanosheets for enhanced visible light CO_2 reduction. *Appl. Catal. B-Environ. 244*: 996-1003.

[10] Li, X.-l.,He, R.-b.,Dai, Y.-j.,Li, S.-s.,Xiao, N.,Wang, A.-x.,Gao, Y.-q.,Li, N.,Gao, J.-f.,Zhang, L.-h.,Ge, L. (2020) Design and fabrication of $Co_9S_8/Zn_{0.5}Cd_{0.5}S$ hollow nanocages with significantly enhanced photocatalytic hydrogen production activity. *Chemical Engineering Journal 400*.

Journal of Physics: Conference Series **2011** (2021) 012090 doi:10.1088/1742-6596/2011/1/012090

Preparation of A_2B_7-type La-Mg-Ni Hydrogen Storage Alloy and Its Catalytic Oxidation Performance for BH_4^-

Liu Xinyu[1,a], Tian Xiao[1,b*], Ji Mingyue[1,c], Tian Lu[1,d]

[1] School of Physics and Electronic Information, Inner Mongolia Normal University, Hohhot, Inner Mongolia, China

[a]email: 1712332846@qq.com, [b]email: nsdtx@imnu.edu.cn, [c]email: 2602637571@qq.com, [d]email: 1648916190@qq.com

*Corresponding author: [b]email: nsdtx@imnu.edu.cn

Abstract： The La-Mg-Ni hydrogen storage alloys with A_2B_7/AB_5 double phase were successfully prepared by vacuum melting combined with ball milling. The phase structure, electrochemical properties and the catalytic oxidation properties for BH_4^- of the $(Mm_{1-x}Mg_x)Ni_{2.485}Co_{0.525}Mn_{0.28}Al_{0.21}$（x=0.1-0.5）alloys were investigated by XRD and electrochemical measurements. The results showed that $(Mm_{1-x}Mg_x)Ni_{2.485}Co_{0.525}Mn_{0.28}Al_{0.21}$ alloys are composed of $LaNi_5$ phase and $(La,Mg)_2Ni_7$ phase. Among the studied alloys, the electrochemical properties of the alloy with x=0.3 is the best, and the electrochemical properties of the alloy with x=0.3 is significantly better than that of the AB_5-type hydrogen storage alloy, which is mainly due to the formation of $(La,Mg)_2Ni_7$ phase in the alloy. At the same time, the alloy with x=0.3 also shows a better catalytic oxidation effect on BH_4^-, which is also better than that of the AB_5-type hydrogen storage alloy. The catalysis of $(Mm_{1-x}Mg_x)Ni_{2.485}Co_{0.525}Mn_{0.28}Al_{0.21}$ alloy catalyst is related to its electrochemical performance. The catalysis of hydrogen storage alloys may have a positive correlation with its own electrochemical performance.

1. Introduction

In this century, the increasingly serious energy crisis and environmental pollution have made clean new energy and new energy technologies gradually become research hotspots in countries around the world. Hydrogen energy has attracted much attention due to its abundant reserves, wide sources, high energy density, and cleanliness [1]. Fuel cells have become a new energy technology actively developed by people due to their high energy conversion efficiency, high energy density, low environmental pollution, low noise, reliable performance and long use time [2]. Direct Borohydride Fuel Cell (DBFC) is a new type of fuel cell that directly converts the chemical energy (hydrogen energy) stored in the borohydride into electrical energy under the action of an anode catalyst [3]. DBFC has higher output voltage and power density than that of ordinary fuel cells. DBFC can use ordinary catalysts to replace precious metals, can work at room temperature, is non-toxic, and is safe in transportation. At the same time, DBFC can overcome the difficulties of hydrogen storage and transportation in the process of hydrogen energy utilization. However, the performance of DBFC largely depends on the electrochemical reaction of borohydride on the anode, and the electrochemical reaction on the anode is directly controlled by the anode catalyst [4]. DBFC generally use precious metals as catalysts. The expensive price and scarce resources of precious metals have restricted the practical application of DBFC. In recent years, many researchers have used hydrogen storage alloys as anode catalysts of DBFC, because hydrogen storage alloys not only have direct catalytic ability, but also can store the generated hydrogen and then release it in the form of electrical energy, which improves fuel utilization through dual functions. Wang et al. [5] used $LmNi_{4.78}Al_{0.22}$ hydrogen storage alloy as the anode catalyst of DBFC. Studies have found that hydrogen storage alloys have good catalytic activity during the discharge process, and can inhibit the hydrolysis reaction, thereby improving fuel utilization. Wang [6] also tried to use AB_5-type $LaNi_{4.5}Al_{0.5}$ hydrogen storage alloy as the anode catalyst of DBFC, and found that the absorption and conversion of hydrogen on the hydrogen storage electrode alleviated the side reaction of hydrolysis. At the same time, Choudhury et al. [7] used AB_5-type $La_{10.5}Ce_{4.3}Pr_{0.5}Nd_{1.4}Ni_{60.0}Co_{12.7}Mn_{5.9}Al_{4.7}$ hydrogen storage alloy as the anode catalyst of DBFC, and found that the maximum power density of DBFC is 81 mW/cm^2, the maximum discharge current density can be up to 148 mA/cm^2. In the early stage, our research group used CNT/AB_5 composite hydrogen storage alloy as a DBFC anode catalyst [8]. Study have found that the catalytic oxidation ability of hydrogen storage alloys in DBFC for BH_4^- is positively correlated with the electrochemical performance of the alloy. In other words, the better the electrochemical performance, the better the catalytic oxidation ability for BH_4^-.

Compared with the commercialized AB_5-type hydrogen storage alloy, the new La-Mg-Ni hydrogen storage alloy with superlattice structure is considered to be the most potential hydrogen storage alloy due to its high discharge capacity (350-410 mAh/g) [9]. If a high-capacity La-Mg-Ni hydrogen storage alloy is used as the anode catalyst of DBFC, it will be more effective. However, the new La-Mg-Ni hydrogen storage alloys are difficult to prepare by conventional smelting methods due to the volatile Mg element in the alloys. In this paper, a new La-Mg-Ni hydrogen storage alloy will be prepared by the method of smelting combined with mechanical ball milling. Then, the new La-Mg-Ni hydrogen storage alloy was used as a DBFC anode catalyst to study the electrochemical hydrogen storage performance of La-Mg-Ni hydrogen storage alloy and its catalytic oxidation performance for BH_4^-.

2. Materials and Methods

Mischmetal Mm, Mg elemental powders and metals Ni, Co, Mn and Al are used as raw materials. Mischmetal Mm is composed of 63.1wt% La, 26.0 wt% Ce, 2.7 wt% Pr and 8.2 wt% Nd. The purity of other metal Ni, Co, Mn, Al and metal Mg powders is greater than 99.5 %. The $(Mm_{1-x}Mg_x)Ni_{2.485}Co_{0.525}Mn_{0.28}Al_{0.21}$ （x=0.1-0.5） master alloys were prepared by the vacuum arc melting combined with vibration ball milling method. Firstly, the $Mm_{1-x}Ni_{2.485}Co_{0.525}Mn_{0.28}Al_{0.21}$ （x=0.1-0.5） as-cast alloy was prepared by the vacuum non-consumable arc melting method and crushed. The powders of $Mm_{1-x}Ni_{2.485}Co_{0.525}Mn_{0.28}Al_{0.21}$ （x=0.1-0.5） were mixed with an appropriate amount of Mg powders in accordance to $(Mm_{1-x}Mg_x)Ni_{2.485}Co_{0.525}Mn_{0.28}Al_{0.21}$ （x=0.1-0.5） . The obtained powder mixtures with the composition of $(Mm_{1-x}Mg_x)Ni_{2.485}Co_{0.525}Mn_{0.28}Al_{0.21}$ （x=0.1-0.5） were subsequently milled in a high-energy vibratory type milling machine for 6 h. In order to prevent the abrasive from oxidizing during the ball milling process, put the ball mill jar into the vacuum chamber of the glove box to vacuum, fill with nitrogen, and vacuum again, so as to repeatedly flush with nitrogen for 2 to 3 times. At the same time, excessive addition of 10% Mg powder to make up for the loss.

The phase structure of the alloy was performed by X-ray diffractometer (XRD, PHILIPS PW1830 type) equipped with Cu Ka radiation, scanned in the 2θ range of 20-80° at room temperature. The scanning method is a step scan method. The collected data were analyzed by the Jade 6.5 software.

The carbonyl nickel powders were mixed with the alloy powders in a mass ratio of 3:1. The mixture having a total mass of about 1.0 g was poured into a steel mold with a diameter of 10 mm and was pressed into a disc. The mixture was pressed with a pressure of 11 tons for 15 minutes. Then, the disc was completely wrapped with foamed nickel and welded the nickel strip. Finally, the experimental electrode to be tested is obtained. The measurement of electrochemical performance and catalytic performance were the same with our previous researches [8, 10].

3. Results & Discussion

The XRD patterns of $(Mm_{1-x}Mg_x)Ni_{2.485}Co_{0.525}Mn_{0.28}Al_{0.21}$ （x=0.1-0.5） hydrogen storage alloys are shown in Fig. 1. It can be found that all alloys are composed of $LaNi_5$ phase and La_2Ni_7 phase. Generally, Mg atoms preferentially exist in the crystal structure of the La_2Ni_7 phase rather than the crystal structure of the $LaNi_5$ phase [11]. Therefore, the La_2Ni_7 phase should be accurately expressed as the (La, Mg)$_2$Ni$_7$ phase. It can also be seen that smelting combined with ball milling is an effective method for preparing La-Mg-Ni alloy. In order to compare the content of (La, Mg)$_2$Ni$_7$ phase in the alloy, the ratio of the main peak intensity of (La, Mg)$_2$Ni$_7$ phase ($2\theta\approx41.91$, expressed by X) to the highest peak intensity ($2\theta\approx42.60$, expressed by Y) in the alloy was roughly calculated. The phase characteristics of the $(Mm_{1-x}Mg_x)Ni_{2.485}Co_{0.525}Mn_{0.28}Al_{0.21}$ alloys are summarized in Table1. It can be seen that the cell volume of $LaNi_5$ phase and (La, Mg)$_2$Ni$_7$ phase in each alloy is not much different. However, with the increase of x, the ratio of X/Y in the alloy shows a trend of first increasing and then decreasing, and the ratio of X/Y in the alloy with x=0.3 has the strongest relative diffraction peak intensity, indicating that the (La, Mg)$_2$Ni$_7$ phase contained in the alloy with x=0.3 was the most. It can be seen that the Mg element content in the alloy is the main reason that affects the formation of (La, Mg)$_2$Ni$_7$ phase in the alloy.

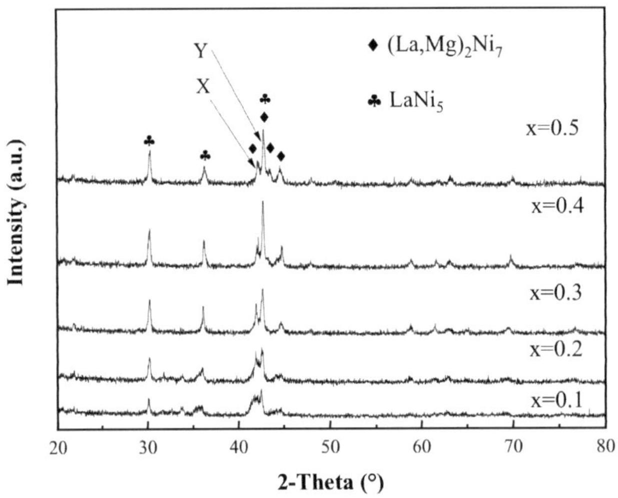

Fig. 1 XRD patterns of $(Mm_{1-x}Mg_x)Ni_{2.485}Co_{0.525}Mn_{0.28}Al_{0.21}$ (x=0.1-0.5) hydrogen storage alloys

Table 1 The phase characteristics of the $(Mm_{1-x}Mg_x)Ni_{2.485}Co_{0.525}Mn_{0.28}Al_{0.21}$ (x=0.1-0.5) alloys

Alloys	Phase structure	lattice parameter (nm)		Cell volume (nm^3)	The ratio of X/Y
		a	c		
x=0.1	$LaNi_5$	0.5065	0.3899	0.0866	0.9318
	La_2Ni_7	0.5086	2.4072	0.5395	
x=0.2	$LaNi_5$	0.4977	0.4056	0.0870	0.9121
	La_2Ni_7	0.5074	2.4144	0.5383	
x=0.3	$LaNi_5$	0.4972	0.4054	0.0868	2.5315
	La_2Ni_7	0.5077	2.4072	0.5373	
x=0.4	$LaNi_5$	0.5054	0.4040	0.0893	0.4955
	La_2Ni_7	0.5054	2.4533	0.5427	
x=0.5	$LaNi_5$	0.4962	0.4035	0.0860	0.4965
	La_2Ni_7	0.5043	2.4563	0.5410	

Fig. 2 shows the variation of the discharge capacity of $(Mm_{1-x}Mg_x)Ni_{2.485}Co_{0.525}Mn_{0.28}Al_{0.21}$ (x=0.1-0.5) hydrogen storage alloys in 6M KOH solution with the number of cycles. For convenience, the main electrochemical properties of the alloys are summarized in Table 2. It can be seen from Fig. 2 and Table 2 that the alloys with x≤0.3 have better activation performance and can be easily activated to reach the maximum capacity within 2-4 cycles, whereas the activation performance of the alloys with x>0.3 become significantly worse. As x in the alloy increases, the maximum discharge capacity of the alloy increases first and then decreases. That is, the maximum discharge capacity of the alloy first increases from 225.22 to 306.62 mAh/g, reaching the maximum value with rise of Mg from x=0.1to x=0.3, then decreases to 84.99 mAh/g with further rise of Mg to x=0.5. This is similar to the variation of the maximum discharge capacity of the $La_xMg_{3-x}Ni_9$ (x =1.6-2.2) alloy with the Mg content in the alloy studied by Liao et al. [12]. In addition, the maximum discharge capacity of the alloy with x=0.3 is significantly higher than that of the AB_5-type $MmNi_{3.55}Co_{0.75}Mn_{0.4}Al_{0.3}$ alloy that we have studied [13]. This is mainly due to the $(La, Mg)_2Ni_7$ phase formed in the alloy. It is generally accepted that, $(La,Mg)_2Ni_7$ phase possesses higher hydrogen storage capacity than $LaNi_5$ phase [14].

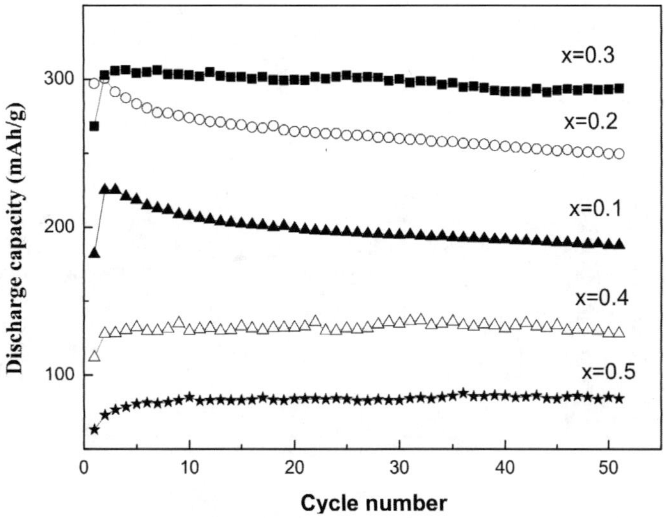

Fig.2. The discharge capacity of $(Mm_{1-x}Mg_x)Ni_{2.485}Co_{0.525}Mn_{0.28}Al_{0.21}$ alloy in 6M KOH solution

The cycle stability of alloy electrodes is generally characterized by capacity retention. It can be seen from Table 2 that the cycle stability of the alloy with x≥ 0.3 is obviously better than that of the alloy with x<0.3. This may be related to the higher content of $LaNi_5$ phase in the alloy. It has been reported in the literature [15] that the higher the abundance of $LaNi_5$ phase in the La-Mg-Ni alloy, the better the cycle stability of the corresponding alloy. On the whole, the alloy with x= 0.3 has the best electrochemical performance among the studied alloys.

Table 2. The main electrochemical properties for the $(Mm_{1-x}Mg_x)Ni_{2.485}Co_{0.525}Mn_{0.28}Al_{0.21}$ alloy

Alloys	*Na*	C_{max}/(mAh/g)	C_{50}/(mAh/g)	S_{50}/ (%)
x=0.1	2	225.22	188.07	83.50
x=0.2	2	300.81	249.74	83.02
x=0.3	4	306.62	293.33	95.66
x=0.4	9	135.00	128.39	95.10
x=0.5	10	84.99	83.81	98.61

Na : The cycle numbers needed to activate the alloy electrode.
C_{max} : The maximum discharge capacity.
C_{50} : The discharge capacity after 50 cycles.
S_{50} : The capacity retaining rate after 50 cycles.

The catalytic oxidation performance of the catalyst for BH_4^- is generally characterized by CV curve. Generally, the lower the corresponding potential of the oxidation peak in the CV curve, the greater the current density of the oxidation peak, and the easier the direct electrocatalytic oxidation of BH_4^- to occur, indicating that the catalytic activity of the catalyst is better [8]. Fig.3 is the CV curve of $(Mm_{1-x}Mg_x)Ni_{2.485}Co_{0.525}Mn_{0.28}Al_{0.21}$ (x= 0.1-0.5) hydrogen storage alloys in 2M KOH+0.1M KBH_4 solution. For the sake of clarity, the peak potential of the oxidation peak and the corresponding current density of the samples are listed in Table 3. It can be seen from Fig.3 that the CV curves of all alloys have similar irreversible oxidation peaks, indicating that all hydrogen storage alloys have a catalytic oxidation effect on BH_4^-. It is found from Table 3 that the corresponding potentials of the oxidation peaks of all hydrogen storage alloy catalysts are relatively low, and the potential spacing is small, between 0.1 V. However, as the Mg content x in the alloy increases, the oxidation peak current density of the hydrogen storage alloy catalyst first increases and then decreases. When x=0.3, the oxidation peak current density of the alloy reaches the maximum, indicating that the alloy has the best catalytic oxidation effect on BH_4^-. The catalytic performance of this alloy on BH_4^- is also significantly better than that of AB_5 alloy [13].

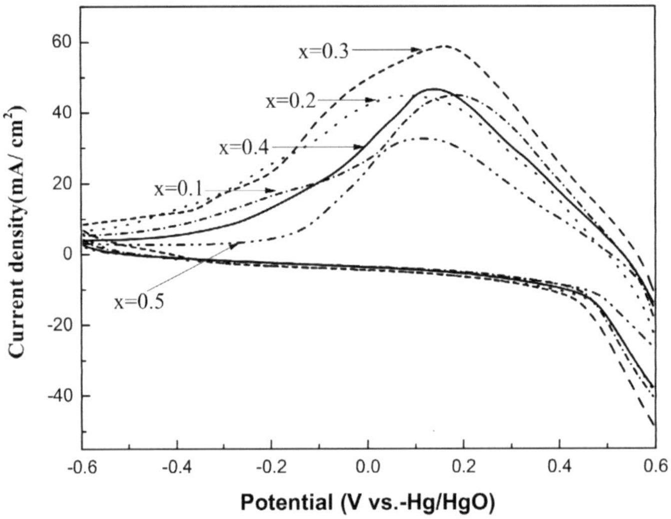

Fig 3. CV curves of $(Mm_{1-x}Mg_x)Ni_{2.485}Co_{0.525}Mn_{0.28}Al_{0.21}$ hydrogen storage alloys

Table 3 Oxidation peak potentials and current densities of the alloys

Alloys	Peak potential (V vs. -Hg/HgO)	Peak current density (mA/cm^2)
x=0.1	0.185	44.970
x=0.2	0.085	44.793
x=0.3	0.165	58.656
x=0.4	0.135	46.627
x=0.5	0.115	32.604

Generally, the chronoamperometry (CA) is a relationship of the oxidation current of borohydride as a function of time. The CA curve can reflect the stability of the catalyst during the catalytic process. The number of transferred electrons during electrocatalytic oxidation can also be calculated from the CA curve and the Cottrell equation [16]. Fig.4 is CA curves of $(Mm_{1-x}Mg_x)Ni_{2.485}Co_{0.525}Mn_{0.28}Al_{0.21}$ (x= 0.1-0.5) alloys in 2 M KOH + 0.1 M KBH_4 with a constant voltage of -0.95 V. Corresponding to Fig.4, the steady current density and the number of transferred electron of the alloys are listed in Table 4. It can be seen from Table 4 that, with the increase of x, the catalytic oxidation performance of the alloy on BH_4^- first becomes better and then becomes worse. The current density of the alloy with x=0.3 is the largest, and the transferred electrons of the alloy with x=0.3 are the most, indicating that the alloy catalyst has the greatest catalytic oxidation effect on BH_4^- and the highest fuel utilization rate. In addition, the catalytic oxidation ability of the alloy catalyst with x=0.3 is significantly better than that of the AB_5-type hydrogen storage alloy catalyst in the previous study [13].

From the above analysis, it can be seen that, among the studied alloys, the electrochemical performance of the alloy with x=0.3 is the best, and its catalytic oxidation ability for BH_4^- is also the best. The catalytic oxidation ability of hydrogen storage alloy to BH_4^- is positively related to the electrochemical performance of the alloy. At the same time, the electrochemical performance and the catalytic oxidation ability on BH_4^- of the alloy are better than that of the AB_5-type hydrogen storage alloy. It can be seen that good electrochemical hydrogen storage is more conducive to the secondary utilization of hydrogen and inhibits the side reaction of hydrolysis, thereby increasing the utilization rate of fuel.

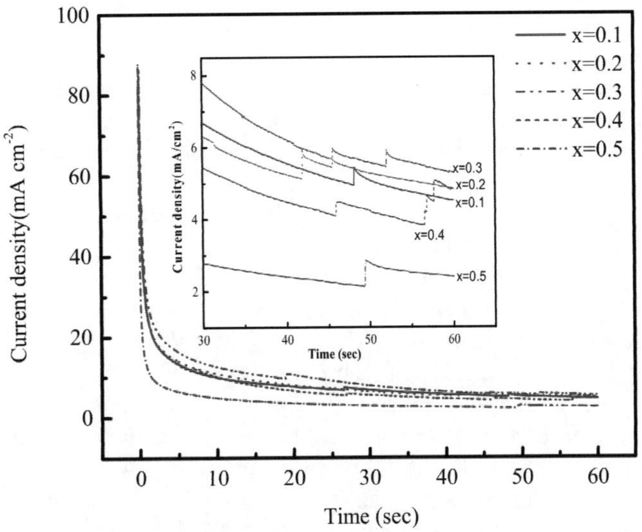

Fig. 4 CA curves of $(Mm_{1-x}Mg_x)Ni_{2.485}Co_{0.525}Mn_{0.28}Al_{0.21}$ hydrogen storage alloys

Table 4 The alloys reache stable current density and number of transferred electrons

Alloys	Current density (mA/cm^2)	Number of electrons (n)
x=0.1	4.5148	1.01
x=0.2	4.8501	1.09
x=0.3	5.2939	1.19
x=0.4	4.8264	1.09
x=0.5	2.4004	0.54

4. Conclusions

1) Melting combined ball milling is an effective method for preparing A_2B_7-type La-Mg-Ni hydrogen storage alloy. $(Mm_{1-x}Mg_x)Ni_{2.485}Co_{0.525}Mn_{0.28}Al_{0.21}$ (x= 0.1-0.5) alloys are composed of $LaNi_5$ main phase and (La, Mg)$_2$Ni$_7$ phase. The ratio of the content of the two phases in the alloy varies with the content of Mg. The content of (La, Mg)$_2$Ni$_7$ phase in the alloys with x=0.2 and x=0.3 is larger.

2) With the increase of Mg content, the electrochemical performance of $(Mm_{1-x}Mg_x)Ni_{2.485}Co_{0.525}Mn_{0.28}Al_{0.21}$ (x= 0.1-0.5) alloy first increases and then decreases. The alloy with x=0.3 shows better electrochemical performance. At the same time, $(Mm_{1-x}Mg_x)Ni_{2.485}Co_{0.525}Mn_{0.28}Al_{0.21}$ (x= 0.1-0.5) alloy as a catalyst for the catalytic oxidation of BH_4^- also shows the same trend. With the increase of Mg content, the catalytic effect first becomes better and then becomes worse. The alloy with x=0.3 shows better catalytic effect.

3) Comprehensive analysis found that the catalysis of $(Mm_{1-x}Mg_x)Ni_{2.485}Co_{0.525}Mn_{0.28}Al_{0.21}$ (x= 0.1-

0.5) alloy catalyst is related to its own electrochemical performance, and there is a positive correlation. In addition, the A_2B_7-type La-Mg-Ni hydrogen storage alloy catalyst has better catalytic oxidation effect on BH_4^- than that of the AB_5-type hydrogen storage alloy catalyst.

Acknowledgments
This work was supported by the Inner Mongolia Autonomous Region Science and Technology Plan Project (2019GG264), Inner Mongolia Normal University Postgraduate Research and Innovation Fund Project (CXJJS20097) and Inner Mongolia Normal University College Student Innovation and Entrepreneurship Training Program (XCY-2001).

References
[1] Meena, P., Singh, R., Sharma, V.K., Jain, I.P. (2018) Role of $NiMn_{9.3}Al_{4.0}Co_{14.1}Fe_{3.6}$ alloy on dehydrogenation kinetics of MgH_2. J. Magnes. Alloy., 6: 318-325.
[2] Wang, S., Jiang, S.P. (2017) Prospects of fuel cell technologies. Natl. Sci. Rev., 4: 163-166.
[3] Yin, X., Wang, Q., Duan, D., Liu, S., Wang, Y. (2019) Amorphous NiB alloy decorated by Cu as the anode catalyst for a direct borohydride fuel cell. Int. J. Hydrogen Energ., 44: 10971-10981.
[4] Tian, X., Duan, R., Zhao, L., NaRen, G. (2017) Anode catalyst for the direct borohydride fuel cell. J. Inorg. Mater., 32: 1233-1242.
[5] Wang, L., Ma, C., Sun, Y., Suda, S. (2005) AB_5-type hydrogen storage alloy used as anodic materials in borohydride fuel cell. J. Alloy. Compd., 391: 318-322.
[6] Wang, L., Ma, C., Mao, X., Sheng, J., Bai, F., Tang, F. (2005) Rare earth hydrogen storage alloy used in borohydride fuel cells. Electrochem. Commun., 7: 1477–1481.
[7] Choudhury, N., Sahai, Y., Buchheit, R. (2011) Chitosan chemical hydrogel electrode binder for direct borohydride fuel cells. Electrochem. Commun., 13: 1-4.
[8] Yan, J., Tian, X., Liu, X., Zhao, X., Wang, R., Zhao, L., Zhang, X. (2020) Correlation between electrochemical properties of the CNTs/AB_5 composite hydrogen storage alloys and their catalytic properties for KBH_4. Int. J. Hydrogen Energ., 45: 452-463.
[9] Wang, W., Zhang, L., Rodríguez-Perez., Ismael, A., Zhao, Y., Liu, X., Zhang, S., Guo, W., Ren, K., Li, Y., Han, S. (2019) A novel AB_4-type RE-Mg-Ni-Al-based hydrogen storage alloy with high power for nickel-metal hydride batteries. Electrochem. Commun., 317: 211-220.
[10] Tian, X., Wei, W., Duan, R., Zheng, X., Zhang, H., Tegus, O., Li, X. (2016) Preparation and electrochemical properties of $La_{0.70}Mg_xNi_{2.45}Co_{0.75}Al_{0.30}$ (x=0, 0.30, 0.33, 0.36, 0.39) hydrogen storage alloys. J. Alloy. Compd., 672: 104-109.
[11] Li, F., Young, K., Ouchi, T., Fetcenko, M. (2009) Annealing effects on structural and electrochemical properties of $(LaPrNdZr)_{0.83}Mg_{0.17}(NiCoAlMn)_{3.3}$ alloy. J. Alloy. Compd., 471:371-377.
[12] Liao, B., Lei, Y., Chen, L., G, L., Pan, H., Wang, Q. (2004) Effect of the La/Mg ratio on the structure and electrochemical properties of $La_xMg_{3-x}Ni_9$ (x=1.6-2.2) hydrogen storage electrode alloys for nickel-metal hydride batteries. J. Power Sources., 129: 358-367.
[13] Yan, J., Tian, X., Yao, Z., Liu, X., Wang, R., Zhao, X., Yang, Y., Li, W. (2020) Study on the Correlation between the catalytic performance of ball mill AB_5 hydrogen storage alloy for BH_4^- and electrochemical performance. J. Rare earth., 41.
[14] Zhang, Y., Ren, H., Cai, Y., Yang, T., Zhang, G., Zhao, D. (2014) Structures and electrochemical hydrogen storage performance of Si added A_2B_7-type alloy electrodes Trans. T. Nonferr. Metal. Soc., 24: 406–414.
[15] Dong, Z., Ma, L., Wu, Y., Wang, L., Shen, X. (2011) Microstructure and electrochemical hydrogen storage characteristics of $(La_{0.7}Mg_{0.3})_{1-x}Ce_xNi_{2.8}Co_{0.5}$ (x=0-0.20) electrode alloys. Int. J. Hydrogen Energ., 36: 3016-3021.
[16] Wei, J., Wang, X., Wang, Y., Chen, Q., Pei, F., Wang, Y. (2009) Investigation of carbon-supported Au hollow nanospheresas electrocatalyst for electrooxidation of sodium borohydride. Int. J. Hydrogen Energ., 34: 3360-3366.

Hyper-spectral Imaging Technology Based on Linear Gradient Bandpass Filter for Electricity Power Equipment Status Monitoring Application

Li Cheng*, Shang Ziwen, Wang Minzhen, Liu Li, Zhang Guangxin, Zhang Jian, Zhao Liyin, Zhang Xiao, Ni Hongxia, Qiao Dongliang

Changchun Institute of Technology, Changchun, Jilin Province, China

*Corresponding author's e-mail: li.cheng@ccit.edu.cn

Abstract. Hyperspectral imaging technology has been applied to the status monitoring of power transmission line and equipment as its continuous and narrow-band induction characteristics, which provides a solution for the operation monitoring of lossless large-area power equipment. However, the optical and electronic equipment is complex, and the related products are large and expensive on the market . This paper describes a preparation method of a linear gradient filter that is cost performance and miniaturization, which can be used to realize the prototype hyper-spectral imaging camera. The high precision preparation process and principle make the camera show good spectral performance, which can scan with precision step length of 2nm between 400nm and 1000nm. The feature extraction and classification algorithm can be used to determine the health conditions of power equipment, such as partial discharge, with an accuracy of about 77%. The equipment collocation algorithm can also be used to identify defects in power equipment, which has been proved to be able to distinguish between aging insulators, icing of conductors and heating of wire clips. This method are promising for an entry-level, low-cost hyper-spectral imaging solution for power detection applications.

1. Introduction

The monitoring of power transmission equipment is of great significance, especially for improving power supply quality and power safety. The method that can accurately understand which areas in the line channel need to be monitored can prevent malfunctions such as partial discharge and equipment overheating from evolving into major accidents. Due to the scale of the power transmission line inspection task, this is difficult to be completed by manual inspection or drone inspection. So, the use of new technologies such as lidar, infrared and ultraviolet cameras, multi-spectral and hyper-spectral imaging has been initially solved this problem. It can be seen that hyper-spectral detection has great potential in power system detection applications, especially partial discharge detection of power equipment under different humidity environments [1-3], insulator fault troubleshooting [4] and large-scale power equipment overheating operation monitoring [5] etc. The application of this technology also reduces the training time and experience accumulation required for operators to perform the same inspection tasks.

Partial discharge is a potential hazard to the safety of power transmission systems. X. Jia, et al. (2013) [1] has used reflectance spectroscopy in the visible light (390-770 nm) to near-infrared region (700-1000 nm) to evaluate it. This analysis has determined the health status of equipment samples under different discharge levels under experimental conditions. L. Jin, Z. Tian, J. Ai, et al. (2018) [4] et al. has used spectral analysis to achieve rapid screening of insulator contamination levels, and demonstrated the use of visible-near-infrared and mid-infrared spectroscopy to distinguish insulators whether flashover was caused by pollution, the classification accuracy was about 90%. Runming Gao, et al. (2019) [5] has adopted hyper-spectral nondestructive monitoring methods and used a real-time monitoring system based on visible light and near-infrared reflectance spectroscopy to study on variation of surface pollution content of insulator electrical equipment in strong electric field. Comparing the survey results with the pollution element index determined by laboratory standard instruments, the reliability of the spectral analysis is verified. The previous studies also showed that reflectivity would be a a sensitive indicator of chlorophyll concentration at 700 nm. Emadi (2012) [6] et al. showed that linear variable filters (LVF) could be used as a miniaturized optical element to replace diffraction gratings. By citing the advantages of LGF small size and wide scanning range, it only need to move the physical distance of the filter component can achieve multi-target capture in the range of 1000 nm, which is particularly important for the optical path design and effective use of space in optical inspection equipment.

Although the capabilities and potential of the spectral analysis test system seem to be huge, the current multispectral and hyperspectral detection systems are generally expensive that priced is more than $10,000 US dollar. Also, it's bulky and heavy, which makes it difficult to expand their application scenarios. This paper solves these problems. By using the new linear gradient band-pass filter (LGF) and new material preparation technology, the size and cost of the spectrometer are significantly reduced. The backstage with a unique algorithm has developed a portable hyper-spectral power detection camera (HETC) and make a prototype.

2. Baffle modelling and thickness distribution

In order to obtain the linear space thickness growth distribution and meet the requirements of the filter, a baffle is designed based on the sputtering angle distribution model, and the growth process of the film is controlled, so as to realize the thickness distribution of the film material.

2.1. Magnetic field modelling

The baffle modelling components are relayed to magnetic field, cathode voltage, and gas pressure, while gas pressure and cathode voltage are assumed constant and uniform across the target surface. Magnetic field tangential to target surface, Btan contributes to electron confinement and plasma generation and as in all magnetrons, and as measured in this example, is highly non-uniform. Goree and Sheridan used a Monto Carlo simulation to show that the ionization efficiency, η, is approximately linear to log **β**, a dimensionless magnetic field, (eqn.1) prior to saturation [7]. This leads to a sputtering yield dependant on Btan, reflected by the erosion track profile physically.

$$\beta = \sqrt{{}^{e}/_{2m}}\left({}^{a}/_{V_{dis}^{1/2}}\right)B_{tan} \tag{1}$$

Where e and m are electron charge and mass, a is radius at Btan, and Vdis is cathode voltage.

The magnetic field can be calculated and this can be fed into a semi empirical model to calculate the etch profile of the target. The etch profile is calculated taking into account an integration of the ionization efficiency along the approximate ion path. Fig.1(a) shows on the left hand side the magnetic field and on the right hand side the etch profile across a section in the middle of the linear magnetron. Measurements are also shown in Fig.1(b) for comparison.

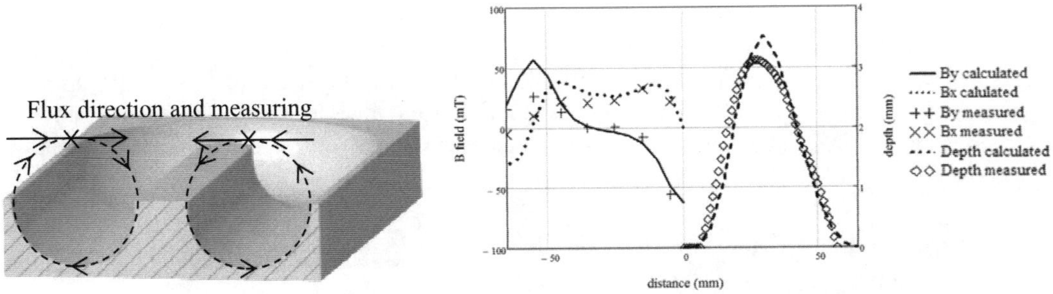

(a) (b)

Figure 1 (a) 3D diagram of target surface showing erosion tracks, direction and location of tangential flux measurements used in the simulation (b) comparison of calculated values to measured values LHS showing Bx & By RHS showing track erosion

2.2. Thickness distribution modelling and filter deposition

In order to realize the linear gradient distribution of the film thickness, a baffle is arranged between the sputtering outlet window and the substrate in the direction perpendicular to the outlet of the ion particles. During ion beam etching, the sample moves back and forth at a constant rate. After a certain amount of etching, a motion perpendicular to the sample is obtained. Through the thickness difference in the direction, the designed wedge - shaped spacer layer is finally obtained. Baffle design based on empirical methods is difficult and time-consuming. In order to design a mask that can produce the desired thickness distribution along the axial direction of the substrate more effectively, a simulation program is written. The parts to consider are the substrate on the rotary fixture, the target, and the mask with a predefined shape function, as shown in Figure 2.

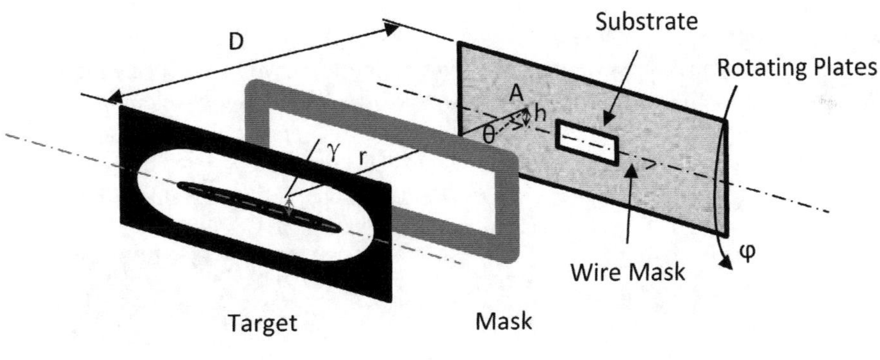

Figure 2 Schematic representation of sputtering geometry

A is the path which atom can reach substrate, and B is the path which atom is blocked by baffle. Substrate is mounted on a rotating fixture , θis the emission angle to the target surface normal of ejected atom flux, γ is the incident angle to the substrate surface normal of deposited atom flux.

3. Results & Discussion

Niobium and silicon are used to deposit high/low refractive index materials for making linear graded filters. The coating characterization is taken by NICOLET6700 spectrometer, which can measure transmission and reflection by using different incidence angle and two different polarization.

3.1. LGF deposition and Results Demonstration

The deposition process of linear gradient filter preparation is produced the first coating performed on the K9 substrate, and the lower film system and the middle cavity layer are plated. Then, the middle layer is etched according to the above method to obtain a certain a wedge-shaped resonant cavity layer with a wedge angle. Finally, bonded with the second coating the preparation of the linear gradient filter can be completed after 32 cycles of repeated cycles.

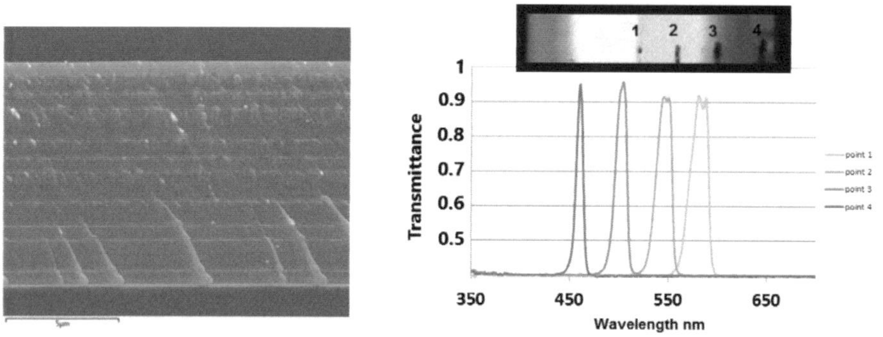

(a) (b)

Figure 3 (a) Cross-sectional SEM images for LGF film (b)The transmittance of several points on LGF has been obtained

The thickness of the film is designed to be 1.6μm at the thickest point , set the deposition working gas parameters as Ar flow rate of 40 sccm (1 sccm = 1 mL/min) and Ar/O2 flow rate of 5/10 sccm. The deposition rates of SiO2and NB2O5 materials are determined under the two conditions to be 1.33A/s and 3.83 A/s.

3.2. Hyper-spectral imaging system HETC

The design of the HETC detection system uses a high-resolution CMOS sensor as the position of the LGF needs to be considered to correspond to the corresponding detection wavelength. A stepper motor with a custom-designed rack and pinion was introduced to achieve linear gradient filter (LGF) translation and accurate positioning in the camera system. This creates discrete vertical bands at linearly spaced wavelengths in each image. Figure 4 shows the basic schematic diagram of the HETC imaging system.

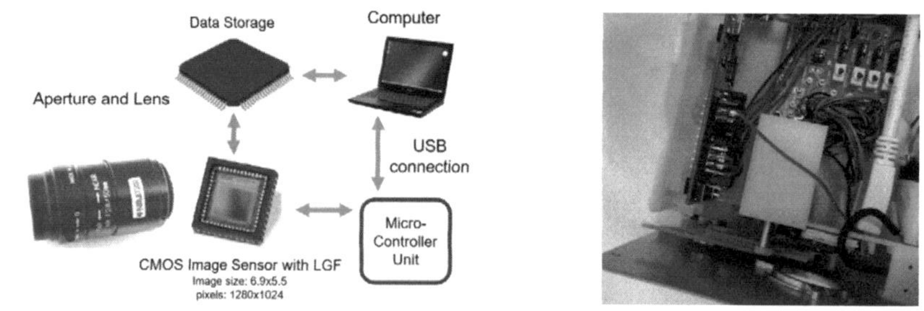

(a) (b)

Figure 4. (a)Basic components and LGF in HETC system set up (b) Prototype HETC assembly with internals exposed

HETC developed for the electricity transmission line monitoring is a low-cost hyper-spectral imaging system based on LGF. As part of the hardware development of the project, HETC is calibrated based on the technical standards of high-end push-broom hyper-spectral systems.

3.3. Classification of Insulator Contamination

By referring to the electric power standard, the on-site insulators are classified into different pollution levels after pollution value detection, and divided into 84 training sets and 36 test sets. The hyperspectral images of the training set samples are collected, and the characteristic bands are extracted after black and white correction and fitting smoothing to reduce redundant information. At the same time, the HSI algorithm is used to extract image features, and the S (color saturation) component that is most relevant to the pollution level is selected to identify the pollution level of the insulator samples in the test set. Then, based on the pollution level classification model and HSI test results, this study realized the level classification and visualization of the insulator surface pollution.

(a) (b) (c) (d) (e) (f)

Figure 5. Visualized image of insulator surface contamination obtained by HETC system

Figure 5 shows the pollution level distribution at different positions on the surface of the on-site insulator string. The gradual colors represent different pollution levels in the picture . Darker color means lower pollution content, and lighter color means higher pollution content. As shown in the figure, contamination on the upper surface of the insulating sheet accumulates more than that on the lower surface, and contamination accumulates in multiple directions on the pillars. The analysis result is basically consistent with the pollution distribution on the surface of the on-site insulation string, which further verifies the feasibility and accuracy of the pollution grade classification model based on the on-site insulator map information to visualize the pollution grade distribution.

4. Conclusions

This paper describes the design and development process of a customized special baffle and provide a linear graded band-pass filter LGF with low cost and wide detection range covering visible and near-infrared wavelengths. The simulation model ensures that LGF with controlled thickness distribution can be mass-produced. Low-cost LGF combined with appropriate detector systems can reduce the cost, weight, and size of the detection equipment. Compared to high-priced multi-lens scanning hyper-spectral imaging systems with similar technical specifications, the hyper-spectral prototype camera has similar performance and the cost save about 85%. The algorithm of the HETC system shows good feature extraction and classification in the process of similar insulator and metal fault judgment, and achieves about 88% accuracy in differentiating the flash-over and aging degree of insulators. These results have broad implications for the future applications of low-cost HETC in power detectionand the potential of remote sensing.

Acknowledgments

This work was funded by Jilin Province Science and Technology Association, Project No. QT202014, and Changchun Institute of Technology, Project No. 04010192020024 and 320200022.

References

[1] X. Jia, B. C. Kuo, M. M. Crawford.(2013) Feature mining for hyperspectral image classification[J]. Proceedings of the IEEE, 101(3):676-697.

[2] G. Lu, B. Fei. (2014) Medical hyperspectral imaging: a review[J]. Journal of Biomedical Optics, 19(01):10901.

[3] SAHOO N C， SALAMA M M A，BARTNIKAS R. (2015) Trends in partial discharge pattern classification: A survey [J]. IEEE Transactions on Dielectrics and Electrical Insulation, 12(2): 248-264
[4] L. Jin, Z. Tian, J. Ai, et al. (2018) Condition evaluation of the contaminated insulators by visible light images assisted with infrared information[J]. IEEE Transactions on Instrumentation and Measurement, 6(67): 1349-1358.
[5] Runming Gao, Xueqin Zhang, Yujun Guo, Chao Fan, Yan Qiu, Yuansheng Li, Guangning Wu. (2019) Study on detection method of composite insulator ageing based on hyperspectral technology[C]. 2019 2nd Conference on Electrical Materials and Power Equipment, 2019: 461-466.
[6] A. Emadi, et al, (2012) "Linear Variable Optical Filter-based Ultraviolet Microspectrometer", Applied Optics, 51,19, 4308-4315
[7] J.Goree, T.Sheridan(1991), Magnetic Field Dependence of Sputtering Magnetron Efficiency, Appl.Phys.Lett., 59,9,1052-1054
[8] Y. Qiu, G. Wu, Z. Xiao, et al. (2019) An Extreme-Learning-Machine-Based Hyperspectral Detection Method of Insulator Pollution Degree. IEEE Access, 07:121156-121164.

Journal of Physics: Conference Series **2011** (2021) 012092 doi:10.1088/1742-6596/2011/1/012092

Sample trapped charge induced signals in tapping-mode atom ic force microscopy

Yu Wang[a*], Chuxiang Zhou[b], Jie Xu [c]

College of Electronic and Optical Engineering, College of Microelectronics, Nanjing Univ ersity of Posts and Telecommunications, Nanjing

[*]Corresponding author: [a]email: 1220024536@njupt.edu.cn

[b]email: 1220024821@njupt.edu.cn, [c]email: jiexu@njupt.edu.cn

Abstract. AFM (Atomic force microscope) has been used to measure the surface topography and electrochemical properties of the sample by the interaction between the tip on the cantilever and the sample. However, when there is a charge in the sample, an induced charge will be generated inside the tip, which will generate a new attractive force between the tip and the sample and reduce the amplitude of the tip. We found that the average amplitude of the tip decreased by about 1 nm in tapping mode AFM. As the amplitude decreasing, AFM keeps the amplitude constant by adjusting z_0 with negative feedback in tapping mode. So, the average charge density will affect the measurement of surface topography.

1. Introduction

AFM is a new type of instrument with high resolution at atomic level invented after scanning tunneling microscope. It can detect the physical properties of various materials and samples in the atmosphere and liquid environment, including morphology, or directly manipulate them. AFM has been widely used in semiconductor, nano-functional materials, biology, chemical industry and food, pharmaceutical research and research institutes of various nanoscale related disciplines, and has become a basic tool for nano-science research [1,2,3].

TM-AFM (Tapping-mode atomic force microscopy) is a dynamic imaging mode in which the tip is oscillated close to its resonance frequency and interacts with the sample once in every oscillation. In this mode of imaging, the cantilever oscillation amplitude is kept constant by the feedback loop and the feedback signal to the piezoelement is recorded as a topography image.

The purpose of this experiment is that when AFM is used to measure the surface topography of the sample, if there are charges on the surface of the sample, the induced charge will be generated inside the tip, and the attractive force will be generated between the charges on the surface of the sample and the induced charge. If the attractive force is ignored, AFM will raise the tip in order to keep the amplitude constant. This will change the measured surface topography. In this simulation experiment, the tip is assumed to be a metal ball, and the charges on the sample surface are distributed in a certain way. In the case of maintaining a relatively stable TM, we find that, as the gradual increase of charge density, the amplitude of the tip will gradually decrease, and the maximum decrease amplitude can be about 1nm.

2. Forces analysis in AFM

2.1 DMT contact mode

In this work, we use DMT (Dejarguin-Muller-Toporov) contact model to model the interaction of tip and sample [4,5]. Regardless of the charge on the sample surface, the tip will be subjected to two kinds of forces in TM-AFM. When the tip is far away from the sample surface, it will be subjected to van der Waals force (attraction force).

As the tip close to the sample, repulsive forces emerge in addition to attractive forces. And the repulsive force is called Hertz [6], by combining these two effects.

The coupled differential equations of motion for the system described above can be written as

$$m_c \frac{\partial^2 Z_c}{\partial t^2} + \gamma_c \frac{\partial Z_c}{\partial t} + k_c Z_c - F_{ts} = F_{dr} \sin(\omega t) \tag{1}$$

In these equations, F_{dr} is the amplitude of cantilever driving force, Qc is the cantilever quality factor and ω_{oc} is the fundamental flexural resonance frequency of the cantilever. In addition, F_{ts} denotes the tip-sample interaction force and is given by

$$F_{ts} = \frac{-HR}{6(Z_0 + Z_c)^2} \qquad Z_0 + Z_c \geq a_0 \tag{2}$$

$$F_{ts} = -\frac{HR}{6a_0^2} + \frac{4}{3} E^* \sqrt{R} (a_0 - Z_0 - Z_c)^{\frac{3}{2}} \qquad Z_0 + Z_c < a_0 \tag{3}$$

The motion amplitude of the tip equation (1) can be calculated by explicit Runge-Kutta methods [7].

2.2 Electrostatic Forces analysis in AFM

When there are charges in the sample, the tip is close to the surface of the sample, and the induced charge will be generated inside the tip. It is assumed that the tip is a metal ball, and the area of charge distribution affected by the sample surface below the tip is a circle with a radius of 3R [8]. When the radius is greater than 3R, the increased charge is far away from the tip, and the tip produces very weak attractive force, which has negligible effect on the amplitude.

First, it is assumed that n charges(electrons) are shared equally on the diameter of a 3R-radius circl e. Each charge(q) in the sample can produce the corresponding induced charge(q'). Then calculate the downward attractive forces of each charge to the tip, and sum them.

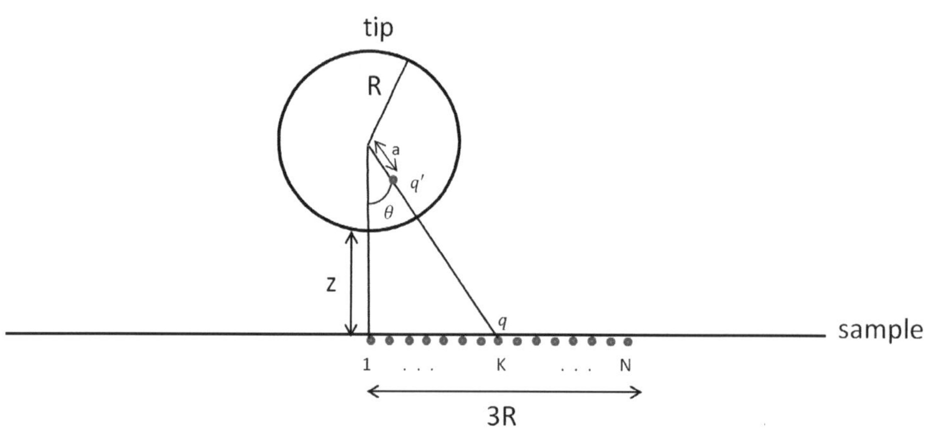

Figure 1. Influence of surface charge distribution on tip

$$\theta = \arctan\left(\frac{\frac{k}{N} \cdot 3R}{R+z}\right) \tag{4}$$

$$a = \frac{R^2}{(R+z)/cos\theta} \tag{5}$$

$$F_{el} = -\sum_{i=1}^{N} \frac{kq^2 \frac{R}{(R+z)/cos\theta}}{\left[\frac{R+z}{cos\theta} - a\right]^2} \cdot cos\theta \tag{6}$$

Then, expanding it to N times, AFM obtains the change of tip amplitude.

$$F_{el} = -n \cdot \sum_{i=1}^{N} \frac{kq^2 \frac{R}{(R+z)/cos\theta}}{\left[\frac{R+z}{cos\theta} - a\right]^2} \cdot cos\theta \tag{7}$$

According to this distribution, the charge distribution is not uniform on the sample with 3R radius under the tip. The closer to the lower part of the tip, the sample has the higher charge density. Therefore, this kind of charge distribution can increase the attraction of the surface charge of the sample to the tip, and the amplitude will be decreased [9]. If the charge distribution is uniform, it is not difficult to find out that the tip will be much less attractive.

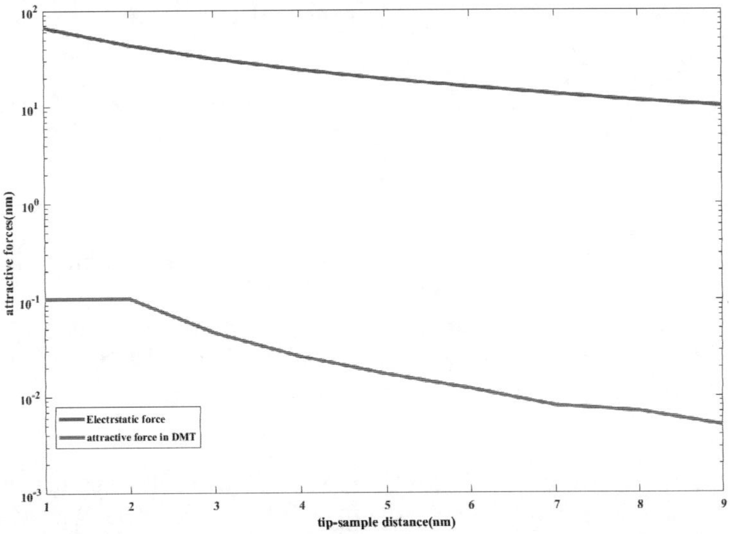

Figure 2. The relationship between the electrostatic force and tip-sample distance

Irrespective of the charges in the sample's surface, the tip is only attracted by van der Waa-ls force. The magnitude of the force is on the order of -10 to -12. When considering the surface charge of the sample, the magnitude of the force is on the order of -8 to -9 [10].When the tip closes to the sample surface, the attractive force increases about 100 times in figure2.

When increasing the value of **n**, the amplitude of the tip will decrease, so we can get the relationship between the average charge density of the sample surface below the tip and the amplitude of the tip.

Table 1. The relationship between average charge density and amplitude of the tip

n	0	500	600	1000	1100	1200
average charge density(e/cm^2)	0	0.707×10^{14}	0.849×10^{14}	0.990×10^{14}	1.556×10^{14}	1.698×10^{14}
amplitude of the tip(nm)	35.10	34.26	34.24	34.09	34.04	33.85

We found that when **n** = 1200, the tip amplitude will be reduced by 1.25nm. The average charge density of the sample is $1.698 \times 10^{14} e/cm^2$.

As the amplitude decreasing, AFM keeps the amplitude constant by adjusting z_0 with negative feedback in tapping mode. When AFM measures the surface morphology of the sample, the charges on the surface of the sample will lead to the AFM false feedback and lead to the tip elevation, which makes surface topography inaccurate. Through simulation, we find out the relationship between the average charges density of sample and the change of $z_0(\Delta z_0)$.

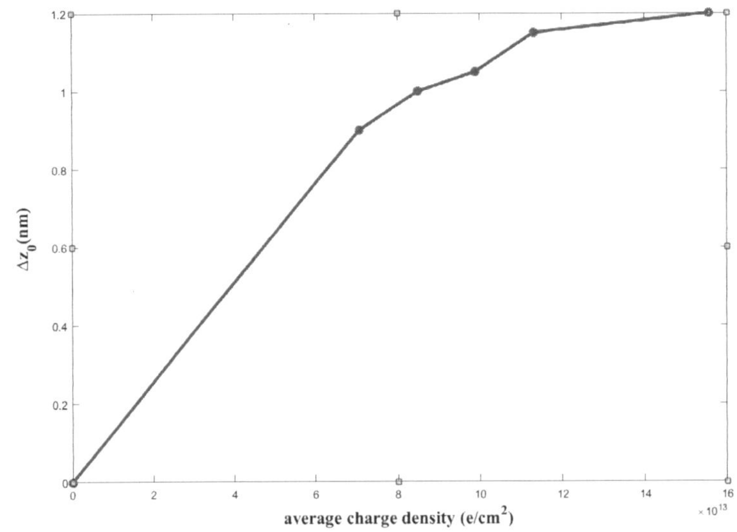

Figure 3. The relationship between the average charges density of sample and the Δz_0

It can be seen from the figure3. that with the increasing of the average charge density of the sample, z_0 is increasing. When the average charge density equals $1.556 \times 10^{14} e/cm^2$, Δz_0 is 1.2nm. In addition, a list of parameter values used in calculations is given in [10].

3. Conclusion

The simulation experiment is realized by MATLAB, we draw the following conclusions: (1) When there are charges on the surface of the sample, the tip will be gravitated and the amplitude of the tip will be reduced; (2) The larger the charge density near the tip is, the greater the attractive force will be generated and the smaller the tip amplitude will be. In TM-AFM, the maximum amplitude will be reduced by about 1nm; (3) In tapping mode, when the average charge density equals $1.556 \times 10^{14} e/cm^2$, the tip will raise 1.2nm by negative feedback in tapping mode, which causes surface topography inaccurate.

Acknowledgment
This work was supported by NSFC (No. 61804080).

References

[1] Belikov, S., Magonov, S. (2019) Simulation of Asymptotic Amplitude-Phase Dyn-amics for AFM Resonant Modes. Proceedings of the American Control Conf-erence., 2477-2482

[2] Patel, AN., Kranz, C. (2018) (Multi)functional Atomic Force Microscopy Imaging. Annual Review of Analytical Chemistry., 11:329-350.

[3] Alexander, S., Hellemans, L., Marti, O., Schneir, J., Elings, V., Hansma, P.K., Matt Longmire, John Gurley. (1989) An atomic-resolution atomic-force microscope implemented using an optical lever. J. Appl. Phys., 164.

[4] Israelachvili, J.(2003) Intermolecular and Surface Forces. In: Academic Press. London.

[5] Hellwing, J., Karlsson, RMP., Wagberg, L., Pettersson, T. (2017) Measuring elasticity of wet cellulose beads with an AFM colloidal probe using a linearized DMT mode. Analytical Methods., 4019-4022.

[6] Park, S. (2017) Asymptotic Approach to Analyze the Mechanical Properties of Bi-ological Cells. Journal of the korean physical society., 70:918-923.

[7] Stark, M., Stark, M., Heckl, M., Guckenberger, R. (2000) Spectroscopy of the anharmonic cantilever oscillations in tapping-mode atomic-force microscopy. Appl. Phys., 77:3293.

[8] Maragliano1, C., Heskes, D., Stefancich, M., Chiesa, M., Souier, T. (2013) Dynamic electrostatic force microscopy technique for the study of electrical properties with improved spatial resolution. Nanotechnology., 24:225703.

[9] Sadewasser, S., Carl, P., Glatzel, T., Lux-Steiner, M.C. (2004) Influence of unco-mpensated electrostatic force on height measurements in non-contact atomic force microscopy. Nanotechnology., 15:S14–S18.

[10] Olav, S., Butrus, T., Khuri-Yakub. (2009) Time-resolved tapping-mode atomic force microscopy. A dissertation submitted to the department of electrical engineering and the committee on graduate studies of STANFORD University.

A transparent photovoltaic device of NiO/ CdS arrays pn junction

Wendi Fu [1], Jiaqi Pan [1] and Chaorong Li [1,*]

[1] Department of Physics and Key Laboratory of ATMMT Ministry of Education

Zhejiang, Sci-Tech University, Hangzhou 310018, China

* Corresponding author: email: crli@zstu.edu.cn

Abstract: The transparent NiO/CdS P-N junction device composite structure was prepared by hydrothermal method and radio frequency magnetron sputtering. The results show that the NiO is orderly arranged on the CdS arrays.In addition, the photoelectric properties of the transparent NiO/CdS device grown at different hydrothermal temperatures are also different. As revealed, the device prepared at 180°Cexhibits a better photoelectric performance.

1. Introduction

Due to the excellent intrinsic properties, the Transparent Conductive Oxide (TCO) film has attracted lots of attentions: (1) good electrical conductivity.(2) high transmittance to visible light.(3) light selectivity. Especially the transparent P-N junction devices, composed of oxide thin films obtained excellent photoelectric conversion performance. Among them, the basic principle of photovoltaic effect is that, the sun irradiates on the semiconductor P-N junction to form a new hole-electron pair. Driven by the electric field of the P-N junction, the holes flow from the N region to the P region, and the electrons flow from the P region to the N region. After the circuit is connected, the current is formed to achieve the photovoltaic conversion, and are widely used in transparent solar cells, flat panel display, thin film solar cells, transparent optoelectronics and other technical fields.

The band gap of CdS is about 2.42eV[1], which can absorb and utilize a large range of visible light, and is regarded as an ideal material to fabricate the photoelectric device. Compared with other semiconductors, CdS obtains more suitable valence band and conduction band location, so which obtains strong oxidation ability and excellent visible light response ability. Particularly the CdS nanomaterials[2], which obtains volume effect, quantum size effect, surface interface effect, macro tunneling effect and other characteristics than other traditional materials, so which can be widely applied in photocatalysis, solar cells, luminescent materials and other fields. However, CdS still have some inherent drawbacks. First of all, the recombination rate of photo-generated electrons and holes is relatively higher. Secondly, the photocorrosion is easy to occur[3]. Aiming at these shortcomings, scholars have developed many ways to curb the compound rate and light corrosion phenomena, such as formation of semiconductor compound, introduction of graphene and amorphous carbon composite, ion doping and precious metal deposition. Especially the formation of pn junction, which can not only enhance the stability, but also increase the photoelectric conversion, thus, can be regarded as a promising method[4].

Undoubtedly, the p-type materials are considered as the core issues. Herein, the nickel oxide (NiO)[5], with a band gap of 3.6~4.0 eV[6] at room temperature is reported as wide band gap p-type semiconductor material and has attracted more and more attention, especially wide application in catalyst materials, magnetic materials, gas sensors and photoelectric fields. Thus, the introduction of NiO to formation of NiO/CdS pn junction is considered as a perfectly method to achieve the photoelectric conversion[7].

In this paper, the CdS/NiO P-N junction device was prepared by hydrothermal and magnetron sputtering methods. The results show that the CdS/NiO P-N junction exhibits a high transparency and fast optical response. In addition, the enhancement mechanism of light response is studied.

2. Experiment

2.1. Materials

All the chemicals are analytical grade and purchased from Aladdin Industrial Corporation (Shanghai, China), the NiO target purchased from Crystal Technical Material (Hefei, China).

2.2. Preparation

2.2.1 Preparation of the CdS nanoarray

CdS arrays were synthesized by hydrothermal method. First, adding 0.15 mmol cadmium nitrate, 0.15 mmol Urea acyl and 0.15 mmol Glutaudio-videoailable onhione into 60 mL deionized water, and stirring for 25 min. After the mixture becomes clear and transparent, the solution is transferred to a 50 mL Teflon lined stainless steel autoclave obtains a substrate with a conductive surface placed facing the bottom of the kettle. The hydrothermal reaction time is 190 min and different reaction temperature(150℃and 180℃) is carried out. After the reactor cool down to room temperature, the samples were washed repeatedly with anhydrous ethanol and dried at 60℃ for 0.5 h. There, the samples with different reaction temperature are labelled as NiO/CdS-1 (150℃) and NiO/CdS-2 (180℃), respectively.

2.2.2 Preparation of NiO film

NiO films were prepared by RF magnetron sputtering using NiO target (the diameter 6cm) on the surface of CdS/glass slide substrate. The background vacuum, sputtering pressure, sputtering power, sputtering time, sputtering temperature were 6×10-4 Pa, 0.9 Pa, 125W, 45 min and 100°C, respectively.The sputtering gas was Ar(40 SCCM).Subsequently, the as-prepared samples were dried at 300 °C in vacuum for 1 h.

2.3 Characterization

Field-emission scanning electron microscope (FESEM, Hitachi S-4800) was used to investigate the surface morphology of the samples. UV-vis absorption spectro photometer (U-3900Hitachi) was employed to investigate transparency of the films. The PL spectra were recorded by a Hitachi F-7000 spectrofluorimeter at the excitation wavelength of 360 nm.The photoresponse such as the dark and illuminated currents was gauged by electrochemical workstation.

3. Results and discussions

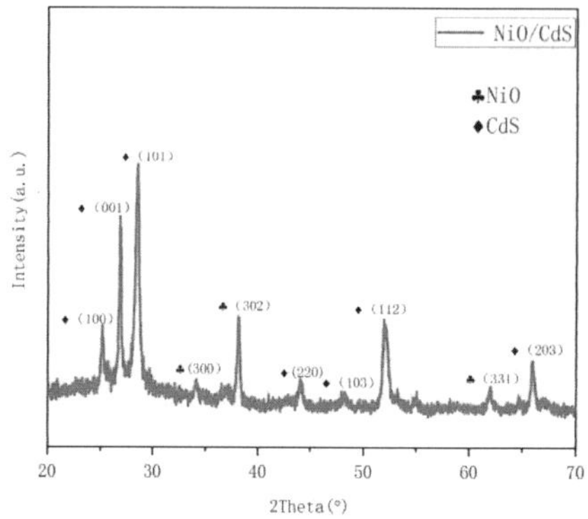

Fig. 1 the XRD of the as-prepared NiO/CdS pn junction device

Fig. 1 is the XRD of as-prepared NiO/CdS pn junction device. As revealed, the peaks at 24.807°, 26.507°, 28.182°, 43.960°, 47.839°, 51.824° and 66.772° are ascribed to the (100), (001),(101),(220),(103),(112) and (203) planes of CdS(PDF#41-1049), the peaks at 34.123°, 38.073° and 62.023° are ascribed to the (300), (301) and (331) planes of NiO(PDF#45-0946).

Fig.2. The SEM of the samples CdS nanoarrays

Fig.2 is the SEM of the surface of the CdS arrays,as revealed,which is uniform and smooth, and could effectively reduce the boundaries between the crystal grains,thereby reducing the boundary reflection and improving the electrical conductivity of the pn junction device.

Journal of Physics: Conference Series **2011** (2021) 012093 doi:10.1088/1742-6596/2011/1/012093

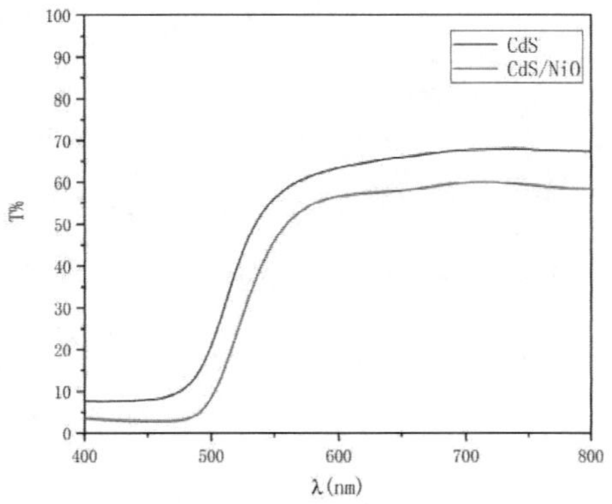

Fig.3. The optical transmittance spectra(T) of the CdS and CdS/NiO

Fig.3 demonstrates the optical transmittance spectra(T) of the CdS nanoarrays and the CdS/NiO pn junction. As shown, all of the samples exhibit a high transparency in the visible light region. Further, due to the wider band gap of NiO, after introduction of NiO film,transparency of as-prepared pn device exhibits no obvious change, and at around ~500 nm.

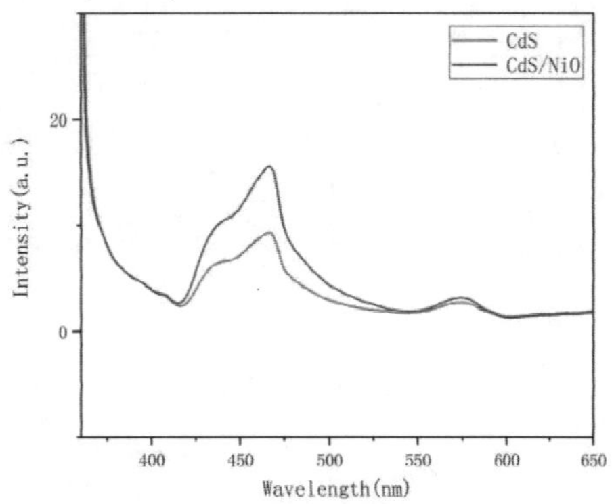

Fig.4. The PL of the CdS and CdS/NiO

Fig.4 demonstrates the PL of CdS film and CdS/NiO pn junction with excitation of 360 nm. With the formation of NiO/CdS heterojunction, the PL exhibits obvious decrease, which is beneficial for photoelectric performance.

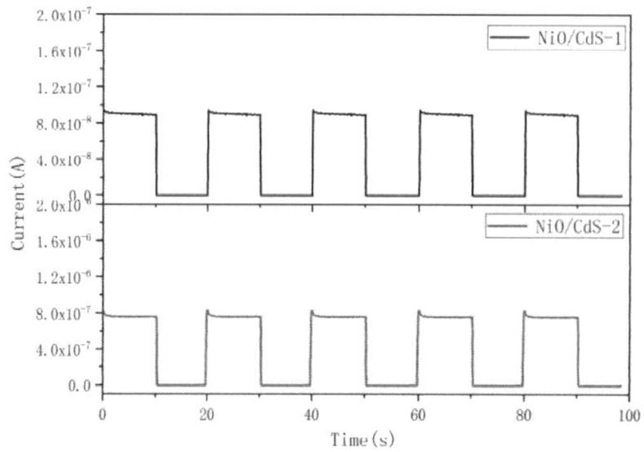

Fig.5. The photosensitivity of the NiO/CdS-1 and NiO/CdS-2

Fig.5 demonstrates the photoelectric properties of the NiO/CdS-1 and NiO/CdS-2. The photoelectric measurement is carried out in room temperature with the illumination and darkness alternate 10 seconds and both of the curves under 0 bias voltage are fitted in uniform coordinate. As revealed, the photoelectric performance of as-prepared pn junction device exhibits an obvious enhancement of about ~10 folds with the increased reaction temperature, which indicate that the suitable reaction temperature is beneficial for the photoelectric performance.

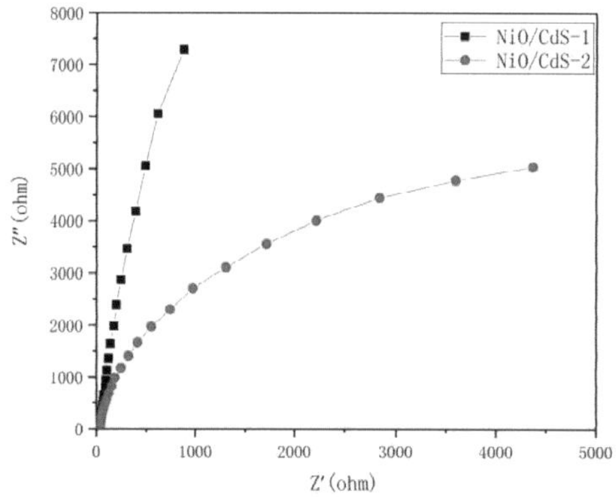

Fig.6.EIS Nyquist plots of different NiO/CdS pn junction transparent photovoltaic device

Fig. 6 demonstrates that the electro chemical impedance interface charge carrier behavior, which can depict the photovoltaic conversion enhancement mechanism. As revealed, the reduction of impedance indicates that the NiO/CdS-2 obtains an optimal interfacial charge carriers transport, which is beneficial for the photovoltaic performance.

4. Conclusions

In this work, we have successfully prepared CdS/NiO pn junction device through a simple process and proved its excellent optical and photoelectric performance, which could be mainly attributed to that the remarkable p-n junction of CdS/NiO could take advantage of the visible light efficiently and obtain a decent carriers transport. With the high transmittance and photoelectric performance, the CdS/NiO pn junction device is regarded as a potential material in transparent-conductive field, and brings us a new vision for designing of transparent devices and solar energy device.

Acknowledgments

This work was supported by Natural Science Foundation of China(Nos. 51802282 and 51672249),

Zhejiang Provincial Natural Science Foundation of China (Nos. LQ17F040004).

References

[1] Zhu, L., Meng, Z.D., Cho, K.Y., et al. (2012) Synthesis of CdS/CNT-TiO_2 with a high photocatalytic activity in photodegradation of methylene blue. New Carbon Mater., 27(3): 166-174.
[2] Pan, J.Q., Wang, P.H., Wang, P.P., Yu, Q., Wang, J.J., Song, C.S., Zheng, Y.Y., Li, C.R. (2021) The photocatalytic overall water splitting hydrogen production of g-C_3N_4/CdS hollow core - shell heterojunction via the HER/OER matching of Pt/MnO_x. J. Chem. Eng., 405: 126622.
[3] Pan, J.Q., Liu, Y.Y., Ou, W., Li, S., Li, H.L., Wang, J.J., Song, C.S., Zheng, Y.Y., Li, C.R. (2020) The photocatalytic hydrogen evolution enhancement of the MoS_2 lamellas modified g-C_3N_4/$SrTiO_3$ core-shell heterojunction. Renew. Energy., 161: 340-349.
[4] Pan, J.Q., Chen, Z.F., Wang, P.H., Wang, P.P., Yu, Q., Zhao, W.J., Wang, J.J., Zhu,M., Zheng, Y.Y., Li, C.R. (2021) The overall water splitting of CdS/Ti^{3+}-$SrTiO_3$ core - shell heterojunction via OER enhancement of MnOx nanoparticles. J. Chem. Eng., 424: 130357.
[5] Pan, J.Q., Li, H.L., Li, S., Ou, W., Liu, Y.Y., Wang, J.J., Song, C.S., Zheng, Y.Y., Li,C.R. (2020) The enhanced photocatalytic hydrogen production of nickel-cobaltbimetals sulfide synergistic modified CdS nanorods with active facets. Renew. Energy., 156: 469-477.
[6] Li, F., Nie, C., You, L., et al. (2018) White light emitting device based on single-phase CdS quantum dots. Nanotechnology., 29(20): 205701-205709.
[7] Pan, J.Q., Li, S., Ou, W., Liu, Y.Y., Li, H.L., Wang, J.J., Song, C.S., Zheng, Y.Y., Li,C.R. (2020) The photovoltaic conversion enhancement of NiO/Tm:CeO_2/SnO_2 transparent p-n junction device with dual-functional Tm:CeO_2 quantum dots. J. Chem. Eng., 393: 124802.

Antisolvent engineering for mixed tin-lead inorganic perovskite solar cells

Hao Chen[1], Ting Zhang[1], Shibin Li[1*]

[1] School of Optoelectronic Science and Engineering, University of Electronic Science and Technology of China, Chengdu, Sichuan, 610054, China

*Corresponding author's e-mail: shibinli@uestc.edu.cn

Abstract. Recently, inorganic perovskite has been a research hotspot due to its thermal stability and promising photovoltaic performance. Here, we demonstrate the antisolvent engineering can assist the fabrication of the mixed tin-lead inorganic perovskite films. The treated devices exhibit enhanced stability in atmosphere air. The champion perovskite solar cells (PSCs) treated with anisole exhibits a power conversion efficiency of 9.3% with an open-circuit voltage over 0.8 V.

1. Introduction

Inorganic perovskite solar cells ($CsPbX_3$, X=I, Br, Cl) are attracting lots of attention due to the promising charge-carrier mobilities and high thermal stability[1,2]. Recently, the power conversion efficiency (PCE) of inorganic perovskite solar cells has surpassed 20%[3]. However, the high fabrication temperature hinders their further application and commercialization[4]. Besides, the bandgap over 1.6 eV limits their photovoltaic performance due to Shockley-Queisser limit, which predicts an ideal bandgap of 1.2-1.4 eV. Therefore, reducing the bandgap of inorganic perovskites to the ideal bandgap and minimizing the fabricating temperature is quite necessary.

Partially replacing Pb with Sn is able to reduce the bandgap of perovskites[5]. In addition, the phase transition temperature of tin-based perovskites is always lower than that of lead-based perovskites[6]. Hence, it is vital to fabricate high-performance low-bandgap inorganic perovskite solar cells. Here, we explore 3 antisolvents and demonstrate that anisole is the best antisolvent to prepare high-quality $CsPb_{0.6}Sn_{0.4}I_3$ films. The antisolvent treated PSC has enhanced stability than the control one. The best inverted planar device shows a PCE of 9.3%.

2. Results and discussion

A 0.8M $CsPb_{0.6}Sn_{0.4}I_3$ precursor solution was prepared by stirring corresponding molar CsI, PbI_2, SnI_2 and SnF_2 (SnI_2 :SnF_2 = 10:1 in molar ratio) in mixed anhydrous solvents of DMSO and DMF for 3 hours. We use a simple one step method combining with an antisolvent engineering to fabricate inorganic perovskite films. To explore the effect of antisolvents on the mixed tin-lead inorganic perovskite films, we choose three different antisolvents chlorobenzene (CB), toluene (Tol) and anisole (Ani). All the antisolvents were dripped onto the film 10 s before the end of the spin-coating and then the films were annealed at 70 °C for 10 min. The scanning electron microscopy (SEM) images of the obtained perovskite films treated with different antisolvents were shown in Figure 1 (a) - 1 (d). After treated with antisolvents, the number of pinholes on the surface is reduced, while the grain sizes remained unchanged. However, all the films had small pinholes, which would lead the decrease of the devices' short-circuit current density (Jsc) and fill factors (FF).

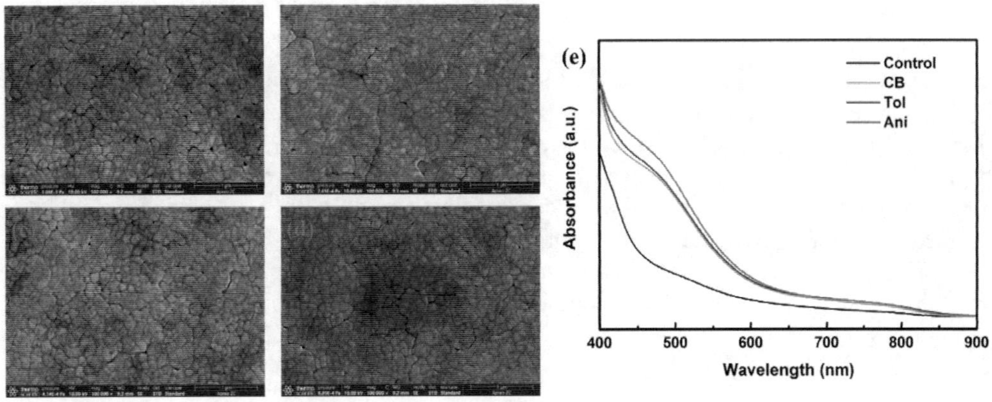

Figure 1. Top-view SEM images of the (a) control film, (b) CB treated film, (c) Tol treated film and (d) Ani treated film. (e) Absorption spectrum of the perovskite films.

The optical property of the films were further tested by Ultraviolet-visible (UV-Vis). All the films were measured without encapsulation. As shown in Figure 1 (e), the obtained film without antisolvent degraded, while the films treated with antisolvents all showed normal absorption across the spectrum. The film treated with Ani showed the highest absorption, which is good for the improvement of the Jsc of the devices. Besides, all the films treated with antisolvents showed similar bandgap (1.39 eV) due to the UV-Vis measurements.

The X-ray diffraction (XRD) patter of the control and ani-treated perovskite films were shown in Figure 2 (a). Obviously, the control film degraded to a nonperovskite phase. In contrast, the ani-treated film showed the characteristic peaks at 14.48 °, 20.36 °, 29.08 ° and 36.02 °, assigning to the (110), (200), (220) and (312) planes of the orthorhombic $CsPb_{0.6}Sn_{0.4}I_3$ structure[7]. The result of the XRD also proved the good quality of the antisolvents treated films. The bad stability of the control film arising from the bad film morphology would restrict the performance of the devices. As shown in Figure 2 (b), the ani-treated device can maintain a good stability in atmosphere air, while the control device degraded in a few hours. The enhanced stability can attribute to the improved film morphology and good crystal quality.

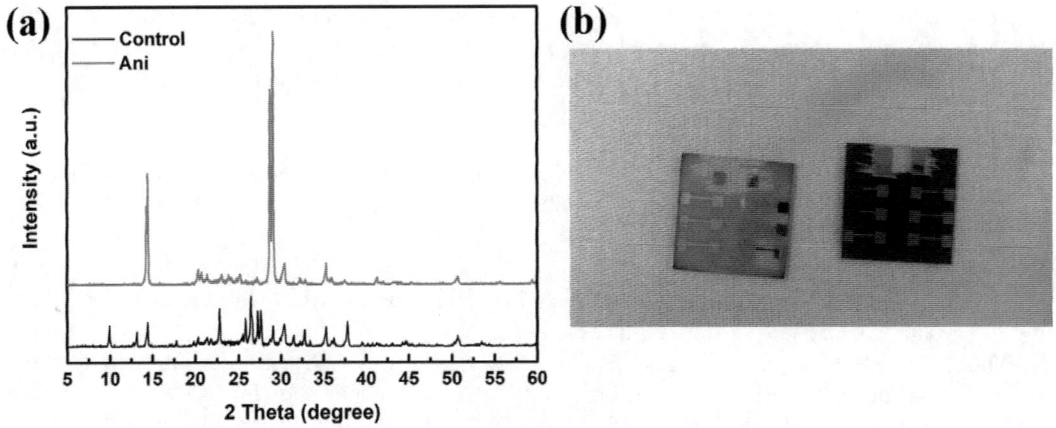

Figure 2. (a) XRD patterns of the control film and anisole treated film. (b) The photograph of the devices based on the perovskite films treated without and with anisole stored in atmosphere air for 2 hours.

Tin-based perovskites prefer the inverted structure to construct solar cells than the normal structure, because traditional metal oxides have oxygen vacancies to accelerate tin oxidation, which will deteriorate the performance of the PSCs[8]. In addition, the hole transport layer (HTL) of the normal devices always need additions to oxidize the material, which will also oxidize the perovskite layer. Therefore, we employ the inverted structure to fabricate PSCs (Figure 3 (a)). Figure 3 (b) presented statistic PCE data of 64 devices. The performance of the devices were measured in air atmosphere without any encapsulation. The PSCs based on the films treated with Ani showed the highest PCE, which matched well with the results of the morphology and absorption. The control devices showed an average PCE less than 4%. This poor performance may result from the bad morphology and instability when measured in air atmosphere. The champion device exhibited a PCE of 9.3% with an open-circuit voltage (Voc) of 0.81 V, Jsc of 18.9 mA/cm^2 and FF of 61.2% (Figure 3 (c)). In contrast, the control device only showed a low PCE of 4.4%.

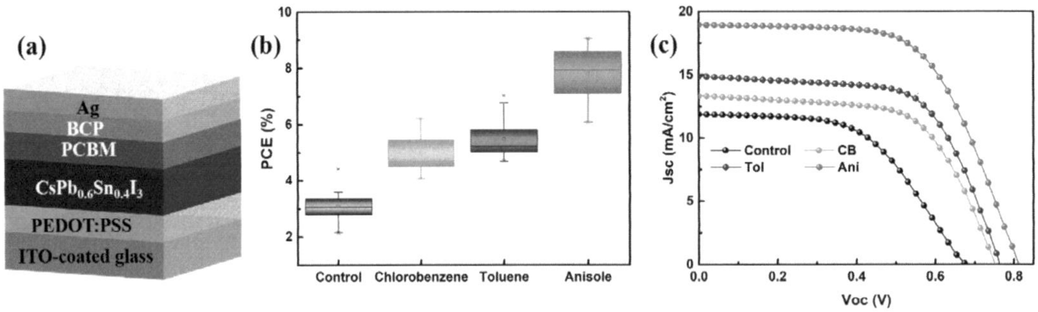

Figure 3. (a) Schematic of the inorganic perovskite solar cells. (b) The PCE statistics data of inorganic perovskite solar cells with different antisolvents. (c) J-V curves of the champion devices based on different antisolvents.

3. Conclusion

In summary, an antisolvent engineering for mixed tin-lead inorganic perovskites is demonstrated. The morphology and absorption of CsPb0.6Sn0.4I3 films treated with 3 different antisolvents are discussed. After treated with antisolvents, the perovskite films show better film morphology and enhanced absorption. Through comparing the performance of the PSCs based on the treated perovskite films, anisole is proved to be an effective antisolvent to fabricate high-quality tin-lead inorganic perovskite films. As a result, the optimized device delivers a PCE of 9.3% with a Voc of 0.81 V. The results provide a great promise for the application of the mixed tin-lead perovskite solar cells.

References

[1] Li, B., Zhang, Y., Fu, L., Yu, T., Zhou, S., Zhang, L., Yin, L., Surface passivation engineering strategy to fully-inorganic cubic $CsPbI_3$ perovskites for high-performance solar cells. Nat. Commun., 9(1): 1-8.

[2] Kajal, S., Kim, G.-H., Myung, C. W., Shin, Y. S., Kim, J., Jeong, J., Jana, A., Kim, J. Y., Kim, K. S., A thermally stable, barium-stabilized α-$CsPbI_3$ perovskite for optoelectronic devices. J. Mater. Chem. A, 7(38): 21740-21746.

[3] Yoon, S. M., Min, H., Kim, J. B., Kim, G., Lee, K. S., Seok, S. I., Surface Engineering of Ambient-Air-Processed Cesium Lead Triiodide Layers for Efficient Solar Cells. Joule, 5(1): 183-196.

[4] Chen, H., Zhang, T., Wang, F., Yang, W., Wang, Y., Zheng, H., Ji, L., Yuan, S., Gu, Y., Liu, D., Peng, X., Chen, L., Li, S., Efficient Stabilization and Passivation for Low-Temperature-Processed gamma-$CsPbI_3$ Solar Cells. ACS Appl. Mater. Inter., 13: 18784-18791.

[5] Hu, M., Chen, M., Guo, P., Zhou, H., Deng, J., Yao, Y., Jiang, Y., Gong, J., Dai, Z., Zhou, Y., Qian, F., Chong, X., Feng, J., Schaller, R. D., Zhu, K., Padture, N. P., Zhou, Y., Sub-1.4eV bandgap inorganic perovskite solar cells with long-term stability. Nat. Commun., 11:1-10.

[6] Yang, Z., Zhang, X., Yang, W., Eperon, G. E., Ginger, D. S., Tin-lead Alloying for Efficient and Stable All-inorganic Perovskite Solar Cells. Chem. Mater., 32: 2782-2794.
[7] Lee, S., Moon, J., Ryu, J., Parida, B., Yoon, S., Lee, D.-G., Cho, J. S., Hayase, S., Kang, D.-W., Inorganic narrow bandgap CsPb0.4Sn0.6I2.4Br0.6 perovskite solar cells with exceptional efficiency. Nano Energy, 77: 105309.
[8] Yao, H., Zhou, F., Li, Z., Ci, Z., Ding, L., Jin, Z., Strategies for Improving the Stability of Tin-Based Perovskite (ASnX3) Solar Cells. Adv. Sci., 7(10): 1903540.

Improved Resistive Switching of Ru: SiO_2/TiO_2 Based Memristive Devices

P. Gu [1,a], I. Nasir [2,b], Z. Q. Yang [1,c], K. X. Chen [1,d], D. Y. Li [3,e], X. D. Jiang [1,f], W. Li[1,g *]

[1]School of Optoelectronic Science and Engineering, University of Electronic Science and Technology of China, Chengdu 611731, China

[2]School of Physics, University of Electronic Science and Technology of China, Chengdu 611731, China

[3]Key Laboratory of Information Materials of Sichuan Province, Southwest Minzu University, Chengdu 610041, P.R. China

[a]email: gupeng18215601875@163.com, [b]email: ilyas_nasir84@yahoo.com, [c]email: 1050332345@qq.com, [d]email: 727362743@qq.com, [e]email: open_please@163.com, [f]email: xdjiang@uestc.edu.cn

*Corresponding author: wli@uestc.edu.cn

Abstract. In this paper, the resistive switching behaviors of Ru:SiO_2/TiO_2 based memristive devices have been investigated. It is found that the random and uncontrolled formation of conductive filaments in the Ru/Ru:SiO_2/p^{++}-Si devices are crucial to realize a filamentary resistive switching. It is also found that the resistive switching behavior of Ru/Ru:SiO_2/p^{++}-Si devices could be significantly improved via inserting a TiO_2 interfacial layer as in the form of Ru/Ru:$SiO_2/TiO_2/p^{++}$-Si device structure. In the modified device, strong and stable conductive filament formation could be realized when the top electrode is positively biased. In addition to nonvolatile memory applications, an analog-type switching behavior has also been realized in our newly proposed resistive switching device. The current obtained analog conductance modulation is essential for simulating synaptic functions in electronic devices for neuromorphic applications.

1. Introduction

Over the past decade, two-terminal resistive switching (RS) devices have gained considerable research interests because of their simple architectures, enormous potential for next-generation nonvolatile memory technology, and are proved to be an attractive candidate for biologically inspired neuromorphic computing systems [1][2]. However, to date, it is still challenging to fabricate an ideal device with excellent performance in terms of reliability and stability as well as low power consumption. Recently, two types of RS devices, such as (i) valence change memory (VCM) and (ii) electro-chemical-metallization (ECM) devices, have been extensively investigated.

It is indicated that oxygen ions or vacancies serve as migratory species during the switching process in VCM devices. These devices have a high data retention capability but consume a lot of power due to the relatively high diffusion energy of oxygen in the oxides-based switching layer [3]. In contrast, the switching mechanism in ECM devices lies in the redox process and subsequent migration of active metal atoms (i.e., Ag [4] and Cu [5]) into the solid electrolyte layer as well as the corresponding formation of conductive filaments (CFs) when a voltage is applied. It is noted that owing to the diffusion coefficient of active metal ions is higher in the dielectric layer [6], there could resulted in a fast migration of metal ions (Ag^+, Cu^+) and a random growth of CFs, as well as the clustering of metal atoms in the electrolyte layer [7]. Therefore, in ECM devices, the repeated switching operation overly expands the area of filaments, which would result in a hard breakdown of devices. These prevailing situations in both VCM and ECM devices are the primary concerns of their industrial applications. Recently, different strategies for improving the RS in ECM devices have been reported, e.g., doping of metal atoms[8] and deliberately inserting an additional layer in the metal/insulator/metal [9] for controlled filament growth/rupture in the solid electrolyte layer.

In this study, the Ru-doped SiO_2 (Ru:SiO_2) based RS devices have been fabricated by a sputtering method using Ru and p^{++}-Si as top and bottom electrodes. The single-layer memristive device structured as Ru/Ru:SiO_2/p^{++}-Si demonstrates a typical filamentary switching with a small on/off ratio and poor data retention. However, after inserting an interfacial layer of TiO_2 between Ru:SiO_2/p^{++}-Si, the device shows a uniform switching behavior with excellent retention properties. Also, the Ru/Ru:SiO_2/TiO_2/p^{++}-Si device exhibits an incremental conductance, which is essential for various neuromorphic applications. Furthermore, the corresponding switching mechanisms both in Ru/Ru:SiO_2/p^{++}-Si and Ru/Ru:SiO_2/TiO_2/P^{++}-Si devices have been discussed in this article.

2. Device Fabrication Methods

Figure 1(a, b) illustrates the configuration of RS devices with the structure of Ru/Ru:SiO_2/p^{++}-Si and Ru/Ru:SiO_2/TiO_2/p^{++}-Si respectively. The following procedure was adopted to fabricate these devices: (1) p^{++}-Si substrates were cleaned with acetone, ethanol, and distilled water for 10 minutes to remove any remaining contaminants; (2) A Ru:SiO_2 layer of 100 nm thickness was deposited via RF sputtering using a SiO_2 target with Ru slices placed on it; (3) A circular hole mask with a diameter of 150 μm was used to form top Ru electrodes directly on the Ru:SiO_2 layer by a DC sputtering process using a pure Ru target. Also, for the improved devices, before the deposition of Ru:SiO_2 layer, a 20 nm thick TiO_2 layer was deposited on the p^{++}-Si substrate first using a TiO_2 target, and the following steps were the same as above mentioned ones. The whole sputtering process for Ru/Ru:SiO_2/p^{++}-Si and Ru/Ru:SiO_2/TiO_2/p^{++}-Si devices was carried out with a deposition pressure of 0.5 P under the flow of Ar gas. The electrical measurements were performed at room temperature with a power source (2636B-Keithley) hooked with a probe station.

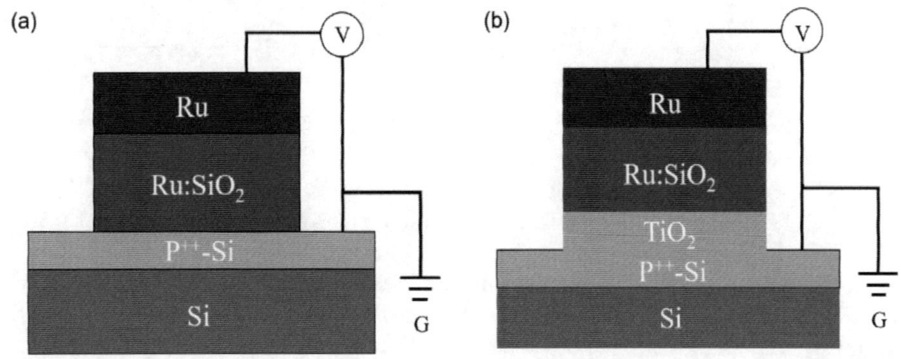

Fig.1. Schematic of (a) Ru/Ru:SiO_2/p^{++}-Si device and (b) Ru/Ru:SiO_2/TiO_2/p^{++}-Si device, respectively.

3. Results and Discussion

The current-voltage (I-V) characteristics of the devices illustrated in Fig. 1 (a, b) are analyzed to investigate the influence of TiO_2 interfacial thin layer in the Ru:SiO_2 based RS device. Fig. 2(a) depicts a typical filamentary RS behavior in Ru/Ru:SiO_2/p^{++}-Si device, where set/reset switching is realized at opposing I-V polarities. Subsequently, a histogram shown in Fig. 1(b) and statistical data illustrated in the inset of Fig. 2(b) present the cycle-to-cycle variations in set/reset voltages. It can be seen that set and reset voltages broadly vary from +0.4 V to +1.7 V and -0.4 to -1.7 V, respectively. Similar to conventional single-layer SiO_2 based RS devices [10], the variations in the operational voltage of Ru/Ru:SiO_2/p^{++}-Si device might originate from the random formation of filaments in the SiO_2 dielectric [11]. In contrast, it can be seen in Fig. 2(d, e) that the stability of the operating voltages is remarkably improved in Ru/Ru:SiO_2/TiO_2/p^{++}-Si device. The reset and set voltages range from +0.7 V to +1.4 V and +1.2 V to +1.9 V, respectively, showing that the set/reset voltage is more tightly distributed in the Ru:SiO_2/TiO_2 bilayer based device than that in the monolayer Ru:SiO_2 based device.

It is known that data retention is also a key factor for memory applications. Therefore, the retention properties of above two devices were measured at a constantly applied reading voltage (V_{read} = 0.5 V), and the data was acquired at certain time intervals (in this study, 100s intervals), as shown in Fig. 2(c, f). It is worthy of noting that the HRS/LRS ratio in the device of Ru/Ru:SiO_2/TiO_2/p^{++}-Si is upheld for more than 10^5 s (Fig. 2(f)) with no discernible degradation, much longer than that (5×10^3s) in the device of Ru/Ru:SiO_2/p^{++}-Si. The resistance value at LRS is kept stable during the test process. From this, it can be concluded that deliberately inserting the TiO_2 layer has improved the RS performance and plays an essential role in the data retention ability of Ru:SiO_2/TiO_2 bilayer based RS memristive devices.

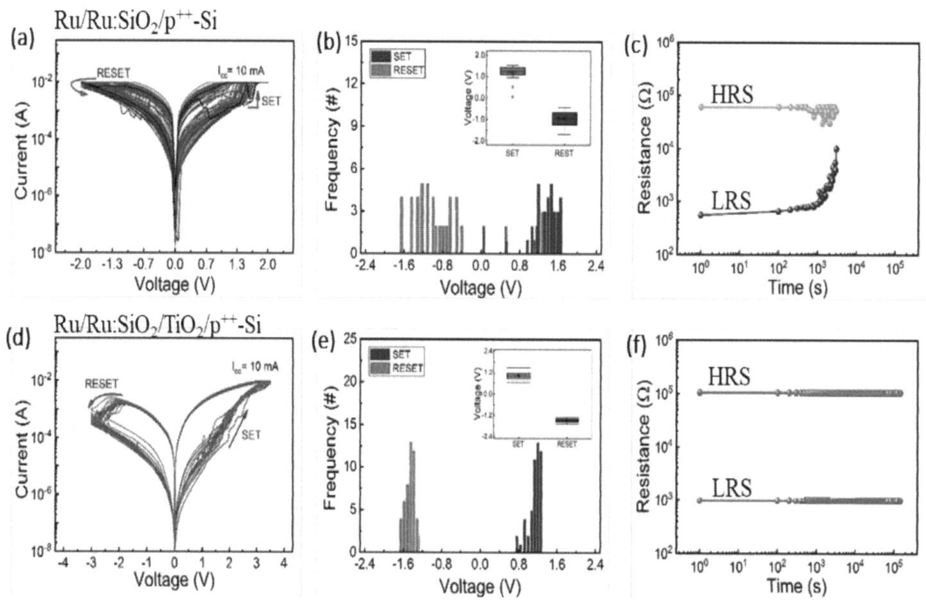

Fig. 2. I-V characteristics of the fabricated memristive Ru/Ru:SiO_2/p^{++}-Si and Ru/Ru:SiO_2/TiO_2/p^{++}-Si devices. (a) and (d) I-V curves of the devices, respectively. (b) and (e) Histograms showing the dispersion of set/reset in the devices, respectively, the insets are the statistical charts. (c, f) Retention abilities for both devices.

In order to investigate the RS mechanism in both Ru/Ru:SiO_2/p^{++}-Si and Ru/Ru:SiO_2/TiO_2/p^{++}-Si devices, I-V characteristics are further analyzed. For the device of Ru/Ru:SiO_2/p^{++}-Si, the fitted results in HRS demonstrate the charge transport behavior following the trap-controlled space charge limited conduction (SCLC) mechanism, which is consist of the Ohmic (slope ~1.1) region, the Mott-Gurney law (slope ~2) region, and the steep current (slope ~8.5) region [12] (shown in Fig. 3(a)). The SCLC, Poole–Frenkel and Ohmic conduction could be realized sequentially in LRS if the trap-filled SCLC dominates the HRS conduction mechanism. A similar phenomenon can be seen for Ru/Ru:SiO_2/TiO_2/p^{++}-Si device in HRS, while a steep drop in current is observed in LRS, as shown in Fig. 3(b). However, the fitted results of Ru/Ru:SiO_2/TiO_2/p^{++}-Si device show that the current decreases linearly with the decreasing voltage, and no steep drop in current is observed, indicating the existence of Ohmic characteristics. Furthermore, the resistances in LRS are also analyzed as a function of temperature (300 to 700 K), as in Fig.3(c) and (d). It is observed that the resistance of the device is increased linearly with temperature, indicating a typical charge transport behavior of metals. For Ru/Ru:SiO_2/TiO_2/p^{++}-Si device, the temperature coefficient (α) is then calculated as 4.13×10^{-4} from the linearly fitted data, which is very close to the α value of Ru ($\approx4.79\times10^{-4}$) [13]. From these results, it can be concluded that a strong Ru filament could be formed when the Ru/Ru:SiO_2/TiO_2/p^{++}-Si device is fully electroformed. Similarly, RS behavior has also been observed in our early research work on Ag:SiO_2 based memristive device via inserting TiO_2 interfacial layer [14].

Moreover, the physical mechanism involved in RS behavior in Ru/Ru:SiO_2/p^{++}-Si and Ru/Ru:SiO_2/TiO_2/p^{++}-Si devices can also well explained by a physical model, as shown in Fig. 4. It can be seen that when the top electrode of the Ru/Ru:SiO_2/p^{++}-Si device is positively biased, multiple tiny conductive filaments of Ru are formed in the SiO_2 dielectric matrix (Fig. 4(a)). The formation of multiple filaments in SiO_2 could have resulted in non-uniform RS behavior, as already discussed. Upon implementation of positive bias on tope electrode, the pre-existence of Ru atoms in SiO_2 matrix may be accumulated, resulting in random growth of Ru filament, and the variation in RS might be due to formation of these weak and multifilament (Fig. 4(b)). In contrast, when the TiO_2 layer is inserted between Ru:SiO_2 and p^{++}-Si, strong filaments could be formed (Fig. 4(d)), which results in uniform RS behavior in Ru/Ru:SiO_2/TiO_2/p^{++}-Si device. Therefore, a uniform set/reset process is realized in the formation and rupture of Ru filaments when positive and negative bias are applied, respectively, to the top electrode of the Ru/Ru:SiO_2/TiO_2/p^{++}-Si device. During the reset process, the Joule heating effect plays an essential role in drifting the Ru atom towards top electrodes, which results in the initial resistance in HRS. Also, the geometry of conductive filaments is important for the data retention ability of the device. If a stable and robust filament is formed across the electrode, the LRS can be sustained for a longer time. From I-V analysis at LRS regions in both kinds of the devices, the physical model can well explain that the formation of strong filaments with high densities has finally induced a stable and a reliable RS behavior in Ru/Ru:SiO_2/TiO_2/p^{++}-Si device.

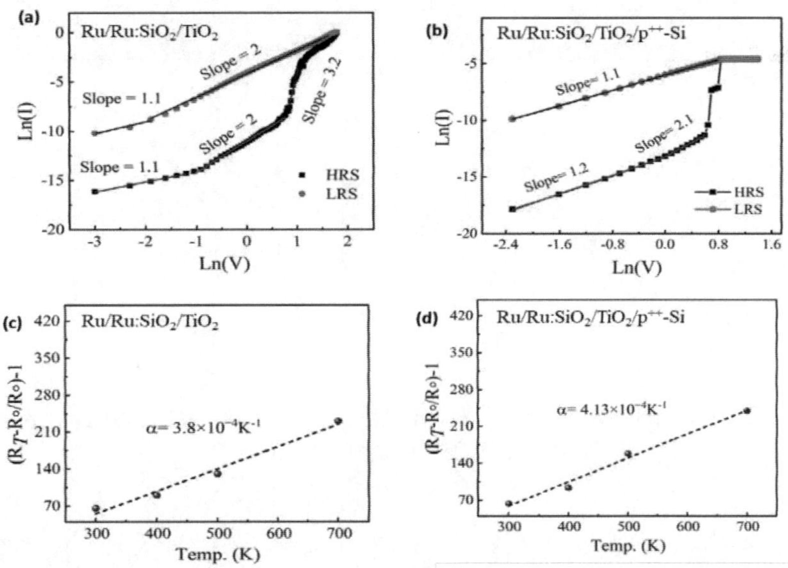

Fig. 3. Fitting results from I-V curves of (a) Ru/Ru:SiO_2/p^{++}-Si and (b) Ru/Ru:SiO_2/TiO_2/p^{++}-Si devices. (c, d)Resistance in LRS v.s. temperature varying from 300 to 700 K in both kinds of devices.

In order to search the possibility of nonvolatile memory applications, the analog conductance modulation in our newly proposed and fabricated memristive device structured as Ru/Ru:SiO_2/TiO_2/p^{++}-Si has also been realized. The gradual set and reset processes on the device using non-identical consecutive sweeps are studied. Fig. 5(a) and (b) show a noticeable increase and decrease in current flow under repeated positive (1st-4th) and negative (5th-8th) voltage sweeps, respectively, indicating the realization of analog conductance modulation. This result is vital for simulating the synaptic properties for neuromorphic applications of electronic devices [15].

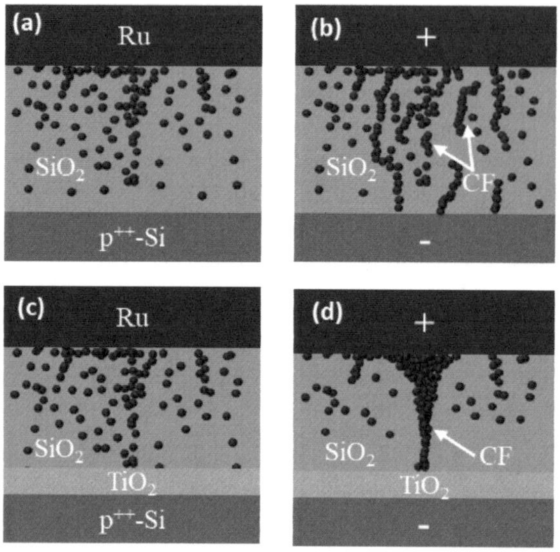

Fig. 4. Growth of highly conductive filaments in Ru/Ru:SiO_2/p++-Si and Ru/Ru:SiO_2/TiO_2/p++-Si devices. (a) As fabricated Ru/Ru:SiO_2/p++-Si structure.(b) Growth of multiple highly conductive filaments for RS behavior in Ru/Ru:SiO_2/p++-Si device. (c) As fabricated Ru/Ru:SiO_2/TiO_2/p++-Si structure. (d) Growth of Ru highly conductive filaments in Ru/Ru:SiO_2/TiO_2/p++-Si device.

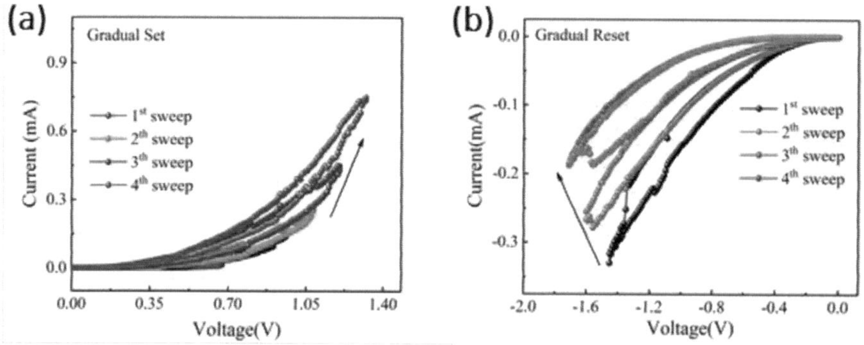

Fig. 5. I-V characteristics under consecutive non-identical positive(a) and negative (b)sweeps applied to Ru/Ru:SiO_2/TiO_2/p++-Si device

4. Conclusion

In conclusion, an effective method is proposed to improve the resistive switching performance of Ru:SiO_2 based memristive device via inserting TiO_2 dielectric film as an interfacial layer between Ru:SiO_2 and p++-Si. A uniform and stable RS behavior is realized in the modified Ru/Ru:SiO_2/TiO_2/p++-Si device. From charge transport behavior and temperature effect on the LRS of the devices, it is concluded that the controlled growth of Ru filaments in Ru/Ru:SiO_2/TiO_2/p++-Si device leads to a reduced variation in switching parameters, i.e., set/reset, HRS/LRS, and improves the retention ability in the memristive devices. Moreover, in the Ru/Ru:SiO_2/TiO_2/p++-Si device, a gradual switching during the set and the reset processes has been achieved for 4 switching windows, indicating the realization of analog switching behavior. This analog conductance modulation is critical for simulating synaptic properties in electronic devices for neuromorphic applications.

References

[1] Kim, H.J., Park, T.H., Yoon, K.J., et al., (2019) Fabrication of a Cu-cone-shaped cation source inserted conductive bridge random access memory and its improved switching reliability. Adv. Funct. Mater., 29: doi:10.1002/adfm.201806278.

[2] Suri, M., Querlioz, D., Bichler, O., et al., (2013) Bio-inspired stochastic computing using binary CBRAM synapses. IEEE Trans. Electron Devices, 60: 2402–2409.

[3] Nakamura, R., Toda, T., Tsukui, S., et al., (2014) Diffusion of oxygen in amorphous Al_2O_3, Ta_2O_5, and Nb_2O_5. J. Appl. Phys., 116: doi:10.1063/1.4889800.

[4] Yuan, F., Zhang, Z., Liu, C., et al., (2017) Real-time observation of the electrode-size-dependent evolution dynamics of the conducting filaments in a SiO_2 layer.ACS Nano, 11: 4097–4104.

[5] Valov, I., (2014) Redox-based resistive switching memories (ReRAMs): Electrochemical systems at the atomic scale.ChemElectroChem., 1: 26–36.

[6] Tsuruoka, T., Valov, I., Tappertzhofen, S., et al., (2015) Redox Reactions at Cu,Ag/Ta_2O_5 Interfaces and the Effects of Ta_2O_5 Film Density on the Forming Process in Atomic Switch Structures. Adv. Funct. Mater., 25: 6374–6381.

[7] Yang, Y., Gao, P., Li, L., et al., (2014) Electrochemical dynamics of nanoscale metallic inclusions in dielectrics. Nat. Commun., 5: doi:10.1038/ncomms5232.

[8] Yan, X., Zhao, J., Liu, S., et al., (2018) Memristor with Ag-Cluster-Doped TiO_2 Films as Artificial Synapse for Neuroinspired Computing. Adv. Funct. Mater., 28: doi:10.1002/adfm.201705320.

[9] Wan, T., Qu, B., Du, H., et al., (2018) Digital to analog resistive switching transition induced by graphene buffer layer in strontium titanate based devices. J. Colloid Interface Sci., doi:10.1016/j.jcis.2017.10.113.

[10] Kim, S.G., Han, J.S., Kim, H., et al., (2018) Recent Advances in Memristive Materials for Artificial Synapses. Adv. Mater. Technol., 3: doi:10.1002/admt.201800457.

[11] Ilyas, N., Li, D., Li, C., et al., (2020) Analog Switching and Artificial Synaptic Behavior of Ag/SiO_x:Ag/TiO_x/p^{++}-Si Memristor Device. Nanoscale Res. Lett. 15: doi:10.1186/s11671-020-3249-7.

[12] Yang, Y.C., Pan, F., Liu, Q., et al., (2009) Fully room-temperature-fabricated nonvolatile resistive memory for ultrafast and high-density memory application. Nano Lett., 9:1636–1643.

[13] Yoon, J.H., Zhang, J., Lin, P., et al., (2020) A Low-Current and Analog Memristor with Ru as Mobile Species. Adv. Mater., 32: doi:10.1002/adma.201904599.

[14] Li, D., Ilyas, N., Li, C., et al., (2020) Synaptic learning and memory functions in SiO_2:Ag/TiO_2 based memristor devices. J. Phys. D. Appl. Phys., 53: doi:10.1088/1361-6463/ab70c9.

[15] Lv, Z., Zhou, Y., Han, S.T., et al., (2018) From biomaterial-based data storage to bio-inspired artificial synapse. Mater. Today, 21: 537–552.

Interlayer ferromagnetic coupling and enhanced magnetism by biaxial strain in $MnBi_2Te_4/VBi_2Te_4$ heterojunction

Zebin Wu[1], Changsheng Song[1*] and Chaorong Li[1*]

[1]Key Laboratory of Optical Field Manipulation of Zhejiang Province, Department of Physics, Zhejiang Sci-Tech University, Hangzhou, 310018, China

[*]Corresponding author's e-mail: cssong@zstu.edu.cn and crli@zstu.edu.cn

Abstract: The pristine anti-ferromagnetic interlayer coupling of even layers $MnBi_2Te_4$ greatly restricts the possibility of realizing quantum anomalous Hall (QAH) effect. In this work, based on first-principles calculations, we find that the interlayer coupling can be transformed into ferromagnetic order from anti-ferromagnetic by substituting cations with V atoms. Furthermore, when applying biaxial strain, a strong band inversion occurs around Fermi level, achieving quantum anomalous Hall effect. A gapless surface states can further confirm it. Meanwhile, the magnetic moment of Mn and V atoms increases with the increase of strain. We attribute this to the electron hopping to magnetic atoms from p orbitals of nonmagnetic Bi and Te atoms. Our work extends the range for designing and applying for spintronic devices with QAH effect.

1. Introduction

$MnBi_2Te_4$(MBT)-family are novel van der Waals topological insulators, which have attracted considerable attention. Since three dimensional MBT has intrinsic magnetism and can display a variety of topological phases, it provides an excellent platform to explore the magnetic topological phases. MBT has a special septuple layer (SL) structure which is composed of Te-Bi-Te-Mn-Te-Bi-Te sequences. And it shows intralayer ferromagnetic (FM) and interlayer anti-ferromagnetic (AFM) exchange coupling.

Different from bulk case, the thin films will exhibit various topological states with various number of SLs. For the even SLs of MBT, the spin magnetic moment was counteracted by another layer due to AFM interlayer coupling. Consequently, there is no Hall current in the band gap. A large external magnetic field is needed to realize the quantum Hall effect[1]. This is also called axion insulator[2]. However, in the Odd SLs, since the net magnetic moment is always greater than zero, this material has enough magnetism to harbor a QAH effect without external magnetic field.

In some recent research, Otrokov et al. [1] reported that a 2 SLs AFM MBT always yields a zero Chern number because of the inversion symmetry. When an artificial FM phase is used in calculation, the Chern number becomes to -1, indicating a QAH insulator. Fu et al.[3] modulated the interlayer coupling of 4 SLs MBT slab from AFM to FM by constructing a sandwiched CrI_3/MBT/CrI_3 heterojunction, where the QAH effect was achieved. Therefore, FM coupling is one of the essential prerequisites for the realization of QAH effect.

In this work, we substituted cation with V to form MBT/VBT heterojunction, which can tune the interlayer coupling from AFM to FM. In addition, we have studied the influence of biaxial strain on band structures of bilayer MBT and MBT/VBT heterojunction by first-principles calculations. The results show that biaxial strain has a great influence on band structure, even inducing MBT/VBT into metallic phase. When the tensile strain increases to 4%, a band inversion phenomenon occurs. As a typical character of QSH insulator, the helical surface states are found in MBT/VBT heterojunction, which connects the bulk conduction and valence bands. Moreover, we calculated the magnetic moment of both MBT-based materials, and gave a reasonable analysis through electron hopping theory. Our work can provide a useful guideline for design of topological spintronics devices with dissipation-less QAH effect.

2. Computational Details

Density functional theory (DFT) calculations are carried out by Vienna ab initio Simulation Package (VASP)[4] with projector augmented wave (PAW) method[5]. The exchange correlation interaction of electrons is treated within the generalized gradient approximation (GGA)[6] by using the Perdew-Burke-Ernzerhof (PBE) functional. The energy cut-off for plane-wave basis was set by 350 eV. The Monkhorst-Pack k mesh of 5×5×1 was adopted for the structural optimization and self-consistent calculations. The convergence criteria for total energy and force on each atom are less than 10^{-5} eV per atom and 0.01V/Å, respectively. Moreover, to avoid unnecessary interactions between adjacent lattices, a vacuum space larger than 15 Å is set alone z-axis. To improve the description of the on-site Coulomb interaction of localized 3*d* states of V and Mn atoms, the GGA + U method[7] was implemented with an effective Hubbard term U set to 3.0eV and 4.0 eV, respectively. And DFT-D2[8] potential was used to describe the interlayer interactions of bilayer MBT and MBT/VBT. The surface states are calculated with the method of surface Green's function[9] by WANNIER90[10] and WANNIERTOOLS package[11].

3. Results and Discussion

We first investigate the total energy of both MBT-based materials. After the atomic position was fully optimized, the lattice constants obtained of bilayer MBT and MBT/VBT was 4.30Å and 4.21Å, respectively. Then we applied biaxial strain on both materials. As can be seen in Fig. 1(b-c), in the absence of strain, both materials have the lowest energy, which indicates that the lattice constant we selected is in the ground state. The blue line presents the energy difference between FM and AFM order (ΔE). In bilayer MBT, ΔE is always greater than zero, which means that bilayer MBT has an AFM interlayer coupling. However, the energy difference of MBT/VBT is negative value, presenting FM order, in agreement with the recent work[12]. Therefore, we conclude that substituted cation with V can turn the interlayer coupling to FM ground state.

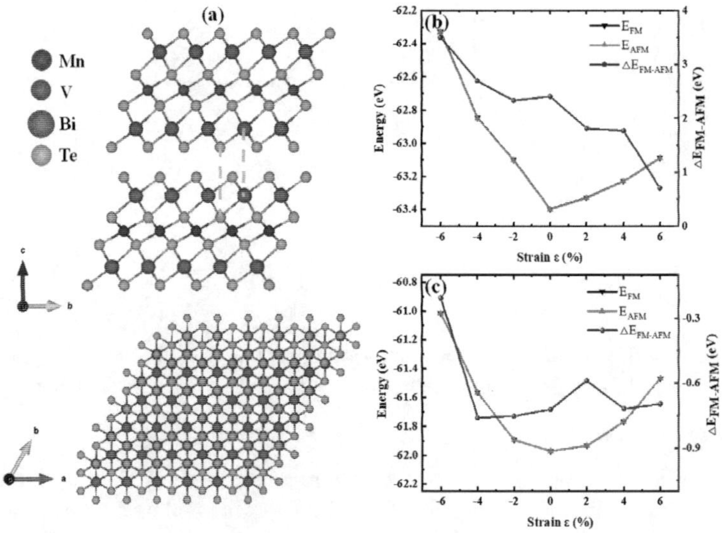

Figure 1. (a) The side and top view of atomic structure of MBT/VBT heterojunction. The violet, red, blue and gray balls represent Mn, V, Bi and Te atoms, respectively. (b-c) Total energy and energy difference of interlayer magnetic coupling under different strain ε of (b) bilayer MBT and (c)MBT/VBT heterojunction.

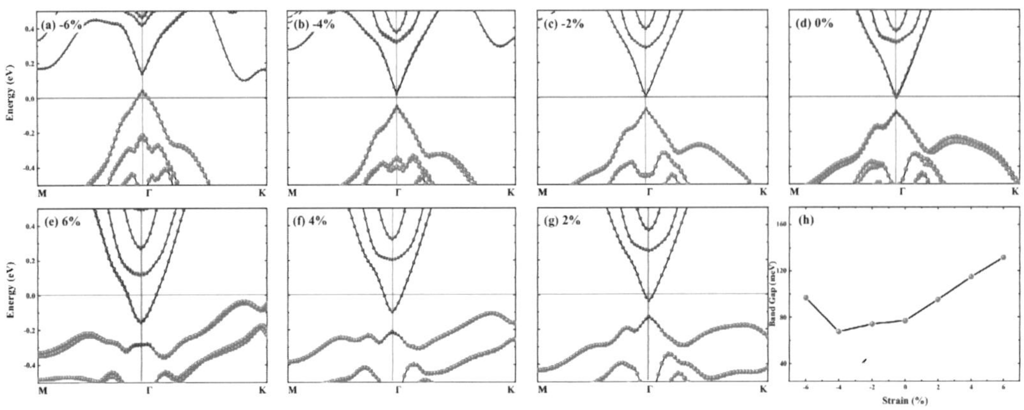

Figure 2. (a-g) Band structure of bilayer MBT under different strain ε. Red and blue ball line represent projectors of Te and Bi atoms. The Fermi level are set to be zero. (h) Band gap at Γ point versus biaxial strain ε.

We next investigate the electronic band structure ang the gap at Γ point. Fig. 2 shows band structures for bilayer MBT under different biaxial strain. We find the valence band maximum (VBM) is mainly composed of Te-p orbitals, while the conduction band minimum (CBM) consists of Bi-p orbitals. One can notice a nontrivial band structure with band inversion, which is consistent with recent works of MBT[2, 13]. When applying biaxial strain, the shape of band structure changes and shifts. A compressive strain can lift both VBM and CBM at Γ point, while tensile strain can push them down. Note that on the path of Γ to K has the opposite moving.

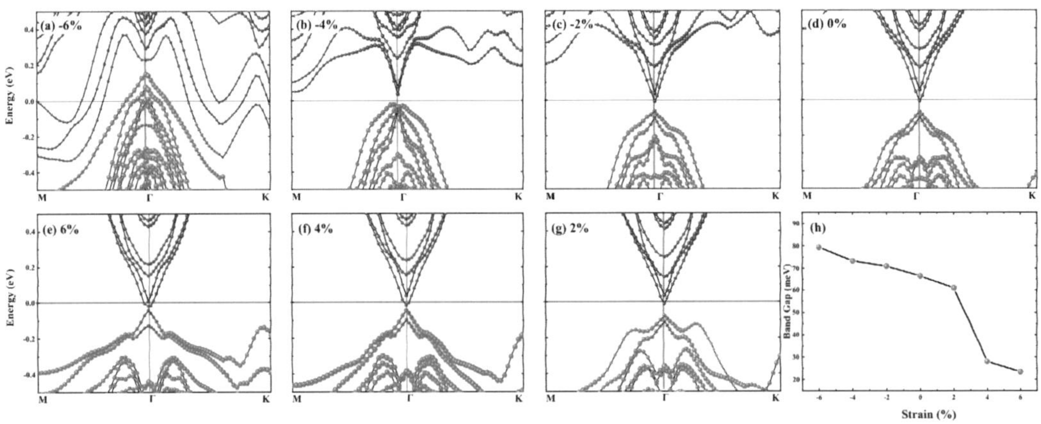

Figure 3. (a-g) Band structure of MBT/VBT heterojunction under different strain ε. Red and blue ball line represent projectors of Te and Bi atoms. The Fermi level are set to be zero. (h) Band gap at Γ point versus biaxial strain ε.

Different with bilayer MBT, in the absence of strain, the MBT/VBT heterojunction does not host band inversion in our calculation. However, Zhu et al.[13] performed calculations and observed the band inversion and gapless surface states at pristine MBT/VBT. We believe this difference can be explained in terms of the differences of lattice constants. In Zhu's work, the selection of lattice constants is closer to experimental values. As illustrated in Fig. 3, band inversion occurs when the tensile strain reaches up to 4%. Although the tendency of MBT/VBT band movement is the same as that of MBT, the band gap variation shows a different trend with MBT. The band gap decreases with increasing tensile strain.

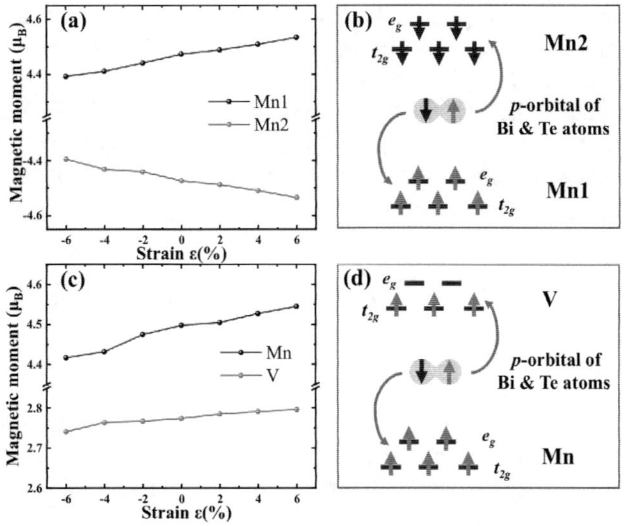

Figure 4. (a)(c) Magnetic moment (μ_B) of Mn and V atoms in (a) bilayer MBT and (b) MBT/VBT as the function of strain ε. (b)(d) Schematics of super-exchange interaction between (b) Mn1-*p*-Mn2 in bilayer MBT and (d) Mn-*p*-V in MBT/VBT.

Fig. 4(a) and (c) exhibits the magnetic moment of Mn and V atoms. One can notice that the magnetic moment increases with the tensile strain in both materials. We attribute this to electron hopping to magnetic atoms. According to Hund's rule, the d orbitals of each spin state will split into a triply degenerate t_{2g} states and a doubly degenerate e_g states. For MBT and VBT, Mn has a d^5 configuration as $t_{2g}^3e_g^2$, while V has a $t_{2g}^3e_g^0$ configuration. With the empty e_g states of V and half-filled e_g t_{2g} of Mn, electrons from p orbital of nonmagnetic atoms hopping to d orbital of magnetic atoms are allowed. Therefore, the magnetic atoms gain more electrons and the magnetic moment grows up.

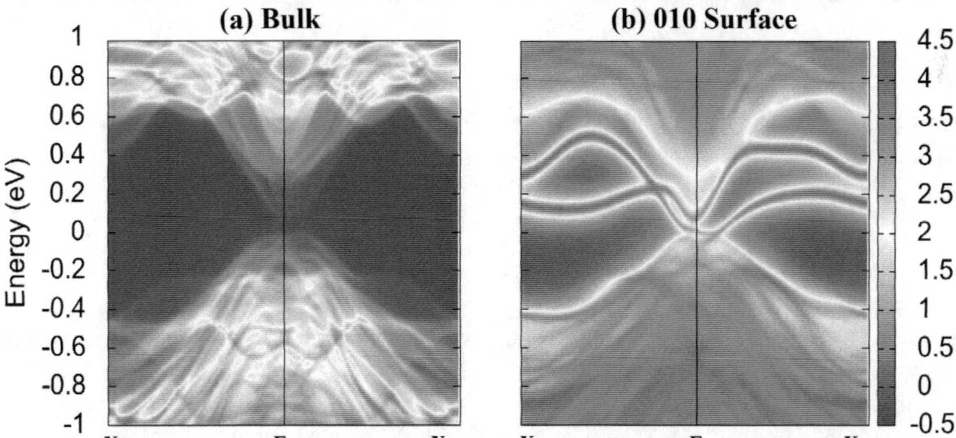

Figure 5. Topological surface states on (010) surface of MBT/VBT heterojunction under 4% tensile strain. The regions with a continuous spectrum correspond to the 2D bulk stats projected onto a 1D Brillouin zone.

We also calculated the surface states of MBT/VBT heterojunction to examine its topological property. As can be shown in Fig.5, (a) presents the band structure on its bulk, while (b) is that on the 010 surface. The former is obviously gapped, but the latter appears gapless surface states, which connects the conduction and valence bands. This further confirms the QAH effect of MBT/VBT heterojunction.

4. Conclusion

Based on first-principles calculations, we have investigated the fundamental properties of bilayer MBT and MBT/VBT heterojunction. Our results show that the interlayer coupling of bilayer MBT is AFM, by contrast, that of MBT/VBT is FM order. Moreover, biaxial strain can effectively modulate the band structure. Remarkably, the QAH state with a gapless surface states in MBT/VBT appears at tensile strain of 4%. The magnetic moment of both bilayer MBT and MBT/VBT are enhanced by applying tensile strain. Our work provides references for experimental control of interlayer coupling to realize QAH effect, especially for MBT family materials.

Acknowledgments

This work was supported by National Natural Science Foundation of China (No. 11804301), the Natural Science Foundation of Zhejiang Province (No. LY21A040008)

References

[1] Otrokov, M. M.; Rusinov, I. P.; Blanco-Rey, M.; Hoffmann, M.; Vyazovskaya, A. Y.; Eremeev, S. V.; Ernst, A.; Echenique, P. M.; Arnau, A.; Chulkov, E. V., (2019) Unique Thickness-Dependent Properties of the Van Der Waals Interlayer Antiferromagnet $MnBi_2Te_4$ Films. Phys. Rev. Lett. 122, 107202.

[2] Li, J.; Li, Y.; Du, S.; Wang, Z.; Gu, B. L.; Zhang, S. C.; He, K.; Duan, W.; Xu, Y., (2019) Intrinsic Magnetic Topological Insulators in Van Der Waals Layered $MnBi_2Te_4$-Family Materials. Sci. Adv. 5, eaaw5685.

[3] Fu, H. X.; Liu, C. X.; Yan, B. H., (2020) Exchange Bias and Quantum Anomalous Hall Effect in the $MnBi_2Te_4/CrI_3$ Heterostructure. Sci. Adv. 6, eaaz0948.

[4] Kresse, G.; Furthmüller, J., (1996) Efficient Iterative Schemes for Ab Initio Total-Energy Calculations Using a Plane-Wave Basis Set. Phys. Rev. B 54, 11169-11186.

[5] Blochl, P. E., (1994) Projector Augmented-Wave Method. Phys. Rev. B: Condens. Matter. 50, 17953-17979.

[6] Perdew, J. P.; Burke, K.; Ernzerhof, M., (1996) Generalized Gradient Approximation Made Simple. Phys. Rev. Lett. 77, 3865-3868.

[7] Dudarev, S. L.; Botton, G. A.; Savrasov, S. Y.; Humphreys, C. J.; Sutton, A. P., (1998) Electron-Energy-Loss Spectra and the Structural Stability of Nickel Oxide: An Lsda+U Study. Phys. Rev. B 57, 1505-1509.

[8] Grimme, S., (2006) Semiempirical Gga-Type Density Functional Constructed with a Long-Range Dispersion Correction. J. Comput. Chem. 27, 1787-1799.

[9] Lee, D. H.; Joannopoulos, J. D., (1981) Simple Scheme for Surface-Band Calculations. Ii. The Green's Function. Phys. Rev. B 23, 4997-5004.

[10] Mostofi, A. A.; Yates, J. R.; Pizzi, G.; Lee, Y.-S.; Souza, I.; Vanderbilt, D.; Marzari, N., (2014) An Updated Version of Wannier90: A Tool for Obtaining Maximally-Localised Wannier Functions. Comput. Phys. Commun. 185, 2309-2310.

[11] Wu, Q.; Zhang, S.; Song, H.-F.; Troyer, M.; Soluyanov, A. A., (2018) Wanniertools: An Open-Source Software Package for Novel Topological Materials. Comput. Phys. Commun. 224, 405-416.

[12] Li, Z.; Li, J. h.; He, K.; Wan, X. g.; Duan, W. h.; Xu, Y., (2020) Tunable Interlayer Magnetism and Band Topology in Van Der Waals Heterostructures of $MnBi_2Te_4$-Family Materials. Phys. Rev. B 102, 081107(R).

[13] Zhu, W. X.; Song, C.; Liao, L. Y.; Zhou, Z. Y.; Bai, H.; Zhou, Y. J.; Pan, F., (2020) Quantum Anomalous Hall Insulator State in Ferromagnetically Ordered $MnBi_2Te_4/VBi_2Te_4$ Heterostructures. Phys. Rev. B 102, 085111.

Preparation of strong green fluorescence carbon quantum dots for cancer cell imaging

Junting Tao[1,a], Hao Zou[1,b], Xinjian Lu[1,c], Xiaokun Liao[1,d], Jiaqi Pan[1,e], Chaorong Li[1,f], Yingying Zheng[1,*]

[1]Department of Physics, and Key Laboratory of ATMMT Ministry of Education, Zhejiang Sci-Tech University, Hangzhou, Zhejiang, People's Republic of China

*Corresponding author: email: zhengyy1718@zstu.edu.cn

Abstract: Carbon quantum dots (CQDs) were prepared throught hydrothermal synthesis method. The TEM and FT-IR spectra showed the successful synthesis of CQDs and the well water solubility of CQDs. After CQDs were co-cultured with human breast cancer cells (MCF-7 cells), the cells were stained and emitted stronge green fluorescence, indicated that CQDs had a good application prospect in cancer cell imaging.

1. Introduction

Cancer is a severe disease and kills many thousands of people every year[1]. Therefore, it is urgent to find a new thoughts for the diagnosis and treatment of cancer. Optical bioimaging is of great significance in the early diagnosis of cancer, which can accurately identify normal cells and cancer cells by using tumor biomarker[2-4]. Carbon quantum dots (CQDs) are a kind of small carbon nanoparticles, which have small size (below 10nm) and excellent optical properties and can be modified or functionalized and then combined with other groups to enhance CQDs' characteristics[5]. CQDs can be used in biological imaging because of its water solubility, low toxicity, biocompatibility and fluorescence properties[6-7]. CQDs contain a large number of carboxyl groups and hydroxyl groups. These carboxyl groups and hydroxyl groups provide a large number of reaction sites and can combine with other functional groups, so that CQDs have the potential ability to target identification of specific cancer cells[8].

Herein, novel fluorescent CQDs were prepared by water soluble nigrosine and betaine hydrochloride through one-step hydrothermal synthesis method. We characterized its microstructure and optical properties and then CQDs were combined with cancer cells to explore its cell imaging applications[9].

2. Materials and Methods

2.1 Chemicals and Materials

Water soluble nigrosine and betaine hydrochloride were purchased from Aladdin Industries Co., Ltd (Shanghai, China). Trypsin-EDTA solution, Fetal bovine serum (FBS), Dulbecco's modified eagle medium (DMEM), Phosphate Buffered Saline (PBS), MTT Cell Proliferation and Cytotoxicity Assay Kit (MTT) were acquired from Lanbao Biotechnology Co., Ltd (Hangzhou, China). All chemicals were of analytical grade and the ultrapure water ($\geq 18\ M\Omega \cdot cm$) water used in the experiment was self-made by purifier of Yjd RO-MB-10T (Hangzhou, China).

2.2 Synthesis of carbon quantum dots (CQDs)

Carbon quantum dots (CQDs) were prepared by one-step hydrothermal synthesis method using water soluble nigrosine and betaine hydrochloride[9]. 0.5g betaine hydrochloride was dissolved in 25ml ultrapure water, and then 0.25g water soluble nigrosine was added into the mixed solution and stirred until uniformity. The above uniformly mixed solution was transferred to a 50ml inner PPL liner, sealed with stainless jacket and hydrothermal synthesis was conducted at 250 °C for 7 h in a box muffle furnace. After the hydrothermal synthesis finished, the supernatant was filtered (slow-speed filter paper and 0.22 μm filter), dialyzed in the dialysis bag for 24 hours (MWCO = 1000D) and then was treated using freeze drying technology.

2.3 Cell viability test

The Cell viability of CQDs was evaluated by MTT assay. MCF-7 cells were incubated in a 96-well plates with DMEM containing 10% FBS, 1% Trypsin-EDTA at 1×10^4 cells density to per well and incubated in an incubator overnight. Then, various different concentrations CQDs solution (0, 0.5, 1.0, 1.5, 2.0, 0.68 mg/ml, respectively) were added to each well and incubated for another 4h and 8h. Microplate reader was used to measure the absorbance at 490 nm after 20 μL MTT was added to each well.

2.4 In vitro cell imaging

MCF-7 cells were incubated in a confocal dish with DMEM containing 10% FBS, 1% Trypsin-EDTA at a density of 1×10^5 cell and incubated in an incubator overnight. 150 μL CQDs solution (0.68 mg/ml) were added to each confocal dish and cocultured for 4 hours. Confocal dish was washed 3 times with 1ml PBS and confocal images were taken by confocal laser scanning microscope.

2.5 Characterization

The CQDs microstructure and surface morphology were characterized by transmission electron microscopy JEM-2100. Fourier transform infrared (FT-IR) spectra was collected by Fourier transform infrared spectrometer Nicolet 5700. Ultraviolet visible (UV–vis) spectra were obtained by Hitachi U-3900 spectrophotometer. Fluorescence characteristics were characterized by Hitachi F-7000. Absorbance were obtained by microplate reader BioTek ELx800. Confocal laser scanning microscope images were taken by confocal laser scanning microscope FV1200-IX81.

3. Results and discussion

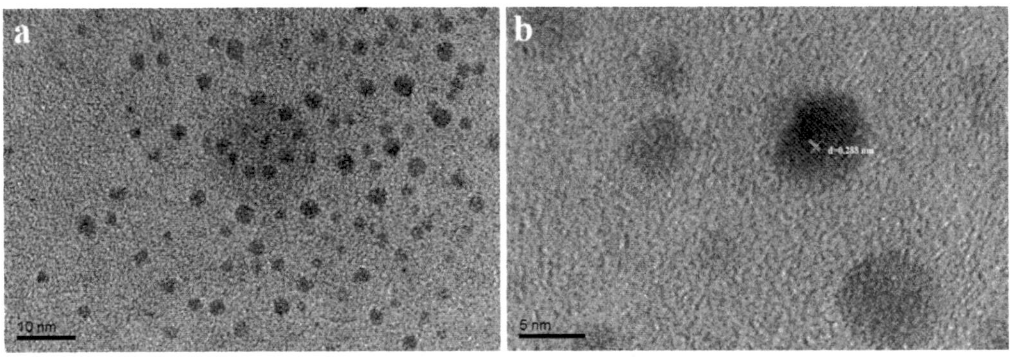

Figure. 1 (a) TEM image and (b) HRTEM image of carbon quantum dots (CQDs)

Figure. 1 showed the microstructure of CQDs. Figure.1 (a) showed that the shape of CQDs was spherical or quasi-spherical, uniformly dispersed, and the size of carbon dots was less than 5 nm. The particle size mainly concentrated around 2-3 nm and the average particle size was 2.69 nm. Figure.1 (b) is HRTEM image, which showed the interplanar spacing was 0.288 nm. TEM diagram proved the successful synthesis of carbon quantum dots.

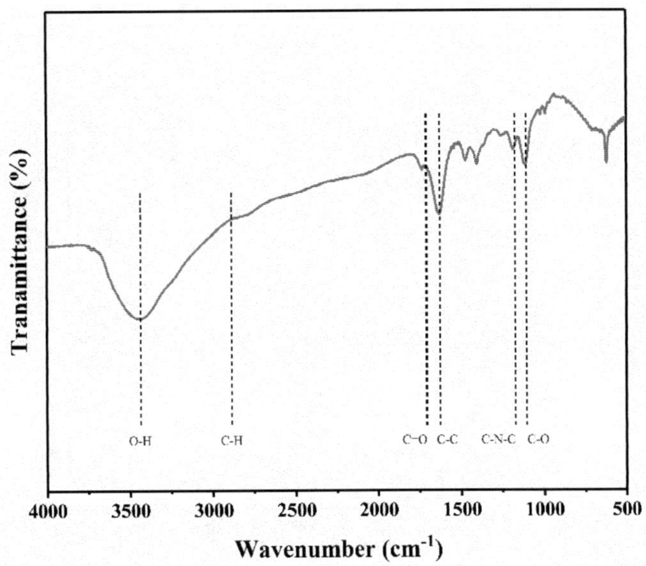

Figure. 2 FT-IR spectrum of carbon quantum dots (CQDs)

FT-IR spectrum showed that there was a wide peak with strong absorption at 3450 cm^{-1}, which could be attributed to the stretching vibration of O-H and N-H. 2800-3000 cm^{-1} was the characteristic absorption band of C-H. The peak 2883 cm^{-1} existed in this range indicated the existence of stretching vibration of C-H. The absorption peak at 1712 cm^{-1} could be attributed to the presence of C=O stretching vibration. C=O at 1712 cm^{-1} combined with the O-H at 3450 cm^{-1} indicated the presence of COOH. The absorption peak at 1622 cm^{-1} was attributable to C= C stretching vibration. The absorption peaks at 1217 and 1090 cm^{-1} could be attributed to the presence of C-N-C and C-O. The FT-IR spectra showed that the surface of the synthesized CQDs mainly contained water-soluble groups such as O-H and COOH so that CQDs had good water solubility.

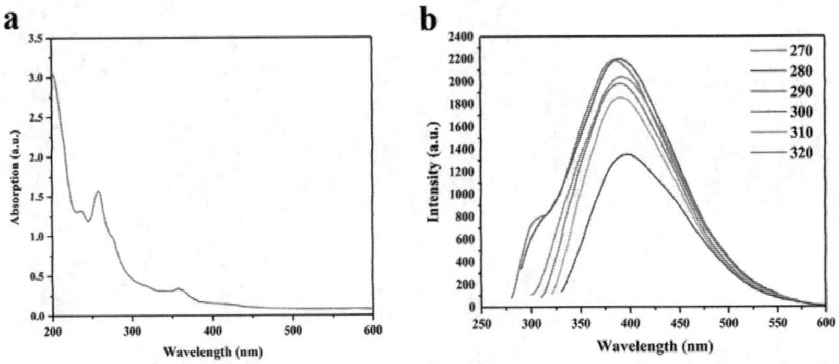

Figure. 3 UV-vis absorption spectra and PL emission spectra of CQDs

At room temperature, the optical properties of CQDs were explored. Figure. 3(a) was the UV-vis absorption spectra of CQDs. The spectra showed that there were two absorption peaks (237 nm and 260 nm) in the range of 200-300 nm, which could be attributed to the C=C bond sp^2 electron π-π^* transfer and C=O bond electron n-π^* transfer, respectively. Absorption peak at 358 nm, which may be due to the electron n-π^* transfer of C-N bond and C-O bond. Figure. 3(b) was the fluorescence spectra of CQDs. The spectra showed the relationship of excitation and emission wavelengths when the excitation wavelength was increased from 270nm to 320nm. It could be seen from the Figure. 2(b) that the best excitation wavelength was at 280 nm, and the corresponding maximum emission wavelength was around 390 nm. The emission wavelength of CQDs hardly changed with the excitation wavelength changed, which might be attributed to the uniform size distribution of CQDs.

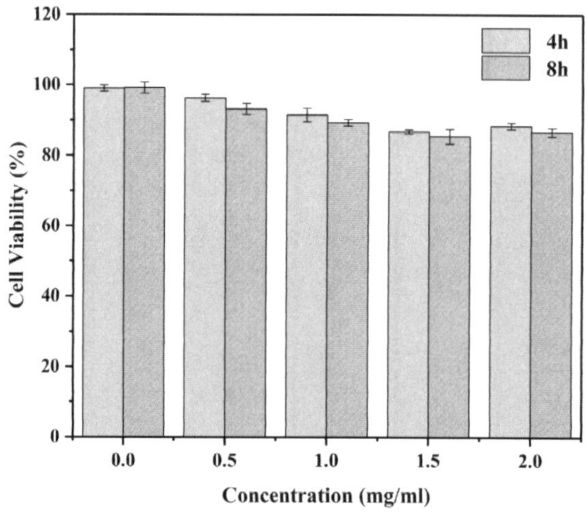

Figure. 4 The cell viability of CQDs with MCF-7 cell for 4h and 8h

The cell viability of CQDs was measured by MTT assay. It could be seen from Figure. 4 that the cell viability value was higher than 85%, which was over than the standard value of 70% according to the ISO 10993-5:2009[10]. This proved that the CQDs had a very low cytotoxic. CQDs and MCF-7 cells were co-cultivated for 4h or 8h and the cell viability value was very high and the gap was small, which proved that CQDs had good cell cytocompatibility and could be potential used in practical biomedical applications. The cell viability results showed that CQDs had the potential application in vitro imaging diagnosis of cancer.

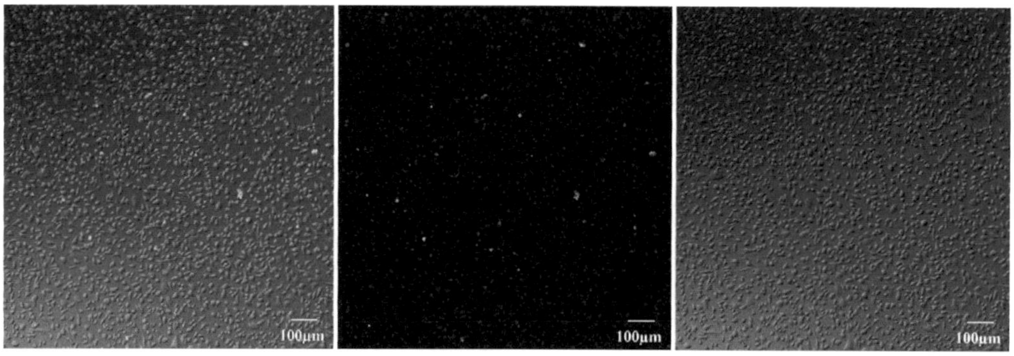

Figure. 5 Confocal laser scanning microscope images of MCF-7 cells after incubation with carbon quantum dots (left column: merged images, middle column: dark field image excited with a 488 nm laser right column: bright field image)

Figure. 5 was a confocal laser scanning microscope images taken after co-cultivate of MCF-7 cells and CQDs for 4 hours. The bright field on the right showed that the cell morphologies were normal and complete without being broken, which also proved that CQDs had good cytocompatibility, low toxicity, and minor impact on cancer cells. MCF-7 cells were excited by 488 nm laser, images showed that the cells had uptaken CQDs and be stained, emitted green fluorescence. Therefore, we could expect that CQDs have great application potential in cancer cell imaging.

4. Conclusions

In this work, we synthesized carbon quantum dots (CQDs) by using water soluble nigrosine and betaine hydrochloride throught hydrothermal synthesis method. CQDs had good water solubility and cell compatibility. MCF-7 cells uptook CQDs and emitted green fluorescence. In addition, CQDs had very low cytotoxicity and small damage to cells with time going by. Therefore, CQDs may be possible combine with other targeted substances to realize the exact recognition of cancer cells, which provide potential feasibility of cancer optical bioimaging diagnosis.

Acknowledgements

This work was supported by Natural Science Foundation of China (Nos. 51802282) and the Zhejiang Provincial Natural Science Foundation of China (No. LQ17F040004).

References

[1] Ferlay, J., Colombet, M., Soerjomataram, I., Mathers, C., Parkin, D. M., Pineros, M., Znaor, A., Bray, F., (2019) Estimating the global cancer incidence and mortality in 2018: GLOBOCAN sources and methods. Int. J. Cancer., 144: 1941-1953.

[2] Xiang, G., Liu, X., Xia, Q., Liu, X., Xu, S., Jiang, S., Zhou, X., Li, L., Wu, D., Ma, L., Wang, X., Zhang, J., (2021) Design of a bi-functional NaScF4: Yb^{3+}/Er^{3+} nanoparticles for deep-tissue bioimaging and optical thermometry through Mn^{2+} doping. Talanta., 224: 121832.

[3] Jafari, S. H., Saadatpour, Z., Salmaninejad, A., Momeni, F., Mokhtari, M., Nahand, J. S., Rahmati, M., Mirzaei, H., Kianmehr, M., (2018) Breast cancer diagnosis: Imaging techniques and biochemical markers. J. Cell. Physiol., 233: 5200-5213

[4] Pegtel, D. M., Gould, S. J., (2019) Exosomes. Annu. Rev. Biochem., 88: 487-514.

[5] Liu, R., Wu, D., Liu, S., Koynov, K., Knoll, W., Li, Q., (2009) An aqueous route to multicolor photoluminescent carbon dots using silica spheres as carriers. Angew. Chem., Int. Ed., 48: 4598-4601.

[6] Sun, Y.-P., Wang, X., Lu, F., Cao, L., Meziani, M. J., Luo, P. G., Gu, L., Veca, L. M., (2008) Doped carbon nanoparticles as a new platform for highly photoluminescent dots. J. Phys. Chem. C., 112: 18295-18298.

[7] Zou, C., Foda, M. F., Tan, X., Shao, K., Wu, L., Lu, Z., Bahlol, H. S., Han, H., (2016) Carbon-dot and quantum-dot-coated dual-emission core–satellite silica nanoparticles for ratiometric intracellular Cu^{2+} imaging. Anal. Chem., 88: 7395-7403.

[8] Javadpour, N., McIntire, K., Waldmann, T., (1978) Human chorionic gonadotropin (HCG) and alpha-fetoprotein (AFP) in sera and tumor cells of patients with testicular seminoma. A prospective study. Cancer., 42: 2768-2772.

[9] Liu, J.-H., Yang, S.-T., Chen, X.-X., Wang, H., (2012) Fluorescent carbon dots and nanodiamonds for biological imaging: preparation, application, pharmacokinetics and toxicity. Curr. Drug Metab., 13: 1046-1056.

[10] Von Byern, J., Mebs, D., Heiss, E., Dicke, U., Wetjen, O., Bakkegard, K., Grunwald, I., Wolbank, S., Muhleder, S., Gugerell, A., Fuchs, H., Nurnberger, S., (2017) Salamanders on the bench - A biocompatibility study of salamander skin secretions in cell cultures. Toxicon., 135: 24-32.

Effects of simulated in-situ electrokinetic remediation on physicochemical properties and phosphorus migration of sediments

Huapeng Wang[1], Jincheng Li[1,2,3*], Xiaoxiao Wang[1], Zuxian Lu[1], Chunman Wei[1], Xuejing Luo[1] and Yijian Zhong[1,2,3]

[1] Guangxi Key Laboratory of Environmental Pollution Control Theory and Technology, Guilin University of Technology, Guilin 541004, China

[2] Collaborative Innovation Center for Water Pollution Control and Water Safety in Karst Area, Guilin University of Technology, Guilin 541004, China

[3] Guangxi Key Laboratory of Environmental Pollution Control Theory and Technology for Science and Education Combined with Science and Technology Innovation Base, Guilin University of Technology, Guilin 541004, China

*Corresponding author's e-mail: 1994025@glut.edu.cn

Abstract. The effect of simulated in-situ electrokinetic remediation (EKR) on phosphorus release from contaminated sediments was investigated. The effects of different anode materials (iron, aluminum, graphite and stainless steel) on the physicochemical properties and phosphorus migration were compared. The results showed that aluminum anode had the best inhibition effect on phosphorus release from sediments, and the highest inhibition rate of phosphorus release was 78.20% compared with non EKR, but the sustainability was poor, and the performance of iron anode is generally stable. The phosphorus in the sediments generally shows the migration trend from cathode to anode. After EKR, the Oxidation-Reduction Potential (ORP) of the sediment decreased with the increase of normalized distance from anode. The anaerobic environment of the sediment on iron anode was the most serious, and the lowest was -616mV. The ORP of the sediment from the anode area to the middle area of other electrodes was effectively improved. The pH of the sediment increases gradually from anode to cathode, and is characterized by acidification in the anode area and alkalization in the cathode area. Iron and stainless steel electrodes have more acidic areas, showing better electrolysis efficiency.

1. Introduction

Sediment is an important part of the aquatic ecosystem. Due to the long-term development and utilization of various chemical elements by human activities, the original biogeochemical cycle of these elements has been broken, and various pollutants enter the water through different channels, leading to the serious deterioration of the aquatic environment, and most pollutants are accumulated in the sediments[1]. sediment can not only contain a variety of pollutants in the water, but also be released into the overlying water under appropriate environmental conditions to pollute the water again[2]. Therefore, research on sediment remediation technology is the key link to solve the aquatic pollution problem, whether it is urban black and odorous water or lake eutrophication.

Electrochemical remediation technology is to realize the electron transfer by constructing a chemical battery to generate the corresponding electrode reaction, thereby completing the oxidation or reduction of the pollutants[3]. According to the different principles of chemical batteries, it can be divided into EKR with the external electric field and galvanic cell remediation with spontaneous reaction. Among them, EKR mainly relies on various electrodynamic processes (electromigration, electroosmotic flow, electrophoresis etc.) under the direct current (DC) electric field to move pollutants to both ends of the electrode to achieve the degradation of pollutants[4]. At present, EKR is mostly used in the field of soil remediation, and the research on sediment remediation is still in its infancy. Secondly, in terms of the types of pollutants, many scholars have focused on the field of organic pollutants and heavy metals[5-6], but less on nutrients[7]. In this study, the effects of four anode materials (iron, aluminum, graphite and stainless steel) on the release of phosphorus from sediments to overlying water were investigated by simulating in-situ EKR. The changes in physical and chemical properties of sediments and the migration of phosphorus were analyzed. The results provide basic data for in-situ EKR of sediment.

2. Materials and methods

2.1. Experimental sediment collections and preparation

Sediment used in the experiment was collected from Qixing Wharf, Huixian Karst National Wetland Park, Guilin, Guangxi (N 30°5′31″, E 110°11′52″). The Qixing Wharf is a gathering point for local residents and tourists. Therefore, the lake is facing a severe environmental crisis, particularly eutrophication caused by large domestic sewage and non-point source pollution.

After removing large rocks and weed, the fresh sediment was screened through an 8-mesh sieve, sealed and frozen for storage. The initial sediment pH was 7.03, the ORP was -349mV, the moisture content was 73.40%, and the TP (Total Phosphorus) concentration was 3106mg/kg.

2.2. Experimental setup configurations and experimental design

Figure 1(a) shows the schematic diagram of the setup used for EKR experiments. The experimental setup is constructed by a Perspex chamber (20 cm×5 cm×20 cm), two cylindrical electrodes (5 cm×φ1cm) with electrode gaps of 15 cm, two copper wire conductors are used to connect the power supply to the electrodes, and a 30 V DC power supply to form 1 V/cm voltage gradient.

A total of 5 experimental setup were designed, one of which was the control check (CK), to assess the influence of 4 anode materials (iron, aluminum, graphite and stainless steel) on the remediation effect (the cathode material is stainless steel), the experimental design parameters are shown in Table 1. The experimental setup was equipped with 5cm of sediment, 10cm of overlying water (RO pure water). After standing for 12h, the setup was started to operate after the sediment-water interface was stabilized. The duration of the experiments was 5 days and TP content in the overlying water was monitored daily. The sediment samples were collected with a medicine spoon after the end of EKR operation. The locations of sampling site are shown in figure 1(b). All experiments were carried out at room temperature (25°C).

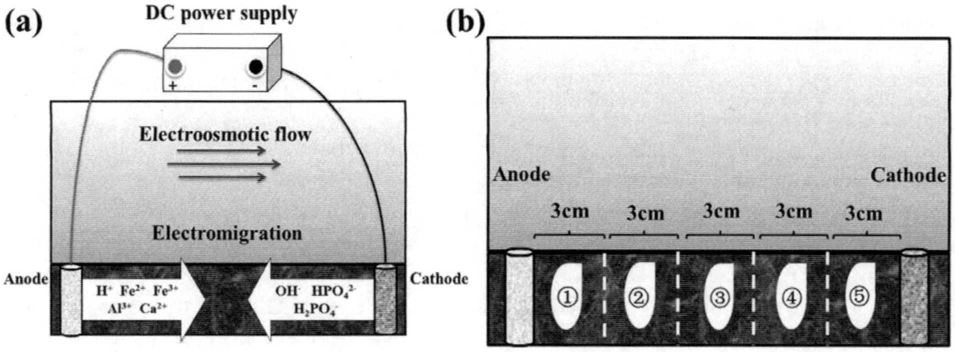

Figure 1 Schematic of the EKR setup (a) and Sampling site locations (b)

2.3. Chemical analyses

The overlying water TP concentration was measured by Lianhua multi-parameter water quality tester [5B-3B(V8), China]. The pH of sediments was measured by E-301F electrode and ORP was measured by 501 composite electrode (PHS-3E, China). The TP concentration within sediment was determined by SMT (The Standards, Measurements and Testing Programme) method[8]. Recorded current changes through DC power supply during EKR process.

Table 1 Experimental design of EKR

Number	Anode materials	Electrode gap/cm	Voltage gradient/($V \cdot cm^{-1}$)	Time/d
EK-0	-	-	-	5
EK-1	Iron	15	1	5
EK-2	Aluminium	15	1	5
EK-3	Graphite	15	1	5
EK-4	Stainless steel	15	1	5

3. Results and discussion

3.1. Changes of TP concentration in overlying water during EKR

It can be seen from figure 2 that TP concentration in the overlying water of Ek-0 increases with time, and reaches the peak value on the second day, increasing by about 66%, and then the growth trend slows down. This is because with the continuous anaerobic of the sedimentary environment, Fe^{3+} is easy to be reduced to Fe^{2+}, which promotes the release of phosphorus adsorbed by $Fe(OH)_3$ or other Fe-P[9], and anaerobic will accelerate the reduction of SO_4^{2-} , and the generated H_2S will combine with Fe^{2+}, It also promotes the release of NaOH-P (phosphorus that can be extracted by NaOH)[10]. Under the EKR of different electrode materials, the TP concentration in the overlying water of each device showed a trend of first increasing and then decreasing. EK-1 and EK-4 gradually inhibited the release of phosphorus in sediments from the second day, and the inhibitory effect of EK-4 was better and stable. Both EK-2 and EK-3 hindered the release of phosphorus from the first day. However, the effect of EK-3 is very small, and the inhibitory effect of EK-2 is the best among the four materials.

The electrolytic reaction of water occurs during the EKR of graphite and stainless steel anodes. The oxygen generated by the anode effectively inhibits the anaerobic environment of the sediment, which significantly weakens the release of phosphorus from the sediment overlying water. The release of phosphorus in EK-3 is higher than that of EK-0 in 3-5 days. This is because the collected sediment samples originally contained a large amount of algae, and the rapid growth of algae absorbed the phosphorus in the overlying water. Figure 3 shows the apparent changes of sediment during the EKR process. It can be seen from figure 3 that there are significantly more algae in EK-0 than in other groups. In the first two days, the phosphorus release from the sediment of EK-1 was even higher than that of Ek-0, and then decreased rapidly, because iron lost two electrons at the anode to form Fe^{2+}. Fe^{2+} had strong reducibility and was easy to combine with the electron acceptor, which made the ORP of sediment very low and promoted the release of phosphorus. However, with the formation of Fe^{3+} and the increase of water pH, Fe^{3+} hydrolyzed to form $Fe(OH)_3$ colloid. The TP concentration of overlying water is greatly reduced by flocculation.

Aluminum as anode will have the following electrolytic reaction:

$2Al\text{-}6e^{-}+3H_2O=Al_2O_3+6H^{+}$ (formed film reaction)

$Al_2O_3+6H^{+}= Al^{3+}+3H_2O$ (dissolution film reaction)

At the initial stage of electrolysis, due to the hydrolysis of Al^{3+}, the generated $Al(OH)_3$ has a good flocculation effect, reducing the TP concentration of overlying water. At the same time, the formed film reaction of aluminum is divided into two steps, the first is the electrolysis reaction of water at the anode (i.e. Releasing O_2), and the second is the oxidation reaction of Al, which also makes the sediment does not release phosphorus due to the low ORP. Therefore, Ek-2 always maintains a relatively low TP concentration in the overlying water. However, as Al_2O_3 is a dense oxide film, it will constantly hinder the electrolytic reaction, and its sustainability remains to be further explored.

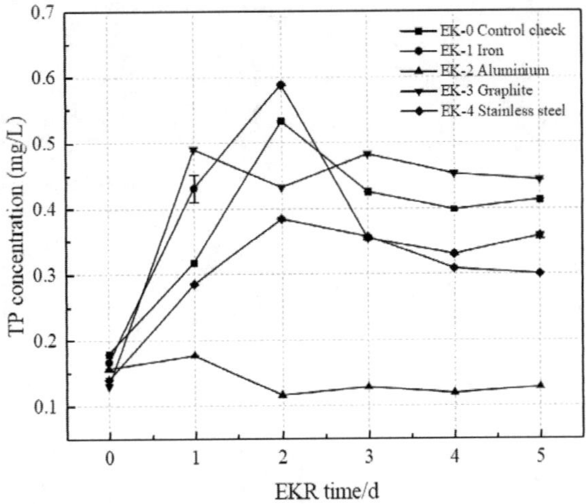

Figure 2 Changes of TP concentration in overlying water during EKR

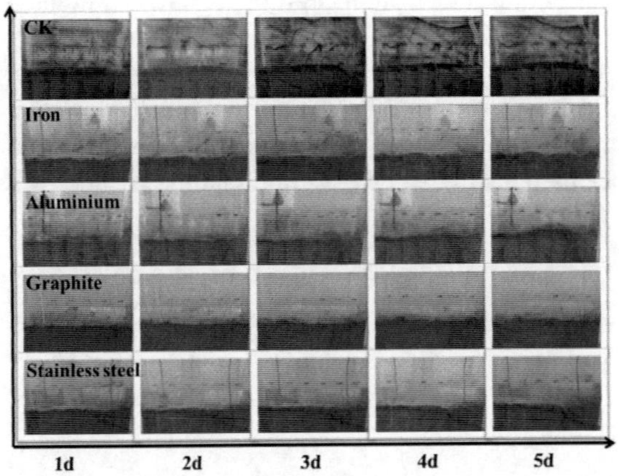

Figure 3 Apparent changes of sediments and overlying water during EKR

3.2. *Current variations during EKR*

Figure 4 shows the changes of current during the EKR. It can be seen from figure 4 that with the extension of EKR time, the current of the four electrode materials gradually decreases, and finally tends to be stable. However, in the initial 12h, the current will increase slightly. This is because the pore water is electrolysis, and the H^+ and OH^- in the solution change more, which promotes the movement of electrons, H^+ promotes the desorption of metal ions in the sediment, increases the number of mobile ions in the system, and increases the current. With the extension of time, the adsorbed metal ions and OH^- form precipitation, blocking the pores of the sediment and increasing the resistance of the sediment. In addition, the accumulation of electrode polarization and concentration polarization will also affect the efficiency of electrode reaction, resulting in the reduction of current. The current drop of EK-2 was the most dramatic. It dropped by 70% in the first 2 days, and then stabilized. Because the hydrolysis of Al^{3+} occurs in the initial stage of EKR, and then with the continuous formation of Al_2O_3 (dense oxide film), it hinders the flow of current and the progress of oxidation reaction, causing the current to drop sharply.

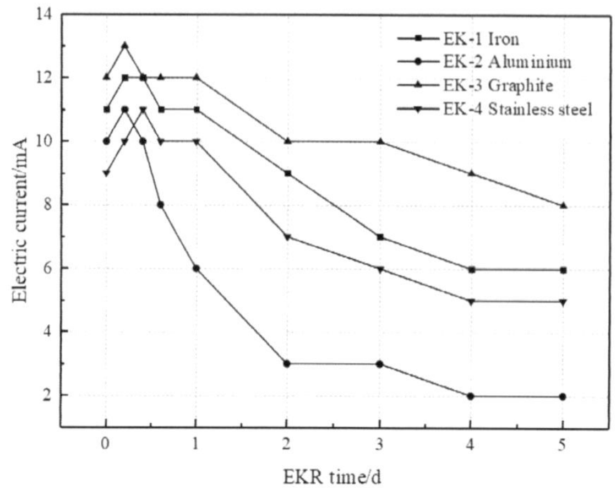

Figure 4 Changes of current during EKR

3.3. Changes in ORP in sediment after EKR

As showing in figure 5, ORP of EK-3 and EK-4 near the anode area is significantly increased, because as inert electrodes, the electrode reaction of the two electrodes at the anode is that OH^- loses electrons to generate O_2, effectively providing electron acceptors for the sediment. EK-1 and EK-2 use active electrodes, which are caused by anode corrosion reaction. Fe loses two electrons to form Fe^{2+}. Fe^{2+} has strong reducibility and it is easy to combine with electron acceptors, which makes the ORP of sediment very low. The first step of formed film reaction of aluminum electrode is the electrolysis of water in the anode. Therefore, aluminum electrode also has the effect of improving ORP of sediment. In general, the EKR under the four electrode materials will show a trend that the farther away from the anode, the lower the ORP of the sediment.

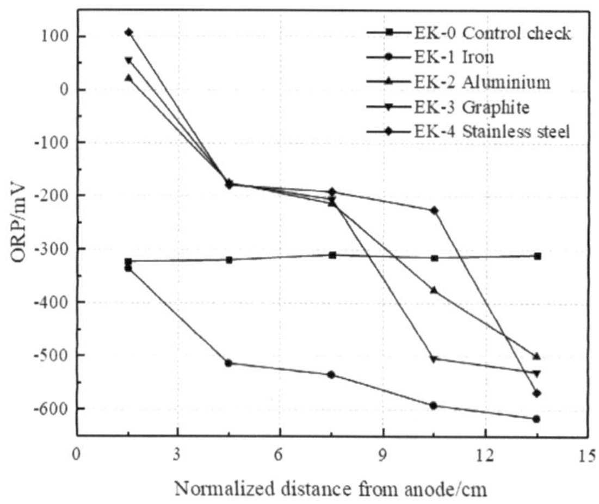

Figure 5 Changes of ORP in sediments after EKR

3.4. Changes of pH in sediment after EKR

Figure 6 shows that the pH of sediment increases gradually from anode to cathode in EKR, and has the characteristics of anodic acidification and cathodic alkalization. Due to the use of micro current in EKR and the buffer property of the sediment, the phenomenon of acid-base is not serious, and the pH value fluctuates in the range of 6 ~ 9. It can be seen that the neutral front of EK-1 and EK-4 migrates further than that of EK-2 and EK-3, that is, iron and stainless steel anode present more acidic area. Generally speaking, effective ionic mobility of H^+ is about 1.8 times that of OH^-, which makes the acid front occupy most of the area[11]. Therefore, it can be considered that the electrolysis efficiency of iron and stainless steel anode is better than that of graphite and aluminum in this experiment. In addition, the pH value of the sediment in the anode area of EK-2 is neutral, which may be due to the formation of Al_2O_3 hindering the electrolysis reaction and reducing the formation of H^+.

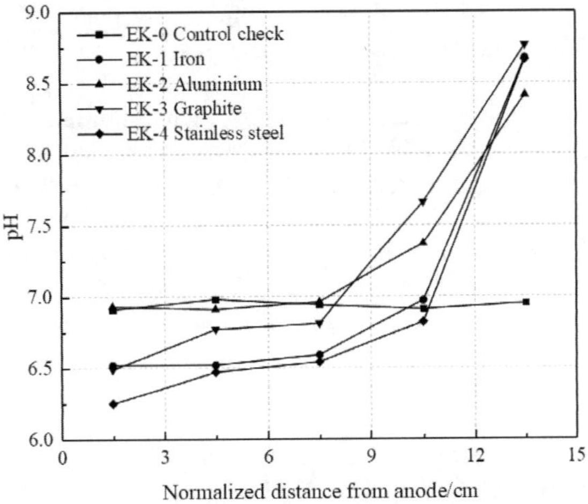

Figure 6 Changes of pH in sediments after EKR

3.5. Phosphorus migrations in sediment after EKR

It can be seen from figure 7 that the TP concentration in the sediment near the cathode area is significantly reduced because H^+ in the cathode is reduced to H_2 and ORP is very low, forming a strong anaerobic environment. Under this condition, microorganisms mineralized the organic phosphorus in the sediments into inorganic phosphate, which moved to the anode under the action of electromigration. Therefore, the concentration of TP in the sediments gradually increased from cathode to anode. The TP concentration of EK-2 in the sediment of the anode area was significantly higher than that of the other three groups. According to the previous analysis of the TP concentration of overlying water, EK-2 fixed more TP, so more phosphorus from the cathode area was accumulated in the sediment of the anode area. Compared with EK-0, the TP concentration of EK-1, EK-3 and EK-4 did not increase significantly, which may be due to the adsorption of migrating phosphorus on the electrode column.

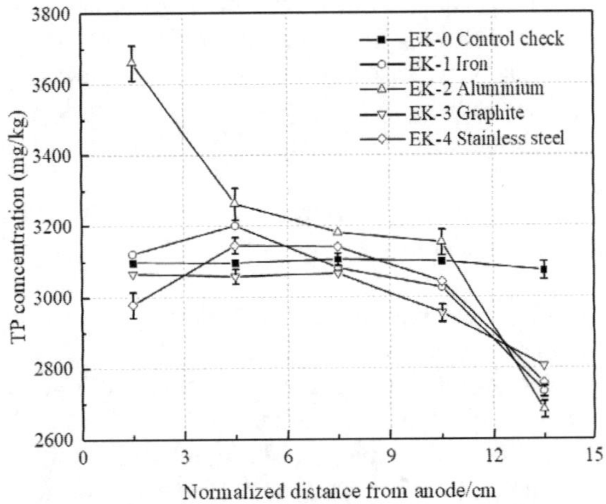

Figure 7 Phosphorus migration in sediments after EKR

4. Conclusion

1) Simulation experiments show that in situ EKR can control the release of phosphorus from sediments. Aluminum anode showed the best control effect in the remediation, and the inhibition rate of phosphorus

release was up to 78.20%. However, its remediation sustainability still needs further study. Stainless steel anode showed moderate performance in the EKR, which is a very suitable electrode material.

2) The current first increases and then decreases, and finally tends to be stable in the EKR. The current of aluminum anode decreases most acutely because of the coating of Al_2O_3 produced by electrolysis.

3) After EKR, ORP of the sediment decreased with the increase of normalized distance from anode, and the iron anode produced reductive Fe^{2+} by electrolysis, which led to the most serious anaerobic environment. The pH of the sediment increases with the increase of normalized distance from anode, and the iron and stainless steel anodes have more acidic area, showing better electrolysis efficiency.

4) After EKR, the phosphorus in the sediment tended to migrate from cathode to anode. The TP concentration of aluminum in the anode area is much higher than that of other electrodes, which may be that aluminum has a more fixation effect on the phosphorus released in the cathode area during electrolysis.

Acknowledgments
This study was supported by the Guangxi 'Bagui Scholar' Construction Project, Guangxi Science and Technology Planning Project (Grant No. GuiKe AD18126018), Xinjiang Major Science and Technology Project (Grant No. 2016A03008-1) and Guangxi Key Research and Development Program Project (Grant No. GuiKe AB16380295).

References

[1] Bouloubassi I, Roussiez V, Azzoug M, et al. Sources, dispersal pathways and mass budget of sedimentary polycyclic aromatic hydrocarbons (PAH) in the NW Mediterranean margin, Gulf of Lions [J]. Marine Chemistry, 2012, 142-144: 18-28.
[2] Liu A, Duodu G O, Mummullage S, et al. Hierarchy of factors which influence polycyclic aromatic hydrocarbons (PAHs) distribution in river sediment [J]. Environmental Pollution, 2017, 223: 81-89.
[3] Wang H P, Li J C, Wei C M, et al. In-situ remediation technology and development of aquatic sediment [J. China Rural Water and Hydropower, 2021(5): 1-7.
[4] Ferro S, Rosestolato D, Bagatin R. Electrokinetic remediation of soils polluted by heavy metals (mercury in particular) [J]. The Chemical Engineering Journal, 2015, 264: 16-23.
[5] Yang G C, Huang S C, Jen Y S, et al. Remediation of phthalates in river sediment by integrated enhanced bioremediation and electrokinetic process [J]. Chemosphere, 2016, 150: 576-585.
[6] Rajic L, Dalmacija B, Perovic S U, et al. Improving the electrokinetic remediation of Nickel-, Cadmium-, and Lead-contaminated sediment [J]. Soil & Sediment Contamination, 2013, 22(2): 199-207.
[7] Tang X Q, Li R, Han D, et al. Impacts of electrokinetic isolation of phosphorus through pore water drainage on sediment phosphorus storage dynamics [J]. Environmental Pollution, 2020, 266(Part2): 115210.
[8] Wang S Y. Sediment-water interface process of lakes—theories and methods [M]. Beijing: Science Press, 2014: 77-78.
[9] Wu F C, Jin X C, Zhang R Y et al. Effects and significance of organic nitrogen and phosphorus in the lake aquatic environment. [J] Lake Science, 2010, 22(1): 1-7.
[10] Xu Y, Chen J A, Wang J F et al. The micro-scale investigation on the effect of redox condition on the release of sediment phosphorus in Lake Hongfeng. [J] Lake Science, 2016, 28(1): 68-74.
[11] Acar Y B, Alshawabkeh A N. Principles of electrokinetic remediation [J]. Environmental Science & Technology, 1993, 27(13): 2638-2647.

Determination of Cr(VI) precipitation in small faucet samples by spectrophotometry and evaluation of the uncertainty

Guo Liping[1*]

[1]Nan'an Quality Measurement and Inspection Institute, Nan'an, Fujian, 362300, China

[*]Corresponding author's e-mail: lpguo815@foxmail.com

Abstract. A spectrophotometric method for the determination of Cr(VI) precipitation in small volume faucet samples which were less than 50 mL was established. The method was verified by precision and standard addition recovery, and the uncertainty of the measurement results was evaluated. After analysis, the precipitation of Cr(VI) in the extracting solution of the faucet samples was measured as (50.7±1.7) μg/L, (k=2, P=95%). Small volume detection method which just needs 5 mL samples solution had the advantages of small sample consumption, simple and quick operation, and can meet the requirements of the product, especially the faucet samples with small internal volume.

1. Introduction

As a common appliance in ordinary families and all kinds of public places, the safety of faucet is directly related to the safety of people's water. Chromate and polychromate are commonly used in the plating layer in the faucet for electroplating, so it is very important to measure the amount of chromium element in the faucet[1]. Especially Cr(VI), it is a carcinogen, which is easily absorbed and accumulated by many tissues in human body, and will harm human health. GB 18145-2014 *Ceramic cartridge faucets* stipulate that Cr(VI) precipitation of the faucets should be detected by the photometric method of the diphenylcarbonyl dihydrazine spectrophotometric according to GB/T 5750.6-2006 *Standard examination methods for drinking water-Metal parametersas*[2][3]. At present, there is a considerable number of kitchen faucets in the inner cavity volume which are less than 50 mL, or even smaller faucet only a few milliliters. The detection of the concentration of Cr(VI) precipitation in the faucet is not applicable to the method specified in GB/T 5750.6-2006. In this paper, according to the amount of reagent in the diphenylcarbonyl dihydrazine spectrophotometric method, the shrinkage experiment was carried out to establish a small volume (5 mL) method to detect the precipitation of Cr(VI) in the faucet immersion solution, and the method was verified. The uncertainty of the measurement results was analyzed.

2. Experiment

2.1. Materials and instruments

- Shimazu Ultraviolet-visible Spectrophotometer (UV-PROBE)
- pipette: 5000 μL,1000 μL,100 μL
- Standard solution: 1000 mg/L Cr(VI) standard solution

2.2. The measurement method

2.2.1. Extracting the faucet samples. The samples were soaked at (23±2) °C, and the preparation requirements of the extracting solution for the test were prepared according to the preparation requirements of Appendix B of GB 18145-2014. The extracting solution which was collected on the 19th day and kept for 16 h was experiment tested.

2.2.2. Preparation and linear range of standard solution. Transferred 100 μL standard solution of Cr(Ⅵ) into a 100 mL volumetric flask, and then got 1 mg/L standard reserve solution of Cr(Ⅵ) at constant volume with pure water, and then used pipette to dilute the standard curve solution step by step to a total volume of 5 mL. Then added 0.25 mL (1+7) sulfuric acid solution and 0.25 mL diphenylcarbonyl dihydrazine acetone solution (2.5 g/L), mixed well and placed for 10 min. The average value of each standard solution was measured twice, and a good linear relationship was obtained between the concentration and the absorption value, Y=0.0008X+0.00025, and the correlation coefficient (R^2) was greater than 0.9999. The standard solution preparation and test results were shown in Table 1, and the standard curve was shown in Figure 1.

Table 1 Cr(VI) standard curve solution configuration table and corresponding measured absorbance values

Concentration (X/(μg/L))	0	4	10	20	40	80	120	160	200
Cr(VI) (1mg/L) add-in (μL)	0	20	50	100	200	400	600	800	1000
Pure Water Addition (mL)	5.00	4.98	4.95	4.90	4.80	4.60	4.40	4.20	4.00
Absorbance (Y)	0.000	0.004	0.008	0.016	0.031	0.064	0.095	0.126	0.159

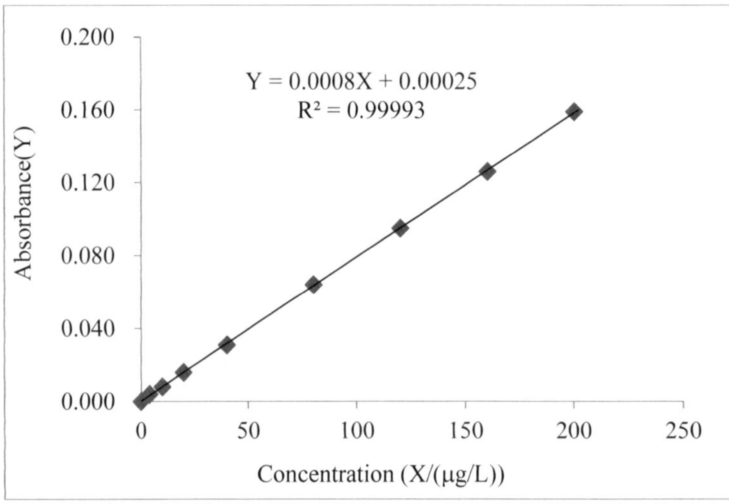

Figure 1 The linear relationship between concentration and absorbance

3. Methods validation

3.1. Standard addition recovery

Each 5 mL sample solution were added with 10 μL, 20 μL, 40 μL Cr(VI) solution with a concentration of 10 μg/mL, respectively, and the standard recovery experiment was carried out. The calculated recovery rate was within the range of deviation[4], and the results were shown in Table 2.

Table 2 Cr(VI) solution recovery rate results

Original (μg/L)	Added (μg/L)	Found (μg/L)	Recovery (%)
39.666	20	60.319	103.27
39.666	40	80.890	103.06
39.666	80	117.780	97.64

3.2. Precision measurement
Under the same conditions, 5 mL of standard Cr(VI) solution (10 μg/L, 80 μg/L, 160 μg/L) were taken separately and repeated for 6 times by the same method, calculating the relative standard deviation of Cr(VI) was 3.53%, 1.54%, 0.59%. It showed that the instrument has good repeatability in this established method, the data results can be seen in Table 3.

Table 3 Precision results

Concentration(μg/L)	10 μg/L	80 μg/L	160 μg/L
1	10.910	81.617	162.135
2	11.455	81.891	159.568
3	10.684	79.813	159.983
4	10.565	79.004	160.545
5	11.494	79.141	159.783
6	10.928	79.951	159.947
Average value	11.006	80.236	160.327
STDEV	0.388	1.235	0.943
RSD	3.53%	1.54%	0.59%

3.3. Method validation results analysis
The results showed that there was a good linear relationship between the concentration of the solution and the absorbance value in the standard curve, and the correlation coefficient was greater than 0.9999. The results showed that the standard addition recoveries of the three groups of Cr(VI) samples were 103.27%, 103.06% and 97.64%, respectively, which were within the range of method recovery deviation. The relative standard deviations of Cr(VI) were calculated to be 3.53%, 1.54% and 0.59%, respectively, for the data of 10 μg/L, 80 μg/L and 160 μg/L standard solution for 6 repeated measurements, which indicated that the analytical data of this method was accurate and reliable.

4. Uncertainty analysis
The uncertainty of measuring Cr(VI) content in faucet samples mainly comes from the following sources: the repeatability of measurement, the uncertainty of measuring instrument, the uncertainty of standard material, the uncertainty of standard curve preparation, the uncertainty of temperature change and the uncertainty of curve fitting.

4.1. Evaluation of uncertain components

4.1.1. Relative standard uncertainty of measurement repeatability $u_{rel}(R)$. The same sample solution was repeated for 9 times and the measured concentrations were as follows: 50.278 μg/L, 49.977 μg/L, 49.273 μg/L, 48.639 μg/L, 52.711 μg/L, 52.039 μg/L, 50.773 μg/L, 50.615 μg/L, 51.885 μg/L, repeatability measurement was evaluated according to Class A standard uncertainty [5].

Average concentration of sample was c= $\bar{X} = \frac{\sum_{i=1}^{9} X}{9} = 50.688\ \mu g/L$

Standard deviation of sample determination results was S= $\sqrt{\frac{\sum_{i=1}^{9}(X_i-\bar{X})^2}{n_1-1}} = 1.334\ \mu g/L$

Standard uncertainty due to measurement repeatability was $u(R) = \frac{S}{\sqrt{n}} = \frac{1.334}{\sqrt{9}} = 0.445\ \mu g/L$

Relative standard uncertainty due to measurement repeatability was $u_{rel}(R) = \frac{u(R)}{\bar{X}} = 0.00877$。

4.1.2. The relative standard uncertainty of the measuring instrument $u_{rel}(Y)$. According to the verification certificate of UV-Vis spectrophotometer, the extended uncertainty of the error of the instrument's transmission ratio was 0.8%, k=2, and its relative standard uncertainty was $u_{rel}(Y) = \frac{U(Y)}{k} = 0.004$.

4.1.3. Relative standard uncertainty $u_{rel}(W)$ for standard reserve solutions. The solution of Cr(VI) reference substance used in this experiment was 1000 mg/L (GBW(E)082820), and the relative extended uncertainty was obtained by inquiry 1%, k=2, and the relative standard uncertainty introduced by the standard solution was calculated as $u_{rel}(W) = \frac{1\%}{2} = 0.005$.

4.1.4. Relative standard uncertainty introduced by standard curve configuration $u_{rel}(Q)$. Three kinds of pipette with nominal capacity of 5000 μL, 1000 μL and 100 μL were used for the preparation of standard solution. The pipette gauges were used to calibrate the introduced standard uncertain devices according to the rectangular distribution. The allowable errors of the pipette capacity of 5000 μL, 1000 μL and 100 μL were obtained as 0.6%, 1% and 2% [6]. The calculation was as follows:

- Standard uncertainty of 5000 μL pipette gun was $u(Q_1) = 0.0173$ mL
- Standard uncertainty of 1000 μL pipette gun was $u(Q_2) = 0.00577$ mL
- Standard uncertainty of 100 μL pipette gun was $u(Q_3) = 0.00116$ mL

Since each uncertainty component was not related to each other, the relative uncertainty introduced by the pipette gun gauge calibration was as follows:

$$u(Q)=\sqrt{9 \times u(Q_1)^2 + 6 \times u(Q_2)^2 + 3 \times u(Q_3)^2}=0.0539\ mL$$

$$u_{rel}(Q)=0.0539/5=0.0108$$

In the formula, "9×", "6×" and "3×" represent the number of times pipette gun was used in the experiment.

4.1.5. Relative standard uncertainty introduced by temperature change $u_{rel}(T)$. The measured sample and the standard curve solution prepared were both 5 mL, the laboratory temperature was controlled at 20±5°C, the expansion coefficient of water was 2.1×10^{-4} mL/℃, distributed in a rectangular way, and the resulting relative standard uncertainty was $u_{rel}(T)=\frac{2.1\times10^{-4}\times5}{5\times\sqrt{3}}=0.000121$.

4.1.6. Relative standard uncertainty introduced by curve fitting $u_{rel}(X)$. In the experiment, 9 concentration points were used as the solution of Cr(VI) standard curve series, and the readings of each concentration point were taken twice to take the average value. The results were shown in Table 1. The regression line equation was obtained by fitting as follows: Y=0.0008X+0.00025, and the correlation coefficient $R^2 = 0.99993$. The residual standard deviation of the standard curve was:

$$S_{(y)} = \sqrt{\frac{\sum_{i=1}^{n_1}(Y_i - (a + bX_i))^2}{n_1 - 2}} = 0.000650\ \mu g/L$$

Standard uncertainty caused by standard curves: $u(X) = \frac{S_{(y)}}{b}\sqrt{\frac{1}{n_1}+\frac{1}{n_2}+\frac{(c-\bar{X})^2}{\sum_{i=1}^{n_1}(X_i-\bar{X})^2}}$ =0.343 μg/L

In the formula, "Y" refers to the average measured punctuation value, "X" refers to the concentration of each punctuation, "n_1" refers to the number of standard curves, "n_2" refers to the total number of standard curve measurements, "a" refers to 0.00025, "b" refers to 0.0008, "c" refers to the average concentration of the sample solution.

The relative standard uncertainty introduced by the standard curve was: $u_{rel}(X)$=0.00677

4.2. The composition of uncertainty of each component

4.2.1. The relative uncertainty of each component. The relative uncertainty of each component in this established method was shown in Table 4.

Table 4 The relative uncertainty of each component

u_{rel} (R)	u_{rel} (Y)	u_{rel} (W)	u_{rel} (Q)	u_{rel} (T)	u_{rel} (X)
0.00877	0.004	0.005	0.0108	0.000121	0.00677

4.2.2. Composition of relative uncertainty u_{rel}. The related combined standard uncertainty of the above uncertainty components was:

$$u_{rel} = \sqrt{u_{rel}(R)^2 + u_{rel}(Y)^2 + u_{rel}(W)^2 + u_{rel}(Q)^2 + u_{rel}(T)^2 + u_{rel}(X)^2} = 0.0167$$

4.3. Extend uncertainty

The extended uncertainty took the inclusion factor k=2, then the extended uncertainty U=0.0167 × 50.7 × 2 = 1.7 μg/L.

4.4. Result representation

The measurement result of Cr(VI) in a small volume sample of 5 mL was (50.7 ± 1.7) μg/L, (k=2, P=95%) by diphenylcarbonyl dihydrazine spectrophotometric method.

5. Conclusions

In this study, a diphenylcarbonyl dihydrazine spectrophotometric spectrophotometric method was established to determine the detection method of Cr(VI) in small-volume faucet samples, which was verified by the method of mark-up recovery rate and repeatability, which showed that the method was feasible. This method just required 5 mL sample solution. At the same time, the method mainly used pipetting gun to prepare standard curve solution and remove sample solution, which was simpler and faster than measuring instruments such as volumetric flask and colorimetrical cylinder.

The uncertainty analysis was shown that the main sources of the uncertainty in the measurement of faucet samples by the compressed volume method were the repeatability of the measurement, the preparation and fitting of the standard curve. Compared with the relative standard uncertainty between the standard curve prepared by the 50 mL large-volume detection method and the sample measurement results, the 5 mL small-volume sample solution detection method can meet the testing requirements of the products, which indicates that the 5 mL small-volume method for measuring the concentration of Cr(VI) was feasible when the sample volume was less than 50 mL.

Acknowledgments

This work was supported by the Science and technology project of Fujian Provincial Bureau of Quality and Technical Supervision (No. FJQI2018053). I would like to thank my colleagues, Mrs. Huang, for her guidance through each stage of the process.

References

[1] Lu, C.Q. (2016) Determination of Chromium (VI) and Chromium (II) Stripped from Stopcock by Liquid Chromatography-Inductively Coupled Plasma Mass Spectrometry. J. Anal. Test., 35(12): 1639-1642.
[2] GB 18145-2014, Ceramic cartridge faucets [S], National Standards of the Peoples Republic of China.
[3] GB 5750.6-2006, Standard examination methods for drinking water-Metal parameters [S] , National Standards of the Peoples Republic of China.
[4] GB/T 27417-2017, Comformity assessment—Guidance on validation and verification of chemical analytical methods [S] , National Standards of the Peoples Republic of China.
[5] JJF 1135-2005, Evaluation of Uncertainty in Chemical Analyswas Measurement [S] , National Standards of the Peoples Republic of China.
[6] JJG 646-2006, Locomotive Pipette [S] , National Standards of the Peoples Republic of China.

Application of New Mxene Based Materials in Electrochemical Sensors

Xiang Zhang [*], **Yu Gu** [—]

Nantong Ecological Environment Monitoring Center, Nantong, Jiangsu, 226000, P.R. China

[*]Corresponding author. E-mail: hbj@nantong.gov.cn

[—] These authors contributed equally to this work.

Abstract. In the current research on the application of new Mxene based materials in electrochemical sensors, the research on the charge storage of new Mxene based materials is not comprehensive enough. The charge storage of new Mxene based materials under different scanning rates is studied in the form of experiments. According to the needs of the experiment, experimental reagents and instruments were selected to prepare Mxene solution and new Mxene based materials. Macroporous Mxene was prepared as the electrode of EES device to store and convert energy. The enzyme was immobilized on the inner surface of Ti3C2Tx to sense biological activity. The experimental results show that the charge storage performance of the new Mxene based materials is the best when the scanning rate is 15mVs-1, which is the most suitable for electrochemical sensors.

1. Introduction

Electrochemical sensors are used to analyze and measure target molecules or substances qualitatively or quantitatively by measuring their electrical and electrochemical properties. Electrochemical sensors originated in the 1950s, with the development of science and technology such as electrochemistry, microelectronics and material processing. In the 1960s, with the advent of ion selective electrode and enzyme electrode, electrochemical sensor entered a stable development stage.[1-3]. Electrochemical sensor consists of two parts: sensor and converter. The sensor has two main functions: selectively interact with the object to be measured, and then convert the measured chemical parameters into conduction module, which can generate response signals. Finally, the response signal of the identification platform is transformed into the signal that can be used for analysis, and the quantity of the object to be measured in the new Mxene based material is detected [4-6]. With the continuous consumption of fossil fuel resources and the increasingly prominent environmental problems, people begin to turn their attention to sustainable and renewable clean energy. Since the beginning of the 21st century, with the in-depth understanding and development of the unique properties of new Mxene based materials, Nano science and technology have developed rapidly in energy, biology, aerospace and other fields [7-9]. Graphene as a representative of two-dimensional materials has some properties that three-dimensional bulk materials do not have, such as the reduction of the size of the surface area is greatly increased than the bulk materials, more atoms are exposed to the surface, thus showing higher electrochemical reaction activity, so two-dimensional nano materials quickly become a research hotspot in the field of energy storage. At present, the literature on the application of new Mxene based materials in electrochemical sensors is not very rich, which needs further study.

2. Experimental preparation

2.1. Selection of experimental reagents and instruments

According to the needs of the experiment, the experimental reagents and instruments are selected as shown in Table 1:

Table 1 Experimental reagents and instruments

Laboratory instruments / reagents	manufacturer	Laboratory instruments / reagents	manufacturer
CHI-660D electrochemical workstation	Shanghai Chenhua Instrument Co., Ltd	glucose	Sinopharm group
KQ218 ultrasonic cleaning instrument	Kunshan Ultrasonic Instrument Co., Ltd	Magnesium chloride	Sinopharm group
Vacuum drying oven	Shanghai Yiheng Scientific Instrument Co., Ltd	potassium chloride	Sinopharm group
Tubular resistance furnace	Shanghai Yafeng Furnace Co., Ltd	hydrofluoric acid	Shanghai McLean Biochemical Technology Co., Ltd
DC high voltage generator	Beijing Institute of mechanical and electrical high voltage Technology Co., Ltd	Restore sweet skin	Sinopharm group
Microinjection pump	Baoding Lange constant flow pump Co., Ltd	Anhydrous ethanol	Sinopharm group
Electronic balance	Shanghai Precision Instrument Factory	Tetramethylbenzidine	Shanghai Yuanye Biotechnology Co., Ltd
DLMAX-2200 X-ray diffractometer	Japan Science Corporation	Fluorination theory	Shanghai prosace group
Disodium hydrogen phosphate dodecahydrate	Sinopharm group	hydrochloric acid	Sinopharm group
Disodium phosphate dihydrate	Sinopharm group	L-cysteine	Sinopharm group

Based on the above experimental instruments and reagents, mxene solution and new mxene based materials were prepared.

2.2. Preparation of Mxene solution

Adding 15 ml 12 m HC and 0.99 g LIF into 70 ml polypropylene plastic bottle, and stirring for 5 min to dissolve LIF in HCl. In order to prevent the system from overheating, 1 g Ti_3AlC_2 was added to the above solution several times. Then, the plastic bottle was transferred to a collector type magnetic stirrer and stirred at 35 °C for 24 h to complete the etching of A1 layer atoms [10]. After the reaction, the product was transferred to an 80 ml centrifuge tube, centrifuged at 6200 rpm and washed with water until the pH of the supernatant was close to neutral. The precipitate centrifuged to neutral was redispersed in water, and then ultrasoniced at 220W for 40 min in hydrogen atmosphere. After centrifugation at 6200rpm for 1h, the supernatant was collected to obtain ti3c2tx solution, which was put into hydrogen for storage.

2.3. Preparation of new Mxene based materials

Diluting the solution containing 100 mg Mxene material to 300ml, add 0.24g, and then ultrasonic wave for 10 minutes to dissolve it. Weigh 0.1g dopamine hydrochloride in the mixed solution and stir it for 24 h at room temperature to complete the self polymerization of dopamine hydrochloride. After centrifugation at 8000 rpm for 5min, the precipitates were collected and freeze-dried for 48 h to obtain three-dimensional polydopamine coated Mxene materials. Then, the orange Mxene was coated with polydopamine. The three-dimensional carbon coated Mxene material, named T-Mxene@C, was prepared by carbonizing the polydopamine layer after being treated at 700 °C for 2h in hydrogen atmosphere Mxene. The specific preparation process. In addition, the Mxene powder obtained by direct freeze-drying of Mxene solution was taken as the reference sample, which was recorded as f-Mxene. The powder of f-Mxene after high temperature treatment at 700 °C was also taken as the reference sample, which was recorded as h-Mxene.

3. Experimental process

3.1. Storing and converting energy

Mxene modules were initially mainly used in the field of supercapacitors, and then extended to other alkali metal ion energy storage fields. Therefore, it is necessary to develop more advanced structures for high performance electrodes based on Mxene. Gogotsi group used PMMA as template to prepare macroporous Mxene as electrode of EES device. Because the ion transport path is significantly shortened, it has ultra-high rate capability. Similarly, the vertical alignment of Mxene nanosheets can also be much shorter than that of 2D structure ion transport path, and show a thickness independent rate capability, up to 200 μ M. Recently, many researches have focused on assembling 2D mxene nanosheets into 3D macrostructures. With the customized 3D network and 3D components, it provides fast ion transport path and high utilization of active sites, so it has excellent electrochemical performance. Therefore, a more comprehensive study of novel 3D assembled Mxene structures is needed to better understand and explore its energy storage applications. Compared with Mxene film, 2D Mxene hybrid film and reduced graphene oxide film electrode, the above Mxene electrode showed a significant increase in capacitance and transmission rate. The formation of 3D interconnection network during gelation can effectively limit the re stacking of MXene nanosheets, thus providing an interconnection network for fast ion transport and providing high electrical conductivity.

3.2. Shielding and absorbing electromagnetic waves

Due to its unique structure and properties, many Mxene materials and Mxene composites have excellent electromagnetic wave attenuation properties. The self-supporting materials etched by HCl + LIF exhibit high conductivity due to low inter chip resistance and in-situ Li ion doping. Further research shows that the shielding effect of electromagnetic wave can reach 92 dB. In addition, the doping of other components in Mxene materials is becoming more and more popular. Similar Fe doped $Ti_3C_2T_x$ Mxene films show an effective reflection loss of - 57.3 dB at 10.1 GHz at a thickness of 1.9 mm, and a broadband reflectivity of less than - 10 dB in the range of 9.1 GHz to 11.1 GHz. In addition, Mxene @ polystyrene microspheres with high conductivity (1081 s / M) were prepared by electrostatic assembly, and they have electromagnetic wave shielding performance of more than 54 dB in the whole X-band. With the load increasing to 1.90% Vo1, the electromagnetic wave absorption can be further improved to - 62 dB. 3D porous Mxene based aerogels with high reflectivity and high conductivity are very suitable for shielding and absorbing electromagnetic waves. The 3D macrostructure generated by the cross-linking of nanosheets also has porous structure, so it is expected to increase the electromagnetic wave loss ability of nanosheets. The hole wall composed of Mxene nanosheets with good dielectric and magnetic properties can increase the energy loss in the electromagnetic wave, so as to achieve the effective attenuation of the electromagnetic wave. All in all, Mxene materials have great potential in the field of electromagnetic wave shielding and absorption.

4. Experimental result

In order to test the electrochemical performance of the new MXenebased materials, according to the mass ratio of 8:1:1, appropriate amount of active material (Ti_3C_2 MXene, 2H-MoSz), binder and conductive agent were respectively weighed and put into the agate mortar, and then appropriate amount of NMP was dripped, and then fully ground to obtain uniform slurry. Finally, the slurry was evenly coated on the carbon paper. The coating area is about 1cm × 5 mg. The electrodes were then dried in a vacuum oven at 45 °C for 12 hours. The Ag / AgCl electrode was selected as counter electrode and reference electrode respectively. The voltage window of cyclic voltammetry test (CV) is - 1.0v-0v, and the voltage window of constant current charge discharge test is - 1.0vo-0v. Electrochemical impedance analysis was performed in the frequency range of 100 kHz ~ 0.01 Hz. The specific capacity of the electrode can be calculated according to the GCD curve, the formula is as follows:

$$W = \frac{hj}{\Delta Q} \tag{1}$$

In formula (1), h is the mass specific capacitance, j is the potential window, and Q is the scanning rate. On the basis of formula (1), according to the CV Curve, the calculation formula of specific capacitance is obtained as follows:

$$R = \frac{1}{\Delta k}\int \frac{h}{t} \tag{2}$$

In formula (2), h is the mass specific capacitance, k is the potential, and t is the point current density. In addition, the coulomb efficiency curve is expressed by GCD curve, the specific expression formula is as follows:

$$\delta = \frac{E}{P} = \frac{r_1}{r_2} \tag{3}$$

In formula (3), E represents discharge capacity, P represents charging capacity, r_1 represents discharge time and r_2 represents charging time. According to the formula (3), the preparation of PVA-H_2SO_4 gel was carried out. When PVA- H_2SO_4 gel is solidified at room temperature, 12h, a symmetrical flexible all solid state supercapacitor based on a new MXene based material can be obtained. The specific capacitance of the device based on the active material area can be calculated by the charge discharge curve:

$$D = \frac{U \times \Delta g}{H \times \Delta y} \tag{4}$$

In formula (4), U is the constant discharge current, H is the opposite area of the active material on the two working electrodes, g is the discharge time, and y is the constant discharge current. The energy density of Mxene based flexible all solid state supercapacitors is calculated

$$F = \frac{1}{(2 \times 3.6)} \times \eta \times \Delta C^2 \tag{5}$$

In formula (5), η represents the voltage window of the charge discharge curve, and C represents the discharge time of the device. Cycle stability is one of the important parameters to evaluate the performance of supercapacitors. In order to study the stability of the material in electrochemical test, the cyclic stability of the electrode was tested at high current density of 50ag-1. After 10000 cycles, the specific capacitance of the electrode decreases slightly, which is% 4% of the initial capacitance, indicating that the three-dimensional network heterostructure of the new Mxene based material is very stable in the process of rapid energy storage. After 20000 long-term cycles, the cycle stability slightly increased to 96.8%, which may be due to the gradual activation of the electrode surface in the electrolyte. A large number of practical results show that the new Mxene based materials have the best application effect in electrochemical sensors when the charge storage is between 81% and 85%. On the basis of the above, the charge storage of the new Mxene based materials at three different scanning rates under the same electrode conditions was tested, and the experimental results are shown in Fig. 1-3

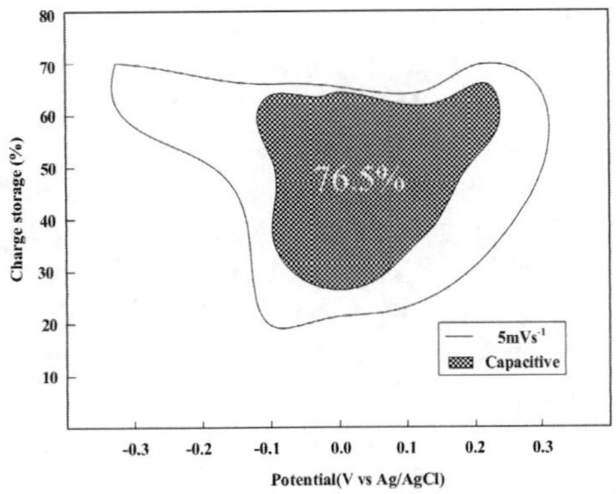

Figure 1 Charge storage capacity of $5mVs^{-1}$

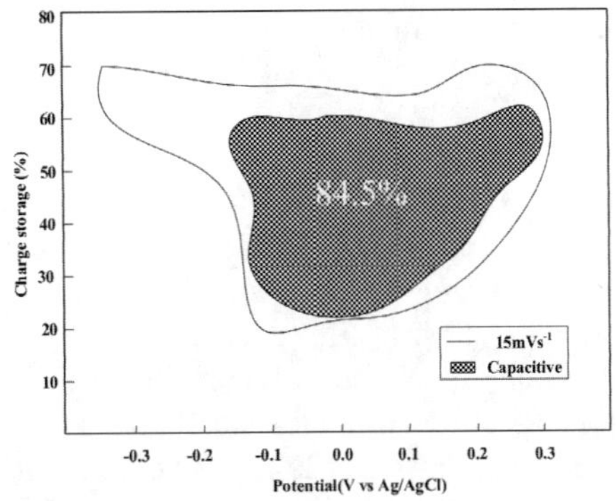

Figure 2 Charge storage capacity of $15mVs^{-1}$

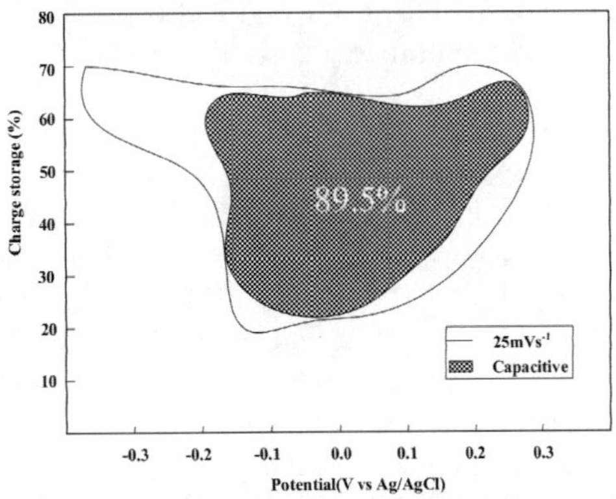

Figure 3 Charge storage capacity of $25mVs^{-1}$

As shown in Figure 1, when the scanning rate is $5mVs^{-1}$, the charge storage capacity is 76.5%; As can be seen from Figure 2, when the scanning rate is $15mVs^{-1}$, the charge storage is 84.5%; According to figure 3, when the scanning rate is$25mVs^{-1}$, the charge storage capacity is 89.5%, which indicates that the charge storage capacity performance of the new Mxene based material is the best when the scanning rate is $15mVs^{-1}$.

5. Conclusions

Through the preparation of Mxene based electrode materials for supercapacitors, the electrochemical performance and energy storage mechanism of these materials were studied. Combined with our understanding of the current research situation of these materials and the existing work basis, we can carry out the next work in the following aspects.1 NiCoAl-LDH and 1T MoS_2 were peeled to prepare a few or even single layer of nanosheets, and then the nanosheets were compounded with new Mxene based materials to prepare heterostructures. This Mxene based heterostructure will have more interfaces, which is beneficial to the improvement of electrochemical performance.2 It can also peel the Mxene micro sheet to obtain a few layers of Mxene sheet, and composite with the stripped NiCoAl-LDH and 1T MoS_2 nano sheet to explore the performance of the material for supercapacitor.3 Both transition metal nitrides and transition metal carbides have ultra-high conductivity, and the former also has high theoretical specific capacity, which belongs to the electrode materials (high power density and energy density) in supercapacitors. It enriches the academic literature on new Mxene based materials, and provides new ideas for the application research of related types of materials. Due to the limited research conditions, there are still some deficiencies in this study, which will be improved in the future.

Acknowledgment
This work was supported by Nantong Social Livelihood Science and Technology Project: Preparation and electrochemical response of functional mxene based conductive composites for rapid detection of contaminants in food.

References
[1] Sasal A, Tyszczuk-Rotko K, M Wójciak, et al. (2020) First Electrochemical Sensor (Screen-Printed Carbon Electrode Modified with Carboxyl Functionalized Multiwalled Carbon Nanotubes) for Ultratrace Determination of Diclofenac[J]. Materials, 13(3):781.
[2] Yang L , Wang J , H Lü, et al. (2021) Electrochemical sensor based on Prussian blue/multi-walled carbon nanotubes functionalized polypyrrole nanowire arrays for hydrogen peroxide and microRNA detection[J]. Microchimica Acta,188(1):1-12.
[3] Bojdi M K , Behbahani M , Ranjbari S . (2020) Developing an Electrochemical Sensor Based on Modified Siliceous Mesocellular Foam for Efficient and Easy Monitoring of Cadmium Ions[J]. ChemistrySelect, 5(22):6617-6625.
[4] Zhou T, WANG L, NIU Q. F. (2020) Research on Electromagnetic Field Simulation and Electrode Position Detection Method of Submerged Arc Furnace[J]. Computer Simulation, 37(5): 206-212.
[5] Pi M, Jiang L, Wang Z, et al. (2021) Robust and ultrasensitive hydrogel sensors enhanced by MXene/cellulose nanocrystals[J]. Journal of Materials Science, 56(14):8871-8886.
[6] Chen Y, Wang Y, Luo Y, et al. (2019) Realization of Artificial Neuron Using MXene Bi-Directional Threshold Switching Memristors[J]. IEEE Electron Device Letters, PP(99):1-1.
[7] Eom W, Shin H, Ambade R B, et al. (2020) Large-scale wet-spinning of highly electroconductive MXene fibers[J]. Nature Communications, 2020, 11(1):2825.
[8] Cui G, Sun X, Zhang G, et al. (2019) Electromagnetic absorption performance of two-dimensional MXene Ti_3C_2T_x exfoliated by HCl + LiF etchant with diverse etching times[J]. Materials Letters, 252(OCT.1):8-10.
[9] Ayman I, Rasheed A, Ajmal S, et al. (2020) CoFe2O4 Nanoparticle-Decorated 2D MXene: A Novel Hybrid Material for Supercapacitor Applications[J]. Energy And Fuels, 34(6):7622-7630.
[10] Liu P , Chen S , Yao M , et al. (2020) Double-layer absorbers based on hierarchical MXene composites for microwave absorption through optimal combination[J]. Journal of Materials Research, 35(11):1-11.

One-step hydrothermal method for preparing carbon dots and its determination of lead (II)

Yongling Shen*, **Hongyi Wu, Jian Li, Guohong Liu, Yanhua Xiao, Zhifei Dai, Hongling Zhen**

[1] Institute of NBC Defence, Beijing, 102205, China.

*Corresponding author's e-mail: yonglingshen@dingtalk.com

Abstract. As a new type of carbon nanomaterials, carbon dots have attracted great interest due to their non-toxic, low preparation cost and unique photoluminescence properties. They have been widely studied and shown great potential in many fields. Here, nitrogen doped fluorescent carbon dots were prepared by one-step hydrothermal method using Epiphyllum leaves as raw materials. Ultraviolet-visible absorption spectrometer and fluorescence spectrometer were used to study the optical properties of carbon dots. The elemental composition and morphology of the surface of carbon dots were analyzed by transmission electron microscope, Fourier transform infrared spectroscopy and X-ray photoelectron spectrometer. The results show that the prepared carbon dots are uniformly dispersed in the aqueous solution, emit blue fluorescence, and the average particle size is 4.2 nm. The carbon dots have good light stability and strong anti-photobleaching ability. Continuous spectral scanning of the carbon dots can keep the fluorescence intensity basically unchanged within 30 minutes. In the concentration of 0～1 $mol \cdot L^{-1}$ NaCl solution, the strength is not affected by the ionic strength. In the pH range of 4～8, the fluorescence intensity of carbon dots is basically stable. The heavy metal ion pb^{2+} has a quenching effect on the fluorescence of carbon dots, and has a good linear relationship in the range of 0.5～200 μm. Therefore, a simple and green method for pb^{2+} has been developed.

1. Introduction

Carbon dots are a new type of fluorescent carbon nanomaterials with a size of less than 10 nm and have the advantages of excellent optical performance, low toxicity, and biocompatibility. The application of carbon dots has been widely studied in biological imaging, sensing, environmental testing, photocatalysis, drug delivery, microfluidics and other fields[1,2].

The preparation methods of carbon dots mainly include arc discharge method [3], laser ablation method [4], electrochemical oxidation method [5], combustion method [6], template method[7], hydrothermal method [8], thermal pyrolysis method [9], microwave method [10] and ultrasonic method [11], etc. So far, the use of cheap, readily available, non-toxic natural product materials as carbon sources, and preparing carbon dots using hydrothermal methods have attracted widespread attention, but these carbon sources have problems such as complex preparation steps [12] and long preparation times [13]. Herein, the epiphyllum leaves were used as the carbon source, and the fluorescent carbon dots with excellent performance were prepared by a simple one-step hydrothermal method. The method is cheap raw materials, mild reaction conditions, simple control and fast. The prepared carbon dots have the advantages of wavelength-dependent luminescence, strong resistance to photobleaching, and high optical stability. Their application in the detection of metal ions pb^{2+} has been further studied.

2. Experimental

2.1. Experimental Materials

Rinse the fresh epiphyllum leaves with water and dry them for later use. Main reagents: $Al_2(SO_4)_3$, $NiSO_4$, $SrSO_4$, $Cd(NO_3)_2$, $Cr(NO_3)_3$, $Pb(NO_3)_2$, $HgCl_2$, $CuCl_2$, KCl, $MnCl_2$, $FeCl_3$, $FeCl_2$, NaCl, $C_8H_5KO_4$, NaOH and quinine sulfate, the reagents used in the experiment are analytically pure. The water used in the experiment was ultrapure water.

2.2. Characterization instrument

Fluorescence spectrometer (F-4600, Japan) was used to test the fluorescence performance, and the excitation and emission gap width was set to 5 nm. Ultraviolet-visible spectrophotometer (TU-1901,China) measured the ultraviolet-visible absorption spectrum. Fourier-infrared spectrometer (Vertex 70, Germany) tested the functional groups on the surface of the carbon dots. Transmission electron microscope (F20U-TWIN, Germany) observed the morphology and size of carbon dots. X-ray photoelectron spectroscopy (Escalab 250xi, USA) was used for surface element analysis.

2.3. Preparation of carbon dots

Put 2.58 g of chopped epiphyllum leaves into a 25 mL PTFE-lined reactor, added 10 mL of water, and heated it in an oven at 180 °C for 9 hours. After cooling, the sand core was filtered to remove large particles of residue. Then centrifuged at a speed of 10,000 rpm for 20 minutes, and then filtered with a 0.22 μm filter membrane. The obtained carbon dots were stored in the refrigerator for later use.

2.4. Pb^{2+} detection

Diluted Pb^{2+} standard solutions of different concentrations with potassium hydrogen phthalate-sodium hydroxide buffer solution (pH=6) to 3 mL, added 30 μL carbon dot solution, mixed well, and placed at room temperature for 1 min. The fluorescence spectrum of the solution was measured under the excitation wavelength of 360 nm, and each sample was measured 3 times in parallel.

3. Results and discussion

3.1. Research on the optical properties of carbon dots

Under high temperature conditions, different organic matter in the leaves of the epiphyllum is carbonized to form carbon dots. As shown in the figure1 (A), the carbon dots are yellow in an aqueous solution and emit blue fluorescence when irradiated with ultraviolet light with a wavelength of 365 nm and the ultraviolet-visible absorption spectrum of carbon dots has an absorption peak at 279nm, which is caused by the absorption of the n-π* transition of C=O[14]. From figure1 (B), it is shows that the emission wavelength of carbon dots gradually redshifts with the change of excitation wavelength, when the excitation wavelength increases from 300 nm to 440 nm, the emission wavelength covers the violet to green part of visible light, while the fluorescence of the emission spectrum first increases and then decreases. This excitation wavelength-dependent characteristics of the carbon dots may be related to the difference in the particle size of the carbon dots [15]and the existence of energy traps on the surface of the carbon dots [16]. When the excitation wavelength is 360 nm, the maximum emission wavelength is 440 nm as shown in the figure1 (C).

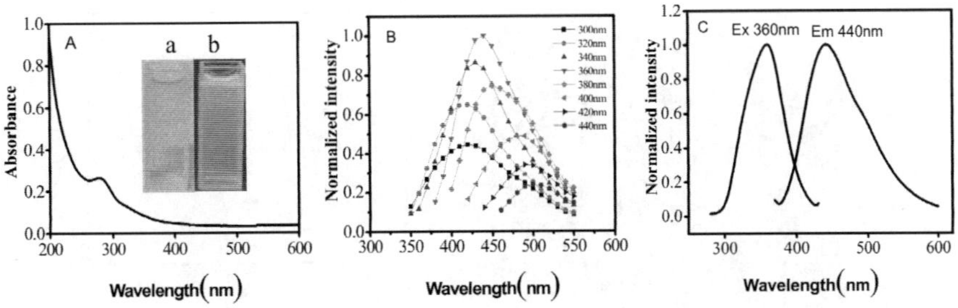

Figure 1. (A) UV absorption spectra of carbon dots, insets a and b are the pictures of carbon dots in aqueous solution under natural light and ultraviolet light, respectively; (B) Emission spectra of carbon dots with the excitation wavelength from 300 nm to 440 nm; (C) Excitation and emission spectra of carbon dots.

The fluorescence stability of carbon dots was further studied. Under continuous and uninterrupted conditions, the carbon dots were spectrally scanned. As shown in the figure2 (A), the intensity remained unchanged within 30 minutes. The results showed that carbon dots had strong resistance to light bleaching strong.

At the same time, the fluorescence stability of carbon dots in high ionic strength solutions was studied. The carbon dots were added to NaCl solutions of different concentrations, and the fluorescence intensity of carbon dots was measured respectively. As shown in the figure2 (B), when the ion concentration of NaCl changes from 0 to 1 $mol \cdot L^{-1}$, the intensity of the carbon dots remained almost unchanged, indicating that the fluorescence stability of the carbon dots was high and not affected by the ionic strength.

In addition, the effect of pH on the fluorescence intensity of carbon dots was also studied. As shown in the figure2 (C), when the pH is very low, the fluorescence is almost quenched. When the pH increases from 4.0 to 7.0, the fluorescence intensity of carbon dots is basically stable. When the pH value is greater than 8.0, the fluorescence intensity of the carbon dots significantly weakened. This is due to protonation reaction in strong acidic condition and deprotonation reaction in strong alkaline condition. Therefore, the -OH, -NH and -COOH groups on the surface of the carbon dots will interact with each other, resulting in the aggregation of carbon dots and fluorescence quenching.

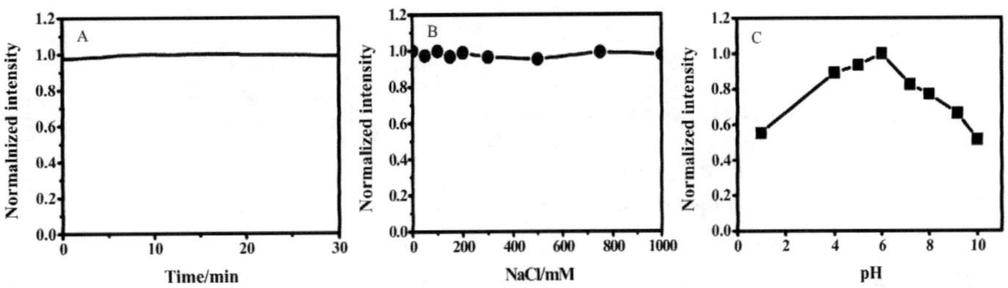

Figure 2. (A) Fluorescence of carbon dots changes with scanning time; (B) Fluorescence changes of carbon dots in different concentrations of NaCl solution; (C)The influence of pH on the fluorescence of carbon dots.

3.2. Morphology and surface analysis of carbon dots

The morphology of the prepared carbon dots was show in figure3 (A). The transmission electron microscope (TEM) image shows that the carbon dots are spherical with a particle size of 4.2 nm (based on statistical analysis of more than 100 dots). Disperse uniformly in the aqueous solution without agglomeration.

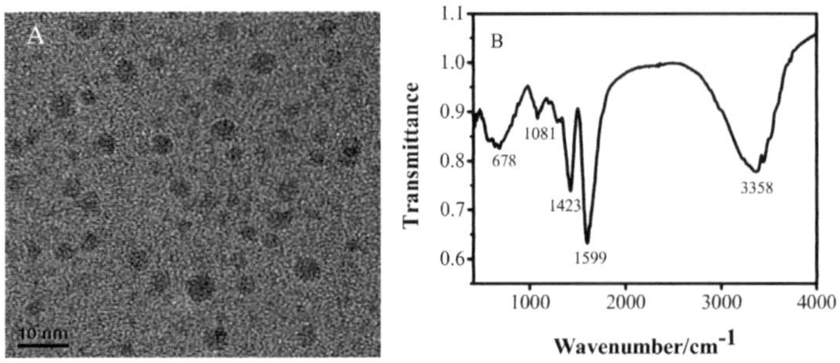

Figure 3. (A) TEM picture of carbon dots; (B) Infrared spectrum of carbon dots.

Fourier transform infrared spectrometer was used to analyze the surface functional groups of carbon dots. As shown in the figure3 (B), the carbon dot has a wide absorption band near 3358 cm^{-1}, which is attributed to the stretching vibration of N-H and O-H; there is a strong peak at 1599 cm^{-1}, which is attributed to the stretching vibration of the C=O bond; 1423 cm^{-1} is the stretching vibration of C-N and C-O; the weak absorption at 1081 cm^{-1} is the bending vibration absorption peak of C-O; the absorption peak at 678 cm^{-1} corresponds to the bending vibration of the O-H bond.

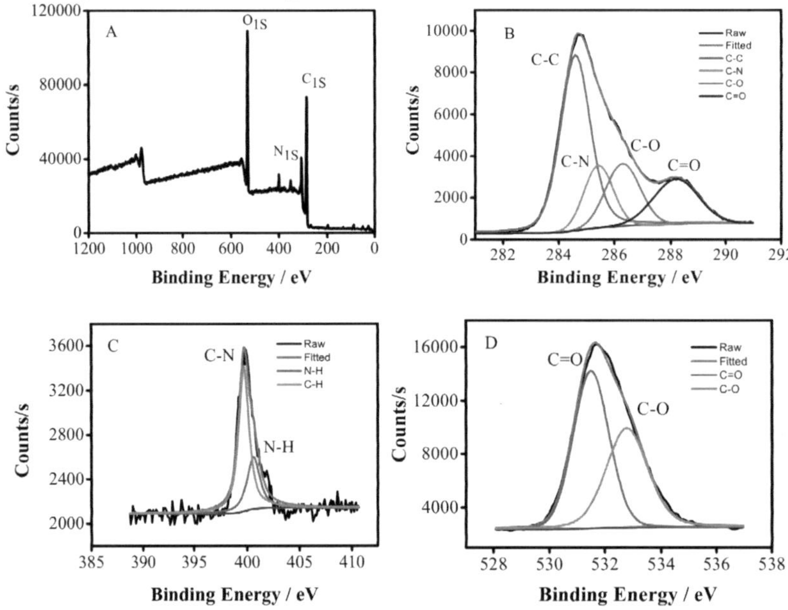

Figure 4. (A) XPS scanning spectrum of carbon dots; (B) C1s spectrum of carbon dots; (C) N1s spectra of carbon dots; (D) O1s spectra of carbon dots.

The XPS test further characterized the chemical state and element composition of the carbon dots surface. The figure4 (A) shows the XPS scan full spectrum of the carbon dots. There are 3 binding energy peaks at 532.08, 400.08, and 286.08 eV, corresponding to the 1s of oxygen, nitrogen and carbon atoms. In which the carbon content is 66.68%, the oxygen content is 28.11%, and the nitrogen content is 5.21%, indicating that the surface of the carbon dots contains oxygen, nitrogen and carbon elements. Figure4 B is the narrow sweep spectrum of C1s, which can be divided into 4 peaks at 284.5, 285.4, 286.3, 288.2 eV, respectively, representing the C element with C=C / C-C, C-N, C-O, $-COO^-$ bonds exist on the surface of carbon dots. The N1s spectrum of carbon dots (Figure4 C) shows that the N element exists in the form of C-N (399.7eV) and N-H (400.59eV). In the O1s spectrum of carbon dots (Figure4 D), the peaks at 531.5eV and 532.8eV are attributed to C=O/C-O-C and C-OH, respectively.

The above results show that the carbon dots are nitrogen-doped fluorescent carbon dots. There are a large number of hydrophilic active groups such as hydroxyl, carboxyl and amino groups on the surface, thus having better dispersal in an aqueous solution. It is conducive to the application of carbon dots in the fields of analysis, sensing and cell imaging.

3.3. Research on the response of Pb^{2+}

The prepared carbon dots are rich in hydroxyl, carboxyl and amino groups on the surface, and these functional groups have the ability to recognize and coordinate metal ions, based on this, the metal ion response of carbon dots was discussed.

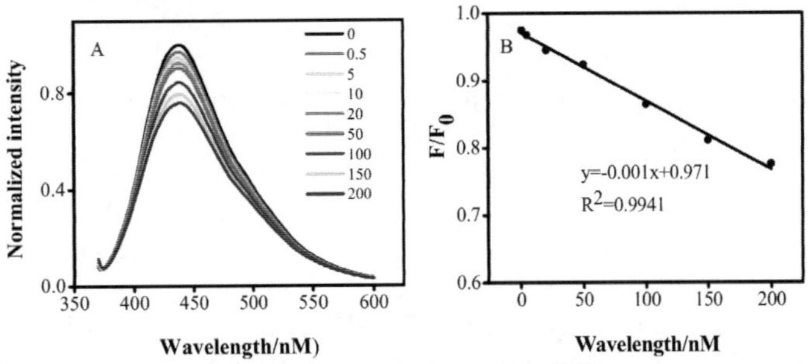

Figure 5. (A) Fluorescence intensity of carbon dots with different concentrations of Pb^{2+} (from 0 to 200 μm); (B) Linear equation of carbon dots for the detection of Pb^{2+}, F and F_0 correspond to fluorescence intensity of carbon dots with and without Pb^{2+}, respectively.

The experimental results (figure5 (A)) show that the fluorescence intensity of carbon dots decreases with the increase of Pb^{2+}, and fluorescence quenching occurs. As shown in the figure5 (B), when the concentration of Pb^{2+} is 0.5~200 μm, the relative fluorescence quenching intensity (F/F_0) of carbon dots is linear with the concentration of Pb^{2+}, and the linear equation is $F/F_0 = -0.001x + 0.971$, a good linear relationship (linear correlation coefficient R^2=0.9941). The detection limit was calculated according to 3s/K (where s is the standard deviation of the blank sample and K is the slope of the fitted curve) by performing 7 tests on the blank samples, and the detection limit is 0.34 μm. It can be seen that the method has a wide linear range and high sensitivity.

The selectivity of carbon dots for the determination of Pb^{2+} was further studied. Under the same experimental conditions, the effects of interfering ions K^+, Mn^{2+}, Sr^{2+}, Cr^{3+}, Ni^{2+}, Al^{3+}, Cd^{2+}, Hg^{2+}, Cu^{2+}, Fe^{2+}, Fe^{3+} were investigated with the concentration of 200 μM, respectively. The experimental results are shown in Figure 6. It is found that only Fe^{3+} has a certain effect, while other metal ions have almost little quenching. The carbon dots have highly selectivity for Pb^{2+}, as the active groups on the surface of the carbon dots are prone to coordination reaction with the Pb^{2+} to form highly stable complexes, leading to the aggregation of carbon dots and fluorescence quenching. However, other metal ions are not prone to forming complexes, so there is almost no quenching effect.

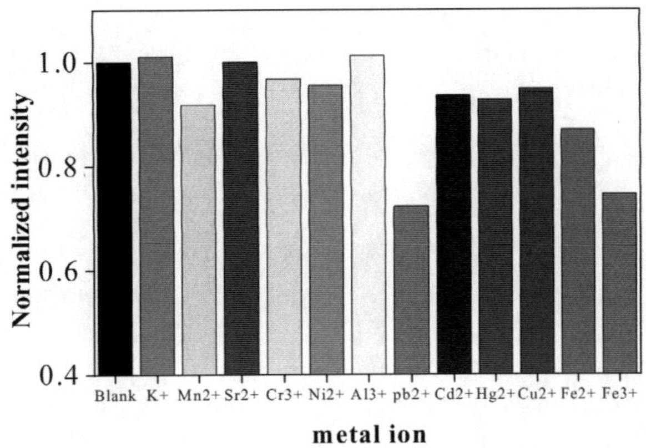

Figure 6. Response of different metal ions to carbon dots

4. In conclusion

In summary, a one-step hydrothermal method was proposed to prepare N-doped carbon dots using epiphyllum leaves as a carbon source. The prepared carbon dots have good fluorescence properties and stability, and their surface contains active groups such as hydroxyl, carboxyl and amino groups. In addition, a method based on fluorescence quenching for the detection of Pb^{2+} has been developed, which provides new possibilities for the development of environmentally friendly, low-cost, high-sensitivity detection methods, and has potential application prospects.

References

[1] Xiao, L., and Sun, H.D., (2018) Novel properties and applications of carbon nanodots. Nanoscale Horiz., 3: 565-597.
[2] Sharma, V., Tiwari, P. and Mobin, S.M. (2017) Sustainable carbon-dots: recent advances in green carbon dots for sensing and bioimaging. J. Mater. Chem. B, 5: 8904-8924.
[3] Xu, X., Ray, R., Gu, Y., Ploehn, H. J., Gearheart, L., Raker, K., and Scrivens, W.A.. (2004) Electrophoretic Analysis and Purification of Fluorescent Single-Walled Carbon Nanotube Fragments. J. Am. Chem. Soc., 126: 12736-12737.
[4] Sun, Y.P., Zhou, B., Lin, Y., et al. (2006) Quantum-sized carbon dots for bright and colorful photoluminescence. J. Am. Chem. Soc., 128: 7756-7757.
[5] Zhou, J., Booker, C., Li, R., et al. (2007) An electrochemical avenue to blue luminescent nanocrystals from multiwalled carbon nanotubes (MWCNTs). J. Am. Chem. Soc., 129:744-745.
[6] Rong, M.C., Zhang, K.X., Wang, Y.R., et al. (2017) The synthesis of B, N-carbon dots by a combustion method and the application of fluorescence detection for Cu^{2+}. Chinese Chem. Lett., 28:207-212.
[7] Yang, Y., Wu, D., Han, S., et al. (2013) Bottom-up fabrication of photoluminescent carbon dots with uniform morphology via a soft-hard template approach. Chem. Commun., 49: 4920-4922.
[8] Zhang, B., Liu, C.Y., Yun, L. (2010) A novel one-step approach to synthesize fluorescent carbon nanoparticles. Eur. J. Inorg. Chem., 2010: 4411-4414.
[9] Liu, X., Li, H. B., Shi, L., Meng, X., Wang, Y., Chen, X., Xu, H., Zhang W., Fang X., Ding T.. (2017) Structure and photoluminescence evolution of nanodots during pyrolysis of citric acid: from molecular nanocluster to carbogenic nanoparticles. J. Mater. Chem. C, 5: 10302-10312.
[10] Zhu H., Wang, X., Li, Y., et al. (2009) Microwave synthesis of fluorescent carbon nanoparticles with electrochemiluminescence properties. Chem. Commun.,45: 5118-5120.
[11] Ma, Z., Ming, H., Huang, H., et al. (2012) One-step ultrasonic synthesis of fluorescent N-doped carbon dots from glucose and their visible-light sensitive photocatalytic ability. New J. Chem., 36:861-864.
[12] Ying, G., Bao, X.L. (2015) Carbon dots-initiated luminol chemiluminescence in the absence of added oxidant. Carbon, 82: 459-469.
[13] Yang, Y., Cui, J., Zheng, M., et al. (2012)One-step synthesis of amino-functionalized fluorescent carbon nanoparticles by hydrothermal carbonization of chitosan. Chem. Commun., 48 :380-382.
[14] Yuan, F.L., Wang, Z., Li, X.D., et al. (2017) Light-emitting diodes: bright multicolor bandgap fluorescent carbon quantum dots for electroluminescent light-emitting diodes. Adv. Mater., 29 :1602236
[15] Sk, A.M., Ananthanarayanan, A., Huang, L., et al. (2014) Revealing the tunable photo-luminescence properties of graphene quantum dots. J. Mater. Chem. C, 2: 6954-6960.
[16] Genc, R., Alas, M.O., Harputlu, E., et al. (2017) High-capacitance hybrid supercapacitor based on multi-colored fluorescent carbon-dots. Sci. Rep., 7(11222).

Crystallization and phase separation of tungsten oxide-bismuth vanadate amorphous film by annealing in air

Chao Li [1, 5], Haili Song [2, *], Yuanyuan Zhang [3], Lei Miao [4], Ruijuan Qi [3], Rong Huang [3], Chengqiang Cui [1, 5, *]

[1]School of Electromechanical Engineering, Guangdong University of Technology, Guangzhou, Guangdong 510006, China

[2]Key Laboratory of Bioinorganic and Synthetic Chemistry of Ministry of Education, School of Chemistry, Sun Yat-Sen University, Guangzhou, Guangdong 510275, China.

[3]Key Laboratory of Polar Materials and Devices (MOE) and Department of Electronics, East China Normal University, Shanghai 200062, China

[4]Guangxi Key Laboratory of Information Material, School of Material Science and Engineering, Guilin University of Electronic Technology, Guilin 541004, China

[5]Ji Hua Laboratory, Foshan, Guangdong 528299, China

[*] Corresponding authors: Haili Song (songhli@mail.sysu.edu.cn)
Chengqiang Cui (cqcui@gdut.edu.cn)

Abstract. A WO_3-$BiVO_4$ amorphous film grown on the (001) yttria-stabilized zirconia (YSZ) substrate was annealed at 600 °C in air. The microstructures evolution was investigated by X-ray diffraction, scanning electron microscopy and high-resolution transmission electron microscopy. After annealed, the amorphous composite film crystalized into monoclinic $BiVO_4$ and orthorhombic WO_3. $BiVO_4$ epitaxially grew on the YSZ, forming the matrix, where part of small irregular WO_3 grains were embedded in. Other large-sized sub-stoichiometric $WO_{3-\delta}$ ($0<\delta<3$) grains with stacking faults lay on the film surface, which may contribute to the compressed strain of $BiVO_4$ matrix vertically. Compared to the typical vertical heteroepitaxial nanocomposite of WO_3-$BiVO_4$ film grown at 600 °C by pulsed laser deposition, in the annealed composite film, the vertical lattice mismatched strain at $WO_3/BiVO_4$ interface decreased and lead to the disappearance of metastable hexagonal WO_3.

1. Introduction

Vertical heteroepitaxial nanocomposite (VHN) thin films have recently stimulated significant research interest for creating promising oxide thin films with optimized and tunable properties [1, 2]. In VHN films, the net strain state has been mainly dominated by the vertical strain rather than the lateral one induced by the substrate, especially in thick films [3,4]. The key of achieving tunable vertical strain and multifunctionalities, is comprehensive understanding of growth mechanisms to realize controlled growth of VHN structure [3]. Up to now, people are still trying various ways to understand the growth of VHN, including synchrotron radiation and transmission electron microscope (TEM), to achieve the desired nanostructures and functionalities [2, 5, 6].

The WO_3-$BiVO_4$ (WO-BVO) VHN film grown shows enhanced photoactivity than that of each component phase [7]. In our previous research [8], the amorphous film has been found in the sample grown at 400 °C resulted by the low driving forces for nucleation and diffusion. When the substrate temperature is higher than 500 °C, the composite film was crystallized and a typical VHN with WO nanopillars embedded in the BVO matrix formed. In continues to our previous works on the microstructure evolution of WO-BVO VHN film, we aim to study the impact of thermal annealing on the structural properties of WO-BVO amorphous composite film. In this work, by annealing the WO-BVO amorphous film at 600 °C in air, the crystallization and phase separation were studied systematically by a combining of X-ray diffraction (XRD) pattern, TEM and X-ray energy dispersive spectrometer (EDS).

2. Experiments

2.1. Annealing of the amorphous WO_3-$BiVO_4$ film
The growth process of WO-BVO films with different substrate temperatures by pulsed laser deposition (PLD) was reported before [8]. The amorphous WO-BVO film grown at 400 °C was annealed using a muffle programmable furnace (LHT02/16, Nabertherm, Germany) in air. The sample was heated at 5 °C/min to the annealing temperature of 600 °C, holding for 1 h, then cooled to room temperature at 10 °C/min in the furnace. The film grown at 400 °C, 600 °C, and the annealed film were abbreviated as 400WB, 600WB, and annealed WB, respectively.

2.2. Microstructure characterizations
XRD (D8 Discover, Bruker, Germany) was performed on the films, to examine the crystal structure and phase composition. mechanical grinding, polishing, precision dimpling, and ion milling ect., which is the standard procedure of TEM specimens, was performed to were prepare cross-section TEM specimens. Transmission electron microscope (JEM-2100F, JEOL, Co., Japan), and energy-dispersive X-ray (MAX-80T, Oxford Instruments, Inc, the United Kingdom) were applied to analyze the microstructure and chemical composition.

3. Results and discussion
In Figure 1(a), the XRD results of WO-BVO composite films prepared at 400 °C and 600 °C were displayed for comparation [8]. It can be clearly seen that after annealed at 600 °C in air, the amorphous WO-BVO film crystallized into monoclinic BVO and orthorhombic WO (o-WO). BVO epitaxially grew along the (001) YSZ substrate, while the o-WO shows polycrystalline growth. Compared to the XRD pattern of WO-BVO film grown at 600 °C, there is no diffraction peak of hexagonal WO (h-WO) in the annealed WB film. To achieving more information, , the 2θform 22° to 32° was enlarged in Figure 1(b). The BVO (004) peak of annealed WB sample is at the right of 600WB sample, and the plane distances were 2.915 Å and 2.925 Å (the bulk value is 2.926 Å), respectively. It shows that the BVO in the annealed sample was subjected to a compressed strain along the vertical direction, whereas the BVO in the 600WB sample was in a relaxed state.

Figure 1(c) shows the low magnification TEM image of the annealed WB sample. After annealed at 600 °C in air, the internal structure of WO-BVO composite film is not very continuous and there are some voids without film compositions, highlighted by green dotted lines. Moreover, the distribution of the two phases is completely different from that of 600WB sample [8]. Part of irregularly-shaped WO nanocrystals were wrapped in the BVO matrix, and other large-sized WO grains directly grew on the film surface. The selected area electron diffraction (SAED) pattern in Figure 2(d) shows that BVO grows epitaxially along the substrate with (001) BVO// (001) YSZ and [100] BVO// [100] YSZ; while WO grains exhibit polycrystalline features.

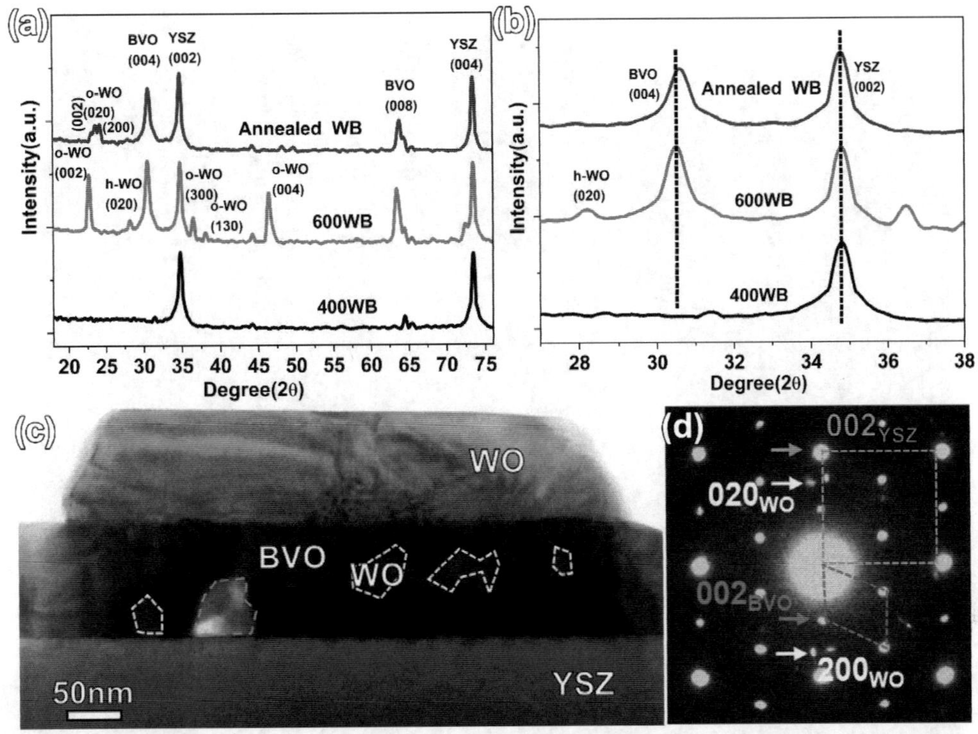

Figure 1. (a) θ-2θ XRD results of 400WB, 600WB and annealed WB samples. (b) XRD spectra of BVO (004) and YSZ (002) peaks from 22° to 32° . (c) The cross-section TEM morphology of annealed WO-BVO sample. (d) The corresponding SAED pattern of the composite film in Figure 1(c).

As shown in Figure 2(a), inside the film, the WO grains randomly grown in the BVO matrix with an average size of 25.6 nm. Figure 2(b) illustrates the interface structure between BVO matrix and YSZ substrate. There is still an amorphous layer with thickness of ~2.9 nm at the film interface. Figure 2(c) shows the stacking faults inside the large-sized WO grains grown on the film surface, which are also revealed by the yellow arrow marked streaking in the SAED pattern, implying that the large-sized WO grains composition is sub-stoichiometric $WO_{3-\delta}$ ($0<\delta<3$)[9]. Figure 2(d) depicts the HRTEM image of the red box marked area in Figure 2(c). Above the stacking faults, there is normal WO structure along the $[01\bar{1}]$ zone axis with the inset Fast Fourier transformed (FFT) pattern calibrated. In the lower FFT pattern of abnormal WO structure, some extinction lattice planes reappeared due to the misalignment of lattice planes, as the yellow circle marked diffraction spots shown.

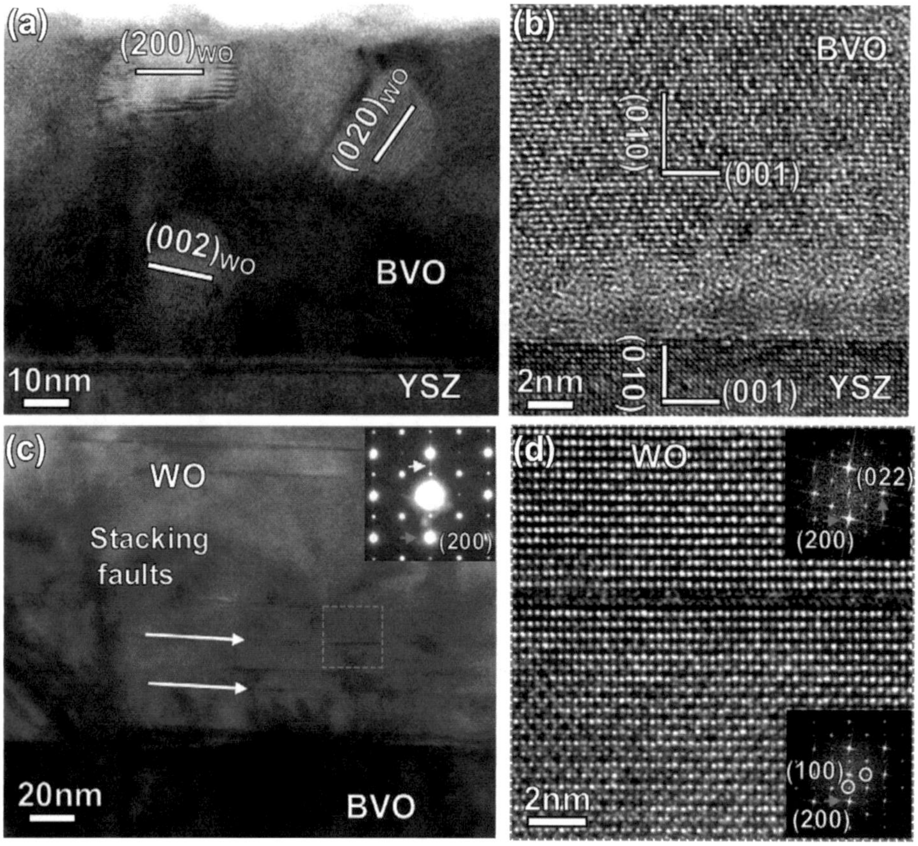

Figure 2. (a) The HRTEM image of the composite film showing WO grains distributed in the BVO matrix. (b) The HRTEM result of the BVO/YSZ interface. (c)The medium magnification TEM image of the WO on the film surface, the inset image is the corresponding SAED pattern. (d) The HRTEM image of the red box marked area in (c), the inset images are the FFT patterns of the areas above and below the stacking faults, respectively.

In Figure 3(a), as the distributions of Bi/V/W elements shown, the composite film grown at 600 °C is a typical VHN nanostructure, in which WO nanopillars inserted in BVO. Nevertheless in the annealed WB sample (Figure 3b), the film is not compact with some voids, the distribution of Bi/V/W elements indicated that some WO grains filled in the BVO matrix and other large-sized WO grains lay on the film surface.

When the WO-BVO amorphous film is annealed at 600 °C in air, the lattice mismatch between BVO and the YSZ substrate is smaller than WO/YSZ [8], BVO preferentially crystallized and grew along the surface of the substrate. Some WO crystal grains were embedded in the BVO matrix and restricted by the surrounding BVO, making it difficult to grow into large size. Other WO grains on the film surface, which were in direct contact with air and less restricted, can grow rapidly under the drive of thermal energy. Distinguished from the WO-BVO film grown at 600 °C by PLD with sufficient oxygen pressure, the composite film annealed in air without enough oxygen can easily form defects, such as the stacking faults in WO, which has been reported elsewhere [10]. In comparison with the WO-BVO film grown at 600°C, there is no h-WO phase in the annealed WB sample. As discussed in our previous papers [8], the driving force of the semi-stable hexagonal WO phase was tensile strain derived from the thermal expansion coefficient mismatch and the crystal lattice mismatch along the WO/BVO epitaxy vertical interface. The growth mode of WO-BVO film annealed at 600°C in air was changed drastically, the irregular WO grains in the BVO matrix decreased the vertical lattice-mismatched strain, thus the metastable h-WO may not be stabilized, similar to the 700WB sample [8]. The density of o-WO (7.149 g/cm^3) is higher than that of BVO (6.25 g/cm^3), hence in the annealed WB film, the large WO grains with the lateral size of ~500nm on the film surface may contribute to the compressed strain state of BVO matrix vertically.

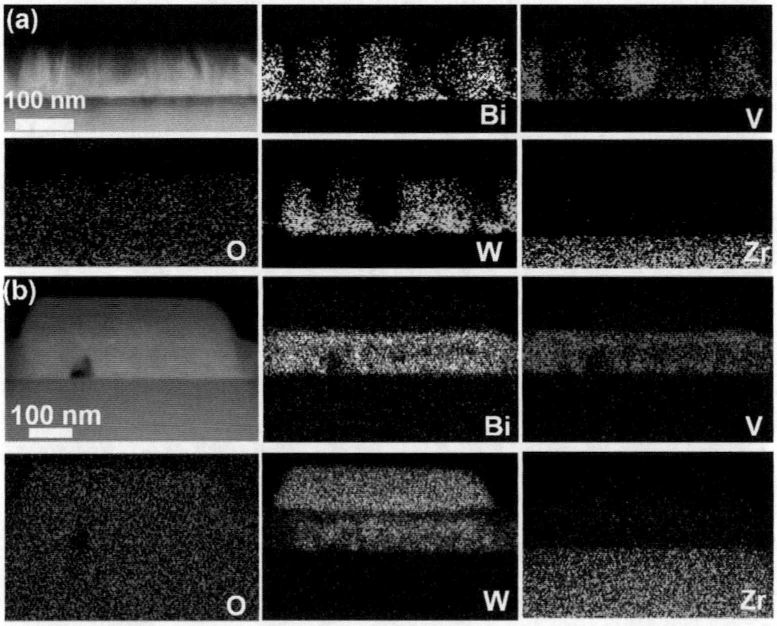

Figure 3. (a) The EDS mapping of (a) 600WB sample;(b) annealed WB sample.

4. Conclusions

The microstructure evolution of amorphous WO-BVO film by annealing at 600°C in air was investigated by XRD, SEM, HRTEM and EDS. Distinguished from the typical VHN composite film grown at 600°C by PLD, after annealed, BVO grew epitaxially along the surface of YSZ substrate, WO formed both irregular polycrystal grains wrapped in BVO matrix and large-sized grains lying on the film surface. The metastable h-WO did not form mainly due to the decreased lattice mismatched strain effect at WO/BVO interface. The large-sized WO grains on the film surface may result in the out of plane compressed state of BVO matrix. In the future, in-situ heating TEM technology will be applied to explore the process of crystallization and phase separation deeply and clearly.

ACKNOWLEDGMENTS

This work was supported by the National Natural Science Foundation of China (Grant No. 61974042, 61874155 and 11774092), Guangdong Basic and Applied Basic Research Foundation (2020A1515110178).

REFERENCES

[1] J. MacManus-Driscoll, A. Suwardi, A. Kursumovic, Z. X. Bi, C. F. Tsai, H. Y. Wang, Q. X. Jia and O. J. Lee. (2015) New strain states and radical property tuning of metal oxides using a nanocomposite thin film approach. APL Mater., 3:062507.

[2] H. J. Liu, W. I. Liang, Y. H. Chu, H. M. Zheng and R. Ramesh. (2014) Self-assembled vertical heteroepitaxial nanostructures: from growth to functionalities. MRS Commun., 4: 31-44.

[3] A. Chen, Z. Bi, Q. Jia, J. L. MacManus-Driscoll and H. Wang. (2013) Microstructure, vertical strain control and tunable functionalities in self-assembled, vertically aligned nanocomposite thin films. Acta Mater., 61: 2783-2792.

[4] J. Huang, A. Gellatly, A. Kauffmann, X. Sun and H. Wang. (2018) Exchange Bias Effect along

Vertical Interfaces in $La_{0.7}Sr_{0.3}MnO_3$:NiO Vertically Aligned Nanocomposite Thin Films Integrated on Silicon Substrates. Cryst. Growth Des., 18: 4388-4394.

[5] J. L. MacManus-Driscoll, M. P. Wells, C. Yun, J.-W. Lee, C.-B. Eom and D. G. Schlom. (2020) New approaches for achieving more perfect transition metal oxide thin films. APL Mater., 8: 040904.

[6] D. H. Kim, S. Ning and C. A. Ross. (2019) Self-assembled multiferroic perovskite–spinel nanocomposite thin films: epitaxial growth, templating and integration on silicon. J. Mater. Chem. C, 7: 9128-9148.

[7] C. Nguyen Van, T. H. Do, J.-W. Chen, W.-Y. Tzeng, K.-A. Tsai, H. Song, H.-J. Liu, Y.-C. Lin, Y.-C. Chen, C.-L. Wu, C.-W. Luo, W.-C. Chou, R. Huang, Y.-J. Hsu and Y.-H. Chu. (2017) WO_3 mesocrystal-assisted photoelectrochemical activity of $BiVO_4$. NPG Asia Mater., 9: e357.

[8] H. Song, C. Li, C. N. Van, W. Dong, R. Qi, Y. Zhang, R. Huang, Y.-H. Chu and C.-G. Duan. (2018) Microstructure evolution determined by the crystalline phases competition in self-assembled WO_3-$BiVO_4$ hetero nanostructures. J. Appl. Phys., 123: 085305.

[9] P. Huang, M. M. Kalyar, R. F. Webster, D. Cherns and M. N. Ashfold. (2014) Tungsten oxide nanorod growth by pulsed laser deposition: influence of substrate and process conditions. Nanoscale, 6: 13586-13597.

[10] A. M. Smith, M. G. Kast, B. A. Nail, S. Aloni and S. W. Boettcher. (2014) A planar-defect-driven growth mechanism of oxygen deficient tungsten oxide nanowires. J. Mater. Chem. A, 2: 6121-6129.

GCMN 2021 IOP Publishing
Journal of Physics: Conference Series **2011** (2021) 012103 doi:10.1088/1742-6596/2011/1/012103

Effects of different solvents on ultraviolet absorption and fluorescence intensity of 2,4-dichlorophenol, 2,4,6-trichlorophenol and Pentachlorophenol

Wenbo Lan[1], Qianru Li[1], Xiaofeng Wang[1*], Liping He[1], Yanbin Meng[2], Yashi Wang[1], Jianhuang Li[3], Can Zeng[1], Liying Wang[1], Mengfen Luo[1], Zhijian Bu[1]

[1]School of Public Health, Xiangnan University, Chenzhou, Hunan, 423043, China

[2]Basic Medical School, Xiangnan University, Chenzhou, Hunan, 423043, China

[3]Office of Network Information Technolog, Xiangnan University, Chenzhou , hunan, 423043, China

*Corresponding author's e-mail: wangxf2016@xnu.edu.cn

Abstract. In order to study the influence of the UV-visible absorption spectra and fluorescence spectrum from 2,4-dichlorophenol (2,4-dcp), 2,4,6-trichlorophenol (2,4,6-tcp) and pentachlorophenol (ptcp) in different solvents. Using methanol, tetrahydrofuran (THF), ethanol, dimethyl sulfoxide (DMSO) and acetone solvent models, computer simulation prediction and analysis of the three chlorophenols were carried out. It provides a reference for the choice of solvents in the determination of chlorophenols by the Ultraviolet-Vis absorption spectrophotometry and fluorescence intensity detection.

1. Introduction

As aromatic organic compounds, chlorophenols generally have strong toxicity and are difficult to explain in nature[1]. At the same time, aromatic organic compounds have good bioaccumulation, environmental stability and biological toxicity[2]. 2,4-dcp, 2,4,6-tcp and ptcp are difficult to degrade, highly toxic and have teratogenic, carcinogenic and mutagenic effects[4, 5]. Therefore, it is very important to determine the properties of its chlorophenols, which are commonly used in Ultraviolet-visible (UV-Vis) absorption spectrum analysis and molecular fluorescence spectrum determination. Due to the close relationship between the spectral method and the solvents used, the UV-Vis Absorption Spectra and fluorescence intensity of chlorophenols in different solvents are different[6-8]. Especially in quantitative determination, the UV-Vis absorption and fluorescence intensity of chlorophenol are directly affected. If the solvents selected are not suitable, it may directly affect the determination, especially affect the detection sensitivity and detection limit[9, 10]. With the rapid development of computer chemistry, computer simulation can replace the conventional experimental results, and the results of computer simulation can provide important basis for experimental research[11-13]. In this study, several solvents were selected, I.E. water, methanol, tetrahydrofuran (THF), ethanol, Dimethyl sulfoxide (DMSO) and acetone, to treat 2,4-dcp, 2,4,6-tcp and ptcp in UV-Vis absorption spectra and fluorescenceintensity spectra were calculated and analyzed, and in order to provide reference for the choice of solvent for the UV-Vis absorption of chlorophenols, the absorption of chlorophenols in vacuum model was compared with that in the same calculation level.

2. Research methods

Quantum view and molecular software Gauss-view were used to construct molecules of 2,4-dcp, 2,4,6-tcp and ptcp. Perform structural optimization calculations in the B3LYP[14-16]/6-311G*[17] basis set in the quantum computing software, and check the stability of the optimized structure through the infrared vibration frequency, and use it in water, methanol, ethanol, propanol, THF and DMSO and other solvents. Calculate the UV-Vis and fluorescence absorption of 2,4-dcp, 2,4,6-tcp and ptcp under the vacuum model with the same calculation level, The UV-Vis absorption value and fluorescence intensity of 2,4-dcp, 2,4,6-tcp and ptcp were compared with each other, and the influence of each solvent model on UV-Vis absorption and fluorescence intensity were analyzed.

3. Research results and analysis

3.1. Structural configuration

The structure optimization calculation was performed on the B3LYP/6-311G* basis set, and the optimized molecular structure diagrams of 2,4-dcp, 2,4,6-tcp and ptcp, there are all the atoms of chlorophenols are in the same plane.

3.2. Infrared vibration

Under the same calculation basis set, the structurally optimized 2,4-dcp, 2,4,6-tcp and ptcp are further calculated for the infrared spectrum vibration frequency, as shown in Figure 1. There was no negative in the value, and it can be seen that there was no false frequency in the vibration, indicating that the optimized structures of 2,4-dcp, 2,4,6-tcp and ptcp exist stably[19]. Due to the difference in the number of Cl atoms in the chlorophenol aromatic ring, it can be seen that the peaks generated by the vibration and swing of the aromatic ring's fixed frame have obvious differences. 2,4,6-tcp and ptcp have obvious oscillation peaks of O-H atoms at 424 cm^{-1}, but do not appear in 2,4-dcp. This is because compared with 2,4,6-tcp and ptcp, the O-H swing peak in 2,4-dcp has a certain red shift, which cannot be shown in figure 1. As for the characteristic absorption peak of each chlorophenol, the O-H stretching vibration value in 2,4-dcp is obviously smaller than that in 2,4,6-tcp and ptcp, and the wave number is larger, and relative to 2,4,6-tcp and ptcp stretching vibration red shift. The reason is that 2,4-dcp has less functional group Cl than 2,4,6-tcp and ptcp, and its total chlorine atom has a weak electron withdrawing effect, which makes the reduction of benzene ring charge density small. The electron transfer in the phenolic hydroxyl group to the aromatic ring is less, the charge density between O-H is higher, and the O-H bond is stronger.

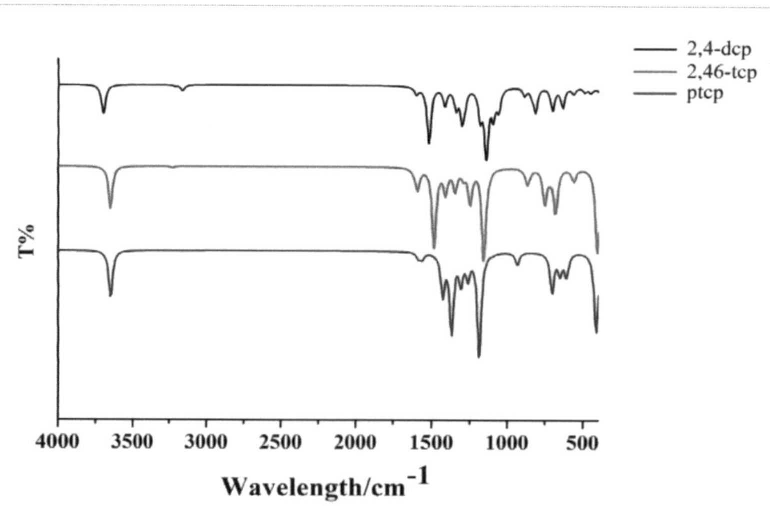

Figure 1. Infrared absorption spectra of the chlorophenols.

3.3. Charge distribution

The Mulliken charge distribution calculations are carried out on the C atoms in the aromatic rings and O atoms and Cl atoms in the phenolic hydroxyl group of 2,4,6-tcp 2,4,6-tcp and ptcp. The results are shown in table 1. With the electron withdrawing effect of 5 Cl atoms in ptcp, the charge density of each C atom in the aromatic ring is increased, and the electronegativity is strengthened. At the same time, the electronegativity of the O atom is weakened, and the force with the H in the phenolic hydroxyl group is reduced, and the OH stretches vibration is stronger, consistent with infrared vibration results.

Table 1. Mulliken charge distribution of some atoms in three phenolic compounds

	2,4-dcp	2,4,6-tcp	ptcp		2,4-dcp	2,4,6-tcp	ptcp
1-C	0.392	0.549	-0.568	O	-0.594	-0.576	-0.571
2-C	-0.362	-0.374	-0.260	2-Cl	0.039	0.057	0.116
3-C	0.064	0.087	-0.128	3-Cl	\	\	0.108
4-C	-0.314	-0.326	-0.140	4-Cl	-0.009	0.010	0.105
5-C	-0.025	0.089	-0.124	5-Cl	\	\	0.110
6-C	-0.149	-0.366	-0.246	6-Cl	\	0.007	0.061

3.4. Ultraviolet visible absorption spectrum

Using water, methanol, THF, ethanol, DMSO and acetone as solvents, respectively, 2,4-dcp, 2,4,6-tcp and ptcp are calculated under the same calculation basis. The corresponding UV absorption spectra are shown in figure 2 *a, b* and *c*. They are 2,4-dcp, 2,4,6-tcp and ptcp' UV absorption spectra in models such as in vacuum, water, methanol, THF, ethanol, DMSO, and acetone, respectively. inand. It can be seen clearly in figure 3 *a* that water, methanol, THF, DMSO and acetone all make the UV-Vis absorption value improved. Among them, the DMSO is the most obvious, and the UV-Vis absorption peak has a more obvious red shift. While in ethanol solvent, the absorption peak is significantly reduced. At the same, it can be seen clearly in figure 3 *b* that water, methanol, THF, DMSO, and acetone all make the absorption peak increase to a certain extent, and the effect of DMSO is the most obvious, each solvent makes the ultraviolet absorption peak red-shift. While the UV-Vis absorption peak in the acetone solvent is significantly reduced. However, it is obvious in figure 3 *c* that in various solvent models, the UV-Vis absorption peak is significantly improved, and there is no obvious difference in solvent increase. This may be affected by the relatively high polarity of 2,4-dcp, while the polarity of 2,4,6-tcp and ptcp is low, and ptcp is simultaneously absorbed electrons by five chlorine atoms[19].

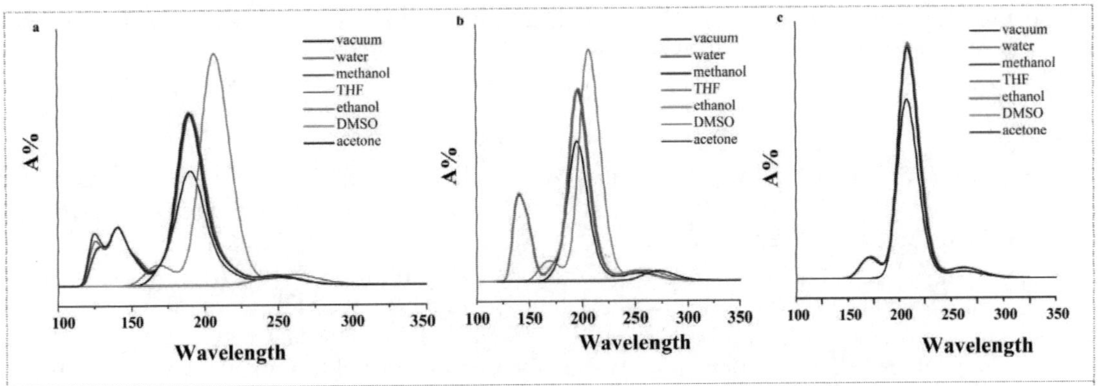

Figure 2. UV-Vis absorption spectrum of the chlorophenols.

3.5. Fluorescence intensity spectrum

Using water, methanol, THF, ethanol, DMSO and acetone as solvents, 2,4-dcp, 2,4,6-tcp and ptcp are calculated under the same calculation basis set. The corresponding fluorescence intensity spectra are shown in Figure 3 *d, e,* and *f,* which represent the fluorescence intensity spectra of 2,4-dcp, 2,4,6-tcp and ptcp in each solvent, respectively. It is obvious from the figure that 2,4,6-tcp and ptcp produce weaker fluorescence in each solvent, and there is no obvious difference in fluorescence wavelength and intensity. However, 2,4-dcp produces strong fluorescence in different solvents. The fluorescence patterns of each solvent are similar, and there are two fluorescence peaks. The stronger fluorescence is at about 226nm, and the second-strong fluorescence value is at about 270nm. And the corresponding solvents for 2,4-dcp to produce fluorescence intensity from strong to weak are water, methanol, ethanol, acetone, DMSO and THF.This may be because trichlorophenol and ptcp have more chlorine atoms than dichlorophenol, which destroys the large π-bond structure of the benzene ring[20].

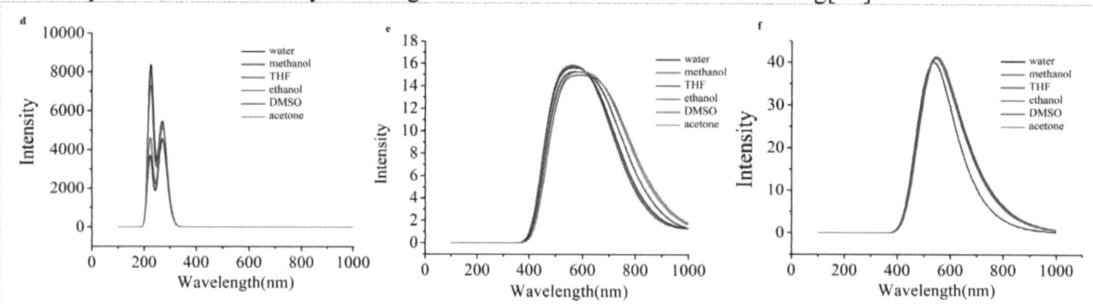

Figure 3. Infrared absorption spectra of the chlorophenols.

4. Conclusion

Among the six solvents of water, methanol, THF, ethanol, DMSO and acetone, the DMSO solvent enhances the UV-Vis absorption effect of 2,4-dcp and 2,4,6-tcp. DMSO is the first of several solvents when doing UV-Vis -visible absorption spectroscopy for the chlorophenol. At the same time, there is no obvious change in the fluorescence spectrum intensity of each chlorophenol in each solvent. The fluorescence intensity of 2,4,6-tcp and ptcp in each solvent is extremely weak, and it is not suitable for quantitative detection of them by fluorescence. However, 2,4-dcp has good fluorescence intensity in various solvents and is suitable for quantitative detection, especially in two solvents, water and methanol.

Acknowledgments

This work was supported by National Natural Science Foundation and National Social Science Foundation Key cultivation projects of Xiannan University [grant number 2020XJ08], the Chenzhou City Heavy Metal Pollution Health Risk Assessment Technology Research and Development Center, the Chenzhou City Science and Technology Innovation Ability Cultivation Project[zdyf201912].

References:

[1] Gupta, S.S., Stadler, M., Noser, C.A., Ghosh, A., Steinhoff, B., Lenoir, D., Horwitz, C.P., Schramm, K.W., Collins,T.J. (2002) Rapid Total Destruction of Chlorophenols by Activated Hydrogen Peroxide. Science., 296: 326-328.

[2] Dwivedi, A.D., Dubey, S.P., Sillanpaa, M., Kwon, V., Lee, C., Varma, R.S. (2015) Fate of engineered nanoparticles: Implications in the environment. Coordin. Chem. Rev., 287: 64-78.

[3] Xing, L., Sun, J., Liu, H., Yu, H., (2012) Combined toxicity of three chlorophenols 2,4-dichlorophenol, 2,4,6-trichlorophenol and pentachlorophenol to Daphnia magna. J. Environ. Monit., 14: 1677-1683.

[4] Zhang, Y., Liu, M., Liu, J., Wang, X., Wang, C., Ai, W., Chen, S., Wang, H. (2018) Combined toxicity of triclosan, 2,4-dichlorophenol and 2,4,6-trichlorophenol to zebrafish (Danio rerio). Environ.Toxicol. Phar., 57: 9-18.

[5] Zamora-Garcia, I.R., Alatorre-Ordaz, A., Ibanez, J.G., Torres-Elguera, J.C., Gutierrez-Granados,

S. (2018) Degradation of pentachlorophenol as a model hazardous and recalcitrant organochlorinated pollutant using AgIII. Environ. Chem., 14.
[6] Liu, X., Cole, J. M.., Low, K. S. (2013) Solvent Effects on the UV-Vis –vis Absorption and Emission of Optoelectronic Coumarins: a Comparison of Three Empirical Solvatochromic Models. J. Phys. Chem. C., 117: 14731–14741.
[7] Díaz, M.S., Freile, M.L., Gutiérrez, M.I. (2009) Solvent effect on the UV-Vis /Vis absorption and fluorescence spectroscopic properties of berberine. Photoch. Photobio. Sci., 8:970-974.
[8] Lv, H.S., Xie, Y.S., Liu, W.Y., Gong, Z.L., Zhao, B.X. (2011) Synthesis, X-ray crystal structures and optical properties of novel substituted pyrazoly 1,3,4-oxadiazole derivatives. J. Fluoresc., 21: 1797-1804.
[9] Fallah, A., Hadjmohammadi, M. R. (2020) Low-density-solvent-based air-assisted liquid - liquid microextraction of azathioprine based on multivariate optimization and its trace determination in biological samples. J. Iran. Chem. Soc., 17: 1945-195217.
[10] Chen. D., Gao, H., Liu, P., Huang, P., Huang, X. (2019) Directly ambient pressure dried robust bridged silsesquioxane and methylsiloxane aerogels: effects of precursors and solvents. Rsc Advances., 9: 8664-8671.
[11] Hung, A.J. (2019) Can machine - learning algorithms replace conventional statistics? BJU International., 123: 1-1.
[12] Mustoe, A.M., Brooks, C.L., Al-Hashimi, H.M. (2014) Topological constraints are major determinants of tRNA tertiary structure and dynamics and provide basis for tertiary folding cooperativity. Nucleic. Acids Research., 42: 11792-804.
[13] Calabro, F.J., Vaina, L.M. (2016)Scale Changes Provide an alternative cue for the discrimination of heading, but not object motion.. J. Exp. Clin. Res, 22: 1782-1791.
[14] Perdew, J.P. (1986) Density-functional approximation for the correlation energy of the inhomogeneous electron gas. Phys. Rev. B. Condens. Matter, 33: 8822-8824.
[15] Pan, Q.J., Schreckenbach, G., Arnold, P.L, Love, J.B. (2011) Theoretical predictions of cofacial bis(actinyl) complexes of a stretched Schiff-base calixpyrrole ligand. Chem. Commun., 47: 5720- 5722.
[16] Pan, Q.J., Schreckenbach, G. (2010) Binuclear hexa- and pentavalent uranium complexes with a polypyrrolic ligand: a density functional study of water- and hydronium-induced reactions. Inorg. Chem.., 49: 6509-17.
[17] Elkechai, A., Mani, Y., Boucekkine, A., Ephritikhine, M. (2012) Density Functional Theory Investigation of the Redox Properties of Tricyclopentadienyl- and Phospholyluranium(IV) Chloride Complexes. Inorg. Chem.., 51: 6943-6952.
[18] Inukai, J., Tryk, D.A., Abe, T., Wakisaka, M., Uchida, H., Watanabe, M. (2013) Direct STM elucidation of the effects of atomic-level structure on Pt(111) electrodes for dissolved CO oxidation. J. Am. Chem. Soc., 135: 1476-1490.
[19] Pradhan, M., Basu, M., Sarkar, S., Pal, T. (2011) Solvent effect on the optical property of uranyl acetylacetonate monohydrate Spectrochim. Acta A., 78: 205-210.
[20] Bharmoria, P., Rao, K.S., Trivedi, T.J., Kumar, A. (2013) Biamphiphilic Ionic Liquid Induced Folding Alterations in the Structure of Bovine Serum Albumin in Aqueous Medium. J. Phys. Chem. B., 118: 115-124.

AUTHOR INDEX